INFRARED AND RAMAN SPECTRA
of
POLYATOMIC MOLECULES

BY

GERHARD HERZBERG, F.R.S.C.

Research Professsor of Physics,
University of Saskatchewan

Forming the second volume of
MOLECULAR SPECTRA AND
MOLECULAR STRUCTURE

NEW YORK
D. VAN NOSTRAND COMPANY, INC.
250 FOURTH AVENUE
1945

PRINTED IN THE UNITED STATES OF AMERICA

LANCASTER PRESS, INC., LANCASTER, PA.

TO MY WIFE

In the manufacture of this book, the publishers have observed the recommendations of the War Production Board with respect to paper, printing and binding in an effort to aid in the conservation of paper and other critical war materials.

CONTENTS

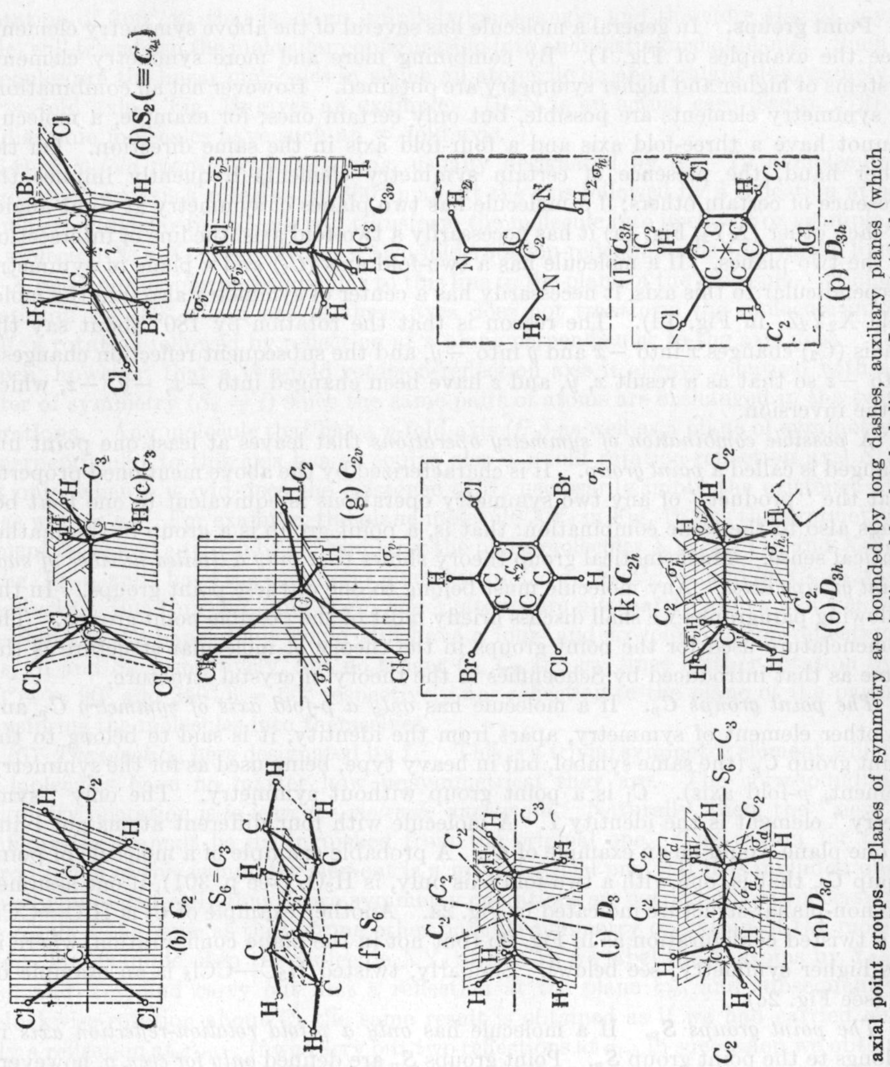

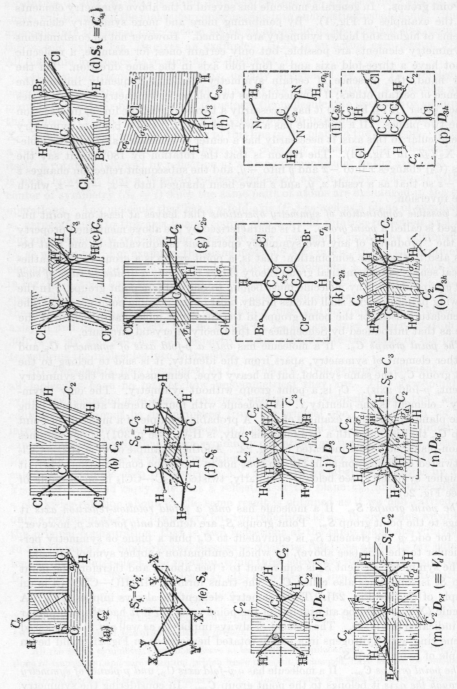

FIG. 2. Illustrations of axial point groups.—Planes of symmetry are bounded by long dashes, axilary planes which serve only to improve the clarity by short dashes. Axes of symmetry are indicated by dot-dash lines. In part (o) the σ_h at the lower right should be σ_v.

is set up vertically. In the present case therefore the planes through the axis are "vertical" planes. That is why they are called σ_v. It can easily be seen that a system with a p-fold axis cannot have just one "vertical" plane of symmetry if $p > 1$. The p planes are symmetrically arranged at angles $360°/p$. The point group $\boldsymbol{C}_{1v}$ is usually written $\boldsymbol{C}_s$ and has one plane of symmetry as the only element of symmetry (apart from I). An example is the non-linear molecule NOCl. There are many molecules belonging to $\boldsymbol{C}_{2v}$, that is, having a two-fold axis and two planes going through that axis and at right angles to each other; some examples are H_2O (see Fig. 1a), H_2CO, NO_2, and CH_2Cl_2 (see Fig. 2g). As mentioned above two planes of symmetry at right angles to each other necessarily imply that the line of intersection is a two-fold axis. There are also many of the simpler polyatomic molecules belonging to point group $\boldsymbol{C}_{3v}$. Molecules like NH_3, CH_3Cl (see Fig. 2h), and others have a three-fold axis and three planes of symmetry through it at mutual angles of 120°. A plane molecule XY_3 (see Fig. 1b) would not, however, be an example of this group, since it has higher symmetry (see below). An example of $\boldsymbol{C}_{4v}$ would be $PtCl_4^{=}$ if it were a square pyramid with the Pt nucleus at the apex. An example of the point group $\boldsymbol{C}_{5v}$ would be the molecule X_5Y_{10} if its structure were as shown in Fig. 1h except that the Y atoms were in a plane different from but parallel to the plane of the X atoms. Similarly, an example of the point group $\boldsymbol{C}_{6v}$ is a benzene molecule (Fig. 1i) in which the planes of the H and C atoms are shifted parallel to each other. The point group $\boldsymbol{C}_{\infty v}$ has an ∞-fold axis and an infinite number of planes through the axis. This is the case for a linear molecule such as HCN (see Fig. 1h) and also for all heteronuclear diatomic molecules.

The point groups $\boldsymbol{D}_p$ *(dihedral groups).* If a molecule has *a p-fold axis* C_p *and p two-fold axes* C_2 *perpendicular to the* C_p at equal angles to one another[3] it belongs to the point group $\boldsymbol{D}_p$. $\boldsymbol{D}_1$ is of course identical with $\boldsymbol{C}_2$. It is not considered as belonging to the groups $\boldsymbol{D}_p$. $\boldsymbol{D}_2$ is frequently called $\boldsymbol{V}$ (from the German "Vierergruppe"). It has three two-fold axes mutually perpendicular to one another (and no other symmetry elements). An example would be C_2H_4 if the two CH_2 groups were rotated with respect to each other by an angle which is not 90° (Fig. 2i). In the point group $\boldsymbol{D}_3$ we have one three-fold axis and three two-fold axes perpendicular to it. An example would be C_2H_6 in which the two CH_3 groups were rotated against each other, as in Fig. 2j, by an angle which is not 60° or 120° (otherwise the symmetry would be higher; see below).

Examples of the point groups D_4, D_5, and D_6 would be the molecules cyclobutane (C_4H_8), cyclopentane (C_5H_{10}), and cyclohexane (C_6H_{12}) if the C nuclei were at the corners of a square, a regular plane pentagon, and a regular plane hexagon respectively, and if the CH_2 planes were to go through the center of the polygons and were all to subtend the same angle (different from 0° and 90°) with the plane of the polygon.

The point groups $\boldsymbol{C}_{ph}$. If a molecule has *a p-fold axis* (C_p) *and a (horizontal) plane* σ_h *perpendicular to it*, it belongs to the point group $\boldsymbol{C}_{ph}$. $\boldsymbol{C}_{1h}$ is obviously identical with $\boldsymbol{C}_s$ (see above); that is, there is only one plane of symmetry. In the point group $\boldsymbol{C}_{2h}$ we have a two-fold axis and a plane of symmetry perpendicular to it. Examples are plane trans $C_2H_2Cl_2$ (see Fig. 1d) and plane trans $C_6H_2Cl_2Br_2$ (see Fig. 2k). The two-fold axes are here perpendicular to the plane of the molecule. The presence of C_2 and σ_h implies a center of symmetry (see p. 5), as is verified by

[3] A molecule that has one C_2 perpendicular to a C_p necessarily has $p - 1$ other C_2's.

the two examples given. An example of point group $\boldsymbol{C_{3h}}$ is probably the guanidinium ion $C^+(NH_2)_3$ (Fig. 2l). An example of point group $\boldsymbol{C_{4h}}$ would be tetrachlorocyclobutane ($C_4H_4Cl_4$), if all atoms were symmetrically arranged in one plane in squares. For $\boldsymbol{C_{4h}}$, $\boldsymbol{C_{6h}}$, $\cdots$ as for $\boldsymbol{C_{2h}}$, there is a center of symmetry i.

The point groups $\boldsymbol{D_{pd}}$. If a molecule has *a p-fold axis and p two-fold axes perpendicular to the* C_p as in the point groups $\boldsymbol{D_p}$ and in addition p (*vertical*) *planes of symmetry* σ_d bisecting the angles between two successive two-fold axes and going through the p-fold axis, it belongs to the point group $\boldsymbol{D_{pd}}$ (d stands for diagonal). $\boldsymbol{D_{1d}}$ does not exist since there would be no angle to bisect. $\boldsymbol{D_{2d}}$ is usually called $\boldsymbol{V_d}$. This point group has three mutually perpendicular two-fold axes, as has the point group $\boldsymbol{V} \equiv \boldsymbol{D_2}$. In addition there are two planes of symmetry bisecting the angles between two of the C_2's. As a consequence the third C_2 is at the same time a four-fold rotation-reflection axis (S_4). An example is the allene molecule ($H_2C{=}C{=}CH_2$), in which the planes of the two CH_2 groups are at right angles to each other (Fig. 2m). The perpendicular (but unstable) form of C_2H_4 is also an example. It is easily seen that these two molecules have all the symmetry elements mentioned. For the point group $\boldsymbol{D_{3d}}$ we have one three-fold axis (C_3), three two-fold axes perpendicular to it, and three bisecting planes going through C_3. As a consequence there is a six-fold rotation-reflection axis (S_6) coinciding with the C_3 and also a center of symmetry (i). An example is the so-called staggered form of ethane, C_2H_6 (Fig. 2n), in which the two CH_3 groups are rotated with respect to each other by 60°.

Examples of $\boldsymbol{D_{4d}}$, $\boldsymbol{D_{5d}}$, $\cdots$ would be molecules X_2Y_8, X_2Y_{10}, $\cdots$, if these consisted of two symmetrical groups rotated by 45°, 36°, $\cdots$, respectively against each other. The sulfur molecule S_8 is probably an actual case of a molecule of point group $\boldsymbol{D_{4d}}$. On account of the presence of a $2p$-fold rotation-reflection axis some authors use the designation $\boldsymbol{S_{2pv}}$ in place of $\boldsymbol{D_{pd}}$.

The point groups $\boldsymbol{D_{ph}}$. If a molecule has *a p-fold axis of symmetry* (C_p) and p (*vertical*) *planes of symmetry* (σ_v) through it at angles $360°/p$ to one another, as in the point group $\boldsymbol{C_{pv}}$, and in addition has *a* (*horizontal*) *plane of symmetry* (σ_h) perpendicular to C_p, it belongs to the point group $\boldsymbol{D_{ph}}$. As a consequence of the presence of these symmetry elements the molecule also has necessarily p two-fold axes (C_2): the lines of intersection of the σ_v's and σ_h (see p. 5). For even p a center of symmetry (i) and a p-fold rotation-reflection axis also follow. $\boldsymbol{D_{1h}}$ is identical with $\boldsymbol{C_{2v}}$ and is therefore not considered as belonging to the groups $\boldsymbol{D_{ph}}$. The point group $\boldsymbol{D_{2h}}$ is usually called $\boldsymbol{V_h}$. It has three mutually perpendicular two-fold axes, three mutually perpendicular planes of symmetry each going through two of the axes, and as a consequence a center of symmetry. Each of the C_2's is also an S_2. An example is ordinary plane ethylene, C_2H_4 (see Fig. 1c). The molecule O_4, if it forms (as is probable) a rectangle, also belongs to this point group. The point group $\boldsymbol{D_{3h}}$ has a three-fold axis, three C_2's at right angles to the former and three planes (σ_v) through it and each C_2, as well as one plane (σ_h) perpendicular to C_3, but no center of symmetry. Examples are all plane and symmetrical molecules XY_3 (see Fig. 1b), such as BF_3 (see p. 298). Other examples are the so-called opposed (eclipsed) form of C_2H_6 (Fig. 2o), 1,3,5-trichlorobenzene, $C_6H_3Cl_3$ (Fig. 2p), and similar molecules. The point group $\boldsymbol{D_{4h}}$ (with one C_4, four C_2, σ_h and four σ_v) again has a center of symmetry and as a consequence a four-fold rotation-reflection axis. Any plane symmetric molecule XY_4 would be an example (see Fig. 1g). An example of $\boldsymbol{D_{5h}}$ would be cyclopentane, if the C atoms were to form a regular pentagon and if the CH_2

planes were symmetrically arranged at right angles to the plane of the pentagon. The X_5Y_{10} molecule in Fig. 1h is another example. The most important example of point group D_{6h} is the benzene molecule, C_6H_6 (see Fig. 1i). The reader should verify that it has all the symmetry elements given above. The point group $D_{\infty h}$ has an ∞-fold axis (C_∞), an infinite number of C_2's perpendicular to the C_∞, an infinite number of planes through the C_∞, and a plane of symmetry perpendicular to the C_∞, which implies a center of symmetry (i). This is the case for symmetrical linear polyatomic molecules such as CO_2 (see p. 272), C_2H_2 (see p. 288), and all homonuclear diatomic molecules.

All the point groups discussed so far have not more than one (if any) three-fold or higher-fold axis. They are also called *axial point groups.* The following point groups of higher symmetry have more than one three-fold or four-fold axis. They are also called *cubic point groups* since they form the basis of the cubic crystal system.

The point group T *(tetrahedral group).* If a molecule has *three mutually perpendicular two-fold axes* (as has the point group $D_2 \equiv V$), and in addition *four three-fold axes* it belongs to the point group T. The two-fold axes bisect the angles between the three-fold axes as in a regular tetrahedron. But the symmetry is less than that of a tetrahedron. An example would be a molecule like neopentane, $C(CH_3)_4$, if the four C atoms of the CH_3 groups occupied the corners of a regular tetrahedron whose center is occupied by the fifth C atom, and if the equilateral triangles formed by the three H atoms of each CH_3 groups were *not* in their most symmetrical positions (see Fig. 3a).

The point group T_d. If a molecule in addition to *three mutually perpendicular two-fold axes* and *four three-fold axes* (point group T) has a *plane of symmetry* σ_d *through each pair of three-fold axes* (that is, two mutually perpendicular planes through each two-fold axis), in all, six planes of symmetry,[4] it belongs to the point group T_d. The presence of these planes implies that the two-fold axes are at the same time four-fold rotation-reflection axes. Since the regular tetrahedron has this symmetry, all tetrahedral molecules are examples of this point group: CH_4 (see Fig. 3b), CCl_4, P_4, and others. The molecule neopentane $C(CH_3)_4$ is also an example if the CH_3 groups in addition to being arranged on a regular tetrahedron have a symmetrical position (that is, if the angle ϕ in Fig. 3a is 0° or 60°).

The point group T_h. If a molecule, in addition to the symmetry properties of point group T, has a *center of symmetry* it belongs to the point group T_h. As a consequence there are also three planes of symmetry through the three (mutually perpendicular) two-fold axes. No actual molecule, even with appropriate distortions, has this point group. But if one could add four CH_3 groups to the $C(CH_3)_4$ of Fig. 3a in positions symmetrical to those already there, such that the central C atom becomes a center of symmetry, one would have an example of the point group T_h.

The point group O *(octahedral group).* If a molecule has *three mutually perpendicular four-fold axes and four three-fold axes* which have the same orientation with respect to one another as the two-fold and three-fold axes of point group T, it belongs to the octahedral point group O. As a consequence of the axes given, the molecule has also six two-fold axes (apart from the three two-fold axes that coincide with the four-fold axes). The regular octahedron and the cube (see Fig. 3d and e) have just these axes of symmetry. But they have in addition a number of planes of symmetry which a molecule of point group O does not have. An example would be a molecule $X(YZ_4)_6$

[4] The presence of one such plane of symmetry has as a necessary consequence the presence of the five others.

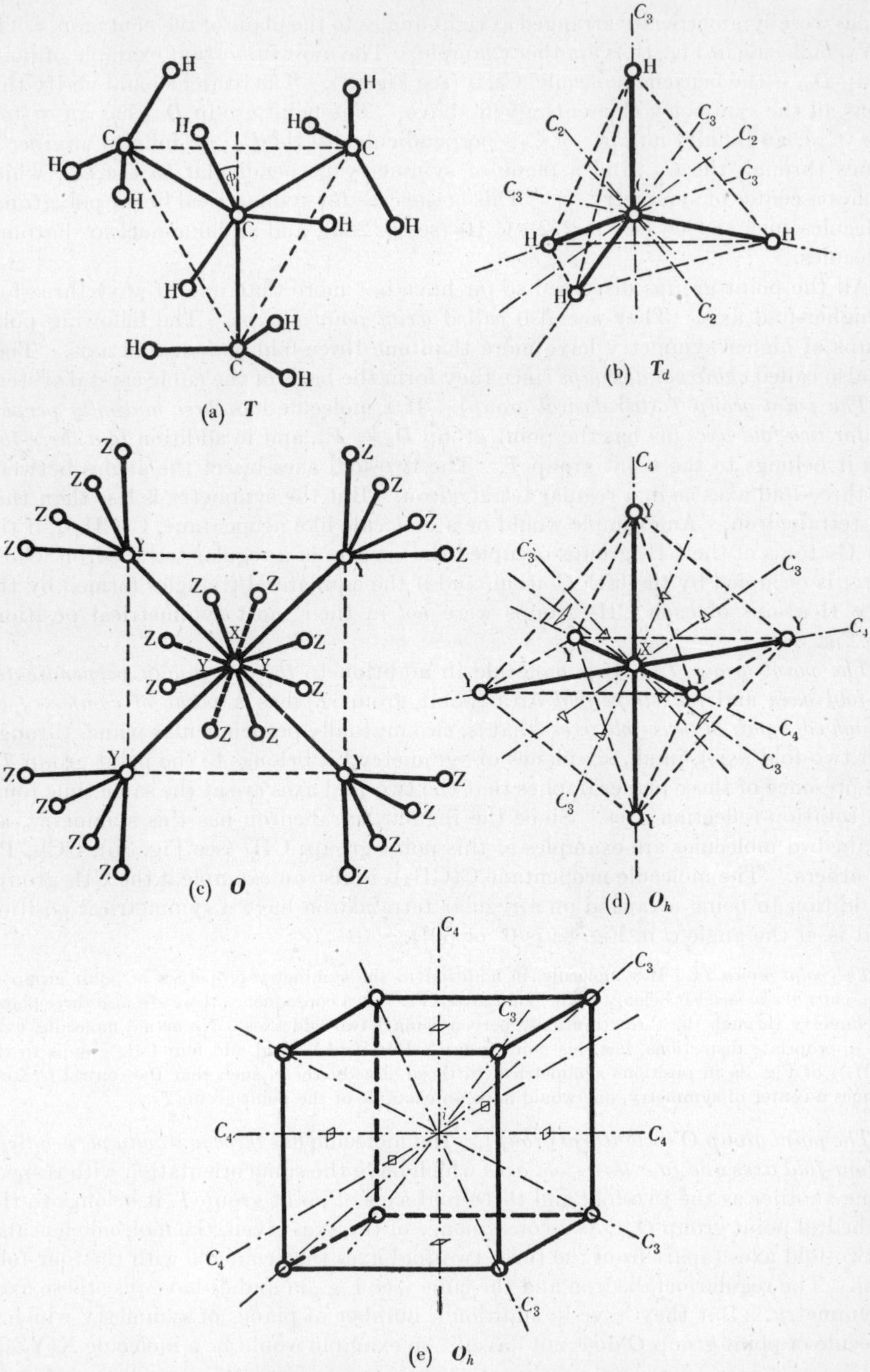

FIG. 3. **Illustrations of cubic point groups.**—(a) and (c) are perpendicular, (b), (d) and (e) are oblique projections. In (d) and (e) the points where the C_3 and C_4 respectively pierce the triangular and square faces are marked by small triangles and squares.

in which symmetrical YZ_4 groups were placed at the corners of a regular octahedron in such a way that the squares formed by the four Z atoms of each group were not in their most symmetrical position but were all rotated about the XY axis by the same amount (see Fig. 3c).

The point group $\boldsymbol{O_h}$. If a molecule has in addition to *three mutually perpendicular four-fold axes* C_4 *and four three-fold axes* C_3 (point group **O**) *a center of symmetry* (i) it belongs to the point group $\boldsymbol{O_h}$. As a consequence it also has six two-fold axes (apart from the three two-fold axes that coincide with the four-fold axes) and nine planes of symmetry. Also, the four-fold axes are at the same time four-fold rotation-reflection axes. The regular octahedron and the cube have this symmetry, as the reader can easily verify from Figs 3d. and e. The molecule SF_6 is very probably an example of point group $\boldsymbol{O_h}$, the F atoms being arranged at the corners of a regular octahedron with the S atom in the center (see p. 336). The molecule S_8 would be another example if the atoms were at the corners of a cube, which, however, is probably not the case.

The only point groups that have *several* higher than four-fold axes are the icosahedral groups $\boldsymbol{I}$ and $\boldsymbol{I_h}$. The former has six five-fold, ten three-fold and fifteen two-fold axes while the latter in ad-

TABLE 1. SYMMETRY ELEMENTS AND EXAMPLES OF THE MORE IMPORTANT POINT GROUPS [5]

Point group	Symmetry elements [6]	Examples [7]
$\boldsymbol{C_1}$	No symmetry	CHFClBr, N_2H_4
$\boldsymbol{C_2}$	One C_2	Non-planar H_2O_2 (Fig. 2a), [partly rotated $H_2C{=}CCl_2$ (Fig. 2b)], $HClC{=}C{=}CHCl$
$\boldsymbol{C_3}$	One C_3	[Partly rotated $H_3C{-}CCl_3$ (Fig. 2c)]
$\boldsymbol{C_i}(\equiv S_2)$	i ($\equiv S_2$)	Trans form of $ClBrHC{-}CHBrCl$ (Fig. 2d)
$\boldsymbol{S_6}$	One S_6, one C_3 (coincident with S_6), i	[Puckered C_6H_6 ring with partly rotated H_6 (Fig. 2f)]
$\boldsymbol{C_s}(\equiv C_{1v} \equiv C_{1h})$	One σ	Non-linear NOCl, plane N_3H
$\boldsymbol{C_{2v}}$	One C_2, two σ_v	H_2O, H_2CO, CH_2Cl_2 (Fig. 2g), plane $H_2C{=}CCl_2$
$\boldsymbol{C_{3v}}$	One C_3, three σ_v	NH_3, CH_3Cl (Fig. 2h), symmetric $H_3C{-}CCl_3$
$\boldsymbol{C_{4v}}$	One C_4, one C_2 (coincident with C_4), four σ_v	Non-planar $(PtCl_4)^{=}$
$\boldsymbol{C_{6v}}$	One C_6, one C_3, one C_2 (both coincident with C_6), six σ_v	[C_6H_6 with C_6 and H_6 in different planes]
$\boldsymbol{C_{\infty v}}$	One C_∞, any C_p, infinite number of σ_v	CN, HCN, COS, $HC{\equiv}CCl$
$\boldsymbol{D_2 \equiv V}$	Three C_2 (mutually perpendicular)	[Partly rotated C_2H_4 (Fig. 2i)]
$\boldsymbol{D_3}$	One C_3, three C_2 ($\perp$ to C_3)	[Partly rotated C_2H_6 (Fig. 2j)]
$\boldsymbol{C_{2h}}$	One C_2, one σ_h, $i \equiv S_2$	Plane trans $C_2H_2Cl_2$ (Fig. 1d), trans $C_6H_2Cl_2Br_2$ (Fig. 2k)
$\boldsymbol{C_{3h}}$	One C_3, one σ_h, one S_3 (coincident with C_3)	$C^+(NH_2)_3$ (Fig. 2l)

[5] For other point groups see the text.

[6] The element I (identity), which is contained in every point group, has been omitted. C_p = p-fold axis, S_p = p-fold rotation-reflection axis, i = center of symmetry, σ_v = vertical plane (see p. 7), σ_h = horizontal plane, σ_d = diagonal plane.

[7] Examples that probably do not correspond to the actual structure of the ground state of the particular molecules are put in square brackets.

TABLE 1—*Continued*

Point group	Symmetry elements[6]	Examples[7]
$D_{2d} \equiv V_d$	Three C_2 (mutually ⊥), one S_4 (coincident with one C_2), two σ_d (through S_4)	$H_2C{=}C{=}CH_2$ (Fig. 2m) [perpendicular (non-planar) C_2H_4]
$D_{3d} (\equiv S_{6v})$	One C_3, three C_2 (⊥ to C_3), S_6 (coincident with C_3), i, three σ_d	Staggered form of C_2H_6 (Fig. 2n), C_6H_{12} (cyclohexane)
$D_{4d} (\equiv S_{8v})$	One C_4, four C_2 (⊥ to C_4), S_8 (coincident with C_4), C_2 (coincident with C_4), four σ_d	Puckered octagon form of S_8 (sulfur)
$D_{2h} \equiv V_h$	Three C_2 (mutually ⊥), three σ (mutually ⊥), i	Plane C_2H_4 (Fig. 1c), plane O_4, plane N_2O_4 (Fig. 1c)
D_{3h}	One C_3, three C_2 (⊥ to C_3), three σ_v, one σ_h	BCl_3 (Fig. 1b) eclipsed form of C_2H_6 (Fig. 2o), 1, 3, 5 $C_6H_3Cl_3$ (Fig. 2p)
D_{4h}	One C_4, four C_2 (⊥ to C_4), four σ_v, one σ_h, one C_2, one S_4 (both coincident with C_4), i	C_4H_8 (cyclobutane) [Plane $(PtCl_4)^{=}$]
D_{5h}	One C_5, five C_2 (⊥ to C_5), five σ_v, one σ_h	[Plane symmetrical cyclopentane]
D_{6h}	One C_6, six C_2 (⊥ to C_6), six σ_v, one σ_h, one C_2, C_3, S_6 (each coincident with C_6), i	Plane symmetrical C_6H_6, C_6Cl_6
$D_{\infty h}$	C_∞, infinite number of C_2 (⊥ to C_∞) and of σ_v, one σ_h, and C_p and S_p (coincident with C_∞), i	O_2, CO_2, C_2H_2, C_3O_2
T	Three C_2 (mutually ⊥), four C_3	[Partly rotated $C(CH_3)_4$ (Fig. 3a)]
T_d	Three C_2 (mutually ⊥), four C_3, six σ, three S_4 (coincident with the C_2's)	CH_4 (Fig. 3b), CCl_4, P_4, symmetrical $C(CH_3)_4$
O_h	Three C_4 (mutually ⊥), four C_3, i, three S_4 and C_2 (coincident with C_4), six C_2, nine σ, four S_6 (coincident with C_3)	SF_6, $(PtCl_6)^{=}$

dition has a center of symmetry which causes numerous planes of symmetry and rotation-reflection axes. The regular icosahedron and the regular pentagon dodecahedron belong to point group I_h. It is not likely that molecules of such a symmetry will ever be found.

Although in principle *molecules* belonging to any one of the mathematically possible point groups may occur, it may be noted that in *crystals* only point groups with one-, two-, three-, four-, and six-fold symmetry axes are possible; hence crystals may have only 32 point groups, which give rise to the 32 crystal classes.

In Table 1 the preceding discussion of point groups is summarized. Only those point groups are included that are likely to be of importance in the study of molecular structure.

CHAPTER I

ROTATION AND ROTATION SPECTRA

As in the case of diatomic molecules, we may in a certain approximation resolve the total energy of a polyatomic molecule into the sum of rotational, vibrational, and electronic energy. However, for polyatomic molecules this approximation is frequently much less accurate than for diatomic molecules, since it often happens that vibrational frequencies are of the same order of magnitude as rotational frequencies, and electronic frequencies of the same order as vibrational frequencies. Consequently the *mutual interactions* of the three types of motion may be much stronger than for diatomic molecules.

In this chapter we shall consider the pure rotation of polyatomic molecules neglecting the interaction with the vibration and with the electronic motion. In other words, we consider the *rotations of a non-vibrating molecule in a fixed* (symmetrical) *electronic state.*

The *moment of inertia* of a rigid body about an axis is defined by

$$I = \Sigma m_i \rho_i^2,$$

where ρ_i is the perpendicular distance of the mass element m_i from the axis. When we determine for a given body this moment of inertia about various axes going through one and the same point, usually the center of mass, we find according to a simple theorem of mechanics that there are three mutually perpendicular directions for which the moment of inertia is a maximum or a minimum. These directions are called the *principal axes* and the corresponding moments of inertia *the principal moments of inertia.* If $1/\sqrt{I}$ is plotted along the respective axes an ellipsoid is obtained which is called the *momental ellipsoid.* The axes of this ellipsoid are the principal axes. If the body (molecule) has symmetry the direction of one or more of the principal axes going through the center of mass can easily be found since *axes of symmetry are always principal axes* and since *a plane of symmetry is always perpendicular to a principal axis.*

If for a molecule the three principal moments of inertia are different, it is called (with respect to its rotations) an *asymmetric top* (or asymmetric rotator). If two of the principal moments of inertia are equal it is called a *symmetric top* (symmetric rotator). In this case the momental ellipsoid is a rotational ellipsoid. If all three principal moments of inertia are equal it is called a *spherical top:* the momental ellipsoid is a sphere. In addition we have the special case of a symmetric top in which one of the principal moments of inertia is zero, or extremely small, while the other two are equal. The momental ellipsoid is a circular cylinder. Such is the case for all *linear polyatomic molecules.* In the following we consider separately the four cases mentioned.

1. Linear Molecules

Linear molecules belong to the point groups $D_{\infty h}$ or $C_{\infty v}$, depending on whether or not they have a plane of symmetry perpendicular to the internuclear axis.

Energy levels. If the angular momentum of the electrons about the internuclear axis is zero, as is the case for the ground states of all known linear polyatomic molecules, the problem can be treated as if the moment of inertia about the internuclear axis were exactly equal to zero, that is, as if we had the *simple rotator*, rigid or nonrigid. (See Molecular Spectra I, p. 122f.). The energy levels are given by the same formula as for diatomic molecules:

$$\frac{E_r}{hc} = F(J) = BJ(J+1) - DJ^2(J+1)^2 + \cdots \qquad \text{(I, 1)}$$

where E_r is the rotational energy (in ergs), $F(J)$ is the rotational *term value* (in cm^{-1}) and J is the *rotational quantum number*. For the rotational constant B we have[1]

$$B = \frac{h}{8\pi^2 c I_B} = \frac{27.994 \times 10^{-40}}{I_B} \qquad \text{(I, 2)}$$

As for diatomic molecules I_B is the *moment of inertia* about an axis perpendicular to the internuclear axis and going through the center of mass. But while for diatomic molecules we have the simple formula $I_B = \mu r^2$ where μ is the reduced mass, for linear polyatomic molecules the general formula

$$I_B = \Sigma m_i r_i^2 \qquad \text{(I, 3)}$$

has to be used, where r_i is the distance of the ith nucleus of mass m_i from the center of mass. For a symmetrical molecule Y—X—Y, for example, I_B is $2m_Y r_Y^2$.

Strictly speaking, in (I, 3) the summation should be over the electrons as well as the nuclei. In view of the smallness of the electron mass this can be taken into account by using for the m_i the masses of the neutral atoms rather than the bare nuclei. The error introduced in this way is much smaller than the present error in the spectroscopic determinations of moments of inertia.

The term $DJ^2(J+1)^2$ in (I, 1), as for diatomic molecules, comes in because of the *non-rigidity* of the molecule. This term is always exceedingly small compared to the term $BJ(J+1)$. It represents the *influence of the centrifugal force* which results in a very slight increase in the internuclear distances when the molecule is rotating. In diatomic molecules the constant D is related to B and the vibrational frequency ω (assuming the harmonic oscillator approximation) by the simple formula

$$D = \frac{4B^3}{\omega^2} \qquad \text{(I, 4)}$$

This formula applies also to linear symmetric XY_2 molecules when ω is the frequency of the one totally symmetric vibration (ν_1 in Fig. 25 p. 66). For XYZ and X_2Y_2 molecules with two totally symmetric vibrations of frequencies ω_1 and ω_2, the constant D is given by

$$D = 4B^3\left(\frac{\delta_1}{\omega_1^2} + \frac{\delta_2}{\omega_2^2}\right) \qquad \text{(I, 4a)}$$

where δ_1 and δ_2 are constants whose values for X_2Y_2 have been given by Shaffer and Nielsen (779), for XYZ by Nielsen (654a).[2] In most cases of linear polyatomic

[1] For the values of the fundamental constants used in this book see the appendix p. 538.
[2] The numbers in brackets refer to the bibliography p. 539–558.

molecules thus far studied experimentally the term $DJ^2(J+1)^2$ in (I, 1) has been neglected. [See, however, Herzberg and Spinks (441) (442)].

Fig. 4 gives a diagram of the rotational energy levels of a linear polyatomic molecule. The spacing of the levels is the larger the smaller the moment of inertia.

As for diatomic molecules the quantum number J corresponds to the angular momentum of the molecule, which has the magnitude

$$\sqrt{J(J+1)}\,\frac{h}{2\pi} \approx J\,\frac{h}{2\pi}. \tag{I, 5}$$

We shall write $\boldsymbol{J}$ for the angular momentum vector.

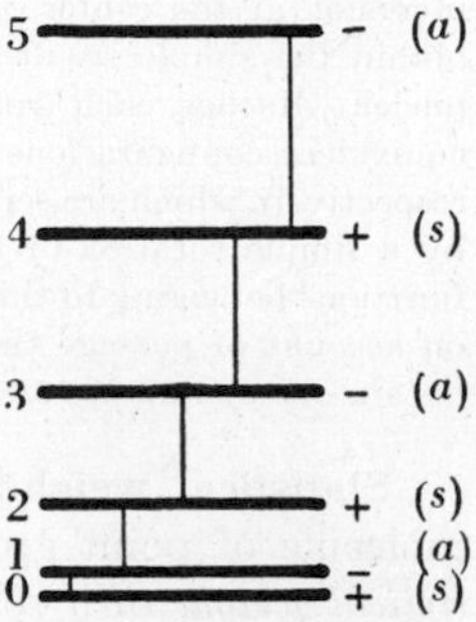

FIG. 4. **Rotational energy levels of a linear molecule.**

Symmetry properties. The *rotational eigenfunctions* ψ_r of linear polyatomic molecules are (like those of diatomic molecules) the surface harmonics represented in Fig. 39 of Molecular Spectra I (p. 74). The total eigenfunction in zero approximation is a product of the electronic, vibrational, and rotational eigenfunctions:

$$\psi = \psi_e \psi_v \psi_r. \tag{I, 6}$$

As in the case of diatomic molecules a rotational level of a linear polyatomic molecule is called *positive or negative depending on whether the total eigenfunction ψ remains unaltered or changes its sign by reflection of all the particles* (electrons and nuclei) *at the origin* (inversion). If the electronic and vibrational eigenfunctions ψ_e and ψ_v are unchanged by all symmetry operations of the molecule (which is the case for the vibrationless ground states of all known linear molecules), the symmetry character positive-negative depends on that of ψ_r only and, as for diatomic molecules, the even rotational levels are positive, the odd ones negative. This dependence is indicated in Fig. 4.

If the linear molecule has the point group $\boldsymbol{D}_{\infty h}$, that is, if it has a center of symmetry (as has C_2H_2, for example) we have in addition to the symmetry property positive-negative, the property *symmetric or antisymmetric with respect to an exchange of the identical nuclei.* The total eigenfunction ψ of the system (apart from the nuclear spin function) remains unchanged or changes sign when all nuclei on one side of the center are *simultaneously* exchanged with the corresponding ones on the other side. We call the corresponding rotational levels *symmetric or antisymmetric in the nuclei.* It will be shown below that, just as for homonuclear diatomic molecules, either the positive rotational levels are symmetric and the negative antisymmetric or the negative levels are symmetric and the positive antisymmetric. The former alternative applies to the vibrationless state of symmetric electronic states (Σ_g^+ states). This is indicated in brackets in Fig. 4.

If any polyatomic molecule has identical nuclei the total eigenfunction (exclusive of nuclear spin) of a (non-degenerate) rotational level must remain unchanged or can only change sign for an exchange of two identical nuclei. In the case of symmetrical linear molecules of point group $\boldsymbol{D}_{\infty h}$ (such as C_2H_2), a simultaneous exchange of all nuclei on one side of the center with those on the other side can be brought about by reflection of all particles at the origin followed by a reflection at the origin of the electrons only. The first operation leaves the total eigenfunction unchanged or only changes its sign for the positive and negative rotational levels respectively, while the second operation leaves the total eigenfunction unchanged (or changes its sign) if the electronic eigenfunction remains unchanged (or changes its sign). Thus in the vibrationless ground state (assuming it to be a Σ_g^+ electronic state) the even rotational levels are symmetric and the odd ones antisymmetric with respect to a simultaneous exchange of all nuclei on one side with those on the other.

In a linear molecule that has two identical nuclei but does not belong to point group $D_{\infty h}$, such as X—Y—Y (for which N_2O is an actual example, see p. 277) or unsymmetrical Y—X——Y, an inversion at the center of mass does not exchange the identical nuclei and therefore one does not obtain the simple result that alternate rotational levels are symmetric and antisymmetric in the nuclei. Rather, each (rotational) level of such a molecule is doubly degenerate since there are two equivalent configurations X—$Y^{(1)}$—$Y^{(2)}$ and X—$Y^{(2)}$—$Y^{(1)}$ or $Y^{(1)}$—X——$Y^{(2)}$ and $Y^{(2)}$—X——$Y^{(1)}$ respectively which are separated by a potential hill and which cannot be transformed into each other by a simple rotation of the whole molecule (see also Chapter II, section 5d). One of the eigenfunctions belonging to this degenerate level is symmetric, the other antisymmetric in the nuclei. If, on account of passage through the potential barrier, a splitting of the degeneracy occurs, then we obtain one symmetric and one antisymmetric level for each originally degenerate level.

Statistical weights and influence of nuclear spin and statistics. For a linear molecule of point group $C_{\infty v}$ (no center of symmetry, for example HCN) the *statistical weight* of a rotational level in a totally symmetric electronic state ($^1\Sigma^+$) is given by the number of possible orientations of J in a magnetic field, that is by $2J + 1$.

Strictly speaking $2J + 1$ has to be multiplied by the total number of possible orientations of the nuclear spins $I_1, I_2, \cdots$ in the magnetic field, that is by $(2I_1 + 1)(2I_2 + 1) \cdots$. But since this factor is the same for all levels of a molecule of point group $C_{\infty v}$, it can usually be omitted.

If the molecule belongs to point group $D_{\infty h}$, that is, if it has a center of symmetry, *alternate rotational levels have different statistical weights*, as in the case of homonuclear diatomic molecules. If the spins of all the nuclei are zero, with the possible exception of the one at the center of symmetry, the antisymmetric rotational levels are missing entirely, that is, for Σ_g^+ electronic states the odd rotational levels are absent.[3] Such is the case for CO_2 and C_3O_2 since they are linear and symmetrical (point group $D_{\infty h}$). If one or more pairs of nuclei outside the center have a nuclear spin $I \neq 0$ all rotational levels are present but the even and odd levels have different statistical weights. If only one pair of identical nuclei has $I \neq 0$ (which is the only case thus far studied experimentally), it is easily seen from the same reasons as for diatomic molecules (Molecular Spectra I, p. 141f.) that the *ratio of the statistical weights of the symmetric and antisymmetric rotational levels is* $(I + 1)/I$ or $I/(I + 1)$, *depending on whether the nuclei follow Bose or Fermi statistics.* One may also say that the statistical weights vary in the same way as in a diatomic molecule containing the same two nuclei with $I \neq 0$. Thus for ordinary acetylene (C_2H_2), which is linear and symmetrical (see p. 288) since $I(C) = 0$, $I(H) = \frac{1}{2}$, the antisymmetric (odd) rotational levels have three times the statistical weight of the symmetric (even) levels, just as for H_2 (see Molecular Spectra I, p. 141). For C_2D_2 and similarly for C_2N_2 the symmetric (even) rotational levels have twice the weight of the antisymmetric (odd) levels, just as for D_2 and N_2.

To the same extremely good approximation as for diatomic molecules intercombinations between the symmetric and antisymmetric rotational levels are prohibited for any type of radiation and even for collisions, that is [4]

$$\textit{symmetric} \not\leftrightarrow \textit{antisymmetric} \tag{I, 7}$$

[3] This statement holds if we assume Bose statistics for the nuclei of spin zero. No nuclei of spin zero that follow Fermi statistics are known.

[4] Here and later $\not\leftrightarrow$ stands for "does not combine with," while $\leftrightarrow$ will be used for "combines with"

Thus we have *ortho and para modifications* also for linear symmetric polyatomic molecules (point group $D_{\infty h}$).

Naturally, if in a linear molecule of point group $D_{\infty h}$ one atom is replaced by an isotopic atom, the distinction between symmetric and antisymmetric rotational levels no longer exists and therefore there is in this case no difference in the statistical weights of the even and odd rotational levels; there are no ortho and para modifications. For example, $O^{16}—C—O^{18}$ has no missing rotational levels, H—C≡C—D has no alternation of the weights of successive rotational levels.

Furthermore, in a molecule such as X—Y—Y (for example N_2O; see p. 277), which has two identical nuclei but not the symmetry $D_{\infty h}$, no difference in the statistical weights of the even and odd levels occurs, except of course for the difference in $2J + 1$.

This conclusion follows immediately from what has been said above (p. 16) about such molecules. It should be noted that in this case the absolute statistical weight is *not* twice $(2J + 1) \times (2I_X + 1)(2I_Y + 1)^2$ as might at first appear from the double degeneracy of the rotational levels (I_X = nuclear spin of atom X, I_Y = nuclear spin of Y). In the case $I_Y = 0$ only the symmetric component of each degenerate level appears and its absolute weight is $(2J + 1)(2I_X + 1)$. In the case $I_Y \neq 0$ the resultant spin due to I_Y is $2I_Y$, or $2I_Y - 1$, or $2I_Y - 2$, $\cdots$, of which the first, third, $\cdots$ values belong to the symmetric (or antisymmetric), the second, fourth, $\cdots$ to the antisymmetric (or symmetric) levels (see Molecular Spectra I, p. 146). This introduces the additional weight factors $[2(2I_Y) + 1] + [2(2I_Y - 2) + 1] + \cdots$ and $[2(2I_Y - 1) + 1] + [2(2I_Y - 3) + 1] + \cdots$ respectively for the symmetric and antisymmetric levels (or the antisymmetric and symmetric ones). The sum of these additional weight factors for a pair of levels of given J is easily seen to be $(2I_Y + 1)^2$. Thus the total statistical weight is $(2J + 1)(2I_X + 1)(2I_Y + 1)^2$, which includes the case $I_Y = 0$. It is the same as would be obtained without considering the identity of the two Y nuclei. The reader may specialize the above proof for N_2O where $I_X = 0$, $I_Y = 1$.

In the case of symmetric linear molecules (point group $D_{\infty h}$) *with several pairs of identical nuclei with* $I \neq 0$ the statistical weight factors of the symmetric and antisymmetric rotational levels can be obtained by an extension of the method used for diatomic molecules as was first done by Placzek and Teller (701).

Assuming the molecule to be brought into a magnetic field that is strong enough to uncouple all nuclear spins from one another it is clear that in a molecule $W(XYZ\cdots)_2$ of point group $D_{\infty h}$ the number of spin configurations of the nuclei on one side of the center (W) is $(2I_X + 1)(2I_Y + 1) \times (2I_Z + 1)\cdots$ and therefore the total number of spin configurations is the square of this quantity (disregarding the contribution of the central atom W if such is present). There are $(2I_X + 1) \times (2I_Y + 1)(2I_Z + 1)\cdots$ configurations for which reflection at the center will leave the configuration unchanged. These correspond to spin functions that are symmetric with respect to a simultaneous exchange of all pairs of identical nuclei. All the other spin configurations occur in pairs (such as $-1, +\frac{1}{2}, 0, +1, +\frac{3}{2}, +1$ and $+1, +\frac{3}{2}, +1, 0, +\frac{1}{2}, -1$) which correspond each to a symmetric and an antisymmetric spin function. Therefore there are

$$\tfrac{1}{2}[(2I_X + 1)^2(2I_Y + 1)^2(2I_Z + 1)^2 \cdots - (2I_X + 1)(2I_Y + 1)(2I_Z + 1) \cdots] \qquad \text{(I, 8)}$$

antisymmetric spin functions and

$$\tfrac{1}{2}[(2I_X + 1)^2(2I_Y + 1)^2(2I_Z + 1)^2 \cdots + (2I_X + 1)(2I_Y + 1)(2I_Z + 1) \cdots] \qquad \text{(I, 9)}$$

symmetric spin functions. The total eigenfunction inclusive of nuclear spin can only be symmetric with respect to a simultaneous exchange of all pairs of identical nuclei for all rotational levels or only antisymmetric and therefore the ratio of (I, 9) to (I, 8) gives the ratio of the statistical weights of the symmetric to the antisymmetric levels or conversely. Which case applies depends on whether the "*resultant*" *statistics* of the group of nuclei XYZ $\cdots$ is Bose or Fermi. The "resultant" statistics is Bose statistics if there is an even number of nuclei following Fermi statistics in the group, it is

Fermi statistics if there is an odd number of nuclei following Fermi statistics.[5] The resultant statistics has to be used since a reflection at the origin exchanges all pairs of identical nuclei *simultaneously.*

As a result, in the case of vibrationless Σ_g^+ electronic states the *even rotational levels have the weight factor* (I, 9), *the odd have the weight factor* (I, 8) *for a resultant Bose statistics while the converse is true for a resultant Fermi statistics.*

It is easily seen that this more elaborate rule gives the same results derived previously in a more elementary way for CO_2, C_2H_2, C_2D_2, C_2N_2. As an illustration of the above considerations and for future use, Table 2 gives the statistical weight

TABLE 2. STATISTICAL WEIGHT FACTORS OF SYMMETRIC AND ANTISYMMETRIC (EVEN AND ODD) ROTATIONAL LEVELS OF SOME LINEAR MOLECULES.[6]

Molecule	Resultant statistics	Statistical weight factors	
		Symmetric (even) levels	Antisymmetric (odd) levels
$C^{12}O_2^{16}$, $C^{13}O_2^{16}$, $C^{12}O_2^{18}$	Bose	1	0[6a]
$C_2^{12}H_2^1$	Fermi	1	3
$C_2^{12}D_2^2$	Bose	6	3
$C_2^{13}H_2^1$	Bose	10	6
$C_2^{13}D_2^2$	Fermi	15	21
$C_2^{12}N_2^{14}$	Bose	6	3
$C_2^{12}N_2^{15}$	Fermi	1	3
$C_3^{12}O_2^{16}$, $C_3^{12}O_2^{18}$	Bose	1	0
$C^{12}C_2^{13}O_2^{16}$, $C_3^{13}O_2^{16}$	Fermi	1	3[6a]
$C_4^{12}H_2^1$	Fermi	1	3
$C_4^{13}H_2^1$	Fermi	28	36
$C_4^{12}D_2^2$	Bose	6	3
$C_2^{12}C_2^{13}H_2^1$	Bose	10	6
$C_4^{13}D_2^2$	Bose	78	66
$C^{12}(N^{14}H^1)_2$	Fermi	15	21
$C^{12}(N^{14}D^2)_2$	Bose	45	36
$C^{12}(N^{15}H^1)_2$	Bose	10	6

factors for a number of polyatomic molecules that are known to be linear or are probably linear, as well as for some of their isotopes. These factors give at the same time the *equilibrium ratios of the ortho and para modifications* of these molecules (the ortho modification corresponding to the larger weight factor). It may be noted that on account of the different resultant statistics, for $C_2^{13}H_2^1$ unlike ordinary $C_2^{12}H_2^1$ the even levels would have the greater statistical weight (ortho modification).

Thermal distribution of rotational levels. The population N_J of the various rotational levels is given by the general formula

$$N_J \sim g_J e^{-(F(J)hc/kT)}, \qquad \text{(I, 10)}$$

[5] This may easily be verified if it is remembered that an exchange of two nuclei following Fermi statistics changes the sign of the total eigenfunction whereas an exchange of two nuclei following Bose statistics does not.

[6] The following values for the nuclear spins have been assumed: $I(C^{13}) = \frac{1}{2}$, $I(N^{15}) = \frac{1}{2}$, $I(O^{18}) = 0$, $I(N^{14}) = 1$, $I(C^{12}) = 0$, $I(H^1) = \frac{1}{2}$, $I(D) = 1$.

[6a] If C^{13} instead of C^{12} is at the center the statistical weights have to be multiplied by 2 [assuming $I(C^{13}) = \frac{1}{2}$].

where g_J is the statistical weight, k the Boltzmann constant and T the absolute temperature. For non-symmetrical linear molecules $g_J = 2J + 1$ (see above), whereas for symmetrical linear molecules the alternation of the statistical weights has to be taken into account. A graphical representation of the distribution function (I, 10) which is the same as for diatomic molecules may be found in Molecular Spectra I, p. 132.

Infrared rotation spectrum. The selection rules for transitions from one rotational level to another due to dipole radiation (without change of electronic or vibrational energy: *pure rotation spectrum*) are exactly the same as for diatomic molecules: A transition can only take place if the molecule has a *permanent dipole moment*—that is, if it has the symmetry $C_{\infty v}$ and not $D_{\infty h}$. Furthermore, we have the rule that positive levels combine only with negative ones, that is

$$+ \leftrightarrow -, \quad + \not\leftrightarrow +, \quad - \not\leftrightarrow -, \tag{I, 11}$$

and the rule

$$\Delta J = \pm 1. \tag{I, 12}$$

That molecules of symmetry $D_{\infty h}$ which have no dipole moment do not exhibit a (dipole) rotation spectrum in the infrared may also be considered as due to the fact that the rules $+ \leftrightarrow -$ and symmetric $\not\leftrightarrow$ antisymmetric cannot be fulfilled at the same time (see Fig. 4).

The formula for the rotation spectrum is obtained by substituting $J' = J'' + 1 \equiv J + 1$ into

$$\nu = F(J') - F(J''), \tag{I, 13}$$

taking $F(J)$ from (I, 1). Here as usual J' is the J value of the upper state, $J'' \equiv J$ that of the lower. One obtains as for diatomic molecules

$$\nu = 2B(J + 1) - 4D(J + 1)^3. \tag{I, 14}$$

Since $D \ll B$, this formula represents a series of very nearly equidistant lines. The transitions are indicated in Fig. 4.

TABLE 3. PREDICTED WAVE NUMBERS AND WAVE LENGTHS OF THE INFRARED ROTATION SPECTRUM OF HCN.

J	ν (cm^{-1})	λ (μ)
0	2.96	3381
1	5.92	1690
2	8.87	1127
3	11.83	845
...	...	...
20	61.98	161.3
21	64.92	154.0
22	67.85	147.4
23 [7]	70.79	141.3

Up to the present time no infrared rotation spectrum of a linear molecule has been observed, since all of them lie very far in the infrared. For HCN, for example,

[7] Last J value observed in the rotation-vibration spectrum.

which has the smallest moment of inertia of all known linear polyatomic molecules, one predicts [with $B = 1.4789$, $D = 3.63 \cdot 10^{-6}$ cm^{-1} as obtained by Herzberg and Spinks (442) from the rotation-vibration spectrum] the wave numbers and wave lengths given in Table 3.

Rotational Raman spectrum. As has been shown in Molecular Spectra I, p. 93, the occurrence of a Raman spectrum depends on *whether the polarizability in a fixed direction changes during the motion.* For a linear molecule the polarizability in the direction of the internuclear axis is always different from that in a direction perpendicular to it (that is, the polarizability ellipsoid is not a sphere) and therefore the polarizability in a fixed direction changes during a rotation of the molecule about an axis perpendicular to the internuclear axis. Thus a linear polyatomic molecule of point group $\boldsymbol{C_{\infty v}}$ or $\boldsymbol{D_{\infty h}}$ always has a rotational Raman spectrum.

Assuming a Σ^+ electronic ground state (which applies to all actual cases) the *selection rules* for Raman transitions are, just as for diatomic molecules with $\Lambda = 0$

$$\Delta J = 0, \pm 2 \tag{I, 15}$$

and

$$+ \leftrightarrow +, \qquad - \leftrightarrow -, \qquad + \not\leftrightarrow -. \tag{I, 16}$$

For molecules of symmetry $\boldsymbol{D_{\infty h}}$ the additional rule *symmetric* $\not\leftrightarrow$ *antisymmetric* here, other than in the case of the infrared spectrum, does not contradict the positive-negative rule (I, 16) so that these molecules, too, exhibit a rotational Raman spectrum.

Substituting $J' = J'' + 2 = J + 2$ and (I, 1) into

$$|\Delta\nu| = F(J') - F(J''),$$

we obtain for the wave-number shifts

$$|\Delta\nu| = (4B - 6D)(J + \tfrac{3}{2}) - 8D(J + \tfrac{3}{2})^3, \tag{I, 17}$$

or, since always $D \ll B$, in very good approximation,

$$|\Delta\nu| = 4B(J + \tfrac{3}{2}). \tag{I, 18}$$

As for diatomic molecules, we have a *series of equidistant lines* (called *S branch* since $\Delta J = +2$) on either side of the exciting line. From the separations of successive lines ($4B$) the rotational constant B and thus the moment of inertia of the molecule may be evaluated. It should be noted that, according to (I, 18), the separation of the first Raman line from the exciting line is $\frac{3}{2}$ times the separation of successive Raman lines.

For molecules of symmetry $\boldsymbol{D_{\infty h}}$, corresponding to the alternation of statistical weights for the odd and even rotational levels (see p. 17f.), an *alternation of intensities* is to be expected. If in such a case the spins of all nuclei with the possible exception of that at the center are zero, alternate lines will be missing. (For a schematic representation see Molecular Spectra I, Fig. 44, p. 95, and Fig. 60, p. 140).

Up to the present time the rotational Raman spectra of only two linear polyatomic molecules, CO_2 and C_2H_2, have been resolved. The Raman shifts observed for CO_2 by Houston and Lewis (458) (mean values of Stokes and anti-Stokes lines) are given in Table 4. Since the distance of the first line from the exciting line is not $\frac{3}{2}$ but $\frac{3}{4}$ of the separation of successive lines (see Molecular Spectra I, p. 140), the observed shifts can be represented by (I, 18) only if it is assumed that the lines with odd J are missing.[8] This is exactly what is to be expected if CO_2 is linear and symmetrical (point group $D_{\infty h}$). Conversely we may conclude from the observed rotational Raman spectrum that *the* CO_2 *molecule is linear and symmetrical*, a conclusion that is corroborated by a great deal of other evidence (see p. 272 and p. 384). The value of the rotational constant B in (I, 18) that best represents all the observed shifts (that is, essentially $\frac{1}{8}$ of the average separation of successive lines) is found to be $B = 0.393_7$ cm^{-1}. The values of the Raman shifts calculated with this B value are given in the last column of Table 4. It is seen that the agreement is within the

TABLE 4. ROTATIONAL RAMAN SHIFTS FOR CO_2 AFTER HOUSTON AND LEWIS (458).

J	Observed shift (in cm^{-1}), mean of Stokes and anti-Stokes lines	Calculated shift, with $B = 0.393_7$ cm^{-1}
0		2.36
2		5.51
4	8.93	8.66
6	11.63	11.81
8	14.84	14.96
10	18.14	18.11
12	21.53	21.26
14	24.60	24.41
16	27.58	27.56
18	30.70	30.71
20	33.60	33.86
22	37.03	37.01
24	40.22	40.16
26	43.39	43.31
28	46.49	46.46
30	49.67	49.61
32	52.96	52.76
34	55.54	55.91

error of measurement. From the rotational constant B, according to (I, 2), one obtains for the moment of inertia $I(CO_2) = 71.1 \times 10^{-40}$ gm cm^2. Since here $I = 2m_O r_{CO}^2$ it follows that the C—O distance in CO_2 is $r_{CO} = 1.157 \times 10^{-8}$ cm (see, however, p. 398 for a more accurate value).

For C_2H_2 Lewis and Houston (576) have found a similar rotational Raman spectrum. However, here alternate lines are not missing but weak, and it is the even lines that are weak in agreement with expectation for a linear and symmetric C_2H_2 (see p. 16). Conversely it follows from the observed Raman spectrum that the C_2H_2 *molecule is symmetric and linear* (see also Chapter IV). The B value obtained is $B = 1.176$ cm^{-1}, from which it follows that the moment of inertia $I(C_2H_2) = 23.80 \times 10^{-40}$ gm cm^2. The internuclear distances cannot be determined from this one figure (see, however, Chapter IV).

The B and I values obtained from rotational Raman spectra are not as accurate as those obtained from infrared rotation-vibration spectra which will be discussed in Chapter IV. Also the values obtained do not refer to the equilibrium position but to the *lowest vibrational state* in which the zero point vibrations take place.

[8] This is the opposite to what is observed for O_2 since for O_2 the electronic ground state is $^3\Sigma_g^-$.

2. Symmetric Top Molecules

As mentioned above, if a molecule has an axis of symmetry this axis coincides with one principal axis of inertia. If a molecule has a three-fold axis (for example a molecule like CH_3Cl), the moments of inertia about any three directions at angles of 120° in a plane perpendicular to the axis of symmetry (for example *aa*, *bb*, *cc* in Fig. 5) are obviously equal. Since the cross section of the momental ellipsoid (see p. 13) with this plane is an ellipse, and since an ellipse has no three equal diameters at angles of 120° except if it degenerates into a circle, it follows that the momental ellipsoid is a rotational ellipsoid, that is, that *a molecule with a three-fold axis is a symmetric top.* In a plane perpendicular to the symmetry axis as in Fig. 5 the moment of inertia about the axis *dd* (and in fact any other axis in the plane) is *exactly* the same as that about the axis *aa*. Similar conclusions apply to molecules with four-fold or higher axes but in general not to molecules with two-fold axes only.

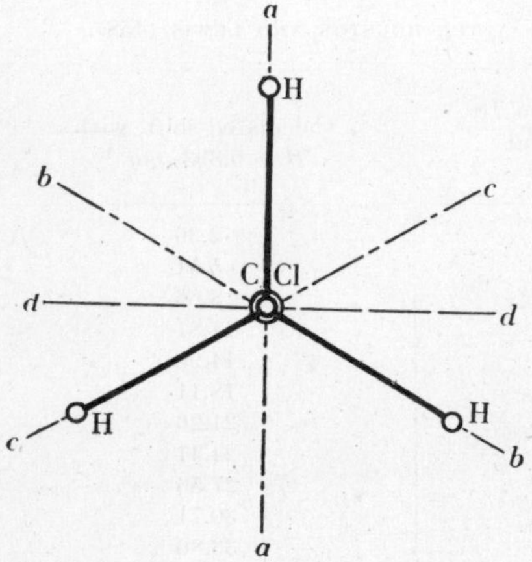

Fig. 5. The CH_3Cl molecule as a symmetric top.—The molecule is projected on a plane perpendicular to the axis of symmetry and through the centre of mass.

In addition to molecules with more than two-fold axes which are symmetric tops because of their symmetry there may be molecules of lower symmetry, or even of no symmetry at all, for which two of the principal moments of inertia happen to have the same value. Such molecules are of course also symmetric tops; but those that are symmetric tops on account of symmetry are more important. In either case we designate the two equal moments of inertia I_B and the third moment of inertia I_A. The axis of this third moment of inertia is usually called the *figure axis* of the symmetric top irrespective of whether it is an axis of symmetry or not.

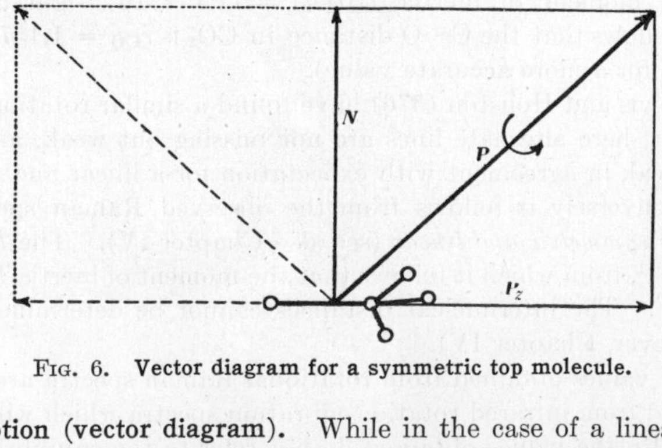

Fig. 6. Vector diagram for a symmetric top molecule.

Classical motion (vector diagram). While in the case of a linear molecule in a Σ electronic state the total angular momentum vector ***P*** (also called ***J*** in the quantum theoretical treatment) is always perpendicular to the internuclear axis, in the case of a symmetric top ***P*** need no longer be perpendicular to the figure axis even if the electronic angular momentum is equal to zero, but has in general *a constant component* $\boldsymbol{P}_z$ *in the direction of the figure axis.* Fig. 6 gives the vector diagram for the angular momenta of a symmetric top. It is essentially the same as for the case of a diatomic molecule in which the rotation of the electrons about the internuclear axis is considered. The only difference is that here $\boldsymbol{P}_z$ is produced by the motion of heavy nuclei and is called $\boldsymbol{K}$, while in diatomic molecules it is produced by the motion of electrons and called $\boldsymbol{\Lambda}$. The figure axis (that is $\boldsymbol{P}_z$) rotates (nutates) about the direction of $\boldsymbol{P}$ which is constant in space. This *nutation* has the frequency $|\boldsymbol{P}|/2\pi I_B$, which is the same as the frequency of rotation of a diatomic molecule of moment of inertia I_B and angular momentum $\boldsymbol{P} \equiv \boldsymbol{J}$ (see Molecular Spectra I, p. 70f.). At the same time the molecule rotates about the figure axis with a frequency

$$\frac{1}{2\pi}\left(\frac{1}{I_A} - \frac{1}{I_B}\right)\boldsymbol{P}_z, \tag{I, 19}$$

not simply $\boldsymbol{P}_z/2\pi I_A$ [for a proof of this see, for example, Teller (836)].

It must be emphasized that the superposition of the two motions, nutation of the figure axis ($\boldsymbol{P}_z$) about $\boldsymbol{P}$ and rotation of the molecule about $\boldsymbol{P}_z$, is of course not simply a rotation of the molecule about the axis of $\boldsymbol{P}$. *$\boldsymbol{P}$ is not fixed in the molecule.* The

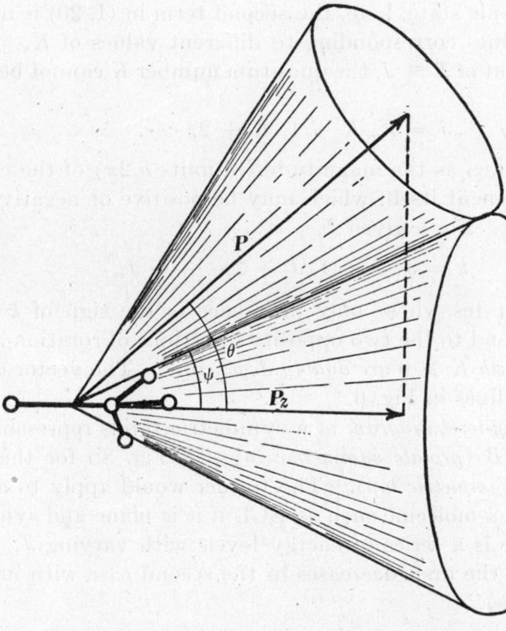

Fig. 7. **Motion of the instantaneous axis of rotation and of the figure axis for a symmetric top.**

molecule rotates about an instantaneous axis whose position in the molecule changes continuously in the following way: Imagine a cone fixed in space with $\boldsymbol{P}$ as axis (Fig. 7) and the center of mass of the molecule as vertex, and with an angle $2(\theta - \psi)$ where θ is the angle between $\boldsymbol{P}$ and $\boldsymbol{P}_z$ and ψ is determined by $\tan\psi = (I_A/I_B)\tan\theta$. Another cone with the figure axis as axis and angle 2ψ may be fixed to the molecule.

If this cone rolls without slipping on the first one with uniform speed it will represent the motion of the molecule.[9] The line of contact is the instantaneous axis of rotation. This axis rotates about the axis of $\boldsymbol{P}$ as does the figure axis and with the same angular velocity. It is seen from Fig. 7 that both this instantaneous axis of rotation and the axis of $\boldsymbol{P}$ (which is fixed in space) continuously change their position with respect to the molecule.

Energy levels. The same formula holds for the quantum theoretical energy levels of a symmetric top as in the case of a diatomic molecule (Molecular Spectra I, p. 125) except that the quantum number Λ has to be replaced by the quantum number K of the component of the angular momentum about the figure axis.[10] Thus we have, for the term values,

$$F(J, K) = BJ(J + 1) + (A - B)K^2, \tag{I, 20}$$

where

$$B = \frac{h}{8\pi^2 c I_B}, \quad A = \frac{h}{8\pi^2 c I_A}. \tag{I, 21}$$

We assume here that there is no electronic angular momentum about the figure axis. Unlike the case of diatomic molecules, A is now of the same order of magnitude as B since both I_A and I_B are moments of inertia produced by heavy nuclei. Furthermore, in a given electronic state, here, the second term in (I, 20) is not constant but can assume various values corresponding to different values of K. However, since $\boldsymbol{P}_z \equiv \boldsymbol{K}$ is the component of $\boldsymbol{P} \equiv \boldsymbol{J}$, the quantum number K cannot be greater than J, or in other words,

$$J = K, K + 1, K + 2, \cdots. \tag{I, 22}$$

K, like Λ, is usually taken as the magnitude (in units $h/2\pi$) of the component of $\boldsymbol{J}$. The value of the component itself, which may be positive or negative, is designated k [see Mulliken (645)]. For a given J,

$$k = J, J - 1, J - 2, \cdots - J. \tag{I, 23}$$

According to (I, 20), states whose only difference is the sign of k have the same energy. They correspond to the two opposite directions of rotation about the figure axis. Thus *all states with $K > 0$ are doubly degenerate.* The vector diagram for $-k$ is indicated by broken lines in Fig. 6.

In Fig. 8a the *energy-level diagram* of a symmetric top is represented for the case $I_A < I_B$, that is $A > B$ (*prolate symmetric top*), in Fig. 8b for the case $I_A > I_B$, that is $A < B$ (*oblate symmetric top*). The former would apply to a molecule such as CH_3Cl, the latter to a molecule such as BCl_3 if it is plane and symmetrical. For every value of K there is a series of energy levels with varying J. For a given J the energy increases in the first, decreases in the second case with increasing K (see the sloping broken lines).

The formula (I, 20) for the energy levels may easily be derived in a semiclassical way [for a more rigorous derivation see Dennison (279) and references quoted there]. In classical mechanics the (kinetic) energy of rotation of a rigid body is

$$E = \tfrac{1}{2}I_x\omega_x^2 + \tfrac{1}{2}I_y\omega_y^2 + \tfrac{1}{2}I_z\omega_z^2 = \frac{P_x^2}{2I_x} + \frac{P_y^2}{2I_y} + \frac{P_z^2}{2I_z}, \tag{I, 24}$$

[9] If $I_A > I_B$ the angle ψ is greater than θ and the moving cone embraces the fixed one.

[10] This K should not be confused with the K used in diatomic molecules as the quantum number of the angular momentum apart from spin.

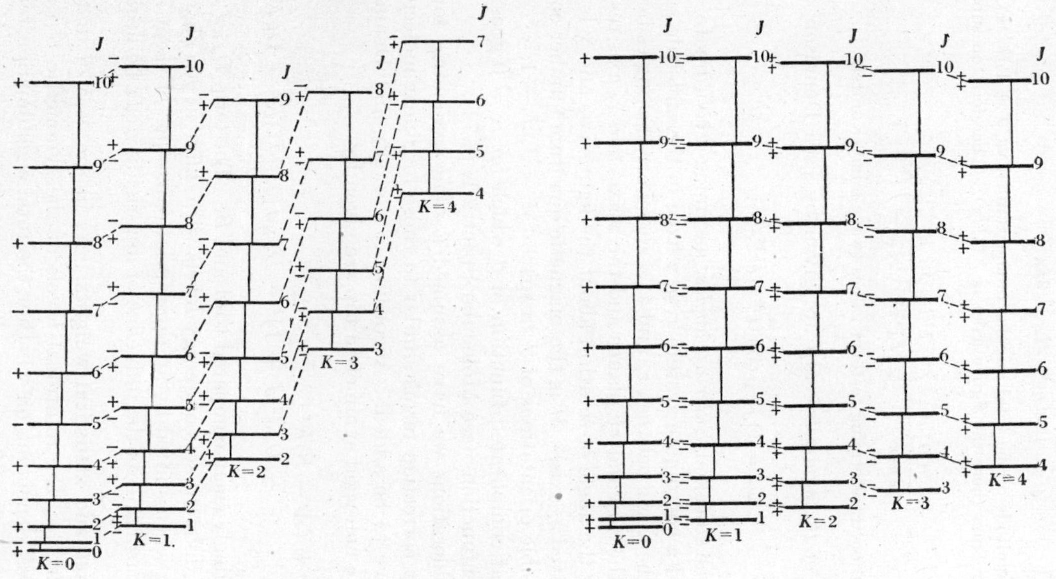

(a) (b)

FIG. 8. **Energy level diagram for symmetric top molecules (schematic); (a) prolate (b) oblate symmetric top.**—In the case of non-planar molecules the signs + and − give the behavior with respect to inversion of only the upper one of each pair of nearly coinciding levels.

where x, y, and z are the directions of the principal axes, and where I_x, ω_x, P_x are moment of inertia, angular velocity, and angular momentum respectively about the x axis and similarly for the other axes. In the present case, $P_z^2 = K^2$, $P_x^2 + P_y^2 = N^2 = J^2 - K^2$, (see Fig. 6) $I_z = I_A$, $I_y = I_x = I_B$. Therefore

$$E = \frac{J^2}{2I_B} - \frac{K^2}{2I_B} + \frac{K^2}{2I_A}. \qquad \text{(I, 25)}$$

In quantum theory the magnitude of the total angular momentum is $\sqrt{J(J+1)}(h/2\pi)$, that of its component in a certain direction (here the figure axis) is $K(h/2\pi)$. Therefore, on substituting in (I, 25), we obtain

$$E = \frac{J(J+1)h^2}{2I_B 4\pi^2} + \left(\frac{h^2}{2I_A 4\pi^2} - \frac{h^2}{2I_B 4\pi^2}\right) K^2,$$

which goes over into (I, 20) on transforming to term values (wave-number units).

The *eigenfunctions of the symmetric top* are given by [see Dennison (279) and Mulliken (645)]

$$\psi_r = \Theta_{JKM}(\vartheta) \cdot e^{ik\varphi} \cdot e^{iM\chi}. \qquad \text{(I, 26)}$$

Here ϑ, φ, and χ are the so-called *Eulerian angles:* assuming a coordinate system x_f, y_f, z_f fixed in space and a coordinate system x, y, z fixed in the molecule such that z is the figure axis, ϑ is the angle between z_f and z, φ and χ are the angles between the line of intersection of the $x_f y_f$ and xy planes and the x axis and x_f axis respectively, that is, φ is essentially the angle of rotation about the figure axis and χ is the angle of rotation about the fixed z_f axis. M is the magnetic quantum number corresponding to the various possible orientations of J in space ($M = J, J-1, \cdots -J$) and $\Theta_{JKM}(\vartheta)$ is a somewhat complicated function of ϑ which for $K = 0$ goes over into the simple rotator eigenfunctions (see Molecular Spectra I, p. 74).

In the above considerations we have assumed a *rigid* symmetric top. For a *non-rigid symmetric top* correction terms similar to those for linear molecules (rotational constant D) have to be added. According to Slawsky and Dennison (795) the energy levels of the non-rigid symmetric top are given by

$$F(J, K) = BJ(J+1) + (A-B)K^2 - D_J J^2(J+1)^2 - D_{JK}J(J+1)K^2 - D_K K^4 \qquad \text{(I, 27)}$$

where the D are exceedingly small compared to A and B. The term $D_{JK}J(J+1)K^2$ has the effect that the different sets of energy levels with different K (Fig. 8) will no longer coincide exactly when shifted by an appropriate amount. Except in cases of extremely accurate measurements the influence of non-rigidity can be neglected.

Symmetry properties and statistical weights. As in the case of diatomic and linear polyatomic molecules, the rotational levels of the symmetric top are either "*positive*" or "*negative*" depending on whether the total eigenfunction remains unchanged or changes sign for a reflection of all particles at the origin. However, in the present case this distinction is much less important as shown by the following considerations. In a *non-planar molecule* reflection of the nuclei at the origin (center of mass) produces a configuration that cannot also be obtained by rotation of the molecule. Therefore there are always two modifications, *a left and a right form* (as in the case of optical isomers), which can be transformed into each other only by passing through a (usually high) potential hill. Each form of the molecule has the same rotational energy levels, since the moments of inertia of the two forms are, of course, the same. Thus each one of the energy levels of the symmetric top, if the molecule is non-planar, is really doubly degenerate. If the potential hill is not infinitely high a slight splitting occurs into two levels whose eigenfunctions contain equal contributions of both the left and the right configuration. One of these levels can be shown to be positive, the other negative. We call this doubling "*inversion doubling*."[11] Because of the fact that *wherever there is a positive level there is, almost coinciding* (and usually not resolved), *also a negative level of the same quantum numbers*, the distinction of positive and negative levels is not very important unless the splitting itself is considered (see Chapter II, section 5d and Chapter IV, section 2a). In Fig. 8 the property + or − for the upper one of the two almost coinciding levels is indicated. The lower one has the opposite symmetry. It should be noted that this property has a different dependence on J and K for the prolate and oblate case. In the former it goes over into that of a diatomic molecule when K is replaced by Λ; but then only the one component set of levels shown occurs. It should also be noted that for $K \neq 0$ there are four sublevels (not drawn separately) for each J value on account of the K degeneracy and the inversion doubling. The two signs given refer to the upper inversion doubling component of each of the two sublevels produced by the K degeneracy.

In the case of a *plane symmetric top molecule* (for example BCl_3) the inversion doubling does not occur, since an inversion of the nuclei can be replaced by a suitable rotation. In this case, which always corresponds to an oblate symmetric top, only one set of energy levels appears with the symmetry properties indicated in Fig. 8b, and here this symmetry property is of greater importance than in the non-planar case.

As long as the symmetric top molecule has *no symmetry*, that is, if two of the principal moments of inertia are only accidentally equal, the nuclear spin increases the statistical weight by the factor $(2I_1 + 1)(2I_2 + 1)(2I_3 + 1)\cdots$, which is the same for all levels. Apart from this constant factor and apart from the inversion doubling, the *statistical weight* is $2J + 1$ for levels with $K = 0$, and $2(2J + 1)$ for levels with $K > 0$.

If the figure axis of the symmetric top is a *p-fold axis of symmetry*, a rotation by $360°/p$ will exchange identical nuclei and therefore *further symmetry properties* of the eigenfunctions arise analogous to the property symmetric or antisymmetric in the nuclei in the case of diatomic and linear polyatomic molecules. This causes differences in the statistical weights of certain levels depending on the spin of the identical nuclei. A more detailed discussion of these relations will be given in Chapter IV. Here we only summarize the results for molecules with a three-fold axis, in so far as they are of importance for the discussion of rotation spectra. Also we shall neglect here the inversion doubling.

For *molecules with a three-fold axis* (point groups $\boldsymbol{C}_3$, $\boldsymbol{C}_{3v}$, $\boldsymbol{C}_{3h}$, $\boldsymbol{D}_3$, $\boldsymbol{D}_{3d}$, $\boldsymbol{D}_{3h}$), in a totally symmetric electronic and vibrational state (ground state) the levels with $K = 0, 3, 6, 9 \cdots$ have a larger statistical weight than those with $K = 1, 2, 4, 5, 7, 8 \cdots$, that is, we have an alternation of the type: *strong, weak, weak, strong, weak, weak, strong,* $\cdots$. This is indicated in Fig. 9, where the "strong" levels are designated by A, the "weak" levels by E (the analogue of a and s for linear molecules). If the spin of the identical nuclei $I = 0$, the levels $K = 1, 2, 4, 5, 7, 8 \cdots$ are entirely missing. For molecules of point group $\boldsymbol{C}_{3v}$, if only three identical atoms are present

[11] This name is not used in the literature but appears descriptive and useful.

and if the nuclear spin $I = \frac{1}{2}$ (for example, NH_3, CH_3Cl) the ratio of the statistical weights of the A and E levels is 2 : 1; if $I = 1$ (for example ND_3, CD_3Cl) the ratio is 11 : 8. This alternation does not depend on the statistics of the nuclei.

For the general case in which the spin of the three identical nuclei is I, Dennison (279) has shown that the weight factors due to the spin are:

For K divisible by 3 (including zero):
$$\frac{1}{3}(2I+1)(4I^2+4I+3)$$
For K not divisible by 3:
$$\frac{1}{3}(2I+1)(4I^2+4I) \qquad \text{(I, 28)}$$

For molecules of point groups D_3, D_{3d}, and D_{3h} there is in addition a *difference of the statistical weights of the levels with even and odd J for* $K = 0$. This alternation does depend on the statistics of the nuclei. For Bose statistics of the identical nuclei,

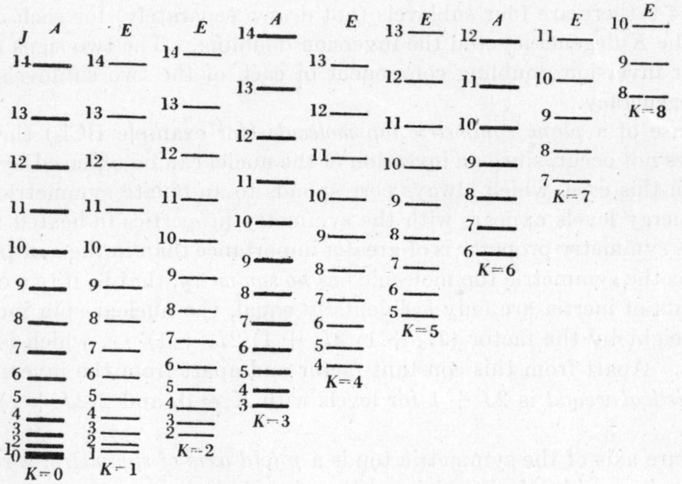

FIG. 9.—**Alternation of statistical weights of the rotational levels for molecules with a three-fold axis.**

the even levels, called A_1 levels (see Fig. 118, p. 408) have a larger statistical weight than the odd levels, called A_2 levels, and the opposite holds for Fermi statistics. If there are only three identical nuclei (for example $CO_3^{=}$, BCl_3), and if $I = 0$ or $I = \frac{1}{2}$, alternate levels are entirely missing; if $I = 1$ the ratio of the statistical weights is 10 : 1; if $I = \frac{3}{2}$ the ratio is 5 : 1; for $I = \frac{5}{2}$ (as for Cl) it is 14 : 5.

In the general case of a molecule of point group D_{3h} with three identical nuclei of spin I, the weight factors due to the nuclear spin for the levels with $K \neq 0$ are exactly the same as those given in (I, 28) for C_{3v}. For $K = 0$ the even rotational levels (A_1 levels) have the weight factor
$$\frac{1}{3}(2I+1)(2I+3)(I+1),$$
the odd levels (A_2 levels)
$$\frac{1}{3}(2I+1)(2I-1)I,$$
if the nuclei follow Bose statistics; the converse holds for Fermi statistics [see Placzek and Teller (701)]. For K divisible by 3 there is for every J value an A_1 and an A_2 level whose weight factors are also given by the preceding expressions. It will be noticed that the sum of these does indeed give the expression for K divisible by 3 in (I, 28).

As in the case of diatomic and linear polyatomic molecules with identical nuclei, here also *levels with different symmetry in the nuclei do not combine with one another;* for example, for $\boldsymbol{C_{3v}}$ we have $A \nleftrightarrow E$. As before, this rule holds very strictly even for collisions. Thus there are two modifications of a gas consisting of molecules of point groups $\boldsymbol{C_3}$, $\boldsymbol{C_{3v}}$, $\boldsymbol{C_{3h}}$, and three modifications (A_1, A_2, and E) for molecules of point groups $\boldsymbol{D_3}$, $\boldsymbol{D_{3h}}$, and $\boldsymbol{D_{3d}}$. In no case, however, have they as yet been separated.

The statistical weights for several more complicated cases of symmetric top molecules have been given by Placzek and Teller (701), Wilson (933) (938) and Schäfer (768) (see also Chapter IV, section 2a).

Thermal distribution of rotational levels. Since the statistical weight and the energy depend now on J and K, the population of the various levels in thermal equilibrium,

$$N_{J,K} \sim g_{J,K}e^{-(E(J,K)/kT)}, \tag{I, 29}$$

cannot be represented as simply as for diatomic or linear polyatomic molecules. For every value of K we have a curve similar to that for diatomic molecules, but because of the increase in energy with increasing K the ordinates are reduced, corresponding to the factor $e^{-[(A-B)K^2hc/kT]}$ and in addition the levels with $J < K$ are missing. This situation is represented in the lower half of Fig. 10a for $B = 2$, $A = 10$ cm^{-1}, $T = 300°$ K., taking account of the factor 2 in the statistical weight for $K \neq 0$ and assuming a very large spin of the nuclei or a molecule which is a symmetric top accidentally and not because of its geometrical symmetry. The curve in the upper half represents (on a different scale) the sum of all curves with different K, that is, it gives the number of molecules with a certain J independent of K. Its maximum is shifted relative to the maximum of the curve for $K = 0$. Figs. 10b and c give similar curves for two actual molecules (NH_3, $B = 9.96$, $A = 6.29$ cm^{-1}, and CH_3Cl, $B = 0.48$, $A = 5.10$ cm^{-1}) but taking account of the difference in statistical weights produced by the spins of the H nuclei (see Fig. 9).

Infrared spectrum. As in the case of linear molecules, an infrared rotation spectrum can appear (as dipole radiation) only if the molecule has a *permanent dipole moment.* If the figure axis of the symmetric top molecule is an axis of symmetry, which is the usual case, the permanent dipole moment of the molecule lies of necessity in this axis. In this case the *selection rules for K and J* are found to be (see below):

$$\Delta K = 0, \qquad \Delta J = 0, \pm 1. \tag{I, 30}$$

That no change of K occurs is, according to the correspondence principle, due to the fact that in the present case the rotation about the figure axis does not change any component of the dipole moment in a fixed direction. In addition, we have the *symmetry selection rules*

$$+ \leftrightarrow -, \qquad + \nleftrightarrow +, \qquad - \nleftrightarrow - \tag{I, 31}$$

and, for point group $\boldsymbol{C_{3v}}$,

$$A \nleftrightarrow E, \qquad A \leftrightarrow A, \qquad E \leftrightarrow E. \tag{I, 32}$$

Selection rules similar to (I, 32) hold for other point groups (see also Chapter IV).

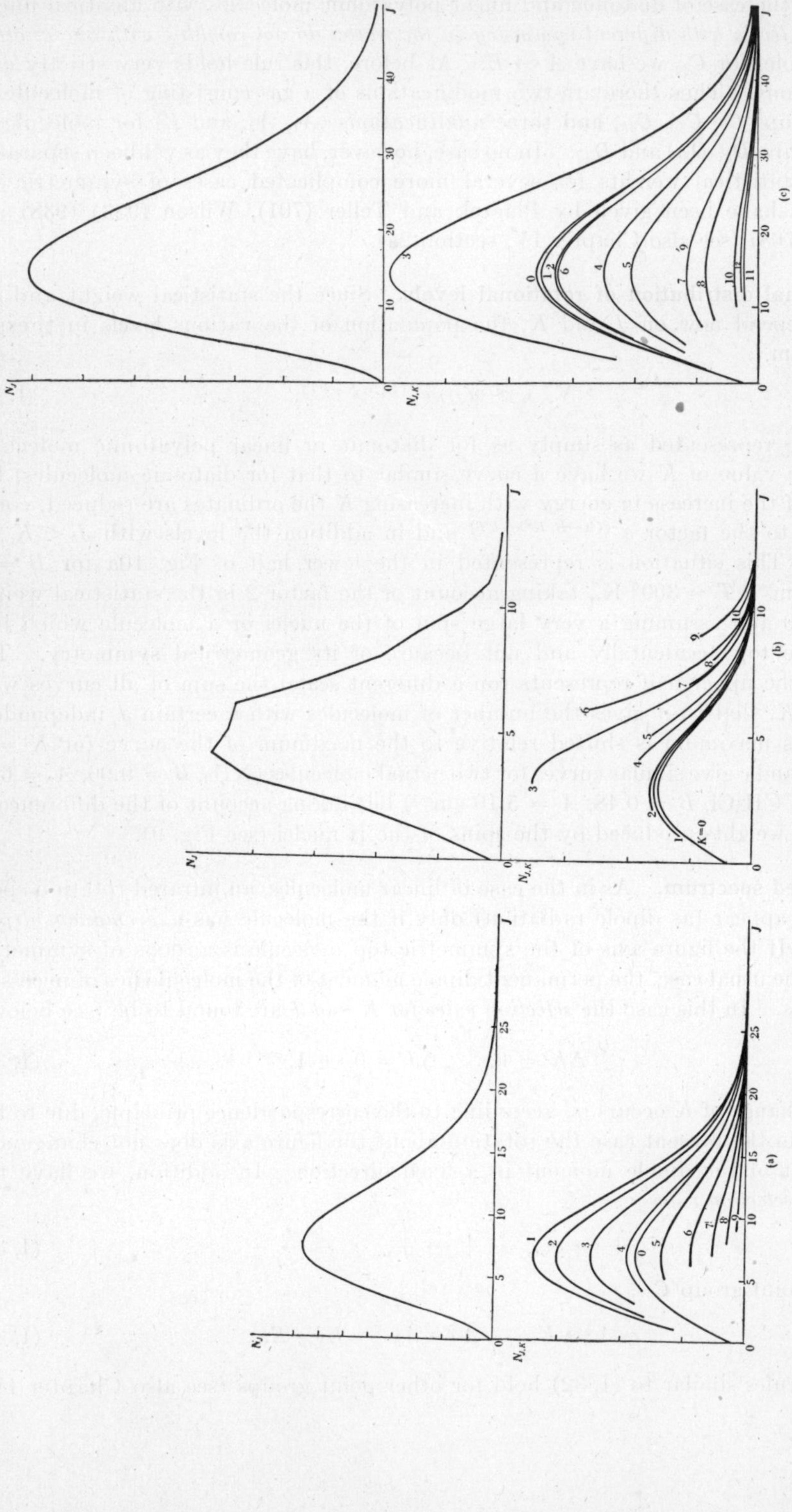

FIG. 10. **Thermal distribution of rotational levels for symmetric top molecules. (a) for $B = 2$, $A = 10\ cm^{-1}$, $T = 300°$ K without alternation of statistical weights. (b) for NH_3: $B = 9.96$, $A = 6.29\ cm^{-1}$, $T = 300°$ K. (c) for CH_3Cl: $B = 0.48$, $A = 5.10\ cm^{-1}$, $T = 300°$ K.—The numbers written on the curves are the K values. The scale of the sum curves above is different from that of the individual curves below.**

The rule (I, 31) can always be fulfilled for non-planar molecules since a + and − level always occur together (inversion doubling; see above).[12]

Since, in the pure rotation spectrum, $\Delta J = 0$ means no transition and $\Delta J = J' - J'' = -1$ does not apply if, as is usual, J' refers to the upper and J'' to the lower state, only $\Delta J = +1$ is of importance; that is, *only neighboring levels with the same K may combine with one another*. In Fig. 8 these transitions are indicated. By using (I, 20) (that is, neglecting centrifugal stretching) we obtain for the positions of the lines in the rotation spectrum, with $K' = K''$ and $J' = J'' + 1 \equiv J + 1$,

$$\nu = F(J', K') - F(J'', K'') = 2B(J + 1). \qquad \text{(I, 33)}$$

This is the same formula as for linear molecules, representing *a simple series of equidistant lines.* The quantum number K drops out entirely. The spectrum is the same as would be obtained for one value of K only; that is, the various sets of levels vertically above one another in Fig. 8 supply the same spectrum. Unlike the case of linear molecules, every line is now obtained in a number of different ways corresponding to the various values of K, the line numbered J in $J + 1$ different ways. Fig. 11a gives a schematic representation of the spectrum. The separation of successive lines is $2B$. If this is measured the *moment of inertia* I_B about an axis perpendicular to the symmetry axis is immediately obtained from (I, 21).

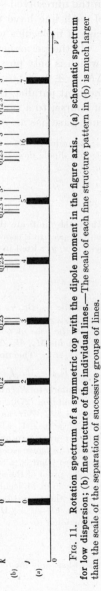

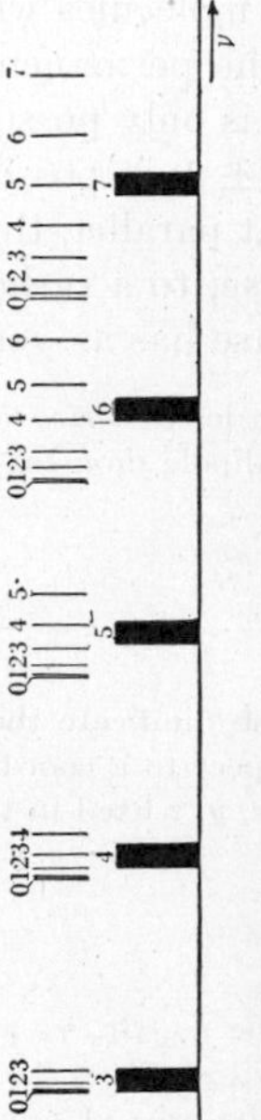

FIG. 11. Rotation spectrum of a symmetric top with the dipole moment in the figure axis. (a) schematic spectrum for low dispersion; (b) fine structure of the individual lines.—The scale of each fine structure pattern in (b) is much larger than the scale of the separation of successive groups of lines.

If *centrifugal stretching* is taken into account, that is, if (I, 27) is used instead of (I, 20), the formula for the spectrum becomes, instead of (I, 33):

$$\nu = 2B(J + 1) - 2D_{KJ}K^2(J + 1) - 4D_J(J + 1)^3. \qquad \text{(I, 34)}$$

According to this formula the rotation lines are no longer exactly equidistant and also the components with different K of each "line" (Fig. 11a) no longer coincide exactly. This splitting is shown in Fig. 11b. However, the splitting would be expected to be exceedingly small and is greatly exaggerated in the figure. If it is not resolved the center of the resultant "line" will show a shift given approximately by $g(J + 1)^3$ where g is not simply $4D_J$. In other words, the unresolved lines of a symmetric top follow the same formula as the rotation lines of linear molecules.

The theoretical *intensity distribution* in the rotation spectrum is different from that for linear molecules since here every "line" consists of a number $(J + 1)$ of components

[12] In the case of plane molecules this rule leads immediately to the result that no transitions with $\Delta K = 0$ are possible. Since according to (I, 30) these would be the only ones possible if the figure axis is a symmetry axis, it follows that no rotation spectrum for a plane symmetric top molecule occurs, in agreement with the fact that it cannot have any permanent dipole moment.

which increases with increasing J. This intensity distribution is essentially given by the upper curves in Fig. 10; the intensity ratio of lines of high J value to lines of low J value is greater than for linear molecules (for which $K = 0$). For low J values the curves have a point of inflexion. There is no intensity alternation in the unresolved series of lines even if there are identical nuclei but the components of each line have the intensity alternation strong, weak, weak, strong, $\cdots$ in the case of molecules with three-fold axes.

If the permanent dipole moment does not lie in the direction of the figure axis (which is only possible for accidentally symmetric tops), in addition to $\Delta K = 0$ also $\Delta K = \pm 1$ is possible, the former corresponding to the component of the dipole moment parallel, the latter to that perpendicular to the figure axis. This gives rise, of course, to a rather more complicated spectrum. We shall not discuss it since no such case has as yet been observed.

In order to *derive the selection rules* for J and K one has to determine the matrix elements of the electric dipole moment $\boldsymbol{M}$ referred to a fixed coordinate system x_f, y_f, z_f (see Molecular Spectra I, p. 16)

$$\begin{aligned} R_{xf} &= \int M_{xf}\psi'\psi''^{*}d\tau, \\ R_{yf} &= \int M_{yf}\psi'\psi''^{*}d\tau, \\ R_{zf} &= \int M_{zf}\psi'\psi''^{*}d\tau, \end{aligned} \tag{I, 35}$$

where the * indicate the conjugate complex eigenfunctions. The components of the dipole moment with respect to a coordinate system fixed in space are related to those with respect to a coordinate system x, y, z fixed in the molecule (z axis = figure axis of the top) by

$$\begin{aligned} M_{xf} &= M_x \cos\alpha_x + M_y \cos\alpha_y + M_z \cos\alpha_z, \\ M_{yf} &= M_x \cos\beta_x + M_y \cos\beta_y + M_z \cos\beta_z \\ M_{zf} &= M_x \cos\gamma_x + M_y \cos\gamma_y + M_z \cos\gamma_z \end{aligned} \tag{I, 36}$$

where the α_x, β_x, γ_x are the angles of the moving x-axis with the three fixed axes and similarly α_y, β_y, γ_y, α_z, β_z, γ_z.

In the case of non-vibrating symmetric top molecules with a permanent dipole moment, the components M_x, M_y, M_z are constant and the eigenfunctions ψ are the symmetric top eigenfunctions ψ_r of (I, 26). Therefore the matrix elements are

$$R_{xf} = M_x \int \cos\alpha_x\psi_r'\psi_r''^{*}d\tau_r + M_y \int \cos\alpha_y\psi_r'\psi_r''^{*}d\tau_r + M_z \int \cos\alpha_z\psi_r'\psi_r''^{*}d\tau_r, \tag{I, 37}$$

and similarly for R_{yf} and R_{zf} with β_x, β_y, β_z, and γ_x, γ_y, γ_z, respectively. Rather involved calculations [see Dennison (278) and Reiche and Rademaker (734)] show that the integrals $\int \cos\alpha_x\psi_r'\psi_r''^{*}d\tau_r$, $\int \cos\alpha_y\psi_r'\psi_r''^{*}d\tau_r$ and similarly those with β_x, β_y, γ_x, γ_y are different from zero only for $\Delta K = \pm 1$ and $\Delta J = 0, \pm 1$, while the integral $\int \cos\alpha_z\psi_r'\psi_r''^{*}d\tau_r$ and similarly those with β_z and γ_z are different from zero only for $\Delta K = 0$ and $\Delta J = 0, \pm 1$. In all practically important cases (dipole moment in figure axis) $M_x = 0$ and $M_y = 0$ and therefore the matrix elements (I, 37) are different from zero (that is, a transition is possible), only when $\Delta K = 0$ and $\Delta J = 0, \pm 1$.

The proof of the symmetry selection rules is similar to the corresponding one for diatomic molecules given in Molecular Spectra I (see also Chapter IV).

For the transitions $\Delta J = +1$, $\Delta K = 0$ the squares of the matrix elements of the transition moment $(M_{xf})^2 + (M_{yf})^2 + (M_{zf})^2$ summed over all possible orientations of J are found to be proportional to

$$\frac{(J+1)^2 - K^2}{(J+1)(2J+1)}.$$

The *intensity of the rotation lines in absorption* is correspondingly given by

$$I(J, K) = C\nu \frac{(J+1)^2 - K^2}{(J+1)(2J+1)} g_{JK} e^{[-F(J,K)hc/kT]}. \tag{I, 38}$$

where g_{JK} is the statistical weight of the lower state. The factor C depends on the permanent dipole moment of the molecule which may thus be determined if the absolute intensity of the absorption lines has been measured [see Foley and Randall (324)].

Of the few *observed* cases of *far infrared absorption spectra*, the molecules NH_3 [Badger and Cartwright (74), Wright and Randall (956), and Barnes (115)], ND_3 [Barnes (115)], and PH_3 [Wright and Randall (956)] have indeed been found to exhibit each a simple series of very nearly equidistant "lines." Fig. 12 shows parts of the spectra of NH_3 and PH_3 under fairly high dispersion.

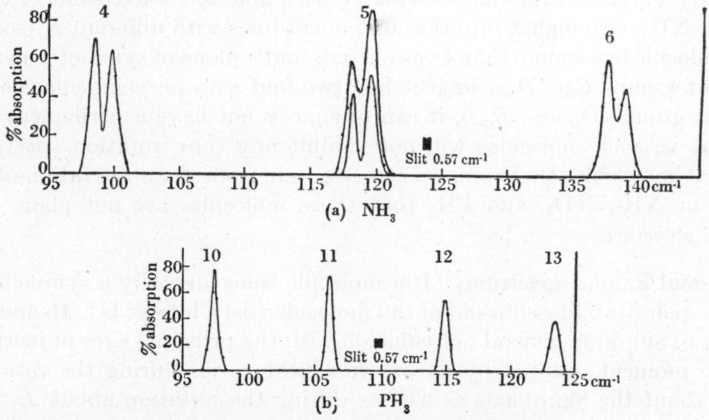

FIG. 12. **Far infra-red absorption spectra of (a) NH_3 and (b) PH_3** [after Wright and Randall (956)].—The absorbing path was 1 cm., the pressure for "line" 4 of (a) was 8 cm., for "line" 5 it was 6.5 and 10 cm. and for "line" 6 it was 7.3 cm., in (b) the gas pressure was 30 cm. throughout.

In Table 5 the observed wave numbers are given for the case of PH_3. In the case of NH_3, Fig. 12a shows that each line is double. This is due to reasons to be discussed in section 5d of the next chapter. The observation of such simple spectra for NH_3, ND_3, and PH_3 proves unambiguously *that these three molecules are symmetric tops* with the dipole moment in the figure axis, that is, that they have a three-fold axis.[13] The series of lines can be represented by the formulae

$$NH_3: \quad \nu = 19.890(J+1) - 0.00178(J+1)^3, \qquad (I, 39)^{14}$$
$$ND_3: \quad \nu = 10.26(J+1) - 0.00045(J+1)^3, \qquad (I, 40)$$
$$PH_3: \quad \nu = 8.892(J+1) - 0.000348(J+1)^3. \qquad (I, 41)$$

TABLE 5. OBSERVED ROTATION SPECTRUM OF PH_3, AFTER WRIGHT AND RANDALL (956).

J[15]	ν (cm^{-1}), observed	ν (cm^{-1}), calculated from (I, 41)
10	97.355	97.35
11	106.10	106.10
12	114.84	114.83
13	123.53	123.53

[13] To be sure, a linear configuration would also give such a simple rotation spectrum. But if such a structure were not already excluded for other reasons, a study of the Raman spectrum and the rotation-vibration spectrum would exclude it.

[14] This is the formula given by Dennison (280). Wright and Randall (956) gave a very slightly different formula.

[15] Wright and Randall's J is that of the upper state, while here, as throughout the book, J refers to the lower state.

The last column of Table 5 gives the values calculated from the formula for PH_3. The coefficient of the linear term, which is very nearly the average distance of successive lines, is 2B. From the coefficients given in (I, 39–41) it follows according to (I, 21) that the *moments of inertia* about axes perpendicular to the symmetry axis are $I_B(NH_3) = 2.815 \times 10^{-40}$, $I_B(ND_3) = 5.457 \times 10^{-40}$, $I_B(PH_3) = 6.296 \times 10^{-40}$ gm cm^2.

Recently Foley and Randall (324) have been able to resolve some of the rotation "lines" of NH_3 with high J into the component lines with different K (see Fig. 11b).

If a molecule has a more than two-fold axis and a plane of symmetry perpendicular to it (point groups $\boldsymbol{C}_{ph}$, $\boldsymbol{D}_{ph}$) or if it has two-fold axes perpendicular to the p-fold axis (point groups $\boldsymbol{D}_p$ and $\boldsymbol{D}_{pd}$), it can obviously not have a permanent dipole moment, that is, such molecules will not exhibit any pure rotation spectrum in the infrared. Conversely, therefore, we can conclude from the observation of a rotation spectrum of NH_3, ND_3, and PH_3 that these molecules are not plane but *have a pyramidal structure.*

Rotational Raman spectrum. If a molecule is accidentally a symmetric top, the axes of the polarizability ellipsoid of the molecule (see Chapter III, 1b and Molecular Spectra I, p. 89) do in general not coincide with the principal axes of inertia; that is, the dipole moment induced by an external field varies during the rotation of the molecule about the figure axis as well as during the nutation about $\boldsymbol{J}$. Therefore, in a light-scattering process (Raman effect) both J and K may change. The selection rules derived by Placzek and Teller (701) are

$$\Delta J = 0, \pm 1, \pm 2; \qquad \Delta K = 0, \pm 1, \pm 2 \qquad \text{(I, 42)}$$

and

$$+ \leftrightarrow +, \qquad - \leftrightarrow -, \qquad + \not\leftrightarrow -. \qquad \text{(I, 43)}$$

The resulting rotational Raman spectrum is rather complicated and will not be discussed here. No actual case of this type is known.

As for the momental ellipsoid (see p. 13), so also for the polarizability ellipsoid the rule holds that an axis of symmetry coincides with one of its axes. Therefore, if the fact that a molecule is a symmetric top is due to its symmetry, one axis of the polarizability ellipsoid coincides with the figure axis, the other two axes being equivalent so that the *polarizability ellipsoid is a rotational ellipsoid* as is the momental ellipsoid. In this case, therefore, a rotation about the figure axis, classically, is not connected with a change of the induced dipole moment and therefore, quantum-theoretically, a change of K cannot be produced by light scattering. Then we have, instead of (I, 42), *the selection rule*

$$\Delta J = 0, \pm 1, \pm 2, \qquad \Delta K = 0, \qquad \text{(I, 44)}$$

with the restriction that $\Delta J = \pm 1$ does not occur for $K = 0$. Of course, (I, 43) remains unchanged. In the pure rotation spectrum $\Delta J = 0$ of course corresponds to the undisplaced line and $\Delta J = J' - J'' = -1, -2$ does not apply (see p. 31). Thus for every value of K we have two series of equidistant lines: $\Delta J = +1$ (R branch) and $\Delta J = +2$ (S branch) on either side of the undisplaced line, with the exception of $K = 0$ for which the R branch cannot occur.[16] However, since for the *rigid* symmetric top the rotational levels for various values of K have exactly the

[16] This corresponds to the results for diatomic molecules with $\Lambda \neq 0$ and $\Lambda = 0$; see Molecular Spectra I, p. 127).

same spacing (see Fig. 8), corresponding lines of the branches for different values of K coincide. Thus only *two branches*, S and R, are to be expected on either side of the undisplaced line. From (I, 20) we obtain for the displacements:

S branches—

$$|\Delta\nu| = F(J+2, K) - F(J, K) = 4BJ + 6B, \qquad J = 0, 1, \cdots \qquad \text{(I, 45)}$$

R branches—

$$|\Delta\nu| = F(J+1, K) - F(J, K) = 2BJ + 2B, \qquad J = 1, 2, \cdots. \qquad \text{(I, 46)}$$

In Fig. 13 the branches are represented schematically. In this figure the Stokes R and S branches have been labeled ^PR and ^OS branches, since they extend to longer wave lengths and thus have the form of P and O branches respectively (see Molecular

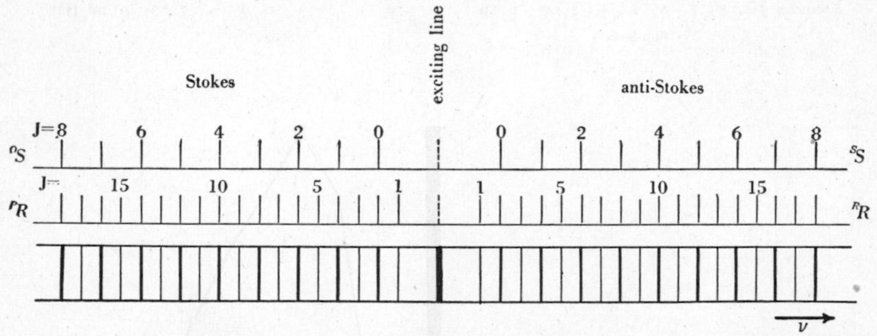

FIG. 13. **Rotational Raman spectrum of a rigid symmetric top (schematic).**—The bottom strip gives the Raman spectrum, the two top strips the identifications of the lines.

Spectra I, p. 273); the anti-Stokes R and S branches have been labeled ^RR and ^SS branches, since they have the form of R and S branches.[17] It is seen from Fig. 13 as well as by comparing equations (I, 45) and (I, 46) that the lines of the R branches with even J coincide with the lines of the S branches. Consequently, there will be an *apparent intensity alternation* which, of course, has nothing to do with the nuclear spin and which is not by any means constant, since the lines of the S branches extend to larger $|\Delta\nu|$ values than do those of the R branches (see below).

The only example of a rotational Raman spectrum of a symmetric top molecule thus far investigated in detail is that of NH_3 [Dickinson, Dillon and Rasetti (287), Amaldi and Placzek (42), and Lewis and Houston (576)]. Fig. 14a is a photometer curve of the Raman spectrum obtained by Lewis and Houston. It shows clearly the strong S branches as well as the weaker R branches (as far as they are not overlapped by the S branches). Table 6 gives the wave numbers of the observed Raman displacements.

If the centrifugal stretching is taken into account, that is, if (I, 27) is used instead of (I, 20), then instead of (I, 45–46) one obtains for the Raman spectrum:

S branches—

$$|\Delta\nu| = (4B - 6D_J)(J + \tfrac{3}{2}) - 4D_{JK}K^2(J + \tfrac{3}{2}) - 8D_J(J + \tfrac{3}{2})^3 \qquad \text{(I, 47)}$$

R branches—

$$|\Delta\nu| = 2B(J+1) - 2D_{JK}K^2(J+1) - 4D_J(J+1)^3 \qquad \text{(I, 48)}$$

[17] The Stokes branches are called P and O branches by many authors. However, this does not appear to be consistent with the international nomenclature as adopted for diatomic molecular spectra (see Molecular Spectra I, p. 96).

The terms with $D_{JK}K^2$ produce a very slight splitting of each "line" into component lines with different K. However, such a splitting has not as yet been resolved. Averaging of the terms with D_{JK} and with D_J produces a slight systematic change of the separations of successive lines and also causes the even R lines no longer to coincide exactly with the S lines. While this also does not lead to an observable splitting, it makes itself felt by the fact that the odd R lines are not exactly half way between adjacent S lines. This can be seen from Table 6, which also shows clearly the systematic

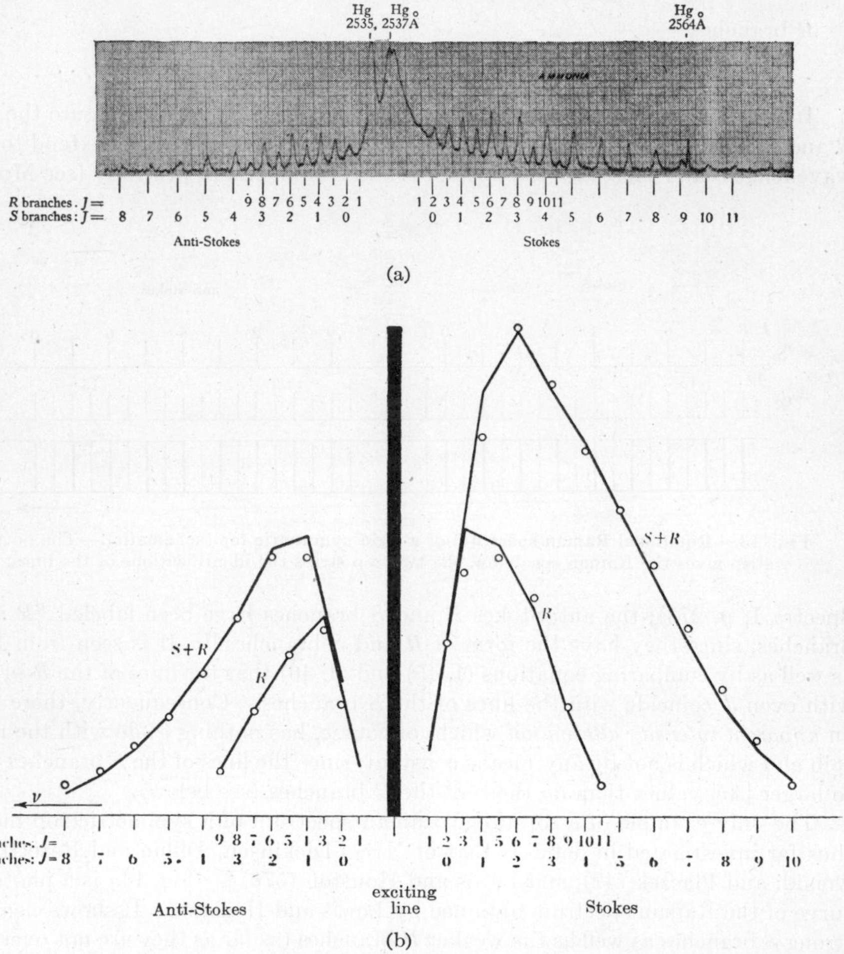

FIG. 14. **Rotational Raman spectrum of gaseous NH_3** [after Lewis and Houston (576)]. **(a) Photometer curve; (b) Intensity distribution.**—The Raman spectrum was excited by the mercury line 2537Å which itself is reduced in intensity on the spectrogram because of absorption by Hg vapor. The circles in (b) represent the observed intensities, the curves represent the calculated values.

change of the separations. Taking account of the correction terms, Lewis and Houston (576) obtained from the observed Raman shifts of Table 6 the value $B = 9.92\ cm^{-1}$, which agrees very satisfactorily with the value $B = 9.945\ cm^{-1}$ obtained from the infrared rotation spectrum (see p. 33). This quantitative agreement as well as the qualitative structure of the spectrum (in particular that only lines with $\Delta K = 0$ occur) shows definitely that NH_3 *is a symmetric top whose figure axis coincides with an axis of symmetry* (three-fold axis).

Placzek and Teller (701) have calculated in detail the *intensity distribution* in the rotational Raman spectrum of a symmetric top molecule on the basis of wave mechanics. According to these calculations, the intensity of the lines of the S branches is given by

$$I_S(J, K) = C\nu^4 \frac{[(J+1)^2 - K^2][(J+2)^2 - K^2]}{(J+1)(J+2)(2J+1)(2J+3)} g_i e^{-E_i/kT}; \tag{I, 49}$$

the intensity of the lines of the R branches is given by

$$I_R(J, K) = C\nu^4 \frac{2K^2[(J+1)^2 - K^2]}{J(J+1)(J+2)(2J+1)} g_i e^{-E_i/kT}; \tag{I, 50}$$

In these equations C is a constant which depends on the difference in polarizability in the direction parallel and perpendicular to the figure axis, ν is the frequency of the particular Raman line, J and K are the quantum numbers of the *lower* state, g_i and E_i are the statistical weight and the energy re-

TABLE 6. OBSERVED ROTATIONAL RAMAN SHIFTS OF GASEOUS NH_3 (AVERAGE OF STOKES AND ANTI-STOKES LINES) AFTER LEWIS AND HOUSTON (576).

J	S branches		R branches	
	$\Delta\nu$ (cm^{-1})	$\Delta(\Delta\nu)$	$\Delta\nu$ (cm^{-1})	$\Delta(\Delta\nu)$
0	59.82			
		39.35		
1	99.17		39.67	
		39.45		20.15
2	138.62		(59.82)[18]	
		39.63		19.66
3	178.25		79.48	
		39.25		19.69
4	217.50		(99.17)	
		39.42		20.08
5	256.92		119.25	
		38.98		19.37
6	295.9		(138.62)	
		38.6		19.76
7	334.5		158.38	
		38.0		19.87
8	372.5[19]		(178.25)	
		39.4		19.13
9	411.9		197.38	
		38.0		20.12
10	449.9		(217.50)	
		36.3		19.0
11	486.2		236.5	

spectively of the *initial* state. Since all lines with different K coincide in the observed Raman spectrum, for a comparison with experiment one has to sum over all values of K for a given J. In doing so one has to take account of the difference in statistical weight introduced by nuclear spin (see p. 27f.). Lewis and Houston (576) have carried out such a comparison. Fig. 14b gives the theoretical curves and the experimental points. In order to evaluate the theoretical intensity distribution, an assumption had to be made about the moment of inertia I_A which can be obtained from other data (see p. 437). However, the intensity distribution is not very sensitive to its exact value. It is seen from Fig. 14b that the intensity distribution reflects in a general way the thermal distribution of the rotational levels (see top curve in the previous Fig. 10b). It should be noted that the R branches fade out at a much smaller distance from the exciting line than the S branches, although of course at about the same J values.

3. Spherical Top Molecules

If a molecule has two or more three-fold or higher-fold axes there are two or more different planes each of which intersects the momental ellipsoid in a circle (see p. 22). Obviously this can only be true if the momental ellipsoid is degenerated into a sphere. Therefore the *moments of inertia about all axes going through the center*

[18] The values in brackets refer to R lines that coincide with S lines.
[19] Overlapped by mercury line.

of mass are exactly equal. The molecule is a *spherical top.* This is the case for all molecules belonging to the cubic point groups, for example CH_4, CCl_4, if they have tetrahedral structure (point group T_d), or SF_6, if it has octahedral structure (point group O_h). Of course a molecule may *accidentally* have all three principal moments of inertia equal, that is, be a spherical top, even when it has lower than cubic symmetry. For example, the NH_3 molecule would be a spherical top, although it has only one three-fold axis (C_{3v}), if the angle between the NH bond and the axis were 52°3′.

Classical motion. For a spherical top, unlike a symmetric top, the instantaneous axis of rotation coincides always with the total angular momentum $\boldsymbol{P}$.[20] Or, in other words, we have a *simple rotation of the molecule* about an axis fixed in space which may have any orientation with respect to the molecule. Any axis fixed in the molecule may be considered as figure axis and describes a simple rotation about $\boldsymbol{P}$. The component of $\boldsymbol{P}$ along any axis fixed in the molecule is a constant. The frequency of rotation about such a "figure axis" is zero according to (I, 19). The cone fixed in space mentioned in discussing the motion of the symmetric top (Fig. 7) shrinks into a line.

Energy levels. The energy levels of the spherical top are obtained from those of the symmetrical top [equation (I, 20)] by putting $I_A = I_B$, that is, $A = B$. We obtain

$$F(J) = BJ(J + 1). \qquad \text{(I, 51)}$$

The energy depends on J only and in exactly the same way as for the simple rotator (linear molecule with $\Lambda = 0$; see Fig. 4). All values of J from 0 up are possible.

The above holds for a *rigid* spherical top. For a *non-rigid* spherical top, in the vibrational ground state, we have to add to (I, 51) a small term $-DJ^2(J + 1)^2$ similar to that for linear molecules (see also Chapter IV, section 3a).

Statistical weights and symmetry properties. Since the spherical top may be considered as a symmetrical top with $A = B$, that is, one in which all levels with the same J but different K (see Fig. 8) coincide, it is clear that, in view of the possible values of K and the double degeneracy for $K \neq 0$ (see above), for the spherical top every level of a given J has a $(2J + 1)$-fold degeneracy in addition to the ordinary $(2J + 1)$-fold space degeneracy. The first degeneracy corresponds to the fact that J may have $2J + 1$ orientations with respect to a fixed direction in the molecule; the second degeneracy corresponds to the fact that J may have $2J + 1$ orientations with respect to a direction fixed in space. Thus the *statistical weight of a level with a given J is* $(2J + 1)^2$, which is quite different from the case of linear molecules.

The factor $(2J + 1)^2$ gives, apart from a constant factor corresponding to the nuclear spin (see p. 27), the complete statistical weight only for a molecule that is accidentally a spherical top or one in which the spins of the identical nuclei are very large. If the molecule is a spherical top by virtue of its symmetry, and if the spins of the identical nuclei are small, the additional factor by which the space degeneracy $2J + 1$ has to be multiplied in order to obtain the total statistical weight is *not* simply $2J + 1$ times the nuclear spin factor. As will be shown in more detail in

[20] This is because $P_x = I_x\omega_x$, $P_y = I_y\omega_y$, $P_z = I_z\omega_z$ (see p. 24), and here $I_x = I_y = I_z$. Therefore the vectors $\boldsymbol{P}$ and $\boldsymbol{\omega}$ have the same direction.

Chapter IV, there are, in the case of tetrahedral molecules (point group T_d) such as CH_4, CD_4, CCl_4, P_4, *three types (species) of rotational levels* called *A*, *E*, and *F* which are analogous to the *a* and *s* levels of linear symmetric molecules and the *A* and *E* levels of molecules with a three-fold axis. It turns out that except for the lowest rotational levels all three species occur for a given *J*.[21] The number of component levels of each type varies in a rather complicated way which can be calculated from group theory, as has been done by Wilson (933) and Maue (605) (see Chapter IV section 3a). Table 7 gives for the first fifteen rotational levels the statistical weights for the cases in which there is only one set of four identical nuclei (as in CH_4, ···) and the spin of these nuclei is $I = 0$ or $I = \frac{1}{2}$ or $I = 1$ or $I = \frac{3}{2}$. In the case of

TABLE 7. STATISTICAL WEIGHTS OF THE ROTATIONAL LEVELS OF TETRAHEDRAL MOLECULES WITH ONE SET OF FOUR IDENTICAL ATOMS OF SPIN *I*.

The statistical weights are given as products whose second factor is $(2J + 1)$.

J	$I = 0$ *A* Total	$I = \frac{1}{2}$ *A* Nuclear quintet	*E* Nuclear singlet	*F* Nuclear triplet	Total	$I = 1$ Total	$I = \frac{3}{2}$ Total
0	1 × 1	5 × 1	0 × 1	0 × 1	5 × 1	15 × 1	36 × 1
1	0 × 3	0 × 3	0 × 3	3 × 3	3 × 3	18 × 3	60 × 3
2	0 × 5	0 × 5	2 × 5	3 × 5	5 × 5	30 × 5	100 × 5
3	1 × 7	5 × 7	0 × 7	6 × 7	11 × 7	51 × 7	156 × 7
4	1 × 9	5 × 9	2 × 9	6 × 9	13 × 9	63 × 9	196 × 9
5	0 × 11	0 × 11	2 × 11	9 × 11	11 × 11	66 × 11	220 × 11
6	2 × 13	10 × 13	2 × 13	9 × 13	21 × 13	96 × 13	292 × 13
7	1 × 15	5 × 15	2 × 15	12 × 15	19 × 15	99 × 15	316 × 15
8	1 × 17	5 × 17	4 × 17	12 × 17	21 × 17	111 × 17	356 × 17
9	2 × 19	10 × 19	2 × 19	15 × 19	27 × 19	132 × 19	412 × 19
10	2 × 21	10 × 21	4 × 21	15 × 21	29 × 21	144 × 21	452 × 21
11	1 × 23	5 × 23	4 × 23	18 × 23	27 × 23	147 × 23	476 × 23
12	3 × 25	15 × 25	4 × 25	18 × 25	37 × 25	177 × 25	548 × 25
13	2 × 27	10 × 27	4 × 27	21 × 27	35 × 27	180 × 27	572 × 27
14	2 × 29	10 × 29	6 × 29	21 × 29	37 × 29	192 × 29	612 × 29
15	3 × 31	15 × 31	4 × 31	24 × 31	43 × 31	213 × 31	668 × 31

$I = 0$ only one species (*A*) of the rotational levels actually occurs (analogous to the case of linear molecules). For $I = \frac{1}{2}$ the three species of rotational levels may be described as *nuclear quintet*, *singlet*, and *triplet* respectively; that is, one species corresponds to a resultant nuclear spin $T = 2$ (all spins parallel), the second to $T = 0$ and the third to $T = 1$. In Table 7 the statistical weights in this case are given separately for the three species. It should be noted that only the levels $J = 0$ and $J = 1$ have but one species of component levels. It can be seen from Table 7 that for large *J* the total statistical weights become approximately proportional to $(2J + 1)^2$.

Just as for linear and symmetric top molecules with identical nuclei, here levels with different symmetry in the nuclei do not combine with one another to any sig-

[21] This is analogous to the fact that for linear molecules with $\Lambda \neq 0$ (Π, Δ ··· states) a symmetric and an antisymmetric level occur for every *J*.

nificant extent; that is,

$$A \leftrightarrow E, \qquad A \leftrightarrow F, \qquad E \leftrightarrow F. \tag{I, 52}$$

Therefore there are *three modifications of tetrahedral molecules with* $I > 0$ analogous to ortho and para modifications of diatomic molecules. They are referred to as the *A, E, and F modifications*, or in the case $I = \frac{1}{2}$ as the *nuclear quintet, singlet, and triplet modifications.*[22] The abundance ratios of these three modifications for high rotational levels approach the values 5 : 2 : 9 for $I = \frac{1}{2}$, 15 : 12 : 54 for $I = 1$, and 36 : 40 : 180 for $I = \frac{3}{2}$. These are independent of the statistics of the nuclei as are also all the individual weights in Table 7.

Thermal distribution of rotational levels. Only for a molecule that is accidentally a spherical top is the population of the rotational levels in thermal equilibrium given by a smooth curve as a function of J:

$$N_J \sim (2J + 1)^2 e^{-[BJ(J+1)hc/kT]}. \tag{I, 53}$$

Such a curve is given in Fig. 15a for $B = 5.25$ cm^{-1} and $T = 300°$ K. It should be noted that unlike the case of diatomic and linear polyatomic molecules the distribu-

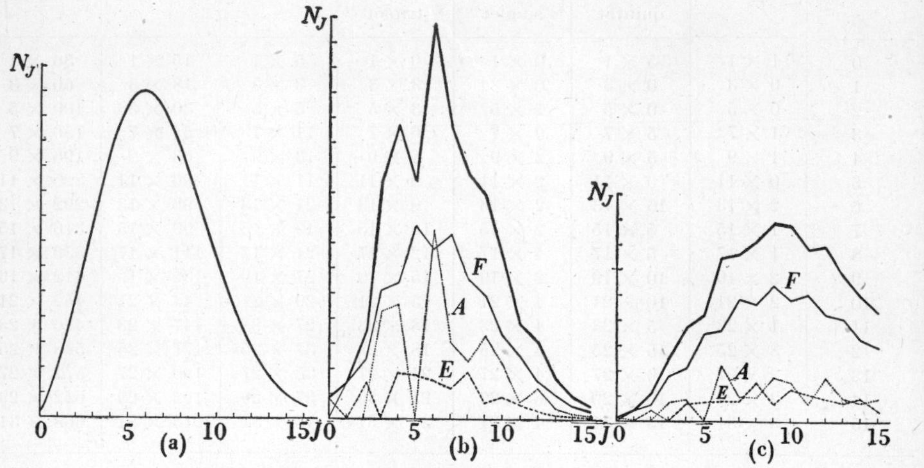

FIG. 15. **Thermal distribution of rotational levels for spherical top molecules (a) accidentally spherical top: B = 5.25 cm.$^{-1}$, T = 300° K; (b) CH_4: B = 5.25 cm.$^{-1}$, T = 300° K; (c) CD_4: B = 2.65 cm.$^{-1}$, T = 300° K.—In (b) and (c) the light curves give the contributions of the three modifications A, E and F of CH_4 and CD_4 separately, the heavy curve gives their sum. The ordinate scale is the same in (a), (b) and (c), the total number of molecules being assumed to be the same.**

tion function does not rise linearly for small J but quadratically, similar to the distribution function for symmetric top molecules (see upper curves in Fig. 10).

If the molecule is a spherical top on account of its symmetry the statistical weights in Table 7 or similar ones in other cases have to be used in order to obtain the thermal distribution. This does not result in a smooth variation. In Fig. 15b the heavy solid "curve" gives the *population of the rotational levels for* CH_4 ($I = \frac{1}{2}$), assuming $B = 5.25$ cm^{-1}, $T = 300°$. The light curves give the contributions of the three

[22] Maue (605) has introduced the names meta-, para-, and ortho-modifications which, however, leads to confusion for $I = \frac{3}{2}$.

modifications. Fig. 15c gives the corresponding curves for CD_4 ($I = 1$, $B = 2.65$ cm^{-1}). For large J values, and therefore always for sufficiently high temperatures, the distribution is very well approximated by (I, 53) (Fig. 15a). Most spherical top molecules other than CH_4, CD_4, SiH_4, SiD_4 and so on, have very small B values, and therefore, at all temperatures at which they are in the gaseous state, essentially only high J values matter, that is, (I, 53) can be used.

If the temperature of CH_4 gas is lowered sufficiently, at first all molecules of modification A go into the lowest A state, which is $J = 0$; all molecules of modification F go into the lowest F state, which is $J = 1$; and all molecules of the modification E go into the lowest E state, which is $J = 2$. Because of the extremely small transition probability between the three types of states [rule (I, 52)], thermal equilibrium, in which practically all molecules are in the state $J = 0$, is established only after a very long time. Once it has been established all molecules belong to the A modification. If now the temperature were raised again, at first the molecules would go only to the higher A states and one would thus have obtained the A modification separately (similar to the production of para-hydrogen). In the A modification (see Table 7) the rotational levels $J = 1$, 2, and 5 do not occur. Up to the present time not even a partial separation of the CH_4 modifications has been obtained experimentally.

Infrared spectrum. As always, a pure rotation spectrum can occur only if the molecule has a permanent dipole moment. If a molecule has an axis of symmetry the permanent dipole moment must necessarily lie in this axis. Therefore, *if a molecule has two or more (non-coinciding) axes of symmetry its permanent dipole moment must be equal to zero.* This is the case for all molecules that are spherical tops on account of their symmetry, that is, molecules that belong to any of the cubic point groups, such as CH_4, SF_6, and others.[23] Therefore they do not exhibit any infrared rotation spectrum. Only if a spherical top molecule is *accidentally* a spherical top can it have a non-zero dipole moment and therefore an *infrared rotation spectrum.* The selection rule for J in this case is simply $\Delta J = 0, \pm 1$ of which only $\Delta J = +1$ is of importance. One obtains a series of equidistant lines as for the symmetric top. The separation of successive lines is $2B$. Actual examples of this case are not known. NH_3 would be such a case if the angle of N—H with the symmetry axis were 52°3′ (which it actually is not).

It should be noted that the structure of the far infrared rotation spectrum alone does not allow a differentiation between a symmetric top molecule and one that is accidentally a spherical top since the position of the rotation lines of the symmetric top does not depend on the moment of inertia about the figure axis. However, in principle a decision between the two alternatives from the rotation spectrum alone would be possible by accurate intensity measurements. A slight difference of intensity distribution in the case of a symmetric top and a spherical top with the same B value arises since for the latter all levels with different K coincide and thus have the same Boltzmann factor, whereas they do not coincide and have different Boltzmann factors for the former.

Rotational Raman spectrum. As mentioned before, an axis of symmetry is always an axis of the polarizability ellipsoid. A three-fold axis of symmetry causes the polarizability ellipsoid to be a rotational ellipsoid and therefore two or more three-fold (or higher-fold) axes cause it to be a *sphere.* This is the case for all molecules that are spherical tops on account of their symmetry (and for no others); that is, for molecules like CH_4, SF_6 if they have cubic symmetry. For any rotation of such a molecule the dipole moment induced by an external field remains unchanged and therefore *no rotational Raman spectrum* appears. Actually a number of authors

[23] They are, of course, not the only molecules with zero dipole moment.

[for example, Bhagavantam (147) and Lewis and Houston (576)] have tried to find the rotational Raman spectrum of CH_4, but without success, as was to be expected from the above consideration. Conversely, the *absence of a rotational Raman spectrum proves that* CH_4 *has tetrahedral symmetry* (point group T_d).

Again, if the molecule is accidentally a spherical top the polarizability ellipsoid is in general *not a sphere* and a rotational Raman spectrum may occur. The selection rule for J is $\Delta J = 0, \pm 1, \pm 2$ and one obtains an R and an S branch of spacing $2B$ and $4B$ respectively on either side of the exciting line, just as for the symmetric top. However, the intensity distribution is slightly different, as in the case of the infrared rotation spectrum (see above). Lewis and Houston (576) in their investigation of the rotational Raman spectrum of NH_3 by careful measurement of the intensities have indeed been able to rule out definitely the possibility that NH_3 is accidentally a spherical top, in agreement with other evidence (see Chapters II and IV).

4. Asymmetric Top Molecules

If a molecule has no three-fold or higher-fold axis, all three principal moments of inertia are in general different and the molecule is an asymmetric top. This is the case for the great majority of polyatomic molecules. For example, the molecules H_2O, C_2H_4, H_2CO are asymmetric tops.

Classical motion. As always, the total angular momentum $\boldsymbol{P}$ of the system remains constant in magnitude and direction during the rotational motion. However there is no longer (as for the symmetric top) a direction in the molecule along which $\boldsymbol{P}$ has a constant component. In other words, there is in general *no axis fixed to the molecule that carries out a simple rotation about* $\boldsymbol{P}$ (as does the figure axis of the symmetric top). The actual (classical) motion can be illustrated in the following way [see Schuler (772)]: By multiplying each radius vector in the momental ellipsoid by $\sqrt{2T}$ where T is the (constant) kinetic energy of rotation, one obtains the so-called *energy ellipsoid*, which is exactly similar to the momental ellipsoid, and like the latter is fixed to the molecule. If this energy ellipsoid, with its center (the center of mass of the molecule) fixed in space, *rolls without slipping on a fixed plane* perpendicular to the total angular momentum $\boldsymbol{P}$, the resultant motion of the molecule (fixed to the ellipsoid) will be one of the possible motions for the particular value of the kinetic energy and the particular direction of $\boldsymbol{P}$. This is illustrated in Fig. 16a in which a is the curve of the successive points of contact on the ellipsoid and b the curve formed by these contact points on the plane. The plane is called the *invariable plane*, since for the free motion of the top it remains unchanged. Which of the infinite number of possible positions the plane has for a given kinetic energy depends on the initial conditions (direction and magnitude of the initial angular velocity).

It is important to realize, as should be clear from Fig. 16a, that the *direction of any one axis of the ellipsoid* does not simply describe a circular cone about $\boldsymbol{P}$ but a more complicated conical surface which is not closed. This same nutation applies to any (two-fold) symmetry axis the molecule might have since such an axis coincides with one of the principal axes. The intersection of the conical surface of nutation, whose vertex is at the center of mass, with the invariable plane is shown in Fig. 16b, c, and d for three different positions of the latter. Only when the ellipsoid is a rotational ellipsoid, that is, when the molecule is a symmetric top, will the axes describe simple circular cones for any position of the invariable plane.

The *instantaneous axis of rotation* is the line (marked **ω** in Fig. 16a)[24] connecting the point of contact of the energy ellipsoid and the invariable plane with the center. It is seen that in the general case it, too, describes a more complicated conical surface both with respect to a fixed coordinate system and one fixed in the molecule.[25]

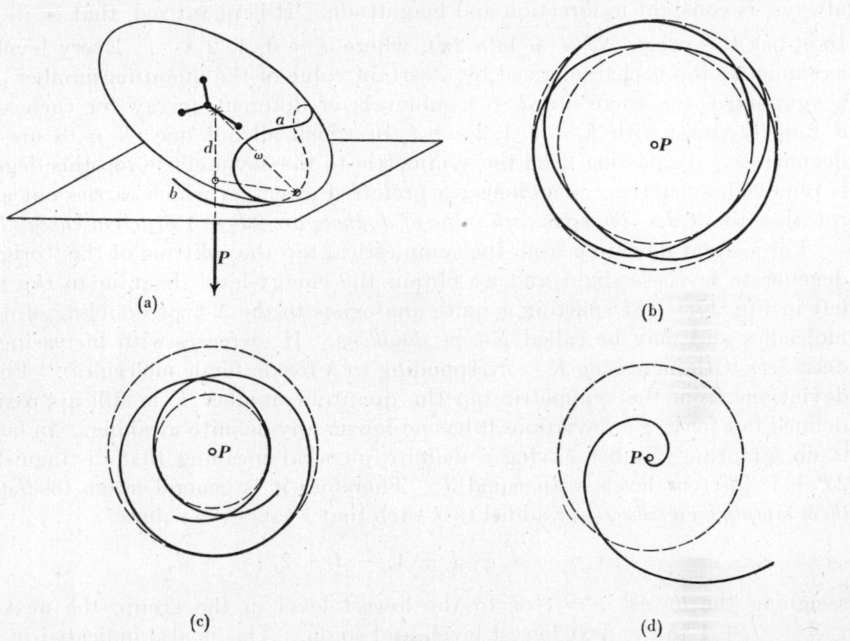

FIG. 16. **Classical motion of the asymmetric top** [after Schuler (772)]. **(a) Energy ellipsoid and invariable plane. (b) Nutation of axis of smallest moment of inertia. (c) Nutation of axis of largest moment of inertia. (d) Nutation of axis of intermediate moment of inertia.**—The three curves (b), (c) and (d) correspond to different positions of the invariable plane such that d is slightly smaller than the particular axis of the energy ellipsoid.

If the distance d of the invariable plane from the center equals the length of any one of the semiaxes of the energy ellipsoid (and only then) we obtain a simple rotation about that principal axis as a special case of the rotational motion of the asymmetric top. If the distance d is somewhat smaller than the largest axis or somewhat larger than the smallest axis of the energy ellipsoid, the motion is somewhat similar to that of the symmetric top: The nutation of the axes takes place between two circular cones of not very different radii as in Fig. 16b and c. But if d is near the length of the intermediate axis the nutation is quite different: the nutation takes place between two opposite circular cones; the point of intersection of each principal axis with the invariable plane moves out in a spiral as in Fig. 16d and periodically comes back; the molecule turns over almost completely during one such "period."

[24] The length of this line is equal to the magnitude of the instantaneous angular velocity which varies in the course of time both in direction and magnitude, whereas for the symmetric top it is constant in magnitude.

[25] This conical surface and the one described by a principal axis, incidentally, in the case of the symmetric top go over into the two circular cones shown in Fig. 7. Or conversely, we may also consider the motion of the asymmetric top as the rolling of a conical surface fixed to the molecule on one fixed in space.

Energy levels. The energy levels of the asymmetric top, according to quantum mechanics, cannot be represented by an explicit formula analogous to that for the symmetric top (I, 20). Therefore let us first try to get a qualitative picture of the energy-level diagram. The total angular momentum $\boldsymbol{J}$ in a given energy level, as always, is constant in direction and magnitude. It is quantized, that is, its magnitude has the values $\sqrt{J(J+1)}(h/2\pi)$, where $J = 0, 1, 2 \cdots$. Every level of the asymmetric top is characterized by a certain value of the quantum number J. For a symmetric top there are $J + 1$ sublevels of different energy for each value of J namely those with $K = 0, 1, 2 \cdots J$, of which all but one ($K = 0$) are doubly degenerate. In passing from the symmetric to the asymmetric top this degeneracy is removed, since there is no longer a preferred direction which carries out a simple rotation about $\boldsymbol{J}$. *Thus for each value of J, there are $2J + 1$ different energy levels.*

For a slight deviation from the symmetrical top the splitting of the "originally" degenerate levels is slight and we obtain the energy-level diagram to the right or left in Fig. 17. The splitting is quite analogous to the Λ-type doubling of diatomic molecules and may be called *K-type doubling*. It increases with increasing J but decreases with increasing K (corresponding to Λ for diatomic molecules). For slight deviations from the symmetric top the quantum number K is still approximately defined but for larger deviations it has no longer any definite meaning. In fact there is no quantum number having a definite physical meaning that distinguishes the $2J + 1$ different levels with equal J. Therefore it is general usage to *distinguish them simply by a subscript τ* added to J such that τ takes the values

$$\tau = -J, -J+1, -J+2, \cdots +J, \qquad \text{(I, 54)}$$

assigning the lowest $\tau = -J$ to the lowest level in the group, the next lowest $\tau = -J + 1$ to the next lowest level, and so on. This is also indicated in Fig. 17 for $J = 0, 1, 2, 3,$ and 4.

Classically, motions with the same total angular momentum are obtained from that described by Fig. 16a if the invariable plane is shifted and the size of the energy ellipsoid changed simultaneously in such a way that $2T/d = |\boldsymbol{P}|$ remains constant. According to quantum theory, of the infinite number of such motions only $2J + 1$ occur, corresponding to $2J + 1$ positions of the invariable plane and $2J + 1$ associated sizes of the energy ellipsoid. For the lowest position of the plane (largest d) and the highest energy (largest $2T$), the energy ellipsoid has its largest axis perpendicular to the plane, that is, we have a simple rotation about the axis of smallest moment of inertia. Even though the highest quantum-theoretical level $\tau = +J$ does not have exactly the highest classical energy, we can conclude that it corresponds approximately to a *rotation about the axis of least moment of inertia* (for the limiting symmetric top that has this axis as figure axis it is the level $K = J$, at the right in Fig. 17). Similarly it can be seen that the lowest level $\tau = -J$ corresponds approximately to a simple *rotation about the axis of largest moment of inertia* ($K = J$ for the limiting symmetric top with this axis as figure axis, at the left in Fig. 17).

Let us call the three principal moments of inertia of an asymmetric top, in order of increasing magnitude, I_A, I_B, and I_C, and let us introduce, similar to the nomenclature for the symmetric top,[26] the quantities

$$A = \frac{h}{8\pi^2 c I_A}, \quad B = \frac{h}{8\pi^2 c I_B}, \quad C = \frac{h}{8\pi^2 c I_C}. \qquad \text{(I, 55)}$$

[26] In reading the literature on the subject it should be noted that some authors use A, B, and C for the moments of inertia, which however is not consistent with the international nomenclature for diatomic molecules.

We may then compare the energy levels of this asymmetric top with two limiting cases, one in which $I_B = I_C$ (prolate symmetric top) and the other in which $I_B = I_A$ (oblate symmetric top). By letting I_B decrease gradually from $I_B = I_C$ to $I_B = I_A$ we can expect to find a continuous change of the energy levels. In the first limiting case ($I_B = I_C$) the energy levels are, according to (I, 20), given by

$$F(J, K) = BJ(J + 1) + (A - B)K^2. \tag{I, 56}$$

These levels are indicated to the extreme right in Fig. 17 but, unlike Fig. 8a, levels with different K are not plotted in separate vertical columns. In the second limiting case ($I_B = I_A$), replacing A by C, we obtain

$$F(J, K) = BJ(J + 1) + (C - B)K^2. \tag{I, 57}$$

These levels are plotted to the extreme left in Fig. 17. This part of Fig. 17 corresponds to Fig. 8b. It should be noted that while at the right for a given J the energy increases with increasing K, at the left it decreases with increasing K since $C \leq B \leq A$.

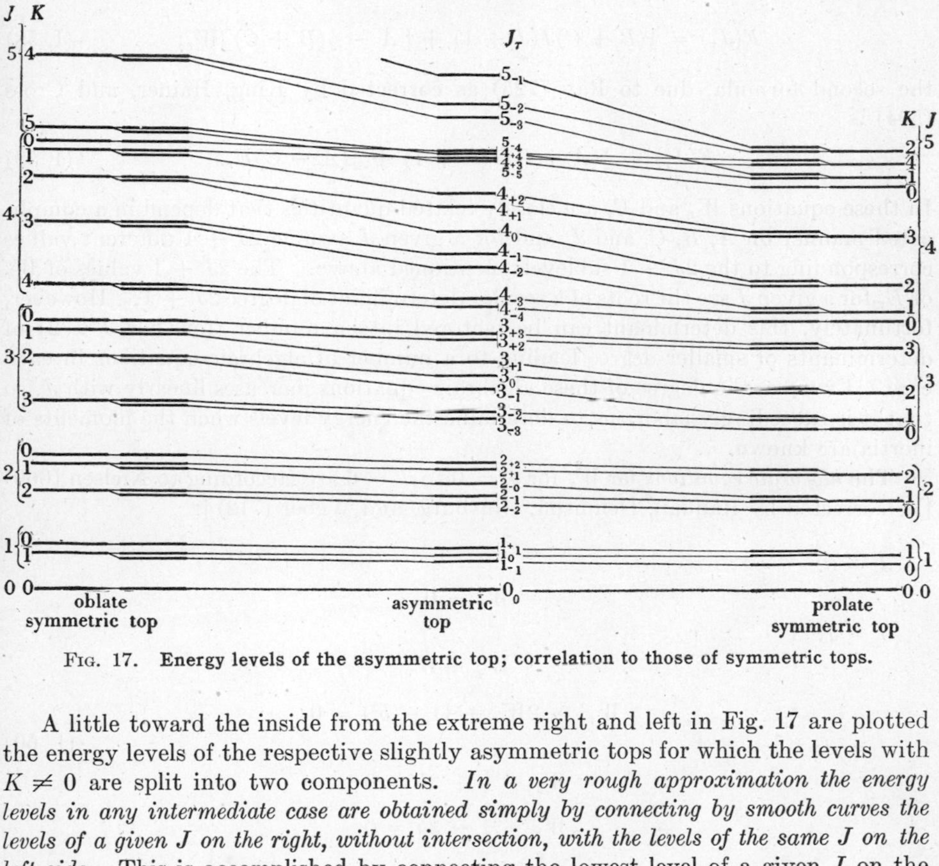

FIG. 17. Energy levels of the asymmetric top; correlation to those of symmetric tops.

A little toward the inside from the extreme right and left in Fig. 17 are plotted the energy levels of the respective slightly asymmetric tops for which the levels with $K \neq 0$ are split into two components. *In a very rough approximation the energy levels in any intermediate case are obtained simply by connecting by smooth curves the levels of a given J on the right, without intersection, with the levels of the same J on the left side.* This is accomplished by connecting the lowest level of a given J on the right with the lowest one of the same J on the left giving the J_{-J} level, the next lowest on the right with the next lowest on the left, giving the J_{-J+1} level, and so on.

It is seen from Fig. 17 that the arrangement of the levels in an intermediate case (obtained as the points of intersection of the connecting lines with a vertical line) is rather irregular.

It should be noted that with increasing J the groups of levels with given J (even in the limiting cases) overlap one another. In fact, in many cases they do so even for very low J values. That they do not overlap in the figure is due only to the fact that $A - C$ has been chosen smaller than either A or C. It can be seen qualitatively from the figure that if B is half-way between A and C [that is, $B = \frac{1}{2}(A + C)$], the energy levels for a given J are symmetrically situated with respect to the central level J_0. This is confirmed by actual calculations.

Fairly elaborate calculations are necessary in order to obtain a *representation of the energy levels* of the asymmetric top *by quantitative formulae.* Such calculations have been carried out by Witmer (946), Wang (912), Kramers and Ittmann (539) (540), Klein (515), Ray (725), and others. The energy formulae may be written in different forms. Two of these have been used in numerical calculations. The first of these energy formulae, due to Wang (912), is

$$F(J_\tau) = \tfrac{1}{2}(B + C)J(J + 1) + [A - \tfrac{1}{2}(B + C)]W_\tau; \tag{I, 58}$$

the second formula, due to Ray (725) as corrected by King, Hainer, and Cross (504) is

$$F(J_\tau) = \tfrac{1}{2}(A + C)J(J + 1) + \tfrac{1}{2}(A - C)E_\tau. \tag{I, 59}$$

In these equations W_τ and E_τ are closely related quantities that depend in a complicated manner on A, B, C, and J, and for a given J assume $2J + 1$ different values corresponding to the $2J + 1$ sublevels mentioned above. The $2J + 1$ values of W_τ or E_τ for a given J are the roots of a secular determinant of degree $2J + 1$. However, fortunately, this determinant can be factored into a number (four for $J > 2$) of determinants of smaller degree leading to a number of algebraic equations in each case. Even so, the degree of these algebraic equations increases linearly with J, so that it is exceedingly laborious to determine the energy levels when the moments of inertia are known.

The *algebraic equations for* W_τ for $J = 0$ to $J = 6$ are, according to Nielsen (660) [as corrected by Randall, Dennison, Ginsburg, and Weber (712)]:

$J = 0$:

$$W_0 = 0$$

$J = 1$:

$$W_\tau = 0$$

$$W_\tau^2 - 2W_\tau + (1 - b^2) = 0$$

$J = 2$: (I, 60)

$$W_\tau - 1 + 3b = 0$$

$$W_\tau - 1 - 3b = 0$$

$$W_\tau - 4 = 0$$

$$W_\tau^2 - 4W_\tau - 12b^2 = 0$$

$J = 3$:

$$W_\tau - 4 = 0$$

$$W_\tau^2 - 4W_\tau - 60b^2 = 0$$

$$W_\tau^2 - (10 - 6b)W_\tau + (9 - 54b - 15b^2) = 0$$

$$W_\tau^2 - (10 + 6b)W_\tau + (9 + 54b - 15b^2) = 0$$

$J = 4$: (I, 60 *continued*)

$$W_\tau^2 - 10(1 - b)W_\tau + (9 - 90b - 63b^2) = 0$$

$$W_\tau^2 - 10(1 + b)W_\tau + (9 + 90b - 63b^2) = 0$$

$$W_\tau^2 - 20W_\tau + (64 - 28b^2) = 0$$

$$W_\tau^3 - 20W_\tau^2 + (64 - 208b^2)W_\tau + 2880b^2 = 0$$

$J = 5$:

$$W_\tau^2 - 20W_\tau + 64 - 108b^2 = 0$$

$$W_\tau^3 - 20W_\tau^2 + (64 - 528b^2)W_\tau + 6720b^2 = 0$$

$$W_\tau^3 - W_\tau^2(35 - 15b) + W_\tau(259 - 510b - 213b^2) - (225 - 3375b - 4245b^2 + 675b^3) = 0$$

$$W_\tau^3 - W_\tau^2(35 + 15b) + W_\tau(259 + 510b - 213b^2) - (225 + 3375b - 4245b^2 - 675b^3) = 0$$

$J = 6$:

$$W_\tau^3 - W_\tau^2(35 - 21b) + W_\tau(259 - 714b - 525b^2) - 225 + 4725b + 9165b^2 - 3465b^3 = 0$$

$$W_\tau^3 - W_\tau^2(35 + 21b) + W_\tau(259 + 714b - 525b^2) - 225 - 4725b + 9165b^2 + 3465b^3 = 0$$

$$W_\tau^3 - 56W_\tau^2 + W_\tau(784 - 336b^2) - 2304 + 9984b^2 = 0$$

$$W_\tau^4 - 56W_\tau^3 + W_\tau^2(784 - 1176b^2) - W_\tau(2304 - 53{,}664b^2) - 483{,}840b^2 + 55{,}440b^4 = 0$$

In these equations the parameter b is an abbreviation for

$$b = \frac{C - B}{2[A - \frac{1}{2}(B + C)]}. \tag{I, 61}$$

In order to find, for example, the energy values for $J = 5$, one would have to solve the quadratic equation and the three cubic equations given under $J = 5$ in (I, 60) and substitute the eleven different W_τ values into (I, 58). The lowest value of W_τ obtained from the set of equations for a given J is W_{-J}, the next lowest W_{-J+1}, and so on. Nielsen (660) has given the equations up to $J = 10$ and Randall, Dennison, Ginsburg, and Weber (712) have given the equations for $J = 11$ as well as a list of typographical errors in Nielsen's equations. Further equations for larger J values have not been worked out but could be derived comparatively easily from Wang's (912) general equations.

The *equations for* E_τ in (I, 59) are obtained from (I, 60) by substituting E_τ/κ for W_τ and $1/\kappa$ for b where

$$\kappa = \frac{2[B - \frac{1}{2}(A + C)]}{A - C}. \tag{I, 62}$$

As B is allowed to vary from C to A, as in Fig. 17—that is, in going from the prolate to the oblate symmetric top—κ goes from -1 to $+1$ whereas b goes from 0 to -1. In the most asymmetric top (for fixed C and A), when B is half-way between A and C, the parameter $\kappa = 0$ while $b = -\frac{1}{3}$. If one wants to calculate the energy levels for a series of asymmetric tops between the two limiting cases, it is preferable to use equation (I, 59), since Ray (725) has shown that, for any given J,

$$E_\tau(\kappa) = E_{-\tau}(-\kappa). \qquad \text{(I, 63)}$$

Thus it is only necessary to calculate the energy values on one side of the most asymmetric case ($\kappa = 0$). Those on the other side are then immediately given by (I, 63) and (I, 59). Recently King, Hainer, and Cross (504) have given an exceedingly useful table of the $E_\tau(\kappa)$ values for $J = 0$ up to $J = 11$ and for $\kappa = 0, -0.1, -0.2, \cdots -1.0$. Fairly accurate energy values for intermediate κ values may be obtained from this table by interpolation.

For a few levels, by combining (I, 58) (I, 60) (I, 61) simple formulae for the energies are obtained:

$$\begin{gathered} F(0_0) = 0; \\ F(1_{-1}) = B + C, \quad F(1_0) = A + C, \quad F(1_{+1}) = A + B; \\ F(2_{-1}) = A + B + 4C, \quad F(2_0) = A + 4B + C, \quad F(2_{+1}) = 4A + B + C; \\ F(3_0) = 4(A + B + C). \end{gathered} \qquad \text{(I, 64)}$$

In the limiting case of the prolate symmetric top ($B = C$, $b = 0$, $\kappa = -1$), as can be verified immediately from the equations (I, 60), W_τ assumes the values, 0, $1^2, 2^2, 3^2 \cdots J^2$, as it should since W_τ is then equivalent to K^2 in (I, 56) for the symmetric top [$\frac{1}{2}(B + C)$ is then equal to B]. A comparison of (I, 59) and (I, 56) shows that E_τ in this case assumes the values $2K^2 - J(J + 1)$ with $K = 0, 1, 2, \cdots J$. In the limiting case of the oblate symmetric top ($B = A$, $b = -1$, $\kappa = +1$), (I, 58) and (I, 59) must go over into (I, 57); that is, W_τ and E_τ become equal to $J(J + 1) - 2K^2$ as can be verified easily for the lowest J values from equations (I, 60).

From Fig. 17 as well as from equations (I, 60) it is seen that for the asymmetric top there are *no simple series of rotational levels* as there are for the symmetric top. However, to a certain approximation, particularly for *slightly asymmetric tops*, simpler formulae giving such *series of levels* may be developed and used to advantage, even though they are not sufficient for an accurate representation [see Mecke (612) (614)]. Since W_τ in the prolate limiting case ($B = C$) is equivalent to K^2, it is to be expected from (I, 58) that for a slightly asymmetric top near this case,

$$F_{\text{prolate}}(J, K) = \tfrac{1}{2}(B + C)J(J + 1) + [A - \tfrac{1}{2}(B + C)]K^2, \qquad \text{(I, 65)}$$

with integral $K \leq J$, will give a fair representation of the energy levels. Similarly, near the oblate limiting case, we would have

$$F_{\text{oblate}}(J, \text{K}) = \tfrac{1}{2}(A + B)J(J + 1) + [C - \tfrac{1}{2}(A + B)]K^2. \qquad \text{(I, 66)}$$

In the first case the two *highest* levels J_{+J} and J_{+J-1} for each J (for which $K = J$; see Fig. 17) are given by

$$\tfrac{1}{2}[F(J_{+J}) + F(J_{+J-1})] = AJ^2 + \tfrac{1}{2}(B + C)J; \qquad \text{(I, 67)}$$

in the second case, the two *lowest* levels ($K = J$) are given by

$$\tfrac{1}{2}[F(J_{-J}) + F(J_{-J+1})] = CJ^2 + \tfrac{1}{2}(A + B)J. \qquad \text{(I, 68)}$$

By taking account of the difference between K^2 and W_τ, Mecke (612) obtained, instead of (I, 67) and (I, 68),

$$\tfrac{1}{2}[F(J_{+J}) + F(J_{+J-1})] = AJ^2 + \tfrac{1}{2}(B + C)J - \tfrac{1}{8}(B - C)J\frac{2J - 1}{J - 1}b\left(1 + \frac{b^2}{4}\right), \quad \text{(I, 69)}$$

$$\tfrac{1}{2}[F(J_{-J}) + F(J_{-J+1})] = CJ^2 + \tfrac{1}{2}(A + B)J - \tfrac{1}{8}(A - B)J\frac{2J - 1}{J - 1}b^*\left(1 + \frac{b^{*2}}{4}\right), \quad \text{(I, 70)}$$

where b is given by (I, 61) and

$$b^* = \frac{B - A}{2[C - \tfrac{1}{2}(A + B)]} \quad \text{(I, 71)}$$

The equations (I, 69) and (I, 70) hold to a fairly good approximation even for strongly asymmetric tops and may then both be applied to the same top.

Wang (912) [see also Mecke (614)] has given formulae for the *splitting of the levels* that are degenerate in the limiting symmetric tops. Instead of reproducing these formulae we give in Fig. 18 a graphical representation of the variation of the energy of the levels with $K = 0$ to 4 as a function of J [after $\tfrac{1}{2}(B + C)J(J + 1)$ has been subtracted] for the case of a very slightly asymmetric (nearly prolate symmetric) top. It is seen that the splitting (K-type doubling) increases with increasing J but much less rapidly for the higher K values. Also it is seen that the average of the levels of the same K deviates from a horizontal, that is, does not follow exactly $\tfrac{1}{2}(B + C)J(J + 1)$.

As is easily verified from equations (I, 60) (without solving them), the *average of all levels with a certain J* follows accurately the formula for the simple rotator with an average rotational constant [see Mecke (612)]:

$$\frac{\Sigma_\tau F(J_\tau)}{2J + 1} = \tfrac{1}{3}(A + B + C)J(J + 1). \quad \text{(I, 72)}$$

This relation is useful in checking calculated levels and the correct assignment of observed levels. Furthermore, by a more careful analysis of the equations (I, 60) Mecke (612) has derived certain *sum rules*, that is, simple formulae for the sum of certain sublevels of a given J. These sum rules are given in Table 8 for values of J up to $J = 6$.[27] They are exceedingly useful in determining the rotational constants from the observed energy levels. The sum rules hold rigorously, just as do equations (I, 60), as long as the asymmetric top is rigid.

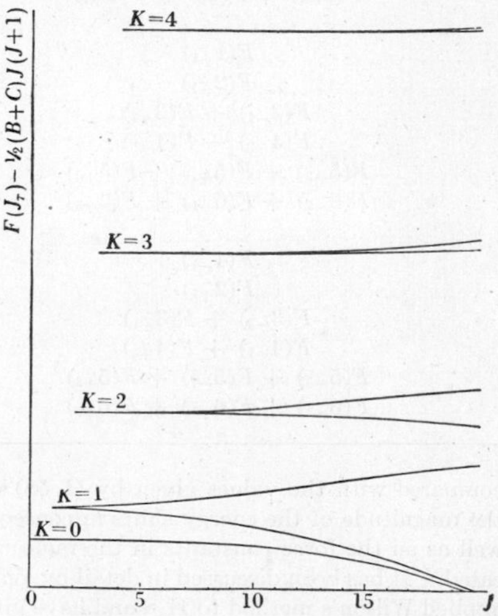

FIG. 18. **Rotational energy of a slightly asymmetric top (b about 0.01) as a function of J [after Dieke and Kistiakowsky (288)].—The term $\tfrac{1}{2}(B + C)J(J + 1)$ is subtracted from the energy, that is, the deviations of the curves from horizontal lines represent the deviations from the levels of the symmetric top.**

Influence of non-rigidity. The actual asymmetric top molecules are not strictly rigid. Therefore, in consequence of the centrifugal forces acting on the nuclei, the molecule is increasingly distorted in the higher rotational levels and this leads to (usually slight) changes of the energy as

[27] They are obtained by applying the rule about the sum of the roots of an algebraic equation to each one of the equations (I, 60).

TABLE 8. SUM RULES FOR THE ASYMMETRIC TOP AFTER MECKE (612).

Designation of levels	Sum	Type of level (see p. 52)
$F(0_0)$	0	$++$
$F(2_{-2}) + F(2_{+2})$	$4(A + B + C)$	
$F(3_0)$	$4(A + B + C)$	
$F(4_{-4}) + F(4_0) + F(4_{+4})$	$20(A + B + C)$	
$F(5_{-2}) + F(5_{+2})$	$20(A + B + C)$	
$F(6_{-6}) + F(6_{-2}) + F(6_{+2}) + F(6_{+6})$	$56(A + B + C)$	
$F(1_0)$	$(A + C)$	$--$
$F(2_0)$	$4B + (A + C)$	
$F(3_{-2}) + F(3_{+2})$	$4B + 10(A + C)$	
$F(4_{-2}) + F(4_{+2})$	$20B + 10(A + C)$	
$F(5_{-4}) + F(5_0) + F(5_{+4})$	$20B + 35(A + C)$	
$F(6_{-4}) + F(6_0) + F(6_{+4})$	$56B + 35(A + C)$	
$F(1_{+1})$	$(A + B)$	$+-$
$F(2_{-1})$	$4C + (A + B)$	
$F(3_{-1}) + F(3_{+3})$	$4C + 10(A + B)$	
$F(4_{-3}) + F(4_{+1})$	$20C + 10(A + B)$	
$F(5_{-3}) + F(5_{+1}) + F(5_{+5})$	$20C + 35(A + B)$	
$F(6_{-5}) + F(6_{-1}) + F(6_{+3})$	$56C + 35(A + B)$	
$F(1_{-1})$	$(B + C)$	$-+$
$F(2_{+1})$	$4A + (B + C)$	
$F(3_{-3}) + F(3_{+1})$	$4A + 10(B + C)$	
$F(4_{-1}) + F(4_{+3})$	$20A + 10(B + C)$	
$F(5_{-5}) + F(5_{-1}) + F(5_{+3})$	$20A + 35(B + C)$	
$F(6_{-3}) + F(6_{+1}) + F(6_{+5})$	$56A + 35(B + C)$	

compared with the values given by (I, 58) or (I, 59). Similarly to the case of the symmetric top, the magnitude of the energy shifts in consequence of the *centrifugal distortion* depends on J and τ as well as on the force constants in the molecule. The exact form of the dependence is quite complicated. It has been discussed in detail by Wilson (936) (937), by Crawford and Cross (242), who have applied Wilson's method to H_2S and have given valuable tables, by Shaffer and Nielsen (780) and by Nielsen (665). Randall, Dennison, Ginsburg, and Weber (712) have given an approximate formula for the deviations, $\delta F(J_{+J})$ and $\delta F(J_{-J})$ of the highest and lowest level of each set with given J from the rigid top levels in the case of a non-linear molecule XY_2 (such as H_2O). They found

$$\delta F = -DJ^4, \tag{I, 73}$$

where the constant D is different for the highest and lowest level and depends on A and C respectively as well as on the force constants in the molecule.

For H_2O the correction for $J = 11$ is as high as 280 cm^{-1} (8.7 per cent of the term value) in the highest level, but is only 4.3 cm^{-1} in the lowest. The large correction for the highest level, in which there is essentially a rotation about the smallest moment of inertia axis (parallel to H—H in H_2O), corresponds to a change of the H—O—H angle from its equilibrium value of 104°27′ to 98°52′ and of the O—H distance from 0.958 to 0.964 Å. The changes introduced by centrifugal distortion are thus quite considerable for the higher rotational levels in light molecules such as H_2O. They are, however, very small for heavier molecules with smaller speeds of rotation.

Symmetry properties and statistical weights. Just as for linear molecules and symmetric top molecules, the total eigenfunction must remain unchanged or can only change sign for an inversion, that is, the rotational levels are *positive* or *negative*. For *non-planar* asymmetric top molecules, just as for symmetric top molecules, each single level of the asymmetric top is really double on account of the possibility of inversion (*inversion doubling*) and always one component is positive, the other negative. For *planar* asymmetric top molecules (H_2O, H_2CO, C_2H_4 $\cdots$) there is no such doubling. It can be shown (see Chapter IV, section 4a) that for them in a totally symmetric vibrational and electronic state the highest level J_{+J} of each set of a given J is $+$, the two next highest are $-$, the two next $+$, and so on (see the first column of signs in Fig. 19).

The above classification according to the symmetry properties of the total eigenfunctions [*over-all species classification* according to Mulliken (645)] is not as frequently used as a classification according to the *symmetry properties of the rotational eigenfunction* only [see Dennison (279)]. For the sake of brevity let us call the three principal axes about which the moments of inertia are I_A, I_B, I_C respectively the *a*, *b*, *c* axes. The rotational eigenfunction ψ_r is a function of the orientation of this system of axes with respect to a fixed coordinate system. $|\psi_r|^2$ gives the probability of finding the various orientations of the axes. Because of the symmetry of the momental ellipsoid, an orientation that differs from a given one by a rotation through 180° about one of the axes must have the same probability. Therefore ψ_r must remain unchanged or only change sign for such a rotation. We call these rotations C_2^a, C_2^b, and C_2^c (the axes are two-fold axes of symmetry of the momental ellipsoid). Thus *the rotational levels of an asymmetric top may be distinguished by their behavior* (+ *or* −) *with respect to the three operations,* C_2^a, C_2^b, C_2^c. Since one of these operations is equivalent to the other two carried out in succession, it is sufficient to determine the behavior with respect to two of them; usually C_2^c and C_2^a are chosen. There are thus *four different types* (*species*) *of levels*, briefly described by $++$, $+-$, $-+$, and $--$, where the first sign refers to the behavior with respect to C_2^c, the second to the behavior with respect to C_2^a.

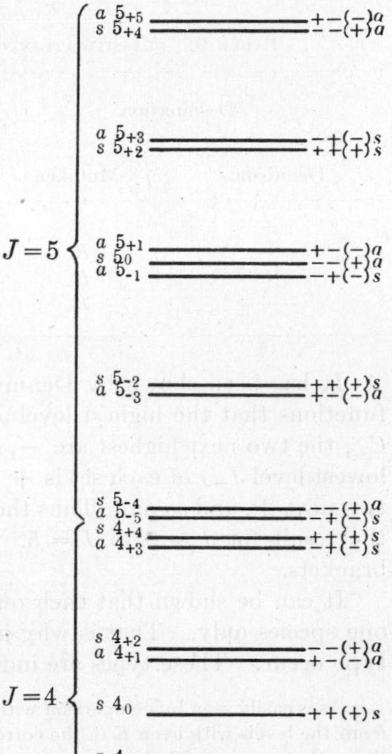

Fig. 19. **Symmetry properties of the rotational levels of asymmetric top molecules for $J = 0$ to 5.**—The designation $++$, $+-$, $\cdots$ applies to any case, the properties *s* (symmetric) and *a* (antisymmetric) given at the right refer to the case that the C_2 lies in the *a* axis (H_2CO, $\cdots$), those at the left to the case that the C_2 lies in the *b* axis (H_2O, $\cdots$). For the sake of clarity the different sets of levels with a given J have been drawn separated. The spacing of the levels within each set corresponds approximately to the most asymmetric case (see p. 48).

The behavior with respect to C_2^b is simply the product of the two signs. The four types are also designated A, B_c, B_a, B_b respectively [Mulliken (645)]. Table 9 summarizes the designation and the behavior of the four types of levels.

TABLE 9. CLASSIFICATION OF THE ENERGY LEVELS OF THE ASYMMETRIC TOP.

Designation		Behavior		
Dennison	Mulliken	C_2^c	C_2^b	C_2^a
$++$	A	$+$	$+$	$+$
$+-$	B_c	$+$	$-$	$-$
$-+$	B_a	$-$	$-$	$+$
$--$	B_b	$-$	$+$	$-$

It has been shown by Dennison (279) from a closer consideration of the eigenfunctions that the highest level J_{+J} for each set of a given J is $+$ with respect to C_2^c, the two next highest are $-$, the two next are $+$, and so on. Furthermore, the lowest level J_{-J} of each set is $+$ with respect to C_2^a, the two next higher are $-$, the two next $+$, and so on. Thus the type of each level can be obtained. Fig. 19 shows the result for $J = 0$ to $J = 5$. The behavior with respect to C_2^b is indicated in brackets.

It can be shown that each one of the algebraic equations (I, 60) gives levels of one species only. That is why in the sum rules of Table 8 only states of the same type occur. These types are indicated in the table.

As is easily seen by comparison with Fig. 17, the levels that are $+$ with respect to C_2^c are derived from the levels with even K in the corresponding oblate symmetric top (to the left in Fig. 17), while those that are $-$ with respect to C_2^c are derived from the levels with odd K. Similarly the levels that are $+$ or $-$ with respect to C_2^a are derived from the levels with even or odd K respectively in the corresponding prolate symmetric top (to the right in Fig. 17). On this basis King, Hainer, and Cross (504) have introduced the designation *ee, oe, eo, oo* for the species $++$, $+-$, $-+$, and $--$ respectively, where the first letter indicates even or odd K_{prolate}, the second even or odd K_{oblate}. It appears that the reverse order would have been preferable, since then $+$ and *e*, $-$ and *o* would be equivalent.

If an asymmetric top molecule has *identical nuclei* the total eigenfunction must be symmetric or antisymmetric with respect to an exchange of any two identical nuclei. However, this leads to a further significant classification only in cases of symmetric molecules in which an exchange of the nuclei can be brought about by a rotation about one of the principal axes, that is, in cases of molecules that have two-fold axes.

We consider first the case of *molecules with one pair of identical nuclei only*, such as H_2O, H_2CO, Cl_2CO and similar molecules of point group C_{2v}. For these molecules, for a totally symmetric vibrational and electronic state (ground state), those rotational levels are *symmetric in the nuclei* that are positive with respect to rotation by 180° about the two-fold axis, and those levels are *antisymmetric* that are negative with respect to the same rotation. For the molecules considered, the two-fold axis of the molecule coincides either with the a or the b axis (least or intermediate moment of inertia). In the first case the levels that are $+$ with respect to C_2^a are symmetric, that is, the $++$ and $-+$ levels, and the levels that are $-$ with respect to C_2^a are antisymmetric, that is, the $+-$ and $--$ levels. This is indicated at the right in Fig. 19. This first case applies, for example, to the molecule H_2CO or to a molecule XY_2 in which the Y—X—Y angle is small (not H_2O). In the second case (two-fold axis coincides with b-axis) the levels that are + or − with respect to C_2^b are symmetric or antisymmetric respectively, that is, the $++$, $--$, and $+-$, $-+$ levels respectively. This is shown at the left in Fig. 19 and applies, for example, to H_2O, NO_2 and other similar molecules with a fairly large angle. It may be noted that in this case, unlike the first, in a set with a given J the levels are alternately a and s, and the lowest (and highest) level is s or a depending on whether J is even or odd.

If the two identical nuclei have *zero spin*, only those levels occur whose total eigenfunction is symmetrical with respect to an exchange of the two nuclei; that is, for a totally symmetric electronic and vibrational state the antisymmetric rotational levels (see Fig. 19) are missing just as for diatomic molecules. If the nuclei have *non-zero spin*, both the symmetric and antisymmetric levels are present but with *different statistical weights*, which again are the same as for the corresponding diatomic molecules and depend on the statistics in the same way. For example, for H_2O, H_2CO the antisymmetric levels have 3 times the statistical weight of the symmetric, for D_2O, D_2CO the statistical weights of the a and s levels are in the ratio 1 : 2. This is, of course, apart from the usual factor $2J + 1$ (which is the same for all $2J + 1$ levels of a given J value). Molecules like HDO, HDCO, of course, do not exhibit any such difference.

TABLE 10. STATISTICAL WEIGHTS OF SYMMETRIC AND ANTISYMMETRIC ROTATIONAL LEVELS IN THE ELECTRONIC AND VIBRATIONAL GROUND STATE OF SOME ASYMMETRIC TOP MOLECULES OF SYMMETRY C_2, C_{2v}, AND C_{2h}.

Molecule	Statistical weight factors [28]	
	Symmetric levels (A)	Antisymmetric levels (B)
H_2O, H_2CO	1	3
D_2O, D_2CO	6	3
cis-, trans-$HDC^{12}{=}C^{12}HD$	15	21
cis-, trans-$HDC^{13}{=}C^{13}HD$	78	66
$D_2C^{12}{=}C^{12}H_2$, $D_2C^{13}{=}C^{13}H_2$, CH_2D_2	15	21
cis-, trans-$HFC^{12}{=}C^{12}FH$, CH_2F_2	10	6
cis-, trans-$HCl^{35}C^{12}{=}C^{12}Cl^{35}H$, $CH_2Cl_2^{35}$	78	66
cis-, trans-$DCl^{35}C^{12}{=}C^{12}Cl^{35}D$, $CD_2Cl_2^{35}$	153	171
NO_2^{16}, $O_2^{16}N{-}NO_2^{18}$	1	0
$O^{16}O^{18}N^{14}{-}N^{14}O^{18}O^{16}$	6	3

As for diatomic molecules, in consequence of the smallness of the nuclear magnetic moments *symmetric and antisymmetric levels do not combine with one another to any significant extent* even by collision, and we have again two modifications of the gases H_2O, H_2CO, and others that may appropriately be called ortho-H_2O, para-H_2O, and so on.

[28] The following values for the nuclear spins have been assumed: $I(\text{H}) = \frac{1}{2}$, $I(\text{D}) = 1$, $I(\text{O}^{16}) = 0$, $I(\text{O}^{18}) = 0$, $I(\text{C}^{12}) = 0$, $I(\text{C}^{13}) = \frac{1}{2}$, $I(\text{F}^{19}) = \frac{1}{2}$, $I(\text{Cl}^{35}) = \frac{5}{2}$, $I(\text{N}^{14}) = 1$.

Since the symmetry of the eigenfunctions is not changed by changing the moments of inertia, the above considerations also supply the symmetry properties *s* and *a* of the levels of a *symmetric* top having two identical nuclei, that is, one that is accidentally symmetric [for example XY_2 with a certain angle; see Mulliken (645)].

For asymmetric top molecules of point groups $\boldsymbol{C_2}$, $\boldsymbol{C_{2v}}$, and $\boldsymbol{C_{2h}}$ with *more than one pair of identical nuclei* that are exchanged by a rotation about the two-fold axis, the same symmetry properties *s* and *a* as in Fig. 19 apply when the C_2 coincides with the smallest or intermediate moment of inertia axis. If it coincides with the largest moment of inertia axis (which is possible in this case but rarely occurs) it is easily seen that the $++$ and $+-$ levels are symmetric, the $-+$ and $--$ levels antisymmetric. Which levels have the greater statistical weight depends on the *resultant statistics* of the set of nuclei that are exchanged by the two-fold rotation (similar to the case of linear molecules with several pairs of identical nuclei) and the magnitudes of the statistical weight factors due to nuclear spin are given by the previous expressions (I, 8) and (I, 9). Table 10 gives the results for a few such molecules. Since in Chapter IV the symmetric and antisymmetric levels of the present case will be called *A* and *B* levels respectively these designations are added in the table.

If a molecule has *three mutually perpendicular two-fold axes* (point groups $\boldsymbol{V}$ and $\boldsymbol{V_h}$) there must be at least *four identical atoms*, and a rotation by 180° about any one of the axes (which coincide with the principal axes of inertia) exchanges at least two pairs of identical nuclei. Since the total eigenfunction can only be symmetric or antisymmetric with respect to such an exchange and since the rotational eigenfunction is positive or negative with respect to the same rotations, we obtain four types of symmetries with respect to the exchanges of nuclei which might be called *ss*, *sa*, *as*, *aa*,[29] where the first letter gives the symmetry with respect to an exchange of nuclei produced by the operation C_2^c, the second with respect to C_2^a. For a totally symmetric electronic and vibrational state it is clear that the $++$ levels are *ss*, the $+-$ levels are *sa*, the $-+$ levels are *as*, and the $--$ levels are *aa*. Wilson (933) calls the *ss*, *sa*, *as*, *aa* levels, *A*, B_1, B_3, B_2 respectively (see Chapter IV, section 4a).

If the spins of the identical nuclei are zero (as for example in O_4 if it forms a rectangle, or in the $C_2O_4^{--}$ ion if it has an ethylene-like structure), the total eigenfunction must be symmetric with respect to an exchange of any two identical nuclei and therefore only the *ss* (*A*) rotational levels occur; that is, the number of rotational levels is reduced very considerably (see the $++$ levels in Fig. 19).

TABLE 11. STATISTICAL WEIGHT FACTORS [30] OF THE ROTATIONAL LEVELS IN THE ELECTRONIC AND VIBRATIONAL GROUND STATES OF SOME MOLECULES OF SYMMETRY $\boldsymbol{V_h}$.

Rotational level	$C_2^{12}H_4$	$C_2^{12}D_4$	$C_2^{13}H_4$	$N_2^{14}O_4^{16}$	$N_2^{15}O_4^{16}$	$N_2^{14}O_4^{17}$ [31]
$++(A)$	7	27	16	6	1	51
$+-(B_1)$	3	18	12	0	0	27
$-+(B_3)$	3	18	24	3	3	39
$--(B_2)$	3	18	12	0	0	27

If the spins of four identical nuclei are different from zero while all other nuclei have zero spin, as in C_2H_4, C_2Cl_4, and similar molecules, the other rotational levels (*sa*, *as*, *aa*) may also occur but with different statistical weights. A group theoretical investigation [Wilson (933); see also Chapter IV] shows that for $I = \frac{1}{2}$ (for example C_2H_4) the weight factors are 7, 3, 3, 3 respectively; for $I = 1$ (for example C_2D_4) they are 27, 18, 18, 18 respectively, independent of the statistics of the nuclei.

If the only identical nuclei with $I \neq 0$ are on one of the symmetry axes, as, for example, in $N_2^{14}O_4^{16}$ if it has symmetry $\boldsymbol{V_h}$, in addition to the *ss* levels only those levels occur that are antisymmetric with respect to those axes on which the nuclei with $I \neq 0$ do not lie. Thus, if the nuclei with $I \neq 0$ lie on the *a*-axis (least moment of inertia) only the *ss* and *as* levels (that is, $++$ and $-+$ in the ground state) occur with weight factors as in the corresponding diatomic molecules. If the spins of the nuclei on as well as those off the axes are different from zero, all four types of levels occur but

[29] These symbols are not used in the literature.

[30] To obtain the complete statistical weight the factors given must be multiplied by $(2J + 1)$.

[31] Assuming $I(O^{17}) = \frac{1}{2}$.

the weight factors of the *sa*, *as*, *aa* levels are no longer equal. The results for two such cases are given in Table 11 which also contains the previously discussed examples.

Of course, there is again *a very strict prohibition of intercombinations between levels of different symmetry in the nuclei.* It holds even for collisions, and therefore, similar to ortho- and para-H_2, there are *four modifications of* C_2H_4 *and* C_2D_4, *and two modifications of* $N_2O_4^{16}$.

Infrared rotation spectrum. As in the other cases discussed before, an infrared rotation spectrum can occur only if the molecule has a *permanent dipole moment.* Therefore molecules of symmetry V_h (such as C_2H_4, N_2O_4) do not exhibit an infrared rotation spectrum but only molecules of symmetry C_{2v}, C_2 such as H_2O, H_2CO, H_2O_2 or molecules of still lower symmetry. If a permanent dipole moment is present we have, as always for dipole radiation, the *selection rule* for J,

$$\Delta J = 0, \pm 1, \tag{I, 74}$$

and the selection rule for the over-all species (+ and −),

$$+ \leftrightarrow -, \quad + \not\leftrightarrow +, \quad - \not\leftrightarrow -. \tag{I, 75}$$

As for symmetric top molecules, the rule (I, 75) is only of importance when the inversion doubling is not negligible. In addition, we have certain symmetry selection rules which depend on the orientation of the permanent dipole moment with respect to the principal axes of inertia:

If the molecule has *no symmetry* the permanent dipole moment will in general not coincide with or be perpendicular to any one of the principal axes. In this case the only restriction is that levels of the same symmetry do not combine with one another:

$$++ \not\leftrightarrow ++, \quad +- \not\leftrightarrow +-, \quad -+ \not\leftrightarrow -+, \quad -- \not\leftrightarrow --. \tag{I, 76}$$

If the molecule has an *axis of symmetry*, the dipole moment lies necessarily in this axis which coincides with one of the principal axes. In this case only those rotational levels can combine with one another whose eigenfunctions have the same symmetry with respect to a rotation by 180° about this axis and opposite symmetry with respect to similar rotations about the other two axes. Therefore, remembering that the symmetry with respect to C_2^b is determined by those for C_2^c and C_2^a, we see that *if the dipole moment lies in the axis of least moment of inertia* (*a* axis), only the transitions

$$++ \leftrightarrow -+ \quad \text{and} \quad +- \leftrightarrow -- \tag{I, 77}$$

can take place. *If the dipole moment lies in the axis of intermediate moment of inertia* (*b* axis), only the transitions

$$++ \leftrightarrow -- \quad \text{and} \quad +- \leftrightarrow -+ \tag{I, 78}$$

can take place. *If the dipole moment lies in the axis of largest moment of inertia* (*c* axis), only the transitions

$$++ \leftrightarrow +- \quad \text{and} \quad -+ \leftrightarrow -- \tag{I, 79}$$

can take place. The prohibition of intercombinations of levels of different symmetry in the nuclei (see above) does not introduce any further restriction of the possible transitions. This can easily be verified if it is noted that the direction of the dipole moment necessarily coincides with that axis about which a rotation exchanges the identical nuclei.

In Fig. 20, for the three orientations of the dipole moment the possible transitions between the rotational levels $J = 3$ and $J = 4$ are indicated. It should be noticed that in addition to transitions between levels belonging to different sets of J values ($\Delta J = \pm 1$) also transitions within one set of given J ($\Delta J = 0$) are possible. It will be seen that in each case the resultant spectrum is quite complicated, particularly since other J values occur at the same time.

In the case of a completely unsymmetrical molecule [selection rule (I, 76)], all three sets of transitions given in Fig. 20 can take place. If the molecule has a plane of symmetry (point group $\boldsymbol{C}_s$) the dipole moment lies in this plane. In this case only two of the three sets of transitions occur, namely those corresponding to the dipole moment in the two principal axes lying in the plane of symmetry.

The proof of the above symmetry selection rules is comparatively simple. The matrix elements R_{xf}, R_{yf}, R_{zf} of the dipole moment are given by the general formula (I, 37) (p. 32)

$$R_{xf} = M_x \int \cos \alpha_x \psi_r' \psi_r''^* d\tau_r + M_y \int \cos \alpha_y \psi_r' \psi_r''^* d\tau_r + M_z \int \cos \alpha_z \psi_r' \psi_r''^* d\tau_r$$

and similarly for R_{yf} and R_{zf}. The moving axes x, y, z may here be taken as the a, b, and c axes respectively. Let us consider the case in which the dipole moment lies in the a-axis ($M_x \neq 0$, $M_y = M_z = 0$). Then

$$\begin{aligned} R_{xf} &= M_x \int \cos \alpha_x \psi_r' \psi_r''^* d\tau_r, \\ R_{yf} &= M_x \int \cos \beta_x \psi_r' \psi_r''^* d\tau_r, \\ R_{zf} &= M_x \int \cos \gamma_x \psi_r' \psi_r''^* d\tau_r, \end{aligned} \qquad \text{(I, 80)}$$

where α_x, β_x, γ_x are the angles of the a-axis with the fixed coordinate axes x_f, y_f, z_f. In order that the transition probability (that is $\boldsymbol{R}$) be different from zero, at least one of the three integrands in (I, 80) must remain unchanged for all transformations of coordinates that transform the system into an indistinguishable one, that is for the three rotations C_2^a, C_2^b, and C_2^c. For the operation C_2^a the angles α_x, β_x, γ_x remain unchanged. Therefore, in order that the integrands remain unchanged for this operation ψ_r' and ψ_r'' must be both + or both −. For the operations C_2^b and C_2^c the a-axis changes its direction into the opposite one and therefore $\cos \alpha_x$, $\cos \beta_x$ and $\cos \gamma_x$ change sign. In order that the integrands remain unchanged for the operations C_2^b and C_2^c the functions ψ_r' and ψ_r'' must therefore have unlike symmetry with respect to these operations. Thus only for transitions $++ \leftrightarrow -+$ and $+- \leftrightarrow --$ can R_{xf}, R_{yf}, R_{zf} and therefore $\boldsymbol{R}$ be different from zero. Only these transitions have a non-zero transition probability. In a similar manner the selection rules for the other cases given above are obtained.

Rigorous *intensity formulae* similar to those for linear and symmetric top molecules have been derived for asymmetric top molecules for J values up to $J = 3$ but not published by Dennison [quoted in (712)]. They become exceedingly complicated for larger J values. The usual way [see Dennison (279) (712)] is to use the formulae for the "nearest" symmetric top, that is, in case of a strongly asymmetric top, for the levels of high τ the prolate, for those of low τ the oblate symmetric top. This approximation is good for all those levels for which the K doubling is fairly small. As a general rule it may be said that large changes of τ are less probable than small changes since in the limiting cases the former would correspond to changes in K of more than one unit which are forbidden.

Very recently Cross, Hainer, and King (249a) have published extensive tables of line strengths based on the rigorous formulae and extended up to $J = 12$. From these tables it appears that the approximations referred to above must be used with caution.

The only asymmetric top molecules whose infrared rotation spectra have been investigated in any detail up to the present time are H_2O and D_2O. Randall, Dennison, Ginsburg, and Weber (712) and Fuson, Randall, and Dennison (343) have measured these spectra with great accuracy and very good resolution. Fig. 21 gives part of the observed H_2O spectrum. It is seen that there are no obvious regularities.

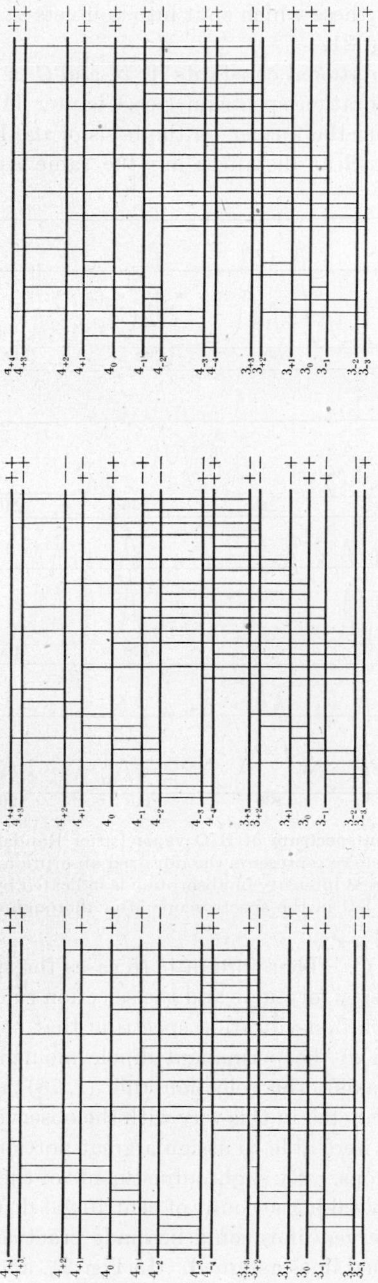

FIG. 20. Allowed transitions between the levels with $J = 3$ and $J = 4$ in the rotation spectrum of asymmetric top molecules.—(a) Dipole moment in a axis. (b) Dipole moment in b axis. (c) Dipole moment in c axis.

On closer examination, however, several series of lines with regularly changing separations can be found. Two of these which split into doublets at long wave lengths are indicated by ○ and × in Fig. 21.

In the case of H_2O the rotational constants A, B, and C in the ground state were known from the rotation-vibration spectrum (see Chapter IV), in the case of D_2O they could be calculated from the former on the basis of the known masses and the assumption that the internuclear distances are the same as in H_2O [see Fuson,

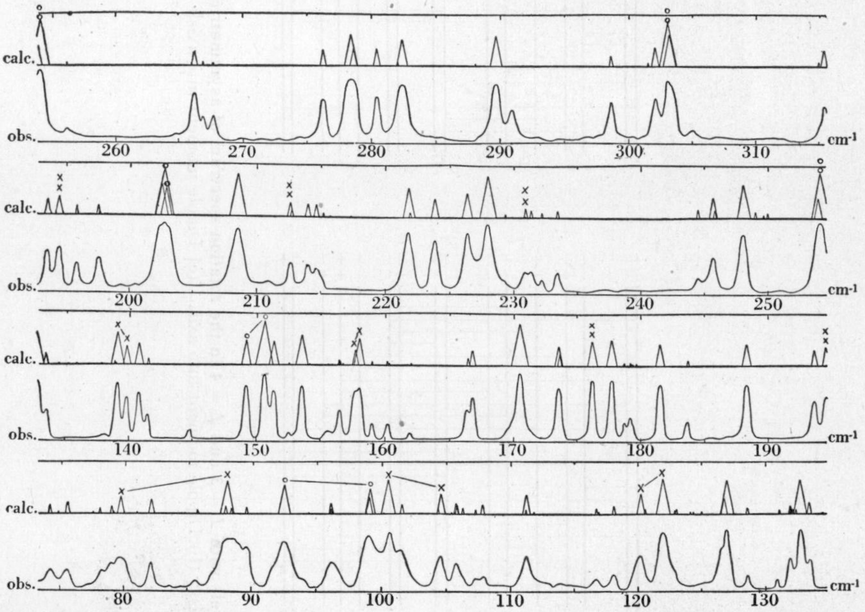

FIG. 21. **Part of the rotation spectrum of H_2O vapor** [after Randall, Dennison Ginsburg and Weber (712)].—The continuous curve represents the infra-red absorption, the small triangles above, the theoretical spectrum. The great intensity of absorption is indicated by the fact that it is entirely due to the small amount of H_2O left in the spectroscope after thorough drying with P_2O_5.

Randall, and Dennison (343)]. Therefore in both cases the spectrum could be predicted on the basis of the energy formulae and the selection rules given above. Since it is also known from the rotation-vibration spectrum that the two-fold axis, which coincides with the direction of the permanent dipole moment, is the intermediate moment of inertia axis (b-axis), the selection rule (I, 78) applies (see Fig. 20b). Comparing the spectrum predicted in this way with the observed, Randall, Dennison, Ginsburg, and Weber (712) were able to assign a great portion of the observed lines to specific rotational transitions. By slight adjustments of the position of the energy levels, that is, essentially, by taking account of centrifugal distortion, the agreement between calculated and observed lines could be made practically perfect and at the same time most of the residual lines assigned. In Fig. 21, above the observed spectrum, the theoretical spectrum derived from the finally adopted energy levels is also shown, where the intensities, indicated by the area of the triangles, are also theoretical (see above). It is seen that the agreement is exceedingly good.

A further very exacting test for the correctness of the assignments is supplied by certain *combination relations*. It can be seen from Fig. 20b that, for example,

$$\nu(4_{+4} - 4_{-2}) - \nu(4_{+4} - 4_{+2}) = \nu(4_{+2} - 3_0) - \nu(4_{-2} - 3_0) = F(4_{+2}) - F(4_{-2}).$$

Similar relations hold for other pairs of levels of the same symmetry. The following is a numerical example from the H_2O spectrum:

$$\nu(7_{-3} - 6_{-5}) = 335.34 \qquad \nu(8_{-3} - 7_{-7}) = 420.10$$
$$\nu(7_{-7} - 6_{-5}) = 139.09 \qquad \nu(8_{-3} - 7_{-3}) = 223.82$$
$$F(7_{-3}) - F(7_{-7}) = 196.25 \qquad F(7_{-3}) - F(7_{-7}) = 196.28$$

Such an excellent agreement has been found for a large number of other pairs of levels, proving the correctness of the analysis. At the same time, *by simply adding up the proper differences of energy levels the energy levels themselves can be determined.*

Of course, an analysis of the H_2O and D_2O rotation spectra would in principle also have been possible if the rotational constants had not been known. One could, for example, have started out from the two series of lines marked in Fig. 21. We have seen above that the two highest and two lowest energy levels of each J follow approximately the formulae (I, 67) and (I, 68) respectively [or more accurately (I, 69) and (I, 70)], and that the doublet separation decreases rapidly with increasing J. Therefore, for the corresponding transitions $J + 1 \leftarrow J$ (for example, $4_{+4} - 3_{+2}$, $4_{+3} - 3_{+3}$ and $4_{-3} - 3_{-3}$, $4_{-4} - 3_{-2}$ in Fig. 20b) we have the approximate formulae

$$\tfrac{1}{2}\{\nu[(J + 1)_{+J+1} - J_{+J-1}] + \nu[(J + 1)_{+J} - J_{+J}]\} = A + \tfrac{1}{2}(B + C) + 2AJ + \cdots, \quad (I, 81)$$
$$\tfrac{1}{2}\{\nu[(J + 1)_{-J-1} - J_{-J+1}] + \nu[(J + 1)_{-J} - J_{-J}]\} = C + \tfrac{1}{2}(A + B) + 2CJ + \cdots; \quad (I, 82)$$

that is, we have two series of doublets, one with an approximately constant spacing $2A$, the other with a spacing $2C$. The doublet splitting should decrease rapidly toward higher frequencies. This is exactly the characteristic of the two series marked in the spectrum Fig. 21. The separations of the two longest wave-length doublets in the two series are 54.0 and 18.8 cm^{-1} respectively. From these one would obtain the approximate A and C values 27.0 and 9.4 cm^{-1} respectively. They agree, within the approximation of the formulae (I, 81) and (I, 82), with the accurate values 27.81 and 9.28 respectively known from the rotation-vibration spectrum. If the latter had not been known one could have used the former values as an initial approximation for calculating the energy levels. The rotational constant B would in that case have been obtained from the relation $I_C = I_B + I_A$, which holds for every plane molecule (see, however, p. 461).

According to the previous discussion, for H_2O the $++$ and $--$ levels (which have even τ) are symmetric, the $+-$ and $-+$ levels (which have odd τ) are antisymmetric in the nuclei; that is, their statistical weights are in the ratio 1 : 3. Therefore in the two series of doublets discussed here (see Fig. 19) *alternately the high- and the low-frequency component should have three times the intensity of the other component.* As far as the doublets are resolved it can be seen from Fig. 21 that this intensity relation is strikingly fulfilled.

The best way to derive the final values of the *rotational constants* A, B, C from the observed spectrum would appear to be by means of Mecke's sum rules (Table 8), after these have been corrected for the influence of centrifugal stretching terms. It is not necessary to know all the energy levels; but it is easily possible on the basis of the sum rules to express certain sums of combination differences in terms of the rotational constants. However, such a re-evaluation of the rotational constants from the far infrared spectrum has not as yet been carried out.

Raman spectrum. Even for an asymmetric top molecule of the highest symmetry (V_h), and *a fortiori* for one of lower symmetry, the polarizability ellipsoid is in general not a sphere, and therefore in general an asymmetric top molecule has a *rotational Raman spectrum.* The selection rule for J is [see Placzek and Teller (701)]

$$\Delta J = 0, \pm 1, \pm 2. \quad (I, 83)$$

If the molecule has no symmetry, and consequently the axes of the polarizability ellipsoid do not coincide with those of the momental ellipsoid, transitions between levels of any of the symmetry types (++, +−, −+, −−) may occur. However, if the molecule has the symmetry C_{2v}, D_2, or V_h, the axes of the two ellipsoids do coincide and only levels of the same symmetry can combine with each other

$$++\leftrightarrow++,\ +-\leftrightarrow+-,\ -+\leftrightarrow-+,\ --\leftrightarrow--. \qquad \text{(I, 84)}$$

Even in this case the rotational Raman spectrum would be very complicated and such a case has not yet been analyzed or even resolved. The only case of an asymmetric top rotational Raman spectrum that has been studied in any detail is that of ethylene (C_2H_4), which is almost a symmetric top [Lewis and Houston (576)]. Here the quantum number K is approximately defined, the selection rule $\Delta K = 0$ holds approximately, and corresponding lines with different K fall nearly together. This explains why an apparently simple S branch is observed on either side of the exciting line. R branches are not observed, apparently because of lack of intensity (compare the intensity of the R branches in the NH_3 Raman spectrum, Fig. 14). An intensity alternation does not occur, as can easily be understood from the previous discussion (p. 52f.) by going over to the limiting case $B \approx C$. The average distance of successive lines in the two branches is found to be 3.68 cm^{-1}, from which it follows [see formulae (I, 45) and (I, 65)] that $\frac{1}{2}(B + C) = 0.92_0$ cm^{-1}, that is, the average moment of inertia about an axis perpendicular to the C—C axis is 30.4×10^{-40} gm cm^2 (for a more accurate determination of this quantity, see Chapter IV, section 2b).

CHAPTER II

VIBRATIONS, VIBRATIONAL ENERGY LEVELS, AND VIBRATIONAL EIGENFUNCTIONS

In the preceding chapter we have considered the rotational motion of polyatomic molecules, assuming that no vibrational motion takes place at the same time, that is, that there are no periodic changes of the internuclear distances. In this chapter we shall consider the *vibrational motion, assuming that no rotation of the whole molecule takes place.* As before, we assume that the molecule is in a fixed electronic state which does not change during the motion.

1. Nature of Normal Vibrations: Classical Theory

We discuss first the vibrational motion of a molecule as it would be according to classical mechanics. We shall see that in this way, as in the case of diatomic molecules, we obtain a fair approximation to the wave mechanical treatment. The classical treatment has the advantage of being more easily visualized.

Vibrational degrees of freedom. If we want to describe the motion of the nuclei in a polyatomic molecule we may choose the ordinary Cartesian coordinates x_k, y_k, z_k of each nucleus k referred to a fixed coordinate system. Then, if there are N nuclei we need $3N$ coordinates to describe their motion: there are $3N$ degrees of freedom. However, if we want to study the vibrational motion of the system we are not interested in the translational motion of the system as a whole, which is described completely by the three coordinates of the center of mass (the three translational degrees of freedom). Therefore $3N - 3$ coordinates are sufficient to fix the *relative* positions of all N nuclei with respect to the center of mass. (The remaining three coordinates may be determined by the condition that the center of mass is at the origin, that is, $\sum m_k x_k = 0$, $\sum m_k y_k = 0$, $\sum m_k z_k = 0$). The motion relative to the center of mass still includes a rotation of the system. The rotation alone, that is, the orientation of the system (considered as rigid) in space, may be described in general by three coordinates (for example, the two angles with two coordinate axes that fix a certain direction in the molecule and the angle of rotation about that direction). Thus $3N - 6$ coordinates are left for describing the relative motion of the nuclei with fixed orientation of the system as a whole,[1] that is, the vibrational motion; or in other words we have *$3N - 6$ vibrational degrees of freedom.* However, for linear molecules two coordinates (for example the two angles of the internuclear axis with two of the coordinate axes) are sufficient to fix the orientation and therefore we have *for linear molecules $3N - 5$ vibrational degrees of freedom.*

[1] More specifically, a fixed orientation of the system as a whole is given by the condition that the angular momentum is zero, that is,

$$\Sigma\, m_k\left(y_k\frac{dz_k}{dt} - z_k\frac{dy_k}{dt}\right) = 0, \qquad \Sigma\, m_k\left(z_k\frac{dx_k}{dt} - x_k\frac{dz_k}{dt}\right) = 0, \qquad \Sigma\, m_k\left(x_k\frac{dy_k}{dt} - y_k\frac{dx_k}{dt}\right) = 0.$$

As a simple example, let us consider a triatomic molecule XYZ. If the molecule is not linear the relative position of the nuclei is given by the three internuclear distances, XY, YZ, XZ; that is, there are $3(=3N-6)$ vibrational degrees of freedom. If the molecule is linear the relative position of the nuclei is given by two internuclear distances XY and YZ and two angles, the angle XYZ and the angle of the plane formed by XYZ in the displaced position against a fixed plane through the (undisplaced) internuclear axis; that is, we have $4(=3N-5)$ vibrational coordinates or degrees of freedom.

As we shall see, the number of vibrational degrees of freedom gives the number of fundamental vibrational frequencies of the molecule, or in other words, the number of different "normal" modes of vibration.

Vibrations of a mass suspended by an elastic bar. Imagine a heavy mass m to be suspended by a homogeneous elastic bar of rectangular cross section as shown in two views in Fig. 22a. If the mass is displaced slightly from its equilibrium position in the x direction and then left to itself it will carry out simple harmonic oscillations in this direction with a frequency

$$\nu_x = \frac{1}{2\pi}\sqrt{\frac{k_x}{m}},$$

where k_x is the force constant in the x direction ($-k_x x$ = restoring force for displacement x). If the mass is displaced in the y direction and released it will similarly carry out simple harmonic oscillations in the y direction with a frequency

$$\nu_y = \frac{1}{2\pi}\sqrt{\frac{k_y}{m}},$$

where k_y is the force constant for the y direction. Unless the rectangular cross section of the bar degenerates into a square, ν_x is different from ν_y. If, however, the mass is displaced in a direction different from x or y, for example to A, it will *not* carry out a simple oscillation in the AOB plane but a very complicated type of motion, a so-called *Lissajous motion*, such as the one given in Fig. 22b. The reason for this is that the restoring force F whose components are $-k_x x$ and $-k_y y$ is not directed toward the origin (see Fig. 22a), since $k_x \neq k_y$. However, the components x and y of the motion are simple harmonic, as before:

$$x = x_0 \cos 2\pi\nu_x t, \qquad y = y_0 \cos 2\pi\nu_y t, \tag{II, 1}$$

where x_0, y_0 are the coordinates of the initial position at A. The complicated *Lissajous motion* (Fig. 22b) *is the superposition of two simple harmonic motions* (of different frequency) *at right angles to each other*. The simple motions into which the complicated motion can be resolved are the so-called *normal vibrations* or *normal modes* of the mass, the x and y coordinates are the *normal coordinates*.

If ν_x/ν_y is irrational, there is no time after which the motion repeats itself. Rather, in the course of time, (assuming that there is no friction) the path of the mass will cover uniformly the whole rectangle whose diagonal is AOB (see Fig. 22b). However, if ν_x/ν_y is rational, after a certain time the path will go back into itself and a definite curve is obtained, such as the one in Fig. 22c, for $\nu_x/\nu_y = 5/3$ This curve is then retraced over and over again.

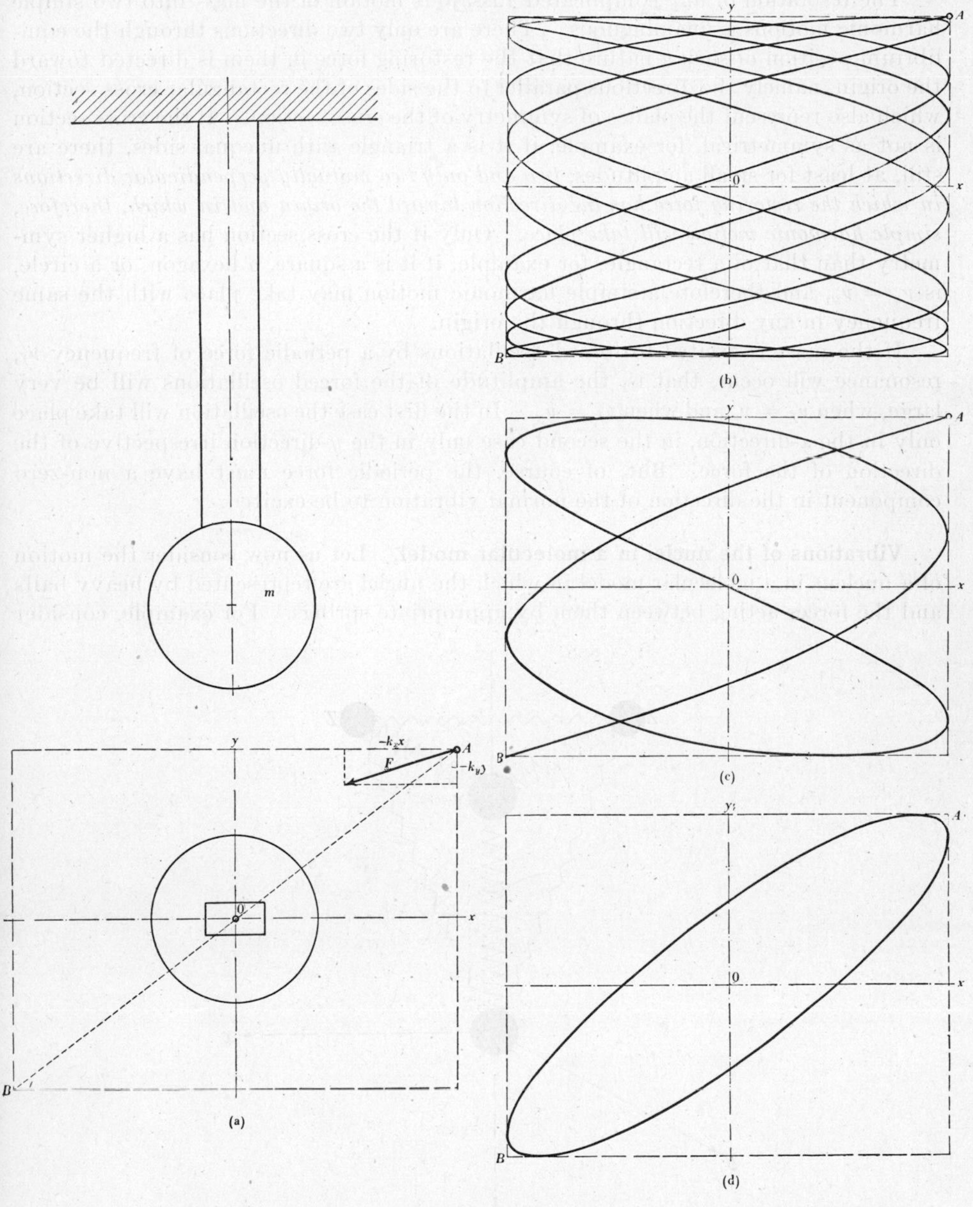

FIG. 22. **Vibrations of a mass suspended by an elastic bar. (a) Front and top view of elastic bar and suspended mass. (b) Lissajous motion for $\nu_x/\nu_y = 2.4461\cdots$. (c) Lissajous motion for $\nu_x/\nu_y = 5/3$. (d) Lissajous motion for $\nu_x = \nu_y$ and phase difference of 30°** (see p. 75).

The resolution of *any* complicated Lissajous motion of the mass into two simple harmonic motions is unambiguous. There are only two directions through the equilibrium position of such a nature that the restoring force in them is directed toward the origin, namely the directions parallel to the sides of the rectangular cross section, which also represent the planes of symmetry of the cross section. If the cross section is not as symmetrical, for example, if it is a triangle with unequal sides, there are still, at least for small amplitudes, *two and only two mutually perpendicular directions in which the restoring force has the direction toward the origin and in which, therefore, simple harmonic motion will take place.* Only if the cross section has a higher symmetry than that of a rectangle, for example, if it is a square, a hexagon, or a circle, is $\nu_x = \nu_y$, and therefore a simple harmonic motion may take place with the same frequency in any direction through the origin.

If the mass is excited to forced oscillations by a periodic force of frequency ν_f, resonance will occur, that is, the amplitude of the forced oscillations will be very large, when $\nu_f = \nu_x$ and when $\nu_f = \nu_y$. In the first case the oscillation will take place only in the x-direction, in the second case only in the y-direction irrespective of the direction of the force. But, of course, the periodic force must have a non-zero component in the direction of the normal vibration to be excited.

Vibrations of the nuclei in a molecular model. Let us now consider the motion of a nucleus in a molecular model in which the nuclei are represented by heavy balls and the forces acting between them by appropriate springs. For example, consider

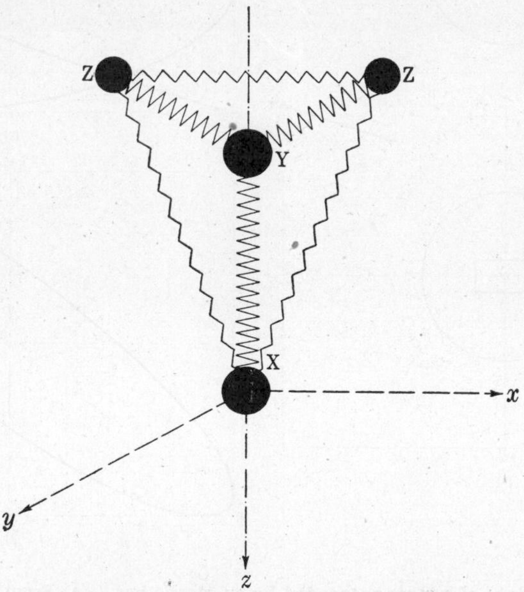

FIG. 23. **Model of an XYZ_2 molecule.**

the plane molecule XYZ_2 (Fig. 23). There are strong restoring forces between X and Y and between Z and Y and weaker forces between X and Z and between the two Z nuclei indicated by strong and weaker springs respectively. Let us assume for a moment that the group YZ_2 is rigidly fixed and that only X can move. Then the forces acting on X if it is displaced from its equilibrium are quite similar to those acting on the mass suspended by an elastic bar considered before. If X is displaced in a direction parallel to Z—Z or in a direction perpendicular to the YZ_2 plane, it

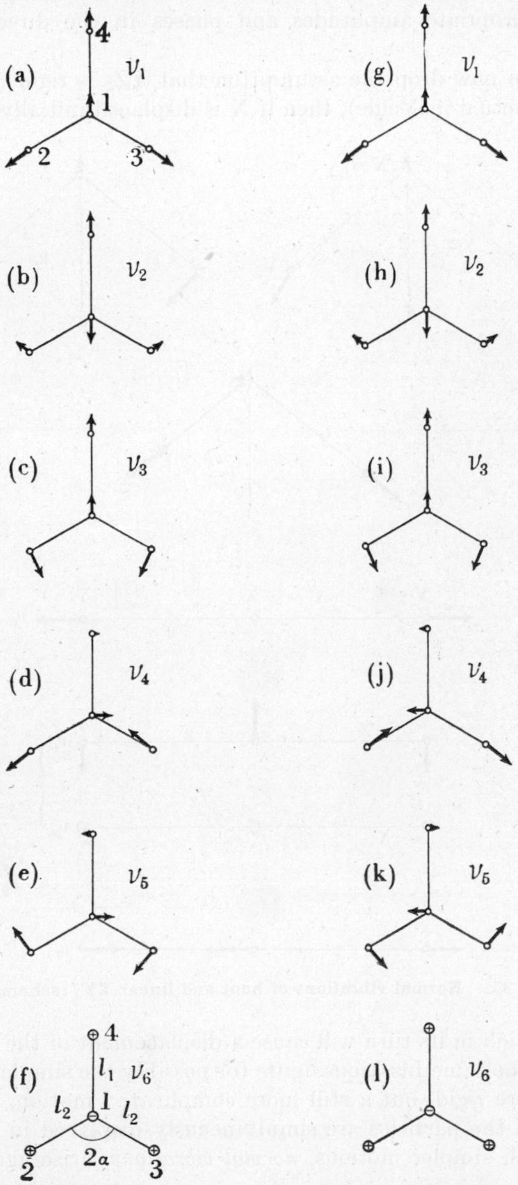

FIG. 24. **Normal vibrations of an XYZ_2 molecule and their behavior for a reflection at the plane of symmetry through XY perpendicular to the plane of the molecule.**—Motions perpendicular to the plane of the paper are indicated by + or − signs in the circles representing the particular nuclei.

will carry out simple harmonic motions of frequency ν_x and $\nu_y(\neq \nu_x)$ respectively. In addition we may also displace the nucleus X in the XY direction, obtaining another simple harmonic motion of frequency ν_z.[2] If X is displaced in any other direction a complicated Lissajous motion results which is now in general in space, not in a plane, and which can be resolved into three simple harmonic motions (*normal vibrations*) of appropriate amplitudes and phases in the directions of the three coordinate axes.

However, if we now drop the assumption that YZ_2 is rigid (which would never be fulfilled in an actual molecule), then if X is displaced initially it will cause a dis-

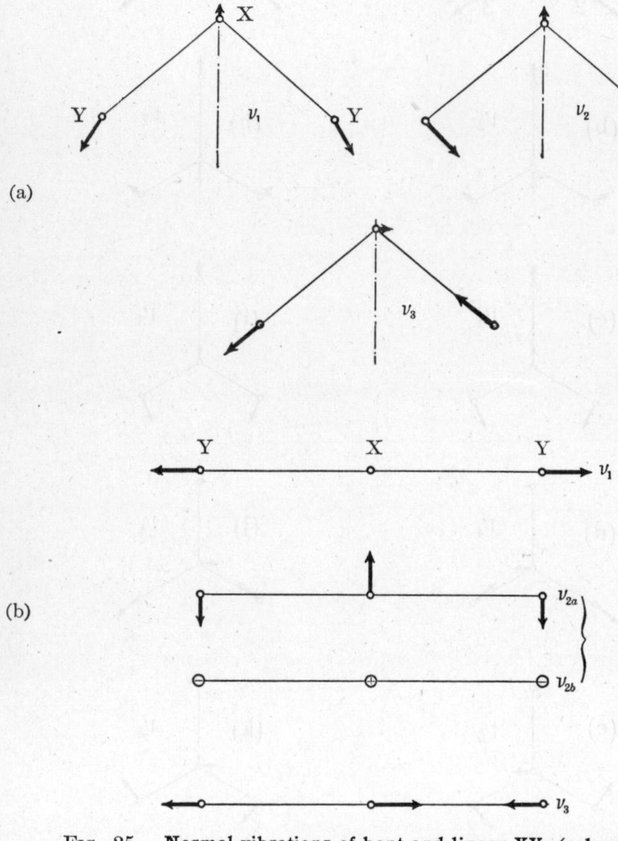

FIG. 25. **Normal vibrations of bent and linear XY_2 (schematic).**

placement of Y which in its turn will cause a displacement of the Z's. Therefore X will not carry out the same Lissajous figure (or possibly the simple harmonic motion) it would if YZ_2 were rigid, but a still more complicated motion, and so will Y and the Z's. Yet if all the particles are simultaneously displaced in a certain way and then released, much simpler motions, *normal vibrations*, arise again, which, similar

[2] There is, of course, an analogue also to this vibration in the case of the elastic bar, but for the bar its frequency is very much larger than those of the other normal vibrations and it is usually not considered.

to the single-particle case, are characterized by the fact that *each particle carries out a simple harmonic motion and that all particles have the same frequency of oscillation and, in general, move in phase.* For example, if the four nuclei are displaced to the respective end points of the solid arrows in Fig. 24a and then released, each one will simply move back and forth about its equilibrium position with the same frequency as all the others, in such a way that after each period the same positions are occupied. The same applies with different frequencies when the initial displacements are as given in Fig. 24b, c, d, e, and f. How these special initial displacements can be found will be taken up in detail later. It may be noted that the relative lengths of the arrows give also the *relative velocities and the amplitudes of the individual nuclei.* They have to be chosen in such a way that there is *no resultant translation or rotation of the molecule as a whole.*

It will be shown later that the extremely complicated motion described above, which arises if one particle is first given a blow and then the system left to itself, or *any* other *motion* of the system can be represented as a *superposition of a number of these normal vibrations.* There are always *as many different normal vibrations as there are vibrational degrees of freedom,* that is, $3N - 6$ or $3N - 5$ respectively. Hence a molecule XYZ_2 has six normal vibrations (Fig. 24). Fig. 25a and b represent as a further example the normal vibrations of bent and linear XY_2 molecules. The second vibration of the latter may occur with equal frequency both in the plane of the paper and perpendicular to it. It is *doubly degenerate* like the normal vibrations of an elastic bar with square or circular cross section (see above).

Mathematical formulation.[3] For any particle i carrying out a simple harmonic motion of frequency ν the *displacement* s_i is given by

$$s_i = s_i^0 \cos (2\pi\nu t + \varphi), \qquad \text{(II, 2)}$$

where s_i^0 is the amplitude, t the time, and φ a phase constant. From (II, 2) follows, for the *acceleration,*

$$a_i = \frac{d^2 s_i}{dt^2} = -4\pi^2\nu^2 s_i^0 \cos (2\pi\nu t + \varphi) = -4\pi^2\nu^2 s_i. \qquad \text{(II, 3)}$$

Therefore we have, for the *restoring force* under whose action the simple harmonic motion is carried out,

$$F^i = m_i a_i = -4\pi^2\nu^2 m_i s_i, \qquad \text{(II, 4)}$$

where m_i is the mass of the particle. Thus the restoring force for any simple harmonic motion of frequency ν is *proportional to* $m_i s_i$ at every moment. This holds also for the component of the motion in any direction.

In a *system of N particles,* such as a molecule, in which every particle acts with a certain force on every other one, in equilibrium the resultant of all forces acting on a given particle is zero. If the particle, say particle 1, is displaced from its equilibrium position by a distance whose components in the direction of the three fixed coordinate axes are x_1, y_1, z_1, there will be a restoring force F^1 which depends on the components of the displacement. In the most general case the components of the force F_x^1, F_y^1, F_z^1 can be developed into a power series in x_1, y_1, z_1. However, if the displacement is sufficiently small only the linear terms need be considered

[3] We are following here the treatment given by Teller (836).

and one has

$$
\begin{aligned}
F_x{}^1 &= -k_{xx}^{11}x_1 - k_{xy}^{11}y_1 - k_{xz}^{11}z_1,\\
F_y{}^1 &= -k_{yx}^{11}x_1 - k_{yy}^{11}y_1 - k_{yz}^{11}z_1,\\
F_z{}^1 &= -k_{zx}^{11}x_1 - k_{zy}^{11}y_1 - k_{zz}^{11}z_1,
\end{aligned}
\qquad \text{(II, 5)}
$$

where the k_{xx}^{11}, k_{yy}^{11}, $\cdots$ are force constants. The negative signs are used since $F_x{}^1$, $F_y{}^1$, $F_z{}^1$ for positive x_1, y_1, z_1 are in general directed in the negative x-, y-, z-direction respectively.

Equations (II, 5) are valid only if all particles except particle 1 remain in their equilibrium positions. If the other particles are displaced as well, the restoring force acting on particle 1 will be changed somewhat. For *small displacements* it will again be a linear function of these other displacements, so that we obtain:

$$
\begin{aligned}
F_x{}^1 &= -k_{xx}^{11}x_1 - k_{xy}^{11}y_1 - k_{xz}^{11}z_1 - k_{xx}^{12}x_2 - k_{xy}^{12}y_2 - k_{xz}^{12}z_2 - \cdots - k_{xz}^{1N}z_N,\\
F_y{}^1 &= -k_{yx}^{11}x_1 - k_{yy}^{11}y_1 - k_{yz}^{11}z_1 - k_{yx}^{12}x_2 - k_{yy}^{12}y_2 - k_{yz}^{12}z_2 - \cdots - k_{yz}^{1N}z_N,\\
F_z{}^1 &= -k_{zx}^{11}x_1 - k_{zy}^{11}y_1 - k_{zz}^{11}z_1 - k_{zx}^{12}x_2 - k_{zy}^{12}y_2 - k_{zz}^{12}z_2 - \cdots - k_{zz}^{1N}z_N.
\end{aligned}
\qquad \text{(II, 6)}
$$

Similar equations are obtained for the forces acting on the other particles

$$
\begin{aligned}
F_x{}^2 &= -k_{xx}^{21}x_1 - k_{xy}^{21}y_1 - k_{xz}^{21}z_1 - k_{xx}^{22}x_2 - k_{xy}^{22}y_2 - k_{xz}^{22}z_2 - \cdots - k_{xz}^{2N}z_N,\\
F_y{}^2 &= -k_{yx}^{21}x_1 - k_{yy}^{21}y_1 - k_{yz}^{21}z_1 - k_{yx}^{22}x_2 - k_{yy}^{22}y_2 - k_{yz}^{22}z_2 - \cdots - k_{yz}^{2N}z_N,\\
F_z{}^2 &= -k_{zx}^{21}x_1 - k_{zy}^{21}y_1 - k_{zz}^{21}z_1 - k_{zx}^{22}x_2 - k_{zy}^{22}y_2 - k_{zz}^{22}z_2 - \cdots - k_{zz}^{2N}z_N,\\
&\cdots\cdots\cdots\cdots\cdots\cdots\cdots\cdots\cdots\cdots\cdots\cdots\cdots\cdots\\
F_z{}^N &= -k_{zx}^{N1}x_1 - k_{zy}^{N1}y_1 - k_{zz}^{N1}z_1 - k_{zx}^{N2}x_2 - k_{zy}^{N2}y_2 - k_{zz}^{N2}z_2 - \cdots - k_{zz}^{NN}z_N.
\end{aligned}
\qquad \text{(II, 7)}
$$

Here it must be realized that the x_i, y_i, z_i are not simply the coordinates of particle i but the *displacement* coordinates, or in other words, the coordinates of particle i with respect to a coordinate system whose origin is at the equilibrium position of particle i and which is therefore different for different particles. The direction of the x_i-, y_i-, z_i-axes is usually but not necessarily the same for all particles i. The coefficients k_{xy}^{il} determine how the x-component of the force on the ith particle depends on the y component of the displacement of the lth particle. It can be shown (see Teller, *loc. cit.*, p. 91) that

$$
k_{xy}^{il} = k_{yx}^{li} \qquad \text{(II, 8)}
$$

which holds for any i and l and where x or y may be any one of x, y, or z.

If we want to find out whether there are any *normal vibrations* of the type described above in the system under consideration, that is, motions in which all particles move with the same frequency according to simple harmonic motion, we have to see whether the above condition (II, 4) for simple harmonic motion can be fulfilled *simultaneously for all particles with the same frequency;* that is, we try to put

$$
F_x{}^i = -4\pi^2\nu^2 m_i x_i, \qquad F_y{}^i = -4\pi^2\nu^2 m_i y_i, \qquad F_z{}^i = -4\pi^2\nu^2 m_i z_i. \qquad \text{(II, 9)}
$$

Substituting this into (II, 6) and (II, 7) we obtain

$$
\begin{aligned}
4\pi^2\nu^2 m_1 x_1 &= k_{xx}^{11}x_1 + k_{xy}^{11}y_1 + k_{xz}^{11}z_1 + k_{xx}^{12}x_2 + \cdots + k_{xz}^{1N}z_N,\\
4\pi^2\nu^2 m_1 y_1 &= k_{yx}^{11}x_1 + k_{yy}^{11}y_1 + k_{yz}^{11}z_1 + k_{yx}^{12}x_2 + \cdots + k_{yz}^{1N}z_N,\\
4\pi^2\nu^2 m_1 z_1 &= k_{zx}^{11}x_1 + k_{zy}^{11}y_1 + k_{zz}^{11}z_1 + k_{zx}^{12}x_2 + \cdots + k_{zz}^{1N}z_N,\\
4\pi^2\nu^2 m_2 x_2 &= k_{xx}^{21}x_1 + k_{xy}^{21}y_1 + k_{xz}^{21}z_1 + k_{xx}^{22}x_2 + \cdots + k_{xz}^{2N}z_N,\\
&\cdots\cdots\cdots\\
4\pi^2\nu^2 m_N z_N &= k_{zx}^{N1}x_1 + k_{zy}^{N1}y_1 + k_{zz}^{N1}z_1 + k_{zx}^{N2}x_2 + \cdots + k_{zz}^{NN}z_N.
\end{aligned}
\qquad \text{(II, 10)}
$$

This system of linear and homogeneous equations for $x_1, y_1, z_1, x_2, y_2, z_2, \cdots z_N$ cannot be solved for arbitrary values of the coefficients occurring therein but, as shown by the theory of linear algebraic equations, only if the determinant of the coefficients is equal to zero. Since the force constants k_{xy}^{il} are fixed for a given system the only way to fulfill this condition is by a suitable choice of the frequency ν. Thus *for certain frequencies defined by the condition*

$$
\begin{vmatrix}
k_{xx}^{11} - 4\pi^2\nu^2 m_1 & k_{xy}^{11} & k_{xz}^{11} & k_{xx}^{12} & \cdots & k_{xz}^{1N}\\
k_{yx}^{11} & k_{yy}^{11} - 4\pi^2\nu^2 m_1 & k_{yz}^{11} & k_{yx}^{12} & \cdots & k_{yz}^{1N}\\
k_{zx}^{11} & k_{zy}^{11} & k_{zz}^{11} - 4\pi^2\nu^2 m_1 & k_{zx}^{12} & \cdots & k_{zz}^{1N}\\
k_{xx}^{21} & k_{xy}^{21} & k_{xz}^{21} & k_{xx}^{22} - 4\pi^2\nu^2 m_2 & \cdots & k_{xz}^{2N}\\
\cdot & \cdot & \cdot & \cdot & \cdot & \cdot\\
k_{zx}^{N1} & k_{zy}^{N1} & k_{zz}^{N1} & k_{zx}^{N2} & \cdots & k_{zz}^{NN} - 4\pi^2\nu^2 m_N
\end{vmatrix} = 0. \qquad \text{(II, 11)}
$$

a simultaneous simple harmonic motion of all particles is possible. The determinant is of the $3N$th degree and therefore has $3N$ roots. Thus in principle the frequencies of the normal vibrations may be determined.

The *form of any one of the normal vibrations* may then be obtained by substituting the corresponding value of ν into the set of equations (II, 10) and solving for $x_1, y_1, z_1, x_2, y_2, z_2, \cdots z_N$. Of course, since these equations are homogeneous only the *ratios* of the $x_1, y_1, z_1, x_2, y_2, z_2, \cdots z_N$ can be determined. The ratio $x_1 : y_1 : z_1 : x_2 : y_2 : z_2 : \cdots : z_N$ is independent of the time for a given ν and gives, therefore, also the ratio of the components of the *amplitudes* of the different particles. It also gives the ratio of the *velocities* at any instant.

A closer examination of the determinantal equation (II, 11), taking account of (II, 8), shows that it has five or six roots that are equal to zero, depending on whether the system (in its equilibrium position) is linear or not. They correspond to *non-genuine normal vibrations* in which simply a translation along any one of the three coordinate axes takes place, or a rotation about two or three suitable axes. Since there is no restoring force for these motions, the "vibrational" frequency is zero.[4] It can be shown further that all the other $3N - 5$ or $3N - 6$ roots are different from zero and real [see Whittaker (25)]. Thus we have $3N - 5$ *or* $3N - 6$ *genuine normal vibrations* in agreement with the previous discussion of vibrational degrees of freedom.

[4] They are therefore also called *null-vibrations*. The zero "vibrational" frequency has, of course, nothing to do with the rotational frequency which does not depend on the restoring forces for small displacements.

Normal coordinates, orthogonality of normal vibrations. Let us denote by

$$\begin{aligned}
&x_1^{(1)}, y_1^{(1)}, z_1^{(1)}, x_2^{(1)}, y_2^{(1)}, z_2^{(1)}, x_3^{(1)}, \cdots z_N^{(1)},\\
&x_1^{(2)}, y_1^{(2)}, z_1^{(2)}, x_2^{(2)}, y_2^{(2)}, z_2^{(2)}, x_3^{(2)}, \cdots z_N^{(2)},\\
&\cdot\quad\cdot\quad\cdot\quad\cdot\quad\cdot\quad\cdot\quad\cdot\quad\cdot\quad\cdot\quad\cdot\quad\cdot\quad\cdot\\
&x_1^{(3N)}, y_1^{(3N)}, z_1^{(3N)}, x_2^{(3N)}, y_2^{(3N)}, z_2^{(3N)}, x_3^{(3N)}, \cdots z_N^{(3N)},
\end{aligned} \tag{II, 12}$$

the displacement coordinates at a certain moment belonging to the first, second, $\cdots$ $3N$th normal vibration of frequency $\nu_1, \nu_2, \cdots \nu_{3N}$ respectively. As mentioned above, these displacement coordinates are determined by (II, 10) only apart from an arbitrary factor. According to a simple theorem about homogeneous linear equations the *ratio of the displacement coordinates* $x_1^{(i)}, y_1^{(i)}, z_1^{(i)}, x_2^{(i)}, \cdots z_N^{(i)}$ is equal to the *ratio of the minors of any one row of the determinant* (II, 11) in which ν_i has been

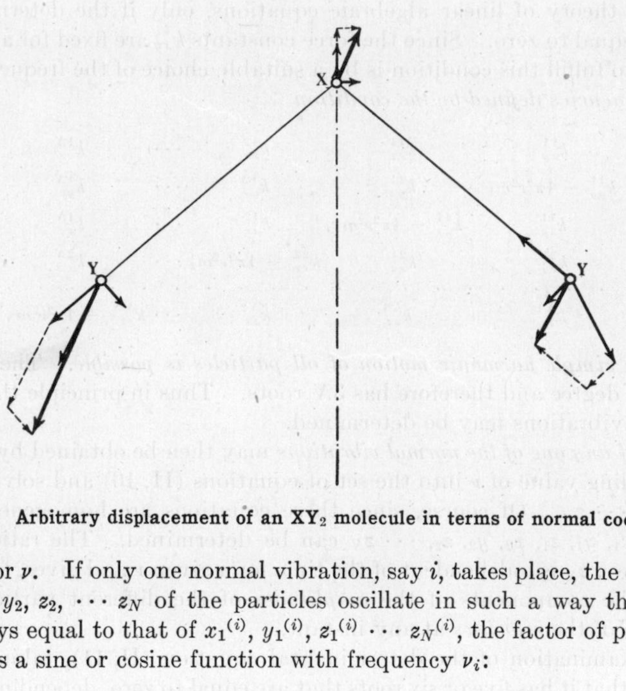

FIG. 26. **Arbitrary displacement of an XY_2 molecule in terms of normal coordinates.**

substituted for ν. If only one normal vibration, say i, takes place, the displacements $x_1, y_1, z_1, x_2, y_2, z_2, \cdots z_N$ of the particles oscillate in such a way that their ratio remains always equal to that of $x_1^{(i)}, y_1^{(i)}, z_1^{(i)} \cdots z_N^{(i)}$, the factor of proportionality ξ_i changing as a sine or cosine function with frequency ν_i:

$$\xi_i = \xi_i^0 \cos(2\pi\nu_i t + \varphi_i). \tag{II, 13}$$

The ξ_i are called *normal coordinates*.

If several normal vibrations are taking place at the same time we have

$$\begin{aligned}
x_1 &= x_1^{(1)}\xi_1 + x_1^{(2)}\xi_2 + x_1^{(3)}\xi_3 + \cdots + x_1^{(3N)}\xi_{3N}\\
y_1 &= y_1^{(1)}\xi_1 + y_1^{(2)}\xi_2 + y_1^{(3)}\xi_3 + \cdots + y_1^{(3N)}\xi_{3N}\\
z_1 &= z_1^{(1)}\xi_1 + z_1^{(2)}\xi_2 + z_1^{(3)}\xi_3 + \cdots + z_1^{(3N)}\xi_{3N}\\
x_2 &= x_2^{(1)}\xi_1 + x_2^{(2)}\xi_2 + x_2^{(3)}\xi_3 + \cdots + x_2^{(3N)}\xi_{3N}\\
&\cdot\quad\cdot\quad\cdot\quad\cdot\quad\cdot\quad\cdot\quad\cdot\quad\cdot\quad\cdot\\
z_N &= z_N^{(1)}\xi_1 + z_N^{(2)}\xi_2 + z_N^{(3)}\xi_3 + \cdots + z_N^{(3N)}\xi_{3N}.
\end{aligned} \tag{II, 14}$$

Conversely, since these $3N$ equations if solved for $\xi_1, \xi_2, \xi_3 \cdots \xi_{3N}$ have one and only one solution, *any vibrational displacement of the system may be expressed in terms of normal coordinates* $\xi_1, \xi_2, \xi_3 \cdots \xi_{3N}$ *instead of in Cartesian coordinates.* For example, the displacement of an XY_2 molecule indicated by heavy arrows in Fig. 26 may be represented by adding 1.2 times the displacements of vibration ν_1 in Fig. 25a, 0.4 times the displacements of ν_2 and 0.7 times the displacements of ν_3. The factors 1.2 0.4, 0.7 are here the values of the normal coordinates ξ_1, ξ_2, and ξ_3.[5] If $x_1, y_1, z_1, x_2, y_2, z_2 \cdots z_N$ are the initial displacements the subsequent motion is determined by (II, 13) and (II, 14), that is, *any vibrational motion of the system may be represented as a superposition of normal vibrations with suitable amplitudes.*

In the example Fig. 26, if the heavy arrows indicate the initial displacements (assuming that the initial velocities are zero), the subsequent Lissajous motion of each particle can be obtained by letting each component normal coordinate change periodically with its characteristic frequency and with the amplitude given. If the initial velocities are not zero, the initial values of the normal coordinates are not the maximum values, since the φ_i in (II, 13) are then not zero. Their values can be obtained from the differentiated equations (II, 13) and (II, 14).

In the case of the elastic bar the normal vibrations are always perpendicular to one another no matter whether the cross section is a rectangle or not. A somewhat similar relation holds for the normal vibrations of a polyatomic molecule although the displacements of one and the same atom in two different normal vibrations need not be perpendicular to each other. (Compare Fig. 24 and 25.) If ν_k and ν_l are the frequencies of two normal vibrations we have, from (II, 10):

$$\begin{aligned}
4\pi^2\nu_k{}^2 m_1 x_1{}^{(k)} &= k^{11}_{xx} x_1{}^{(k)} + k^{11}_{xy} y_1{}^{(k)} + \cdots \quad \Big| \; x_1{}^{(l)} \\
4\pi^2\nu_k{}^2 m_1 y_1{}^{(k)} &= k^{11}_{yx} x_1{}^{(k)} + k^{11}_{yy} y_1{}^{(k)} + \cdots \quad \Big| \; y_1{}^{(l)} \\
4\pi^2\nu_k{}^2 m_1 z_1{}^{(k)} &= k^{11}_{zx} x_1{}^{(k)} + k^{11}_{zy} y_1{}^{(k)} + \cdots \quad \Big| \; z_1{}^{(l)} \\
4\pi^2\nu_k{}^2 m_2 x_2{}^{(k)} &= k^{21}_{xx} x_1{}^{(k)} + k^{21}_{xy} y_1{}^{(k)} + \cdots \quad \Big| \; x_2{}^{(l)} \\
&\cdots\cdots\cdots
\end{aligned} \qquad \text{(II, 15)}$$

and

$$\begin{aligned}
4\pi^2\nu_l{}^2 m_1 x_1{}^{(l)} &= k^{11}_{xx} x_1{}^{(l)} + k^{11}_{xy} y_1{}^{(l)} + \cdots \quad \Big| \; x_1{}^{(k)} \\
4\pi^2\nu_l{}^2 m_1 y_1{}^{(l)} &= k^{11}_{yx} x_1{}^{(l)} + k^{11}_{yy} y_1{}^{(l)} + \cdots \quad \Big| \; y_1{}^{(k)} \\
4\pi^2\nu_l{}^2 m_1 z_1{}^{(l)} &= k^{11}_{zx} x_1{}^{(l)} + k^{11}_{zy} y_1{}^{(l)} + \cdots \quad \Big| \; z_1{}^{(k)} \\
4\pi^2\nu_l{}^2 m_2 x_2{}^{(l)} &= k^{21}_{xx} x_1{}^{(l)} + k^{21}_{xy} y_1{}^{(l)} + \cdots \quad \Big| \; x_2{}^{(k)}. \\
&\cdots\cdots\cdots
\end{aligned} \qquad \text{(II, 16)}$$

If we multiply the individual equations by the displacements indicated on the right, add the resulting equations (II, 15), and subtract from the sum all the resulting equations (II, 16), the right-hand sides will cancel and we obtain

$$4\pi(\nu_k{}^2 - \nu_l{}^2)[\sum_i m_i x_i{}^{(k)} x_i{}^{(l)} + \sum_i m_i y_i{}^{(k)} y_i{}^{(l)} + \sum_i m_i z_i{}^{(k)} z_i{}^{(l)}] = 0. \qquad \text{(II, 17)}$$

[5] The displacement is one in which no motion of the center of mass and no rotation about it takes place. Therefore, normal coordinates corresponding to non-genuine vibrations need not be considered.

Therefore, if the two normal vibrations have different frequency we have

$$\sum_i m_i(x_i^{(k)}x_i^{(l)} + y_i^{(k)}y_i^{(l)} + z_i^{(k)}z_i^{(l)}) = 0, \qquad \text{(II, 18)}$$

where we have to sum over all particles. (II, 18) is the *relation of orthogonality. Two normal vibrations of different frequencies are orthogonal to each other.* It is easily seen that (II, 18) is fulfilled when all displacement vectors of the two normal vibrations are mutually perpendicular to one another. But (II, 18) does not require them to be perpendicular.

By the equations (II, 10), only the ratios of the amplitudes of the individual particles for a given normal vibration are defined. It is sometimes convenient to choose their actual magnitude in such a way that the expression

$$M_k = \sum_i m_i(x_i^{(k)2} + y_i^{(k)2} + z_i^{(k)2}) \qquad \text{(II, 19)}$$

is the same for all normal vibrations ν_k. The normal vibrations are then said to be *normalized.* M_k is the expression in the bracket in (II, 17) if $k = l$. It is of course *not* equal to zero.

If we apply the relation of orthogonality (II, 18) to a genuine normal vibration and the non-genuine normal vibration consisting of a translation in the x-direction ($x_1^{(l)} = x_2^{(l)} = x_3^{(l)} = \cdots x_N^{(l)}$, $y_1^{(l)} = y_2^{(l)} = \cdots = 0$, $z_1^{(l)} = z_2^{(l)} = \cdots = 0$), we obtain

$$x_1^{(l)} \sum_i m_i x_i^{(k)} = 0,$$

and therefore

$$\sum_i m_i x_i^{(k)} = 0. \qquad \text{(II, 20)}$$

This equation means that there is *no displacement of the center of mass* in the x direction (and similarly in the y and z direction) for any genuine normal vibration, a result that, of course, also follows from the fact that there are no external forces acting on the system. In a similar manner it can be shown that in a genuine normal vibration (if it is not degenerate), there is *no resultant angular momentum* of the system.

Potential energy and kinetic energy. The normal vibrations and normal coordinates of a molecule can also be introduced by using, instead of Newton's second law (force = mass × acceleration), the law of conservation of energy (total energy = kinetic + potential energy). This is the method most frequently used in actual calculations.

Considering that the force is the negative derivative of the potential energy with respect to the displacement, the potential energy for a simple harmonic oscillator is found from (II, 4) to be

$$V_i = 2\pi^2\nu_i^2 m_i s_i^2 = \tfrac{1}{2}k_i s_i^2, \qquad \text{(II, 21)}$$

where

$$k_i = 4\pi^2\nu_i^2 m_i \qquad \text{(II, 22)}$$

is the force constant. The potential energy is taken to be zero at the equilibrium position $s_i = 0$. The kinetic energy is

$$T_i = \tfrac{1}{2}m_i v_i^2 = \tfrac{1}{2}m_i\dot{s}_i^2, \qquad \text{(II, 23)}$$

where $\dot{s}_i$ as usual stands for ds_i/dt. The total energy is therefore

$$H_i = V_i + T_i = \tfrac{1}{2}(k_i s_i^2 + m_i\dot{s}_i^2). \qquad \text{(II, 24)}$$

The *potential energy of the nuclei in the molecule* referred to the equilibrium position as $V = 0$ is given in first approximation, that is, as long as the displacements are sufficiently small, by

$$V = \tfrac{1}{2}\sum_{ij} (k_{xx}^{ij}x_ix_j + k_{yy}^{ij}y_iy_j + k_{zz}^{ij}z_iz_j) + \sum_{ij} (k_{xy}^{ij}x_iy_j + k_{xz}^{ij}x_iz_j + k_{yz}^{ij}y_iz_j). \qquad \text{(II, 25)}$$

This can be easily verified by forming $\frac{\partial V}{\partial x_1}, \frac{\partial V}{\partial x_2}, \cdots \frac{\partial V}{\partial y_1}, \frac{\partial V}{\partial y_2}, \cdots$, which according to the definition of the potential energy must be equal to $-F_x^1, -F_x^2 \cdots -F_y^1, -F_y^2 \cdots$ respectively [compare equations (II, 6) and (II, 7)]. In order to facilitate writing let us denote the coordinates $x_1, y_1, z_1, x_2, y_2, z_2 \cdots$ by $q_1, q_2, q_3, q_4, q_5, q_6 \cdots$ in this order. Then the potential energy may be written (with $k_{ij} = k_{ji}$)

$$V = \tfrac{1}{2}\sum_{ij} k_{ij}q_iq_j = \tfrac{1}{2}k_{11}q_1^2 + \tfrac{1}{2}k_{22}q_2^2 + \cdots + k_{12}q_1q_2 + k_{13}q_1q_3 + \cdots. \qquad \text{(II, 26)}$$

This holds, as long as the displacements are small compared to the internuclear distances, even if $q_1, q_2, q_3 \cdots$ are displacement coordinates other than those chosen above. In the present case

$$k_{11} = k_{xx}^{11},\ k_{22} = k_{yy}^{11},\ \cdots\ k_{12} = k_{xy}^{11},\ \cdots \qquad \text{(II, 27)}$$

The *kinetic energy* is given by

$$T = \sum_i \tfrac{1}{2}m_i(\dot{x}_i^2 + \dot{y}_i^2 + \dot{z}_i^2), \qquad \text{(II, 28)}$$

or, with the new notation,

$$T = \tfrac{1}{2}\sum_{ij} b_{ij}\dot{q}_i\dot{q}_j, \qquad \text{(II, 29)}$$

where in the present case

$$b_{ij} = 0 \text{ for } i \neq j, \text{ and } b_{11} = b_{22} = b_{33} = m_1,\ b_{44} = b_{55} = b_{66} = m_2,\ \cdots \qquad \text{(II, 30)}$$

When the q_i are not simply Cartesian coordinates, b_{ij} is in general different from zero also for $i \neq j$, but (II, 29) still holds.

Let us now introduce new coordinates $\eta_1, \eta_2 \cdots \eta_{3N}$ by means of the linear equations:

$$\begin{aligned}
[x_1 =]q_1 &= c_{11}\eta_1 + c_{12}\eta_2 + c_{13}\eta_3 + \cdots \\
[y_1 =]q_2 &= c_{21}\eta_1 + c_{22}\eta_2 + c_{23}\eta_3 + \cdots \\
[z_1 =]q_3 &= c_{31}\eta_1 + c_{32}\eta_2 + c_{33}\eta_3 + \cdots \\
&\cdots \\
q_i &= c_{i1}\eta_1 + c_{i2}\eta_2 + c_{i3}\eta_3 + \cdots. \\
&\cdots
\end{aligned} \qquad \text{(II, 31)}$$

The theory of quadratic forms (see (25)) shows that by appropriate choice of the coefficients c_{ik} of this *linear transformation* we can bring simultaneously both V and T to a simpler form in terms of the new coordinates, namely, to

$$V = \tfrac{1}{2}(\lambda_1\eta_1^2 + \lambda_2\eta_2^2 + \cdots + \lambda_i\eta_i^2 + \cdots + \lambda_{3N}\eta_{3N}^2), \qquad \text{(II, 32)}$$

$$T = \tfrac{1}{2}(\dot{\eta}_1^2 + \dot{\eta}_2^2 + \cdots + \dot{\eta}_i^2 + \cdots + \dot{\eta}_{3N}^2). \qquad \text{(II, 33)}$$

The λ_i in (II, 32) can be shown to be the roots of the determinantal equation which is also called the *secular equation* of the problem:

$$\begin{vmatrix} k_{11}-b_{11}\lambda & k_{12}-b_{12}\lambda & k_{13}-b_{13}\lambda \cdots \\ k_{21}-b_{21}\lambda & k_{22}-b_{22}\lambda & k_{23}-b_{23}\lambda \cdots \\ k_{31}-b_{31}\lambda & k_{32}-b_{32}\lambda & k_{33}-b_{33}\lambda \cdots \\ \cdot\ \cdot\ \cdot & \cdot\ \cdot\ \cdot & \cdot\ \cdot\ \cdot \end{vmatrix} = 0. \qquad \text{(II, 34)}$$

where the k_{ij} and b_{ij} are from (II, 26) and (II, 29) respectively.

The *total energy* is now

$$H = V + T = \tfrac{1}{2}(\lambda_1\eta_1^2 + \dot{\eta}_1^2) + \tfrac{1}{2}(\lambda_2\eta_2^2 + \dot{\eta}_2^2) + \cdots, \qquad \text{(II, 35)}$$

that is, it is the sum of $3N$ mutually independent terms each of which has the form of the total energy of a simple harmonic oscillator (II, 24) of mass 1. That is, the motion of the system of N particles may be considered as a *superposition of* $3N$ *independent simple harmonic motions in the new coordinates*

$$\eta_i = \eta_i^0 \cos(2\pi\nu_i t + \varphi_i), \qquad \text{(II, 36)}$$

where the frequencies ν_i are, according to (II, 21), (II, 22), and (II, 32), related to the constants λ_i by

$$\lambda_i = 4\pi^2\nu_i^2. \qquad \text{(II, 37)}$$

In the present case if we substitute (II, 27) and (II, 30) into (II, 34) we obtain as equation for the λ_i's:

$$\begin{vmatrix} k_{xx}^{11}-m_1\lambda & k_{xy}^{11} & k_{xz}^{11} & k_{xx}^{12} & \cdots & k_{xz}^{1N} \\ k_{yx}^{11} & k_{yy}^{11}-m_1\lambda & k_{yz}^{11} & k_{yx}^{12} & \cdots & k_{yz}^{1N} \\ k_{zx}^{11} & k_{zy}^{11} & k_{zz}^{11}-m_1\lambda & k_{zx}^{12} & \cdots & k_{zz}^{1N} \\ k_{xx}^{21} & k_{xy}^{21} & k_{xz}^{21} & k_{xx}^{22}-m_2\lambda & \cdots & k_{xz}^{2N} \\ \cdot\ \cdot\ \cdot & \cdot\ \cdot\ \cdot & \cdot\ \cdot\ \cdot & \cdot\ \cdot\ \cdot & \cdot\ \cdot & \cdot\ \cdot\ \cdot \\ k_{zx}^{N1} & k_{zy}^{N1} & k_{zz}^{N1} & k_{zx}^{N2} & \cdots & k_{zz}^{NN}\ \ m_N\lambda \end{vmatrix} = 0. \qquad \text{(II, 38)}$$

It is seen that this equation is identical with (II, 11) if we take account of (II, 37). Thus, as was to be expected, both methods lead to the same frequencies for the simple harmonic oscillations (normal vibrations) as whose superposition any vibrational motion may be considered.

The *form of a given normal vibration*, say η_j, is obtained by putting all other η_i's in (II, 31) equal to zero. One obtains

$$\begin{aligned} x_1^{(j)} &= c_{1j}\eta_j, & y_1^{(j)} &= c_{2j}\eta_j, & z_1^{(j)} &= c_{3j}\eta_j, \\ x_2^{(j)} &= c_{4j}\eta_j, & y_2^{(j)} &= c_{5j}\eta_j, & z_2^{(j)} &= c_{6j}\eta_j. \\ &\cdot\ \cdot\ \cdot & &\cdot\ \cdot\ \cdot & &\cdot\ \cdot\ \cdot \end{aligned} \qquad \text{(II, 39)}$$

that is, considering (II, 36), all displacement coordinates vary with the same frequency ν_j. The ratios of the displacement coordinates $c_{1j}:c_{2j}:c_{3j}:\cdots$ (which is all that matters) are the same as obtained by the first method (p. 69f.) since it can be shown that c_{1j}, c_{2j}, c_{3j}, $\cdots$ are in the ratio of the minors of any one row of the determinant (II, 34) or (II, 38) with $\lambda = \lambda_j$ just as were the previous displacement coordinates. Thus the normal coordinates η_i introduced by (II, 31) are essentially the same as the ξ_i introduced by (II, 14) except for constant factors. It is easily seen by substituting (II, 14) into (II, 28), taking account of (II, 18) and (II, 19), and comparing with (II, 33) that the ξ_i would be identical with the η_i if all M_k were put equal to 1, that is, the normal coordinates η_i are *normalized to unity.*

Degenerate vibrations, generalization of the definition of a normal vibration. It may happen that two (or more) roots of the determinantal equation (II, 11) [or (II, 34) or (II, 38)] coincide, that is, that two (or more) normal vibrations have the same frequency. The two (or more) vibrations are then called *degenerate with one another*. There are then two (or more) sets of solutions of (II, 10) for the degenerate frequency, say $x_1^{(\iota)}$, $y_1^{(\iota)}$, $z_1^{(\iota)}$, $x_2^{(\iota)}$, $y_2^{(\iota)}$, $z_2^{(\iota)}$, $\cdots$ and $x_1^{(\kappa)}$, $y_1^{(\kappa)}$, $z_1^{(\kappa)}$, $x_2^{(\kappa)}$, $y_2^{(\kappa)}$, $z_2^{(\kappa)}$, $\cdots$. Because of the homogeneity of the equations (II, 10) any *linear combination* $ax_1^{(\iota)} + bx_1^{(\kappa)}$, $ay_1^{(\iota)} + by_1^{(\kappa)}$, $\cdots$ with any values of the constants a and b is *also a solution of* (II, 10) *for the same frequency*. The corresponding motion (unlike the motion resulting from the composition of two normal vibrations of *different* frequencies) is again a simple motion since all atoms move with the same frequency, and may also be called a normal vibration. Thus we have really an infinite number of different simple vibrational motions for the same frequency which, however, can be represented as a superposition of two (or more, if the degeneracy is higher than two-fold) linearly independent vibrations.

A good example is supplied by the elastic bar discussed above if its cross section is a square (or a circle), since then its two normal vibrations have the same frequency. Consequently the mass suspended by the bar may carry out a simple harmonic vibration with the same frequency in any direction through the equilibrium position. If the two "original" simple harmonic motions are superimposed with different phase, a motion of the mass on an ellipse (or circle if the phase shift is 90° and the amplitudes of the two component motions are the same) will result (see Fig. 22d), and this ellipse will be traversed with the frequency of the degenerate vibration.

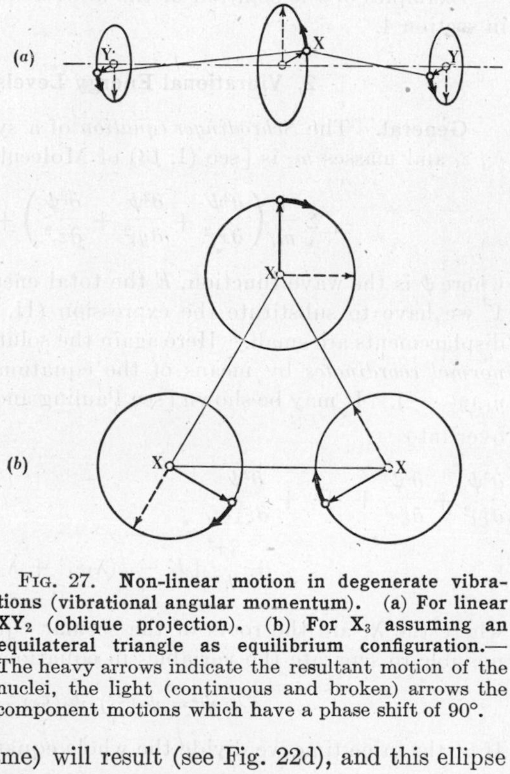

FIG. 27. **Non-linear motion in degenerate vibrations (vibrational angular momentum). (a) For linear XY_2 (oblique projection). (b) For X_3 assuming an equilateral triangle as equilibrium configuration.—The heavy arrows indicate the resultant motion of the nuclei, the light (continuous and broken) arrows the component motions which have a phase shift of 90°.**

Similarly, in the molecule we may superimpose the two components of a degenerate vibration *with different phase* and obtain again a simple motion of the same frequency in which, however, not all atoms move in phase and in straight lines although they do move with the same frequency. For example, if we superimpose the two perpendicular vibrations ν_{2a} and ν_{2b} of a linear XY_2 molecule (Fig. 25b) with a phase shift of 90°, each atom will swing around the axis in a circle as indicated in Fig. 27a, giving rise to a (constant) *vibrational angular momentum about the axis.* Conversely, by superimposing this motion and the opposite one, ν_{2a} or ν_{2b} or any other linear perpendicular vibration can be obtained. As a second example, Fig. 27b gives the superposition of two mutually degenerate vibrations of an X_3 molecule (forming an equilateral triangle) with a phase difference of 90°. Again *each nucleus traverses a circle* in the same sense, giving rise to a *vibrational angular momentum;* but here the internuclear distances do not remain constant during the "vibration." For phase differences other than 90° elliptical motions are obtained (see Fig. 22d).

In order to include such motions under the term "normal vibrations," it is necessary in the definition of a normal vibration to drop the condition that all atoms move in phase and in straight lines and go through their equilibrium positions at the same time. It is sufficient to state that *in a normal vibration all atoms move with the same frequency in such a way that the Cartesian components of the displacements change according to sine curves.*

Examples of the solution of the above equations for specific cases will be given in section 4.

2. Vibrational Energy Levels and Eigenfunctions

General. The *Schrödinger equation* of a system of N particles of coordinates x_i, y_i, z_i and masses m_i is [see (I, 13) of Molecular Spectra I]

$$\sum_i \frac{1}{m_i}\left(\frac{\partial^2\psi}{\partial x_i^2} + \frac{\partial^2\psi}{\partial y_i^2} + \frac{\partial^2\psi}{\partial z_i^2}\right) + \frac{8\pi^2}{h^2}(E - V)\psi = 0, \qquad \text{(II, 40)}$$

where ψ is the wave function, E the total energy, and V the potential energy. For V we have to substitute the expression (II, 25), in which it is assumed that the displacements are small. Here again the solution is greatly simplified if we *introduce normal coordinates* by means of the equations (II, 31) (using $\xi_1, \xi_2 \cdots$ instead of $\eta_1, \eta_2 \cdots$). It may be shown [see Pauling and Wilson (18)] that (II, 40) then goes over into

$$\frac{\partial^2\psi}{\partial \xi_1^2} + \frac{\partial^2\psi}{\partial \xi_2^2} + \cdots + \frac{\partial^2\psi}{\partial \xi_{3N}^2} + \frac{8\pi^2}{h^2}[E - \tfrac{1}{2}(\lambda_1\xi_1^2 + \lambda_2\xi_2^2 + \cdots + \lambda_{3N}\xi_{3N}^2)]\psi = 0, \qquad \text{(II, 41)}$$

where the λ_i are the roots of the secular equation (II, 34) or (II, 38). It is now possible to separate the variables in equation (II, 41) by means of the substitution

$$\psi = \psi_1(\xi_1)\cdot\psi_2(\xi_2) \cdots \psi_{3N}(\xi_{3N}). \qquad \text{(II, 42)}$$

If at the same time we divide the whole equation (II, 41) by ψ, we obtain

$$\frac{1}{\psi_1}\frac{\partial^2\psi_1}{\partial \xi_1^2} + \frac{1}{\psi_2}\frac{\partial^2\psi_2}{\partial \xi_2^2} + \cdots + \frac{1}{\psi_{3N}}\frac{\partial^2\psi_{3N}}{\partial \xi_{3N}^2} + \frac{8\pi^2}{h^2}[E - \tfrac{1}{2}(\lambda_1\xi_1^2 + \lambda_2\xi_2^2 + \cdots + \lambda_{3N}\xi_{3N}^2)] = 0. \qquad \text{(II, 43)}$$

This equation may be resolved into a sum of $3N$ equations:

$$\frac{1}{\psi_i}\frac{d^2\psi_i}{d\xi_i^2} + \frac{8\pi^2}{h^2}(E_i - \tfrac{1}{2}\lambda_i\xi_i^2) = 0, \qquad \text{(II, 44)}$$

with

$$E = E_1 + E_2 + \cdots + E_{3N}. \qquad \text{(II, 45)}$$

The equation (II, 44) is the wave equation of a single simple harmonic oscillator of potential energy $\frac{1}{2}\lambda_i\xi_i^2$ and mass 1 whose coordinate is the normal coordinate ξ_i [see equation (III, 28) of Molecular Spectra I]. Thus *in wave mechanics as in classical mechanics the vibrational motion of the molecule may be considered,* in a first good approximation, *as a superposition of 3N simple harmonic motions in the 3N normal coordinates.*

Energy levels. The eigenvalues of equation (II, 44), that is, the *energy values of the harmonic oscillator i,* are given by

$$E_i = h\nu_i(v_i + \tfrac{1}{2}), \qquad v_i = 0, 1, 2 \cdots \qquad \text{(II, 46)}$$

where

$$\nu_i = \frac{1}{2\pi}\sqrt{\lambda_i} \qquad \text{(II, 47)}$$

is the classical oscillation frequency of the normal vibration i, and v_i is the *vibrational quantum number.* Therefore, according to equation (II, 45), the *total vibrational energy* of the system can assume only the values

$$E(v_1, v_2, v_3, \cdots) = h\nu_1(v_1 + \tfrac{1}{2}) + h\nu_2(v_2 + \tfrac{1}{2}) + h\nu_3(v_3 + \tfrac{1}{2}) + \cdots, \qquad \text{(II, 48)}$$

or, if we go over to the *term values,*

$$G(v_1, v_2, v_3, \cdots) = \frac{E(v_1, v_2, v_3, \cdots)}{hc}$$
$$= \omega_1(v_1 + \tfrac{1}{2}) + \omega_2(v_2 + \tfrac{1}{2}) + \omega_3(v_3 + \tfrac{1}{2}) + \cdots. \qquad \text{(II, 49)}$$

Here we have put

$$\omega_1 = \frac{\nu_1}{c}, \ \omega_2 = \frac{\nu_2}{c}, \ \omega_3 = \frac{\nu_3}{c}, \cdots. \qquad \text{(II, 50)}$$

This designation is in agreement with the nomenclature accepted for diatomic molecules but is not always followed by writers on polyatomic molecules. The ω_i are the (classical) *vibrational frequencies measured in cm^{-1} units.* It is thus seen that by solving the classical vibration problem, that is, determining the classical vibration frequencies ν_i from the secular equation (II, 34) or (II, 38), the quantum theoretical energy values are immediately found.

In (II, 45), (II, 48), and (II, 49), the *non-genuine vibrations* (translations and rotations) are still included. However, since for them $\nu = 0$, they do not give a contribution to the vibrational energy and we shall therefore in future disregard them and consider in (II, 49) the summation over the $3N - 6$ or $3N - 5$ genuine normal vibrations only.

Part of the *energy level diagram* for the simplest case of three genuine normal vibrations (triatomic molecules) is represented in Fig. 28. It consists of a large number of (overlapping) series of equidistant levels: one series with spacing ω_1 corresponding to various values of v_1 for $v_2 = 0$, $v_3 = 0$, one series with spacing ω_2 corresponding to various values of v_2 for $v_1 = 0$, $v_3 = 0$, one series with spacing ω_3 corresponding to various values of v_3 for $v_1 = 0$, $v_2 = 0$; in addition, further v_1 series occur with other (all possible) fixed values of v_2 and v_3 (for example $v_2 = 1$, $v_3 = 0$), further v_2 series with other fixed v_1 and v_3 values, and further v_3 series with other fixed v_1 and v_2 values. Only some of these different series of levels could be drawn in Fig. 28.

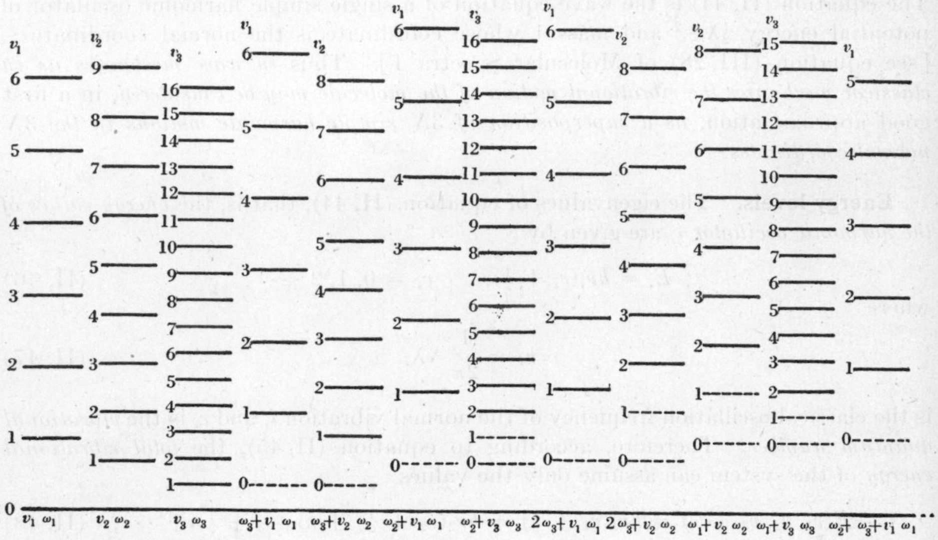

FIG. 28. **Vibrational energy level diagram of a triatomic molecule.**—Some levels occur in several of the series of levels shown. On all but one of these occurrences they are indicated by broken lines.

It is thus seen that the vibrational energy-level diagram even of a triatomic molecule and even assuming harmonic oscillations is much more complicated than that of a diatomic molecule. In a molecule having more than three atoms there are correspondingly more series of energy levels.

According to (II, 48), for $v_1 = 0, v_2 = 0, v_3 = 0, \cdots$, that is, in the lowest possible state, the vibrational energy is not zero but we have a *zero-point vibrational energy:*

$$G(0, 0, 0 \cdots) = \tfrac{1}{2}\omega_1 + \tfrac{1}{2}\omega_2 + \tfrac{1}{2}\omega_3 \cdots. \qquad \text{(II, 51)}$$

For a molecule with several atoms this zero-point energy may have quite a considerable magnitude. Frequently it is convenient to refer the vibrational energy to the lowest possible state as zero (as for diatomic molecules). For this we have

$$G_0(v_1, v_2, v_3, \cdots) = G(v_1, v_2, v_3, \cdots) - G(0, 0, 0, \cdots)$$
$$= \omega_1 v_1 + \omega_2 v_2 + \omega_3 v_3 + \cdots. \qquad \text{(II, 52)}$$

Eigenfunctions. The eigenfunctions $\psi_i(\xi_i)$ of equation (II, 44) are the ordinary *harmonic oscillator eigenfunctions* as pictured by the broken curves in Fig. 29 (which is identical with Fig. 41 of Molecular Spectra I) where the abscissa apart from a constant factor is the normal coordinate ξ_i. The mathematical form of the function is

$$\psi_i(\xi_i) = N_{v_i} e^{-(\alpha_i/2)\xi_i^2} H_{v_i}(\sqrt{\alpha_i}\xi_i), \qquad \text{(II, 53)}$$

where N_{v_i} is a normalization constant, $\alpha_i = 2\pi\nu_i/h$, and $H_{v_i}(\sqrt{\alpha_i}\xi_i)$ is a so-called *Hermite polynomial* of the v_ith degree. The full curves in Fig. 29 give $[\psi_i(\xi_i)]^2$ which is proportional to the *probability* of finding the oscillator with coordinate ξ_i.

The *total vibrational eigenfunction*, according to (II, 42), is the product of $3N - 6$ or $3N - 5$ harmonic oscillator functions (II, 53). It is not very easy to visualize this function. It should be understood that it is a function in the $3N - 6$- (or $3N - 5$)-dimensional space of the $3N - 6$ (or $3N - 5$) normal coordinates, which are not simply the displacements of the individual atoms. In order to get ψ in terms of the Cartesian coordinates of the displacements one would have to express the ξ_i in terms of the Cartesian coordinates from (II, 14) and substitute into (II, 42) and (II, 53). Even then the rather more complicated function obtained is in a $3N$-dimensional space.

Let us inquire into the dependence of ψ on the displacements of one particular atom if only one normal vibration is excited, for example ν_1 in Fig. 24. The eigenfunction is then given by

$$\psi = NH_{v_1}(\sqrt{\alpha_1}\xi_1) \times e^{-(\alpha_1/2)\xi_1^2 - (\alpha_2/2)\xi_2^2 - (\alpha_3/2)\xi_3^2 - \cdots} \quad \text{(II, 54)}$$

If, for a moment, we neglect the zero-point motion of all the other normal vibrations, that is, if we put $\xi_2 = 0$, $\xi_3 = 0$, $\cdots$, we have $\psi = \psi_1(\xi_1)$; and since in this case, according to (II, 14), the displacement components of all atoms are proportional to ξ_1, the eigenfunction ψ as a function of every one of the displacement components has the same course as given in Fig. 29 except for an appropriate change of scale of the abscissa axis. The probability density of a particular atom would correspondingly be different from zero only along the line of classical motion of the atom and would vary in this line according to the full curves of Fig. 29 with an appropriate abscissa scale. However, actually we can never neglect the zero-point motion of the other vibrations; that is, ξ_2, ξ_3, $\cdots$ are different from zero even if $v_2 = v_3 = \cdots = 0$. In this case (II, 54) does not simplify to (II, 53), and (II, 14) does in general not lead to such simple expressions for the ξ_i. The general consequence of this is that the

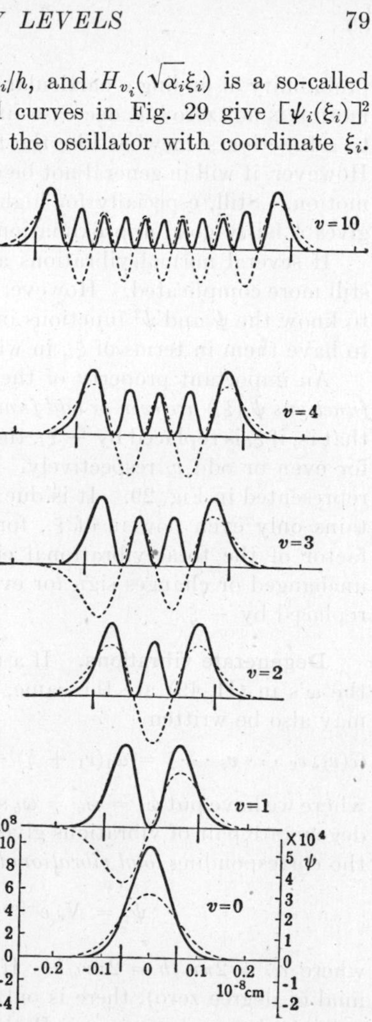

FIG. 29. **Eigenfunctions and probability distributions of the harmonic oscillator for $v_i = 0, 1, 2, 3, 4$, and 10.—The eigenfunctions are represented by the broken line curves, the probability distributions by the full line curves. All curves are drawn to the same scale. The vibrational frequency has been assumed to be 1000 cm^{-1}. The abscissa is proportional to the normal coordinate ξ. It is the actual displacement from the equilibrium position in a diatomic molecule of reduced mass 10.**

probability of finding a particular atom outside the line corresponding to the classical motion is not zero but decreases in any direction perpendicular to this line according to a function somewhat like the full curve $v = 0$ in Fig. 29 (Gauss error function). However, it will in general not be cylindrically symmetrical about the line of classical motion. Still, especially for higher v_1 values, the classical picture (Fig. 24 and 25) gives a fairly good approximation to the wave-mechanical probability distribution.

If several normal vibrations are excited simultaneously, the resulting ψ will be still more complicated. However, for most practical purposes it is quite unnecessary to know the ψ and ψ^2 functions in terms of $x_i, y_i, z_i, \cdots$, but it is entirely sufficient to have them in terms of ξ_i, in which form they are very simple [see (II, 42)].

An important property of the vibrational eigenfunctions should be noted. *The functions $\psi_i(\xi_i)$ are even or odd functions of the ξ_i depending on whether v_i is even or odd;* that is, if ξ_i is replaced by $-\xi_i$, the function $\psi_i(\xi_i)$ remains unchanged or changes sign for even or odd v_i respectively. This can easily be verified for the eigenfunctions represented in Fig. 29. It is due to the fact that for even v_i the function $\psi_i(\xi_i)$ contains only even powers of ξ_i, for odd v_i only odd powers of ξ_i. Since $\psi_i(\xi_i)$ is a factor of the total vibrational eigenfunction ψ, this latter function, too, remains unchanged or changes sign for even or odd v_i respectively if the corresponding ξ_i is replaced by $-\xi_i$.

Degenerate vibrations. If a molecule has a doubly degenerate vibration, two of the ω's in (II, 49) are the same, say $\omega_a = \omega_b$, and the formula for the *term values* may also be written

$$G(v_1, v_2 \cdots v_i \cdots) = \omega_1(v_1 + \tfrac{1}{2}) + \omega_2(v_2 + \tfrac{1}{2}) + \cdots + \omega_i(v_i + 1) + \cdots, \qquad \text{(II, 55)}$$

where we have put $\omega_i = \omega_a = \omega_b$ and $v_i = v_a + v_b$. Naturally each one of a mutually degenerate pair of vibrations gives its contribution $\frac{1}{2}\omega_i$ to the zero-point energy. In the corresponding *total vibrational eigenfunction* we have the factor

$$\psi_i = N_{v_i} e^{-(\alpha_i/2)(\xi_a^2+\xi_b^2)} H_{v_a}(\sqrt{\alpha_i}\xi_a) H_{v_b}(\sqrt{\alpha_i}\xi_b), \qquad \text{(II, 56)}$$

where $\alpha_i = 2\pi\nu_a/h = 2\pi\nu_b/h$. If $v_a = v_b = 0$, since $H_0(\sqrt{\alpha}\xi) =$ constant (polynomial of degree zero), there is only one function, that is, the *zero-point vibration does not introduce a degeneracy.* If the degenerate vibration is excited by one quantum we have either $v_a = 1, v_b = 0$ or $v_a = 0, v_b = 1$; that is, there are two eigenfunctions for the state $v_i = v_a + v_b = 1$ of energy $G = \cdots \omega_i(1 + 1) \cdots$. It is *doubly degenerate.* In this case any linear combination of the two eigenfunctions (II, 56) is also an eigenfunction of the same energy level. If two quanta are excited [$v_i = 2$, $G = \cdots \omega_i(2 + 1) \cdots$] we may have $v_a = 2, v_b = 0$ or $v_a = 0, v_b = 2$ or $v_a = 1$, $v_b = 1$; that is, there is a triple degeneracy. Quite generally the degree of degeneracy if v_i quanta of the doubly degenerate vibration are excited is equal to the number of different ways in which v_i can be written as a sum of two positive integers (where the order of the integers matters), that is, it is $v_i + 1$. This is indicated for the eight lowest vibrational levels in Fig. 30a. However, this high degeneracy exists only as long as strictly harmonic vibrations are assumed. As will be shown later the anharmonocity that is always present produces a partial splitting of this degeneracy.

By introducing *polar normal coordinates* by

$$\xi_a = \rho_i \cos \varphi_i, \qquad \xi_b = \rho_i \sin \varphi_i, \qquad \text{(II, 57)}$$

and choosing proper linear combinations of (II, 56), it can be shown [see Pauling-Wilson (18)] that the eigenfunctions of a doubly degenerate vibration may also be written

$$\psi_i = e^{(-\alpha_i/2)\rho_i^2} F_{v_i}^{l_i}(\sqrt{\alpha_i}\rho_i) e^{\pm j l_i \varphi_i}, \tag{II, 58}$$

where $F_{v_i}^{l_i}(\sqrt{\alpha_i}\rho_i)$ is a polynomial of degree v_i in ρ_i, where $j = +\sqrt{-1}$, and where l_i can take the values

$$l_i = v_i,\ v_i - 2,\ v_i - 4,\ \cdots 1 \text{ or } 0, \tag{II, 59}$$

depending on whether v_i is odd or even. In Fig. 30a the l_i values and degeneracies are indicated for the lower vibrational levels. The polynomials $F_{v_i}^{l_i}(\sqrt{\alpha_i}\rho_i)$ are related to the associated Laguerre polynomials [see, for example, Shaffer (776)]. For the lowest v_i and l_i values they are [see Kemble (12)]:

$$F_0{}^0 = 1, \qquad F_1{}^1 = -\sqrt{\alpha_i}\rho_i, \qquad F_2{}^2 = 2\alpha_i\rho_i{}^2, \qquad F_2{}^0 = 1 - \alpha_i\rho_i{}^2.$$

The factor $e^{\pm j l_i \varphi_i}$ in the eigenfunction (II, 58) indicates that there is in general an angular momentum of the vibrational motion. If in the two mutually degenerate vibrations the displacement vectors for each atom can be chosen at right angles to each other and therefore the angle φ_i is an angle in actual space (for example for XY_2 and X_3 in Fig. 27) the angular momentum would be $l_i(h/2\pi)$, where l_i is given by (II, 59) (for other cases see Chapter IV, section 2a).

Classically the transformation (II, 57) corresponds to the superposition of the degenerate linear oscillations with a phase difference of 90° (see above, p. 75), which, unless ξ_a and ξ_b are motions in the same line, results in circular or elliptical motions of the nuclei.

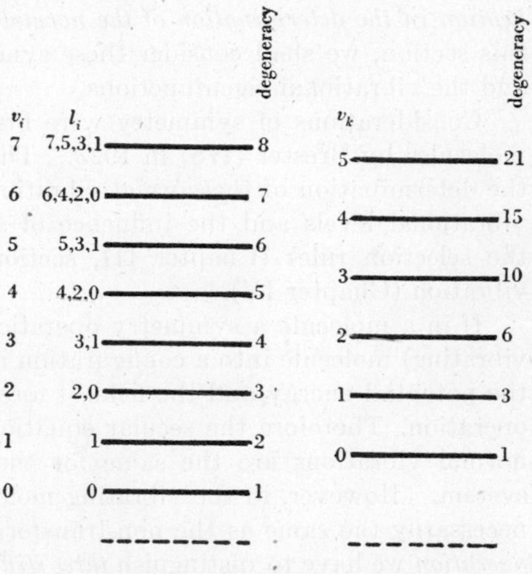

FIG. 30. Vibrational levels of (a) a doubly degenerate, (b) a triply degenerate vibration and their degrees of degeneracy.—The broken lines indicate the zero of energy. It should be noted that the lowest vibrational level $v = 0$ is $\frac{2}{2}\omega_i$ and $\frac{3}{2}\omega_k$ above this zero.

For *triply degenerate vibrations* three of the ω's in (II, 49), say ω_a, ω_b, ω_c, are the same, and we can write

$$G(v_1, v_2 \cdots v_k \cdots) = \omega_1(v_1 + \tfrac{1}{2}) + \omega_2(v_2 + \tfrac{1}{2}) + \cdots \omega_k(v_k + \tfrac{3}{2}) + \cdots, \tag{II, 60}$$

where $\omega_k = \omega_a = \omega_b = \omega_c$ and $v_k = v_a + v_b + v_c$. The corresponding factor in the eigenfunction is similar to (II, 56) except that there are now three terms in the exponential and three factors H_v. As before, for $v_k = 0$ there is only one eigenfunction, that is, *no degeneracy*. If the triply degenerate vibration is excited by one quantum ($v_k = 1$), there are three eigenfunctions ($v_a = 1$ or $v_b = 1$ or $v_c = 1$); that is, this state is *triply degenerate*. If the triply degenerate vibration is excited by two quanta we may have $v_a = 2, v_b = 0, v_c = 0$; or $v_a = 0, v_b = 2, v_c = 0$; or $v_a = 0, v_b = 0, v_c = 2$; or $v_a = 1, v_b = 1, v_c = 0$; or $v_a = 1, v_b = 0, v_c = 1$; or $v_a = 0, v_b = 1, v_c = 1$—that is, we have a *six-fold degeneracy*. In general, if the triply degenerate vibration is excited by v_k quanta, we have a $\frac{1}{2}(v_k + 1)(v_k + 2)$-fold degeneracy (again only if the anharmonicity is neglected). In Fig. 30b the degrees degeneracy for the lower vibrational levels are indicated.

In the most general *case of various degeneracies of the normal vibrations* the vibrational term values may be written conveniently

$$G(v_1, v_2, v_3 \cdots) = \Sigma\omega_i\left(v_i + \frac{d_i}{2}\right), \tag{II, 61}$$

where d_i is the degree of degeneracy of the vibration ω_i ($d_i = 1$ for non-degenerate vibration). It is seen that this formula includes (II, 49), (II, 55), and (II, 60).

3. Symmetry of Normal Vibrations and Vibrational Eigenfunctions

The degree of the secular equation (II, 38) from which the normal vibrations are obtained is $3N$ where N is the number of atoms in the molecule. Therefore, even for a moderate number N, the secular equation is by no means easy to solve. However, if a molecule has symmetry, also the normal vibrations and vibrational eigenfunctions have certain symmetry properties and in consequence considerable *simplification in the determination of the normal vibrations* is brought about. Therefore, in this section, we shall consider these symmetry properties of the normal vibrations and the vibrational eigenfunctions.

Considerations of symmetry were first applied to the vibrations of polyatomic molecules by Brester (178) in 1923. They are of greatest importance not only for the determination of the normal vibrations but also for the discussion of the higher vibrational levels and the influence of anharmonicity (section 5 of this chapter), the selection rules (Chapter III, section 2) and the interaction of rotation and vibration (Chapter IV).

If in a molecule a symmetry operation is carried out that transforms the (non-vibrating) molecule into a configuration indistinguishable from the original one, also the potential energy and the field of force will be the same as before the symmetry operation. Therefore the secular equation and consequently the frequencies of the normal vibrations are the same for the transformed as for the non-transformed system. However, in the vibrating molecule the transformed *displacements* are not necessarily the same as the non-transformed ones. *With respect to a given symmetry operation* we have to distinguish *three different behaviors of a normal vibration: It may remain unchanged, it may change sign, or it may change by more than just the sign.*

Mathematically there are two equivalent ways of carrying out a symmetry operation. We may either keep the coordinate system fixed and rotate or reflect the molecule, that is, change the position of the nuclei (*position transformation*), or we may keep the molecule fixed and refer it to different, rotated, or reflected coordinate systems (*coordinate transformation*). In what follows we shall always use the first method.

(a) *Effect of symmetry operations on non-degenerate normal vibrations*

For a given non-degenerate normal vibration ν_i there is only one possible ratio for the displacement coordinates of the various atoms (see p. 69f.). If a symmetry operation is carried out this ratio remains unchanged since the frequency ν_i substituted into the equations (II, 10) is the same. Therefore since the displacements are defined apart from a constant factor only (which is defined by the normalization, see p. 72), a symmetry operation can at most bring about a simultaneous change of sign of all displacement coordinates belonging to a given non-degenerate vibration, that is, a change of sign of the normal coordinate. The only other possibility is that it will leave them unchanged. *Thus a non-degenerate vibration can only be symmetric or antisymmetric with respect to any symmetry operation that is permitted by the symmetry of the molecule.*

As an example, consider the normal vibrations of the plane XYZ_2 molecule represented in the previous Fig. 24a-f. When these vibrations are reflected at the plane of symmetry of the molecule $\sigma_v(xz)$ which is perpendicular to the plane of the molecule, the diagrams in Fig. 24g-l are obtained. It is seen that this reflection leads to identical pictures for ν_1, ν_2, ν_3, ν_6, whereas for ν_4 and ν_5 the direction of all displacement vectors has been inverted; that is, the corresponding normal coordinates ξ_4 and ξ_5 have been changed into their negatives $-\xi_4$ and $-\xi_5$ (phase shift by 180°). These vibrations are antisymmetric with respect to the plane $\sigma_v(xz)$. In a similar way it can be seen that all vibrations but ν_6 are symmetric with respect to the plane of the molecule, while ν_6 is antisymmetric. Finally ν_4, ν_5, and ν_6 are antisymmetric with respect to a rotation by 180° about the X—Y axis (two-fold axis).

It is an immediate result of the above rule (and is verified by the example of Fig. 24) that a nucleus that has its equilibrium position on a plane of symmetry in a non-degenerate vibration can only move *in the plane* (if the vibration is symmetric to the plane) *or perpendicular to the plane* (if it is antisymmetric). Similarly, a nucleus that has its equilibrium position on an axis of symmetry can only move *along this axis* (if the vibration is symmetric with respect to this axis) *or perpendicular to it* (if it is antisymmetric).

The following restriction to the above rule may be noted. If the molecule has a p-fold axis of symmetry and p is odd, a non-degenerate vibration can only be symmetric with respect to a rotation by $2\pi/p$ about this axis since, if it were antisymmetric, after p such rotations, that is, a rotation by 2π, it would not transform into itself as it must. However, a non-degenerate vibration may be antisymmetric as well as symmetric with respect to an even-fold axis, since then after p such rotations it will transform into the original configuration. The non-degenerate vibrations of the symmetrical molecules X_3, X_4, X_5, X_6 represented in Figs. 32a, 37, 38a, 40 exemplify this restriction.

(*b*) *Effect of symmetry operations on degenerate normal vibrations*

Two simple examples. Whereas non-degenerate vibrations can only be symmetric or antisymmetric with respect to any symmetry operation, degenerate normal vibrations may also change by more than just the sign. Before we discuss the reason for this, let us consider two examples. In Fig. 25b the *normal vibrations of a linear symmetric triatomic molecule* XY_2 (for instance CO_2) are represented. The two vibrations ν_{2a} and ν_{2b} are obviously degenerate with each other. They are antisymmetric with respect to an inversion at the center of symmetry as is the vibration ν_3. Another symmetry operation is the rotation C_∞^φ by an arbitrary angle φ about the internuclear axis. This leaves ν_1 and ν_3 unchanged; they are symmetric with respect to this symmetry operation. But both ν_{2a} and ν_{2b} are changed by more than just the sign. This is represented in Fig. 31a, in which a side view of the vibrations before and after the symmetry operation is given. Before the transformation, ν_{2a} takes place in the x-direction, nucleus i having an amplitude $x_i^{(a)}$. After the transformation the amplitude in the x-direction is $x_i^{(a)} \cos\varphi$, and in addition there is now also a component of the motion in the y-direction with amplitude $x_i^{(a)} \sin\varphi$. Both components change periodically in simple harmonic motion whose frequency is obviously the same as before the symmetry operation. Similar considerations apply to ν_{2b}.

As a second example, consider *the normal vibrations of a molecule* X_3 *forming an equilateral triangle* (three-fold axis of symmetry) in Fig. 32a. The diagrams obtained

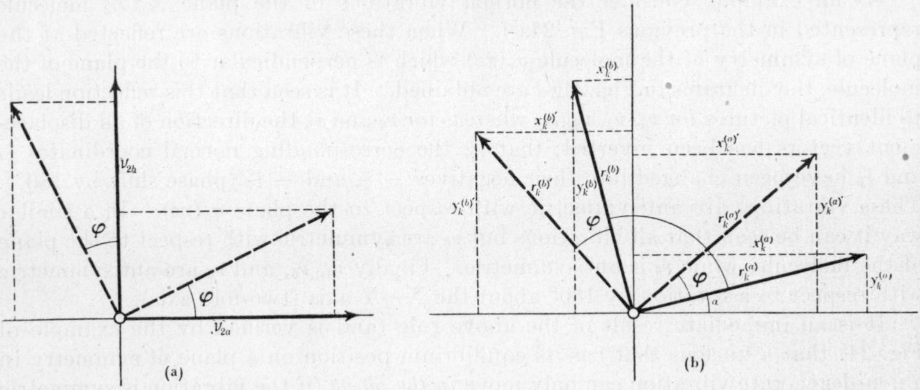

FIG. 31. Effect of the symmetry operation C_∞^φ on the degenerate vibration of linear XY_2.

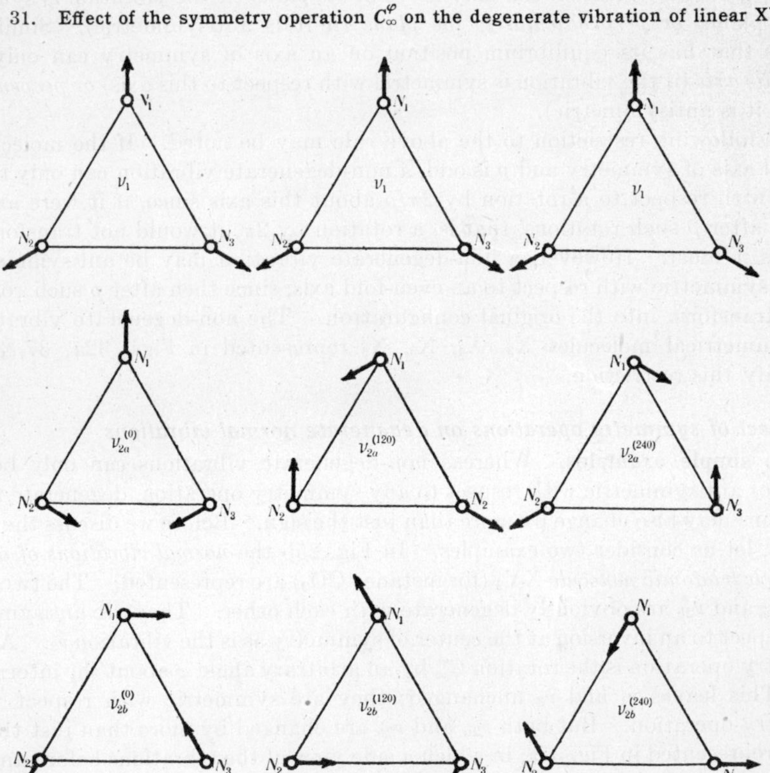

FIG. 32. Effect of the symmetry operations C_3 and C_3^2 on the normal vibrations of an X_3 molecule.

by rotating the molecule with its displacement vectors in a clockwise direction by 120° and 240° (or − 120°) are shown in Fig. 32b and c. It is seen that while ν_1 remains unchanged by these rotations, that is, is symmetric with respect to the three-fold axis, the two other vibrations ν_{2a} and ν_{2b}, which are degenerate with each other (see below), are neither symmetric nor antisymmetric but change into different vibrations which, however, have obviously the same frequency. They differ only in that the nucleus N_1 at the top, for example, instead of moving up and down in $\nu_{2a}^{(0)}$, after the rotation, in $\nu_{2a}^{(120)}$, moves at an angle of 120° to the vertical. Similar changes occur for the other two atoms N_2 and N_3.

By superposition of $\nu_{2a}^{(0)}$ and $\nu_{2a}^{(120)}$ or $\nu_{2a}^{(240)}$ with suitable amplitudes, we can obtain a simple harmonic motion of the nucleus N_1 in any other direction, and similarly of N_2 and N_3. For instance, we may obtain a motion of N_1 in a horizontal direction by superimposing on $\nu_{2a}^{(0)}$ the vibration $\nu_{2a}^{(240)}$ with double the amplitude. This is shown in Fig. 33a. It is seen that in this way, apart from a constant factor, the

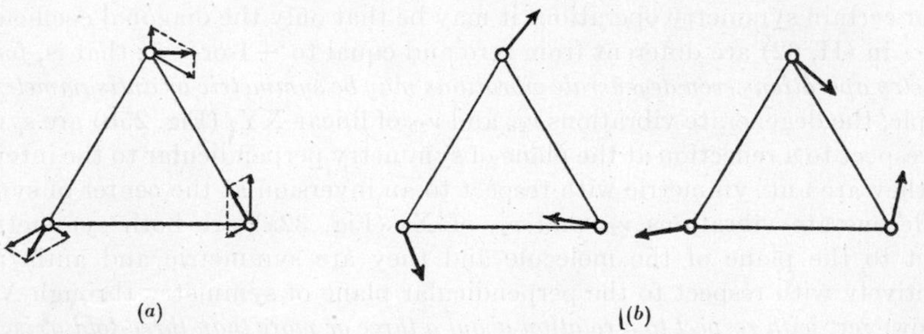

FIG. 33. **Degenerate vibration ν_2 of X_3. (a) Superposition of $\nu_{2a}^{(0)}$ and $\nu_{2a}^{(240)}$ to give $\nu_{2b}^{(0)}$. (b) Orthogonal pair different from ν_{2a} and ν_{2b} in Fig. 32.**

vibration $\nu_{2b}^{(0)}$ is obtained. Thus ν_{2b} is a linear combination of $\nu_{2a}^{(0)}$ and $\nu_{2a}^{(240)}$. Since $\nu_{2a}^{(0)}$ and $\nu_{2a}^{(240)}$ have the same frequency, ν_{2b} also has this frequency. That is why ν_{2a} and ν_{2b} are degenerate with each other. Conversely, of course, we may obtain $\nu_{2a}^{(240)}$ as a linear combination of $\nu_{2a}^{(0)}$ and $\nu_{2b}^{(0)}$. Any other oscillation of the same frequency may be represented as a linear combination of $\nu_{2a}^{(0)}$ and $\nu_{2b}^{(0)}$ (for example, also, $\nu_{2a}^{(120)}$, $\nu_{2b}^{(120)}$, $\nu_{2b}^{(240)}$). The vibrations $\nu_{2a}^{(0)}$ and $\nu_{2b}^{(0)}$ are *mutually orthogonal*: in $\nu_{2b}^{(0)}$ all nuclei move at right angles to the paths they traverse in $\nu_{2a}^{(0)}$. However, $\nu_{2a}^{(0)}$ and $\nu_{2b}^{(0)}$ are not the only orthogonal pair by whose superposition all other vibrations of the same frequency may be represented. There is an infinite number of such pairs. Fig. 33b gives another example. $\nu_{2a}^{(0)}$ and $\nu_{2b}^{(0)}$ are distinguished by being symmetric and antisymmetric respectively with respect to the plane of symmetry though N_1.

As we have seen before (p. 75), we have a degenerate vibration when two or more roots of the secular equation coincide. There are then two or more sets of solutions of the equations (II, 10) for the same frequency ν_i:

$$x_1{}^{(a)}y_1{}^{(a)}z_1{}^{(a)}x_2{}^{(a)}y_2{}^{(a)}z_2{}^{(a)}\cdots,$$
$$x_1{}^{(b)}y_1{}^{(b)}z_1{}^{(b)}x_2{}^{(b)}y_2{}^{(b)}z_2{}^{(b)}\cdots,$$
$$\cdots\cdots\cdots\cdots$$

That is to say, there are *two or more normal coordinates* ξ_{ia}, ξ_{ib}, $\cdots$ that differ by more than just a constant factor. A symmetry operation which does not change the field of force may therefore change each of the degenerate normal coordinates into *a linear combination* of these normal coordinates, since such a linear combination is also a solution of equations (II, 10) (see p. 75). Thus we have

$$\begin{aligned} \xi_{ia} &\rightarrow \xi'_{ia} = d_{aa}\xi_{ia} + d_{ab}\xi_{ib} + \cdots; \\ \xi_{ib} &\rightarrow \xi'_{ib} = d_{ba}\xi_{ia} + d_{bb}\xi_{ib} + \cdots, \\ &\cdot \quad \cdot \quad \cdot \quad \cdot \quad \cdot \quad \cdot \quad \cdot \quad \cdot \quad \cdot \quad \cdot \end{aligned} \tag{II, 62}$$

where the $\rightarrow$ indicates "goes over, by a symmetry operation, into" and the d_{aa}, d_{ab} $\cdots$ are constant coefficients to be determined below. The $\cdots$ should be dropped for doubly degenerate vibrations. Since ξ_{ia}, ξ_{ib}, $\cdots$ may be taken to be the displacement vectors of any nucleus k in the degenerate vibration ν_i, it is immediately clear that for a doubly degenerate vibration the transformed displacement vectors of a given nucleus all lie in the plane determined by the two "original" displacement vectors, while for a triply degenerate vibration they are not restricted to a plane.

For certain symmetry operations it may be that only the diagonal coefficients d_{aa}, d_{bb}, $\cdots$ in (II, 62) are different from zero and equal to $+1$ or -1; that is, *for certain symmetry operations even degenerate vibrations may be symmetric or antisymmetric.* For example, the degenerate vibrations ν_{2a} and ν_{2b} of linear XY_2 (Fig. 25b) are symmetric with respect to a reflection at the plane of symmetry perpendicular to the internuclear axis; they are antisymmetric with respect to an inversion at the center of symmetry. The degenerate vibrations ν_{2a} and ν_{2b} of X_3 (Fig. 32a) are both symmetric with respect to the plane of the molecule and they are symmetric and antisymmetric respectively with respect to the perpendicular plane of symmetry through N_1.

However, *with respect to a rotation about a three or more than three-fold axis, degenerate vibrations are in general neither symmetric nor antisymmetric but change according to* (II, 62) *with non-vanishing* d_{ab}, d_{ba}, $\cdots$.[6] For example, in the case of the *linear molecule* XY_2 (see Fig. 25b), if we take the normal coordinates ξ_{2a} and ξ_{2b} of the two degenerate vibrations orthogonal to each other and normalized, that is, if the displacement vectors $r_k^{(a)}$ and $r_k^{(b)}$ of each atom k are perpendicular to each other and of equal magnitude, we have, for a simultaneous rotation of the two displacement vectors by an angle φ (see Fig. 31b),

$$x_k^{(a)\prime} = r_k^{(a)} \cos(\alpha + \varphi) = x_k^{(a)} \cos\varphi - y_k^{(a)} \sin\varphi,$$

and similarly

$$y_k^{(a)\prime} = x_k^{(a)} \sin\varphi + y_k^{(a)} \cos\varphi,$$

where the primed coordinates are the ones after the rotation. Since, according to Fig. 31b, $y_k^{(a)} = -x_k^{(b)}$ and $x_k^{(a)} = y_k^{(b)}$, we may also write

$$\begin{aligned} x_k^{(a)\prime} &= x_k^{(a)} \cos\varphi + x_k^{(b)} \sin\varphi, \\ y_k^{(a)\prime} &= y_k^{(a)} \cos\varphi + y_k^{(b)} \sin\varphi. \end{aligned} \tag{II, 63}$$

Similarly we obtain

$$\begin{aligned} x_k^{(b)\prime} &= -x_k^{(a)} \sin\varphi + x_k^{(b)} \cos\varphi, \\ y_k^{(b)\prime} &= -y_k^{(a)} \sin\varphi + y_k^{(b)} \cos\varphi; \end{aligned} \tag{II, 64}$$

[6] This holds as long as only real normal coordinates and coefficients are admitted. For complex normal coordinates, see below.

or, since (II, 63) and (II, 64) hold for any k,

$$\xi'_{2a} = \xi_{2a} \cos \varphi + \xi_{2b} \sin \varphi$$
$$\xi'_{2b} = - \xi_{2a} \sin \varphi + \xi_{2b} \cos \varphi; \qquad \text{(II, 65)}$$

that is, for a rotation by an angle φ about the symmetry axis (C_∞^φ), the coefficients in (II, 62) are in the case of a degenerate vibration of a linear (triatomic) molecule:

$$d_{aa} = d_{bb} = \cos \varphi, \qquad d_{ab} = - d_{ba} = \sin \varphi, \qquad \text{(II, 66)}$$

Equation (II, 65) is independent of the angle α, that is, of the orientation of the fixed coordinate system, x, y.

In a similar manner, in the *molecule* X_3, if ξ_{2a} and ξ_{2b} are the mutually orthogonal normal coordinates belonging to the vibrations ν_{2a} and ν_{2b} of Fig. 32, and if we assume that they have the same amplitude, then we have for a (clockwise) rotation of the whole diagram by an angle of 120° (which is a symmetry operation),

$$\xi_{2a} \rightarrow \xi'_{2a} = \xi_{2a} \cos 120° + \xi_{2b} \sin 120°,$$
$$\xi_{2b} \rightarrow \xi'_{2b} = - \xi_{2a} \sin 120° + \xi_{2b} \cos 120°; \qquad \text{(II, 67)}$$

that is, again we have a linear transformation of the type (II, 62). More explicitly, the equations (II, 67) mean that for every nucleus k,

$$x_k^{(a)} \rightarrow x_k^{(a)'} = x_k^{(a)} \cos 120° + x_k^{(b)} \sin 120°,$$
$$y_k^{(a)} \rightarrow y_k^{(a)'} = y_k^{(a)} \cos 120° + y_k^{(b)} \sin 120°,$$
$$x_k^{(b)} \rightarrow x_k^{(b)'} = - x_k^{(a)} \sin 120° + x_k^{(b)} \cos 120°, \qquad \text{(II, 68)}$$
$$y_k^{(b)} \rightarrow y_k^{(b)'} = - y_k^{(a)} \sin 120° + y_k^{(b)} \cos 120°.$$

Here it is to be remembered (see p. 82) that the primed coordinates are referred to the same coordinate system as the unprimed and that we number the nuclei according to their position (the nucleus at the top in Fig. 32 is always N_1, and so on) and not according to what they had been before the transformation. With this in mind, the equations (II, 68) can easily be verified by the reader from Fig. 32. They hold irrespective of which orthogonal pair is chosen, whether $\nu_{2a}^{(0)}$, $\nu_{2b}^{(0)}$ or $\nu_{2a}^{(120)}$, $\nu_{2b}^{(120)}$ or $\nu_{2a}^{(240)}$, $\nu_{2b}^{(240)}$ in Fig. 32, or ν_{2a} and ν_{2b} in Fig. 33b, or any other orthogonal pair. Always *the displacement vectors of a nucleus of a given position* (for example, the one at the top) *rotate by* 120° *in a counter-clockwise direction if the molecule is rotated by* 120° *in the clockwise direction.* This rule has here been derived from the assumed form of the degenerate normal vibrations. Since we shall see that it is a necessary consequence of the theory of normal vibrations, it can conversely be used to determine the form of the degenerate vibrations.

When the symmetry operation is the counter-clockwise rotation by 120°, one has to replace 120° by −120° in (II, 67) and (II, 68). The result is obviously the same as when the clockwise rotation is carried out twice in succession, that is, when 120° is replaced by 240°.

Plane doubly degenerate vibrations. We shall now discuss the general case of doubly degenerate normal vibrations in which the nuclei move in planes perpendicular to a p-fold axis of symmetry [see Cabannes (189)]. We shall also show how the form of these degenerate vibrations can be determined. In Fig. 34, N_k and N_{k+1} are two identical nuclei that are transformed into each other by rotation about the p-fold axis C_p (assumed to be perpendicular to the plane of the paper) through the angle $2\pi/p$. The displacements of these nuclei during the two mutually degenerate and orthogonal vibrations ν_{ia} and ν_{ib} of normal coordinates ξ_{ia} and ξ_{ib} are indicated by the heavy arrows $r_k^{(a)}$, $r_k^{(b)}$ and $r_{k+1}^{(a)}$, $r_{k+1}^{(b)}$. We assume for the present that all amplitudes r of

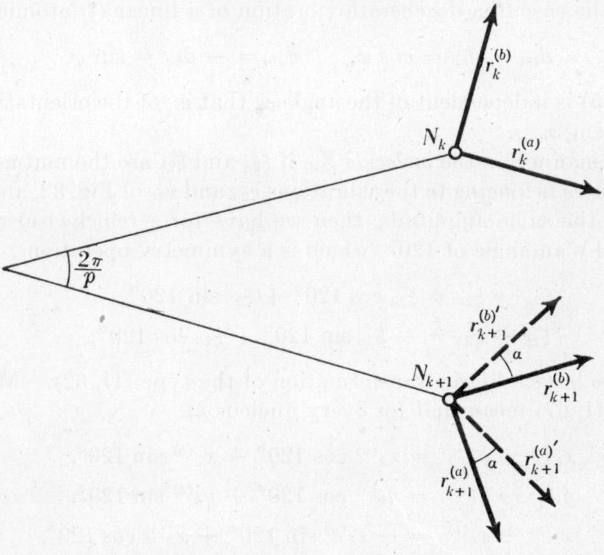

FIG. 34. **Doubly degenerate vibrations in a molecule with a p-fold axis.**

the p identical nuclei have the same magnitude. If a clockwise rotation by $2\pi/p$ is carried out the vectors $r_k^{(a)}$, $r_k^{(b)}$ go over into the dotted displacements $r_{k+1}^{(a)'}$ and $r_{k+1}^{(b)'}$ of N_{k+1}, making an angle α with $r_{k+1}^{(a)}$ and $r_{k+1}^{(b)}$ respectively. Applying (II, 63) and (II, 64) to the present case, we have

$$\begin{aligned} x_{k+1}^{(a)'} &= x_{k+1}^{(a)} \cos \alpha + x_{k+1}^{(b)} \sin \alpha, \\ y_{k+1}^{(a)'} &= y_{k+1}^{(a)} \cos \alpha + y_{k+1}^{(b)} \sin \alpha, \\ x_{k+1}^{(b)'} &= -\, x_{k+1}^{(a)} \sin \alpha + x_{k+1}^{(b)} \cos \alpha, \\ y_{k+1}^{(b)'} &= -\, y_{k+1}^{(a)} \sin \alpha + y_{k+1}^{(b)} \cos \alpha. \end{aligned} \qquad \text{(II, 69)}$$

If we were to choose the directions of $r_k^{(a)}$ and $r_k^{(b)}$ arbitrarily for every k, the angle α would be different for different k. But in order to obtain a linear transformation as in (II, 62) for the normal coordinates, all displacement components must transform in the same way; that is, we have to choose the displacements so that α is independent of k. Then we have

$$\begin{aligned} \xi'_{ia} &= \xi_{ia} \cos \alpha + \xi_{ib} \sin \alpha, \\ \xi'_{ib} &= -\, \xi_{ia} \sin \alpha + \xi_{ib} \cos \alpha, \end{aligned} \qquad \text{(II, 70)}$$

as *the law of transformation of the two degenerate* (*orthogonal*) *normal coordinates for a rotation by* $2\pi/p$. It might at first appear that the angle α can have any fixed value. However, after p rotations by $2\pi/p$ we must obviously obtain the original diagram.

Therefore, since for each rotation of the molecule a displacement vector turns by α, $p\alpha$ must be $\pm 2\pi$ or $\pm 2\cdot 2\pi$ or $\pm 3\cdot 2\pi \cdots$ or $\pm (p-1)2\pi$. The value $\pm p\cdot 2\pi$ would mean $\alpha = \pm 2\pi$, that is, give the same transformation as $\alpha = 0$, namely $\xi'_{ia} = \xi_{ia}$, $\xi'_{ib} = \xi_{ib}$. The vibration would be symmetric with respect to the rotation about the symmetry axis. Obviously the values $\pm (p+1)2\pi$, $\pm (p+2)2\pi \cdots$ for $p\alpha$ would be equivalent to $\pm 2\pi$, $\pm 2\cdot 2\pi \cdots$. Thus we have

$$\alpha = \pm \frac{2\pi}{p}\cdot l, \qquad 0 < l < p, \tag{II, 71}$$

where l is an integral number.

When a degenerate vibration transforms according to (II, 70) with $0 < |\alpha| < 2\pi$ (and $\alpha \neq \pi$) for a rotation by $2\pi/p$ about a p-fold axis, we say, for brevity, that it is *degenerate with respect to this axis*. It may or may not be symmetric or antisymmetric with respect to other symmetry elements if such are present. It might appear from (II, 71) that in a molecule with a p-fold axis there are $p-1$ different types of vibrations that are degenerate with respect to the p-fold axis, namely those with $l = 1, l = 2 \cdots l = p-1$. However, the types $l = 1$ and $l = p-1$, $l = 2$ and $l = p-2$, $l = 3$ and $l = p-3 \cdots$, are equivalent, since $\alpha = \pm (p-j)\dfrac{2\pi}{p}$ is equivalent to $\alpha = \mp j\cdot\dfrac{2\pi}{p}$ (the difference of the two being 2π). Furthermore, if p is even, $l = p/2$ is one of the possible l values; that is, $\alpha = \pi$. In this case, according to (II, 70), $\xi'_{ia} = -\xi_{ia}$ and $\xi'_{ib} = -\xi_{ib}$, which means that this type of vibration is antisymmetric and not degenerate with respect to the p-fold axis considered. Thus if we put $p = 2q - 2$ if p is even, and if we put $p = 2q - 1$ if p is odd, there are q *different types* (*species*) *of degenerate vibrations distinguished by the value of l, that is, by the angle α by which the displacement vectors rotate for a rotation of the molecule by* $2\pi/p$. Of course there may be several degenerate vibrations of the same type (see section 4a). Degenerate vibrations that differ only by the sign of α are considered as of the same type for the following reason: Suppose ξ_{ia} and ξ_{ib} transform according to (II, 70) with $\alpha = -\dfrac{2\pi}{p}l$, then it is immediately seen that ξ_{ia} and $-\xi_{ib}$ transform according to (II, 70) with $\alpha = +\dfrac{2\pi}{p}l$. But the pair ξ_{ia}, ξ_{ib} is obviously perfectly equivalent to ξ_{ia}, $-\xi_{ib}$.

For a *molecule with a three-fold axis* C_3 the number l introduced above may have the values 1 and 2. But since $2 = p - 1 = 3 - 1$ there is *only one type* (*species*) *of degenerate vibrations*, namely the one in which the displacement vectors in a plane perpendicular to the symmetry axis rotate by $\pm 120°$ for a clockwise rotation of the molecule by 120°. We have already seen (Figs. 32 and 33b) that this is actually fulfilled for the degenerate pair ν_{2a} and ν_{2b} of X_3, with $\alpha = +120°$ (counted positive in a counter-clockwise direction). Conversely, if we did not know the form of the degenerate normal vibration we could determine it from the above condition. This is shown in Fig. 35. For example, starting out from a vertical vector at N_1 [Fig. 35a (1)], we rotate this vector counter-clockwise by 120° (dotted vector), then rotate the whole diagram by 120° counter-clockwise. The dotted vector then goes over into the heavy vector at N_2 [Fig. 35a (2)], which represents the actual displacement vector of N_2 when the vertical vector at N_1 is the displacement vector of N_1, since by clockwise rotation of the whole molecule by 120° the displacement vector of N_1 would rotate by 120° in the counter-clockwise direction. Similarly rotating the displacement vector of N_2 again by 120°, obtaining a dotted vector at N_2, and rotating the whole diagram by 120°, we obtain the displacement vector of N_3 [Fig. 35a (3)]. Thus we have obtained ν_{2a} of Fig. 32. In the same way, by starting out from

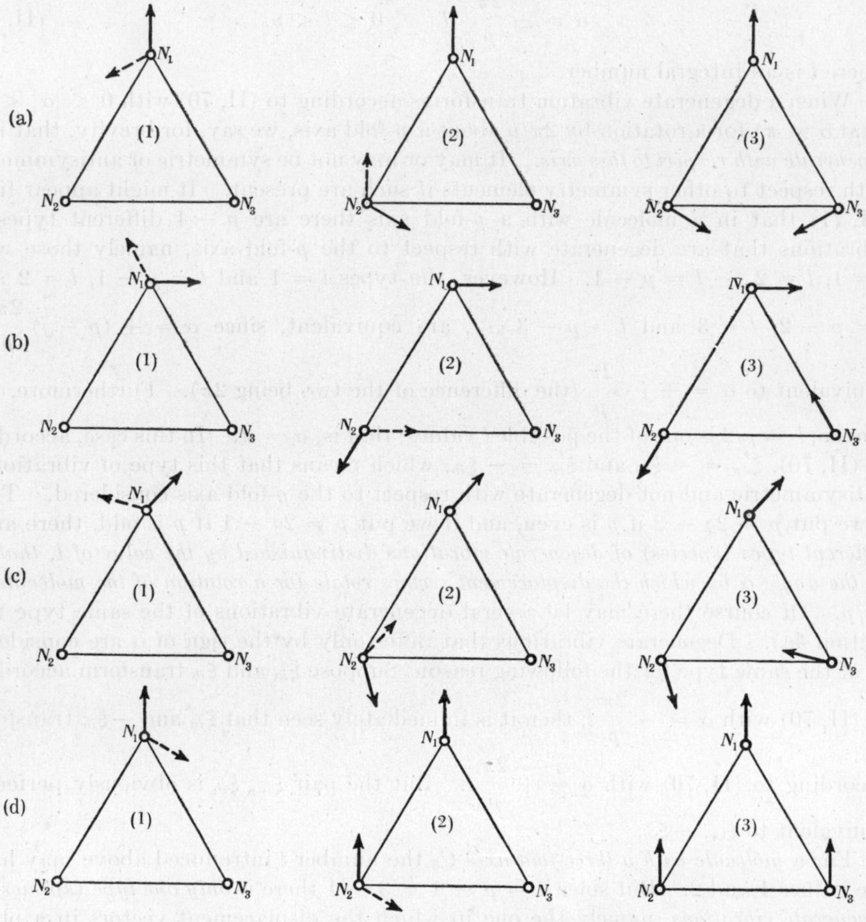

FIG. 35. **Derivation of degenerate normal vibrations of X_3.**

a horizontal displacement at N_1 (Fig. 35b), the vibration ν_{2b} is obtained, which, as shown before, is degenerate with ν_{2a}. Fig. 35c shows the same for a displacement of N_1 of arbitrary direction, giving a vibration that is a linear combination of ν_{2a} and ν_{2b} (see Fig. 33b). If instead of $\alpha = +120°$ we choose $\alpha = -120°$ we obtain, as shown by Fig. 35d, simply a *translation* of the whole molecule in the direction of the initially assumed displacement vector. Thus the two translations perpendicular to the three-fold axis are a *non-genuine degenerate vibration* of the type $l = 1$.

As an example of a somewhat more complicated case of a molecule with a three-fold axis, Fig. 36 gives the *normal vibrations of a plane symmetrical* X_3Y_3 *molecule of point group* $\boldsymbol{D_{3h}}$. Since here a Y nucleus may move either in the same direction as the adjacent X nucleus or in the opposite direction, there are now *two* totally symmetric vibrations, ν_1 and ν_2; there are two vibrations that are symmetric with respect to the C_3, but antisymmetric with respect to the three σ_v: the rotation about the C_3 (which is a non-genuine vibration and therefore not shown) and the vibration ν_3; there are *two*

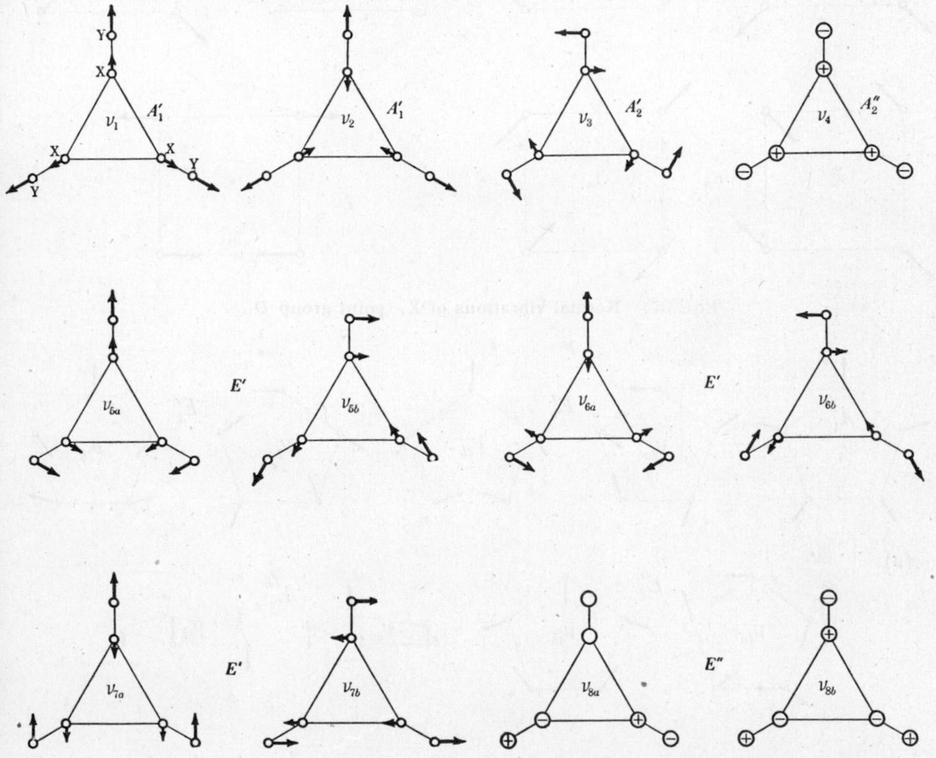

FIG. 36. **Normal vibrations of an X_3Y_3 molecule (point group D_{3h}).**— Motions perpendicular to the plane of the paper are indicated by + and − signs in the circles representing the nuclei.

degenerate vibrations with $\alpha = +120$, namely ν_5 and ν_6; and there are two degenerate vibrations with $\alpha = -120$, namely the translations perpendicular to C_3 which represent a non-genuine vibration (not shown) and the genuine vibration ν_7. This is in addition to the vibrations parallel to the three-fold axis, which will be discussed later. It should be noted that the vibrations ν_{6a}, ν_{6b}, ν_{7a}, and ν_{7b} cannot be obtained as linear combinations of ν_{5a} and ν_{5b}; that is, ν_6 and ν_7 have in general a frequency different from that of ν_5.

For a *molecule with a four-fold axis of symmetry*, l may have the values 1, 2, and 3. But $l = 2 = p/2$ corresponds to vibrations that are antisymmetric with respect to the axis and $l = 3 = p - 1$ is equivalent $l = 1$, so that there is again only one type (species) of degenerate vibrations. As an example, Fig. 37 gives the normal vibrations of a molecule X_4 with a four-fold axis. ν_1 is symmetric, ν_2, ν_3, and ν_4 are antisymmetric with respect to a rotation by $2\pi/4 = 90°$ about the axis, whereas ν_{5a} and ν_{5b} are a degenerate pair. Any linear combination of ν_{5a} and ν_{5b}, for example ν_5', is also a vibration of the same frequency ($\xi_5' = \xi_{5b} - \xi_{5a}$). It is seen that by a clock-

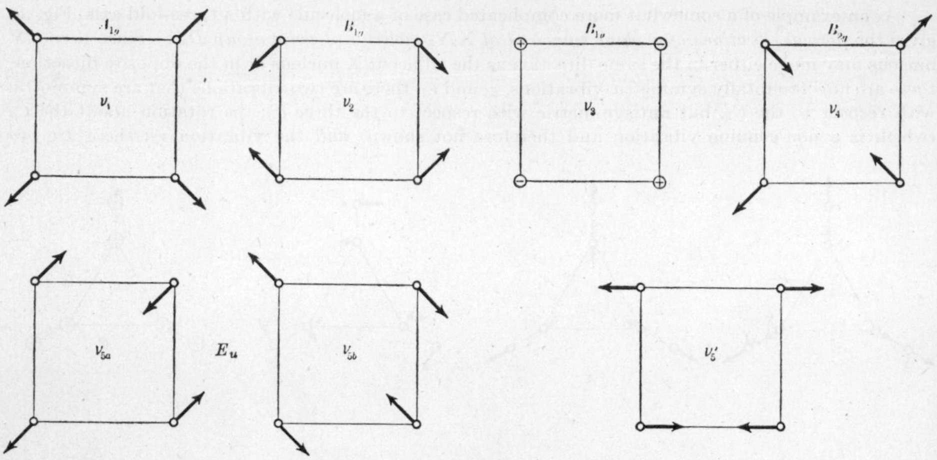

FIG. 37. **Normal vibrations of X_4 (point group D_{4h}).**

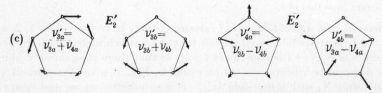

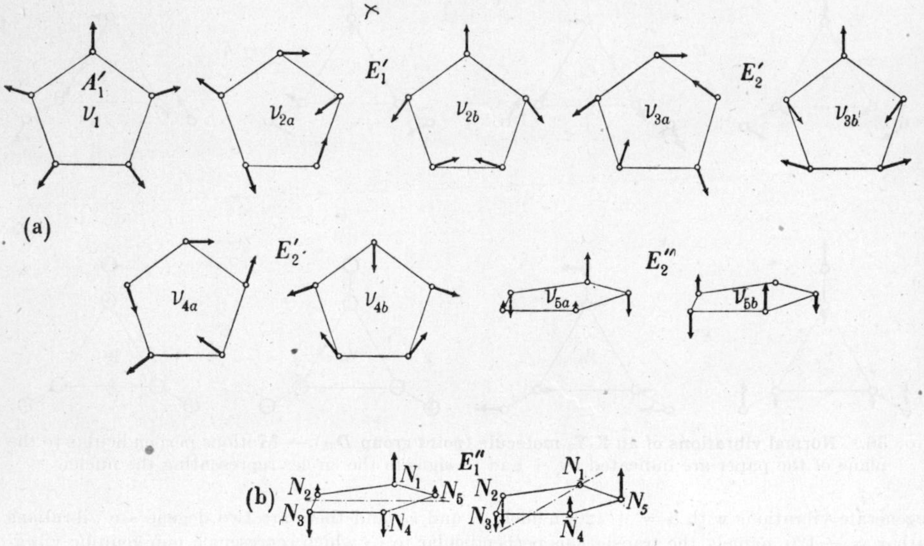

FIG. 38. **Normal vibrations of an X_5 molecule (point group D_{5h}). (a) Genuine normal vibrations. (b) A non-genuine degenerate vibration (see p. 96). (c) Alternative degenerate vibrations $\nu_{3a}' = \nu_{3a} + \nu_{4a}$, $\nu_{3b}' = \nu_{3b} + \nu_{4b}$, $\nu_{4a}' = \nu_{3b} - \nu_{4b}$, $\nu_{4b}' = \nu_{3a} - \nu_{4a}$.—Vibrations perpendicular to the plane of the molecule are shown in oblique projection.**

wise rotation by 90° of the molecule in these degenerate vibrations every displacement vector rotates counter-clockwise by +90°.

For a *molecule with a five-fold axis of symmetry*, l may have the values 1, 2, 3, and 4, but according to the above $l = 1$ is equivalent to $l = 4$ and $l = 2$ is equivalent to $l = 3$, so that we have *two types (species) of degenerate vibrations* corresponding to $\alpha = \pm 2\pi/5$ and $\alpha = \pm 2\cdot 2\pi/5$ respectively. Let us derive as an example the form of the (plane) normal vibrations of a molecule X_5 whose nuclei are arranged at the corners of a regular pentagon. The only vibration that is symmetric with respect to the axis C_5 is evidently the pulsation vibration ν_1 in Fig. 38a. If we want to

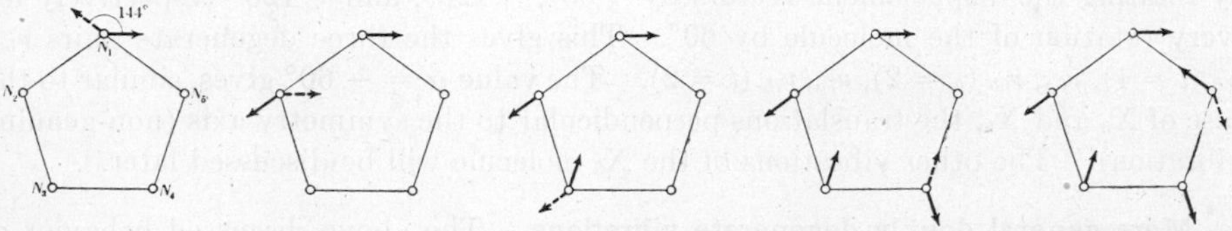

FIG. 39. **Derivation of the vibration ν_{3a} of X_5.**—From one diagram to the next a rotation of the molecule by 72° is assumed to have taken place. In this rotation a broken line vector goes over into the full-line displacement vector of the next nucleus (rotation by 144°).

determine a vibration of the type $l = 1$ we have to use $\alpha = \pm 360°/5 = \pm 72°$ and proceed in the same way as indicated in Fig. 35 for an X_3 molecule. With $\alpha = +72°$ we obtain thus the degenerate pair ν_{2a} and ν_{2b} in Fig. 38a; with $\alpha = -72°$ we obtain (similar to Fig. 35d) the two translations perpendicular to the five-fold axis which thus represent a non-genuine degenerate vibration with $l = 1$. If now we want to determine a vibration of the type $l = 2$, that is, $\alpha = \pm 2\cdot 360°/5 = \pm 144°$, we have to rotate the displacement vectors by 144° for every rotation of the molecule by 72°. This is shown for $\alpha = +144°$ in Fig. 39 and gives the vibration ν_{3a} of Fig. 38a. ν_{3b} is obtained in the same way but using a vertical instead of a horizontal displacement

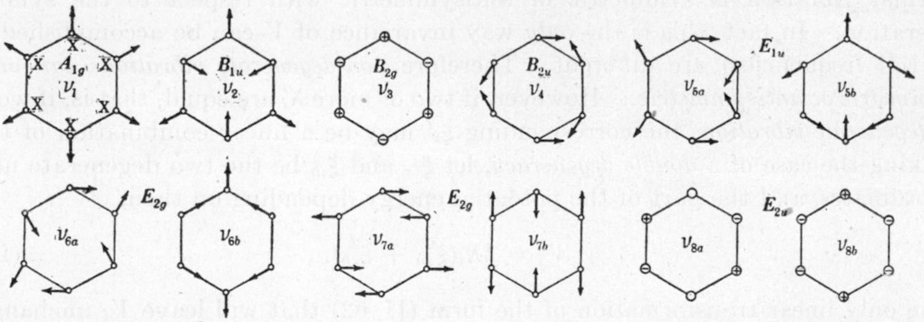

FIG. 40. **Normal vibrations of an X_6 molecule (point group D_{6h}).**—Displacement vectors at right angles to the plane of the paper are indicated by + and − signs. In ν_{8b} these vectors have not all the same length.

vector at the top. If $\alpha = -144°$, another degenerate pair of genuine normal vibrations is obtained, ν_{4a} and ν_{4b} of Fig. 38a.[7] Thus there are in this example two plane degenerate vibrations of type $l = 2$.

[7] The displacement vector of the nucleus at the top in ν_{4b} has been chosen opposite to that in ν_{3b} in order to have the pair ν_{4a} and ν_{4b} transform according to (II, 70) with $\alpha = +144°$, just as ν_{3a} and ν_{3b} (see p. 89).

For *a molecule with a six-fold axis of symmetry*, l may have the values 1, 2, 3, 4, and 5; but according to the above $l = 3 = p/2$ corresponds to vibrations that are antisymmetric with respect to the axis C_6, and $l = 4$ and 5 are equivalent to $l = 2$ and $l = 1$ respectively so that we have again only *two types of degenerate vibrations*. Fig. 40 gives as an example the normal vibrations of a molecule X_6 whose nuclei in their equilibrium positions lie at the corners of a regular hexagon. There is again only one vibration that is symmetric with respect to a rotation by 60° about the axis, but there are three vibrations that are antisymmetric with respect to this rotation. The plane degenerate vibrations are obtained in a manner similar to that used for X_5 by rotating the displacement vectors by +60°, +120°, and −120° respectively for every rotation of the molecule by 60°. This gives the three degenerate pairs ν_{5a}, ν_{5b} ($l = 1$), ν_{6a}, ν_{6b} ($l = 2$), ν_{7a}, ν_{7b} ($l = 2$). The value $\alpha = -60°$ gives, similar to the case of X_3 and X_5, the translations perpendicular to the symmetry axis (non-genuine vibration). The other vibrations of the X_6 molecule will be discussed later.

More general doubly degenerate vibrations. The above discussed behavior of the normal coordinates with respect to symmetry operations may also be derived from the requirement that *the potential energy* (II, 32) *must be invariant with respect to all symmetry operations permitted by the molecule in its equilibrium position.* That is, if the potential energy is

$$V = \tfrac{1}{2}(\lambda_1\xi_1^2 + \lambda_2\xi_2^2 + \lambda_3\xi_3^2 + \cdots), \qquad \text{(II, 72)}$$

after a symmetry operation has been carried out, it must be

$$V = \tfrac{1}{2}(\lambda_1\xi_1'^2 + \lambda_2\xi_2'^2 + \lambda_3\xi_3'^2 + \cdots) \qquad \text{(II, 73)}$$

where the ξ_i' are the normal coordinates in the transformed molecule.[8]

Invariance of V accordingly exists if $\xi_i' = \xi_i$ or $\xi_i' = -\xi_i$, that is, when the normal vibration is symmetric or antisymmetric with respect to the symmetry operation. In fact, this is the *only* way invariance of V can be accomplished if all λ_i (all frequencies) are different. Therefore *non-degenerate vibrations can only be symmetric or antisymmetric.* However, if two or more λ_i are equal, that is, if we have a *degenerate vibration*, the corresponding ξ_i' may be a linear combination of the ξ_i. Taking the case of a *double degeneracy*, let ξ_{ia} and ξ_{ib} be the two degenerate normal coordinates, and the part of the potential energy depending on them

$$V_i = \tfrac{1}{2}\lambda_i(\xi_{ia}^2 + \xi_{ib}^2). \qquad \text{(II, 74)}$$

The only linear transformation of the form (II, 62) that will leave V_i unchanged is an *orthogonal transformation*, that is, a transformation that transforms one Cartesian coordinate system into another with the same origin. Such a transformation, in the present case of two dimensions, is performed only either by

$$\begin{aligned} \xi_{ia}' &= +\,\xi_{ia}\cos\beta + \xi_{ib}\sin\beta, \\ \xi_{ib}' &= -\,\xi_{ia}\sin\beta + \xi_{ib}\cos\beta \end{aligned} \qquad \text{(II, 75)}$$

[8] It should be noted that it is not sufficient that the numerical value of V for a given set of ξ_i values remains the same; it must remain unchanged for any set of ξ_i values and therefore must have the same functional dependence on the transformed normal coordinates as on the non-transformed.

or by

$$\begin{aligned} \xi'_{ia} &= -\ \xi_{ia}\cos\beta + \xi_{ib}\sin\beta, \\ \xi'_{ib} &= +\ \xi_{ia}\sin\beta + \xi_{ib}\cos\beta. \end{aligned} \qquad \text{(II, 76)}$$

The first transformation means a simple rotation of a rectangular coordinate system formed by ξ_{ia} and ξ_{ib} by an angle β [compare equation (II, 65) and the accompanying discussion]. The second transformation means a rotation plus a reflection at the origin. It is easily verified, by forming $\xi'^2_{ia} + \xi'^2_{ib}$ from (II, 75) or (II, 76), that $\xi'^2_{ia} + \xi'^2_{ib} = \xi^2_{ia} + \xi^2_{ib}$; that is, that the transformations (II, 75) and (II, 76) leave the potential energy V_i in (II, 74) invariant. It is also easily seen, by substituting (II, 62) in (II, 74) and requiring that V_i be invariant, that the above two transformations are the only ones giving this invariance.

The transformation (II, 76), if carried out twice in succession, has the property of leading to the original normal coordinates for any value of β, as can immediately be verified. It can therefore correspond only to those symmetry operations which if applied twice bring the system back to itself, such as a reflection at a plane. Only the transformation (II, 75) can represent the change of degenerate normal coordinates produced by a rotation about a p-fold axis with $p > 2$. It is of exactly the same form as (II, 70). However, the proof for (II, 75) is more general and we can now drop a number of restrictions under which (II, 70) was derived: Equation (II, 70) was derived for the x and y components of the displacements only [see (II, 69)], whereas (II, 75) holds generally *also for the z components* (that is, the components in the direction of the axis of symmetry). In deriving (II, 69) and (II, 70) we assumed that each displacement vector of ξ_{ia} was perpendicular to the corresponding displacement vector of ξ_{ib} (see Fig. 34) and had the same magnitude, whereas now we need only to assume that ξ_{ia} and ξ_{ib} are *orthogonal and normalized* in the sense of equations (II, 18) and (II, 19).

The value of β in (II, 75) is not entirely arbitrary but, in the same way as above for α, it follows that for a rotation by $2\pi/p$ about a p-fold axis,

$$\beta = \pm\frac{2\pi}{p}l, \qquad l = 1, 2 \cdots p - 1. \qquad \text{(II, 77)}$$

Thus for those vibrations that fulfill the above-mentioned restrictions the considerations are exactly the same as before. However, we can now also discuss those degenerate normal vibrations that do not fulfill these restrictions.

Let us consider, as an example, those *normal vibrations of an* X_5 *molecule* (Fig. 38) *that are perpendicular (antisymmetric) to the plane of the molecule.* The only one such vibration that is symmetric with respect to the axis is the non-genuine vibration consisting of a translation in the z direction (not shown). The other vibrations in this group are degenerate with respect to the axis. It is easily seen that a vibration parallel to the axis can be degenerate only with a vibration that is also parallel to the axis [since otherwise a rotation by $2\pi/p$ could not transform the one into a linear combination (II, 75) of the two original ones]. Thus the displacement vectors of the individual atoms in the two mutually degenerate vibrations are not perpendicular but *parallel* to each other. In order that they shall be orthogonal we must choose them so that [see (II, 18)]

$$\Sigma m_k z_k^{(a)} z_k^{(b)} = 0. \qquad \text{(II, 78)}$$

One way of accomplishing this is to choose the displacements of N_1 as $z_1^{(a)} = s$, $z_1^{(b)} = 0$ (then at least the contribution of N_1 to the sum is zero). The displacements of N_2 are now obtained by carrying out the transformation (II, 75) and then rotating the molecule by 72° (similar to Fig. 39). The latter operation here does not change the direction of the displacement vectors. Thus we have

$$\begin{aligned} z_2^{(a)} &= z_1^{(a)\prime} = s\cos\beta + 0, \\ z_2^{(b)} &= z_1^{(b)\prime} = -\,s\sin\beta + 0. \end{aligned} \tag{II, 79}$$

Similarly we obtain

$$\begin{aligned} z_3^{(a)} &= z_2^{(a)\prime} = s\cos 2\beta, \\ z_3^{(b)} &= z_2^{(b)\prime} = -\,s\sin 2\beta. \\ &\quad . \quad . \quad . \quad . \quad . \quad . \quad . \end{aligned} \tag{II, 80}$$

Taking first $\beta = +(360°/p)\cdot 1 = +72°$ we see that the displacements z_k of the successive nuclei $N_1, N_2, N_3, \cdots$ in the first vibration are

$$s, \qquad s\cos 72°, \qquad s\cos 144°, \qquad s\cos 216°, \qquad s\cos 288°,$$

and in the second vibration, which is degenerate with the first,

$$0, \qquad -s\sin 72°, \qquad -s\sin 144°, \qquad -s\sin 216°, \qquad -s\sin 288°.$$

It is easily verified that these two sets of displacements fulfill the relation of orthogonality (II, 78). They are illustrated in Fig. 38b, from which it can be seen that they represent simple rotations about two mutually perpendicular axes in the plane of the molecule. These two vibrations are thus non-genuine.

If we now take $\beta = +(360°/p)\cdot 2 = +144°$ we see that the z displacements in the two mutually degenerate vibrations are, according to (II, 79) and (II, 80),

$$s, \qquad s\cos 144°, \qquad s\cos 288°, \qquad s\cos 72°, \qquad s\cos 216°$$

and

$$0, \qquad -s\sin 144°, \qquad -s\sin 288°, \qquad -s\sin 72°, \qquad -s\sin 216°.$$

These displacements are shown as ν_{5a} and ν_{5b} in Fig. 38a. They represent evidently genuine normal vibrations (which are mutually orthogonal). It is immediately seen that the values $\beta = -72°$ and $-144°$ lead to the same vibrations as $\beta = +72°$ and $+144°$ respectively. As previously, $\beta = \pm 2\pi/5\cdot 3$ and $2\pi/5\cdot 4$ lead to the same vibrations as $\beta = \pm 2\pi/5\cdot 2$ and $\pm 2\pi/5\cdot 1$ respectively.

Doubly degenerate vibrations with the same relative amplitudes as for ν_{5a} and ν_{5b} may also take place in the plane of the molecule, either in such a way that all atoms move radially or in such a way that they all move tangentially. These are represented in Fig. 38c. It is easily seen that these vibrations fulfill the transformation law (II, 75), even though the displacement vectors are not all the same and are not at right angles to each other in a degenerate pair. The displacement vectors in these vibrations can therefore not be obtained by the simple rotation method illustrated by Fig. 39. However, these two degenerate vibrations are not independent of those already given. Their normal coordinates are the linear combinations $\xi_{3a} + \xi_{4a}$, $\xi_{3b} + \xi_{4b}$ and $\xi_{3b} - \xi_{4b}$, $\xi_{3a} - \xi_{4a}$ of the previous ν_3 and ν_4 (Fig. 38a). There is an infinite number of other pairs of linear combinations ($\xi_{3a} + c\xi_{4a}$, $\xi_{3b} + c\xi_{4b}$ with c an arbitrary number) that would also fulfill the transformation law (II, 75).

But only two of these pairs are independent of one another. Which two are the actual vibrations ν_3 and ν_4 depends on the forces acting between the atoms (see section 4). Only for certain special force fields would the vibrations be as given for ν_3 and ν_4 in Fig. 38a, and for certain other special force fields they would be as given in Fig. 38c. On the other hand, the other vibrations, ν_1, ν_2, and ν_5 of X_5, are unambiguously given for any force field since there is only one of each species (see section 4a).

It should be realized that linear combinations $\xi_2 + c\xi_3$ or $\xi_2 + c\xi_4$ do not fulfill the transformation laws (II, 75) and are therefore not possible normal vibrations, since $\beta = +72°$ for ν_2 but $\beta = \pm 144°$ for ν_3 and ν_4.

For a molecule with a three-fold axis only $\beta = \pm 120°$ is possible. In the case of a triatomic molecule X_3 this leads (just like $\beta = \pm 72°$ for X_5) only to a non-genuine vibration perpendicular to the plane of the molecule.

However in the molecule X_3Y_3 both a non-genuine and a genuine vibration (ν_8 in Fig. 36) result, with amplitudes

$$s, \quad s\cos 120°, \quad s\cos 240° \qquad \text{and} \qquad 0, \quad -s\sin 120°, \quad -s\sin 240°.$$

For this molecule, unlike X_3, also plane degenerate vibrations with the amplitudes given are possible in a radial or tangential direction, analogous to those discussed for X_5 (Fig. 38c). However, they are again simply certain linear combinations of the plane degenerate vibrations already given (ν_5, ν_6, ν_7 in Fig. 36).

The *vibrations perpendicular to the plane of the molecule* X_6 (see Fig. 40) are obtained in a way entirely similar to that for X_5. Again $\beta = \pm 2\pi/p \cdot 1$, which here equals $\pm 60°$, gives two non-genuine vibrations, while $\beta = 2\pi/p \cdot 2 = 120°$ gives the degenerate genuine vibrations ν_{8a} and ν_{8b}, shown in Fig. 40. $\beta = 2\pi/p \cdot 3$ gives the non-degenerate vibration that is antisymmetric with respect to the six-fold axis (ν_3 in Fig. 40). Again, the plane vibrations ν_6 and ν_7 are not unambiguous. The linear combinations $\xi_{6a} + \xi_{7a}$, $\xi_{6b} + \xi_{7b}$ and $\xi_{6a} - \xi_{7a}$, $\xi_{6b} - \xi_{7b}$ would be tangential and radial vibrations similar to those of X_5 in Fig. 38c.

Further examples of doubly degenerate vibrations are those of XY_3, XY_4, X_2Y_6, X_6Y_6, and XYZ_3 given in Figs. 45, 48, 49, 50, and 91 respectively.

It remains now to discuss briefly the *behavior of doubly degenerate vibrations with respect to reflection at a plane, rotation about a two-fold axis, and inversion.* It can be shown (see below) that two mutually degenerate vibrations are either both symmetric or both *antisymmetric* with respect to a center of symmetry i, a plane of symmetry σ_h perpendicular to the axis of symmetry, and a two-fold axis coinciding with the p-fold axis, with respect to which the vibrations are degenerate, if such symmetry elements are present. Examples of this rule are supplied by the degenerate vibrations in Figs. 25b, 32a, 33, 36, 37, 38, 40.

With respect to planes through the p-fold axis or to two-fold axes perpendicular to it, degenerate vibrations *may or may not be symmetric or antisymmetric.* For example, the vibration ν_{2a} of X_3 (Fig. 32a) is symmetric, ν_{2b} is antisymmetric with respect to a reflection at the plane of symmetry perpendicular to the plane of the molecule through N_1, and also with respect to a two-fold rotation about the two-fold axis through N_1, whereas they change by more than just the sign for a reflection at the planes through N_2 and N_3 as well as for the rotations about the two-fold axes through N_2 and N_3. However, one can always find two linear combinations of the two mutually degenerate vibrations that are symmetric and antisymmetric respectively with respect to the particular plane or two-fold axis. In the example of X_3, a vibration that is antisymmetric with respect to the plane σ_v through N_2 is $\nu_{2b}^{(240)}$ of Fig. 32c and one that is symmetric is $\nu_{2a}^{(240)}$. However, the vibrations $\nu_{2b}^{(240)}$ and $\nu_{2a}^{(240)}$ are, as we have seen before, linear combinations of the original ν_{2a} and ν_{2b}. Similarly, for a linear molecule we can always choose two mutually degenerate vibrations so that one is in the plane considered and the other perpendicular to it; that is, so that one is symmetric and the other antisymmetric with respect to that plane.

According to the preceding discussion a more-than-two-fold axis in a molecule necessarily involves the existence of degenerate vibrations, that is, vibrations that change by more than just the sign for a rotation about that axis. On the other hand, since normal vibrations can always be so chosen that they remain unchanged or only change sign for a reflection at a plane, for a two-fold rotation and for an inversion, a molecule with no more-than-two-fold axes need not have any degenerate vibration although accidentally two (or more) of its vibrations may be degenerate with one another. *Only if a molecule has at least one more-than-two-fold axis does it necessarily have degenerate vibrations.*

The fact mentioned above, that both vibrations of a degenerate pair behave in the same way with respect to an inversion, can be seen in the following way. If the substitutions $x_k \rightarrow -x_k$, $y_k \rightarrow -y_k$, $z_k \rightarrow -z_k$ are made in equations (II, 10) the same equations are obtained, since in a molecule with a center of symmetry the force constants k_{xy}^{il} are invariant to inversion. Therefore the ratio of the displacements (given by a row of minors of the determinant of the equations) remains unchanged; that is, any degenerate normal vibration can only be symmetric or antisymmetric with respect to inversion. This holds also for a linear combination of two mutually degenerate vibrations and therefore both components of a pair must show the same behavior. In a similar though somewhat more complicated way it can be shown that they must also show the same behavior with respect to a plane σ_h perpendicular to C_p and for a C_2 coinciding with C_p. In all these cases, then, $\xi'_{ia} = \xi_{ia}$, $\xi'_{ib} = \xi_{ib}$ or $\xi'_{ia} = -\xi_{ia}$, $\xi'_{ib} = -\xi_{ib}$, which is a special case of the transformation (II, 75) with $\beta = 0$ and 180° respectively.

For all other reflections and two-fold rotations a degenerate vibration does not necessarily remain unchanged or only change sign, and therefore the transformation (II, 76) applies, since it also fulfills the requirement that two successive reflections or rotations lead back to the original normal coordinates, which (II, 75) does not except for $\beta = 0$ and 180°.[9] For two special values of β, $\beta = 0$ and $\beta = 180°$, the transformation (II, 76) leads to a simple result, namely $\xi'_{ia} = -\xi_{ia}$, $\xi'_{ib} = +\xi_{ib}$ and $\xi'_{ia} = +\xi_{ia}$, $\xi'_{ib} = -\xi_{ib}$ respectively, that is, for these values of β, one component of the degenerate pair is symmetric, the other is antisymmetric with respect to reflection or two-fold rotation of the type discussed. The important point is now that if two mutually degenerate normal coordinates ξ_{ia} and ξ_{ib} are not symmetric or antisymmetric with respect to a reflection or two-fold rotation, two mutually orthogonal linear combinations of them, $\bar{\xi}_{ia}$ and $\bar{\xi}_{ib}$, can always be found that are symmetric and antisymmetric respectively. This is immediately seen if it is realized that (II, 76) represents a rotation by an angle β in the ξ_{ia}, ξ_{ib} plane plus an inversion. Therefore by applying the opposite rotation to ξ_{ia} and ξ_{ib} by means of transformation (II, 75), normal coordinates $\bar{\xi}_{ia}$ and $\bar{\xi}_{ib}$ must be obtained that are transformed according to (II, 76) with $\beta = 0$ or $\beta = 180°$; that is, one of them is symmetric, the other antisymmetric with respect to the operation in question. A good illustration is the case of ν_2 of a molecule X_3 reflected at a plane through N_2 (see above and Fig. 32).

Complex normal coordinates. Sometimes, instead of using two *real* (orthogonal) mutually degenerate normal coordinates ξ_a, ξ_b, it is convenient to introduce complex normal coordinates. Since any linear combination of ξ_a and ξ_b is a solution of equations (II, 10), then

$$\eta_a = \xi_a + i\xi_b, \qquad \eta_b = \xi_a - i\xi_b \tag{II, 81}$$

[9] That (II, 76) applies for a reflection at a plane through the symmetry axis can easily be verified, for example, for the case of linear XY_2 (see p. 86f.).

are formally also solutions (where $i = +\sqrt{-1}$).[10] If we now apply transformation (II, 75), we obtain

$$\eta_a' = \xi_a' + i\xi_b' = \xi_a \cos\beta + \xi_b \sin\beta + i(-\xi_a \sin\beta + \xi_b \cos\beta),$$
$$\eta_b' = \xi_a' - i\xi_b' = \xi_a \cos\beta + \xi_b \sin\beta - i(-\xi_a \sin\beta + \xi_b \cos\beta);$$

or, with $\cos\beta \pm i\sin\beta = e^{\pm i\beta}$,

$$\eta_a' = \eta_a e^{-i\beta}, \quad \eta_b' = \eta_b e^{+i\beta}. \tag{II, 82}$$

On the other hand, transformation (II, 76) leads to

$$\eta_a' = \xi_a' + i\xi_b' = -\eta_b e^{-i\beta}, \quad \eta_b' = \xi_a' - i\xi_b' = -\eta_a e^{+i\beta}. \tag{II, 82a}$$

The first transformation, therefore, changes the complex normal coordinates only by a (complex) factor while the second changes one into the other with a (complex) factor.[11] The second transformation (II, 82a) applies as previously only to planes through, or two-fold axes perpendicular to, the more-than-two-fold axes causing the degeneracy. Therefore, if there are no such planes or axes, only (II, 82) applies. In that case, the vibrations are said to be *separably degenerate* [see Placzek (700)] since a pair can be found, namely the complex η_a and η_b in (II, 81), such that each one of them transforms into itself, possibly apart from a constant factor, for any symmetry operation permitted by the system. In the previous examples, however, the degenerate vibrations are not separable since there are planes through the axis of symmetry as well as two-fold axes perpendicular to it for which transformation (II, 82a) applies. A system in which there were only separably degenerate vibrations would, for instance, be the molecule X_3Y_3 if the X_3 triangle were rotated with respect to the Y_3 triangle.

Triply degenerate vibrations. For triply degenerate vibrations the contribution to the potential energy is

$$V_i = \tfrac{1}{2}\lambda_i(\xi_{ia}^2 + \xi_{ib}^2 + \xi_{ic}^2), \tag{II, 83}$$

which must remain unchanged for any symmetry operation. This condition is fulfilled by a *three-dimensional orthogonal transformation* of the form (II, 62). The d_{ab} are the direction cosines of the new axes as compared to the old ones. The normal coordinates transform in the same way as Cartesian coordinates when one goes over from one system of axes to another with the same origin.

As an illustration, Fig. 41 shows the *normal vibrations of a tetrahedral* XY_4 *molecule*. The three component vibrations of each of the two triply degenerate vibrations ν_3 and ν_4 are shown. By carrying out the symmetry operations (for example, rotation about one of the three-fold axes), any one of the vibrations is transformed into a vibration that is, in general, a linear combination of all three mutually degenerate vibrations. The various transformed displacement vectors of a particular atom no longer lie in a plane.

We shall not discuss a general method of obtaining the form of the normal vibrations. But it is easily seen that the vibrations shown in Fig. 41 do fulfill the necessary requirements: the three vibrations ν_{3a}, ν_{3b}, ν_{3c} have obviously the same frequency and they are orthogonal to one another; that is, they are linearly independent. The same holds for ν_{4a}, ν_{4b}, ν_{4c}. Since there are in XY_4 two triply degenerate vibrations,

[10] We omit here the subscripts i for ξ and η in order to avoid confusion with $i = +\sqrt{-1}$.

[11] The factor in this second case can be made equal to 1 (that is $\beta = 0$) if one starts out from the proper normal coordinates $\bar{\xi}_a$ and $\bar{\xi}_b$ that are symmetric and antisymmetric with respect to the plane or two-fold axis considered (see above).

their actual forms are not uniquely determined by symmetry. They are linear combinations of the two sets given. In a tetrahedral molecule Y_4 (without a central atom) only one triply degenerate vibration occurs whose form is uniquely determined by symmetry. It is represented by ν_4 of Fig. 41 if the central atom is omitted.

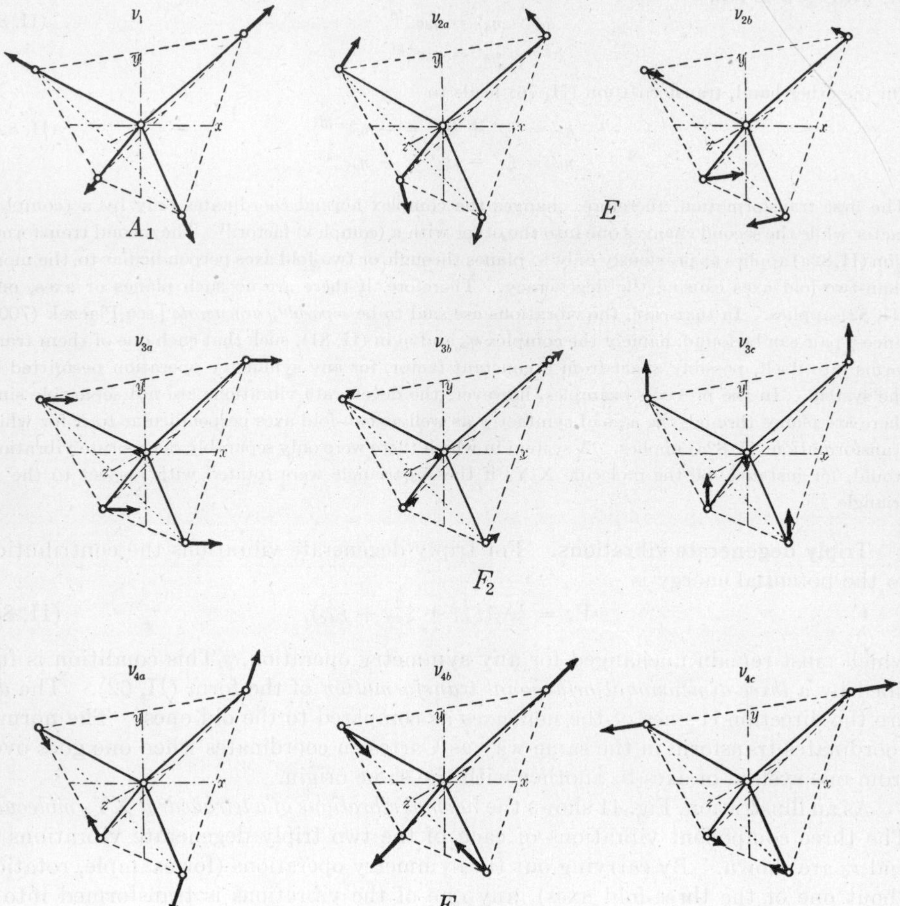

FIG. 41. **Normal vibrations of a tetrahedral XY_4 molecule.**—The three two-fold axes (dot-dash lines) are chosen as x, y, and z axes.

While in general triply degenerate vibrations (just as doubly degenerate vibrations) change by more than the sign for a reflection at a plane of symmetry or for a rotation about a two-fold axis, it is always possible to find a set of linear combinations each of which is either symmetric or antisymmetric with respect to a particular plane or two-fold axis. Thus ν_{3a} and ν_{3c} in Fig. 41 are symmetric, ν_{3b} is antisymmetric with respect to the xy plane. In the same way as previously for doubly degenerate vibrations it can be shown that three mutually degenerate vibrations are either all symmetric or all antisymmetric with respect to a center of symmetry. But unlike the case of doubly degenerate vibrations there are no other elements of symmetry with respect to which all three vibrations behave in the same way.

It can be shown that it is always possible to find one (and only one) linear combination of three mutually degenerate vibrations that is symmetric with respect to a particular three-fold (or four-fold) axis. (Thus for ν_3 of XY_4 in Fig. 41 the combination $\nu_{3a} + \nu_{3b} + \nu_{3c}$, for ν_4 the combination $\nu_{4a} + \nu_{4b} + \nu_{4c}$, are seen to be symmetric with respect to one of the three-fold axes.) Two other combinations behave, then, in the same way as a doubly degenerate pair with respect to rotation about this axis. However, with respect to another three-fold axis the first linear combination is not symmetric (compare the above example). Thus it is the presence of more than one three-fold axis that leads necessarily to the occurrence of triply degenerate vibrations.

(c) *Effect of symmetry operations on the vibrational eigenfunctions*

Since the vibrational eigenfunction is a function of the normal coordinates, its behavior with respect to symmetry operations depends on the behavior of the normal coordinates with respect to them.

Molecules with non-degenerate vibrations only. If a non-degenerate vibration, say ν_i, is symmetric with respect to a certain symmetry element [for example, the vibrations ν_1, ν_2, ν_3, ν_6 of XYZ_2 in Fig. 24 with respect to the plane $\sigma_v(xz)$], that is, if the corresponding normal coordinate ξ_i is symmetric, it follows that its contribution $\psi_i(\xi_i)$ to the vibrational eigenfunction is *symmetric* (remains unchanged) with respect to the particular symmetry operation *for all values of* v_i. If a normal vibration, say ν_k, is antisymmetric with respect to a symmetry element [for example, the vibrations ν_4 and ν_5 of XYZ_2 in Fig. 24 with respect to the plane $\sigma_v(xz)$], that is, if the corresponding normal coordinate ξ_k is antisymmetric, it follows that $\psi_k(\xi_k)$, since it is an odd (even) function of ξ_k for odd (even) v_k, changes sign—*is antisymmetric—for odd* v_k but remains unchanged—*is symmetric—for even* v_k if the particular symmetry operation is carried out. This behavior is represented in Fig 42a and b. It is very important for the following considerations.

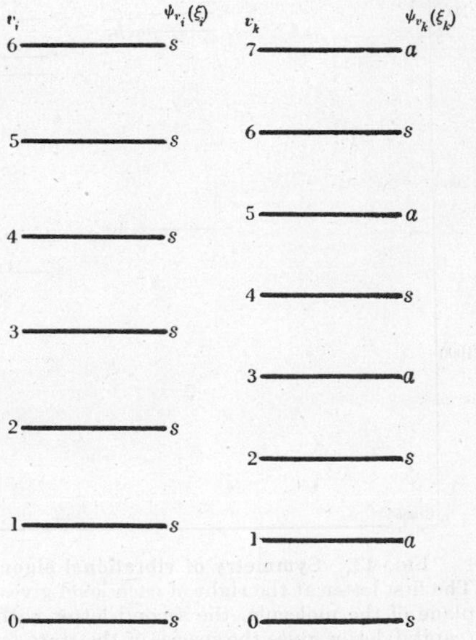

FIG. 42. **Symmetry of vibrational eigenfunctions of non-degenerate vibrations. (a) Symmetrical (b) antisymmetrical vibration.**

The total vibrational eigenfunction ψ, according to (II, 42), is a product of harmonic oscillator eigenfunctions $\psi_1(\xi_1), \psi_2(\xi_2), \cdots$, corresponding to the $3N - 6$ or $3N - 5$ normal coordinates. Therefore, if there are only non-degenerate normal vibrations, the total eigenfunction will be symmetric with respect to a given symmetry operation if there is an even number of component functions $\psi_k(\xi_k)$ that are antisymmetric with respect to that same symmetry operation; the total eigenfunction will be antisymmetric if there is an odd number of antisymmetric component functions. Its behavior with respect to the given symmetry operation is independent of the number of symmetric component functions. In other words, since, for antisymmetric ξ_k, the function $\psi_k(\xi_k)$ is antisymmetric for odd v_k, *the total vibrational eigenfunction is symmetric with respect to a certain symmetry operation if the sum* $\sum_a v_k$, *extended over all normal vibrations that are antisymmetric with respect to that symmetry operation, is even; the total eigenfunction is antisymmetric with respect to the same symmetry operation if the sum* $\sum_a v_k$ *is odd.*

As an illustration of this rule Fig. 43 gives the symmetry properties of the total vibrational eigenfunction in the lower vibrational levels of the H_2CO molecule which

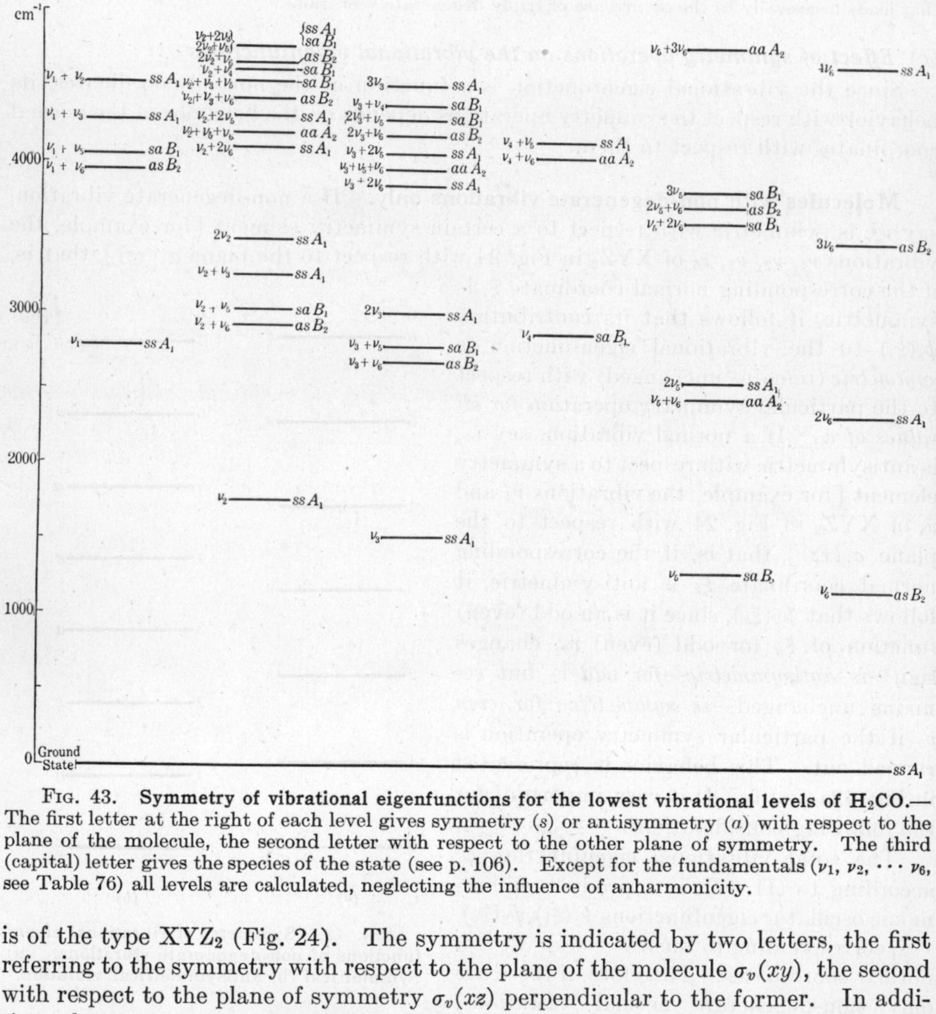

FIG. 43. **Symmetry of vibrational eigenfunctions for the lowest vibrational levels of H_2CO.—The first letter at the right of each level gives symmetry (s) or antisymmetry (a) with respect to the plane of the molecule, the second letter with respect to the other plane of symmetry. The third (capital) letter gives the species of the state (see p. 106). Except for the fundamentals (ν_1, ν_2, $\cdots$ ν_6, see Table 76) all levels are calculated, neglecting the influence of anharmonicity.**

is of the type XYZ_2 (Fig. 24). The symmetry is indicated by two letters, the first referring to the symmetry with respect to the plane of the molecule $\sigma_v(xy)$, the second with respect to the plane of symmetry $\sigma_v(xz)$ perpendicular to the former. In addition, the more systematic symbols for the symmetry to be explained in the next subsection are given. For the designation and form of the normal vibrations, see Fig. 24. When, for example, two quanta of the oscillation ν_6, which is antisymmetric with respect to $\sigma_v(xy)$, and three quanta each of the two oscillations ν_4 and ν_5, which are antisymmetric with respect to $\sigma_v(xz)$, are excited, then the total vibrational function is symmetric with respect to both planes of symmetry since $\sum_a v_k$ is even for both.

Molecules with degenerate vibrations. When a molecule has degenerate vibrations as well as non-degenerate vibrations, the same relations apply as for the preceding case if the degenerate vibrations ν_j are not excited by more than the zero-point vibration, that is, if all have $v_j = 0$. This is because for $v_j = 0$ the contribution of a doubly degenerate vibration to the total vibrational eigenfunction is, according to equation (II, 56),

$$e^{-(\alpha_j/2)(\xi_{ja}^2+\xi_{jb}^2)},$$

which remains unchanged for any symmetry operation just as does the potential energy (II, 74). In such a case, therefore, and similarly for higher degeneracies, the total vibrational eigenfunction can only be symmetric or antisymmetric with respect to any symmetry operation permitted by the molecule (including p-fold rotations); it is symmetric when $\sum_a v_k$ formed as above is even, antisymmetric when it is odd. The same holds also for the case in which accidentally degenerate vibrations are excited with $v_j > 0$ if the correct linear combinations are used. However, it does not hold if necessarily degenerate vibrations which occur in molecules with more-than-two-fold axes are excited with $v_j > 0$.

If a *degenerate vibration is excited with one quantum* ($v_j = 1$), there are two (or three) different total eigenfunctions belonging to the same energy value and the total eigenfunction will be no longer only symmetric or antisymmetric with respect to all symmetry operations but will change into a linear combination of the two (or three) degenerate functions. In the case of doubly degenerate vibrations, the eigenfunction is, for any v_j, given by (II, 56). In this equation, for $v_j = 1$ we have either $v_a = 1$, $v_b = 0$, or $v_a = 0$, $v_b = 1$. Since the Hermite polynomial H_v is of the vth degree, the two eigenfunctions for $v_j = 1$ are

$$\psi_{ja} = Ce^{-(\alpha_j/2)(\xi_{ja}^2+\xi_{jb}^2)} \cdot \xi_{ja}$$

and

$$\psi_{jb} = Ce^{-(\alpha_j/2)(\xi_{ja}^2+\xi_{jb}^2)} \cdot \xi_{jb} \qquad \text{(II, 84)}$$

where C is a constant. The exponential factor, as mentioned previously, remains unchanged for any symmetry operation just as does the potential energy. Therefore ψ_{ja} and ψ_{jb} must have *the same behavior with respect to all symmetry operations as the normal coordinates* ξ_{ja} and ξ_{jb} respectively. In particular, for those symmetry operations (for example rotation about a C_p) for which ξ_{ja} and ξ_{jb} go over into linear combinations [(II, 75) or (II, 76)], ψ_{ja} and ψ_{jb} go over into the corresponding linear combinations *with the same coefficients.* Similar considerations apply also to triply degenerate vibrations.

The symmetry of the *total vibrational eigenfunction* is, of course, again determined by the behavior of its factors with respect to the symmetry operations. For example, if in a linear triatomic molecule XY_2 one quantum each of the three normal vibrations (Fig. 25b) is excited, the total eigenfunction will be antisymmetric with respect to a reflection at a plane through X perpendicular to the internuclear axis, but it will be "*degenerate*" with respect to a rotation by an arbitrary angle about the internuclear axis.

If a degenerate vibration is excited with $v_j > 1$ the degree of degeneracy is increased (see p. 80f.). It is immediately clear from (II, 56) since the factor $H_{v_a}(\sqrt{\alpha_j}\xi_{ja})H_{v_b}(\sqrt{\alpha_j}\xi_{jb})$ is no longer simply proportional to one normal coordinate, that the transformation properties of the eigenfunctions are rather more complicated for $v_j > 1$ than for $v_j = 1$. But, as for $v_j = 1$, *each one of the degenerate eigenfunctions, when a symmetry operation is carried out, goes over into a linear combination of the eigenfunctions belonging to the same energy level.* As we shall see in more detail in subsection 3e, one set of eigenfunctions can be found consisting of independent groups which for a symmetry operation transform only among themselves. It may happen that one or the other of these groups has one eigenfunction only, which then is symmetric or antisymmetric with respect to the symmetry element that causes the degeneracy. Thus, for example (see p. 128), if $v = 2$ for the degenerate perpendicular oscillation of linear XY_2, we have a triply degenerate state. But it is possible to choose the three eigenfunctions in such a way that one is symmetric with respect to the ∞-fold axis and the other two form a pair that is "degenerate" with respect to this axis. If the anharmonicity of the vibrations is taken into account, a splitting of the triply degenerate level occurs into two levels, a non-degenerate (symmetric) and a doubly degenerate level, with just the eigenfunctions mentioned (see section 5 of this chapter).

Generalization. In the preceding discussion we have derived the symmetry properties of the vibrational eigenfunctions from those of the normal coordinates. Actually the symmetry properties of the eigenfunctions are much more general and do not depend on the assumption of harmonic oscillations. The potential energy, even if it is not simply a quadratic function of the displacement coordinates [as in (II, 25)], must be invariant to all symmetry operations permitted by the point group of the molecule. Therefore the *Schrödinger equation* (II, 40) *is invariant to these symmetry operations* and consequently the *eigenfunction can only be symmetric or antisymmetric* with respect to these symmetry operations if the state is *non-degenerate,* while it can also *transform into a linear combination of mutually degenerate eigenfunctions for a degenerate state* (see also Molecular Spectra I, p. 239). It can be shown that in the latter case the transformation is an *orthogonal transformation,* which for double degeneracy is given by (II, 75) or (II, 76).

The same reasoning shows also that *the rotational, electronic and total eigenfunctions can only be symmetric, antisymmetric, or degenerate with respect to any of the symmetry operations.*

(*d*) *Symmetry types* (*species*) *of normal vibrations and eigenfunctions*

Thus far we have only considered the behavior of the normal vibrations and vibrational eigenfunctions with respect to individual symmetry operations. However, since only certain combinations of symmetry elements occur in the various point groups (see p. 5 f.), and since some of their symmetry elements are a necessary consequence of others, *only certain combinations of symmetry properties of the normal vibrations and vibrational* (and electronic) *eigenfunctions are possible,* as was first shown by Brester (178). We call such combinations of symmetry properties *symmetry types* or *species* [following Mulliken (643)]. In *group theory* they are the so-called *irreducible representations* of the point group considered, a name that some authors prefer. On the basis of the discussions in the preceding subsection, the symmetry types can be derived fairly easily without the explicit use of group theory for all molecules except those belonging to cubic point groups [see also Placzek (700)][11a] This will be done in the following pages. For the cubic point groups we shall accept the results of group theory without proof.

In every molecule there are normal vibrations and eigenfunctions that are symmetric with respect to all symmetry operations permitted by the system. These are called *totally symmetric* and this symmetry type is designated by A or A_1 or A' or by similar symbols (see below).

Point groups C_1, C_2, C_s, and C_i. In the case of the point group C_1 there is no symmetry and consequently there is only one species A for the normal vibrations and eigenfunctions which may also be said to be symmetrical with respect to the identical symmetry operation I.

If there is *one element of symmetry* only, as in the point groups C_2 (one two-fold axis), C_s (one plane of symmetry), and C_i (center of symmetry only), the vibrations and eigenfunctions may be symmetric or antisymmetric with respect to the one element of symmetry. Thus there are *two species* for each point group, the symmetric one being called A, A', and A_g for C_2, C_s, and C_i respectively, the antisymmetric one B, A'', and A_u respectively.[12] This is represented in Table 12 where $+1$

TABLE 12. SYMMETRY TYPES (SPECIES) FOR THE POINT GROUPS C_2, C_s, $C_i \equiv S_2$.

C_2	I	$C_2(z)$		C_s	I	$\sigma(xy)$		$C_i \equiv S_2$	I	i	
A	$+1$	$+1$	T_z, R_z	A'	$+1$	$+1$	T_x, T_y, R_z	A_g	$+1$	$+1$	R_x, R_y, R_z
B	$+1$	-1	T_x, T_y, R_x, R_y	A''	$+1$	-1	T_z, R_x, R_y	A_u	$+1$	-1	T_x, T_y, T_z

and -1 is used to indicate symmetric and antisymmetric. In the first line is given the point group (heavy type) and the symmetry operations including the identity I. Below are the symmetry types and the behavior of the vibrations or eigenfunctions having these symmetry types for the symmetry operations given in the top row. In the last column of each subtable are given the non-genuine vibrations, *translation* in the x, y, or z directions (T_x, T_y, T_z), and *rotation* about the x, y, and z axes (R_x, R_y, R_z) that belong to the particular species (see also the next subsection). It is clear, for example, that for point group C_2 a translation in the direction of the C_2 and a rotation about the C_2 is symmetric with respect to the operation C_2 while the other translations and rotations are antisymmetric with respect to it.

As an example, consider the plane but non-linear N_3H molecule. It will have, according to the above, normal vibrations that are symmetric and normal vibrations that are antisymmetric with respect to the plane of the molecule. During the former all atoms remain always in the plane, during the latter they move in lines perpendicular to the plane. It may be noted that in an asymmetric triatomic molecule such as NOCl which belongs also to the point group C_s the vibrational motion can take place only in the plane; that is, there are no vibrations of type A''.

[11a] For a discussion using group theory throughout see Rosenthal and Murphy (750) and Meister, Cleveland, and Murray (620a).

[12] Following Placzek (700) the letter A is used for all species that are symmetric, the letter B for those that are antisymmetric with respect to an axis of symmetry. As for diatomic molecules, species that are symmetric or antisymmetric with respect to a center of symmetry are distinguished by the subscript g and u respectively.

Point groups C_{2v}, C_{2h}, and $D_2 \equiv V$. If a molecule (point group) has *two necessary elements of symmetry* it is obvious that there are *four different possible symmetry types* (always assuming that there are no more than two-fold axes) which may be characterized briefly by $++$, $+-$, $-+$, $--$: a vibration (eigenfunction) may be symmetric or antisymmetric with respect to either element of symmetry. Point groups of this type are C_{2v}, C_{2h}, and $D_2 \equiv V$. Each of these has three elements of symmetry (see Table 1, p. 11), but one of them is in each case a consequence of the other two; that is, by carrying out two of the symmetry operations in succession the same result is obtained as by carrying out the third symmetry operation on the original system. Therefore the behavior of the vibrations and eigenfunctions with respect to the third operation is given by their behavior with respect to the other two. For example, if a vibration is antisymmetric with respect to two of the symmetry operations, carrying them out in succession will bring the vibration back to the original form; that is, the vibration is symmetric with respect to the third symmetry operation. Table 13

TABLE 13. SYMMETRY TYPES (SPECIES) FOR THE POINT GROUPS C_{2v}, C_{2h}, AND $D_2 \equiv V$.

C_{2v}	I	$C_2(z)$	$\sigma_v(xz)$	$\sigma_v(yz)$	
A_1	+1	+1	+1	+1	T_z
A_2	+1	+1	−1	−1	R_z
B_1	+1	−1	+1	−1	T_x, R_y
B_2	+1	−1	−1	+1	T_y, R_x

C_{2h}	I	$C_2(z)$	$\sigma_h(xy)$	i	
A_g	+1	+1	+1	+1	R_z
A_u	+1	+1	−1	−1	T_z
B_g	+1	−1	−1	+1	R_x, R_y
B_u	+1	−1	+1	−1	T_x, T_y

$D_2 \equiv V$	I	$C_2(z)$	$C_2(y)$	$C_2(x)$	
A	+1	+1	+1	+1	—
B_1	+1	+1	−1	−1	T_z, R_z
B_2	+1	−1	+1	−1	T_y, R_y
B_3	+1	−1	−1	+1	T_x, R_x

gives the designation of the symmetry types and the behavior of the vibrations (eigenfunctions) having these symmetry types for the point groups C_{2v}, C_{2h}, and $D_2 \equiv V$. It should be noted that for these point groups, in consequence of the above connection between the symmetry operations, symmetry types in which an eigenfunction would be antisymmetric with respect to all three elements of symmetry or antisymmetric with respect to one element of symmetry only do not exist.

As an example, consider the plane Y type molecule XYZ_2 (for instance, H_2CO), whose normal vibrations were given in Fig. 24 and which belongs to point group C_{2v}. It is seen that the three normal vibrations ν_1, ν_2, ν_3 are totally symmetric, that is, belong to species A_1, the vibrations ν_4, ν_5 belong to species B_1 (if we call the plane of the molecule the xz plane) and ν_6 belongs to species B_2 (vibration antisymmetric with respect to the plane of the molecule). There is no genuine normal vibration of species A_2. But there may be vibrational (and electronic) eigenfunctions of species A_2; for example, if a vibration of species B_1 and one of species B_2 are each excited by one quantum the resultant eigenfunction will be antisymmetric with respect to both planes of symmetry (see p. 102), that is, will be of species A_2. In more complicated molecules belonging to point group C_{2v} there may also be normal vibrations belonging to species A_2. It is immediately seen that for all molecules of point group C_{2v} the translation in the direction of the C_2 is totally symmetric (species A_1) while the rotation about this axis is antisymmetric with respect to both planes of symmetry (species A_2). Similarly the species of the other *non-genuine vibrations* follow, as indicated in the last column of each subtable in Table 13.

Point group $V_h \equiv D_{2h}$. If there are *three necessary symmetry elements* (but no more-than-two-fold axis) there may be *eight different symmetry types* of the normal vibrations and eigenfunctions, that is, just as many as there are possible combinations of + and − in sets of three: +++, ++−, +−+, −++, +−−, −+−, −−+, −−−. The point group $V_h \equiv D_{2h}$ to which the plane molecule C_2H_4 belongs is the only one with three necessary symmetry elements that are not more-than-two-fold axes. As such necessary symmetry elements we may choose any three independent ones of the seven elements of symmetry (see Table 1), for example, the three mutually perpendicular planes of symmetry $\sigma(xy)$, $\sigma(xz)$, $\sigma(yz)$. The possible behavior of normal vibrations (eigenfunctions) with respect to these three elements is given in columns 3, 4, and 5 of Table 14. All the eight possibilities mentioned

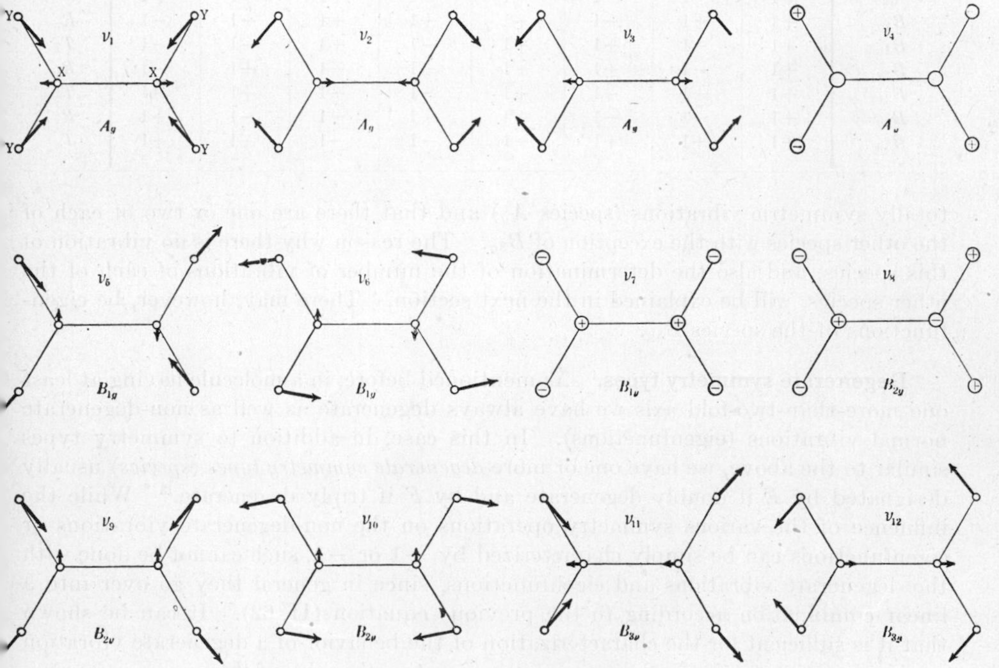

FIG. 44. **Normal vibrations of an X_2Y_4 molecule of point group V_h.**—It is assumed that the mass of X is larger than that of Y as in C_2H_4 or C_2D_4.

above are given. The designations of these species (symmetry types) are indicated in the first column.

Since an inversion may be replaced by successive reflections at three mutually perpendicular planes, the behavior of the normal vibrations of point group V_h with respect to an inversion (column 6) may be obtained simply by multiplying together columns 3, 4, and 5 in Table 14. Since a rotation by 180° about a two-fold axis may be replaced by an inversion followed by a reflection at a plane perpendicular to the two-fold axis (see p. 5), the behavior with respect to the three C_2's (columns 7, 8, and 9) is obtained by multiplying columns 3, 4, and 5 respectively by column 6.

As an example, in Fig. 44 the normal vibrations of a plane molecule X_2Y_4 of point group V_h (such as ethylene, C_2H_4, see p. 325) are given. For each vibration the species to which it belongs is indicated. The correctness of this assignment can easily be verified by the reader with the help of Table 14. Unfortunately the designation in this case is not unambiguous, since any one of the three C_2's may be chosen as z axis. In the figures the z axis has been assumed to be perpendicular to the plane of the molecule and the x axis in the line X—X. It is seen that there are three

TABLE 14. SYMMETRY TYPES (SPECIES) FOR THE POINT GROUP $D_{2h} \equiv V_h$.

$D_{2h} \equiv V_h$	I	$\sigma(xy)$	$\sigma(xz)$	$\sigma(yz)$	i	$C_2(z)$	$C_2(y)$	$C_2(x)$	
A_g	+1	+1	+1	+1	+1	+1	+1	+1	
A_u	+1	−1	−1	−1	−1	+1	+1	+1	
B_{1g}	+1	+1	−1	−1	+1	+1	−1	−1	R_z
B_{1u}	+1	−1	+1	+1	−1	+1	−1	−1	T_z
B_{2g}	+1	−1	+1	−1	+1	−1	+1	−1	R_y
B_{2u}	+1	+1	−1	+1	−1	−1	+1	−1	T_y
B_{3g}	+1	−1	−1	+1	+1	−1	−1	+1	R_x
B_{3u}	+1	+1	+1	−1	−1	−1	−1	+1	T_x

totally symmetric vibrations (species A_g) and that there are one or two of each of the other species with the exception of B_{3g}. The reason why there is no vibration of this species, and also the determination of the number of vibrations of each of the other species, will be explained in the next section. There may, however, be eigenfunctions of the species B_{3g}.

Degenerate symmetry types. As mentioned before, in a molecule having at least one more-than-two-fold axis we have always degenerate as well as non-degenerate normal vibrations (eigenfunctions). In this case, in addition to symmetry types similar to the above, we have one or more *degenerate symmetry types* (*species*) usually designated by E if doubly degenerate and by F if triply degenerate.[13] While the influence of the various symmetry operations on the non-degenerate vibrations or eigenfunctions can be simply characterized by +1 or −1, such cannot be done with the degenerate vibrations and eigenfunctions, since in general they go over into a linear combination according to the previous equation (II, 62). It can be shown that it is sufficient for the characterization of the behavior of a degenerate vibration or eigenfunction *to give for every symmetry operation the value of the sum*

$$\chi = d_{aa} + d_{bb} + \cdots \tag{II, 85}$$

of the coefficients with two equal subscripts in the equations (II, 62). In group theory these sums are called *characters* of the *irreducible representation* (species) considered (see p. 104). In forming these characters it is assumed that the normal coordinates or eigenfunctions are mutually orthogonal. The characters are independent of which particular orthogonal pair is chosen (for example, for X_3 the pair in Fig. 32a or that in Fig. 33b) and which coordinate system is used.

For doubly degenerate vibrations the sums in (II, 85) consist of two terms only ($d_{aa} + d_{bb}$) and can easily be obtained on the basis of our previous discussion (see below). For triply degenerate vibrations or eigenfunctions we shall accept without proof the characters given by group theory [Wigner (923)]. For non-degenerate vibrations the sums (II, 85) consist of one term only, which is +1 or −1 since

[13] Some authors use T in place of F.

$\xi_i' = \xi_i$ or $\xi_i' = -\xi_i$. The values +1 and −1 given in Tables 12–14 are these characters for the point groups C_2, C_s, C_i, C_{2v}, C_{2h}, V, and V_h.

For the identical transformation I we have $\xi_a' = \xi_a$, $\xi_b' = \xi_b$, $\xi_c' = \xi_c$, if ξ_a, ξ_b, ξ_c are three mutually degenerate vibrations or eigenfunctions. Therefore, for a triply degenerate vibration, $\chi^{(I)} = +3$. Similarly, for a doubly degenerate vibration, $\chi^{(I)} = +2$ and of course for a non-degenerate vibration, $\chi^{(I)} = +1$. For a rotation by an angle $2\pi/p$ about a p-fold axis the character of a doubly degenerate vibration according to (II, 75) and (II, 77) is $2\cos(2\pi/p)l$. The same character corresponds to a rotation by an angle $-2\pi/p$, as can immediately be seen from (II, 75). The operations *rotation by* $+2\pi/p$ and *rotation by* $-2\pi/p$ are said to belong to the *same class*.[14] Similarly the rotations by $\pm 2\cdot 2\pi/p$ have the same character $2\cos(2\pi/p)2l$ and belong to the same class (which is different from the former class). Similarly the characters of a given (degenerate or non-degenerate) species for a reflection at any one of the p vertical planes of point group C_{pv} or for a rotation about any one of the p two-fold axes of point groups D_p and D_{ph} are the same. These symmetry operations therefore belong to the same class.

For non-degenerate symmetry types it is easily seen that different behavior (different characters) for two of the planes σ_v (or two of the C_2's $\perp C_p$) would lead to a contradiction to the fact that the vibration or eigenfunction must be symmetric or antisymmetric with respect to the C_p. For doubly degenerate symmetry types it was shown on p. 98 that the reflections at σ_v or rotations about the C_2's are represented by the transformation (II, 76). Therefore for them the character $\chi = d_{aa} + d_{bb} = 0$, independently of the angle β.

As we have seen previously, mutually degenerate vibrations (eigenfunctions) always have the same behavior with respect to an inversion. Therefore the character of a doubly degenerate species for the inversion is either +2 (when both components are symmetric) or −2 (when they are antisymmetric); for a triply degenerate species it is either +3 or −3. Similarly for a reflection at a plane σ_h perpendicular to C_p (see p. 97f.), the character of a doubly degenerate vibration is either +2 or −2.

On the basis of the preceding discussion, we can now proceed to discuss the symmetry types (species) of normal vibrations and eigenfunctions for the more important point groups with more-than-two-fold axes.

Point groups C_{3v} and D_3. According to our previous discussion (p. 89), a vibration or eigenfunction can only be symmetric or degenerate with respect to a three-fold axis, but not antisymmetric, since $p = 3$ is odd. Consequently there are for the point groups C_{3v} and D_3 *only two species of non-degenerate vibrations*, both symmetric with respect to C_3: one is symmetric with respect to the three planes σ_v or three C_2's, the other is antisymmetric. They are called A_1 and A_2. There can be no vibrations or eigenfunctions that are symmetric with respect to one and antisymmetric with respect to another of the planes σ_v or the C_2's (see above).

For both point groups *only one degenerate species* occurs, since according to p. 89 only $l = 1$ is possible and since there is only one possible behavior with respect

[14] In group theory a class is formed by those elements of a group that are conjugate to one another, that is, can be obtained from one element s by forming tst^{-1} where t may be any element of the group [see van der Waerden (23)]. The number of irreducible representations (our symmetry types or species) equals the number of classes of the group (in our case classes of symmetry elements of the point group).

to the three σ_v's or the three C_2's represented by the equations (II, 76) and leading to the character $\chi = 0$. The character for the operation C_3, according to the above, is

$$\chi^{(C_3)} = 2 \cos \frac{2\pi}{p} \cdot 1 = 2 \cos 120° = -1.$$

Altogether we have thus three different types (species) of vibrations or eigenfunctions in the point groups C_{3v} and D_3. These are summarized together with their

TABLE 15. SYMMETRY TYPES (SPECIES) AND CHARACTERS FOR THE POINT GROUPS C_{3v} AND D_3.

C_{3v}	I	$2C_3(z)$	$3\sigma_v$		D_3	I	$2C_3(z)$	$3C_2$	
A_1	+1	+1	+1	T_z	A_1	+1	+1	+1	
A_2	+1	+1	−1	R_z	A_2	+1	+1	−1	T_z, R_z
E	+2	−1	0	T_x, T_y, R_x, R_y	E	+2	−1	0	T_x, T_y, R_x, R_y

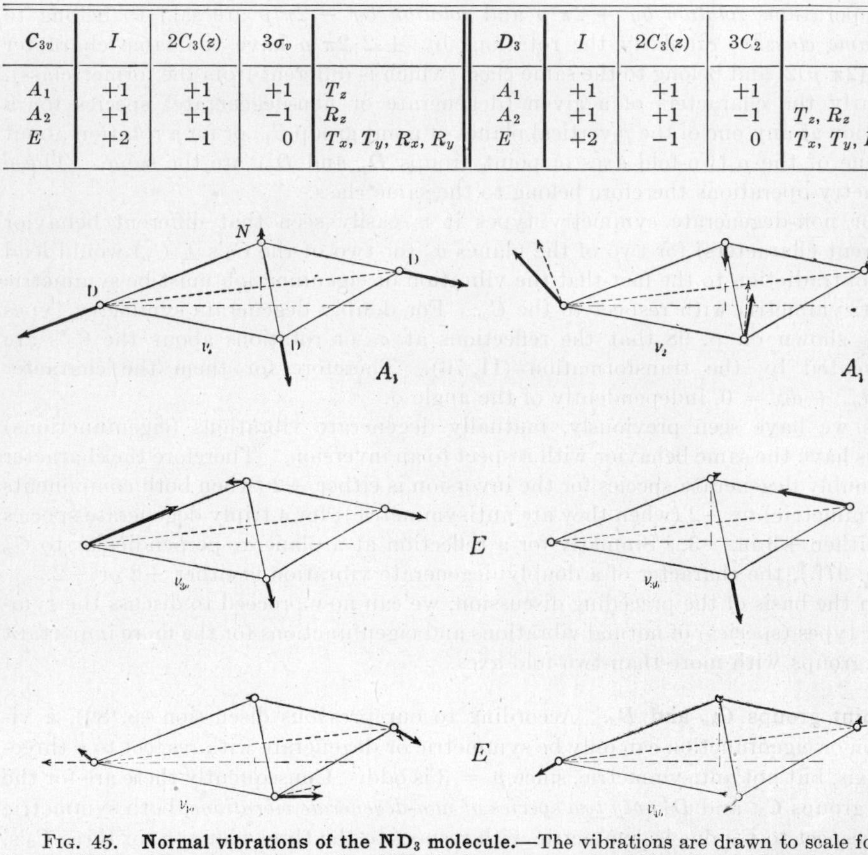

FIG. 45. **Normal vibrations of the ND_3 molecule.**—The vibrations are drawn to scale for ND_3 (see p. 177) in oblique projection. (For NH_3 the large mass ratio of N to H would not have allowed the displacement vectors of N to be drawn to the same scale as those of H). Both components of the degenerate vibrations are shown. The broken-line arrows in ν_2 and ν_4 give the symmetry coordinates of Fig. 58 (see p. 155). They are added so that the form of the vibrations can be more clearly visualized. In ν_{3b} there is a very small displacement (too small to show in the scale of the diagram) of the left D nucleus parallel to the line connecting the two other D nuclei (see also the discussion of Fig. 60 on p. 171). It should be noted that ν_{3a} and ν_{4a} are symmetric, ν_{3b} and ν_{4b} antisymmetric with respect to the plane of symmetry through the left D nucleus, that is, the plane of the paper.

characters in Table 15. In this table the numbers 2 and 3 respectively in front of C_3, σ_v, C_2 indicate the number of operations of the particular class (see above). We have seen previously (p. 90 and p. 97) that the translations in the x and y direction and the rotations about the x and y axes are degenerate non-genuine vibrations (last column of each subtable).

As an example, in Fig. 45 the normal vibrations of the ND_3 molecule which belongs to point group C_{3v} are represented to scale (see p. 177). There are two totally symmetric vibrations (species A_1) and two degenerate vibrations (species E). During the former oscillations the molecule remains always a symmetric pyramid, but it does not during the latter. There are no genuine vibrations of species A_2, but the rotation about the z axis has this type. Also some eigenfunctions of the higher vibrational levels of the degenerate vibrations ν_3 and ν_4 may have this species. Such an eigenfunction is represented schematically by Fig. 46. It is antisymmetric with respect to all three planes σ_v. The molecule $H_3C—CCl_3$, if it belongs to point group C_{3v} (as is very likely), will have a genuine normal vibration of species A_2—the torsion oscillation of the two groups CH_3 and CCl_3 with respect to each other about the C—C axis.

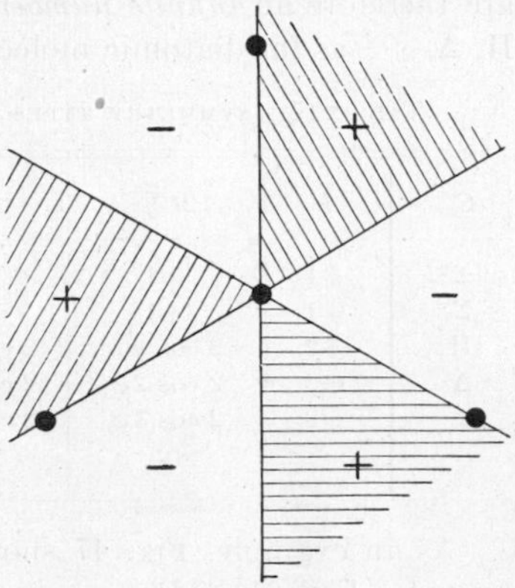

Fig. 46. Eigenfunction of symmetry type A_2 in a molecule of point group C_{3v}.

Point group C_{5v}. The symmetry types for the point group C_{5v} are in all respects similar to those of C_{3v} except that now we have *two types of degenerate vibrations*, namely, those with $l = 1$ and those with $l = 2$ (see p. 93) which are distinguished as E_1 and E_2. The characters are given in Table 16. In this table C_5^2 means a rotation by $\pm 2\pi/p \cdot 2 = \pm 144°$ about the five-fold axis which is a symmetry operation distinct from the rotation by $\pm 2\pi/p = 72°$ (a rotation by $\pm 2\pi/p \cdot 3$, however, leads to the same result as one by $\mp 2\pi/p \cdot 2$ and therefore is not a separate symmetry element). The characters of the degenerate species with respect to C_5 are $2 \cos (2\pi/5)l$ (see p. 109); with respect to C_5^2 they are $2 \cos (2\pi/5)2l$, which gives the values in Table 16.

Table 16. Symmetry types (species) and characters for the point group C_{5v}.

C_{5v}	I	$2C_5$	$2C_5^2$	$5\sigma_v$	
A_1	+1	+1	+1	+1	T_z
A_2	+1	+1	+1	−1	R_z
E_1	+2	2 cos 72°	2 cos 144°	0	T_x, T_y, R_x, R_y
E_2	+2	2 cos 144°	2 cos 72°	0	

The same symmetry types and characters as for C_{5v} apply to the point group D_5 if only $5\sigma_v$ in Table 16 is replaced by $5C_2$. The symmetry types and characters for point groups C_{7v} and D_7 would be similar to those for C_{5v} and D_5 except that there would be *three* degenerate species ($l = 1, 2, 3$) which would be designated E_1, E_2, E_3.

Point group $C_{\infty v}$. Just as for the point groups C_{3v} and C_{5v}, for $C_{\infty v}$ there can be no symmetry types that are antisymmetric with respect to rotation about the axis of symmetry (here by an arbitrary angle φ). Thus there are the same *two non-degenerate symmetry types*. They are, however, here designated Σ^+ and Σ^- in order to have agreement with the accepted nomenclature for electronic states of (heteronuclear) diatomic molecules which also belong to the point group $C_{\infty v}$. According to (II, 77) the number l can assume the values 1, 2, 3, $\cdots$ up to infinity, and there are therefore an *infinite number of degenerate symmetry types*. They are designated Π, Δ, $\cdots$ as for diatomic molecules. Table 17 gives the characters.

TABLE 17. SYMMETRY TYPES (SPECIES) AND CHARACTERS FOR THE POINT GROUP $C_{\infty v}$.

$C_{\infty v}$	I	$2C_\infty^\varphi$	$2C_\infty^{2\varphi}$	$2C_\infty^{3\varphi}$	$\cdots$	$\infty\sigma_v$	
Σ^+	+1	+1	+1	+1	$\cdots$	+1	T_z
Σ^-	+1	+1	+1	+1	$\cdots$	−1	R_z
Π	+2	$2\cos\varphi$	$2\cos 2\varphi$	$2\cos 3\varphi$	$\cdots$	0	T_x, T_y, R_x, R_y
Δ	+2	$2\cos 2\varphi$	$2\cos 2\cdot 2\varphi$	$2\cos 3\cdot 2\varphi$	$\cdots$	0	
Φ	+2	$2\cos 3\varphi$	$2\cos 2\cdot 3\varphi$	$2\cos 3\cdot 3\varphi$	$\cdots$	0	
$\cdots$	$\cdots$	$\cdots$	$\cdots$	$\cdots$	$\cdots$		

As an example, Fig. 47 shows the normal vibrations of a linear XYZ molecule (see also Fig. 61). For any number of atoms the normal vibrations belong to the species Σ^+ and Π (see section 4 of this chapter), but the eigenfunctions of the higher vibrational levels of the perpendicular vibrations (ν_2 of XYZ in Fig. 47) may have species Σ^-, Δ, Φ, $\cdots$ (see the next subsection).

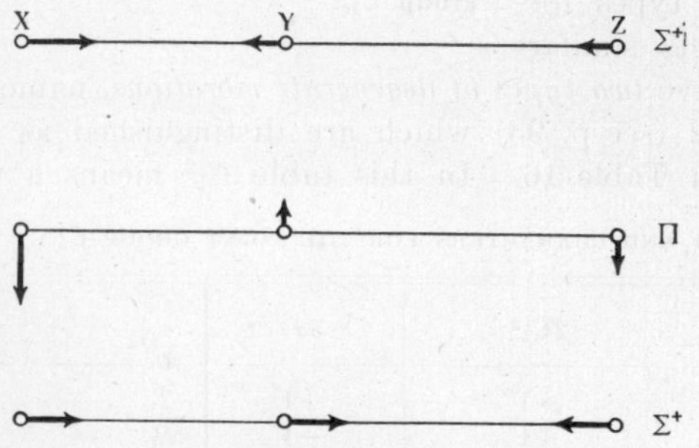

FIG. 47. **Normal vibrations of a linear XYZ molecule.**

That the designations of the species of $C_{\infty v}$ used here are really *equivalent to those used for the electronic states of diatomic molecules* is obvious for Σ^+ and Σ^-, since the symmetry properties are the same (see Molecular Spectra I, p. 238). It is also easily seen for Π, Δ, $\cdots$: The electronic eigenfunctions of a diatomic molecule in a Π, Δ, $\cdots$ state are given by (see Molecular Spectra I, p. 233)

$$\chi e^{i\Lambda\varphi} \quad \text{and} \quad \bar{\chi} e^{-i\Lambda\varphi}, \qquad \text{(II, 86)}$$

where χ and $\bar{\chi}$ do not contain the azimuthal angle φ. Rotation by the angle φ will therefore multiply the first function by the factor $e^{i\Lambda\varphi}$ and the second function by $e^{-i\Lambda\varphi}$ so that the character is $e^{i\Lambda\varphi} + e^{-i\Lambda\varphi} = 2\cos\Lambda\varphi$ as for Π, Δ, $\cdots$ in Table 17 if Λ is identified with l.

One may also say that in C_{3v} (and C_{4v}; see below) the type E corresponds to Π; but there are no analogues of Δ, Φ, $\cdots$. Similarly, for C_{5v} (and C_{6v}) the types E_1 and E_2 correspond to Π and Δ of linear or diatomic molecules.

Point groups C_{4v}, D_4, and $D_{2d} \equiv V_d$. If a molecule has a p-fold axis (C_p or S_p) with even p, an oscillation or eigenfunction may also be antisymmetric with respect to this axis (see p. 83). Therefore there are twice as many non-degenerate symmetry types as for odd p. In the case of C_{pv} the p planes σ_v must be divided into two classes, $p/2$ planes called σ_v and $p/2$ planes called σ_d (the latter being diagonal to the former), since the transformation properties (the characters) of these two sets are different. It is immediately seen (compare for example Fig. 1g and i) that a reflection of the molecule at σ_d can be replaced by a reflection at σ_v and a subsequent rotation by $2\pi/p$ about the C_p. Only the C_p and the $(p/2)\sigma_v$'s are necessary symmetry elements and the *four non-degenerate symmetry types* correspond to the four combinations $++$, $+-$, $-+$, $--$ of behavior with respect to these two operations. The behavior with respect to σ_d, which is not always the same as that with respect to σ_v, follows by multiplication of the behavior (character) with respect to C_p by that with respect to σ_v.

TABLE 18. SYMMETRY TYPES AND CHARACTERS FOR THE POINT GROUPS C_{4v}, D_4, AND $D_{2d} \equiv V_d$.

C_{4v}	I	$2C_4(z)$	$C_4^2 \equiv C_2''$	$2\sigma_v$	$2\sigma_d$	
D_4	I	$2C_4(z)$	$C_4^2 \equiv C_2''$	$2C_2$	$2C_2'$	
$D_{2d} \equiv V_d$	I	$2S_4(z)$	$S_4^2 \equiv C_2''$	$2C_2$	$2\sigma_d$	
A_1	$+1$	$+1$	$+1$	$+1$	$+1$	T_z for C_{4v}
A_2	$+1$	$+1$	$+1$	-1	-1	T_z for D_4; R_z
B_1	$+1$	-1	$+1$	$+1$	-1	
B_2	$+1$	-1	$+1$	-1	$+1$	T_z for V_d
E	$+2$	0	-2	0	0	T_x, T_y, R_x, R_y

The characters for C_{4v} that result in this way are given in Table 18. It includes also the *degenerate species* E which is the only one possible, since only $l = 1$ can occur (see p. 91). The character of this degenerate species with respect to C_4 is $2 \cos \beta = 2 \cos 2\pi/4 = 0$. The rotation about the two-fold axis C_2'' that is coincident with C_4 is identical with two successive rotations by $2\pi/4$ about C_4. Therefore all non-degenerate vibrations or eigenfunctions are symmetric with respect to it. But both

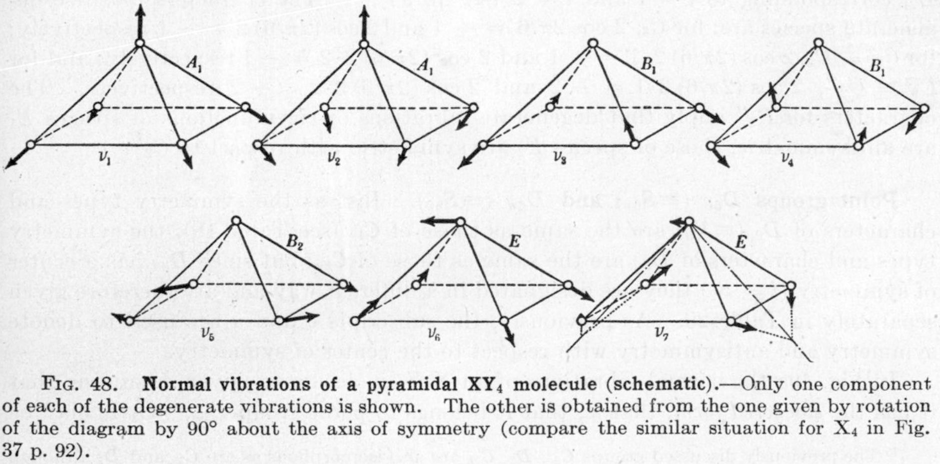

FIG. 48. Normal vibrations of a pyramidal XY_4 molecule (schematic).—Only one component of each of the degenerate vibrations is shown. The other is obtained from the one given by rotation of the diagram by 90° about the axis of symmetry (compare the similar situation for X_4 in Fig. 37 p. 92).

components of a degenerate pair are antisymmetric with respect to this symmetry operation, as can immediately be seen from (II, 75). Therefore the character is -2. The characters of E for σ_v and σ_d are 0 from the same reasons as for the point groups C_{3v} and C_{5v}. In Fig. 48, as an example, are illustrated the normal vibrations of a pyramidal XY_4 molecule. There are genuine vibrations of every species except A_2.

The *point groups* D_4 and $D_{2d} \equiv V_d$ have symmetry elements that are entirely similar to those of C_{4v} and with respect to which the vibrations and eigenfunctions behave in the same way: The C_4 of D_4 and the S_4 of D_{2d} correspond to the C_4 of C_{4v}, the C_2's perpendicular to C_4 or S_4 correspond to the σ_v's and σ_d's of C_{4v}. Therefore the point groups have the *same symmetry types and characters as* C_{4v}. The symmetry types are also designated in the same way. Thus it is not necessary to give special tables for them. Instead, their symmetry elements have been added at the top of Table 18. Point groups that have the same number and characters of their symmetry types such as C_{4v}, D_4, D_{2d}, are called *isomorphous*.[15]

Point groups C_{6v} and D_6. The symmetry types and characters of the point groups C_{6v} and D_6 given in Table 19 are similar in all respects to those of C_{4v} and D_4 except that now there are *two types of degenerate vibrations* (*eigenfunctions*) E_1 and

TABLE 19. SYMMETRY TYPES (SPECIES) AND CHARACTERS FOR THE POINT GROUPS C_{6v} AND D_6.

C_{6v}	I	$2C_6(z)$	$2C_6^2 \equiv 2C_3$	$C_6^3 \equiv C_2''$	$3\sigma_v$	$3\sigma_d$	
D_6	I	$2C_6(z)$	$2C_6^2 \equiv 2C_3$	$C_6^3 \equiv C_2''$	$3C_2$	$3C_2'$	
A_1	+1	+1	+1	+1	+1	+1	T_z for C_{6v}
A_2	+1	+1	+1	+1	−1	−1	T_z for D_6; R_z
B_1	+1	−1	+1	−1	+1	−1	
B_2	+1	−1	+1	−1	−1	+1	
E_1	+2	+1	−1	−2	0	0	T_x, T_y, R_x, R_y
E_2	+2	−1	−1	+2	0	0	

E_2, corresponding to $l = 1$ and $l = 2$ (see p. 94).[15a] The characters of these degenerate species are: for C_6, $2 \cos 2\pi/6 = +1$ and $2 \cos (2\pi/6)\, 2 = -1$ respectively; for $C_6^2 \equiv C_3$, $2 \cos (2\pi/6)\, 2 \cdot 1 = -1$ and $2 \cos (2\pi/6)\, 2 \cdot 2 = -1$ respectively; and for $C_6^3 \equiv C_2''$, $2 \cos (2\pi/6)\, 3 \cdot 1 = -2$ and $2 \cos (2\pi/6)\, 3 \cdot 2 = +2$ respectively. The characters for C_2'' imply that degenerate vibrations or eigenfunctions of species E_1 are antisymmetric, those of species E_2 are symmetric with respect to C_2''.

Point groups D_{3d} ($\equiv S_{6v}$) and D_{4d} ($\equiv S_{8v}$). Just as the symmetry types and characters of $D_{2d}(\equiv V_d)$ are the same as those of C_{4v} (see Table 18), the symmetry types and characters of D_{3d} are the same as those of C_{6v}; but since D_{3d} has a center of symmetry ($i \equiv S_2$) they are designated in a different way and are therefore given separately in Table 20. As previously, the subscripts g and u are used to denote symmetry and antisymmetry with respect to the center of symmetry.

In Fig. 49a the normal vibrations of an X_2Y_6 molecule are given, assuming that it has the *staggered form*, that is, that it belongs to point group D_{3d}. C_2H_6 may be

[15] The previously discussed groups C_{2v}, D_2, C_{2h} are also isomorphous as are C_{3v} and D_3 although for the former the designation of the species is different (see Table 13).

[15a] Some authors [see for example Sponer and Teller (802)] use E^- and E^+ for our E_1 and E_2 where the − and + signs indicate antisymmetry and symmetry with respect to C_2''. However this nomenclature is not applicable to point groups with five and higher than six fold axes (see C_{5v} above) and has therefore not been adopted here. The designation E_1, E_2, $\cdots$ is due to Tisza (867) [see also Jahn and Teller (471)]. Mulliken (641) used E^*, E^{**}, $\cdots$ in place of E_1, E_2, $\cdots$ but now also favors the latter.

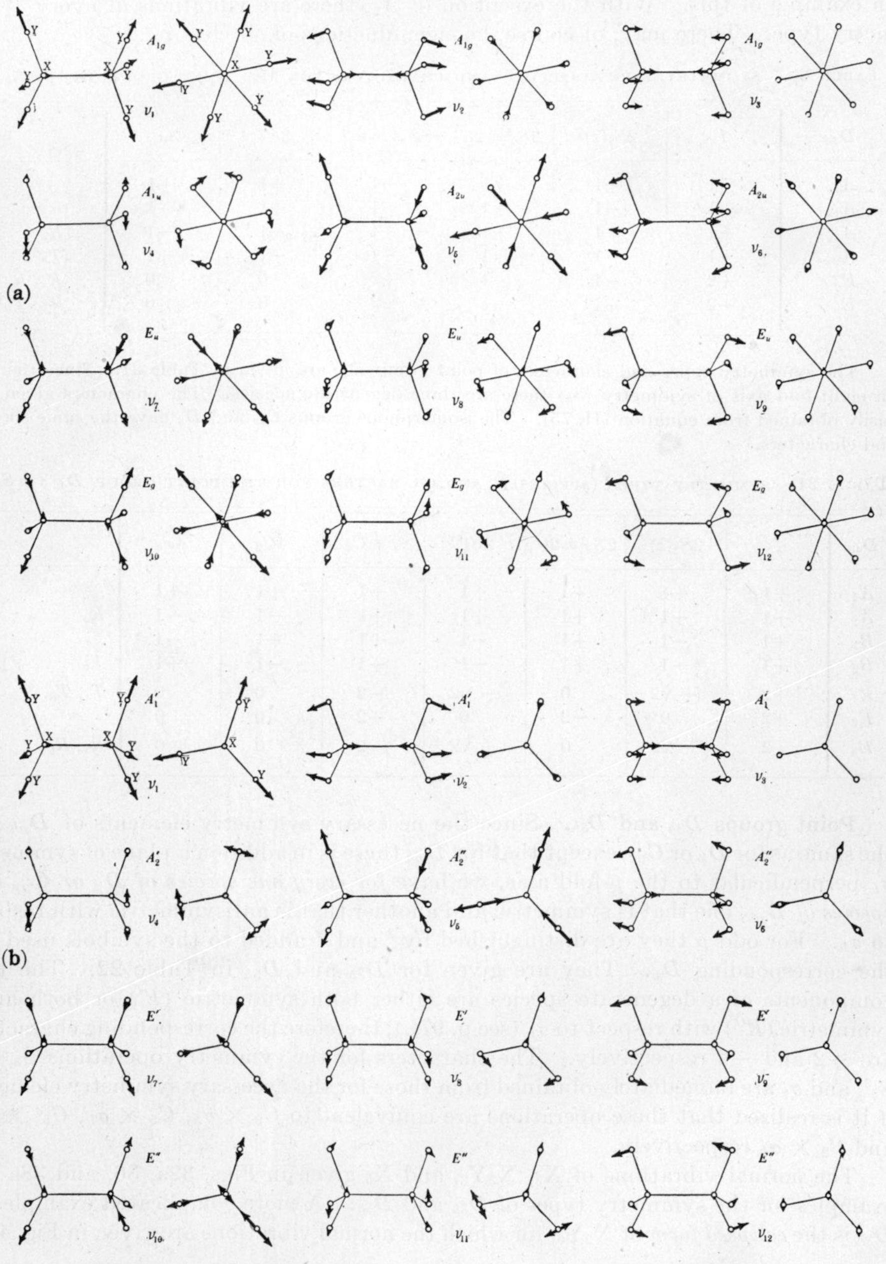

FIG. 49. **Normal vibrations of an X_2Y_6 molecule (schematic). (a) Assuming point group D_{3d}. (b) Assuming point group D_{3h}.—Only one component of each degenerate vibration is given in side and front view. For the other components of ν_7, ν_8, ν_{10}, ν_{11} see Fig. 45. The other components of ν_9 and ν_{12} are obtained from those given by rotating all vectors by 90°.**

an example of this. With the exception of A_{2g} there are vibrations of every symmetry type. There may, of course, be eigenfunctions of species A_{2g}.

TABLE 20. SYMMETRY TYPES (SPECIES) AND CHARACTERS FOR THE POINT GROUP D_{3d} ($\equiv S_{6v}$).

D_{3d}	I	$2S_6(z)$	$2S_6^2 \equiv 2C_3$	$S_6^3 \equiv S_2 \equiv i$	$3C_2$	$3\sigma_d$	
A_{1g}	+1	+1	+1	+1	+1	+1	
A_{1u}	+1	−1	+1	−1	+1	−1	
A_{2g}	+1	+1	+1	+1	−1	−1	R_z
A_{2u}	+1	−1	+1	−1	−1	+1	T_z
E_g	+2	−1	−1	+2	0	0	R_x, R_y
E_u	+2	+1	−1	−2	0	0	T_x, T_y

The symmetry types and characters of point group D_{4d} are given in Table 21. Since there is an eight-fold axis of symmetry (S_8) there are three degenerate species. The characters given are easily obtained from equation (II, 75). The isomorphous groups C_{8v} and D_8 have the same species and characters.

TABLE 21. SYMMETRY TYPES (SPECIES)[15b] AND CHARACTERS FOR THE POINT GROUP D_{4d} ($\equiv S_{8v}$).

D_{4d}	I	$2S_8(z)$	$2S_8^2 \equiv 2C_4$	$2S_8^3$	$S_8^4 \equiv C_2''$	$4C_2$	$4\sigma_d$	
A_1	+1	+1	+1	+1	+1	+1	+1	
A_2	+1	+1	+1	+1	+1	−1	−1	R_z
B_1	+1	−1	+1	−1	+1	+1	−1	
B_2	+1	−1	+1	−1	+1	−1	+1	T_z
E_1	+2	$+\sqrt{2}$	0	$-\sqrt{2}$	−2	0	0	$T_x, T_y,$
E_2	+2	0	−2	0	+2	0	0	
E_3	+2	$-\sqrt{2}$	0	$+\sqrt{2}$	−2	0	0	R_x, R_y

Point groups D_{3h} and D_{5h}. Since the necessary symmetry elements of D_{ph} are the same as for D_p or C_{pv}, except that for D_{ph} there is in addition a plane of symmetry σ_h perpendicular to the p-fold axis, we have *for every one species of D_p or C_{pv} two species of D_{ph}*, one that is symmetric and another that is antisymmetric with respect to σ_h. For odd p they are distinguished by ′ and ″ added to the symbols used for the corresponding D_p. They are given for D_{3h} and D_{5h} in Table 22. The two components of a degenerate species are either both symmetric (E') or both antisymmetric (E'') with respect to σ_h (see p. 97f.); therefore the corresponding characters are +2 and −2 respectively. The characters for the symmetry operations S_3, S_5, S_5^3, and σ_v are immediately obtained from those for the necessary symmetry elements if it is realized that these operations are equivalent to $C_3 \times \sigma_h$, $C_5 \times \sigma_h$, $C_5^2 \times \sigma_h$, and $C_2 \times \sigma_h$ respectively.

The normal vibrations of X_3, X_3Y_3, and X_5 given in Figs. 32a, 36, and 38a are examples for the symmetry types of D_{3h} and D_{5h}. A more complicated example for D_{3h} is the *eclipsed form* of X_2Y_6, for which the normal vibrations are given in Fig. 49b.

Point groups D_{4h} and D_{6h}. In the point groups D_{ph} with even p, the necessary elements of symmetry imply a center of symmetry i, and therefore the symmetry types, which are in the same relation to D_p as for odd p, are not distinguished by ′

[15b] Jahn and Teller (471) interchange B_1 and B_2.

TABLE 22. SYMMETRY TYPES (SPECIES) AND CHARACTERS FOR THE POINT GROUPS D_{3h} AND D_{5h}.

D_{3h}	I	$2C_3(z)$	$3C_2$	σ_h	$2S_3$	$3\sigma_v$	
A_1'	+1	+1	+1	+1	+1	+1	
A_1''	+1	+1	+1	−1	−1	−1	
A_2'	+1	+1	−1	+1	+1	−1	R_z
A_2''	+1	+1	−1	−1	−1	+1	T_z
E'	+2	−1	0	+2	−1	0	T_x, T_y
E''	+2	−1	0	−2	+1	0	R_x, R_y

D_{5h}	I	$2C_5$	$2C_5^2$	σ_h	$5C_2$	$5\sigma_v$	$2S_5$	$2S_5^3$	
A_1'	+1	+1	+1	+1	+1	+1	+1	+1	
A_1''	+1	+1	+1	−1	+1	−1	−1	−1	
A_2'	+1	+1	+1	+1	−1	−1	+1	+1	R_z
A_2''	+1	+1	+1	−1	−1	+1	−1	−1	T_z
E_1'	+2	2 cos 72°	2 cos 144°	+2	0	0	+2 cos 72°	+2 cos 144°	T_x, T_y
E_1''	+2	2 cos 72°	2 cos 144°	−2	0	0	−2 cos 72°	−2 cos 144°	R_x, R_y
E_2'	+2	2 cos 144°	2 cos 72°	+2	0	0	+2 cos 144°	+2 cos 72°	
E_2''	+2	2 cos 144°	2 cos 72°	−2	0	0	−2 cos 144°	−2 cos 72°	

TABLE 23. SYMMETRY TYPES (SPECIES) AND CHARACTERS FOR THE POINT GROUPS D_{4h} AND D_{6h}.

D_{4h}	I	$2C_4(z)$	$C_4^2 \equiv C_2''$	$2C_2$	$2C_2'$	σ_h	$2\sigma_v$	$2\sigma_d$	$2S_4$	$S_2 \equiv i$	
A_{1g}	+1	+1	+1	+1	+1	+1	+1	+1	+1	+1	
A_{1u}	+1	+1	+1	+1	+1	−1	−1	−1	−1	−1	
A_{2g}	+1	+1	+1	−1	−1	+1	−1	−1	+1	+1	R_z
A_{2u}	+1	+1	+1	−1	−1	−1	+1	+1	−1	−1	T_z
B_{1g}	+1	−1	+1	+1	−1	+1	+1	−1	−1	+1	
B_{1u}	+1	−1	+1	+1	−1	−1	−1	+1	+1	−1	
B_{2g}	+1	−1	+1	−1	+1	+1	−1	+1	−1	+1	
B_{2u}	+1	−1	+1	−1	+1	−1	+1	−1	+1	−1	
E_g	+2	0	−2	0	0	−2	0	0	0	+2	R_x, R_y
E_u	+2	0	−2	0	0	+2	0	0	0	−2	T_x, T_y

D_{6h}	I	$2C_6(z)$	$2C_6^2 \equiv 2C_3$	$C_6^3 \equiv C_2''$	$3C_2$	$3C_2'$	σ_h	$3\sigma_v$	$3\sigma_d$	$2S_6$	$2S_3$	$S_6^3 \equiv S_2 \equiv i$	
A_{1g}	+1	+1	+1	+1	+1	+1	+1	+1	+1	+1	+1	+1	
A_{1u}	+1	+1	+1	+1	+1	+1	−1	−1	−1	−1	−1	−1	
A_{2g}	+1	+1	+1	+1	−1	−1	+1	−1	−1	+1	+1	+1	R_z
A_{2u}	+1	+1	+1	+1	−1	−1	−1	+1	+1	−1	−1	−1	T_z
B_{1g}	+1	−1	+1	−1	+1	−1	−1	−1	+1	+1	−1	+1	
B_{1u}	+1	−1	+1	−1	+1	−1	+1	+1	−1	−1	+1	−1	
B_{2g}	+1	−1	+1	−1	−1	+1	−1	+1	−1	+1	−1	+1	
B_{2u}	+1	−1	+1	−1	−1	+1	+1	−1	+1	−1	+1	−1	
E_{1g}	+2	+1	−1	−2	0	0	−2	0	0	−1	+1	+2	R_x, R_y
E_{1u}	+2	+1	−1	−2	0	0	+2	0	0	+1	−1	−2	T_x, T_y
E_{2g}	+2	−1	−1	+2	0	0	+2	0	0	−1	−1	+2	
E_{2u}	+2	−1	−1	+2	0	0	−2	0	0	+1	+1	−2	

and $''$ but by a subscript g or u depending on whether they are symmetric or antisymmetric with respect to the center of symmetry. Table 23 gives the symmetry types and characters of D_{4h} and D_{6h} as obtained from those of D_4 and D_6 of Tables 18 and 19. Again the characters for i, σ_v, σ_d, S_4, S_6, S_3 are obtained in a way anal-

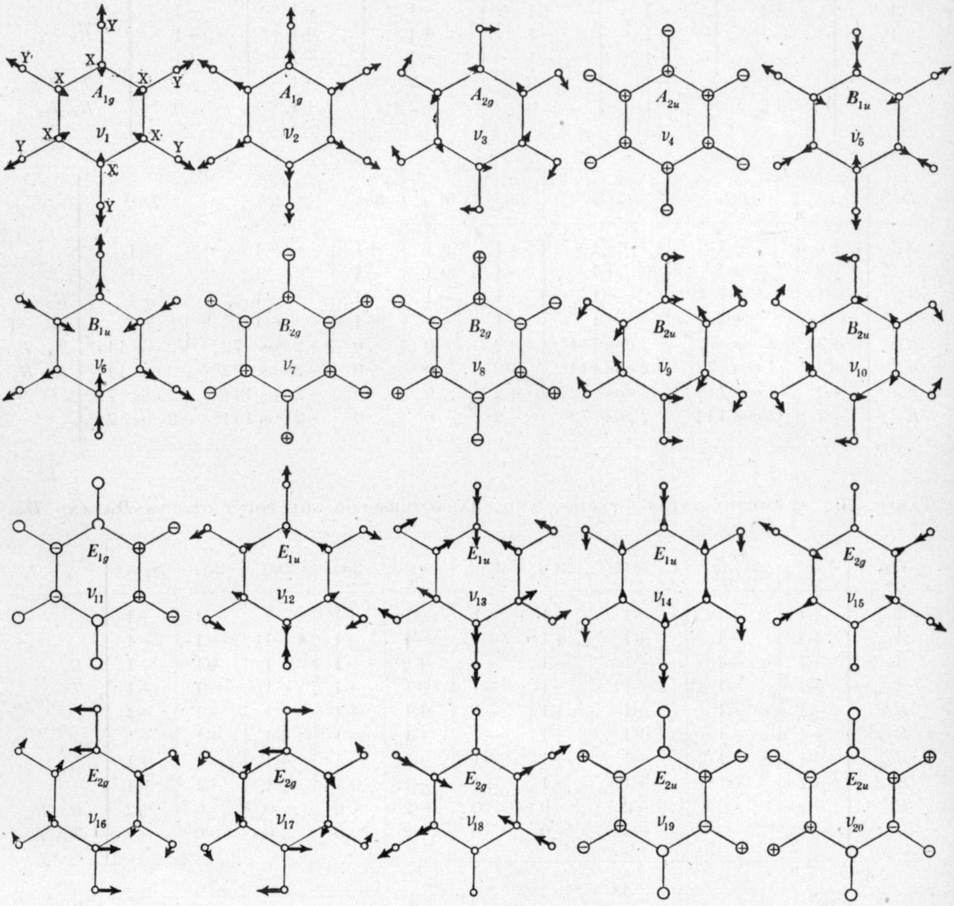

FIG. 50. **Normal vibrations of an X_6Y_6 molecule (point group D_{6h}).—Only one component of each degenerate vibration is given. For the other components compare Fig. 40 and 38c.**

ogous to the one indicated above for D_{3h} and D_{5h}. The normal vibrations of X_4 and X_6 given in Fig. 37 and Fig. 40 are examples for the symmetry types of D_{4h} and D_{6h} respectively. As a more complicated example, Fig. 50 gives the normal vibrations of a plane X_6Y_6 molecule (see C_6H_6, Chapter III, p. 362).

Point group $D_{\infty h}$. *Linear symmetric molecules* belong to point group $D_{\infty h}$. The symmetry types of $D_{\infty h}$ are quite analogous to those of D_{ph} with odd p except that there are now an infinite number of degenerate species corresponding to $l = 1, 2, 3 \cdots$. The designations are chosen the same as for the electronic states of homonuclear diatomic molecules. They are given together with the characters in Table 24.

In linear symmetric molecules ($\boldsymbol{D}_{\infty h}$) only normal vibrations of the species $\Sigma_g{}^+$, $\Sigma_u{}^+$, Π_g, and Π_u occur (see section 4a). This is illustrated by the normal vibrations of linear X_2Y_2 (for example C_2H_2, see p. 288) in Fig. 64 p. 181 (compare also XY_2 in

TABLE 24. SYMMETRY TYPES (SPECIES) AND CHARACTERS FOR THE POINT GROUP $\boldsymbol{D}_{\infty h}$.

$D_{\infty h}$	I	$2C_\infty^\varphi$	$2C_\infty^{2\varphi}$	$2C_\infty^{3\varphi}$	$\cdots$	σ_h	∞C_2	$\infty\sigma_v$	$2S_\infty^\varphi$	$2S_\infty^{2\varphi}$	$\cdots$	$S_2\equiv i$	
$\Sigma_g{}^+$	+1	+1	+1	+1	$\cdots$	+1	+1	+1	+1	+1	$\cdots$	+1	
$\Sigma_u{}^+$	+1	+1	+1	+1	$\cdots$	−1	−1	+1	−1	−1	$\cdots$	−1	T_z
$\Sigma_g{}^-$	+1	+1	+1	+1	$\cdots$	+1	−1	−1	+1	+1	$\cdots$	+1	R_z
$\Sigma_u{}^-$	+1	+1	+1	+1	$\cdots$	−1	+1	−1	−1	−1	$\cdots$	−1	
Π_g	+2	$2\cos\varphi$	$2\cos 2\varphi$	$2\cos 3\varphi$	$\cdots$	−2	0	0	$-2\cos\varphi$	$-2\cos 2\varphi$	$\cdots$	+2	R_x, R_y
Π_u	+2	$2\cos\varphi$	$2\cos 2\varphi$	$2\cos 3\varphi$	$\cdots$	+2	0	0	$+2\cos\varphi$	$+2\cos 2\varphi$	$\cdots$	−2	T_x, T_y
Δ_g	+2	$2\cos 2\varphi$	$2\cos 4\varphi$	$2\cos 6\varphi$	$\cdots$	+2	0	0	$+2\cos 2\varphi$	$+2\cos 4\varphi$	$\cdots$	+2	
Δ_u	+2	$2\cos 2\varphi$	$2\cos 4\varphi$	$2\cos 6\varphi$	$\cdots$	−2	0	0	$-2\cos 2\varphi$	$-2\cos 4\varphi$	$\cdots$	−2	
Φ_g	+2	$2\cos 3\varphi$	$2\cos 6\varphi$	$2\cos 9\varphi$	$\cdots$	−2	0	0	$-2\cos 3\varphi$	$-2\cos 4\varphi$	$\cdots$	+2	
Φ_u	+2	$2\cos 3\varphi$	$2\cos 6\varphi$	$2\cos 9\varphi$	$\cdots$	+2	0	0	$+2\cos 3\varphi$	$+2\cos 4\varphi$	$\cdots$	−2	
$\cdots$	$\cdots$	$\cdots$	$\cdots$	$\cdots$	$\cdots$	$\cdots$	$\cdots$	$\cdots$	$\cdots$	$\cdots$	$\cdots$	$\cdots$	

Fig. 25b). However the eigenfunctions of the higher vibrational levels and of the electronic states may also be of any one of the other species.

Point groups $\boldsymbol{C}_p$. The symmetry types of the (not very important) point groups $\boldsymbol{C}_p$, which have a p-fold axis only ($p > 2$), are obtained from those of the point groups $\boldsymbol{C}_{pv}$ (Tables 15–19) by dropping the symmetry elements σ_v and σ_d and therefore by dropping the distinction between A_1 and A_2, between B_1 and B_2, and between Σ^+ and Σ^-. Thus there is only one non-degenerate species (A or Σ) for $\boldsymbol{C}_3$, $\boldsymbol{C}_5$, and $\boldsymbol{C}_\infty$,

TABLE 25. SYMMETRY TYPES (SPECIES) AND CHARACTERS FOR THE POINT GROUPS $\boldsymbol{C}_3$, $\boldsymbol{C}_6$, AND $\boldsymbol{S}_6$.

$\boldsymbol{C}_3$	I	$2C_3$	
A	+1	+1	T_z, R_z
E	+2	$2\cos 120° = -1$	T_x, T_y, R_x, R_y

$\boldsymbol{C}_6$		I	C_6	$C_6{}^2\equiv C_3$	$C_6{}^3\equiv C_2''$	
	$\boldsymbol{S}_6$	I	S_6	$S_6{}^2\equiv C_3$	$S_6{}^3\equiv S_2\equiv i$	
A	A_g	+1	+1	+1	+1	T_z for $\boldsymbol{C}_6$, R_z
B	B_u	+1	−1	+1	−1	T_z for $\boldsymbol{S}_6$
E_1	E_{1u}	+2	+1	−1	−2	R_x, R_y for $\boldsymbol{C}_6$, T_x, T_y
E_2	E_{2g}	+2	−1	−1	+2	R_x, R_y for $\boldsymbol{S}_6$

and only two (A and B) for $\boldsymbol{C}_4$ and $\boldsymbol{C}_6$. The number of degenerate species is the same as for $\boldsymbol{C}_{pv}$. In Table 25 the symmetry types and characters for $\boldsymbol{C}_3$ and $\boldsymbol{C}_6$ are given. The reader can easily construct similar tables for $\boldsymbol{C}_5$, $\boldsymbol{C}_\infty$, $\boldsymbol{C}_4$, from the corresponding tables for $\boldsymbol{C}_{5v}$, $\boldsymbol{C}_{\infty v}$, $\boldsymbol{C}_{4v}$.

For some considerations it is necessary to take account of the fact that the degenerate vibrations and eigenfunctions of the point groups $\boldsymbol{C}_p$ are separably degenerate (see p. 99). The complex normal coordinates or eigenfunctions given by (II, 81) are not mixed by any of the symmetry operations. Therefore the characters are frequently given for each component separately. They are simply the (complex) factors by which the normal coordinates (or eigenfunctions) are multiplied for the respective symmetry operations. These characters are given in Table 26. If the two

TABLE 26. SYMMETRY TYPES AND CHARACTERS FOR THE POINT GROUP C_3, TAKING ACCOUNT OF SEPARABILITY.

C_3	I	C_3	C_3^2	
A	+1	+1	+1	T_z, R_z
E	+1	$e^{2\pi i/3}$	$e^{-2\pi i/3}$	T_x+iT_y, R_x+iR_y
	+1	$e^{-2\pi i/3}$	$e^{2\pi i/3}$	T_x-iT_y, R_x-iR_y

vibrations of a pair are considered together, the character is the same for C_3 and $C_3^2 \equiv C_3$; but for each individual one of the separated degenerate normal coordinates they are not the same. For one of them the characters are (see p. 99) $e^{2\pi i/3}$ and $e^{2\cdot 2\pi i/3} = e^{-2\pi i/3}$ for C_3 and C_3^2 respectively, and for the other they are the conjugate complex of these values. This is why there are now two columns C_3, C_3^2 instead of the one column $2C_3$ in the table of characters. It may be noted that the sums of the characters of the pair of separably degenerate vibrations are +2, 2 cos 120°, 2 cos 120°, that is, they are the characters given in Table 25.

For the other point groups C_p the characters are similar. In particular, for the operation C_p they are $e^{2\pi i/p}$ and $e^{-2\pi i/p}$ for the two components of E_1, they are $e^{2\cdot 2\pi i/p}$ and $e^{-2\cdot 2\pi i/p}$ for the two components of E_2.

TABLE 27. SYMMETRY TYPES AND CHARACTERS FOR THE POINT GROUPS C_{3h}, C_{4h}, AND C_{6h}.

C_{3h}	I	C_3	σ_h	S_3	
A'	+1	+1	+1	+1	R_z
A''	+1	+1	−1	−1	T_z
E'	+2	−1	+2	−1	T_x, T_y
E''	+2	−1	−2	+1	R_x, R_y

C_{4h}	I	C_4	$C_4^2 \equiv C_2''$	σ_h	S_4	$S_2 \equiv i$	
A_g	+1	+1	+1	+1	+1	+1	R_z
A_u	+1	+1	+1	−1	−1	−1	T_z
B_g	+1	−1	+1	+1	−1	+1	
B_u	+1	−1	+1	−1	+1	−1	
E_g	+2	0	−2	−2	0	+2	R_x, R_y
E_u	+2	0	−2	+2	0	−2	T_x, T_y

C_{6h}	I	C_6	$C_6^2 \equiv C_3$	$C_6^3 \equiv C_2''$	σ_h	S_6	S_3	$S_2 \equiv i$	
A_g	+1	+1	+1	+1	+1	+1	+1	+1	R_z
A_u	+1	+1	+1	+1	−1	−1	−1	−1	T_z
B_g	+1	−1	+1	−1	−1	+1	−1	+1	
B_u	+1	−1	+1	−1	+1	−1	+1	−1	
E_{1g}	+2	+1	−1	−2	−2	−1	+1	+2	R_x, R_y
E_{1u}	+2	+1	−1	−2	+2	+1	−1	−2	T_x, T_y
E_{2g}	+2	−1	−1	+2	+2	−1	−1	+2	
E_{2u}	+2	−1	−1	+2	−2	+1	+1	−2	

Point groups S_4 and S_6. The point groups S_4 and S_6 are isomorphous with C_4 and C_6 respectively. The symmetry types and characters of S_6 have been included in Table 25 for C_6. It should be noted that on account of the presence of a center of symmetry for S_6 the designation of the symmetry types is different from those of C_6.

Point groups C_{ph}. In the point groups C_{ph} there is a plane of symmetry σ_h perpendicular to the axis of symmetry C_p. For every one symmetry type of C_p there are therefore two for C_{ph}, one that is symmetric and one that is antisymmetric with respect to σ_h. In particular, the two components of a degenerate pair are either both symmetric or both antisymmetric with respect to σ_h. For odd p the species that are symmetric or antisymmetric with respect to σ_h are distinguished by $'$ and $''$, while for even p, since there is a center of symmetry i, they are distinguished by subscripts g or u depending on the behavior with respect to i, which is determined by that with respect to σ_h and C_2''. Table 27 gives the symmetry types and characters for C_{3h}, C_{4h}, and C_{6h}. The reader can easily obtain those for C_{5h} from those of C_5 if required.

As for the point groups C_p, the degenerate species of C_{ph} are separably degenerate and therefore it is sometimes convenient to give the characters of the degenerate components separately. For C_{3h} they are exactly like those in Table 26 for C_3. For C_{4h} and C_{6h} they can easily be obtained on the basis of the above discussion for C_4 and C_6.

Point groups T_d and O. The cubic point group T_d (to which molecules like CH_4 and others belong) has four three-fold axes. The non-degenerate vibrations or eigenfunctions can only be symmetric with respect to these axes (see p. 83), but, as for point group C_{3v}, they may be symmetric or antisymmetric with respect to the six planes of symmetry σ_d through the C_3's, and in consequence of that to the three four-fold rotation-reflection axes S_4. Thus there are *two types* (A_1 and A_2) *of non-degenerate vibrations or eigenfunctions*. A closer examination on the basis of group theory [see Wigner (923)] shows that there is just *one doubly degenerate species* E as for C_{3v} and *two triply degenerate species* F_1 and F_2. Their characters are given without further proof in Table 28.

TABLE 28. SYMMETRY TYPES (SPECIES) AND CHARACTERS FOR THE POINT GROUPS T_d AND O.[16]

T_d	I	$8C_3$	$6\sigma_d$	$6S_4$	$3S_4^2 \equiv 3C_2$	
O	I	$8C_3$	$6C_2$	$6C_4$	$3C_4^2 \equiv 3C_2''$	
A_1	+1	+1	+1	+1	+1	
A_2	+1	+1	−1	−1	+1	
E	+2	−1	0	0	+2	
F_1	+3	0	−1	+1	−1	T_x, T_y, T_z for O, R_x, R_y, R_z
F_2	+3	0	+1	−1	−1	T_x, T_y, T_z for T_d

As mentioned before, degenerate vibrations can always be chosen in such a way that they are symmetric or antisymmetric with respect to planes, two-fold axes, and a center of symmetry. In the present case one vibration of a doubly degenerate pair can be chosen symmetric, the other antisymmetric, with respect to σ_d, and therefore the corresponding character is 0. All doubly degenerate vibrations are symmetric with respect to the two-fold axes C_2. The two triply degenerate symmetry types are distinguished by the fact that in one of them two of the mutually degenerate vibrations can be made antisymmetric and one symmetric with respect to a σ_d plane while in the other type two vibrations are symmetric and one antisymmetric. That is why the corresponding characters ($\chi = d_{aa} + d_{bb} + d_{cc}$) are -1 and $+1$ respectively. The behavior of these triply degenerate vibrations is perhaps best visualized

[16] There are four three-fold axes (see p. 9) but there are eight symmetry operations C_3, namely the rotations by $\pm 2\pi/3$ about the four axes. Similarly there are three four-fold rotation-reflection axes and therefore six operations S_4.

if it is noted that the three rotations R_x, R_y, R_z form a non-genuine vibration of species F_1, while the three translations T_x, T_y, T_z form a non-genuine vibration of species F_2. If a σ_d plane is the xy plane, the rotations R_x, R_y are antisymmetric and the rotation R_z is symmetric with respect to this σ_d, that is, $\chi_{F_1} = -1$; while the translations T_x, T_y are symmetric, the translation T_z is antisymmetric, that is, $\chi_{F_2} = +1$.

The normal vibrations of a tetrahedral XY_4 molecule given in the previous Fig. 41 represent examples of the species A_1, E, and F_2. It is easily verified that they do have the required symmetry properties. For XY_4 there are no genuine normal vibrations of the species A_2 and F_1, but the eigenfunctions of higher vibrational levels may belong to these species.

There is a one-to-one correspondence between the symmetry elements of point groups T_d and point group O: they are isomorphous. Consequently the symmetry types and characters are the same. Therefore, in order to give the symmetry types and characters of O it was only necessary to add at the top of Table 28 the symmetry elements of O. For the point group O other than for T_d the three translations T_x, T_y, T_z form a non-genuine vibration of species F_1 as do the three rotations.

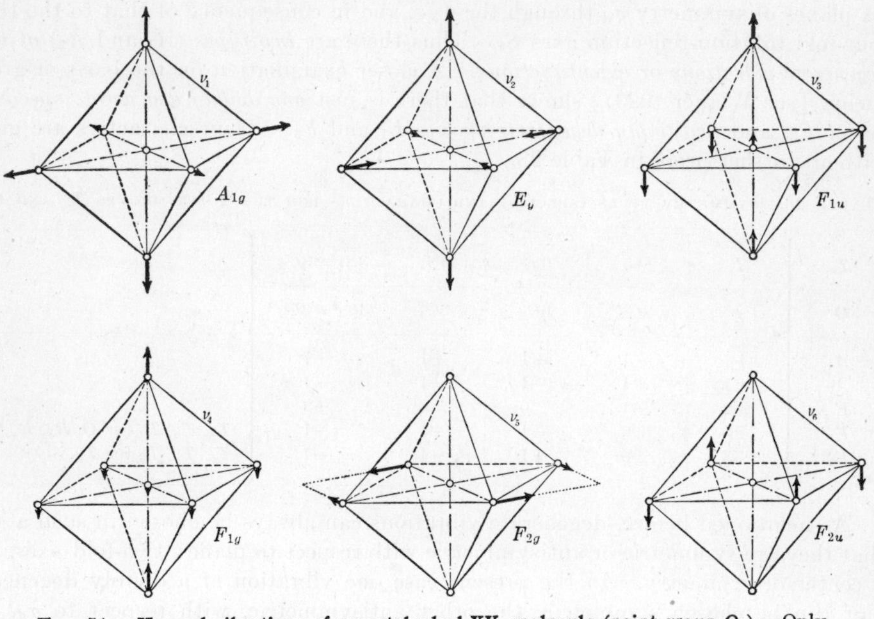

FIG. 51. **Normal vibrations of an octahedral XY_6 molecule (point group O_h).—Only one component of each degenerate vibration is shown.**

Point group O_h. The point group O_h has in addition to the symmetry elements of O a center of symmetry i, as well as several other symmetry elements necessitated by it. Therefore, in place of each symmetry type of O there are two in O_h, one that is symmetric and one that is antisymmetric with respect to i. Thus we obtain the species and characters given in Table 29.[16a] As an example, in Fig. 51 the normal

[16a] It may be noted that the characters in columns 8, 9, 10, and 11 are obtained from those in columns 5, 3, 6, and 4 respectively (which are identical with those of point group O) by multiplication with those in column 7.

vibrations of an octahedral XY_6 molecule (such as SF_6; see p. 336) are illustrated. Normal vibrations of the species A_{1u}, A_{2g}, A_{2u}, E_u, F_{1g} do not occur (see Table 36).

TABLE 29. SYMMETRY TYPES (SPECIES) AND CHARACTERS FOR THE POINT GROUP O_h.

O_h	I	$8C_3$	$6C_2$	$6C_4$	$3C_4^2 \equiv 3C_2''$	$S_2 \equiv i$	$6S_4$	$8S_6$	$3\sigma_h$	$6\sigma_d$	
A_{1g}	+1	+1	+1	+1	+1	+1	+1	+1	+1	+1	
A_{1u}	+1	+1	+1	+1	+1	−1	−1	−1	−1	−1	
A_{2g}	+1	+1	−1	−1	+1	+1	−1	+1	+1	−1	
A_{2u}	+1	+1	−1	−1	+1	−1	+1	−1	−1	+1	
E_g	+2	−1	0	0	+2	+2	0	−1	+2	0	
E_u	+2	−1	0	0	+2	−2	0	+1	−2	0	
F_{1g}	+3	0	−1	+1	−1	+3	+1	0	−1	−1	R_x, R_y, R_z
F_{1u}	+3	0	−1	+1	−1	−3	−1	0	+1	+1	T_x, T_y, T_z
F_{2g}	+3	0	+1	−1	−1	+3	−1	0	−1	+1	
F_{2u}	+3	0	+1	−1	−1	−3	+1	0	+1	−1	

Point group *T*. Since the point group T has no planes of symmetry σ_d, and consequently no S_4, but otherwise the same symmetry elements as the point group T_d (Table 28), the symmetry types are similar except that the distinction between A_1 and A_2 and between F_1 and F_2 has to be dropped. Thus the symmetry types and characters in Table 30 are obtained.

TABLE 30. SYMMETRY TYPES (SPECIES) AND CHARACTERS FOR THE POINT GROUP T.

T	I	$8C_3$	$3C_2$	
A	+1	+1	+1	
E	+2	−1	+2	
F	+3	0	−1	T, R

Since there are no planes through the three-fold axes, the doubly degenerate vibrations and eigenfunctions are separably degenerate (see p. 99). The characters of the separated (complex) normal coordinates (II, 81) for the operation C_3 are the same as for point group C_3 (Table 26), for C_2 they are +1 for both components.

Since molecules belonging to the point groups T_h, I, I_h are not likely to be found we omit a discussion of their symmetry types and characters [see Tisza (867)]. However, it may perhaps be mentioned that the point groups I and I_h, in addition to triply degenerate species, also have species of four-fold and five-fold degeneracy.

(e) *Symmetry types (species) of the higher vibrational levels*

From the considerations at the end of section 3c, it follows that any eigenfunction of a polyatomic molecule (whether electronic, vibrational, rotational, or total) must belong to one of the symmetry types of the particular point group discussed in the preceding subsection. Thus also the vibrational eigenfunctions of states in which several normal vibrations of different symmetry types are excited with one or more quanta must belong to one of the possible symmetry types. This holds irrespective of whether or not the vibrations may be considered as strictly harmonic (see also section 5). The question arises, therefore: What is the resultant symmetry type (species) of a state in which several vibrations are excited, or in which one or more vibrations are excited by more than one quantum?

Non-degenerate vibrations. The answer to the above question is very easy to find in the case of non-degenerate vibrations on the basis of our previous discussion (p. 101f.). Since the total vibrational eigenfunction is symmetric or antisymmetric with respect to a certain symmetry element depending on whether $\sum_a v_k$ (that is, the sum of the vibrational quantum numbers of all vibrations that are antisymmetric with respect to this symmetry element) is even or odd, we can immediately obtain *the behavior of the total vibrational eigenfunction with respect to all symmetry elements and thus its symmetry type.* It is sufficient to restrict this consideration to the *necessary symmetry elements.* For example, if in C_2H_4 (assuming that it belongs to the point group $\boldsymbol{D_{2h}} \equiv \boldsymbol{V_h}$) the vibration ν_4 (see Fig. 44) of species A_u is doubly excited, ν_8 of species B_{2g} is triply excited, and ν_{12} of species B_{3u} is triply excited, then it follows from Table 14 (p. 108) and from the above rule that the resultant eigenfunction is antisymmetric with respect to $\sigma(xy)$, symmetric with respect to $\sigma(xz)$ and symmetric with respect to $\sigma(yz)$, that is, is of type B_{1u}. Symbolically we may write this result as

$$(a_u)^2 \cdot (b_{2g})^3 \cdot (b_{3u})^3 = B_{1u},$$

where we have used *small letters for the species symbols of the individual vibrations* and a *capital letter for that of the resultant state.*[17] In a similar manner the resultant species in the previous Fig. 43 for H_2CO have been obtained.

It is at once evident that the above rule is equivalent to saying that *the characters of the resultant species are obtained by multiplying for each symmetry element the characters of the species of the component normal vibrations taken to the v_kth power if v_k is the vibrational quantum number of the particular vibration.* This simple method is also applicable to the non-degenerate vibrations of point groups with more-than-two-fold axes. It is immediately clear from this rule that the vibrational levels in which a non-totally symmetric vibration is excited with even v_k are totally symmetric, while those with odd v_k have the symmetry of the normal vibration. Thus in Fig. 42b, if the vibration is of type B_{1u} (point group $\boldsymbol{V_h}$) the levels designated s and a are A_{1g} and B_{1u} respectively. Similarly, if two non-totally symmetric vibrations of the same species are each singly excited, the resultant state is totally symmetric; if a totally symmetric and non-totally symmetric vibration are each singly excited, the resultant state has the species of the latter. Finally, a rule concerning the behavior with respect to a center of symmetry i (if such is present) is very useful, since this behavior is indicated in the species symbol by the same subscripts g or u for all point groups: *If two vibrations with the same behavior with respect to* i (that is, both g or both u) *are each singly excited, the resultant state is symmetric* (g) *with respect to* i; *if the two vibrations have opposite symmetry, the resultant state is antisymmetric* (u). This *g, u rule* is, of course, simply a special case of the above rule concerning $\sum_a v_k$. However, it holds quite generally even for degenerate vibrations, since even they can only be symmetric or antisymmetric with respect to i (see p. 97f.). The *g, u rule* is therefore valid in all cases in which a center of symmetry is present.

For the convenience of the reader, Table 31 includes those cases of *binary combinations of non-degenerate vibrations* that are not covered by one of the above special

[17] This usage has been suggested by Mulliken (643), but it is not observed by all authors. Sponer and Teller (802) have suggested the use of corresponding Greek letters for vibrational species, light-faced for the individual vibrations, bold-faced for the resultant state. They reserve the roman letters for electronic states.

rules. This table, therefore, should make it possible, if necessary by successive application, to find the resultant species without looking up the character tables and multiplying the characters. Thus in the previous example the above rules tell us that $(a_u)^2 = A_g$, $(b_{2g})^3 = B_{2g}$, and $(b_{3u})^3 = B_{3u}$. Furthermore $(a_u)^2(b_{2g})^3 = A_g \cdot B_{2g} = B_{2g}$, and from Table 31 and the g, u rule $(a_u)^2(b_{2g})^3(b_{3u})^3 = B_{2g} \cdot B_{3u} = B_{1u}$. (The table holds, of course, equally for small and capital letters).

Binary combinations of a non-degenerate and a degenerate vibration. If a non-degenerate vibration is singly excited at the same time that a degenerate vibration is singly excited (that is, if we have a binary combination of these two), the resultant state has, of course, a species of the same degree of degeneracy as the one degenerate vibration. However, if for the point group considered there are several degenerate species, the species of the resulting state need not be the same as that of the degenerate vibration. Group theory shows that the species of the resulting state is obtained in the same way as for two non-degenerate vibrations by taking the *product of the characters of the two species for each symmetry operation.* The numbers obtained in this way are the characters of the resulting state.

For example, if in a molecule of point group $\boldsymbol{C}_{3v}$ (for instance H_3C—CCl_3) two vibrations, one of species A_2, the other of species E, are singly excited, then the characters of the resulting state are, according to the above rule and Table 15: $+1 \times +2 = +2$, $+1 \times -1 = -1$, and $-1 \times 0 = 0$. These are the characters of species E, which is therefore the species of the resulting state. If in a molecule of point group $\boldsymbol{D}_{6h}$ one vibration each of species B_{1u} and E_{2g} is singly excited, the characters of the resulting state are $+2$, $+1$, -1, -2, 0, 0, $+2 \cdots$ (see Table 23); that is, it is an E_{1u} state. Symbolically we write in the two examples:

$$a_2 \cdot e = E, \qquad b_{1u} \cdot e_{2g} = E_{1u}.$$

For the convenience of the reader, Table 31 gives the results for all binary combinations of a degenerate and a non-degenerate vibration (species) for all important point groups.

Multiple excitation of a single, degenerate vibration. If a degenerate vibration is excited to higher vibrational states of quantum number v_j, the resultant species are not as easily obtained. They have been derived by Tisza (867) with the aid of group theory. We shall describe here only the results. As we have seen previously (p. 103), *the eigenfunction is totally symmetric if* $v_j = 0$, *it has the same (degenerate) species as the normal vibration if* $v_j = 1$. If $v_j > 1$, the resulting state has a degree of degeneracy greater than that of the singly excited vibration as long as the vibration is strictly harmonic (see p. 80f.). However, it can be shown [see Tisza (867)] that this more highly degenerate state may be considered as a superposition of a number of less highly degenerate and possibly non-degenerate states which belong to the various symmetry types of the point group of the molecule, and which are accidentally degenerate with one another. In fact, slight perturbations, such as the anharmonicity usually present, cause a splitting of the accidental degeneracies but leave, of course, the necessary degeneracies of the component states.

For example, in the case of non-symmetrical linear molecules (point group $\boldsymbol{C}_{\infty v}$, example HCN), if a vibration of species Π (to which all perpendicular vibrations belong; see Fig. 47) is excited by three quanta ($v_j = 3$) the resultant vibrational state is four-foldly degenerate (see p. 80), consisting of two substates, one of species Π and one of species Φ. The reason for this is immediately clear if it is remembered that a Π vibration may be considered as having an angular momentum of one unit ($l = 1$) about the internuclear axis and that there are just four ways of adding three

TABLE 31. SYMMETRY TYPES (SPECIES) FOR THOSE LEVELS IN WHICH TWO DIFFERENT VIBRATIONS, AT LEAST ONE OF WHICH IS NON-DEGENERATE, ARE SINGLY EXCITED.

Point group	Vibrations excited	Resultant state	Vibrations excited	Resultant state	Vibrations excited	Resultant state	Vibrations excited	Resultant state
C_{2v}	$a_2 \cdot b_1$	B_2	$a_2 \cdot b_2$	B_1	$b_1 \cdot b_2$	A_2		
C_{2h}	$a_u \cdot b_g$	B_u	$a_u \cdot b_u$	B_g	$b_g \cdot b_u$	A_u		
$D_2 \equiv V[D_{2h} \equiv V_h]$[18]	$a \cdot b_i$	B_i	$b_1 \cdot b_2$	B_3	$b_1 \cdot b_3$	B_2	$b_2 \cdot b_3$	B_1
$D_{3h}[C_{3v}, D_3, C_{3h}, C_3]$[19]	$a_1'' \cdot a_2'$	A_2''	$a_1'' \cdot a_2''$	A_2'	$a_2' \cdot a_2''$	A_1''		
	$a_1' \cdot e'$	E'	$a_1' \cdot e''$	E''	$a_1'' \cdot e'$	E''	$a_1'' \cdot e''$	E'
	$a_2' \cdot e'$	E'	$a_2' \cdot e''$	E''	$a_2'' \cdot e'$	E''	$a_2'' \cdot e''$	E'
$C_{4v}, D_4, D_{2d} \equiv V_d[D_{4h}, C_{4h}, C_4, S_4]$[20]	$a_2 \cdot b_1$	B_2	$a_2 \cdot b_2$	B_1	$b_1 \cdot b_2$	A_2		
	$a_1 \cdot e$	E	$a_2 \cdot e$	E	$b_1 \cdot e$	E	$b_2 \cdot e$	E
$D_{5h}[C_{5v}, D_5, C_{5h}, C_5]$[21]	$a_1'' \cdot a_2'$	A_2''	$a_1'' \cdot a_2''$	A_2'	$a_2' \cdot a_2''$	A_1''		
	$a_1' \cdot e_1'$	E_1'	$a_1' \cdot e_1''$	E_1''	$a_1' \cdot e_2'$	E_2'	$a_1' \cdot e_2''$	E_2''
	$a_1'' \cdot e_1'$	E_1''	$a_1'' \cdot e_1''$	E_1'	$a_1'' \cdot e_2'$	E_2''	$a_1'' \cdot e_2''$	E_2'
	$a_2' \cdot e_1'$	E_1'	$a_2' \cdot e_1''$	E_1''	$a_2' \cdot e_2'$	E_2'	$a_2' \cdot e_2''$	E_2''
	$a_2'' \cdot e_1'$	E_1''	$a_2'' \cdot e_1''$	E_1'	$a_2'' \cdot e_2'$	E_2''	$a_2'' \cdot e_2''$	E_2'
$C_{6v}, D_6[D_{6h}, D_{3d}, C_{6h}, C_6, S_6]$[22]	$a_2 \cdot b_1$	B_2	$a_2 \cdot b_2$	B_1	$b_1 \cdot b_2$	A_2		
	$a_1 \cdot e_1$	E_1	$a_1 \cdot e_2$	E_2	$a_2 \cdot e_1$	E_1	$a_2 \cdot e_2$	E_2
	$b_1 \cdot e_1$	E_2	$b_1 \cdot e_2$	E_1	$b_2 \cdot e_1$	E_2	$b_2 \cdot e_2$	E_1
D_{4d}, C_{8v}, D_8	$a_2 \cdot b_1$	B_2	$a_2 \cdot b_2$	B_1	$b_1 \cdot b_2$	A_2		
	$a_1 \cdot e_1$	E_1	$a_2 \cdot e_1$	E_1	$b_1 \cdot e_1$	E_3	$b_2 \cdot e_1$	E_3
	$a_1 \cdot e_2$	E_2	$a_2 \cdot e_2$	E_2	$b_1 \cdot e_2$	E_2	$b_2 \cdot e_2$	E_2
	$a_1 \cdot e_3$	E_3	$a_2 \cdot e_3$	E_3	$b_1 \cdot e_3$	E_1	$b_2 \cdot e_3$	E_1
$C_{\infty v}[D_{\infty h}]$[23]	$\sigma^+ \cdot \pi$	Π	$\sigma^+ \cdot \delta$	Δ	$\sigma^- \cdot \pi$	Π	$\sigma^- \cdot \delta$	Δ
$T_d, O[O_h, T]$[24]	$a_1 \cdot e$	E	$a_1 \cdot f_1$	F_1	$a_1 \cdot f_2$	F_2		
	$a_2 \cdot e$	E	$a_2 \cdot f_1$	F_2	$a_2 \cdot f_2$	F_1		

[18] For D_{2h} the g, u rule has to be taken into account. The subscript i may be 1, 2, or 3.

[19] For C_{3v} and D_3 the $'$ and $''$, for C_{3h} the subscripts 1 and 2, and for C_3 both the $'$ and $''$ and the subscripts 1 and 2 should be omitted.

[20] For D_{4h} and C_{4h} the g, u rule has to be taken into account; for C_{4h}, C_4, and S_4, the subscripts 1 and 2 should be omitted.

[21] For C_{5v} and D_5 the $'$ and $''$, for C_{5h} the subscripts 1 and 2 of A, and for C_5 both should be omitted.

[22] For D_{6h}, D_{3d}, and S_6 the g, u rule must be taken into account; for C_{6h}, C_6, and S_6 the subscripts 1 and 2 of A and B should be omitted; for D_{3d}, B should be put equal to A and the subscripts 1 and 2 of E should be omitted.

[23] For $D_{\infty h}$ the g, u rule must be taken into account.

[24] For O_h the g, u rule must be taken into account; for T the subscripts 1 and 2 should be omitted.

TABLE 32. SYMMETRY TYPES (SPECIES) OF THE HIGHER VIBRATIONAL LEVELS OF DEGENERATE VIBRATIONS.

The numbers in front of some symbols (for example, $2E'$) indicate how many sublevels of that particular species occur if this number is greater than one.

Point group	Vibrational level	Resulting states	Vibrational level	Resulting states
$D_{3h}[C_{3v}, D_3, C_{3h}, C_3]$[19]	$(e')^2$	$A_1'+E'$	$(e'')^2$	$A_1'+E'$
	$(e')^3$	$A_1'+A_2'+E'$	$(e'')^3$	$A_1''+A_2''+E''$
	$(e')^4$	$A_1'+2E'$	$(e'')^4$	$A_1'+2E'$
	$(e')^5$	$A_1'+A_2'+2E'$	$(e'')^5$	$A_1''+A_2''+2E''$
	$(e')^6$	$2A_1'+A_2'+2E'$	$(e'')^6$	$2A_1'+A_2'+2E'$
$D_{4h}[C_{4v}, D_4, D_{2d} \equiv V_d, C_{4h}, C_4, S_4]$[25]	$(e_g)^2$	$A_{1g}+B_{1g}+B_{2g}$	$(e_u)^2$	$A_{1g}+B_{1g}+B_{2g}$
	$(e_g)^3$	$2E_g$	$(e_u)^3$	$2E_u$
	$(e_g)^4$	$2A_{1g}+A_{2g}+B_{1g}+B_{2g}$	$(e_u)^4$	$2A_{1g}+A_{2g}+B_{1g}+B_{2g}$
	$(e_g)^5$	$3E_g$	$(e_u)^5$	$3E_u$
	$(e_g)^6$	$2A_{1g}+A_{2g}+2B_{1g}+2B_{2g}$	$(e_u)^6$	$2A_{1g}+A_{2g}+2B_{1g}+2B_{2g}$
$D_{5h}[C_{5v}, D_5, C_{5h}, C_5]$[21]	$(e_1')^2$	$A_1'+E_2'$	$(e_1'')^2$	$A_1'+E_2'$
	$(e_1')^3$	$E_1'+E_2'$	$(e_1'')^3$	$E_1''+E_2''$
	$(e_1')^4$	$A_1'+E_1'+E_2'$	$(e_1'')^4$	$A_1'+E_1'+E_2'$
	$(e_2')^2$	$A_1'+E_1'$	$(e_2'')^2$	$A_1'+E_1'$
	$(e_2')^3$	$E_1'+E_2'$	$(e_2'')^3$	$E_1''+E_2''$
	$(e_2')^4$	$A_1'+E_1'+E_2'$	$(e_2'')^4$	$A_1'+E_1'+E_2'$
$C_{6v}, D_6[D_{6h}, D_{3d}, C_{6h}, C_6, S_6]$[22]	$(e_1)^2$	A_1+E_2	$(e_2)^2$	A_1+E_2
	$(e_1)^3$	$B_1+B_2+E_1$	$(e_2)^3$	$A_1+A_2+E_2$
	$(e_1)^4$	A_1+2E_2	$(e_2)^4$	A_1+2E_2
	$(e_1)^5$	$B_1+B_2+2E_1$	$(e_2)^5$	$A_1+A_2+2E_2$
	$(e_1)^6$	$2A_1+A_2+2E_2$	$(e_2)^6$	$2A_1+A_2+2E_2$
D_{4d}, C_{8v}, D_8	$(e_1)^2$	A_1+E_2	$(e_2)^2$	$A_1+B_1+B_2$
	$(e_1)^3$	E_1+E_3	$(e_2)^3$	$2E_2$
	$(e_1)^4$	$A_1+B_1+B_2+E_2$	$(e_2)^4$	$2A_1+A_2+B_1+B_2$
	$(e_3)^2$	A_1+E_2	$(e_3)^4$	$A_1+B_1+B_2+E_2$
	$(e_3)^3$	E_1+E_3		
$D_{\infty h}[C_{\infty v}]$[26]	$(\pi_g)^2$	$\Sigma_g^+ + \Delta_g$	$(\pi_u)^2$	$\Sigma_g^+ + \Delta_g$
	$(\pi_g)^3$	$\Pi_g + \Phi_g$	$(\pi_u)^3$	$\Pi_u + \Phi_u$
	$(\pi_g)^4$	$\Sigma_g^+ + \Delta_g + \Gamma_g$	$(\pi_u)^4$	$\Sigma_g^+ + \Delta_g + \Gamma_g$
	$(\pi_g)^5$	$\Pi_g + \Phi_g + \mathrm{H}_g$	$(\pi_u)^5$	$\Pi_u + \Phi_u + \mathrm{H}_u$
	$(\pi_g)^6$	$\Sigma_g^+ + \Delta_g + \Gamma_g + \mathrm{I}_g$	$(\pi_u)^6$	$\Sigma_g^+ + \Delta_g + \Gamma_g + \mathrm{I}_g$
$T_d, O[O_h, T]$[24]	$(e)^2$	A_1+E	$(e)^5$	A_1+A_2+2E
	$(e)^3$	A_1+A_2+E	$(e)^6$	$2A_1+A_2+2E$
	$(e)^4$	A_1+2E	$(e)^7$	A_1+A_2+3E
	$(f_1)^2$	A_1+E+F_2	$(f_2)^2$	A_1+E+F_2
	$(f_1)^3$	$A_2+2F_1+F_2$	$(f_2)^3$	$A_1+F_1+2F_2$
	$(f_1)^4$	$2A_1+2E+F_1+2F_2$	$(f_2)^4$	$2A_1+2E+F_1+2F_2$
	$(f_1)^5$	$A_2+E+4F_1+2F_2$	$(f_2)^5$	$A_1+E+2F_1+4F_2$
	$(f_1)^6$	$3A_1+A_2+3E+2F_1+4F_2$	$(f_2)^6$	$3A_i+A_2+3E+2F_1+4F_2$
	$(f_1)^7$	$2A_2+2E+6F_1+4F_2$	$(f_2)^7$	$2A_1+2E+4F_1+6F_2$

[25] For C_{4v}, D_4, $D_{2d} \equiv V_d$ the subscripts g and u, for C_{4h} the subscripts 1 and 2, and for C_4 and S_4 all subscripts should be omitted.

[26] For $C_{\infty v}$ the subscripts g and u should be dropped. The higher levels of π_g and π_u only are given since π vibrations are the only vibrations that occur (see section 4a).

vectors $l = 1$ in the internuclear axis: $\substack{\leftarrow\\\rightarrow\\\rightarrow}$, $\substack{\rightarrow\\\leftarrow\\\leftarrow}$, and $\substack{\leftarrow\\\leftarrow\\\leftarrow}$, $\substack{\rightarrow\\\rightarrow\\\rightarrow}$, corresponding to the species Π (resultant $l = 1$) and Φ (resultant $l = 3$). Symbolically we write

$$(\pi)^3 = \Pi + \Phi.$$

If the anharmonicity is taken into account these two states have slightly different energies but each one remains doubly degenerate. If the π vibration is excited by two quanta ($v_j = 2$) a triply degenerate state is obtained which splits into one non-degenerate state (Σ^+) and one doubly degenerate state (Δ). They correspond to $\rightleftarrows$ and $\leftleftarrows$, $\rightrightarrows$. Symbolically we write

$$(\pi)^2 = \Sigma^+ + \Delta.$$

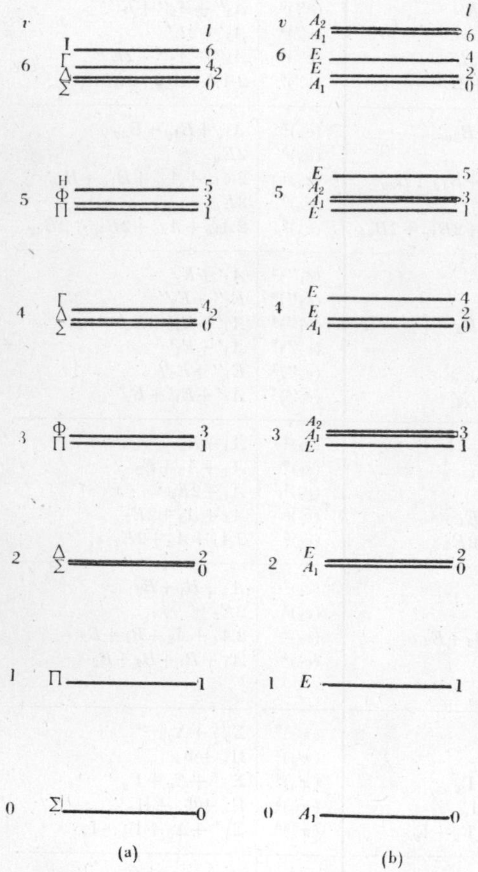

FIG. 52. Splitting and species designation of higher vibrational levels of π and e vibrations of molecules of point groups $C_{\infty v}$ and C_{3v} respectively.

In a similar manner, if in a molecule of point group C_{3v} (for example, NH_3 or CH_3Cl) a degenerate vibration of species E is doubly excited ($v_j = 2$), a triply degenerate state arises which splits into one with $l = 0$ and one with $l = 2$. But here, unlike the linear case, l no longer represents the angular momentum and, as we have seen previously, $l = 2$ is equivalent to $l = 1$; thus we obtain

$$(e)^2 = A_1 + E.$$

If the vibration of species E is triply excited, two doubly degenerate states with $l = 1$ and $l = 3$ arise as in the linear case. But now $l = 3$ is equivalent to $l = 0$ and therefore the double degeneracy is split into two non-degenerate levels with $l = 0$, which group theory shows to be A_1 and A_2. Thus we have

$$(e)^3 = A_1 + A_2 + E.$$

These and similar data for higher vibrational levels are given for all important point groups in Table 32 in a somewhat condensed form. In Fig. 52 the species of the higher vibrational levels of a π vibration of a linear molecule and of an e vibration of a molecule of point group C_{3v} are indicated in an energy-level diagram similar to Fig. 42 for non-degenerate vibrations. The different levels of the same v_j are shown with somewhat different energy corresponding to the splitting that actually occurs in consequence of anharmonicity. The resultant l values are also indicated. They are identical with the l_i values given by equation (II, 59).

Binary combinations of two different degenerate vibrations. Just as in the preceding case, *several degenerate or non-degenerate species result if two degenerate vibrations are singly excited.* For example, if in a linear molecule two different vibrations of species Π are singly excited, the three states Σ^+, Σ^-, and Δ result. They correspond to the four vector diagrams ⇄, ⇆, ⇇, and ⇉.[27] The last two are degenerate with each other forming the Δ state, while one linear combination of the first two gives Σ^+ and the other (orthogonal) one gives Σ^-. It must be noted that here three states arise whereas by double excitation of a single π vibration only the two states Σ^+ and Δ result. This is because in the latter case the diagrams ⇄ and ⇆ are indistinguishable and therefore count as one state only. Table 33 gives the similar results for all binary combinations of the important point groups.

TABLE 33. SYMMETRY TYPES (SPECIES) OF THOSE LEVELS IN WHICH TWO DIFFERENT DEGENERATE VIBRATIONS ARE SINGLY EXCITED.

Point group	Vibrational configuration	Resulting states	Vibrational configuration	Resulting states
$D_{3h}[C_{3v}, D_3, C_{3h}, C_3]$[19]	$e' \cdot e'$	$A_1' + A_2' + E'$	$e'' \cdot e''$	$A_1' + A_2' + E'$
	$e' \cdot e''$	$A_1'' + A_2'' + E''$		
$D_{4h}[C_{4v}, D_4, D_{2d} \equiv V_d, C_{4h}, C_4, S_4]$[25]	$e_g \cdot e_g$	$A_{1g} + A_{2g} + B_{1g} + B_{2g}$	$e_u \cdot e_u$	$A_{1g} + A_{2g} + B_{1g} + B_{2g}$
	$e_g \cdot e_u$	$A_{1u} + A_{2u} + B_{1u} + B_{2u}$		
$D_{5h}[C_{5v}, D_5, C_{5h}, C_5]$[21]	$e_1' \cdot e_1'$	$A_1' + A_2' + E_2'$	$e_1'' \cdot e_2'$	$E_1'' + E_2''$
	$e_1' \cdot e_1''$	$A_1'' + A_2'' + E_2''$	$e_1'' \cdot e_2''$	$E_1' + E_2'$
	$e_1' \cdot e_2'$	$E_1' + E_2'$	$e_2' \cdot e_2'$	$A_1' + A_2' + E_1'$
	$e_1' \cdot e_2''$	$E_1'' + E_2''$	$e_2' \cdot e_2''$	$A_1'' + A_2'' + E_1''$
	$e_1'' \cdot e_1''$	$A_1' + A_2' + E_2'$	$e_2'' \cdot e_2''$	$A_1' + A_2' + E_1'$
$C_{6v}, D_6[D_{6h}, D_{3d}, C_{6h}, C_6, S_6]$[22]	$e_1 \cdot e_1$	$A_1 + A_2 + E_2$	$e_1 \cdot e_2$	$B_1 + B_2 + E_1$
	$e_2 \cdot e_2$	$A_1 + A_2 + E_2$		
D_{4d}, C_{8v}, D_8	$e_1 \cdot e_1$	$A_1 + A_2 + E_2$	$e_2 \cdot e_2$	$A_1 + A_2 + B_1 + B_2$
	$e_1 \cdot e_2$	$E_1 + E_3$	$e_2 \cdot e_3$	$E_1 + E_3$
	$e_1 \cdot e_3$	$B_1 + B_2 + E_2$	$e_3 \cdot e_3$	$A_1 + A_2 + E_2$
$C_{\infty v}[D_{\infty h}]$[23]	$\pi \cdot \pi$	$\Sigma^+ + \Sigma^- + \Delta$	$\delta \cdot \delta$	$\Sigma^+ + \Sigma^- + \Gamma$
	$\pi \cdot \delta$	$\Pi + \Phi$	$\delta \cdot \varphi$	$\Pi + H$
	$\pi \cdot \varphi$	$\Delta + \Gamma$	$\varphi \cdot \varphi$	$\Sigma^+ + \Sigma^- + I$
$T_d, O[O_h, T]$[24]	$e \cdot e$	$A_1 + A_2 + E$	$f_1 \cdot f_1$	$A_1 + E + F_1 + F_2$
	$e \cdot f_1$	$F_1 + F_2$	$f_1 \cdot f_2$	$A_2 + E + F_1 + F_2$
	$e \cdot f_2$	$F_1 + F_2$	$f_2 \cdot f_2$	$A_1 + E + F_1 + F_2$

The way in which the data of Table 33 have been obtained is similar to the way in which Table 31 has been derived: The characters of the resulting state are again obtained by *multiplication of the corresponding characters of the component states.* But now the characters so obtained do not directly occur in the species table of the particular point group. However, they can be reduced to a sum of characters of species that do occur and these are the component states formed; that is, if C and D

[27] The situation is exactly analogous to the determination of the resultant states of an electron configuration $\pi\pi$ of a diatomic molecule (see Molecular Spectra I, p. 359, Fig. 140).

are the species of the two singly excited vibrations and $G, H \cdots$ the resulting species, then we must have for every symmetry operation k (as proved by group theory),

$$\chi_C^{(k)} \cdot \chi_D^{(k)} = \chi_G^{(k)} + \chi_H^{(k)} + \cdots \tag{II, 87}$$

where $\chi_C^{(k)} \cdots$ are the characters for the operation k. From this condition the resulting states are uniquely determined. For example, if in a molecule of point group $\boldsymbol{C}_{3v}$ two different vibrations of species E are singly excited the resulting characters are, according to Table 15, +4, +1, 0. It is easily seen that this can only be obtained as the sum of corresponding characters of A_1, A_2, and E. These latter species are the species of the resulting states. In group-theory language, A_1, A_2, and E are the *irreducible representations* into which the reducible representation with characters +4, +1, 0 can be reduced, or A_1, A_2, and E are the irreducible components of the *direct product* $E \times E$.

More general cases. If more than two normal vibrations are singly excited, one has first to form the resulting states of two of them according to Table 31 or 33, then combine every one of these resulting states with the third vibration again according to Table 31 or 33, and so on. For example, if in XY_4 one quantum each of the four normal vibrations of species A_1, E, F_2, and F_2 are excited (see Fig. 41), we obtain for the symmetry type of the eigenfunctions of the resulting state

$$\begin{aligned}(a_1 \cdot e) \cdot (f_2 \cdot f_2) &= E \cdot (A_1 + E + F_1 + F_2)\\ &= E \cdot A_1 + E \cdot E + E \cdot F_1 + E \cdot F_2\\ &= E + A_1 + A_2 + E + F_1 + F_2 + F_1 + F_2;\end{aligned}$$

that is, the resultant state consists of eight substates, one each of species A_1 and A_2 and two each of species E, F_1, and F_2.

If finally several normal vibrations are multiply excited, one has first to find the resultant species for each multiply excited vibration according to Table 32 (or, for non-degenerate vibrations, according to the rules given on p. 124), and then to combine the results by means of Table 31 or Table 33. As an example, let us consider an excited vibrational state of C_6H_6 (see Fig. 50) in which (assuming point group $\boldsymbol{D}_{6h}$) two quanta of ν_3 (a_{2g}), three quanta of $\nu_8(b_{2g})$, two quanta of $\nu_{13}(e_{1u})$, and three quanta of $\nu_{19}(e_{2u})$ are excited; that is, we consider the configuration

$$(a_{2g})^2(b_{2g})^3(e_{1u})^2(e_{2u})^3.$$

From the previous rule (p. 124), we obtain

$$(a_{2g})^2 = A_{1g}; \qquad (b_{2g})^3 = B_{2g};$$

$$(a_{2g})^2 \cdot (b_{2g})^3 = A_{1g} \cdot B_{2g} = B_{2g};$$

and from Table 32,

$$(e_{1u})^2 = A_{1g} + E_{2g}, \qquad (e_{2u})^3 = A_{1u} + A_{2u} + E_{2u}.$$

Multiplying according to Table 31, we obtain

$$(a_{2g})^2(b_{2g})^3(e_{1u})^2 = B_{2g} + E_{1g};$$

and finally, from Tables 31 and 33,

$$\begin{aligned}(a_{2g})^2(b_{2g})^3(e_{1u})^2(e_{2u})^3 &= (B_{2g} + E_{1g})(A_{1u} + A_{2u} + E_{2u})\\ &= B_{2u} + E_{1u} + B_{1u} + E_{1u} + E_{1u} + B_{1u} + B_{2u} + E_{1u}.\end{aligned}$$

Thus the resulting state has eight substates, two each of symmetry B_{1u} and B_{2u} and four of symmetry E_{1u}. Naturally, vibrational states in which several vibrations are multiply excited lie rather high and are therefore not often observed in ordinary infrared and Raman spectra; but they may be of importance in photographic infrared and electronic band spectra.

4. Determination of Normal Modes of Vibration

As we have seen in section 1, the $3N - 6$ (or $3N - 5$) normal modes of vibration of a polyatomic molecule are unambiguously determined by the secular equation (II, 11) if the force constants are known. But even if that be the case, the actual calculation of the normal modes is a very tedious and slow process since the secular equation is usually of a rather high degree. However, if the molecule has symmetry elements we know from the preceding section what types of normal vibrations there may be. If there is only one vibration of a given species its form is completely determined without any detailed solution of the secular equation, and even if there are two vibrations of a given type it is in general not difficult to obtain a fair idea of what the two vibrations will be like. Therefore, we derive first the number of vibrations of each species in various molecules.[28]

(*a*) *Number of normal vibrations of a given symmetry type (species)*

Sets of equivalent nuclei. The nuclei in a polyatomic molecule may be divided into certain *sets of identical nuclei that can be transformed into one another by the symmetry operations permitted by the molecule* (we may call them equivalent nuclei). For example, the three H atoms of $H_3C—CCl_3$ (whether it belongs to point group $\boldsymbol{C}_3$ or $\boldsymbol{C}_{3v}$) form one such set, the three Cl atoms another, since they can be transformed into each other by three-fold rotations about the symmetry axis. However, the two C atoms do not belong to one set (are not equivalent) but form two sets, since they cannot be transformed into one another by symmetry operations. On the other hand, in C_2H_6 (point group $\boldsymbol{D}_3$, $\boldsymbol{D}_{3d}$, or $\boldsymbol{D}_{3h}$) the six H atoms belong to one set, the two C atoms to another. *The position of all nuclei belonging to one set is fixed if the position of one nucleus of the set is given.* The representative nucleus of the set may have a general position (not on any symmetry element), or it may be on one of the symmetry elements, or it may lie on two or on more symmetry elements. The number of nuclei in a set depends on the position of the representative nucleus. It is largest if the representative nucleus has a general position, since then, starting out from this nucleus, every one of the necessary symmetry operations of the molecule will "produce" another nucleus of this set. On the other hand, the number of nuclei in a set is smallest, namely equal to one, if the representative nucleus lies on all symmetry elements (for example, the C atoms in $H_3C—CCl_3$).

Non-degenerate vibrations. In the case of non-degenerate vibrations, for a given symmetry type the displacements of all nuclei of a set are fixed by the displacements of one of them. Therefore *the nuclei of a set can at most contribute three degrees of freedom to each non-degenerate symmetry type.* If the representative nucleus does not lie on any symmetry element, the set will actually contribute these three degrees of freedom, since then there are no restrictions on the motion of the nucleus. But if the representative nucleus lies on one or more elements of symmetry it may not

[28] A very clear exposition of this derivation, which we partly follow here, has been given by Cabannes (189).

have all three degrees of freedom if the motion is to be in conformity with the symmetry type. Thus if there are m sets of nuclei not on any symmetry elements there will be $3m$ degrees of freedom of each non-degenerate symmetry type contributed by these sets; if there are h sets of nuclei lying on certain symmetry elements there will be $2h$, or h, or 0 degrees of freedom contributed to a given symmetry type depending on the symmetry type and on the symmetry elements of the sets (see Table 35 and the examples below). In this way the total number of degrees of freedom can be determined for each species. If then the number of non-genuine vibrations of this species is subtracted one obtains the *number of genuine vibrations* of this species.

As an illustration let us consider a molecule of point group C_{2v}, having a two-fold axis $C_2(z)$ and two planes of symmetry $\sigma_v(xz)$ and $\sigma_v(yz)$ through it. There are four symmetry types A_1, A_2, B_1, B_2 (see Table 13, p. 106). A nucleus not in one of the planes of symmetry will produce three other nuclei symmetrically placed. According to the above, if there are m such sets of four nuclei they will contribute $3m$ degrees of freedom to each symmetry type. If there is a nucleus on the $\sigma_v(xz)$ plane there must also be another one obtained by reflecting at the $\sigma_v(yz)$ plane. Such a set of two nuclei will not contribute three degrees of freedom but less. If the motion of one such nucleus is to be symmetric with respect to both planes of symmetry (species A_1), it must necessarily take place in the $\sigma_v(xz)$ plane and therefore it contributes only two degrees of freedom to A_1; if the motion is to be antisymmetric with respect to both planes (species A_2) it must necessarily take place in a line perpendicular to the $\sigma_v(xz)$ plane, that is, the set contributes only one degree of freedom to the species A_2. Similarly it contributes only one degree of freedom to B_2 and two degrees of freedom to B_1. If there are m_{xz} such sets there will thus be the contributions to the degrees of freedom of each symmetry type given in the third column of Table 34.

TABLE 34. DETERMINATION OF THE NUMBER OF VIBRATIONS OF EACH SPECIES FOR THE POINT GROUP C_{2v}.

Species	Degrees of freedom contributed by sets of nuclei				Number of normal vibrations		
					Non-genuine		Genuine
	On no symmetry element	On $\sigma_v(xz)$	On $\sigma_v(yz)$	On C_2, $\sigma_v(xz)$, $\sigma_v(yz)$	T	R	
A_1	$3m$	$2m_{xz}$	$2m_{yz}$	$1m_0$	1		$3m+2m_{xz}+2m_{yz}+m_0-1$
A_2	$3m$	$1m_{xz}$	$1m_{yz}$	0		1	$3m+m_{xz}+m_{yz}-1$
B_1	$3m$	$2m_{xz}$	$1m_{yz}$	$1m_0$	1	1	$3m+2m_{xz}+m_{yz}+m_0-2$
B_2	$3m$	$1m_{xz}$	$2m_{yz}$	$1m_0$	1	1	$3m+m_{xz}+2m_{yz}+m_0-2$

Similar considerations apply to sets of nuclei in the $\sigma_v(yz)$ plane. The degrees of freedom contributed by m_{yz} such sets are given in the fourth column of Table 34. Finally, each nucleus on the C_2 axis lies on the two planes of symmetry as well, that is, on all symmetry elements, and therefore forms a set of its own. If its motion is to be symmetrical with respect to both planes it can move only in the axis of symmetry, that is, it contributes one degree of freedom to A_1; it cannot move antisymmetrically with respect to both planes of symmetry and therefore does not contribute to A_2. In order to be antisymmetric with respect to $\sigma_v(yz)$ and symmetric with respect to

$\sigma_v(xz)$ it must move perpendicular to $\sigma_v(yz)$, and therefore contributes only one degree of freedom to B_1; similarly it is found that it contributes one degree of freedom to B_2. Thus if there are m_0 nuclei on the axis they give the contributions indicated in the fifth column of Table 34.

Since there are four nuclei in each set of the first type, two in each set of the second and third types, and one in each set of the fourth type, the total number of nuclei is

$$N = 4m + 2m_{xz} + 2m_{yz} + m_0.$$

It is easily verified from Table 34 that the total number of degrees of freedom is $3N$.

In order to get the number of *genuine* normal vibrations we have now to subtract the non-genuine vibrations. Their species are given in Table 13. There are one each of species A_1 and A_2 and two each of species B_1 and B_2. Subtracting these from the number of degrees of freedom obtained before for each species, we obtain for the four species the number of genuine normal vibrations given in the last column of Table 34.

If we apply the above results to the non-linear molecule XY_2 we find, since $m = 0$, $m_{xz} = 1$, $m_{yz} = 0$, $m_0 = 1$, that there are two vibrations of species A_1, no vibrations of species A_2, one vibration of species B_1, and no vibration of species B_2.[29] The three normal vibrations have been given in the previous Fig. 25a. Since there is only one vibration of species B_1 which is antisymmetric with respect to the plane of symmetry $\sigma_v(yz)$ (which is perpendicular to the plane of the molecule), its form is unambiguously determined without further calculation: nucleus X in this species must necessarily move perpendicular to $\sigma_v(yz)$, say with amplitude s_X. The antisymmetry with respect to $\sigma_v(yz)$ and the symmetry with respect to $\sigma_v(xz)$ further require that the atoms Y move in the $\sigma_v(xz)$ plane in lines at the same angle β to the z axis and in opposite phase. Furthermore, the total linear momentum must be zero, that is, the center of mass must remain at rest, and the total moment of momentum (angular momentum) about any point must remain constant (usually zero) during the vibration. If the displacements were the broken arrows in Fig. 53 which satisfy the symmetry requirements, the total moment of momentum about the point c would obviously not be constant but vary from positive to negative values as the X atom moves back and forth. Only if the point c is made to coincide with X, that is, only if the

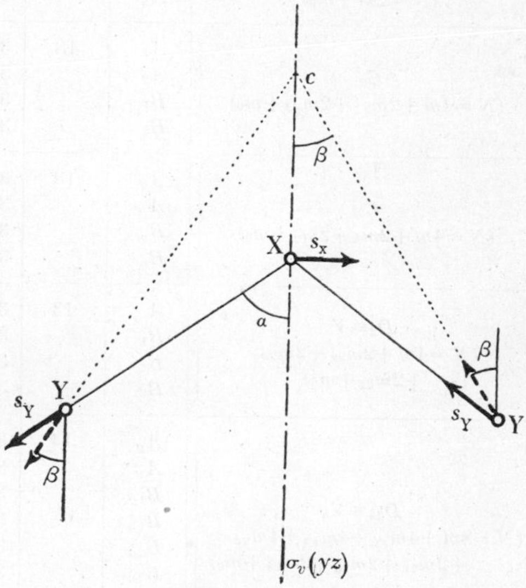

FIG. 53. **Determination of the b_1 vibration of bent XY_2.**

[29] This result could also have been easily obtained directly by applying the same reasoning as above for the general case.

displacements of the Y nuclei are in the direction XY, is the condition of constant (zero) moment of momentum fulfilled. The magnitude s_Y of the displacements of the Y nuclei is obtained from the condition that the component of the total linear momentum perpendicular to the plane $\sigma_v(yz)$ is zero; that is, since the velocities are proportional to the amplitudes of the displacements, $2m_Ys_Y \sin\alpha = m_Xs_X$, where α is half the angle at the top of the triangle formed by the molecule, s_X is the displacement of the X nucleus and m_X and m_Y are the masses of X and Y. Thus the form

TABLE 35. NUMBER OF VIBRATIONS OF EACH SPECIES FOR THE POINT GROUPS HAVING NON-DEGENERATE VIBRATIONS ONLY.

Point group, total number of atoms	Species of vibration	Explained in Table	Number of vibrations[30]
C_2 ($N = 2m + m_0$)	A	12	$3m + m_0 - 2$
	B		$3m + 2m_0 - 4$
$C_s \equiv C_{1h}$ ($N = 2m + m_0$)	A'	12	$3m + 2m_0 - 3$
	A''		$3m + m_0 - 3$
$C_i \equiv S_2$ ($N = 2m + m_0$)	A_g	12	$3m - 3$
	A_u		$3m + 3m_0 - 3$
C_{2v} ($N = 4m + 2m_{xz} + 2m_{yz} + m_0$)	A_1	13	$3m + 2m_{xz} + 2m_{yz} + m_0 - 1$
	A_2		$3m + m_{xz} + m_{yz} - 1$
	B_1		$3m + 2m_{xz} + m_{yz} + m_0 - 2$
	B_2		$3m + m_{xz} + 2m_{yz} + m_0 - 2$
C_{2h} ($N = 4m + 2m_h + 2m_2 + m_0$)	A_g	13	$3m + 2m_h + m_2 - 1$
	A_u		$3m + m_h + m_2 + m_0 - 1$
	B_g		$3m + m_h + 2m_2 - 2$
	B_u		$3m + 2m_h + 2m_2 + 2m_0 - 2$
$D_2 \equiv V$ ($N = 4m + 2m_{2x} + 2m_{2y} + 2m_{2z} + m_0$)	A	13	$3m + m_{2x} + m_{2y} + m_{2z}$
	B_1		$3m + 2m_{2x} + 2m_{2y} + m_{2z} + m_0 - 2$
	B_2		$3m + 2m_{2x} + m_{2y} + 2m_{2z} + m_0 - 2$
	B_3		$3m + m_{2x} + 2m_{2y} + 2m_{2z} + m_0 - 2$
$D_{2h} \equiv V_h$ ($N = 8m + 4m_{xy} + 4m_{xz} + 4m_{yz} + 2m_{2x} + 2m_{2y} + 2m_{2z} + m_0$)	A_g	14	$3m + 2m_{xy} + 2m_{xz} + 2m_{yz} + m_{2x} + m_{2y} + m_{2z}$
	A_u		$3m + m_{xy} + m_{xz} + m_{yz}$
	B_{1g}		$3m + 2m_{xy} + m_{xz} + m_{yz} + m_{2x} + m_{2y} - 1$
	B_{1u}		$3m + m_{xy} + 2m_{xz} + 2m_{yz} + m_{2x} + m_{2y} + m_{2z} + m_0 - 1$
	B_{2g}		$3m + m_{xy} + 2m_{xz} + m_{yz} + m_{2x} + m_{2z} - 1$
	B_{2u}		$3m + 2m_{xy} + m_{xz} + 2m_{yz} + m_{2x} + m_{2y} + m_{2z} + m_0 - 1$
	B_{3g}		$3m + m_{xy} + m_{xz} + 2m_{yz} + m_{2y} + m_{2z} - 1$
	B_{3u}		$3m + 2m_{xy} + 2m_{xz} + m_{yz} + m_{2x} + m_{2y} + m_{2z} + m_0 - 1$

[30] m is always the number of sets of equivalent nuclei not on any element of symmetry; m_0 is the number of nuclei lying on all symmetry elements present; m_{xy}, m_{xz}, m_{yz} are the numbers of sets of nuclei lying on the xy, xz, yz plane respectively but not on any axes going through these planes; m_2 is the number of sets of nuclei on a two-fold axis but not at the point of intersection with another element of symmetry; m_{2x}, m_{2y}, m_{2z} are the numbers of sets of nuclei lying on the x, y, or z axis if they are two-fold axes, but not on all of them; m_h is the number of sets of nuclei on a plane σ_h but not on the axis perpendicular to this plane.

of the normal vibration of species B_1 is completely determined. For the two A_1 vibrations the form is not unambiguously determined from symmetry alone, but one can say immediately that the X nucleus must move in the z axis for both, and the two Y nuclei must move in a symmetrical fashion such that the component of their linear momentum in the z direction is opposite and equal to that of the X nucleus. The direction of motion of the Y nuclei is not fixed by symmetry and the condition of zero linear and angular momentum; that is why there are two vibrations of species A_1.

If, as a second example, we apply the results of Table 34 to the H_2CO molecule, we find, since now $m = 0$, $m_{xz} = 1$, $m_{yz} = 0$, $m_0 = 2$, that there are three vibrations of species A_1, no vibrations of species A_2, two vibrations of species B_1, and one vibration of species B_2. These vibrations are represented in Fig. 24a–f. The one vibration of species B_2 is again unambiguously determined by symmetry.

In a similar way, as indicated above for $\boldsymbol{C_{2v}}$, the number of vibrations of each species can be derived for the other point groups having non-degenerate species only. The results [first derived by Brester (178)] are given in the last column of Table 35. In the first column for every point group the total number N of nuclei is expressed in terms of the number of sets of nuclei; that is, the factors in front of the m values in this first column indicate the number of nuclei in each set. This can serve as a check on the correct selection of the sets.

Degenerate vibrations. For molecules with more than two-fold axes the number of vibrations of the non-degenerate symmetry types can be determined in exactly the same way as described above. However, as to the number of degenerate vibrations some further considerations are necessary. Let us consider as an example the point group $\boldsymbol{C_{3v}}$ to which molecules like NH_3, CH_3Cl, H_3C—CCl_3 belong. *A set of nuclei lying on the three planes* σ_v (but not on the axis) will contribute six degrees of freedom, that is, *three vibrations to the degenerate species E*, since to every one displacement (which may be in any one of the three coordinate directions) of one nucleus of the set correspond two different displacements of each of the other nuclei of the set (compare the previous discussion of the form of degenerate vibrations, p. 89f.). Thus, if there are m_v such sets they will contribute $3m_v$ degenerate vibrations (genuine or non-genuine).

If there is a set of nuclei *in a general position* (none of these occur in the examples mentioned above) they contribute twelve (rather than six) degrees of freedom of type E, that is, six degenerate vibrations. This is because, first, every one displacement of one nucleus of a set has corresponding to it two different displacements of each of those nuclei of the set that are produced from the representative nucleus by rotations about C_3. This accounts for six degrees of freedom. But the three nuclei thus produced comprise only half the set, since for every one of them there is another one placed symmetrically to it with respect to σ_v. Reflection at the plane σ_v will in general change the vibration into something else, but if suitable linear combinations of the degenerate vibrations are chosen one will be symmetric, the other antisymmetric with respect to this plane; that is, the number of degrees of freedom contributed is doubled. Thus, in all, one such set of nuclei in a general position contributes twelve degrees of freedom to the degenerate species or, in other words, *six degenerate vibrations*. If there are m such sets of nuclei they will contribute $6m$ degenerate vibrations. This holds not only for species E of point group $\boldsymbol{C_{3v}}$ but also *for the doubly degenerate*

Table 36. number of vibrations of each species for the point groups with degenerate vibrations.

Point group, total number of atoms	Species of vibration	Explained in Table	Number of vibrations[31]
C_3 $(N=3m+m_0)$	A	25	$3m+m_0-2$
	E		$3m+m_0-2$
C_4 $(N=4m+m_0)$	A		$3m+m_0-2$
	B		$3m$
	E		$3m+m_0-2$
C_6 $(N=6m+m_0)$	A	25	$3m+m_0-2$
	B		$3m$
	E_1		$3m+m_0-2$
	E_2		$3m$
S_4 $(N=4m+2m_2+m_0)$	A		$3m+m_2-1$
	B		$3m+m_2+m_0-1$
	E		$3m+2m_2+m_0-2$
S_6 $(N=6m+2m_3+m_0)$	A_g	25	$3m+m_3-1$
	B_u		$3m+m_3+m_0-1$
	E_{1u}		$3m+m_3+m_0-1$
	E_{2g}		$3m+m_3-1$
D_3 $(N=6m+3m_2+2m_3+m_0)$	A_1	15	$3m+m_2+m_3$
	A_2		$3m+2m_2+m_3+m_0-2$
	E		$6m+3m_2+2m_3+m_0-2$
D_4 $(N=8m+4m_2+4m_2'+2m_4+m_0)$	A_1	18	$3m+m_2+m_2'+m_4$
	A_2		$3m+2m_2+2m_2'+m_4+m_0-2$
	B_1		$3m+m_2+2m_2'$
	B_2		$3m+2m_2+m_2'$
	E		$6m+3m_2+3m_2'+2m_4+m_0-2$
D_6 $(N=12m+6m_2+6m_2'+2m_6+m_0)$	A_1	19	$3m+m_2+m_2'+m_6$
	A_2		$3m+2m_2+2m_2'+m_6+m_0-2$
	B_1		$3m+m_2+2m_2'$
	B_2		$3m+2m_2+m_2'$
	E_1		$6m+3m_2+3m_2'+2m_6+m_0-2$
	E_2		$6m+3m_2+3m_2'$
C_{3v} $(N=6m+3m_v+m_0)$	A_1	15	$3m+2m_v+m_0-1$
	A_2		$3m+m_v-1$
	E		$6m+3m_v+m_0-2$
C_{4v} $(N=8m+4m_v+4m_d+m_0)$	A_1	18	$3m+2m_v+2m_d+m_0-1$
	A_2		$3m+m_v+m_d-1$
	B_1		$3m+2m_v+m_d$
	B_2		$3m+m_v+2m_d$
	E		$6m+3m_v+3m_d+m_0-2$

[31] See footnote, p. 139.

TABLE 36.—*Continued*

Point group, total number of atoms	Species of vibration	Explained in Table	Number of vibrations[31]
C_{5v} ($N = 10m + 5m_v + m_0$)	A_1	16	$3m + 2m_v + m_0 - 1$
	A_2		$3m + m_v - 1$
	E_1		$6m + 3m_v + m_0 - 2$
	E_2		$6m + 3m_v$
C_{6v} ($N = 12m + 6m_v + 6m_d + m_0$)	A_1	19	$3m + 2m_v + 2m_d + m_0 - 1$
	A_2		$3m + m_v + m_d - 1$
	B_1		$3m + 2m_v + m_d$
	B_2		$3m + m_v + 2m_d$
	E_1		$6m + 3m_v + 3m_d + m_0 - 2$
	E_2		$6m + 3m_v + 3m_d$
$C_{\infty v}$ ($N = m_0$)	Σ^+	17	$m_0 - 1$
	Σ^-		0
	Π		$m_0 - 2$
	$\Delta, \Phi \cdots$		0
C_{3h} ($N = 6m + 3m_h + 2m_3 + m_0$)	A'	27	$3m + 2m_h + m_3 - 1$
	A''		$3m + m_h + m_3 + m_0 - 1$
	E'		$3m + 2m_h + m_3 + m_0 - 1$
	E''		$3m + m_h + m_3 - 1$
C_{4h} ($N = 8m + 4m_h + 2m_4 + m_0$)	A_g	27	$3m + 2m_h + m_4 - 1$
	A_u		$3m + m_h + m_4 + m_0 - 1$
	B_g		$3m + 2m_h$
	B_u		$3m + m_h$
	E_g		$3m + m_h + m_4 - 1$
	E_u		$3m + 2m_h + m_4 + m_0 - 1$
C_{6h} ($N = 12m + 6m_h + 2m_6 + m_0$)	A_g	27	$3m + 2m_h + m_6 - 1$
	A_u		$3m + m_h + m_6 + m_0 - 1$
	B_g		$3m + m_h$
	B_u		$3m + 2m_h$
	E_{1g}		$3m + m_h + m_6 - 1$
	E_{1u}		$3m + 2m_h + m_6 + m_0 - 1$
	E_{2g}		$3m + 2m_h$
	E_{2u}		$3m + m_h$
$D_{2d} \equiv V_d (\equiv S_{4v})$ ($N = 8m + 4m_d + 4m_2 + 2m_4 + m_0$)	A_1	18	$3m + 2m_d + m_2 + m_4$
	A_2		$3m + m_d + 2m_2 - 1$
	B_1		$3m + m_d + m_2$
	B_2		$3m + 2m_d + 2m_2 + m_4 + m_0 - 1$
	E		$6m + 3m_d + 3m_2 + 2m_4 + m_0 - 2$
$D_{3d} (\equiv S_{6v})$ ($N = 12m + 6m_d + 6m_2 + 2m_6 + m_0$)	A_{1g}	20	$3m + 2m_d + m_2 + m_6$
	A_{1u}		$3m + m_d + m_2$
	A_{2g}		$3m + m_d + 2m_2 - 1$
	A_{2u}		$3m + 2m_d + 2m_2 + m_6 + m_0 - 1$
	E_g		$6m + 3m_d + 3m_2 + m_6 - 1$
	E_u		$6m + 3m_d + 3m_2 + m_6 + m_0 - 1$

[31] See footnote, p. 139.

TABLE 36.—*Continued*

Point group, total number of atoms	Species of vibration	Explained in Table	Number of vibrations[31]
$D_{4d}(\equiv S_{8v})$ $(N=16m+8m_d+8m_2+2m_8+m_0)$	A_1	21	$3m+2m_d+m_2+m_8$
	A_2		$3m+m_d+2m_2-1$
	B_1		$3m+m_d+m_2$
	B_2		$3m+2m_d+2m_2+m_8+m_0-1$
	E_1		$6m+3m_d+3m_2+m_8+m_0-1$
	E_2		$6m+3m_d+3m_2$
	E_3		$6m+3m_d+3m_2+m_8-1$
D_{3h} $(N=12m+6m_v+6m_h+3m_2+2m_3+m_0)$	A_1'	22	$3m+2m_v+2m_h+m_2+m_3$
	A_1''		$3m+m_v+m_h$
	A_2'		$3m+m_v+2m_h+m_2-1$
	A_2''		$3m+2m_v+m_h+m_2+m_3+m_0-1$
	E'		$6m+3m_v+4m_h+2m_2+m_3+m_0-1$
	E''		$6m+3m_v+2m_h+m_2+m_3-1$
D_{4h} $(N=16m+8m_v+8m_d+8m_h+4m_2+4m_2'+2m_4+m_0)$	A_{1g}	23	$3m+2m_v+2m_d+2m_h+m_2+m_2'+m_4$
	A_{1u}		$3m+m_v+m_d+m_h$
	A_{2g}		$3m+m_v+m_d+2m_h+m_2+m_2'-1$
	A_{2u}		$3m+2m_v+2m_d+m_h+m_2+m_2'+m_4+m_0-1$
	B_{1g}		$3m+2m_v+m_d+2m_h+m_2+m_2'$
	B_{1u}		$3m+m_v+2m_d+m_h+m_2'$
	B_{2g}		$3m+m_v+2m_d+2m_h+m_2+m_2'$
	B_{2u}		$3m+2m_v+m_d+m_h+m_2$
	E_g		$6m+3m_v+3m_d+2m_h+m_2+m_2'+m_4-1$
	E_u		$6m+3m_v+3m_d+4m_h+2m_2+2m_2'+m_4+m_0-1$
D_{5h} $(N=20m+10m_v+10m_h+5m_2+2m_5+m_0)$	A_1'	22	$3m+2m_v+2m_h+m_2+m_5$
	A_1''		$3m+m_v+m_h$
	A_2'		$3m+m_v+2m_h+m_2-1$
	A_2''		$3m+2m_v+m_h+m_2+m_5+m_0-1$
	E_1'		$6m+3m_v+4m_h+2m_2+m_5+m_0-1$
	E_1''		$6m+3m_v+2m_h+m_2+m_5-1$
	E_2'		$6m+3m_v+4m_h+2m_2$
	E_2''		$6m+3m_v+2m_h+m_2$
D_{6h} $(N=24m+12m_v+12m_d+12m_h+6m_2+6m_2'+2m_6+m_0)$	A_{1g}	23	$3m+2m_v+2m_d+2m_h+m_2+m_2'+m_6$
	A_{1u}		$3m+m_v+m_d+m_h$
	A_{2g}		$3m+m_v+m_d+2m_h+m_2+m_2'-1$
	A_{2u}		$3m+2m_v+2m_d+m_h+m_2+m_2'+m_6+m_0-1$
	B_{1g}		$3m+m_v+2m_d+m_h+m_2'$
	B_{1u}		$3m+2m_v+m_d+2m_h+m_2+m_2'$
	B_{2g}		$3m+2m_v+m_d+m_h+m_2$
	B_{2u}		$3m+m_v+2m_d+2m_h+m_2+m_2'$
	E_{1g}		$6m+3m_v+3m_d+2m_h+m_2+m_2'+m_6-1$
	E_{1u}		$6m+3m_v+3m_d+4m_h+2m_2+2m_2'+m_6+m_0-1$
	E_{2g}		$6m+3m_v+3m_d+4m_h+2m_2+2m_2'$
	E_{2u}		$6m+3m_v+3m_d+2m_h+m_2+m_2'$

[31] See footnote, p. 139.

TABLE 36.—*Continued*

Point group, total number of atoms	Species of vibration	Explained in Table	Number of vibrations[31]
$D_{\infty h}$ ($N = 2m_\infty + m_0$)	Σ_g^+	24	m_∞
	Σ_u^+		$m_\infty + m_0 - 1$
	Σ_g^-, Σ_u^-		0
	Π_g		$m_\infty - 1$
	Π_u		$m_\infty + m_0 - 1$
	$\Delta_g, \Delta_u, \Phi_g, \Phi_u, \cdots$		0
T ($N = 12m + 6m_2 + 4m_3 + m_0$)	A	30	$3m + m_2 + m_3$
	E		$3m + m_2 + m_3$
	F		$9m + 5m_2 + 3m_3 + m_0 - 2$
T_d ($N = 24m + 12m_d + 6m_2 + 4m_3 + m_0$)	A_1	28	$3m + 2m_d + m_2 + m_3$
	A_2		$3m + m_d$
	E		$6m + 3m_d + m_2 + m_3$
	F_1		$9m + 4m_d + 2m_2 + m_3 - 1$
	F_2		$9m + 5m_d + 3m_2 + 2m_3 + m_0 - 1$
O_h ($N = 48m + 24m_h + 24m_d + 12m_2 + 8m_3 + 6m_4 + m_0$)	A_{1g}	29	$3m + 2m_h + 2m_d + m_2 + m_3 + m_4$
	A_{1u}		$3m + m_h + m_d$
	A_{2g}		$3m + 2m_h + m_d + m_2$
	A_{2u}		$3m + m_h + 2m_d + m_2 + m_3$
	E_g		$6m + 4m_h + 3m_d + 2m_2 + m_3 + m_4$
	E_u		$6m + 2m_h + 3m_d + m_2 + m_3$
	F_{1g}		$9m + 4m_h + 4m_d + 2m_2 + m_3 + m_4 - 1$
	F_{1u}		$9m + 5m_h + 5m_d + 3m_2 + 2m_3 + 2m_4 + m_0 - 1$
	F_{2g}		$9m + 4m_h + 5m_d + 2m_2 + 2m_3 + m_4$
	F_{2u}		$9m + 5m_h + 4m_d + 2m_2 + m_3 + m_4$

species of all point groups with the exception of C_p, C_{ph}, and T. For the latter point groups, there are no σ_v's through C_p, or C_2's perpendicular to C_p, and therefore the reason for the doubling of the number of degrees of freedom contributed disappears. For them, m sets of nuclei in a general position contribute only $3m$ doubly degenerate vibrations.

Finally, a nucleus *on the axis of symmetry* can contribute to the degenerate degrees of freedom only if it moves perpendicular to the axis. It will then contribute two degrees of freedom, that is, *one degenerate vibration.* Thus the total number of degenerate vibrations (species E) of point group C_{3v} is $6m + 3m_v + m_0$ including the non-genuine vibrations (m_0 = number of nuclei on the axis). There are two non-genuine degenerate vibrations (see Table 15), and therefore the number of genuine degenerate vibrations is $6m + 3m_v + m_0 - 2$. For example, for non-planar XY_3

[31] m is the number of sets of nuclei not on any element of symmetry; m_0 is the number of nuclei on all elements of symmetry; m_2, m_3, m_4, $\cdots$ are the numbers of sets of nuclei on a two-fold, three-fold, four-fold, $\cdots$ axis but not on any other element of symmetry that does not wholly coincide with that axis; m_2' is the number of sets of nuclei on a two-fold axis called C_2' in the previous character tables; m_v, m_d, m_h are the numbers of sets of nuclei on planes σ_v, σ_d, σ_h, respectively but not on any other element of symmetry.

(NH_3 and others), $m = 0$, $m_v = 1$, $m_0 = 1$, and therefore there are two degenerate vibrations (see Fig. 45).

In a similar way the number of degenerate vibrations for the other *axial point groups* (with one more-than-two-fold axis) can be obtained. The results, together with those for the non-degenerate symmetry types of these point groups are given in Table 36. It may be noted that nuclei on an axis of symmetry C_p contribute only to the degenerate vibrations of the first kind E_1 with $l = 1$ but not to E_2 and higher ones if these occur. Therefore since for linear molecules all nuclei are on the axis they have no normal vibrations of species Δ, Φ, $\cdots$. It is also easily seen that they can have no vibrations of species Σ^-.

In the case of the *cubic point groups* the same considerations apply to the non-degenerate and doubly degenerate vibrations as given above. For the triply degenerate vibrations further considerations are necessary. However, we give in Table 36 only the results without further proof [see Jahn and Teller (471)]. As an example consider the tetrahedral XY_4 molecule of point group $\boldsymbol{T_d}$ (CH_4 and others). Here there are no nuclei in a general position ($m = 0$), no nuclei on the planes σ_d but on no other element of symmetry ($m_d = 0$), one set of nuclei on the three-fold axes ($m_3 = 1$), and one set of nuclei on all elements of symmetry ($m_0 = 1$). Therefore, according to Table 36, there is one vibration of species A_1, one of species E, and two of species F_2. As a somewhat more complicated example consider the $C(CH_3)_4$ molecule, assuming point group $\boldsymbol{T_d}$. Here $m = 0$, $m_d = 1$, $m_2 = 0$, $m_3 = 1$, $m_0 = 1$. Therefore we have, according to Table 36, three vibrations of species A_1, one of species A_2, four of species E, four of species F_1 and seven of species F_2.

For the rare point groups $\boldsymbol{T_h}$, $\boldsymbol{O}$, $\boldsymbol{I}$, $\boldsymbol{I_h}$ not given here we refer to the paper by Jahn and Teller (471).

(b) Methods for the general solution of the secular equation

In order to determine the vibrational frequencies from the force constants and at the same time the form of the normal vibrations in cases where it is not determined by the symmetry alone, it is necessary to solve the secular equation (II, 11) or (II, 38). Actually, of course, the force constants are in general not known, but the frequencies of the normal vibrations are observed by means of the spectra. The relations between the force constants and the frequencies obtained from the secular equation may then be used *to obtain the force constants* or, in other words, *the potential function of the molecule in terms of the observed frequencies.* In fact, this determination of the forces holding the nuclei in a molecule to the equilibrium positions is one of the main objects of the study of the vibrational structure of polyatomic molecular spectra.

One may attempt to solve the secular equation (II, 11) directly in terms of Cartesian coordinates or one may try to introduce other coordinates in terms of which the secular equation becomes simpler, or finally one may try to solve the problem by mechanical models.

Solution in Cartesian coordinates. The secular determinant in Cartesian coordinates (II, 11) or (II, 38) has $3N$ rows and $3N$ columns. The resulting equation is therefore of the $3N$th degree in $\lambda(= 4\pi^2\nu^2)$; that is, even in the case of a triatomic molecule of the ninth degree. We know that the secular equation has six (or, for linear molecules, five) *zero roots* corresponding to the six (or five) non-genuine vibrations (the translations and rotations of the whole molecule). Therefore the secular equation must have the factor λ^6 (or λ^5). However, this factor cannot immediately be separated from the secular determinant (II, 38).

We illustrate a possible procedure by the example of a *non-linear* XYZ *molecule*. If we assume the z-axis to be perpendicular to the plane of the molecule and the x and y axes as indicated in Fig. 54, all force constants k_{xz}, k_{yz}, and k_{zz} (with any superscripts) must vanish since no restoring force is produced by (small) displacements in the z direction. Therefore the secular determinant (II, 38) simplifies to

$$-\lambda^3 m_1 m_2 m_3 \begin{vmatrix} k_{xx}^{11}-\lambda m_1 & k_{xy}^{11} & k_{xx}^{12} & k_{xy}^{12} & k_{xx}^{13} & k_{xy}^{13} \\ k_{yx}^{11} & k_{yy}^{11}-\lambda m_1 & k_{yx}^{12} & k_{yy}^{12} & k_{yx}^{13} & k_{yy}^{13} \\ k_{xx}^{21} & k_{xy}^{21} & k_{xx}^{22}-\lambda m_2 & k_{xy}^{22} & k_{xx}^{23} & k_{xy}^{23} \\ k_{yx}^{21} & k_{yy}^{21} & k_{yx}^{22} & k_{yy}^{22}-\lambda m_2 & k_{yx}^{23} & k_{yy}^{23} \\ k_{xx}^{31} & k_{xy}^{31} & k_{xx}^{32} & k_{xy}^{32} & k_{xx}^{33}-\lambda m_3 & k_{xy}^{33} \\ k_{yx}^{31} & k_{yy}^{31} & k_{yx}^{32} & k_{yy}^{32} & k_{yx}^{33} & k_{yy}^{33}-\lambda m_3 \end{vmatrix} = 0, \quad \text{(II, 88)}$$

where the subscripts 1, 2, 3 refer to the X, Y, Z nuclei respectively. Thus in this special case three of the zero roots are immediately eliminated. (They correspond to the translation in the z direction and the rotations about the x and y axes).

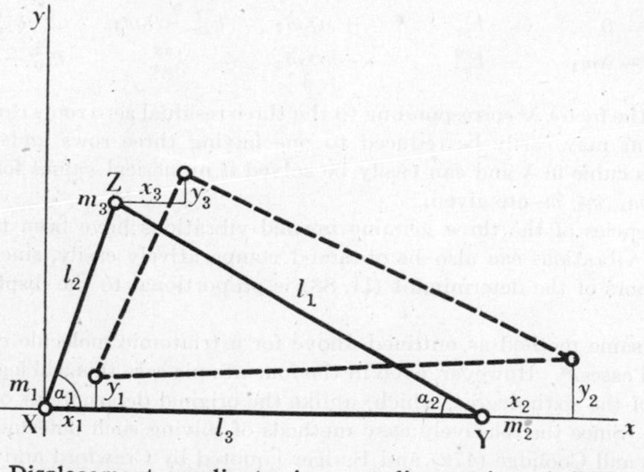

FIG. 54. **Displacement coordinates in a non-linear triatomic XYZ molecule.**

If it is now considered that for a translation of the whole molecule in the x direction ($x_1 = x_2 = x_3$, $y_1 = y_2 = y_3 = 0$, $z_1 = z_2 = z_3 = 0$) the restoring force is zero, it follows from (II, 6) and (II, 7) that

$$\begin{aligned} k_{xx}^{11} + k_{xx}^{12} + k_{xx}^{13} &= 0, \\ k_{yx}^{11} + k_{yx}^{12} + k_{yx}^{13} &= 0, \\ \cdots\cdots & \\ k_{yx}^{31} + k_{yx}^{32} + k_{yx}^{33} &= 0. \end{aligned} \quad \text{(II, 89)}$$

Similarly it follows, by using a displacement of the whole molecule in the y direction, that

$$\begin{aligned} k_{xy}^{11} + k_{xy}^{12} + k_{xy}^{13} &= 0, \\ k_{yy}^{11} + k_{yy}^{12} + k_{yy}^{13} &= 0, \\ \cdots\cdots & \\ k_{yy}^{31} + k_{yy}^{32} + k_{yy}^{33} &= 0. \end{aligned} \quad \text{(II, 90)}$$

Finally, for a rotation of the whole molecule about any point, for example X in Fig. 54, the restoring force must be zero. But for such a rotation the displacement components of X, Y, Z are: 0, 0; 0, c; $-c(l_2/l_3)\sin\alpha_1$, $c(l_2/l_3)\cos\alpha_1$ where c is the arbitrary displacement of Y perpendicular to XY.

If we substitute these displacements into (II, 6) and (II, 7) and consider that for them the restoring force is zero, we obtain

$$\begin{aligned} k_{xy}^{12} - k_{xx}^{13}a + k_{xy}^{13}b &= 0, \\ k_{yy}^{12} - k_{yx}^{13}a + k_{yy}^{13}b &= 0, \\ \cdots\cdots\cdots& \\ k_{yy}^{32} - k_{yx}^{33}a + k_{yy}^{33}b &= 0, \end{aligned} \qquad \text{(II, 91)}$$

where for abbreviation we have put

$$a = \frac{l_2}{l_3}\sin\alpha_1, \qquad b = \frac{l_2}{l_3}\cos\alpha_1. \qquad \text{(II, 92)}$$

If now the third and fifth column of the determinant (II, 88) are added to the first, the fourth and sixth to the second, and if the fifth column multiplied by $-a$, and the sixth multiplied by b, are added to the fourth column we obtain, considering the equations (II, 89), (II, 90), and (II, 91),

$$\begin{vmatrix} -\lambda m_1 & 0 & k_{xx}^{12} & 0 & k_{xx}^{13} & k_{xy}^{13} \\ 0 & -\lambda m_1 & k_{yx}^{12} & 0 & k_{yx}^{13} & k_{yy}^{13} \\ -\lambda m_2 & 0 & k_{xx}^{22} - \lambda m_2 & 0 & k_{xx}^{23} & k_{xy}^{23} \\ 0 & -\lambda m_2 & k_{yx}^{22} & -\lambda m_2 & k_{yx}^{23} & k_{yy}^{23} \\ -\lambda m_3 & 0 & k_{xx}^{32} & +a\lambda m_3 & k_{xx}^{33} - \lambda m_3 & k_{xy}^{33} \\ 0 & -\lambda m_3 & k_{yx}^{32} & -b\lambda m_3 & k_{yx}^{33} & k_{yy}^{33} - \lambda m_3 \end{vmatrix} = 0. \qquad \text{(II, 93)}$$

From this equation the factor λ^3 corresponding to the three residual zero roots can be taken out. The resulting determinant may easily be reduced to one having three rows and columns only. The resulting equation is cubic in λ and can easily be solved if numerical values for the force constants k_{xy}^{ij} and the masses m_1, m_2, m_3 are given.

Once the frequencies of the three genuine normal vibrations have been thus determined, the form of the normal vibrations can also be obtained comparatively easily, since according to p. 70 any row of first minors of the determinant (II, 88) is proportional to the displacement coordinates $x_1, y_1, x_2, y_2, x_3, y_3$.

Essentially the same method as outlined above for a triatomic molecule could also be applied to more complicated cases.[32] However, even in the four-atomic case this will lead to a rather complicated determinant of the sixth degree, which, unlike the original determinant of the twelfth degree, is not symmetrical. Since the relatively easy methods of solving such determinants of high degree suggested by James and Coolidge (472) and Badger [quoted by Crawford and Cross (242)] require symmetrical determinants, it may be advantageous to solve directly the original secular determinant (II, 38) without eliminating the zero roots first.

Solution in "internal" coordinates. The relative position of the nuclei is fixed by $3N - 6$ (or $3N - 5$) coordinates. Instead of following the above procedure, one may express the potential energy and the kinetic energy in terms of these $3N - 6$ "internal" coordinates and thus obtain directly a secular equation of the $(3N - 6)$th degree which does not contain the zero roots. Various choices of the internal coordinates are possible [see Wilson and Crawford (943)]. Perhaps the most obvious choice in the case of an unsymmetrical molecule is to use $3N - 6$ internuclear distances, or rather *deviations Q_i from $3N - 6$ equilibrium internuclear distances.* Such coordinates are also called *central force coordinates* [see for example Shaffer and Newton (778)] since they are particularly adapted to the central force system (see p. 159). Since for small amplitudes these Q_i are linear functions of the rectangular

[32] Bauermeister and Weizel (129) have applied a somewhat similar procedure to linear triatomic and four-atomic molecules.

displacement coordinates, the *potential energy* is a quadratic function of the Q_i (see p. 73) and may be written

$$V = \tfrac{1}{2}\sum_{ij} a_{ij}Q_iQ_j, \tag{II, 94}$$

where $a_{ij} = a_{ji}$. The values of a_{ij} depend on the force constants k_{xy}^{ij} of (II, 25) or k_{ij} of (II, 26). Actually the a_{ij} are more significant than the k_{ij} (see below): whereas there are $\frac{1}{2}(3N)(3N + 1)$ force constants k_{ij} which are not independent of one another [compare equations (II, 89–91) for the triatomic case], there are only $\frac{1}{2}(3N - 6)(3N - 5)$ potential constants a_{ij} which are mutually independent and all different from zero as long as the molecule has no symmetry.[33] Because of the linear relationship between the Q_i and the x_i, y_i, z_i also the *kinetic energy* remains a quadratic function of the $\dot{Q}_i$:

$$T = \tfrac{1}{2}\sum_{ij} b_{ij}\dot{Q}_i\dot{Q}_j, \tag{II, 95}$$

where $b_{ij} = b_{ji}$. However now, unlike previously, the b_{ij} for $i \neq j$ are in general not zero, that is, the kinetic energy as well as the potential energy contains cross terms.[33a]

The *secular determinant* which determines the normal vibrations is now of the form [see equation (II, 34)].

$$\begin{vmatrix} b_{11}\lambda - a_{11} & b_{12}\lambda - a_{12} & b_{13}\lambda - a_{13} & \cdots \\ b_{21}\lambda - a_{21} & b_{22}\lambda - a_{22} & b_{23}\lambda - a_{23} & \cdots \\ b_{31}\lambda - a_{31} & b_{32}\lambda - a_{32} & b_{33}\lambda - a_{33} & \cdots \\ \cdot & \cdot & \cdot & \cdot \end{vmatrix} = 0, \tag{II, 96}$$

which is of degree $3N - 6$ only. It can immediately be set up if V and T have been expressed in terms of Q_i and $\dot{Q}_i$, respectively, according to (II, 94) and (II, 95). However, it is usually not easy to express the kinetic energy in terms of these new coordinates except when the molecule has symmetry and in that case other methods are even more convenient.

As an example, let us consider again the unsymmetrical non-linear triatomic XYZ molecule. For it,

$$V = \tfrac{1}{2}a_{11}Q_1^2 + \tfrac{1}{2}a_{22}Q_2^2 + \tfrac{1}{2}a_{33}Q_3^2 + a_{12}Q_1Q_2 + a_{23}Q_2Q_3 + a_{13}Q_1Q_3, \tag{II, 97}$$

where Q_1, Q_2, and Q_3 are the changes of the lengths of the three sides of the triangle from the equilibrium values, l_1, l_2, l_3 (Fig. 54). There are in the general case six potential constants.

In order to write the kinetic energy in terms of the $\dot{Q}_i$, we have first to find the transformation by which the rectangular coordinates go over into the Q_i. From Fig. 54 it is easily seen that if x_1, y_1, x_2, y_2, x_3, y_3 are, as before, the Cartesian displacement coordinates of the three atoms, then for small amplitudes, we have

$$\begin{aligned} Q_1 &= (x_2 - x_3)\cos\alpha_2 + (y_3 - y_2)\sin\alpha_2, \\ Q_2 &= (x_3 - x_1)\cos\alpha_1 + (y_3 - y_1)\sin\alpha_1, \\ Q_3 &= (x_2 - x_1). \end{aligned} \tag{II, 98}$$

[33] If i and j can assume n different values there are n combinations of the form ii and $\frac{1}{2}n(n - 1)$ of the form ij with $i \neq j$.

[33a] The b_{ij} used here are not the same as those used in equations (II, 29) and (II, 30).

While these three equations are sufficient to determine the Q_i from the x_i, y_i the latter are not unambiguously determined by the former. For that purpose we have to introduce the condition that the center of mass remains at rest, that is,

$$\begin{aligned} m_1x_1 + m_2x_2 + m_3x_3 &= 0, \\ m_1y_1 + m_2y_2 + m_3y_3 &= 0, \end{aligned} \tag{II, 99}$$

and the condition that the angular momentum of the molecule vanishes (see p. 133); that is (since the velocity components are in the ratio of the displacement components),

$$m_3x_3l_2 \sin \alpha_1 - m_3y_3l_2 \cos \alpha_1 - m_2y_2l_3 = 0. \tag{II, 100}$$

From the six equations (II, 98–100) the x_i, y_i can now be obtained in terms of the Q_i. The result is

$$\begin{aligned} x_1 &= \frac{A_{11}}{D}Q_1 + \frac{A_{21}}{D}Q_2 + \frac{A_{31}}{D}Q_3, \\ x_2 &= \frac{A_{12}}{D}Q_1 + \frac{A_{22}}{D}Q_2 + \frac{A_{32}}{D}Q_3, \\ x_3 &= \frac{A_{13}}{D}Q_1 + \frac{A_{23}}{D}Q_2 + \frac{A_{33}}{D}Q_3, \\ y_1 &= \frac{A_{14}}{D}Q_1 + \frac{A_{24}}{D}Q_2 + \frac{A_{34}}{D}Q_3, \\ y_2 &= \frac{A_{15}}{D}Q_1 + \frac{A_{25}}{D}Q_2 + \frac{A_{35}}{D}Q_3, \\ y_3 &= \frac{A_{16}}{D}Q_1 + \frac{A_{26}}{D}Q_2 + \frac{A_{36}}{D}Q_3, \end{aligned} \tag{II, 101}$$

where the A_{ik} are the first minors of the determinant D formed from the coefficients of x_1, x_2, x_3, y_1, y_2, y_3 of the equations (II, 98–100). It is obvious that the actual evaluation of the A_{ik} and of D, although quite elementary, will be rather tedious. If it is accomplished the kinetic energy can be expressed in terms of the $\dot{Q}_i$ since (II, 101) holds also for the time derivatives. We obtain

$$\begin{aligned} 2T &= m_1\dot{x}_1^2 + m_2\dot{x}_2^2 + m_3\dot{x}_3^2 + m_1\dot{y}_1^2 + m_2\dot{y}_2^2 + m_3\dot{y}_3^2 \\ &= \frac{1}{D^2}[\dot{Q}_1^2(m_1A_{11}^2 + m_2A_{12}^2 + m_3A_{13}^2 + m_1A_{14}^2 + m_2A_{15}^2 + m_3A_{16}^2) \\ &\quad + \dot{Q}_2^2(m_1A_{21}^2 + \cdots + m_3A_{26}^2) + \dot{Q}_3^2(m_1A_{31}^2 + \cdots + m_3A_{36}^2) \\ &\quad + 2\dot{Q}_1\dot{Q}_2(m_1A_{11}A_{21} + \cdots + m_3A_{16}A_{26}) + 2\dot{Q}_1\dot{Q}_3(m_1A_{11}A_{31} + \cdots \\ &\quad + m_3A_{16}A_{36}) + 2\dot{Q}_2\dot{Q}_3(m_1A_{21}A_{31} + \cdots + m_3A_{26}A_{36})]. \end{aligned} \tag{II, 102}$$

From this the values for the coefficients b_{ij} in (II, 95) are immediately given and the secular determinant can be set up. It has now only three rows and three columns [just the three rows and columns written out in full in (II, 96)]. This secular equation has no zero roots. Its three roots λ_1, λ_2, λ_3 give the *frequencies of the three genuine normal vibrations.* They depend through the b_{ij} on the masses of the nuclei and the dimensions of the triangle formed by them, and of course on the potential constants a_{ij}.

The *form of the normal vibrations* is again determined by the ratio of the first minors, C_{ik}, in any one row of the determinant (II, 96) in which for λ the specific value for the particular normal vibration has been substituted. For example, in the present case, for the normal vibration λ_1,

$$Q_1^{(1)} : Q_2^{(1)} : Q_3^{(1)} = C_{1k}^{(1)} : C_{2k}^{(1)} : C_{3k}^{(1)}; \qquad \text{(II, 103)}$$

that is, the changes of the lengths of the three sides of the triangle for this normal vibration are in the (constant) ratio $C_{1k}^{(1)} : C_{2k}^{(1)} : C_{3k}^{(1)}$. If one wants to have the form of the normal vibrations in terms of Cartesian coordinates, one has to substitute (II, 103) or similar equations for λ_2 and λ_3 into (II, 101).

It is apparent that the calculations in this method, at least for the example considered, are by no means less cumbersome than in the preceding method (using Cartesian coordinates throughout). However, the advantage is that the *force constants*, a_{ij}, used in the second method are more easily visualized. For example, a_{11}, a_{22}, and a_{33} are the force constants corresponding to a change of the bond lengths Y—Z only, X—Z only, and X—Y only, respectively. The force constants k_{xy}^{ij} in (II, 88) can be expressed in terms of the a_{ij} by substituting (II, 98) into (II, 97) and comparing the coefficients of x_1^2, x_1x_2, $\cdots$ with those in (II, 25).

Considerable simplification in most of the above formulae arises when the nuclei X and Y are identical ($m_1 = m_2$) and the sides l_1 and l_2 of the triangle are equal, that is, when we have a symmetrical bent molecule. In this case the potential energy must remain unchanged for a reflection at the plane of symmetry through Z (which exchanges X and Y). Therefore we have, in equation (II, 97),

$$a_{11} = a_{22}, \qquad a_{13} = a_{23}, \qquad \text{(II, 104)}$$

and in (II, 88) and (II, 93),

$$\begin{aligned} &k_{xx}^{11} = k_{xx}^{22}, \quad &&k_{yy}^{11} = k_{yy}^{22}, \\ &k_{xx}^{13} = k_{xx}^{23}, \quad &&k_{yy}^{13} = k_{yy}^{23}, \\ &k_{xy}^{11} = -k_{xy}^{22}, \quad &&k_{xy}^{13} = -k_{xy}^{23}, \quad k_{yx}^{13} = -k_{yx}^{23}. \end{aligned} \qquad \text{(II, 105)}$$

While the substitution of these relations into (II, 93) and (II, 96–102) simplifies them greatly, the resulting equation for the frequencies is still cubic in λ, which is rather awkward if one wants to have an analytical expression for the frequencies in terms of the force constants. A further reduction of the degree of the secular equation can be obtained in this case if still other coordinates are introduced (see below).

Solution by the use of symmetry coordinates. The best method of determining the normal vibrations in the case of symmetrical molecules has proved to be the method of "*symmetry coordinates*" first introduced by Howard and Wilson (462). While somewhat different symmetry coordinates have been introduced by Rosenthal and Murphy (750) and by Redlich and Tompa (733), we shall discuss here only the symmetry coordinates introduced by Howard and Wilson [see also Wilson and Crawford (943)], since they seem to be the most useful for actual calculations.

From our previous considerations we know how many genuine vibrations of each symmetry type there are for a given molecule. Suppose there are f_1, f_2, $\cdots$ of the first, second, $\cdots$ species. Then let us introduce, instead of the $3N$ Cartesian displacement coordinates, $3N - 6$ (or $3N - 5$) new internal coordinates (symmetry coordinates), f_1 of which have the first symmetry type, f_2 the second, and so on. That is, *to a given value of one of these coordinates correspond displacements of the nuclei in agreement with one of the symmetry types.* At the same time we choose them so that *no motion of the center of mass* and *no rotation of the molecule as a whole* takes place (zero angular momentum). These symmetry coordinates are therefore conceivable normal coordinates which would be *the* normal coordinates for certain very special and usually simple values of the force constants.[34]

For example, in the case of the symmetrical non-linear XY_2 molecule, Fig. 55a shows such symmetry coordinates of species A_1 and B_1 (see Table 13), the only species for which genuine normal vibrations occur (see p. 133). While there is only one normal vibration of species B_1 and therefore only one symmetry coordinate of

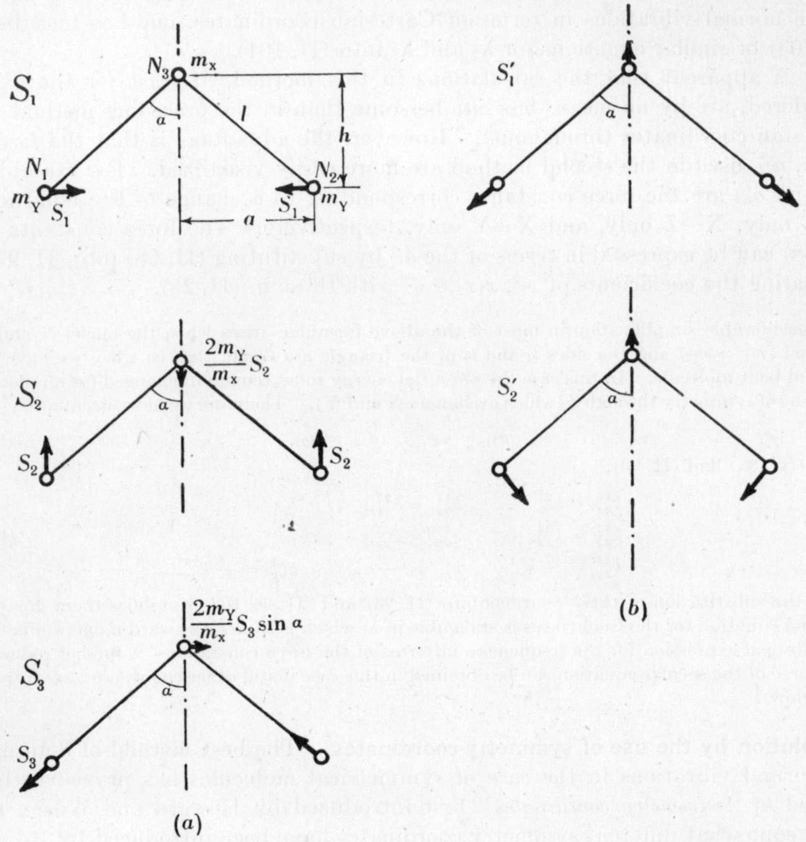

FIG. 55. **Symmetry coordinates of non-linear symmetric XY_2.**

this species, which is identical with the normal coordinate (see Fig. 53), there is an infinite number of possible symmetry coordinates of species A_1 of which two mutually orthogonal ones were selected in Fig. 55a. The actual normal coordinates are linear combinations of these two symmetry coordinates. Of course one might just as well have selected another pair of symmetry coordinates of this species such as those given in Fig. 55b. The latter are valence type symmetry coordinates [see

[34] Conversely, Dennison (279), before the development of this theory, used the choice of special "limiting" force fields in order to determine the number and type of the normal vibrations of simple molecules.

Wilson (942) and p. 168f.] since the nuclei move as much as possible along the chemical bonds or perpendicular to them.

Since the symmetry coordinates are linearly related to the rectangular coordinates, as long as infinitesimal displacements are considered, the potential energy is a quadratic function of the symmetry coordinates as well as of the Cartesian coordinates. That is, if $S_1, S_2 \cdots$ are the symmetry coordinates we obtain, from (II, 25),

$$2V = \sum_{ik} c_{ik} S_i S_k, \tag{II, 106}$$

and similarly for the kinetic energy, from (II, 28),

$$2T = \sum_{ik} d_{ik} \dot{S}_i \dot{S}_k. \tag{II, 107}$$

However, if S_i and S_k belong to different symmetry types (species) there will be at least one symmetry operation with respect to which S_i and S_k behave differently. For example, for non-degenerate symmetry types there will be one operation for which say $S_i \rightarrow S_i$ and $S_k \rightarrow -S_k$ (or $S_i \rightarrow -S_i$ and $S_k \rightarrow +S_k$); that is, there will be at least one symmetry operation for which $S_i S_k$ (and similarly $\dot{S}_i \dot{S}_k$) changes sign. Since, however, the potential energy (as well as the kinetic energy) must be invariant with respect to all symmetry operations, it follows that *the c_{ik} and d_{ik} are zero whenever S_i and S_k belong to different species.* Therefore the secular determinant assumes the form indicated in Fig. 56, where all elements outside the shaded area are zero since they correspond to S_i and S_k of different species. Each shaded square corresponds to one species. The determinant can now be written as a *product of factors corresponding to each species.* Each of these factors put equal to zero gives the normal frequencies of this species. The degree of the resulting algebraic equation is equal to the number of genuine vibrations of the particular species and is thus in general (if the molecule is symmetrical) considerably reduced as compared to the degree of the secular equation obtained according to the two previous methods.

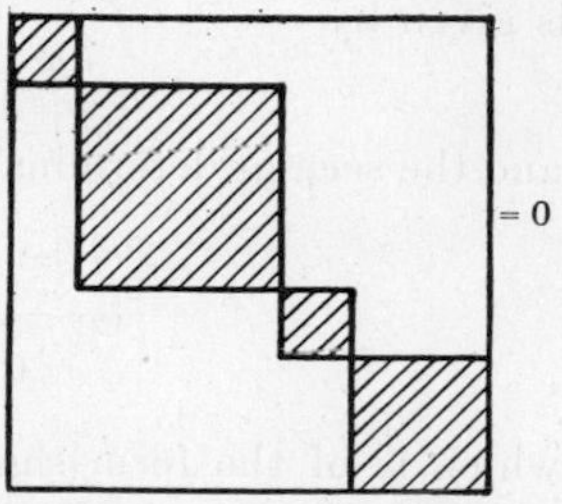

FIG. 56. Form of the secular determinant set up in terms of symmetry coordinates.

While the factoring of the secular equation has here only been proved for non-degenerate symmetry types, it holds also for the *degenerate symmetry types* [see, for example, Rosenthal and Murphy (750)]. Furthermore it is found that if S_{ia} and S_{ib} are two mutually orthogonal degenerate symmetry coordinates of a certain species the potential energy depends in exactly the same way on S_{ia} as it does on S_{ib}, and the product $S_{ia}S_{ib}$ does not occur. A corresponding result applies to the kinetic energy. Therefore there are two identical factors in the factored secular determinant (two identical shaded squares in Fig. 56) for each doubly degenerate species.[34a] In the case of triply degenerate vibrations there would be three such identical factors, the solution of one of which would give the frequencies of the degenerate vibrations of that particular species.

[34a] Here it is assumed that if several vibrations of the degenerate species exist the symmetry coordinates are all chosen in such a way that α in the transformation (II, 75) has not only the same magnitude but also the same sign (see p. 89).

We can now also give the *number of independent potential constants for a molecule.* If there are f_j vibrations (that is, symmetry coordinates) of a given species j, the number of coefficients in (II, 106) contributed by this species is $\frac{1}{2}f_j(f_j + 1)$, and since there are no cross terms between coordinates of different species the total number of independent potential constants is simply

$$\tfrac{1}{2}\sum f_j(f_j + 1), \tag{II, 108}$$

where the f_j are given by the Tables 35 and 36.

The fact that the secular determinant factors is independent of whether or not the different symmetry coordinates of a given species have been chosen orthogonal to one another (see p. 72). But if they are chosen orthogonal to one another it can be shown that the kinetic energy contains only diagonal terms $d_{ii}\dot{S}_i^2$ (see the examples below).

Application to non-linear XY_2. For a non-linear symmetrical XY_2 molecule which has two normal vibrations of species A_1 and one of species B_1, the *potential energy* expressed in terms of any symmetry coordinates (for example those of Fig. 55a) is, according to the above,

$$2V = c_{11}S_1^2 + 2c_{12}S_1S_2 + c_{22}S_2^2 + c_{33}S_3^2; \tag{II, 109}$$

that is, there are four independent potential constants. Similarly, the *kinetic energy* is given by

$$2T = d_{11}\dot{S}_1^2 + 2d_{12}\dot{S}_1\dot{S}_2 + d_{22}\dot{S}_2^2 + d_{33}\dot{S}_3^2, \tag{II, 110}$$

and the secular determinant is, with $c_{12} = c_{21}$, $d_{12} = d_{21}$ [compare equation (II, 34)],

$$\begin{vmatrix} c_{11} - \lambda d_{11} & c_{12} - \lambda d_{12} & 0 \\ c_{21} - \lambda d_{21} & c_{22} - \lambda d_{22} & 0 \\ 0 & 0 & c_{33} - \lambda d_{33} \end{vmatrix} = 0, \tag{II, 111}$$

which is of the form shown in Fig. 56 and can immediately be factored into the two equations

$$c_{33} - \lambda d_{33} = 0 \tag{II, 112}$$

and

$$(c_{11} - \lambda d_{11})(c_{22} - \lambda d_{22}) - (c_{12} - \lambda d_{12})^2 = 0, \tag{II, 113}$$

or

$$\lambda^2 - \frac{c_{22}d_{11} + c_{11}d_{22} - 2c_{12}d_{12}}{d_{11}d_{22} - d_{12}^2}\lambda + \frac{c_{11}c_{22} - c_{12}^2}{d_{11}d_{22} - d_{12}^2} = 0. \tag{II, 114}$$

Equation (II, 112) yields directly the frequency of the one antisymmetric normal vibration (species B_1):

$$\lambda_3 = 4\pi^2\nu_3^2 = \frac{c_{33}}{d_{33}}. \tag{II, 115}$$

Equation (II, 114) gives the frequencies of the two symmetric normal vibrations:

$$\left.\begin{matrix}\lambda_1\\ \lambda_2\end{matrix}\right\} = \frac{c_{22}d_{11} + c_{11}d_{22} - 2c_{12}d_{12}}{2(d_{11}d_{22} - d_{12}^2)}$$
$$\pm \frac{\sqrt{(c_{22}d_{11} + c_{11}d_{22} - 2c_{12}d_{12})^2 - 4(c_{11}c_{22} - c_{12}^2)(d_{11}d_{22} - d_{12}^2)}}{2(d_{11}d_{22} - d_{12}^2)}. \tag{II, 116}$$

We may also write, according to a well-known theorem concerning algebraic equations,

$$\lambda_1 + \lambda_2 = 4\pi^2(\nu_1^2 + \nu_2^2) = \frac{c_{22}d_{11} + c_{11}d_{22} - 2c_{12}d_{12}}{d_{11}d_{22} - d_{12}^2}, \qquad \text{(II, 117)}$$

$$\lambda_1\lambda_2 = 16\pi^4\nu_1^2\nu_2^2 = \frac{c_{11}c_{22} - c_{12}^2}{d_{11}d_{22} - d_{12}^2}. \qquad \text{(II, 118)}$$

We must now express the c_{ik} and d_{ik} in terms of more familiar quantities. In order to obtain the d_{ik} we have to express the Cartesian displacement coordinates, x_i, y_i in terms of the symmetry coordinates S_1, S_2, S_3, and then substitute in the familiar expression (II, 28) for the kinetic energy. If we consider in Fig. 55a S_1, S_2, and S_3 more specifically as the displacements of nucleus N_1 in the first, second, and third symmetry coordinates, respectively, the displacement coordinates of the other nuclei for each symmetry coordinate are easily expressed in terms of S_1, S_2, S_3, on the basis of Fig. 55a and using the law of conservation of linear momentum (see also p. 134). Considering further that the most general displacement is simply a superposition of the three symmetry coordinates, we obtain

$$\begin{aligned} x_1 &= S_1 - S_3 \sin\alpha, & x_2 &= -S_1 - S_3 \sin\alpha, & x_3 &= \frac{2m_Y}{m_X} S_3 \sin\alpha, \\ y_1 &= S_2 - S_3 \cos\alpha, & y_2 &= S_2 + S_3 \cos\alpha, & y_3 &= -\frac{2m_Y}{m_X} S_2. \end{aligned} \qquad \text{(II, 119)}$$

Substituting into (II, 28) and comparing with (II, 110), we obtain

$$d_{11} = 2m_Y, \quad d_{22} = 2m_Y p, \quad d_{33} = 2m_Y r, \quad d_{12} = d_{13} = d_{23} = 0, \qquad \text{(II, 120)}$$

where for abbreviation

$$p = 1 + \frac{2m_Y}{m_X}, \quad r = 1 + \frac{2m_Y}{m_X}\sin^2\alpha. \qquad \text{(II, 121)}$$

The coefficients d_{ik} with $i \neq k$ vanish since we have chosen S_1 and S_2 orthogonal to each other (see p. 148).

The potential energy is usually expressed in terms of the changes of $3N - 6$ internuclear distances or of fewer internuclear distances and a number of angles (*central force and valence force coordinates;* see below). In the case of a triatomic molecule we may use the changes Q_i of the three internuclear distances and the corresponding force constants a_{ik} of equation (II, 97). In order to find the relation of the symmetry coordinates and the Q_i we substitute (II, 119) into (II, 98) and obtain, considering that here $\alpha_1 = \alpha_2 = (\pi/2) - \alpha$,

$$\begin{aligned} Q_1 &= -\sin\alpha\, S_1 - p\cos\alpha\, S_2 - rS_3, \\ Q_2 &= -\sin\alpha\, S_1 - p\cos\alpha\, S_2 + rS_3, \\ Q_3 &= -2S_1. \end{aligned} \qquad \text{(II, 122)}$$

Substituting this into (II, 97) we obtain, since here $a_{13} = a_{23}$ and $a_{11} = a_{22}$,

$$\begin{aligned} c_{11} &= 2(a_{11} + a_{12})\sin^2\alpha + 4a_{33} + 8a_{13}\sin\alpha, \\ c_{12} &= 2p(a_{11} + a_{12})\sin\alpha\cos\alpha + 4a_{13}p\cos\alpha, \\ c_{22} &= 2p^2(a_{11} + a_{12})\cos^2\alpha, \\ c_{33} &= 2r^2(a_{11} - a_{12}), \end{aligned} \qquad \text{(II, 123)}$$

and of course, as was to be expected, $c_{13} = c_{23} = 0$. Finally, substituting (II, 120) and (II, 123) into (II, 115), (II, 117), and (II, 118), we obtain *the frequencies in terms of the potential constants a_{ik}, the masses, and the angle α:*

$$\lambda_3 = 4\pi^2\nu_3^2 = \left(\frac{1}{m_Y} + \frac{2}{m_X}\sin^2\alpha\right)(a_{11} - a_{12}), \qquad \text{(II, 124)}$$

$$\begin{aligned}\lambda_1 + \lambda_2 &= 4\pi^2(\nu_1^2 + \nu_2^2)\\ &= \left(\frac{1}{m_Y} + \frac{2}{m_X}\cos^2\alpha\right)(a_{11} + a_{12}) + \frac{2a_{33}}{m_Y} + \frac{4a_{13}}{m_Y}\sin\alpha,\end{aligned} \qquad \text{(II, 125)}$$

$$\lambda_1 \cdot \lambda_2 = 16\pi^4\nu_1^2\nu_2^2 = \frac{2(2m_Y + m_X)}{m_Y^2 m_X}\cos^2\alpha[(a_{11} + a_{12})a_{33} - 2a_{13}^2]. \qquad \text{(II, 126)}$$

In order to obtain λ_1 and λ_2 separately one would have to solve the quadratic equation [corresponding to (II, 114)]

$$\lambda^2 - (\lambda_1 + \lambda_2)\lambda + \lambda_1 \cdot \lambda_2 = 0. \qquad \text{(II, 127)}$$

As before (see p. 145), the *normal coordinate* belonging to a certain normal frequency is given by any row of first minors of the secular determinant after substitution of the particular value of λ. In the present case, substituting λ_3 into (II, 111) and taking the last row of minors, we obtain for the normal coordinate ξ_3

$$S_1 : S_2 : S_3 = 0 : 0 : \begin{vmatrix} c_{11} - \lambda_3 d_{11} & c_{12} - \lambda_3 d_{12} \\ c_{21} - \lambda_3 d_{21} & c_{22} - \lambda_3 d_{22} \end{vmatrix}. \qquad \text{(II, 128)}$$

In other words, the normal coordinate ξ_3, apart from a constant factor, is identical with the symmetry coordinate S_3 in agreement with our previous discussion (p. 133). Substituting λ_1 into (II, 111) and taking the first row of minors, we obtain for ξ_1,

$$\begin{aligned} S_1 : S_2 : S_3 &= (c_{22} - \lambda_1 d_{22}) : -(c_{21} - \lambda_1 d_{21}) : 0 \\ &= [2p(a_{11} + a_{12})\cos^2\alpha - 2\lambda_1 m_Y] : \\ &\quad -[2(a_{11} + a_{12})\sin\alpha\cos\alpha + 4a_{13}\cos\alpha] : 0, \end{aligned} \qquad \text{(II, 129)}$$

and the same expression with λ_2 for ξ_2. Thus ξ_1 as well as ξ_2 is a superposition of S_1 and S_2 of Fig. 55a in the ratio given by (II, 129). It is seen at the same time that one needs only to take the ratio of the minors of that factor of the secular determinant that corresponds to the species considered.

Application to symmetrical planar X_2Y_4. In the ethylene-like molecule X_2Y_4 of point group V_h (see Table 14), if we choose as previously the z axis perpendicular to the plane of the molecule and the x axis in the line X—X, we have for the number of the various sets of identical nuclei (see Table 35): $m = 0$, $m_{xy} = 1$, $m_{xz} = 0$, $m_{yz} = 0$, $m_{2x} = 1$, $m_{2y} = 0$, $m_{2z} = 0$, $m_i = 0$. Therefore there are three vibrations of species A_g, one of species A_u, two of species B_{1g}, one of species B_{1u}, one of species B_{2g}, two of species B_{2u}, none of species B_{3g}, and two of species B_{3u}; or, symbolically, the resolution into the various species is given by

$$3A_g + A_u + 2B_{1g} + B_{1u} + B_{2g} + 2B_{2u} + 2B_{3u}. \qquad \text{(II, 130)}$$

A possible set of *symmetry coordinates* $S_1, \cdots, S_{12}$ is given in Fig. 57. In view of a later application (p. 189f.), they have been so chosen that wherever possible they are symmetric and antisymmetric combinations of the symmetry coördinates of XY_2 (Fig. 55a). Otherwise all symmetry coordinates could have been so chosen that the nuclei move only in the x, y, and z directions. Of course an infinite number of other possibilities exist for those species for which more than one genuine vibration occurs.

The *potential and kinetic energies* in terms of the symmetry coordinates of Fig. 57 are

$$2V = \sum_{i,k=1,2,3} c_{ik}S_iS_k + c_{44}S_4^2 + \sum_{i,k=5,6} c_{ik}S_iS_k + c_{77}S_7^2 + c_{88}S_8^2 + \sum_{i,k=9,10} c_{ik}S_iS_k + \sum_{i,k=11,12} c_{ik}S_iS_k, \qquad \text{(II, 131)}$$

$$2T = \sum_{i,k=1,2,3} d_{ik}\dot{S}_i\dot{S}_k + d_{44}\dot{S}_4^2 + \sum_{i,k=5,6} d_{ik}\dot{S}_i\dot{S}_k + d_{77}\dot{S}_7^2 + d_{88}\dot{S}_8^2 + \sum_{i,k=9,10} d_{ik}\dot{S}_i\dot{S}_k + \sum_{i,k=11,12} d_{ik}\dot{S}_i\dot{S}_k, \qquad \text{(II, 132)}$$

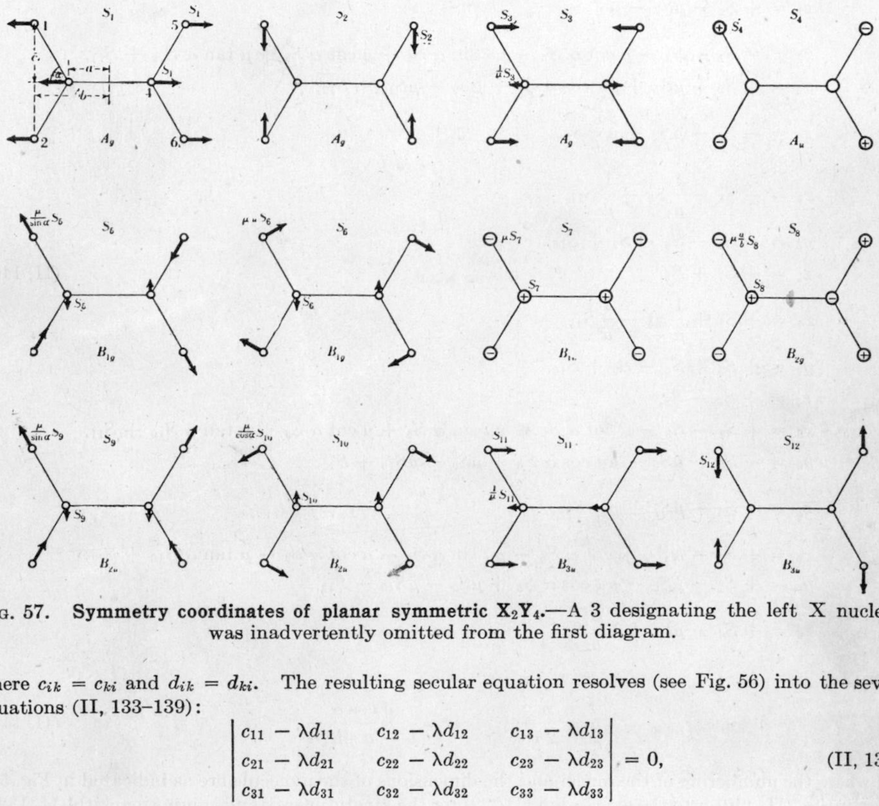

FIG. 57. **Symmetry coordinates of planar symmetric X_2Y_4.**—A 3 designating the left X nucleus was inadvertently omitted from the first diagram.

where $c_{ik} = c_{ki}$ and $d_{ik} = d_{ki}$. The resulting secular equation resolves (see Fig. 56) into the seven equations (II, 133–139):

$$\begin{vmatrix} c_{11} - \lambda d_{11} & c_{12} - \lambda d_{12} & c_{13} - \lambda d_{13} \\ c_{21} - \lambda d_{21} & c_{22} - \lambda d_{22} & c_{23} - \lambda d_{23} \\ c_{31} - \lambda d_{31} & c_{32} - \lambda d_{32} & c_{33} - \lambda d_{33} \end{vmatrix} = 0, \qquad \text{(II, 133)}$$

$$c_{44} - \lambda d_{44} = 0, \qquad \text{(II, 134)}$$

$$\begin{vmatrix} c_{55} - \lambda d_{55} & c_{56} - \lambda d_{56} \\ c_{65} - \lambda d_{65} & c_{66} - \lambda d_{66} \end{vmatrix} = 0, \qquad \text{(II, 135)}$$

$$c_{77} - \lambda d_{77} = 0, \qquad \text{(II, 136)}$$

$$c_{88} - \lambda d_{88} = 0, \qquad \text{(II, 137)}$$

$$\begin{vmatrix} c_{99} - \lambda d_{99} & c_{9\,10} - \lambda d_{9\,10} \\ c_{10\,9} - \lambda d_{10\,9} & c_{10\,10} - \lambda d_{10\,10} \end{vmatrix} = 0, \qquad \text{(II, 138)}$$

$$\begin{vmatrix} c_{11\,11} - \lambda d_{11\,11} & c_{11\,12} - \lambda d_{11\,12} \\ c_{12\,11} - \lambda d_{12\,11} & c_{12\,12} - \lambda d_{12\,12} \end{vmatrix} = 0. \qquad \text{(II, 139)}$$

None of these equations is of a degree higher than three, whereas without using symmetry coordinates we should have had to solve a secular determinant of 12 rows and 12 columns, even after the zero roots had been eliminated.

From equations (II, 133–139) the *frequencies of the normal vibrations* can immediately be obtained in terms of the c_{ik} and d_{ik}. In order to obtain the d_{ik} in terms of the masses of the nuclei and the dimensions of the molecule we have to express the Cartesian displacement coordinates in terms of the symmetry coordinates. We see readily from Fig. 57, that

$$\begin{aligned}
x_1 &= -S_1 + S_3 - \mu \cot\alpha\, S_5 + \mu u \sin\alpha\, S_6 - \mu\cot\alpha\, S_9 - \mu\tan\alpha\, S_{10} + S_{11},\\
y_1 &= -S_2 + \mu\, S_5 + \mu u \cos\alpha\, S_6 + \mu S_9 - \mu S_{10} - S_{12},\\
z_1 &= +S_4 - \mu S_7 - \mu\frac{a}{b} S_8,\\
x_2 &= -S_1 + S_3 + \mu\cot\alpha\, S_5 - \mu u\sin\alpha\, S_6 + \mu\cot\alpha\, S_9 + \mu\tan\alpha\, S_{10} + S_{11},\\
y_2 &= +S_2 + \mu S_5 + \mu u\cos\alpha\, S_6 + \mu S_9 - \mu S_{10} + S_{12},\\
z_2 &= -S_4 - \mu S_7 - \mu\frac{a}{b} S_8,\\
x_3 &= -S_1 - \frac{1}{\mu} S_3 - \frac{1}{\mu} S_{11},\\
y_3 &= -S_5 - S_6 - S_9 + S_{10},\\
z_3 &= +S_7 + S_8,\\
x_4 &= +S_1 + \frac{1}{\mu} S_3 - \frac{1}{\mu} S_{11},\\
y_4 &= +S_5 + S_6 - S_9 + S_{10},\\
z_4 &= +S_7 - S_8,\\
x_5 &= +S_1 - S_3 - \mu\cot\alpha\, S_5 + \mu u\sin\alpha\, S_6 + \mu\cot\alpha\, S_9 + \mu\tan\alpha\, S_{10} + S_{11},\\
y_5 &= -S_2 - \mu S_5 - \mu u\cos\alpha\, S_6 + \mu S_9 - \mu S_{10} + S_{12},\\
z_5 &= -S_4 - \mu S_7 + \mu\frac{a}{b} S_8,\\
x_6 &= +S_1 - S_3 + \mu\cot\alpha\, S_5 - \mu u\sin\alpha\, S_6 - \mu\cot\alpha\, S_9 - \mu\tan\alpha\, S_{10} + S_{11},\\
y_6 &= +S_2 - \mu S_5 - \mu u\cos\alpha\, S_6 + \mu S_9 - \mu S_{10} - S_{12},\\
z_6 &= +S_4 - \mu S_7 + \mu\frac{a}{b} S_8,
\end{aligned} \qquad \text{(II, 140)}$$

where

$$\mu = \frac{m_X}{2m_Y}, \qquad u = \frac{a\cos\alpha}{b - a\sin^2\alpha}, \qquad \text{(II, 141)}$$

and where the numbering of the nuclei and the dimensions of the molecule are as indicated in Fig. 57. Substituting (II, 140) into the expression (II, 28) for the kinetic energy and comparing with (II, 132), we obtain, with $M = 2m_X + 4m_Y$,

$$\begin{aligned}
&d_{11} = M, \quad d_{22} = 4m_Y, \quad d_{33} = \frac{M}{\mu}, \quad d_{12} = d_{13} = d_{23} = 0, \quad d_{44} = 4m_Y,\\
&d_{55} = \mu(M + 2m_X\cot^2\alpha), \quad d_{66} = 4m_Y\mu^2u^2 + 2m_X, \quad d_{56} = 2m_X,\\
&d_{77} = \frac{m_X^2}{m_Y} + 2m_X, \quad d_{88} = \frac{m_X^2}{m_Y}\frac{a^2}{b^2} + 2m_X, \quad d_{99} = \mu(M + 2m_X\cot^2\alpha),\\
&d_{1010} = \mu(M + 2m_X\tan^2\alpha), \quad d_{910} = -2m_X, \quad d_{1111} = \frac{M}{\mu}, \quad d_{1212} = 4m_Y, \quad d_{1112} = 0.
\end{aligned} \qquad \text{(II, 142)}$$

It may be noted that for the two species A_g and B_{3u} for which the symmetry coordinates have been chosen orthogonal to one another the d_{ik} with $i \neq k$ vanish.

It would be fairly simple though somewhat tedious to express the *potential constants* c_{ik} in terms of a_{ik} formed in a way similar to that described above for the molecule XY_2.[35] Instead of doing this at the present stage we shall later (p. 183) express the c_{ik} in terms of the potential constants of a somewhat more specialized force field.

While for the vibrations ν_4, ν_7, and ν_8 the normal coordinates are, apart from a factor, identical with the symmetry coordinates (Fig. 57), for the three totally symmetric vibrations ν_1, ν_2, ν_3 the actual form of the vibrations is obtained by superimposing S_1, S_2, and S_3 in the ratio of one row of minors of (II, 133) with $\lambda = \lambda_1$, λ_2, or λ_3, respectively, that is, in the ratio of

$$S_1 : S_2 : S_3 = \begin{vmatrix} c_{22} - \lambda_i d_{22} & c_{23} - \lambda_i d_{23} \\ c_{32} - \lambda_i d_{32} & c_{33} - \lambda_i d_{33} \end{vmatrix} : \begin{vmatrix} c_{23} - \lambda_i d_{23} & c_{21} - \lambda_i d_{21} \\ c_{33} - \lambda_i d_{33} & c_{31} - \lambda_i d_{31} \end{vmatrix} : \begin{vmatrix} c_{21} - \lambda_i d_{21} & c_{22} - \lambda_i d_{22} \\ c_{31} - \lambda_i d_{31} & c_{32} - \lambda_i d_{32} \end{vmatrix}.$$

Similar relations hold for the vibrations ν_5, ν_6, ν_9, ν_{10}, ν_{11}, and ν_{12}.

Application to linear symmetrical XY_2. In a linear symmetric XY_2 molecule of point group $D_{\infty h}$ (for example CO_2) there is only one vibration of each of the species Σ_g^+, Σ_u^+, Π_u (see Table 36). The symmetry coordinates are identical with the normal coordinates given in Fig. 25b. According to the above, the potential and kinetic energies expressed in these coordinates are

$$2V = c_{11}S_1^2 + c_{22}(S_{2a}^2 + S_{2b}^2) + c_{33}S_3^2, \qquad \text{(II, 143)}$$

$$2T = d_{11}\dot{S}_1^2 + d_{22}(\dot{S}_{2a}^2 + \dot{S}_{2b}^2) + d_{33}\dot{S}_3^2, \qquad \text{(II, 144)}$$

where S_{2a} and S_{2b} are the mutually degenerate and orthogonal symmetry coordinates of species Π_u. The secular equation resolves into three linear equations:

$$c_{11} - \lambda_1 d_{11} = 0, \qquad c_{22} - \lambda_2 d_{22} = 0, \qquad c_{33} - \lambda_3 d_{33} = 0. \qquad \text{(II, 145)}$$

From Fig. 25b, the Cartesian displacement coordinates may easily be expressed in terms of the symmetry coordinates. Numbering the nuclei 1, 2, 3, from left to right and taking the x axis in the direction of the internuclear axis, we obtain

$$\begin{aligned} x_1 &= -S_1 - \frac{m_X}{2m_Y}S_3, & y_1 &= -\frac{m_X}{2m_Y}S_{2a}, & z_1 &= -\frac{m_X}{2m_Y}S_{2b}, \\ x_2 &= +S_3, & y_2 &= +S_{2a}, & z_2 &= +S_{2b} \\ x_3 &= +S_1 - \frac{m_X}{2m_Y}S_3, & y_3 &= -\frac{m_X}{2m_Y}S_{2a}, & z_3 &= -\frac{m_X}{2m_Y}S_{2b}, \end{aligned} \qquad \text{(II, 146)}$$

where S_1 is the displacement of the right nucleus Y in ν_1 while S_{2a}, S_{2b}, and S_3 are the displacements of the X nucleus in ν_{2a}, ν_{2b}, and ν_3, respectively. Substituting (II, 146) into the expression (II, 28) for the *kinetic energy* and comparing with (II, 144), we obtain

$$d_{11} = 2m_Y, \qquad d_{22} = m_X + \frac{m_X^2}{2m_Y}, \qquad d_{33} = m_X + \frac{m_X^2}{2m_Y}. \qquad \text{(II, 147)}$$

In the present case it is customary to express the potential energy in terms of the changes Q_1 and Q_2 of the two internuclear distances and the angle of deviation from a straight line either in the xy or in the xz plane, which we designate by φ_a

[35] This calculation is somewhat simplified if it is noted that the relation between the c_{ik} of a given symmetry type and the a_{ik} can be obtained by putting in (II, 140) those S_i equal to zero that do not belong to that symmetry type.

and φ_b, respectively. These are valence force coordinates (see below). It is immediately seen that

$$Q_1 = x_2 - x_1, \qquad Q_2 = x_3 - x_2,$$
$$\varphi_a = \frac{1}{l}(y_1 + y_3 - 2y_2), \qquad \varphi_b = \frac{1}{l}(z_1 + z_3 - 2z_2), \tag{II, 148}$$

where l is the equilibrium internuclear distance between X and Y. Substituting (II, 146) into (II, 148), we have

$$Q_1 = S_1 + (1 + \mu)S_3, \qquad Q_2 = S_1 - (1 + \mu)S_3,$$
$$\varphi_a = \frac{2}{l}(1 + \mu)S_{2a}, \qquad \varphi_b = \frac{2}{l}(1 + \mu)S_{2b}, \tag{II, 149}$$

where $\mu = m_X/2m_Y$. In terms of Q_1, Q_2, φ_a, φ_b the *potential energy* is

$$2V = a_{11}Q_1^2 + 2a_{12}Q_1Q_2 + a_{22}Q_2^2 + a_{33}\varphi_a^2 + a_{44}\varphi_b^2, \tag{II, 150}$$

where no cross terms $Q_1\varphi_a$, $Q_1\varphi_b$, $Q_2\varphi_a$, $Q_2\varphi_b$, $\varphi_a\varphi_b$ occur because φ_a and φ_b are, apart from a constant factor, the symmetry coordinates S_{2a} and S_{2b}. Since the potential energy must remain unchanged for a reflection at the center of symmetry and for a rotation by 90° about the axis, we must have

$$a_{11} = a_{22} \quad \text{and} \quad a_{33} = a_{44}.$$

Substituting (II, 149) into (II, 150) and comparing with (II, 143), we obtain

$$c_{11} = 2(a_{11} + a_{12}), \qquad c_{22} = a_{33}\frac{4(1 + \mu)^2}{l^2}, \qquad c_{33} = 2(1 + \mu)^2(a_{11} - a_{12}). \tag{II, 151}$$

From this and from equations (II, 145) and (II, 147), the *frequencies of the normal vibrations* follow:

$$4\pi^2\nu_1^2 = \lambda_1 = \frac{a_{11} + a_{12}}{m_Y},$$
$$4\pi^2\nu_2^2 = \lambda_2 = \frac{2(m_X + 2m_Y)}{m_X m_Y}\frac{a_{33}}{l^2}, \tag{II, 152}$$
$$4\pi^2\nu_3^2 = \lambda_3 = \frac{m_X + 2m_Y}{m_X m_Y}(a_{11} - a_{12}).$$

Here a_{11} is the force constant of the X—Y bond, a_{12} is the force constant that gives the interaction of the two bonds and a_{33} is the force constant for the bending of the molecule. It is seen that the frequency of the degenerate vibration ν_2 of the molecule depends only on a_{33} as was to be expected, while the frequencies of the non-degenerate vibrations ν_1 and ν_3 depend only on a_{11} and a_{12}.

Application to pyramidal XY_3. As a somewhat more complicated example of a molecule with degenerate vibrations we consider the pyramidal XY_3 molecule of point group $\boldsymbol{C_{3v}}$. According to Table 36 there are two vibrations each of the species A_1 and E. *Symmetry coordinates* S_1, S_2, S_{3a}, S_{3b}, S_{4a}, S_{4b}, as suggested by Howard and Wilson (462), are given in Fig. 58.[36] In terms of these symmetry coordinates the *potential and kinetic energies* are [see (II, 106) and (II, 107)]

$$2V = c_{11}S_1^2 + 2c_{12}S_1S_2 + c_{22}S_2^2 + c_{33}(S_{3a}^2 + S_{3b}^2) + 2c_{34}(S_{3a}S_{4a} + S_{3b}S_{4b}) + c_{44}(S_{4a}^2 + S_{4b}^2), \quad \text{(II, 153)}$$

$$2T = d_{11}\dot{S}_1^2 + 2d_{12}\dot{S}_1\dot{S}_2 + d_{22}\dot{S}_2^2 + d_{33}(\dot{S}_{3a}^2 + \dot{S}_{3b}^2) + 2d_{34}(\dot{S}_{3a}\dot{S}_{4a} + \dot{S}_{3b}\dot{S}_{4b}) + d_{44}(\dot{S}_{4a}^2 + \dot{S}_{4b}^2). \quad \text{(II, 154)}$$

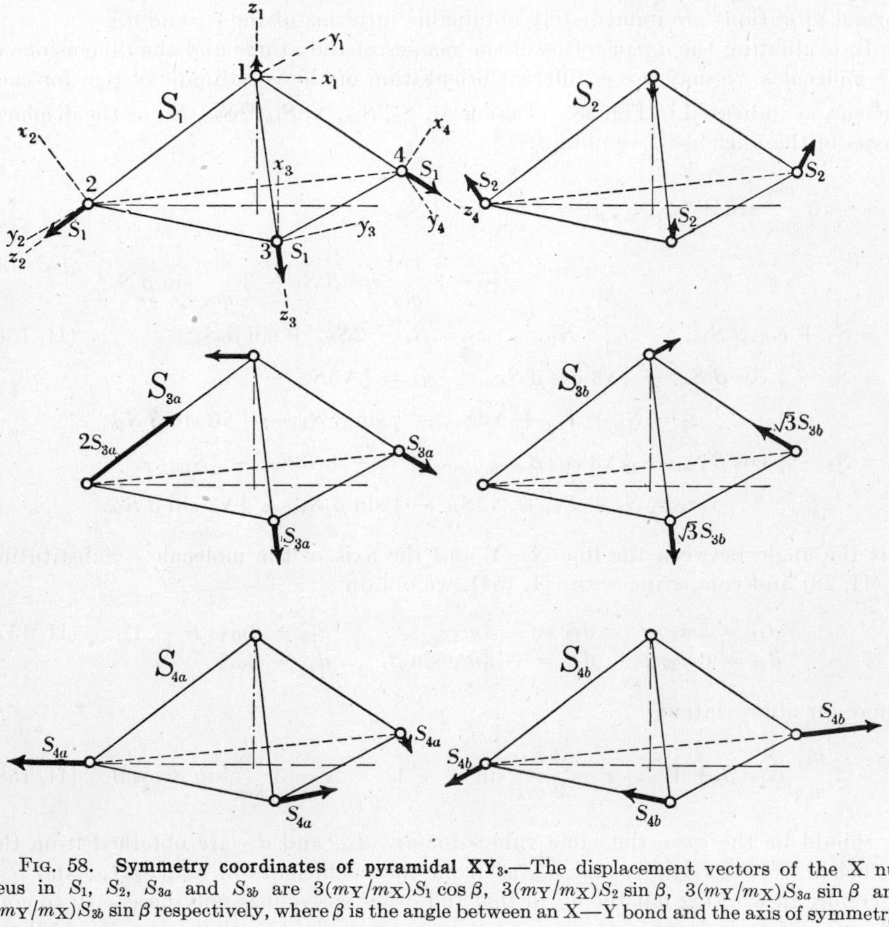

FIG. 58. **Symmetry coordinates of pyramidal XY_3.**—The displacement vectors of the X nucleus in S_1, S_2, S_{3a} and S_{3b} are $3(m_Y/m_X)S_1 \cos\beta$, $3(m_Y/m_X)S_2 \sin\beta$, $3(m_Y/m_X)S_{3a} \sin\beta$ and $3(m_Y/m_X)S_{3b} \sin\beta$ respectively, where β is the angle between an X—Y bond and the axis of symmetry

Here account has been taken of the requirement (see p 147) that the dependence on each pair of mutually degenerate coordinates must be identical and that cross terms between coordinates of different species do not occur. There are six independent potential constants c_{ik}. The *secular determinant* breaks up into three

[36] Naturally we could also have chosen symmetry coordinates S_1, S_2, and S_3 in which the Y atoms move parallel or perpendicular to the axis of symmetry similar to those of XY_2 (Fig. 55a). This would simplify the calculation of the d_{ik} but complicate the calculation of the c_{ik}.

factors,

$$\begin{vmatrix} c_{11} - \lambda d_{11} & c_{12} - \lambda d_{12} \\ c_{21} - \lambda d_{21} & c_{22} - \lambda d_{22} \end{vmatrix} = 0, \quad \begin{vmatrix} c_{33} - \lambda d_{33} & c_{34} - \lambda d_{34} \\ c_{43} - \lambda d_{43} & c_{44} - \lambda d_{44} \end{vmatrix} = 0,$$
$$\begin{vmatrix} c_{33} - \lambda d_{33} & c_{34} - \lambda d_{34} \\ c_{43} - \lambda d_{43} & c_{44} - \lambda d_{44} \end{vmatrix} = 0, \qquad \text{(II, 155)}$$

of which the last two, corresponding to the degenerate vibrations, are identical and where, as always, $c_{ik} = c_{ki}$ and $d_{ik} = d_{ki}$. From (II, 155) the frequencies of the normal vibrations are immediately obtainable in terms of the c_{ik} and d_{ik}.

In evaluating the d_{ik} in terms of the masses of the atoms and the dimensions of the molecules we may use a different orientation of the coordinate system for each nucleus, as indicated in Fig. 58. Taking S_1, S_2, S_{3a}, $\sqrt{3}S_{3b}$,[37] S_{4a}, S_{4b} as the displacements of the nucleus 3, we obtain:

$$x_1 = -3\frac{m_Y}{m_X}\sin\beta\, S_{3a}, \qquad y_1 = 3\frac{m_Y}{m_X}\sin\beta\, S_{3b},$$
$$z_1 = 3\frac{m_Y}{m_X}\cos\beta\, S_1 - 3\frac{m_Y}{m_X}\sin\beta\, S_2;$$
$$x_2 = S_2 + \cos\beta\, S_{4a}, \qquad y_2 = S_{4b}, \qquad z_2 = S_1 - 2S_{3a} + \sin\beta\, S_{4a}; \qquad \text{(II, 156)}$$
$$x_3 = S_2 - \tfrac{1}{2}\cos\beta\, S_{4a} - \tfrac{1}{2}\sqrt{3}\cos\beta\, S_{4b}, \qquad y_3 = \tfrac{1}{2}\sqrt{3}S_{4a} - \tfrac{1}{2}S_{4b},$$
$$z_3 = S_1 + S_{3a} + \sqrt{3}S_{3b} - \tfrac{1}{2}\sin\beta\, S_{4a} - \tfrac{1}{2}\sqrt{3}\sin\beta\, S_{4b};$$
$$x_4 = S_2 - \tfrac{1}{2}\cos\beta\, S_{4a} + \tfrac{1}{2}\sqrt{3}\cos\beta\, S_{4b}, \qquad y_4 = -\tfrac{1}{2}\sqrt{3}S_{4a} - \tfrac{1}{2}S_{4b},$$
$$z_4 = S_1 + S_{3a} - \sqrt{3}S_{3b} - \tfrac{1}{2}\sin\beta\, S_{4a} + \tfrac{1}{2}\sqrt{3}\sin\beta\, S_{4b}.$$

β is the angle between the line X—Y and the axis of the molecule. Substituting in (II, 28) and comparing with (II, 154), we obtain

$$\begin{aligned} d_{11} &= 3m_Y p, & d_{12} &= -3m_Y s, & d_{22} &= 3m_Y(2r - 1), \\ d_{33} &= 6m_Y r, & d_{34} &= -3m_Y\sin\beta, & d_{44} &= 3m_Y, \end{aligned} \qquad \text{(II, 157)}$$

where for abbreviation

$$p = 3\frac{m_Y}{m_X}\cos^2\beta + 1, \qquad r = \frac{3}{2}\frac{m_Y}{m_X}\sin^2\beta + 1, \qquad s = 3\frac{m_Y}{m_X}\sin\beta\cos\beta. \qquad \text{(II, 158)}$$

As should be the case, the same values for d_{33}, d_{34}, and d_{44} are obtained from the coefficients of $\dot{S}_{3a}^2$, $\dot{S}_{3a}\dot{S}_{4a}$, and $\dot{S}_{4a}^2$ as from the coefficients of $\dot{S}_{3b}^2$, $\dot{S}_{3b}\dot{S}_{4b}$, and $\dot{S}_{4b}^2$. Therefore in working out the d_{ik} in this and other cases, it is only necessary to consider one of each degenerate pair of symmetry coordinates; that is, in (II, 156) S_{3b} and S_{4b} could have been put equal to zero. It may be noted that here d_{12} and d_{34} are not equal to zero since S_1 has not been chosen orthogonal to S_2, and S_3 not orthogonal to S_4.

At this stage we shall not attempt to express the potential constants c_{ik} in terms of more conventional quantities. This will be done later (p. 162 and p. 175) in terms of the constants of more specialized force fields.

[37] The factor $\sqrt{3}$ is introduced in order to have the same normalization for S_{3b} as for S_{3a}.

In a manner similar to the previous cases (XY_2, X_2Y_4), the actual form of the normal vibrations is again obtained by superimposing the symmetry coordinates in the ratio of the minors of the secular determinant, as below.

For ν_1:

$$S_1 : S_2 = (c_{22} - \lambda_1 d_{22}) : -(c_{21} - \lambda_1 d_{21}),$$

For ν_2:

$$S_1 : S_2 = (c_{22} - \lambda_2 d_{22}) : -(c_{21} - \lambda_2 d_{21}),$$

For ν_3: (II, 159)

$$S_3 : S_4 = (c_{44} - \lambda_3 d_{44}) : -(c_{43} - \lambda_3 d_{43}),$$

For ν_4:

$$S_3 : S_4 = (c_{44} - \lambda_4 d_{44}) : -(c_{43} - \lambda_4 d_{43}).$$

All the above examples were molecules with comparatively few atoms and of rather high symmetry. For molecules with more atoms or less symmetry, the secular equations frequently cannot be resolved into factors of an easily manageable degree even by using symmetry coordinates. More elaborate methods of dealing with such cases have been worked out, particularly by Wilson (940) (942), who has also developed a method to obtain directly the expanded secular equation (in algebraic rather than determinantal form).

Solution by the use of mechanical models. In view of the complexity of the mathematical calculations in many cases, Kettering, Shutts, and Andrews (501) first introduced the *experimental study of the vibrations of molecular models.* Steel balls representing the nuclei are connected by springs representing the forces between them; these models, suspended by rubber bands, are set into vibration by means of an eccentric disk driven by a variable-speed motor. At certain speeds of the motor resonance occurs, whereas otherwise the model remains quiet. *The resonance frequencies are the normal frequencies of the model.* At the same time, by a stroboscopic method or photographically [Andrews and Murray (53)], the type of motion belonging to each particular normal frequency can be determined. If the ratios of the linear dimensions, the masses and the force constants are the same in the model as in the actual molecule, the frequencies of the model are in a constant ratio to those of the actual molecule. Thus, if the force constants and the geometrical structure of a molecule are known the *fundamental frequencies can be predicted* from those of the model rather than by calculation; or, conversely, by using a number of different models and comparing the model frequencies with the observed frequencies of the molecule one may obtain *conclusions about the geometrical structure as well as the ratio of the force constants.*

The method has been successfully applied to CCl_4, $CHCl_3$, C_2H_6, C_2H_4, C_2H_2, CH_3OH, C_2H_5OH by Kettering, Shutts, and Andrews (501); to a number of metal carbonyls and allyls by Duncan and Murray (293); to C_6H_6, $C_6H_5CH_3$, and the phenyl halides by Teets and Andrews (833); and to the various chlorobenzenes, C_6H_5Cl to C_6Cl_6 by Murray, Deitz, and Andrews (649). Trenkler, in a series of papers (870) (871) (872) has considerably improved on the method for the case of the plane oscillations of plane molecules. He investigated in detail the dependence of the frequencies of three- and four-particle systems on the arrangement of the masses as well as on the ratio of the masses and the strength of the bonds. He also studied the benzene ring and other cyclic molecules with six atoms as well as some benzene derivatives.

There are a number of *limitations* to the method of molecular models as used by the authors above mentioned. It is difficult to approach the ideal of mass points connected by weightless springs in these models, since the springs must have considerable strength. Furthermore, in order to study the influence of a change of force constant or geometrical configuration one has to build a different model for each set of trial values. Finally, while the models are well adapted to the valence force system (see section 4d) it would not be easy to study more complicated force systems with their aid.

MacDougall and Wilson (593) have overcome these difficulties in a very ingenious way. They gave up the geometrical similarity between the model and the molecule but built a *system of coupled harmonic oscillators* having the same kinetic and potential energies (except for a scale factor) as the molecule under consideration. Such a system will have, apart from a constant factor, the same frequencies as the molecule. As oscillators MacDougall and Wilson use rods with heavy masses attached to them supported by an axis through the center of mass and connected by springs to a solid base. A number of such units are placed side by side and interconnected by a number of springs representing the coupling of the oscillators. Since the springs are only extended or compressed and do not need to resist bending, they can easily be made of negligible weight as compared to the heavy masses that give each oscillator its moment of inertia.

If ϑ_i is the angular displacement of the ith unit from its equilibrium position and if I_i is its moment of inertia, the total kinetic energy T of the set is given by:

$$2T = \sum_{i=1}^{n} I_i \dot{\vartheta}_i^2, \qquad \text{(II, 160)}$$

where the summation is over all n units; the potential energy V is given by

$$2V = \sum_{i,j} a_{ij}\vartheta_i\vartheta_j, \qquad \text{(II, 161)}$$

where the a_{ij} depend on the position and stiffness of the springs. Mathematically, the frequencies, $\nu = 1/2\pi\sqrt{\lambda}$, of the system are obtained from the corresponding secular equation

$$\begin{vmatrix} a_{11} - I_1\lambda & a_{12} & a_{13} & \cdots & a_{1n} \\ a_{21} & a_{22} - I_2\lambda & a_{23} & \cdots & a_{2n} \\ \cdot & \cdot & \cdot & \cdot & \cdot \\ a_{n1} & a_{n2} & a_{n3} & \cdots & a_{nn} - I_n\lambda \end{vmatrix} = 0,$$

which is seen to be of the same form as the secular equation for the molecular vibrations. Therefore by adjusting the springs in the model, that is, the a_{ij}, and the moments of inertia, I_i, in such a way that they are proportional to the corresponding quantities in the molecular problem, the molecular vibration frequencies can be determined from the observed resonance frequencies of the mechanical model. In fact, any secular equation may be solved in this way. There is no restriction on the potential energy except that it is to be a quadratic function of some coordinates. But these coordinates have to be chosen in such a way that the kinetic energy does not contain any cross terms. One may, for example, use simply Cartesian coordinates for each atom. However, that requires $3N$ units of the machine if there are

N atoms since each unit corresponds to one coordinate. For symmetrical molecules the necessary number of units of the machine can be considerably reduced by using the symmetry coordinates introduced above and solving mechanically the secular equation for each symmetry type. The number of units is then equal to the number of genuine vibrations of that symmetry type. The springs have to be so adjusted that the a_{ik} are proportional to the previous c_{ik} and the moments of inertia so that they are proportional to the d_{ik}. Also, the symmetry coordinates have to be chosen in such a way that the kinetic energy contains no cross terms. This is the case if the symmetry coordinates are chosen orthogonal to one another (see p. 148).

(c) *Assumption of central forces*

General considerations. Only in very few cases such as H_2O [Van Vleck and Cross (885)] and H_3 and H_3^+ [Hirschfelder (453)] is it feasible to obtain the values of the potential constants of a polyatomic molecule on a purely theoretical basis and then predict the vibrational frequencies by solving the secular equation in the above described way and compare them with the experimental values.[38] In all other cases one must try *to determine the potential constants (or force constants) from the experimentally observed fundamental frequencies of the particular molecule.* However, in all except a very few cases the number of potential constants [$\sum f_j(f_j + 1)$ summed over all symmetry types (see p. 148)] is larger than the number of normal vibrations ($3N - 6$) and therefore the former cannot all be determined from the latter. One way out of this difficulty is the investigation of isotopic molecules for which the potential constants are the same (at least to a very high approximation) but the frequencies different, thus supplying additional equations for the potential constants (see section 6). Another way is to make certain more specific assumptions about the forces in the molecules such that the number of force constants to be determined is reduced.

The first such assumption that we shall have to discuss in more detail is *the assumption that only central forces act* between the atoms in a molecule; that is, we assume that *the force acting on a given atom in a molecule is the resultant of the attractions and repulsions by all the other atoms,* and that these attractions and repulsions depend only on the distances from these other atoms and lie in the lines connecting them with the one considered. This is equivalent to assuming that the *potential energy is a purely quadratic function of the changes Q_i of the distances l_i between the nuclei* (the central force coordinates) without any cross products:

$$2V = \sum a_{ii}Q_i^2. \tag{II, 162}$$

The equivalence of the two assumptions is immediately seen if the derivatives of V, which give the force components, are formed; for example,

$$F_x^k = -\frac{\partial V}{\partial x_k} = -\sum a_{ii}Q_i \frac{\partial Q_i}{\partial x_k}.$$

Here $\partial Q_i/\partial x_k$ vanishes for all but those Q_i that involve the nucleus k, and for these it is equal to $\cos(l_i, x)$. Thus the contributions to the force F^k on k have the magnitudes $a_{ii}Q_i$ and have the directions of the l_i.

[38] Even for diatomic molecules there are only two or three cases where the vibrational frequencies have been predicted from theory.

It is obvious that the assumption of the potential function (II, 162) leads to a considerable simplification, particularly when in consequence of symmetry some of the a_{ii} are identical. Frequently the number of force constants under this assumption is smaller than the number of normal frequencies. Therefore not all normal frequencies need be known to evaluate the force constants and those not used may be predicted or their assignment checked by the calculations.

Application to non-linear symmetric XY_2. In the case of non-linear symmetrical XY_2 molecules the assumption of central forces means that a_{12}, a_{13}, and a_{23} in the previous equation (II, 97) are zero and we have, since $a_{22} = a_{11}$,

$$2V = a_{11}(Q_1^2 + Q_2^2) + a_{33}Q_3^2; \tag{II, 163}$$

that is, there are only *two potential constants* which are indeed overdetermined by the three normal frequencies. Putting $a_{12} = 0$ and $a_{13} = 0$ in the previous more general equations (II, 124–126) we obtain for the *frequencies:*

$$\lambda_3 = 4\pi^2\nu_3^2 = \left(1 + \frac{2m_Y}{m_X}\sin^2\alpha\right)\frac{a_{11}}{m_Y} \tag{II, 164}$$

$$\lambda_1 + \lambda_2 = 4\pi^2(\nu_1^2 + \nu_2^2) = 2\frac{a_{33}}{m_Y} + \frac{a_{11}}{m_Y}\left(1 + \frac{2m_Y}{m_X}\cos^2\alpha\right) \tag{II, 165}$$

$$\lambda_1\lambda_2 = 16\pi^4\nu_1^2\nu_2^2 = 2\left(1 + \frac{2m_Y}{m_X}\right)\cos^2\alpha\,\frac{a_{11}a_{33}}{m_Y^2} \tag{II, 166}$$

These equations were first derived by Dennison (277). If the frequencies ν_1, ν_2, ν_3 as well as the angle α are known one can determine a_{11} from (II, 164) and a_{33} from either (II, 165) or (II, 166). The third equation may then be used as a check on the consistency of the assumption of central forces.

In Table 37 are given for a number of triatomic molecules of the type XY_2 the constants a_{11} and a_{33} as determined from the observed frequencies by means of (II, 164) and (II, 166) assuming the values of α given in the fifth column. These α values have been obtained from the rotational structure of infrared bands (see Chapter IV) or from electron diffraction measurements. Instead of the actual frequencies, in Table 37 the *wave numbers of the fundamental bands* (see Chapter III) are given, which in agreement with common practice are also denoted ν_1, ν_2, ν_3.[40] In substituting them into the equations (II, 164–166) the values given must therefore be multiplied by $c = 2.99776 \times 10^{10}$. If the ν_i are substituted in cm^{-1} and the masses in atomic-weight units the ν_i^2 on the left side of these and similar equations have to be multiplied, instead of by $4\pi^2$, by

$$4\pi^2c^2M_1 = 5.8894 \times 10^{-2}, \tag{II, 167}$$

where M_1 is $\frac{1}{16}$ of the mass of the O^{16} atom ($= 1.6600 \times 10^{-24}$ gm).[40a]

It is seen from Table 37 that, as was to be expected, the force constant a_{33} for the Y—Y "bond" comes out considerably smaller than the force constant a_{11} for the X—Y bonds. Also in the case of the pairs H_2O and D_2O, H_2S and D_2S, H_2Se and

[40] In the nomenclature used for diatomic molecules they would have to be called $\Delta G_{\frac{1}{2}}$.

[40a] Compare the values of the fundamental constants in the appendix p. 538.

D_2Se, the force constants are nearly the same, as they should be for isotopic molecules (see section 6).

The last two columns of Table 37 give the left- and right-hand sides of equation (II, 165) multiplied by m_Y. If there were actually central forces only, these two columns should agree. It is seen that there is agreement within about 25 per cent. While this agreement is not very good, one may say that at least very roughly the forces between atoms may be considered as central. It is noteworthy that the differences between the last two columns of Table 37 are all in the same direction.

TABLE 37. FORCE CONSTANTS OF NON-LINEAR XY_2 MOLECULES AS OBTAINED FROM THE OBSERVED FREQUENCIES, ASSUMING CENTRAL FORCES. [REVISION AND EXTENSION OF A SIMILAR TABLE BY PENNEY AND SUTHERLAND (692)].

Molecule	ν_1 (cm^{-1})	ν_2 (cm^{-1})	ν_3 (cm^{-1})	2α	a_{11}	a_{33}	$4\pi^2 m_Y \times (\nu_1^2 + \nu_2^2)$	$2a_{33} + a_{11}\left(1 + \frac{2m_Y}{m_X}\cos^2\alpha\right)$
H_2O	3652	1595	3756	105°	7.76	1.85	9.44	11.82
D_2O[39a]	2666	1179	2784	105°	7.94	1.89	10.08	12.46
H_2S	2611	1290	2684	92°	4.14	0.940	5.04	6.15
D_2S	1892	934	1999	92°	4.46	0.908	5.28	6.54
H_2Se	2260	1074	2350	90°	3.24	0.625	3.72	4.53
D_2Se	1630	745	1696	90°	3.33	0.594	3.81	4.60
CH_2[39b]	2968	1444	3000	115°	4.77	2.01	6.48	9.03
SO_2[39a]	1151	524	1361	120°	9.97	3.24	15.33	18.94
NO_2	1320	648	1621	120°[39]	9.13	4.34	20.40	23.02
F_2O	830	490	1110	104°	5.57	1.45	10.40	13.49
Cl_2O	680	330	973	111°	4.93	1.28	11.94	14.51
					$\times 10^5$ dynes/cm		$\times 10^5$ dynes/cm	

One might be tempted, in cases in which the angle α is not known from other data, to determine from the three equations (II, 164–166) not only the constants a_{11} and a_{33} but also the angle α. However, if this is done for the examples of Table 37 no agreement with the known angles is obtained; in fact, in most cases imaginary values for α result ($\sin\alpha > 1$). While only slight adjustments of the frequencies would be necessary to produce the correct values of α, it is obviously impossible to use this method for a determination of α in other cases.

Linear triatomic and plane more-than-triatomic molecules. In the case of a linear triatomic molecule (symmetric or non-symmetric), the assumption of central forces would lead to the frequency zero for the perpendicular (degenerate) vibration since in this vibration the distances of the atoms do not change except in higher order; or, in other words if $\alpha = 90°$ is substituted in equations (II, 165 and 166) the observed non-zero frequency of the perpendicular vibration leads to an infinite

[39] Since the above table was calculated, Maxwell and Mosley (607) have obtained an angle of 130° ± 2° from electron diffraction, while Harris and King (413) have obtained an angle of 154° from the ultraviolet absorption spectrum.

[39a] The frequencies given here are those available at the time this table was calculated. The most recent values are given in Chapter III, section 3a.

[39b] The frequencies of CH_2 are those obtained by Sutherland and Dennison (828) from a study of the frequencies of H_2CO.

value for a_{33}. Hence the assumption of central forces cannot be made for linear molecules and one would expect it to be a very poor approximation for very wide-angled triatomic molecules.

Similarly, for plane molecules with more than three atoms such as H_2CO, C_2H_4, and others, the assumption of central forces would lead to zero frequencies for the vibrations perpendicular to the plane.

Application to pyramidal XY_3 molecules. The assumption of central forces leads again to reasonable results for non-planar four-atomic molecules such as the pyramidal molecules XY_3 discussed previously. The *potential energy* in the latter case is

$$2V = a_1(Q_{12}^2 + Q_{13}^2 + Q_{14}^2) + a_2(Q_{23}^2 + Q_{24}^2 + Q_{34}^2); \qquad \text{(II, 168)}$$

the Q_{ik} are the changes of distance between atoms i and k (for the numbering of the atoms see Fig. 58), a_1 is the force constant for the central force between atoms X and Y, and a_2 is the force constant for the central force between two Y atoms. As always for central forces there are no cross terms between different Q_{ik} in V.

Expressing the Q_{ik} in terms of the displacement coordinates x_i, y_i, z_i introduced above (see Fig. 58), we obtain

$$\begin{aligned}
Q_{12} &= x_1 \sin\beta + z_1 \cos\beta + z_2,\\
Q_{13} &= -\tfrac{1}{2}x_1 \sin\beta + \tfrac{1}{2}\sqrt{3}y_1 \sin\beta + z_1\cos\beta + z_3,\\
Q_{14} &= -\tfrac{1}{2}x_1 \sin\beta - \tfrac{1}{2}\sqrt{3}y_1 \sin\beta + z_1\cos\beta + z_4,\\
Q_{23} &= \tfrac{1}{2}\sqrt{3}x_2\cos\beta - \tfrac{1}{2}y_2 + \tfrac{1}{2}\sqrt{3}z_2\sin\beta + \tfrac{1}{2}\sqrt{3}x_3\cos\beta + \tfrac{1}{2}y_3 + \tfrac{1}{2}\sqrt{3}z_3\sin\beta,\\
Q_{24} &= \tfrac{1}{2}\sqrt{3}x_2\cos\beta + \tfrac{1}{2}y_2 + \tfrac{1}{2}\sqrt{3}z_2\sin\beta + \tfrac{1}{2}\sqrt{3}x_4\cos\beta - \tfrac{1}{2}y_4 + \tfrac{1}{2}\sqrt{3}z_4\sin\beta,\\
Q_{34} &= \tfrac{1}{2}\sqrt{3}x_3\cos\beta - \tfrac{1}{2}y_3 + \tfrac{1}{2}\sqrt{3}z_3\sin\beta + \tfrac{1}{2}\sqrt{3}x_4\cos\beta + \tfrac{1}{2}y_4 + \tfrac{1}{2}\sqrt{3}z_4\sin\beta.
\end{aligned}$$

Substituting the x_i, y_i, z_i from (II, 156) into these equations but putting $S_{3b} = 0$ and $S_{4b} = 0$ (since we know beforehand that the dependence on S_{3b} and S_{4b} will be the same as that on S_{3a} and S_{4a}), we obtain, *in terms of symmetry coordinates*,

$$\begin{aligned}
Q_{12} &= pS_1 - sS_2 - 2rS_{3a} + \sin\beta S_{4a},\\
Q_{13} &= pS_1 - sS_2 + rS_{3a} - \tfrac{1}{2}\sin\beta S_{4a},\\
Q_{14} &= pS_1 - sS_2 + rS_{3a} - \tfrac{1}{2}\sin\beta S_{4a},\\
Q_{23} &= \sqrt{3}\sin\beta S_1 + \sqrt{3}\cos\beta S_2 - \tfrac{1}{2}\sqrt{3}\sin\beta S_{3a} + \tfrac{1}{2}\sqrt{3}S_{4a},\\
Q_{24} &= \sqrt{3}\sin\beta S_1 + \sqrt{3}\cos\beta S_2 - \tfrac{1}{2}\sqrt{3}\sin\beta S_{3a} + \tfrac{1}{2}\sqrt{3}S_{4a};\\
Q_{34} &= \sqrt{3}\sin\beta S_1 + \sqrt{3}\cos\beta S_2 + \sqrt{3}\sin\beta S_{3a} - \sqrt{3}S_{4a},
\end{aligned} \qquad \text{(II, 169)}$$

where p, r, and s are given by (II, 158).

Substituting (II, 169) into (II, 168), we obtain for the *potential constants* c_{ik} in the general equation (II, 153),

$$\begin{aligned}
c_{11} &= 3p^2a_1 + 9a_2\sin^2\beta, & c_{12} &= -3psa_1 + 9a_2\sin\beta\cos\beta,\\
c_{22} &= 3s^2a_1 + 9a_2\cos^2\beta, & c_{33} &= 6r^2a_1 + \tfrac{9}{2}a_2\sin^2\beta,\\
c_{34} &= -3ra_1\sin\beta - \tfrac{9}{2}a_2\sin\beta, & c_{44} &= \tfrac{3}{2}a_1\sin^2\beta + \tfrac{9}{2}a_2.
\end{aligned} \qquad \text{(II, 170)}$$

The other c_{ik} are of course zero. We can now substitute these c_{ik} and the d_{ik} from (II, 157) into (II, 155) in order to obtain expressions for the *normal frequencies*. We obtain, for the two non-degenerate vibrations

$$\lambda_1 + \lambda_2 = 4\pi^2(\nu_1^2 + \nu_2^2) = \left(1 + 3\frac{m_Y}{m_X}\cos^2\beta\right)\frac{a_1}{m_Y} + 3\frac{a_2}{m_Y}, \quad \text{(II, 171)}$$

$$\lambda_1\lambda_2 = 16\pi^4\nu_1^2\nu_2^2 = 3\left(1 + 3\frac{m_Y}{m_X}\right)\cos^2\beta\frac{a_1a_2}{m_Y^2}, \quad \text{(II, 172)}$$

and similarly for the degenerate vibrations,

$$\lambda_3 + \lambda_4 = 4\pi^2(\nu_3^2 + \nu_4^2) = \left(1 + \frac{3}{2}\frac{m_Y}{m_X}\sin^2\beta\right)\frac{a_1}{m_Y} + \frac{3}{2}\frac{a_2}{m_Y}, \quad \text{(II, 173)}$$

$$\lambda_3\lambda_4 = 16\pi^4\nu_3^2\nu_4^2 = \frac{3}{4}\left(2 - \sin^2\beta + \frac{3m_Y}{m_X}\sin^2\beta\right)\frac{a_1a_2}{m_Y^2}. \quad \text{(II, 174)}$$

These solutions, although in a slightly different form, were first obtained by Dennison (277).

Since there are four frequencies and only two potential constants, one may use (II, 171) and (II, 172) to determine the latter from the non-degenerate frequencies, and use (II, 173) and (II, 174), that is, the degenerate frequencies, as a check on the validity of the assumption of central forces, or vice versa. If the angle is not known from other data one may also use three of the equations to determine the three unknowns a_1, a_2, and β and use the fourth equation as a check. For the calculation of the *angle* β it is best to divide (II, 174) by (II, 172), since then the force constants drop out. One obtains

$$\cos^2\beta = \frac{1}{4\dfrac{\nu_3^2\nu_4^2}{\nu_1^2\nu_2^2} + \dfrac{3m_Y - m_X}{3m_Y + m_X}}. \quad \text{(II, 175)}$$

Table 38, which is similar to Table 37, gives for a number of XY_3 molecules the observed fundamental frequencies ν_1, ν_2, ν_3, ν_4 (in cm^{-1}),[41] the angle β as obtained from other evidence [see Howard and Wilson (462)], the angle β as obtained from (II, 175), the force constants a_1 and a_2 as obtained from equations (II, 173) and (II, 174) (using the observed ν_3, ν_4, and β values), and the frequencies ν_1 and ν_2 as calculated from equations (II, 171) and (II, 172) with the a_1 and a_2 values obtained from ν_3 and ν_4. It is seen that with one exception the angles β calculated from the observed frequencies according to (II, 175) agree within less than 15° with the observed β values. The agreement of the observed and calculated ν_1 and ν_2 is fair in most cases although a few big discrepancies occur. It should be mentioned that if a_1 and a_2 had been calculated from ν_1 and ν_2 complex values would have been obtained in most cases. On account of the quadratic nature of the equations (II, 173) and (II, 174), two sets of values for a_1 and a_2 are obtained. Of these, that one has been chosen for which $a_2 < a_1$. The fact that in most cases a_2 is much smaller than a_1

[41] The question which observed fundamentals are degenerate, that is, which are ν_3 and ν_4, and which not, will be taken up in more detail in Chapter III. Which non-degenerate vibration is called ν_1, which ν_2, and which degenerate vibration is called ν_3, which ν_4 is of course without influence on the results. *Within one species we shall in general designate the higher frequency by the lower subscript.*

is reasonable and corresponds to the fact that ν_4 is much smaller than ν_3. On the whole, one can say that the assumption of central forces is a better approximation for pyramidal XY_3 molecules than for non-linear XY_2 molecules, even though it is by no means entirely satisfactory.

It is interesting to apply the equations for the frequencies of the XY_3 molecule to the special case in which X = Y, that is, a *pyramidal molecule consisting of four equal atoms.* Let us assume that in this case all atoms are in equivalent positions, that is, that they lie at the corners of a regular tetrahedron. Such a case seems to be provided by the molecules P_4 and As_4 [see Chapter III, p. 299]. In this case it is easily seen [42] that $\sin\beta = 1/\sqrt{3}$, and also that $a_1 = a_2$ in (II, 168), since there is no difference between X and Y. Substituting this and $m_X = m_Y$ into (II, 171–174), we obtain

$$\lambda_1 = \frac{4a_1}{m_X}, \qquad \lambda_2 = \lambda_3 = \frac{2a_1}{m_X}, \qquad \lambda_4 = \frac{a_1}{m_X}. \tag{II, 176}$$

Two of the frequencies one non-degenerate the other doubly degenerate coincide, thus giving rise to a triply degenerate normal vibration. This is in agreement with what one would expect for a mole-

TABLE 38. FUNDAMENTAL FREQUENCIES, FORCE CONSTANTS, AND ANGLES FOR THE PYRAMIDAL MOLECULES XY_3, ASSUMING CENTRAL FORCES.[43]

Molecule	Observed frequencies (cm^{-1})				$\beta_{observed}$[44]	$\beta_{calculated}$	Force constants from ν_3, ν_4, $\beta_{observed}$, and equations (II, 173) and (II, 174)		Frequencies ν_1 and ν_2 from a_1, a_2, $\beta_{observed}$, and equations (II, 171) and (II, 172)	
	ν_1	ν_2	ν_3	ν_4			a_1	a_2	ν_1 (cm^{-1})	ν_2 (cm^{-1})
NH_3	3337	950[45]	3414	1628	69°	73°	3.89	2.83	4473	859
ND_3	2419	749[45]	2555	1191	69°	72°	5.05	2.29	3133	679
PH_3	2327	991[45]	2421	1121	62°[46]	63°	2.57	1.04	2991	784
PD_3	1694	730[45]	—	806	62°[46]	—	—	—	—	—
PF_3	890	531	840	486	62°	56°	3.58	2.93	1045	380
PCl_3	510	257	480	190	64°	51°	1.94	0.63	459	184
PBr_3	380	162	400	116	65°	55°	1.79	0.47	203	203
AsF_3	707	341	644	274	60°	45°	3.23	0.88	723	240
$AsCl_3$	410	193	370	159	59°	50°	1.73	0.51	404	154
$SbCl_3$	360	165	320	134	57°	45°	1.49	0.38	357	130
$BiCl_3$	288	130	242	96	57°	25°	0.94	0.20	270	90
							$\times 10^5$ dynes/cm	$\times 10^5$ dynes/cm		

[42] The angle α between any two X—X bonds is 60° and therefore according to footnote 44 the above result follows.

[43] For the frequencies of NH_3 and ND_3 see Table 72, p. 295. The frequencies of PH_3 and PD_3 have been taken from Lee and Wu (572), those of the other molecules from Kohlrausch (14).

[44] β is related to the angle α between each two X—Y bonds by $\sin\beta = (2/\sqrt{3})\sin\alpha/2$. The β values for the halides were obtained from the α values given by Howard and Wilson (462). Pauling in his book (17) gives slightly different values based on more recent electron diffraction work. However it did not appear worth while to repeat the calculations embodied in the table with these new β values.

[45] Average of the doublets due to passage through potential barrier (see p. 221f.).

[46] Recently Stevenson (806), assuming a PH distance of 1.415 Å (obtained from Pauling's table of covalent radii), has obtained a bond angle of 93° which corresponds to $\beta = 57°$.

cule X_4 of symmetry T_d (see Table 36), and is independent of the assumption of central forces. According to (II, 176), the ratio of the three frequencies, assuming central forces, should be

$$\nu_1 : \nu_2 : \nu_4 = 2 : \sqrt{2} : 1. \tag{II, 177}$$

For P_4 the observed ratio is 2.00 : 1.54 : 1.20.

Application to tetrahedral XY_4 molecules. For five-atomic XY_4 molecules in which the X atom occupies the center and the Y atoms the corners of a tetrahedron, the considerations are somewhat more complicated because the assumption of central forces between each pair of atoms no longer leads to vanishing forces in the equilibrium position. Whereas in the case of the molecule XY_3 in the equilibrium position the angle will adjust itself in such a way that there is no force in the X—Y or Y—Y direction, in the case of the tetrahedral XY_4 molecule the angle is fixed and the

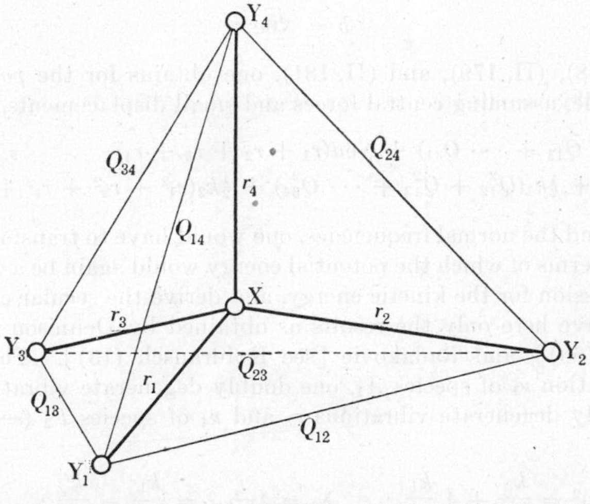

FIG. 59. **Displacement coordinates in tetrahedral XY_4.**—The Q_{ik} and r_i are not the lengths of the lines on which they are written but the changes of these lengths.

resultant force on a Y atom, for example, may be zero even though the forces between the Y atom considered and the other atoms are not individually zero. For example, a repulsion between the Y atoms could, in the equilibrium position, be balanced by an attraction between X and Y.

The expression for the potential energy in such a case was first derived by Dennison (276). If we designate the relative displacements of the Y nuclei along the edges of the tetrahedron by Q_{12}, Q_{13}, Q_{14}, Q_{23}, Q_{24}, Q_{34} (see Fig. 59) and the displacement of the Y atoms relative to X by r_1, r_2, r_3, r_4, then we have, in the case of small displacements, for the magnitude of the forces acting on Y_1 due to Y_2, Y_3, and Y_4,

$$\begin{aligned} F_{Y_1Y_2} &= a - k_1Q_{12}, \\ F_{Y_1Y_3} &= a - k_1Q_{13}, \\ F_{Y_1Y_4} &= a - k_1Q_{14}, \end{aligned} \tag{II, 178}$$

and for the force due to X,

$$F_{Y_1X} = -b - k_2r_1. \tag{II, 179}$$

Similar equations hold for the forces on Y_2, Y_3, Y_4. The force on X due to Y_1 is of course equal and opposite to F_{Y_1X}. The constants a and b represent the constant part of the force that does not vanish in the equilibrium position. We have chosen a positive sign for a and a negative sign for b in (II, 178) and (II, 179) in order to indicate repulsion and attraction respectively as the normally expected state of affairs. In the equilibrium position ($Q_{ik} = 0$ and $r_i = 0$), the three forces a acting on Y_1 must be in equilibrium with the force b, that is,

$$3a \cos \varphi = b,$$

where φ is the angle between Y—Y and Y—X. This angle in a tetrahedron is given by

$$\sin \varphi = \tfrac{1}{3}\sqrt{3}. \tag{II, 180}$$

It therefore follows that

$$b = \sqrt{6}a. \tag{II, 181}$$

From (II, 178), (II, 179), and (II, 181), one obtains for the *potential energy* of the XY_4 molecule, assuming central forces and small displacements,[47]

$$V = -a(Q_{12} + Q_{13} + \cdots Q_{34}) + \sqrt{6}a(r_1 + r_2 + r_3 + r_4) + \tfrac{1}{2}k_1(Q_{12}^2 + Q_{13}^2 + \cdots Q_{34}^2) + \tfrac{1}{2}k_2(r_1^2 + r_2^2 + r_3^2 + r_4^2). \tag{II, 182}$$

In order to find the normal frequencies, one would have to transform to symmetry coordinates (in terms of which the potential energy would again be a quadratic form), set up the expression for the kinetic energy, and derive the secular equation. However, we shall give here only the results as obtained by Dennison (276), Jaumann [see Schaefer (763)], and Radakovic [see Kohlrausch (13)]. There is one non-degenerate vibration ν_1 of species A_1, one doubly degenerate vibration ν_2 of species E, and two triply degenerate vibrations ν_3 and ν_4 of species F_2 (see p. 140). One obtains

$$\lambda_1 = 4\pi^2\nu_1^2 = \frac{k_2}{m_Y} + 4\frac{k_1}{m_Y}; \quad \lambda_2 = 4\pi^2\nu_2^2 = \frac{k_1}{m_Y} - \frac{k'}{m_Y};$$

$$\lambda_3 + \lambda_4 = 4\pi^2(\nu_3^2 + \nu_4^2) = \frac{2k_1}{m_Y} + \frac{4m_Y + 3m_X}{3m_Xm_Y}k_2 - \frac{2(3m_X + 16m_Y)}{3m_Xm_Y}k'; \tag{II, 183}$$

$$\lambda_3\lambda_4 = 16\pi^4\nu_3^2\nu_4^2 = \frac{2(4m_Y + m_X)}{3m_Xm_Y^2}(k_1k_2 - 8k_1k' - 5k_2k' - 8k'^2).$$

Here we have used, as an abbreviation,

$$k' = \frac{a}{l}, \tag{II, 184}$$

where l is the length of the edge of the tetrahedron in the equilibrium position. Thus k' is of the dimension of a force constant, whereas a was not.[48]

While there are four frequencies, there are only three force constants k_1, k_2, k' to determine. Thus again one of the equations may serve as a check on the assump-

[47] It is easy to verify this if it is remembered that $F_{Y_1Y_2} = -\dfrac{\partial V}{\partial Q_{12}}$, $F_{Y_1X} = -\dfrac{\partial V}{\partial r_1}$, and so on.

[48] Our k' is $\frac{1}{4}$ of Dennison's and Jaumann's K'.

tion of central forces. The force constant k_1 is most easily determined by the following combination of the equations (II, 183):

$$\lambda_3 + \lambda_4 - \frac{4m_Y + 3m_X}{3m_X}\lambda_1 - \frac{2(3m_X + 16m_Y)}{3m_X}\lambda_2 = -\frac{4(m_X + 4m_Y)}{m_X m_Y}k_1. \quad \text{(II, 185)}$$

k_2 and k' then follow immediately from the first two equations (II, 183) and the fourth may serve as a check.

Table 39 gives the observed fundamental frequencies and the force constants determined in the above described way for a number of tetrahedral five-atomic

TABLE 39. FUNDAMENTAL FREQUENCIES AND FORCE CONSTANTS OF TETRAHEDRAL XY_4 MOLECULES ASSUMING CENTRAL FORCES.

Molecule	Observed frequencies (cm^{-1})[49]				Force constants			Fourth equation (II, 183)	
	ν_1	ν_2	ν_3	ν_4	k_1	k_2	k'	L.H.S.[49a]	R.H.S.[49a]
CH_4	2914	1526	3020	1306	0.60	2.66	−0.788	5.42	14.27
CD_4	2085	1054	2258	996	0.61	2.72	−0.708	1.75	2.94
SiH_4	2187	978	2183	910	0.22	1.96	−0.348	1.37	2.61
NH_4^+[50]	3033	1685	3134	1397	0.71	2.62	−0.975	6.65	10.62
CF_4	904	437	1265	630	1.58	2.81	−0.553	0.220	0.227
SiF_4	800	260	1022	420	0.47	5.29	−0.288	0.064	0.072
CCl_4	458	218	776[51]	314	0.79	1.24	−0.207	0.021	0.022
$SiCl_4$	424	150	608	221	0.36	2.33	−0.115	0.0063	0.0077
$TiCl_4$	386	120	495	141	0.23	2.19	−0.072	0.0017	0.0029
$GeCl_4$	396	134	453	172	0.27	2.19	−0.104	0.0021	0.0029
$SnCl_4$	366	104	403	134	0.15	2.20	−0.076	0.0010	0.0024
CBr_4	267	123	672	183	0.56	1.11	−0.148	0.0052	0.0056
$SiBr_4$	249	90	487	137	0.29	1.76	−0.093	0.0016	0.0019
$GeBr_4$	234	78	328	111	0.22	1.72	−0.070	0.00046	0.00060
$SnBr_4$	220	64	279	88	0.14	1.72	−0.053	0.00021	0.00028
SO_4^{--}	981	451	1104	613	1.18	4.36	−0.739	0.159	0.093
ClO_4^-	935	462	1102	628	1.05	4.05	−0.965	0.166	0.089
PO_4^{---}	980	363	1082	515	0.90	5.44	−0.339	0.108	0.125
					$\times 10^5$ dynes/cm			$\times 10^5$ dynes/cm	

molecules. As was to be expected, k_1 is much smaller than k_2 in all cases, corresponding to a much smaller resistance to deformation without change of the X—Y distance than to stretching of the X—Y bond. In most cases k' comes out negative, which means that, contrary to our expectation, in the equilibrium position there are (weak) attractive forces between each pair of Y atoms balanced by repulsive forces between X and Y. The last two columns of Table 39 give the left- and right-hand sides of the fourth equation (II, 183). It is seen that the agreement is fairly good for the

[49] With a few exceptions the observed frequencies are taken from Kohlrausch (14).

[49a] Using atomic weights for the masses and $4\pi^2c^2\nu_i^2M_1 = 5.8894 \times 10^{-2}\nu_i^2$ for λ_i (see p. 160).

[50] The evidence that NH_4^+ is tetrahedral is inconclusive [see Ananthakrishnan (47)].

[51] This is the average of the two observed frequencies which correspond to a Fermi resonance (see section 5).

halides of C and Si but poor for the other molecules. Trumpy (873) has given a method of obtaining a set of force constants k_1, k_2, k' which gives the best possible representation of the observed *four* frequencies [see also Kohlrausch (13)]. The force constants in Table 39 are not this best set. Rather they are evaluated in such a way as to give a perfect representation of ν_1, ν_2, and $\nu_3^2 + \nu_4^2$ but a poorer representation of the individual ν_3 and ν_4.

We refrain from giving any further examples for the application of the assumption of central forces since in general the assumption of valence forces (see the following subsection) has proved to be somewhat more satisfactory in representing the observed fundamental frequencies. It has the added advantage of being applicable also to linear and plane molecules.

(*d*) *Assumption of valence forces*

The assumption of valence forces was first made by Bjerrum (155) in 1914. He assumed that there is a *strong restoring force in the line of every valence bond* if the distance of the two atoms bound by this bond is changed. In addition he assumed that there is a *restoring force opposing a change of the angle between two valence bonds* connecting one atom with two others. As for the central force system, here likewise often not all normal frequencies are necessary for a calculation of the force constants and therefore a valuable check on the assignment of the frequencies is possible.

Application to non-linear symmetric XY_2. In the case of the non-linear symmetric molecule XY_2 there is, according to the assumption of valence forces, a restoring force between X and each Y as well as a force stabilizing the Y—X—Y angle. The latter replaces the central force between the two Y atoms in the central force system. Thus we have for the *potential energy* (neglecting higher powers),

$$2V = k_1(Q_1^2 + Q_2^2) + k_\delta \delta^2, \qquad \text{(II, 186)}$$

where Q_1 and Q_2 as previously are the changes of the X—Y distances and where δ is the change of the angle 2α (Fig. 55a). It is assumed in the valence force field that there are *no cross terms in the potential energy if it is expressed in terms of the Q_i and δ.* Q_1, Q_2, and δ are also called *valence force coordinates* as contrasted to the central force coordinates Q_1, Q_2, and Q_3 used previously.

We have now to express δ in terms of the symmetry coordinates (Fig. 55a) just as previously we expressed Q_1, Q_2, Q_3 in terms of these coordinates. In the equilibrium position we have

$$\cot \alpha = \frac{h}{a},$$

where $2a$ and h are the base and the height of the triangle formed by the nuclei. In the displaced position of the symmetry coordinate S_2 the angle α has changed to $\alpha + \delta_2/2$ and we have, with (II, 121),

$$\cot\left(\alpha + \frac{\delta_2}{2}\right) = \frac{h - S_2 - \dfrac{2m_Y}{m_X} S_2}{a} = \frac{h - pS_2}{a};$$

and therefore, for small displacements and with $a = l \sin\alpha$,

$$\cot\left(\alpha + \frac{\delta_2}{2}\right) - \cot\alpha = -\frac{\delta_2/2}{\sin^2\alpha} = -\frac{pS_2}{l\sin\alpha}$$

or

$$\delta_2 = \frac{2p}{l} S_2 \sin\alpha,$$

where l is the equilibrium X—Y distance. Similarly δ_1, the change of 2α during the displacement S_1, comes out to be

$$\delta_1 = -\frac{2}{l} S_1 \cos\alpha,$$

and therefore, since S_3 does not produce any change of 2α (see Fig. 55a), the total change δ of 2α when all oscillations are excited is

$$\delta = \delta_1 + \delta_2 = \frac{2p}{l} S_2 \sin\alpha - \frac{2}{l} S_1 \cos\alpha. \qquad \text{(II, 187)}$$

Substituting (II, 187) and Q_1 and Q_2 from (II, 122) into the expression (II, 186) for the potential energy and comparing with (II, 109), one obtains for the *potential constants* c_{ik},

$$c_{11} = 2k_1 \sin^2\alpha + \frac{4}{l^2} k_\delta \cos^2\alpha;$$

$$c_{12} = 2pk_1 \sin\alpha\cos\alpha - \frac{4}{l^2} pk_\delta \sin\alpha\cos\alpha; \qquad \text{(II, 188)}$$

$$c_{22} = 2p^2k_1 \cos^2\alpha + \frac{4p^2}{l^2} k_\delta \sin^2\alpha;$$

$$c_{33} = 2r^2k_1.$$

The constants d_{ik} of the kinetic energy are, as before, given by (II, 120). If these d_{ik} as well as the new c_{ik} are substituted in (II, 115), (II, 117), and (II, 118), one obtains for the *normal frequencies*:

$$\lambda_3 = 4\pi^2\nu_3^2 = \left(1 + \frac{2m_Y}{m_X}\sin^2\alpha\right)\frac{k_1}{m_Y}, \qquad \text{(II, 189)}$$

$$\lambda_1 + \lambda_2 = 4\pi^2(\nu_1^2 + \nu_2^2)$$
$$= \left(1 + \frac{2m_Y}{m_X}\cos^2\alpha\right)\frac{k_1}{m_Y} + \frac{2}{m_Y}\left(1 + \frac{2m_Y}{m_X}\sin^2\alpha\right)\frac{k_\delta}{l^2}, \qquad \text{(II, 190)}$$

$$\lambda_1\lambda_2 = 16\pi^4\nu_1^2\nu_2^2 = 2\left(1 + \frac{2m_Y}{m_X}\right)\frac{k_1}{m_Y^2}\frac{k_\delta}{l^2}. \qquad \text{(II, 191)}$$

It should be noted that k_δ/l^2 has the dimension of a force constant, dynes per centimeter, as has k_1. Again, as in the case of the central force field, if ν_1, ν_2, ν_3 and the angle α are known, the force constants k_1 and k_δ/l^2 may be obtained from the equations (II, 189) and (II, 191),[52] and (II, 190) may then serve as a check on the consistency of the assumption of valence forces.

Table 40 gives, for the same molecules as Table 37, the force constants k_1 and k_δ/l^2 obtained in this way and the left- and right-hand sides of equation (II, 190) multiplied by m_Y. It is seen that the agreement between these two columns is

TABLE 40. FORCE CONSTANTS AND VALENCE ANGLES OF NON-LINEAR XY_2 MOLECULES OBTAINED UNDER THE ASSUMPTION OF VALENCE FORCES.[53] [REVISION AND EXTENSION OF A SIMILAR TABLE BY PENNEY AND SUTHERLAND (692)].[53a]

Molecule	k_1	k_δ/l^2	$4\pi^2 m_Y(\nu_1^2+\nu_2^2)$	$m_Y\times$[R.H.S. of (II, 190)]	2α Observed	2α Calculated
H_2O	7.76	0.69	9.44	9.60	105°	120°
D_2O	7.94	0.70	10.08	10.30	105°	113°
H_2S	4.14	0.45	5.04	5.21	92°	156°
D_2S	4.46	0.44	5.28	5.66	92°	159°
H_2Se	3.24	0.31	3.72	3.91	90°	imaginary
D_2Se	3.33	0.30	3.81	4.02	90°	imaginary
CH_2	4.77	0.58	6.48	6.31	115°	98°
SO_2	9.97	0.81	15.33	15.30	120°	120°
NO_2	9.13	1.52	20.40	22.58	120°[39]	119°
F_2O	5.57	0.55	10.40	13.32	104°	87°
Cl_2O	4.93	0.41	11.94	15.25	111°	81°
	$\times 10^5$ dynes/cm		$\times 10^5$ dynes/cm			

decidedly better than the agreement between the corresponding two columns of Table 37. It is noteworthy that in all cases k_δ/l^2 is much smaller than k_1, indicating a much smaller resistance to the bending than to the stretching of the particular bond.

Instead of using equation (II, 190) as a check one may use it together with the others in order to determine the angle 2α. By successive eliminations the following equation for $w = 1 + (2m_Y/m_X)\sin^2\alpha$ is obtained:

$$w^3 - \left(1 + \frac{2m_Y}{m_X}\right)\frac{(\nu_1^2 + \nu_2^2 + \nu_3^2)\nu_3^2}{\nu_1^2\nu_2^2}\,w + 2\left(1 + \frac{m_Y}{m_X}\right)\left(1 + \frac{2m_Y}{m_X}\right)\frac{\nu_3^4}{\nu_1^2\nu_2^2} = 0. \quad \text{(II, 192)}$$

The values of 2α calculated from this equation are given in the last column of Table 40 and compared to the observed angles given in the preceding column. While from the central force system in almost all cases imaginary values for the angles result, here with two exceptions (H_2Se, D_2Se) real values are obtained which except for H_2S and D_2S are fairly close to the observed angles. One may therefore, with some reservation, even attempt to obtain this angle from the observed frequencies in cases where it is not known from other data.

[52] k_δ itself can only be obtained when in addition the internuclear distance l is known, but usually one is satisfied to determine k_δ/l^2.

[53] The frequencies used are those given in Table 37. About the units see equation (II, 167) and the accompanying discussion.

[53a] Our k_δ is one half of Penney and Sutherland's k_α.

From the above it appears that the *valence force field is much superior to the central force field in the case of* XY_2 *molecules.* Conversely the former may be used with greater confidence for the correlation of the observed frequencies to the three normal vibrations of XY_2 molecules.

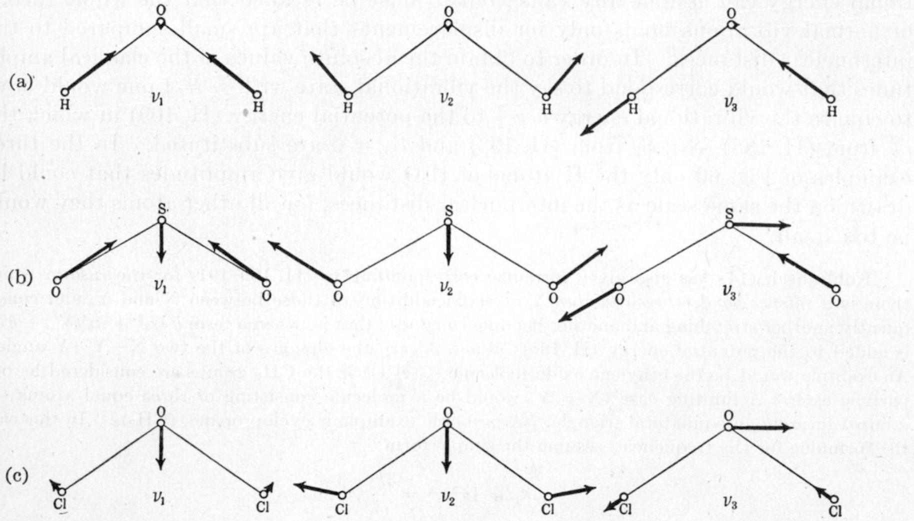

FIG. 60. Actual form of the normal vibrations of H_2O, SO_2 and Cl_2O.—The diagrams give the correct relative amplitudes of the nuclei. The scale of amplitudes is much larger than the scale of internuclear distances if the state $v_i = 1$ is considered. No attempt is made to have the same scale in different diagrams, that is, the vibrations are not normalized.

In order to obtain the *actual form of the normal vibrations* ν_1 and ν_2, which are not uniquely determined by symmetry, we have to superimpose the symmetry coordinates S_1 and S_2 of Fig. 55a in the ratio $(c_{22} - \lambda_i d_{22}) : -(c_{21} - \lambda_i d_{21})$ (see p. 150), where $\lambda_i = \lambda_1$ or λ_2. Substituting the c_{ik} and d_{ik} from (II, 188) and (II, 120) respectively, we obtain the ratio

$$S_1 : S_2 = \left(2pk_1 \cos^2\alpha + 4p\frac{k_\delta}{l^2}\sin^2\alpha - 2\lambda_i m_Y \right) : \\ -2\sin\alpha\cos\alpha\left(k_1 - 2\frac{k_\delta}{l^2}\right), \quad \text{(II, 193)}$$

where $p = 1 + (2m_Y/m_X)$. The following numerical values result from this formula and the data of Table 40.

For H_2O:

$$\nu_1, \quad S_1 : S_2 = 1.20 : 1; \qquad \nu_2, \quad S_1 : S_2 = -0.875 : 1.$$

For SO_2:

$$\nu_1, \quad S_1 : S_2 = 1.40 : 1; \qquad \nu_2, \quad S_1 : S_2 = -1.34 : 1.$$

For Cl_2O:

$$\nu_1, \quad S_1 : S_2 = -1.025 : 1; \qquad \nu_2, \quad S_1 : S_2 = -4.88 : 1.$$

In Fig. 60 the superposition is shown for these three cases. The vibrations ν_3 are also given to scale. Essentially the same form for ν_1 and ν_2 would have been obtained on the basis of the central force system, as can be seen from equation (II, 129) in which $a_{12} = a_{13} = 0$.

Fig. 60 gives only the correct *relative* amplitudes of the nuclei. Their absolute values depend on the magnitude of the vibrational energy. Classically this vibrational energy can assume any value, but it must be recalled that the whole theory of normal vibrations holds only for displacements that are small compared to the internuclear distances. In order to obtain the absolute values of the classical amplitudes that would correspond to say the vibrational state with $v_i = 1$ one would have to equate the vibrational energy $hc\nu_i\frac{3}{2}$ to the potential energy (II, 109) in which the c_{ik} from (II, 188), $S_1 : S_2$ from (II, 193) and $S_3 = 0$ are substituted. In the three examples of Fig. 60 only the H atoms of H_2O would give amplitudes that could be drawn on the same scale as the internuclear distances, for all other atoms they would be too small.

Kohlrausch (14) has also given formulae corresponding to (II, 189–191) for the case in which there is a *valence bond between the two* Y *atoms* in addition to those between X and Y, and consequently another stretching and another bending constant; that is, when a term $k_2Q_3^2 + k_\delta'(\delta'^2 + \delta''^2)$ is added to the potential energy (II, 186); δ' and δ'' are the changes of the two X—Y—Y angles. An example would be the ethylene oxide molecule (C_2H_4O) if the CH_2 groups are considered as one particle each. A limiting case (X = Y) would be a molecule consisting of three equal atoms (or groups) forming an equilateral triangle. A probable example is cyclopropane, $(CH_2)_3$. In this case the formulae for the frequencies assume the simple form:

$$\lambda_1 = 4\pi^2\nu_1^2 = \frac{3k_1}{m},$$

$$\lambda_2 = \lambda_3 = 4\pi^2\nu_2^2 = 4\pi^2\nu_3^2 = \frac{3}{2m}\left(k_1 + 3\frac{k_\delta}{l^2}\right). \qquad \text{(II, 194)}$$

Two of the frequencies coincide (see p. 85). These formulae could, of course, easily be obtained directly by using the symmetry coordinates (normal coordinates) given in Fig. 32a.

Application to linear XY_2 molecules. A great advantage of the valence force field is that it leads to reasonable results even for linear molecules. Unlike (II, 166), equation (II, 191) does not contain the factor $\cos^2\alpha$ and therefore does not require one of the two frequencies ν_1 or ν_2 to be equal to zero for $\alpha = 90°$. One obtains in this case for the frequencies of the three normal vibrations (see Fig. 25b) from (II, 189–191)

$$\lambda_1 = 4\pi^2\nu_1^2 = \frac{k_1}{m_Y}, \qquad \text{(II, 195)}$$

$$\lambda_2 = 4\pi^2\nu_2^2 = \frac{2}{m_Y}\left(1 + \frac{2m_Y}{m_X}\right)\frac{k_\delta}{l^2}, \qquad \text{(II, 196)}$$

$$\lambda_3 = 4\pi^2\nu_3^2 = \left(1 + \frac{2m_Y}{m_X}\right)\frac{k_1}{m_Y}. \qquad \text{(II, 197)}$$

These formulae are, of course, identical with those obtained from the previous general formulae (II, 152) if one puts

$$a_{11} = k_1, \qquad a_{12} = 0, \qquad a_{33} = k_\delta.$$

In Table 41 are given the observed frequencies for the linear molecules CO_2 and CS_2 and the values for k_1 and k_δ/l^2 derived from (II, 195–197). k_1 may be obtained from ν_1 as well as from ν_3. The quality of the agreement between the two gives an indication of the quality of the approximation supplied by the valence force field for these molecules.

If it were not known that CO_2 and CS_2 are linear, one would obtain from (II, 192) the angles 156° and 164° respectively.

TABLE 41. OBSERVED FUNDAMENTAL FREQUENCIES AND FORCE CONSTANTS OF LINEAR SYMMETRIC MOLECULES XY_2.

Molecule	ν_1 (cm^{-1})	ν_2 (cm^{-1})	ν_3 (cm^{-1})	k_1 from ν_1	k_1 from ν_3	k_δ/l^2
CO_2	1337[54]	667	2349	16.8	14.2	0.57
CS_2	657	397	1523	8.1	6.9	0.234
					$\times 10^5$ dynes/cm	

Application to linear XYZ molecules. In the case of an unsymmetric linear triatomic molecule (XYZ) such as HCN, ClCN, N_2O, there are *two different force constants* k_1 and k_2 corresponding to the *stretching* of the two different bonds (H—C and C≡N for HCN) as well as *one constant* k_δ corresponding to the *bending* of the molecule. A calculation similar to the one described above for non-linear symmetric molecules leads to the following *formulae for the frequencies* [see Lechner (562), Bauermeister and Weizel (129), and Rosenthal (749)]:

$$\lambda_1 + \lambda_3 = 4\pi^2(\nu_1^2 + \nu_3^2) = k_1\left(\frac{1}{m_X} + \frac{1}{m_Y}\right) + k_2\left(\frac{1}{m_Y} + \frac{1}{m_Z}\right), \quad \text{(II, 198)}$$

$$\lambda_1\lambda_3 = 16\pi^4\nu_1^2\nu_3^2 = \frac{m_X + m_Y + m_Z}{m_X m_Y m_Z} k_1 k_2, \quad \text{(II, 199)}$$

$$\lambda_2 = 4\pi^2\nu_2^2 = \frac{1}{l_1^2 l_2^2}\left(\frac{l_1^2}{m_Z} + \frac{l_2^2}{m_X} + \frac{(l_1 + l_2)^2}{m_Y}\right) k_\delta. \quad \text{(II, 200)}$$

Here l_1 and l_2 are the distances of X and Z respectively from Y. In order to evaluate the three force constants all three frequencies are necessary and no check is possible. The bending constant k_δ can be evaluated only when the internuclear distances are known.

Table 42 gives for a number of such linear unsymmetrical molecules the observed frequencies and the force constants calculated from them according to (II, 198–200), assuming the internuclear distances given in the columns headed l_1 and l_2, which are obtained from other data. In agreement with expectation, $k_\delta/(l_1 l_2)$ is always much smaller than either k_1 or k_2, indicating a much smaller resistance to bending than to stretching.

Fig. 61 gives as an example the *form of the normal vibrations* for HCN and ClCN. It should be noted that the two outer atoms oscillate in opposite directions for the

[54] This is the average of the two observed frequencies which are due to Fermi resonance (see p. 217).

lower frequency ν_1 and in the same direction for the higher frequency ν_3. The direction of motion of the central atom in ν_1 is opposite in the two cases because of the difference in the ratio of the masses. Correspondingly in HCN it is ν_1 in which mainly the C—N bond is stretched, in ClCN it is ν_3.

TABLE 42. FUNDAMENTAL FREQUENCIES AND FORCE CONSTANTS OF LINEAR UNSYMMETRIC MOLECULES, AFTER PENNEY AND SUTHERLAND (692).[55a]

Molecule	ν_1	ν_2	ν_3	l_1	l_2	k_1	k_2	k_δ/l_1l_2	k_δ
HCN	2089	712	3312	1.06	1.15	5.8	17.9	0.20	0.25
ClCN	729	397	2201	1.76	1.15	5.2	16.7	0.20	0.40
BrCN	580	368	2187	1.93	1.15	4.2	16.9	0.17	0.37
ICN	470	321	2158	2.12	1.15	3.0	16.7	0.12	0.30
SCN^-	750	398	2066	1.54	1.15	5.4	14.4	0.20	0.36
SCO	859	527	2079	1.54	1.16	8.0	14.2	0.37	0.66
NNO[55]	1285	589	2224	1.15	1.23	14.6	13.7	0.49	0.69
	cm^{-1}			$\times 10^{-8}$ cm		$\times 10^5$ dynes/cm			$\times 10^{-11}$ dynes · cm / radian

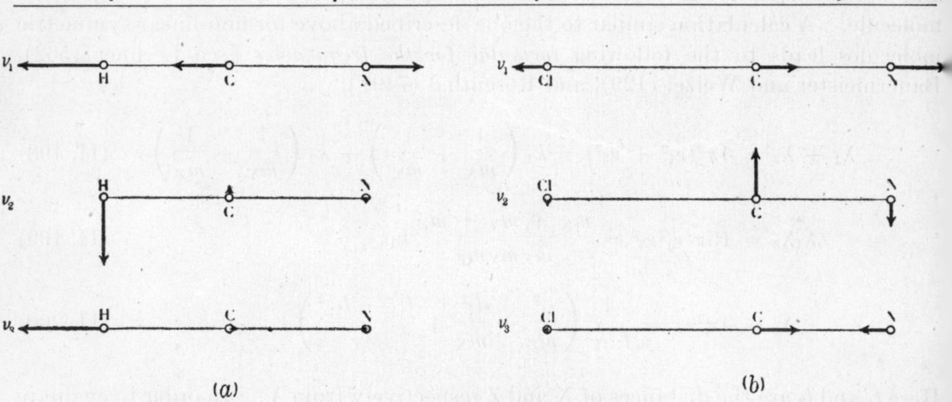

(a) (b)

FIG. 61. **Actual form of the normal vibrations of (a) HCN and (b) ClCN.**—See caption of Fig. 60. Some displacements are so small compared to the others that the corresponding arrows fall within the circle representing the particular nucleus.

The numbering of the vibrations used in Table 42 and Fig. 61 is the one used by most authors. For various reasons it was not changed to a numbering consistent with our usual scheme (see p. 271).

Application to non-linear XYZ molecules. Lechner (562) and Cross and Van Vleck (250) have also calculated the frequencies for an unsymmetrical non-linear triatomic molecule under the assumption of valence forces [see also Wilson (940)]. We refrain from giving their formulae [see

[55] The equations (II, 198–199) lead to complex values for k_1 and k_2 in this case. The values given have been obtained by a slight adjustment of the observed frequencies.

[55a] Our k_δ is one fourth of Penney and Sutherland's k_α even though their k_α is defined in the same way as our k_δ. This appears to be due to an error in their paper.

Kohlrausch (14)] but give in Fig. 62 the normal modes of ethyl chloride (C_2H_5Cl) as calculated by Cross and Van Vleck considering the CH_3 and CH_2 groups as mass points, that is, the C_2H_5Cl as a three-particle system.

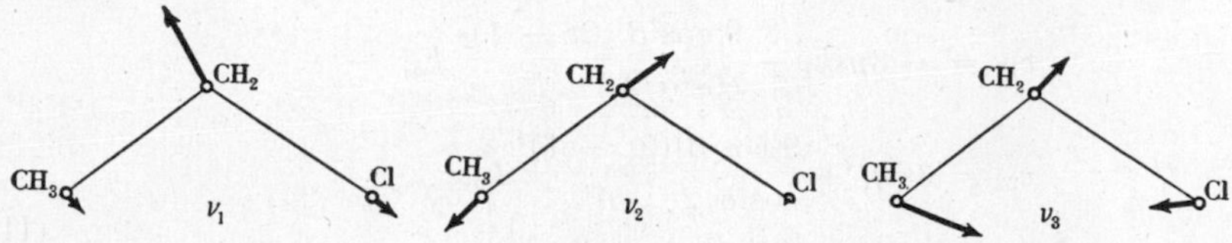

FIG. 62. **Actual form of normal vibrations of C_2H_5Cl considered as a three-particle system** [after Cross and Van Vleck (250)].

Application to pyramidal XY_3 molecules. In the valence force treatment of pyramidal XY_3 molecules, instead of assuming a central force between each pair of Y atoms we now assume a force that tends to restore the angle α between each pair of XY bonds. Thus the *potential energy* is

$$2V = k_1(Q_{12}^2 + Q_{13}^2 + Q_{14}^2) + k_\delta(\delta_{23}^2 + \delta_{24}^2 + \delta_{34}^2), \qquad \text{(II, 201)}$$

where, as in the central-force treatment, Q_{12}, Q_{13}, and Q_{14} are the changes of the XY distances and where the δ_{ik} are the changes of the angles between the lines XY_i and XY_k. k_1 and k_δ are the force constants.

We have now to express the Q_{ik} and δ_{ik} in terms of the symmetry coordinates introduced previously. For the Q_{ik} the previous formulae (II, 169) hold. A calculation similar to the one given above for the XY_2 molecule leads to the following expressions for the δ_{ik}:

$$\begin{aligned}\delta_{23} = \delta_{24} = &-\frac{\sqrt{3}\cos\beta}{\cos\alpha/2}\frac{s}{l}S_1 + \frac{\sqrt{3}(2r-1)\cos\beta}{l\cos\alpha/2}S_2 \\ &+ \frac{3}{4}\frac{\sqrt{3}}{l}\frac{m_Y}{m_X}\frac{\sin^3\beta}{\cos\alpha/2}S_{3a} + \frac{\sqrt{3}(1+\cos^2\beta)}{4l\cos\alpha/2}S_{4a}, \\ \delta_{34} = &-\frac{\sqrt{3}\cos\beta}{\cos\alpha/2}\frac{s}{l}S_1 + \frac{\sqrt{3}(2r-1)\cos\beta}{l\cos\alpha/2}S_2 \\ &- \frac{3}{2}\frac{\sqrt{3}}{l}\frac{m_Y}{m_X}\frac{\sin^3\beta}{\cos\alpha/2}S_{3a} - \frac{\sqrt{3}(1+\cos^2\beta)}{2l\cos\alpha/2}S_{4a}.\end{aligned} \qquad \text{(II, 202)}$$

Here r and s are given by (II, 158), l is the equilibrium distance between X and Y, α is the equilibrium angle between two XY bonds, and β is the equilibrium angle between an X—Y bond and the axis of symmetry of the molecule [$\sin\beta = (2/\sqrt{3})\sin\alpha/2$].

Substituting the Q_{ik} from (II, 169) and the δ_{ik} from (II, 202) into the expression (II, 201) for the potential energy, and comparing with (II, 153), the following formulae for the *potential constants* c_{ik} are obtained:

$$c_{11} = 3p^2k_1 + \frac{9\cos^2\beta}{\cos^2\alpha/2}\frac{s^2}{l^2}k_\delta,$$

$$c_{12} = -3psk_1 - \frac{9\cos^2\beta}{\cos^2\alpha/2}\frac{(2r-1)s}{l^2}k_\delta,$$

$$c_{22} = 3s^2k_1 + \frac{9\cos^2\beta}{\cos^2\alpha/2}\frac{(2r-1)^2}{l^2}k_\delta,$$

$$c_{33} = 6r^2k_1 + \frac{81}{8}\frac{m_Y^2}{m_X^2}\frac{\sin^6\beta}{\cos^2\alpha/2}\frac{k_\delta}{l^2}, \tag{II, 203}$$

$$c_{34} = -3rk_1\sin\beta + \frac{27}{8}\frac{m_Y}{m_X}\frac{(1+\cos^2\beta)\sin^3\beta}{\cos^2\alpha/2}\frac{k_\delta}{l^2},$$

$$c_{44} = \tfrac{3}{2}k_1\sin^2\beta + \frac{9}{8}\frac{(1+\cos^2\beta)^2}{\cos^2\alpha/2}\frac{k_\delta}{l^2}.$$

The terms depending on k_1 in these equations are, of course, the same as those with a_1 in the previous equations (II, 170) based on the central force field.

If now the c_{ik} of (II, 203) and the d_{ik} of (II, 157) are substituted into the factored secular equation (II, 155), the following equations for the *normal frequencies* are obtained [noting that $\cos^2\alpha/2 = \frac{1}{4}(1+3\cos^2\beta)$]:

$$\lambda_1 + \lambda_2 = \left(1 + 3\frac{m_Y}{m_X}\cos^2\beta\right)\frac{k_1}{m_Y} + \left(1 + 3\frac{m_Y}{m_X}\sin^2\beta\right)\frac{12\cos^2\beta}{(1+3\cos^2\beta)}\frac{k_\delta}{m_Y l^2}, \tag{II, 204}$$

$$\lambda_1\lambda_2 = \frac{12\cos^2\beta}{1+3\cos^2\beta}\left(3\frac{m_Y}{m_X}+1\right)\frac{k_1}{m_Y^2}\cdot\frac{k_\delta}{l^2}, \tag{II, 205}$$

$$\lambda_3 + \lambda_4 = \left(1 + \frac{3}{2}\frac{m_Y}{m_X}\sin^2\beta\right)\frac{k_1}{m_Y} + \frac{3\left(1+\cos^2\beta+\frac{3}{2}\frac{m_Y}{m_X}\sin^4\beta\right)}{1+3\cos^2\beta}\frac{k_\delta}{m_Y l^2}, \tag{II, 206}$$

$$\lambda_3\lambda_4 = \frac{3\left(1+\cos^2\beta+3\frac{m_Y}{m_X}\sin^2\beta\right)}{1+3\cos^2\beta}\frac{k_1}{m_Y^2}\frac{k_\delta}{l^2}. \tag{II, 207}$$

These equations were first derived by Lechner (563). If all four frequencies are observed we have thus four equations for the two unknown force constants k_1 and k_δ/l^2, just as in the case of the assumption of central forces

In Table 43 are given for the same molecules as in Table 38 the force constants k_1 and k_δ/l^2, first as evaluated from ν_1 and ν_2 [equations (II, 204 and 205)] and then as evaluated from ν_3 and ν_4 [equations (II, 206 and 207)]. It is seen that the agreement between the two sets of values is quite satisfactory, showing that the assumption of valence forces is a fairly good approximation for the pyramidal XY_3 molecules. As in previous cases k_δ/l^2 is small compared to k_1 in all cases, in agreement with expectation.

If the *angle* β is not known, instead of using the equations (II, 206) and (II, 207) as a check on (II, 204) and (II, 205) one may use (II, 204) and (II, 206) to determine

TABLE 43. FORCE CONSTANTS (IN 10^5 DYNES/CM) FOR THE PYRAMIDAL XY_3 MOLECULES ASSUMING VALENCE FORCES.[56]

Molecule	From ν_1 and ν_2		From ν_3 and ν_4		From $\nu_1, \nu_2, \nu_3, \nu_4$ according to (II, 248)			
	k_1	k_δ/l^2	k_1	k_δ/l^2	k_1	k_δ/l^2	k_1'	k_δ'/l^2
NH_3	6.42	0.41	6.32	0.61	6.35	0.54	+0.03	−0.07
ND_3	6.58	0.44	6.48	0.62	6.51	0.56	+0.03	−0.06
PH_3	3.09	0.33	3.32	0.33	3.24	0.33	−0.07	0
PD_3	3.16	0.35	—					
PF_3	5.38	1.15	4.20	1.04	4.59	1.07	+0.39	+0.04
PCl_3	2.66	0.43	1.85	0.26	2.12	0.32	+0.27	+0.06
PBr_3	1.69	0.41	1.60	0.20	1.63	0.27	+0.03	+0.07
AsF_3	4.61	0.52	3.58	0.35	3.92	0.41	+0.34	+0.06
$AsCl_3$	2.42	0.26	1.84	0.21	2.03	0.23	+0.19	+0.02
$SbCl_3$	2.10	0.21	1.62	0.16	1.78	0.18	+0.16	+0.02
$BiCl_3$	1.49	0.14	1.03	0.09	1.19	0.10	+0.15	+0.02

β. It is easily seen that the formula for β thus obtained is identical with (II, 175) which was derived on the basis of the assumption of central forces.

If the *actual form of the normal vibrations* is calculated according to the previous general equations (II, 159) one obtains for example for NH_3 and ND_3 the result that the symmetry coordinates of Fig. 58 have to be superimposed in the following ratios in order to obtain the normal coordinates:

		NH_3	ND_3
$\nu_1,$	$S_1 : S_2 =$	$1 : -0.0066$	$1 : -0.0135$
$\nu_2,$	$S_1 : S_2 =$	$1 : \ 1.98$	$1 : \ 2.54$
$\nu_3,$	$S_3 : S_4 =$	$1 : \ 0.027$	$1 : \ 0.061$
$\nu_3,$	$S_3 : S_4 =$	$1 : \ 2.54$	$1 : \ 2.81.$

For ND_3 these normal vibrations are the ones represented in the previous Fig. 45. It is seen from the above ratios, as well as by comparing Fig. 58 with Fig. 45, that the symmetry coordinates introduced above are fairly close to the normal coordinates. This is also the reason why these symmetry coordinates were chosen in preference to simpler ones (see p. 155).

Application to plane XY_3 molecules. If we were to apply the equations (II, 204–207) to a plane symmetrical molecule XY_3 (point group D_{3h}), we would obtain $\lambda_1\lambda_2 = 0$ since $\cos\beta = 0$; that is, one of the normal frequencies would be zero. This is obviously the vibration perpendicular to the plane of the molecule (see Fig. 63), since for such a vibration the XY distance and the YXY angles change only in higher order, so that in the present approximation the potential energy (II, 201) would not

[56] For the fundamental frequencies and the values of the angles assumed, see Table 38, p. 164.

change. In order to obtain a non-zero frequency for this vibration one has to introduce an *additional restoring force* that tends to bring the angle between X—Y and the plane back to zero. The addition of such a force is in conformity with the general assumption of valence forces. The *potential energy* then is

$$2V = k_1(Q_{12}^2 + Q_{13}^2 + Q_{14}^2) + k_\delta(\delta_{23}^2 + \delta_{24}^2 + \delta_{34}^2) + k_\Delta(\Delta_{12}^2 + \Delta_{13}^2 + \Delta_{14}^2), \quad \text{(II, 208)}$$

where Δ_{12}, Δ_{13}, Δ_{14} are the deviations of the lines XY from the Y_3 plane.

We can choose the same symmetry coordinates as for the pyramidal molecules (Fig. 58) except that now $\beta = 90°$ and consequently S_1 and S_2 are of different species (A_1' and A_2'' respectively; see Table 22). While the relations for the Q_{ik} and δ_{ik} remain the same as before, we have for the Δ_{ik}

$$\Delta_{12} = \Delta_{13} = \Delta_{14} = \left(\frac{3m_Y}{m_X} + 1\right)\frac{S_2}{l}. \quad \text{(II, 209)}$$

Substituting this and the previous relations (II, 169) and (II, 202) for the Q_{ik} and δ_{ik} into (II, 208), we obtain the same potential constants c_{ik} as before [equations (II, 203)], except that for c_{22} the expression $3[(3m_Y/m_X) + 1]^2 k_\Delta/l^2$ has to be added and, of course, everywhere $\beta = 90°$ has to be substituted. It should be noted that in consequence of the latter substitution we have $c_{12} = 0$, as it should be since S_1 and S_2 now belong to a different symmetry type (see p. 147). Because of this we have simply $\lambda_1 = c_{11}/d_{11}$ and $\lambda_2 = c_{22}/d_{22}$, and thus, using the d_{ik} of (II, 157) with $\beta = 90°$, we obtain for the *frequencies* the simple formulae

TABLE 44. FUNDAMENTAL FREQUENCIES AND FORCE CONSTANTS OF PLANE SYMMETRICAL XY_3 MOLECULES [57] ASSUMING VALENCE FORCES.

Molecule	ν_1	ν_2	ν_3	ν_4	k_1	k_δ/l^2	k_Δ/l^2	$\lambda_3+\lambda_4$ [58]	R.H.S. of (II, 211) [58]
$B^{11}F_3$	888	691	1446	480	8.83	0.37	0.87	1.37	1.88
$B^{11}Cl_3$	471	462	958	243	4.63	0.16	0.42	0.58	0.84
$B^{11}Br_3$	279	372	806	151	3.66	0.13	0.29	0.40	0.55
CO_3^{--}	1063	879	1415	680	10.65	0.51	1.46	1.45	2.29
NO_3^-	1050	831	1390	720	10.39	0.64	1.47	1.44	2.09
SO_3	1069	652	1330	532	10.77	0.46	1.60	1.21	1.33
	cm^{-1}				$\times 10^5$ dynes/cm			$\times 10^5$ dynes/cm	

$$\lambda_1 = \frac{k_1}{m_Y}, \quad \lambda_2 = \left(1 + \frac{3m_Y}{m_X}\right)\frac{k_\Delta}{m_Y l^2}, \quad \text{(II, 210)}$$

$$\lambda_3 + \lambda_4 = \left(1 + \frac{3}{2}\frac{m_Y}{m_X}\right)\left(\frac{k_1}{m_Y} + \frac{3k_\delta}{m_Y l^2}\right), \quad \text{(II, 211)}$$

$$\lambda_3\lambda_4 = 3\left(1 + \frac{3m_Y}{m_X}\right)\frac{k_1}{m_Y^2}\frac{k_\delta}{l^2}, \quad \text{(II, 212)}$$

[57] For the sources of the BF_3 frequencies see Chapter III. The frequencies of BCl_3 and BBr_3 were taken from Anderson, Lasettre, and Yost (52) and those of CO_3^{--} and NO_3^- from Kujumzelis (545), Schaefer, Bormuth, and Matossi (765), and Schaefer and Bormuth (764), those of SO_3 from Gerding and Lecomte (348).

[58] Here λ_i is taken as $4\pi^2c^2M_1\nu_i^2 = 5.8894 \times 10^{-2}\nu_i^2$ (see p. 160) and correspondingly the masses at the right of equation (II, 211) are taken in atomic weight units.

where the last two equations are obtained from (II, 206) and (II, 207) by substituting $\beta = 0$ since (II, 209) does not contain S_3 and S_4. As would be expected, according to (II, 210), ν_1 depends on k_1 only, ν_2 on k_Δ only.

In Table 44 are given the frequencies of a number of plane symmetrical XY_3 molecules and the force constants determined from (II, 210) and (II, 212). The last two columns give the left and right hand side of equation (II, 211) as evaluated with the k_1 and k_δ/l^2 values given. The agreement is fair, but there is a systematic difference. In Fig. 63 as an example the actual form of the normal vibrations of BF_3 is represented.

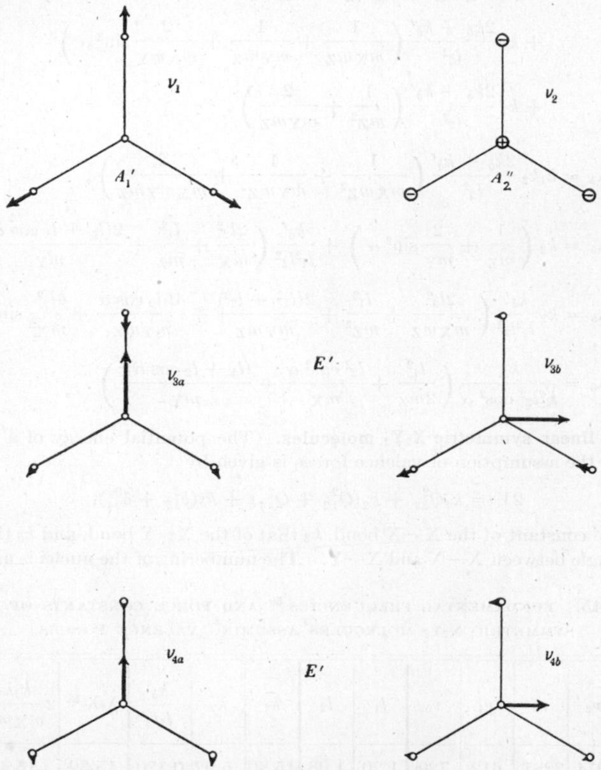

FIG. 63. **Actual form of the normal vibrations of the *BF*₃ molecule.**—See caption of Fig. 60. The forms for ν_3 and ν_4 have been calculated from force constants $k_1 = 5.45$ and $k_2 = 0.60 \times 10^5$ dynes/cm. obtained from the observed frequencies ν_3 and ν_4 according to (II, 211) and (II, 212) and not the k_1 and k_2 values in Table 44 which are obtained from (II, 210) and (II, 212).

Application to plane XYZ_2 molecules. For the plane XYZ_2 molecules, of which formaldehyde (H_2CO) and phosgene (Cl_2CO) are well-known examples, there are six non-degenerate vibrations, five of which are in the plane of the molecule (see Fig. 24a-f). The potential energy, under the assumption of valence forces, is

$$2V = k_1Q_{14}^2 + k_2(Q_{12}^2 + Q_{13}^2) + k_\delta\delta_{23}^2 + k_\delta'(\delta_{24}^2 + \delta_{34}^2) + k_\Delta\Delta^2, \qquad \text{(II, 213)}$$

where the nuclei are numbered as indicated in Fig. 24a. The force constants k_1 and k_2 represent the resistance to stretching of the X—Y and Y—Z bonds respectively, k_δ refers to the change of the angle δ_{23} between the two Y—Z bonds, k_δ' to the change (δ_{24} or δ_{34}) of the angle between the Y—X and a Y—Z bond and k_Δ to the change Δ of the angle between X—Y and the plane of YZ_2. The

Q_{ik} have the same meaning as for pyramidal XY_3 above. A calculation very similar to those carried out above leads to the following results [see Lechner (563), Kohlrausch (14), and Burkard (185) for the oscillations in the plane of the molecule and Fig. 24a-f for the numbering of the oscillations]:

$$\lambda_1 + \lambda_2 + \lambda_3 = k_1\left(\frac{1}{m_X} + \frac{1}{m_Y}\right) + k_2\left(\frac{1}{m_Z} + \frac{2}{m_Y}\cos^2\alpha\right) + \frac{2k_\delta + k_\delta'}{l_2^2}\left(\frac{1}{m_Z} + \frac{2}{m_Y}\sin^2\alpha\right), \quad \text{(II, 214)}$$

$$\lambda_1\lambda_2 + \lambda_2\lambda_3 + \lambda_1\lambda_3 = k_1k_2\left(\frac{1}{m_Xm_Z} + \frac{1}{m_Ym_Z} + \frac{2}{m_Xm_Y}\cos^2\alpha\right) + k_1\frac{2k_\delta + k_\delta'}{l_2^2}\left(\frac{1}{m_Xm_Z} + \frac{1}{m_Ym_Z} + \frac{2}{m_Xm_Y}\sin^2\alpha\right) + k_2\frac{2k_\delta + k_\delta'}{l_2^2}\left(\frac{1}{m_Z^2} + \frac{2}{m_Ym_Z}\right), \quad \text{(II, 215)}$$

$$\lambda_1\lambda_2\lambda_3 = k_1k_2\frac{2k_\delta + k_\delta'}{l_2^2}\left(\frac{1}{m_Xm_Z^2} + \frac{1}{m_Ym_Z^2} + \frac{2}{m_Xm_Ym_Z}\right), \quad \text{(II, 216)}$$

$$\lambda_4 + \lambda_5 = k_2\left(\frac{1}{m_Z} + \frac{2}{m_Y}\sin^2\alpha\right) + \frac{k_\delta'}{l_1^2l_2^2}\left(\frac{2l_2^2}{m_X} + \frac{l_1^2}{m_Z} + \frac{2(l_2 + l_1\cos\alpha)^2}{m_Y}\right), \quad \text{(II, 217)}$$

$$\lambda_4\lambda_5 = k_2\frac{k_\delta'}{l_1^2l_2^2}\left(\frac{2l_2^2}{m_Xm_Z} + \frac{l_1^2}{m_Z^2} + \frac{2(l_1^2 + l_2^2)}{m_Ym_Z} + \frac{4l_1l_2\cos\alpha}{m_Ym_Z} + \frac{4l_2^2}{m_X^2}\sin^2\alpha\right), \quad \text{(II, 218)}$$

$$\lambda_6 = \frac{k_\Delta}{l_1^2l_2^2\cos^2\alpha}\left(\frac{l_1^2}{2m_Z} + \frac{l_2^2\cos^2\alpha}{m_X} + \frac{(l_1 + l_2\cos\alpha)^2}{m_Y}\right). \quad \text{(II, 219)}$$

Application to linear symmetric X_2Y_2 molecules. The potential energy of a linear symmetric X_2Y_2 molecule on the assumption of valence forces is given by

$$2V = k_1Q_{23}^2 + k_2(Q_{12}^2 + Q_{34}^2) + k_\delta(\delta_{13}^2 + \delta_{24}^2). \quad \text{(II, 220)}$$

Here k_1 is the force constant of the X—X bond, k_2 that of the X—Y bond, and k_δ that corresponding to the change of angle between X—X and X—Y. The numbering of the nuclei is as indicated in Fig. 64b. The Q_{ik} are defined as before; δ_{13} is the deviation of the angle between 1–2 and 2–3 from 180° and δ_{24} the corresponding deviation for 2–3–4. Since there is only one each of the vibrations of species Σ_u^+, Π_g, and Π_u (see Table 36), there is for them only one possible choice of symmetry coordinates which are identical with the normal coordinates in Fig. 64a (see p. 150). Possible symmetry coordinates S_1 and S_2 for the two totally symmetric vibrations are given in Fig. 64b.

TABLE 45. FUNDAMENTAL FREQUENCIES [59] AND FORCE CONSTANTS OF LINEAR SYMMETRIC X_2Y_2 MOLECULES ASSUMING VALENCE FORCES.

Molecule	ν_1	ν_2	ν_3	ν_4	ν_5	l_1	l_2	k_1	k_2	$\frac{k_\delta}{l_1l_2}$	$\lambda_1\lambda_2$ [58]	$2\frac{k_1k_2}{m_Xm_Y}$ [60]	$\nu_5$calculated
C_2H_2	3374	1974	3287	612	729	1.20	1.06	15.80	5.92	0.120	15.39	15.46	500
C_2D_2	2700	1762	2427	505	539	1.20	1.06	15.90	5.99	0.116	7.85	7.87	360
C_2N_2	2322	842	2149	506	226	1.46	1.15	5.22	17.60	0.195	1.32	1.09	254
	cm^{-1}					$\times 10^{-8}$ cm		$\times 10^5$ dynes/cm			$\times 10^{10}$ (dynes/cm)^2		cm^{-1}

[58] On p. 178.

[59] For a more detailed discussion of the observations see Chapter III.

[60] m in atomic-weight units.

It is very easy in this case to find the relation between the coordinates Q_{ik} and δ_{ik} and the symmetry coordinates, and to set up the secular equation. We give only the results for the normal frequencies:

$$\lambda_1 + \lambda_2 = 2\frac{k_1}{m_X} + \left(1 + \frac{m_Y}{m_X}\right)\frac{k_2}{m_Y}, \quad \text{(II, 221)}$$

$$\lambda_1\lambda_2 = 2\frac{k_1}{m_X}\frac{k_2}{m_Y}, \quad \text{(II, 222)}$$

$$\lambda_3 = \left(1 + \frac{m_Y}{m_X}\right)\frac{k_2}{m_Y}, \quad \text{(II, 223)}$$

$$\lambda_4 = \frac{k_\delta}{l_1^2 l_2^2}\left[\frac{l_1^2}{m_Y} + \frac{(l_1 + 2l_2)^2}{m_X}\right], \quad \text{(II, 224)}$$

$$\lambda_5 = \left(1 + \frac{m_Y}{m_X}\right)\frac{k_\delta}{m_Y l_2^2}. \quad \text{(II, 225)}$$

In Table 45 are given the observed fundamental frequencies of the three linear molecules C_2H_2, C_2D_2, and C_2N_2, and the *force constants* k_1, k_2, k_δ/l_1l_2 for these molecules as obtained from (II, 221), (II, 223), and (II, 224), assuming the internuclear distances given. The last three columns give the left- and right-hand sides of (II, 222) and the value of ν_5 as calculated from (II, 225), assuming the k_δ/l_1l_2 obtained from ν_4. It is seen that for the acetylenes equation (II, 222) is quite well fulfilled but the calculated values of ν_5 are in very poor agreement with the observed, while for cyanogen (II, 222) is poorly fulfilled but the agreement for ν_5 is fairly good.

The *form of the normal vibrations* ν_1 *and* ν_2 which is not determined by symmetry can be obtained in the same way as described earlier for non-linear XY_2 (see p. 171). The result for C_2N_2 is illustrated in Fig. 64a, in which also the other three normal vibrations are drawn to scale for this case. It is seen that ν_2 is essentially a C—C vibration whereas ν_1 is a C—N vibration.

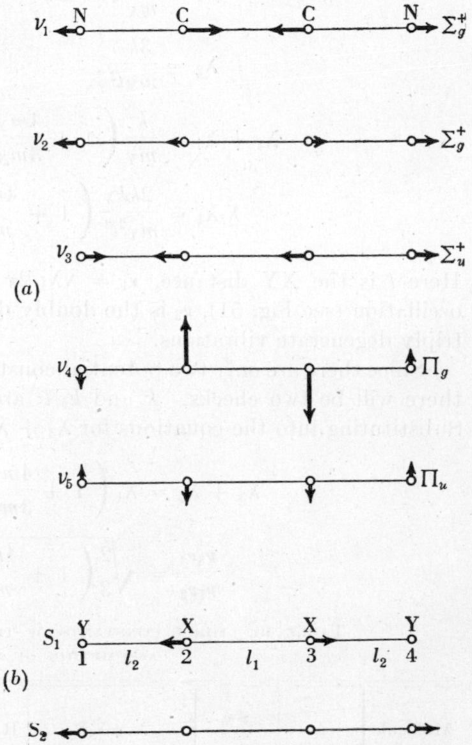

FIG. 64. **Normal coordinates and symmetry coordinates of linear symmetric X_2Y_2.**—The normal vibrations in (a) are drawn to scale for C_2N_2 (see Table 45). The two displacement vectors of X in ν_3 and ν_5 are m_Y/m_X times the displacement vectors of Y, those of X in ν_4 are $m_Y/m_X[2(l_2/l_1) + 1]$ times those of Y. In (b) two convenient symmetry coordinates of species Σ_g^+ are shown. The symmetry coordinates for the other species are identical with the normal coordinates in (a).

Application to tetrahedral XY_4 molecules. The potential energy of tetrahedral XY_4 molecules on the assumption of valence forces is given by

$$2V = k(r_1^2 + r_2^2 + r_3^2 + r_4^2) + k_\delta(\delta_{12}^2 + \delta_{13}^2 + \delta_{14}^2 + \delta_{23}^2 + \delta_{24}^2 + \delta_{34}^2), \quad \text{(II, 226)}$$

where the force constant k corresponds to the change of length r_i of the XY bond and k_δ corresponds to the change δ_{ik} of the angle YXY. Unlike the case of central forces, here all individual forces are zero in the equilibrium position. Correspondingly there are only two rather than three force constants. Introduction of symmetry coordinates and setting up of the secular equation leads to the following equations for the *four fundamental frequencies* [Lechner (564) quoted by

Kohlrausch (14)]:

$$\lambda_1 = \frac{k}{m_Y}, \tag{II, 227}$$

$$\lambda_2 = \frac{3k_\delta}{m_Y l^2}, \tag{II, 228}$$

$$\lambda_3 + \lambda_4 = \frac{k}{m_Y}\left(1 + \frac{4m_Y}{3m_X}\right) + \frac{2k_\delta}{m_Y l^2}\left(1 + \frac{8m_Y}{3m_X}\right), \tag{II, 229}$$

$$\lambda_3\lambda_4 = \frac{2kk_\delta}{m_Y{}^2 l^2}\left(1 + \frac{4m_Y}{m_X}\right). \tag{II, 230}$$

Here l is the XY distance, $\nu_1 = \sqrt{\lambda_1}/2\pi$ is the totally symmetric non-degenerate oscillation (see Fig. 51), ν_2 is the doubly degenerate vibration, and ν_3 and ν_4 are the triply degenerate vibrations.

Since there are only two potential constants but four equations for the frequencies, there will be two checks. k and k_δ/l^2 are most easily determined from λ_1 and λ_2. Substituting into the equations for $\lambda_3 + \lambda_4$ and $\lambda_3\lambda_4$, we obtain

$$\lambda_3 + \lambda_4 = \lambda_1\left(1 + \frac{4m_Y}{3m_X}\right) + \tfrac{2}{3}\lambda_2\left(1 + \frac{8m_Y}{3m_X}\right) \tag{II, 231}$$

$$\frac{\nu_3\nu_4}{\nu_1\nu_2} = \sqrt{\frac{2}{3}\left(1 + \frac{4m_Y}{m_X}\right)}. \tag{II, 232}$$

TABLE 46. FORCE CONSTANTS OF TETRAHEDRAL MOLECULES XY_4 ON THE ASSUMPTION OF VALENCE FORCES.

Molecule	k	$\frac{k_\delta}{l^2}$	$\lambda_3+\lambda_4$[58]	R.H.S. (II, 231)	$\frac{\nu_3\nu_4}{\nu_1\nu_2}$	$\sqrt{\frac{2}{3}\left(1+\frac{4m_Y}{m_X}\right)}$
CH_4	5.04	0.461	6.38	6.69	0.89	0.94
CD_4	5.16	0.439	3.59	3.76	1.02	1.06
SiH_4	2.84	0.189	3.29	3.36	0.93	0.87
NH_4^+	5.46	0.562	6.93	7.27	0.86	0.93
CF_4	9.14	0.713	1.18	1.89	2.02	2.21
SiF_4	7.16	0.252	0.72	0.79	2.06	1.57
CCl_4	4.38	0.331	0.41	0.78	2.44	2.92
$SiCl_4$	3.75	0.157	0.25	0.32	2.11	2.01
$TiCl_4$	3.11	0.100	0.156	0.191	1.51	1.63
$GeCl_4$	3.27	0.125	0.138	0.169	1.47	1.40
$SnCl_4$	2.80	0.075	0.106	0.118	1.42	1.21
CBr_4	3.36	0.237	0.29	0.53	3.74	4.29
$SiBr_4$	2.92	0.127	0.151	0.20	2.98	2.87
$GeBr_4$	2.58	0.095	0.071	0.089	1.99	1.90
$SnBr_4$	2.28	0.064	0.050	0.059	1.74	1.57
SO_4^{--}	9.07	0.639	0.94	1.13	1.53	1.41
ClO_4^-	8.24	0.670	0.95	1.01	1.60	1.37
PO_4^{---}	9.05	0.414	0.85	1.08	1.57	1.43
	$\times 10^5$ dynes/cm		$\times 10^5$ dynes/cm			

[58] On p. 178.

Table 46 gives for the same molecules as in Table 39 the force constants k and k_δ/l^2 and the left- and right-hand sides of equations (II, 231) and (II, 232). The agreement is good for the hydrides but very poor for CF_4, CCl_4, and CBr_4, and fair for the other molecules given. On the whole, the consistency for the valence force system in the case of tetrahedral molecules is at least as good as for the central force system (Table 39), even though the former uses only two adjustable force constants as compared to three for the latter. But a refinement of the valence force field (see below) leads to a still better agreement.

Application to plane X_2Y_4 (point group V_h). For ethylene-like molecules X_2Y_4 with bonds between X—X and X—Y, the potential energy, under the assumption of valence forces, is given by

$$2V = k_1Q_{34}^2 + k_2(Q_{13}^2 + Q_{23}^2 + Q_{45}^2 + Q_{46}^2) + k_\delta(\delta_{12}^2 + \delta_{56}^2) + k_\delta'(\delta_{14}^2 + \delta_{24}^2 + \delta_{35}^2 + \delta_{36}^2) + k_\beta(\beta_{12}^2 + \beta_{56}^2) + k_\gamma\gamma^2. \quad \text{(II, 233)}$$

Here the Q_{ik}, as previously, are the changes of the distances between nuclei i and k (with the numbering used in Fig. 57), δ_{ik} are the changes of the valence angles involving nucleus i and k, β_{12} and β_{56} are the angles between the XY_2 planes and the XX axis, and γ is the angle of torsion of the two XY_2 about the XX axis. k_1 and k_2 are the stretching force constants of the X—X and X—Y bonds respectively, k_δ and k_δ' are the bending force constants corresponding to the Y—X—Y and Y—X—X angles respectively, k_β is the force constant corresponding to the bending of the plane of the molecule and k_γ is the torsional force constant.

If the Q_{ik}, δ_{ik}, β_{ik}, and γ are expressed in terms of the symmetry coordinates of Fig. 57, the potential constants c_{ik} in (II, 131) can be obtained. Substituting these and the d_{ik} from (II, 142) into (II, 133), one finds for the three vibrations of species A_g:

$$\lambda_1 + \lambda_2 + \lambda_3 = \frac{2k_1}{m_X} + \frac{k_2}{m_Y}\left(1 + \frac{2m_Y}{m_X}\cos^2\alpha\right) + \frac{k_\delta' + 2k_\delta}{m_Y \cdot l_2^2}\left(1 + \frac{2m_Y}{m_X}\sin^2\alpha\right),$$

$$\lambda_1\lambda_2 + \lambda_2\lambda_3 + \lambda_1\lambda_3 = \frac{2k_1k_2}{m_Xm_Y} + \frac{k_2(k_\delta' + 2k_\delta)}{m_Y^2l_2^2}\left(1 + \frac{2m_Y}{m_X}\right) + \frac{2k_1(k_\delta' + 2k_\delta)}{m_Xm_Yl_2^2}, \quad \text{(II, 234)}$$

$$\lambda_1\lambda_2\lambda_3 = \frac{2k_1k_2(k_\delta' + 2k_\delta)}{m_Xm_Y^2l_2^2}.$$

Similarly, from (II, 134) one obtains for the vibration of species A_u (torsional oscillation):

$$\lambda_4 = \frac{1}{\sin^2\alpha}\frac{k_\gamma}{m_Yl_2^2}. \quad \text{(II, 235)}$$

From (II, 135) one obtains for the two vibrations of species B_{1g}:

$$\lambda_5 + \lambda_6 = \frac{k_2}{m_Y}\left(1 + \frac{2m_Y}{m_X}\sin^2\alpha\right) + \frac{k_\delta'}{m_Yl_2^2}\left[1 + \frac{2m_Y}{m_X}\left(2\frac{l_2}{l_1} - \cos\alpha\right)^2\right],$$

$$\lambda_5\lambda_6 = \frac{k_2k_\delta'}{m_Y^2l_2^2}\left[1 + \frac{2m_Y}{m_X}\left(1 - 4\frac{l_2}{l_1}\cos\alpha + 4\frac{l_2^2}{l_1^2}\right)\right]. \quad \text{(II, 236)}$$

From (II, 136) and (II, 137) one obtains for the vibrations of species B_{1u} and B_{2g}:

$$\lambda_7 = \frac{1}{2\cos^2\alpha m_Y}\frac{k_\beta}{l_2^2}\left(1 + \frac{2m_Y}{m_X}\right), \quad \text{(II, 237)}$$

$$\lambda_8 = \frac{1}{\cos^2\alpha}\frac{k_\beta}{l_1^2l_2^2}\left[\frac{l_1^2}{m_Y} + \frac{(l_1 + 2l_2\cos\alpha)^2}{m_X}\right]. \quad \text{(II, 238)}$$

From (II, 138) one obtains for the two vibrations of species B_{2u}:

$$\lambda_9 + \lambda_{10} = \frac{k_2}{m_Y}\left(1 + \frac{2m_Y}{m_X}\sin^2\alpha\right) + \frac{k_\delta'}{m_Yl_2^2}\left(1 + \frac{2m_Y}{m_X}\cos^2\alpha\right),$$

$$\lambda_9\lambda_{10} = \frac{k_2k_\delta'}{m_Y^2l_2^2}\left(1 + \frac{2m_Y}{m_X}\right). \quad \text{(II, 239)}$$

And finally, from (II, 139), for the two vibrations of species B_{3u}:

$$\lambda_{11} + \lambda_{12} = \frac{k_2}{m_Y}\left(1 + \frac{2m_Y}{m_X}\cos^2\alpha\right) + \frac{k_\delta' + 2k_\delta}{m_Y l_2^2}\left(1 + \frac{2m_Y}{m_X}\sin^2\alpha\right),$$
$$\lambda_{11}\lambda_{12} = \frac{k_2(k_\delta' + 2k_\delta)}{m_Y^2 l_2^2}\left(1 + \frac{2m_Y}{m_X}\right). \qquad \text{(II, 240)}$$

TABLE 47. OBSERVED FUNDAMENTAL FREQUENCIES [61] AND FORCE CONSTANTS OF PLANE SYMMETRIC X_2Y_4 MOLECULES.

	C_2H_4	C_2D_4	C_2Cl_4	N_2O_4	
$\nu_1(a_g)$	3019	2251	1571	1360	
$\nu_2(a_g)$	1623	1515	447	813	
$\nu_3(a_g)$	1342	981	237	283	
$\nu_4(a_u)$	825	—	—	—	
$\nu_5(b_{1g})$	3272[62]	2304	1000	1724	
$\nu_6(b_{1g})$	1050	883	347	500	cm^{-1}
$\nu_7(b_{1u})$	949	720	332?	680	
$\nu_8(b_{2g})$	943	780	512	—	
$\nu_9(b_{2u})$	3106	2345	913	1749	
$\nu_{10}(b_{2u})$	995			380	
$\nu_{11}(b_{3u})$	2990	2200	782	1265	
$\nu_{12}(b_{3u})$	1444	1078	387?	752	
$\lambda_1+\lambda_2+\lambda_3-\lambda_{11}-\lambda_{12}$ [58]	1.50	1.37	1.16	0.25	
$\left(1+\frac{2m_Y}{m_X}\right)\frac{\lambda_1\lambda_2\lambda_3}{\lambda_{11}\lambda_{12}}$ [58]	1.60	1.56	1.24	0.21	
$k_1=\frac{m_X}{2}\left(1+\frac{2m_Y}{m_X}\right)\frac{\lambda_1\lambda_2\lambda_3}{\lambda_{11}\lambda_{12}}$ [58]	9.57	9.39	7.47	1.47	
k_2 [63]	5.08	5.27	3.21	11.06	$\times 10^5$ dynes/cm
$\frac{k_\delta}{l_2^2}$ [64]	0.269	0.338		0.623	
$\frac{k_\delta'}{l_2^2}$ [65]	0.567	0.448		0.108	
$\frac{k_\beta}{l_2^2}$ (from ν_7)	0.229	0.231	1.67	0.663	
$\frac{k_\gamma}{l_2^2}$	0.303				

[58] On p. 178.

[61] For a more detailed discussion of the vibrational assignments for C_2H_4, C_2D_4, and C_2Cl_4, see Chapter III, section 3d. The assignments for N_2O_4 are those of Sutherland (822).

[62] If for ν_5 instead of 3272 the value 3075 cm^{-1} is used (as is done by several investigators), and k_δ'/l_2^2 calculated a much better agreement with the k_δ'/l_2^2 value derived from ν_9 and ν_{10} is obtained. However, this does not seem to be sufficient reason to change the assignment (see Chapter III, section 3d).

[63] For C_2H_4, C_2D_4, and C_2Cl_4 from ν_{11} and ν_{12}, for N_2O_4 from ν_9 and ν_{10}. For C_2Cl_4 a slight adjustment of ν_{11} and ν_{12} was necessary in order to prevent k_2 from becoming complex. For C_2H_4 the k_2 value obtained from ν_9 and ν_{10} agrees exactly with that obtained from ν_{11} and ν_{12}.

[64] For C_2H_4 and C_2D_4 from ν_{11}, ν_{12} and k_δ', for N_2O_4 from ν_1, ν_2, ν_3 and k_δ'.

[65] For C_2H_4 and N_2O_4 from ν_9 and ν_{10}, for C_2D_4 from ν_5 and ν_6.

In the above equations l_1 is the X—X, l_2 the X—Y distance in the equilibrium position and α is half the angle at X in the XY_2 triangle (see Fig. 57). The equations for the vibrations in the plane of the molecule have been given by Kohlrausch (14), for the out-of-plane vibrations by Bernard and Manneback (139); but it should be noted that some of their force constants are defined in a slightly different way.

TABLE 48. MOLECULES TREATED ON THE BASIS OF THE SIMPLE VALENCE FORCE SYSTEM.

Molecule	Point group	References
Non-linear XY_2	C_{2v}	See the preceding text
Linear XY_2	$D_{\infty h}$	
Non-linear XYZ	C_s	
Linear XYZ	$C_{\infty v}$	
Pyramidal XY_3	C_{3v}	
Plane XY_3	D_{3h}	
Plane XYZ_2	C_{2v}	
Linear X_2Y_2	$D_{\infty h}$	
Non-linear X_2Y_2	C_{2v} and C_{2h}	Kohlrausch (14), Trenkler (872)
Linear WXYZ	$C_{\infty v}$	Kohlrausch (14), Lechner (563)
Tetrahedral X_4	T_d	Bhagavantam and Venkatarayudu (153)
Square X_4	D_{4h}	Kohlrausch (14)
Plane X_4	V_h	Trenkler (872)
Tetrahedral XY_4	T_d	See the preceding text
Plane XY_4	D_{4h}	Kohlrausch (14), Wilson (932)
Linear XY_2Z_2	$D_{\infty h}$	Kohlrausch (14)
Non-linear XY_2Z_2	C_{2v} or C_{2h}	Thompson and Linnett (858)
Tetrahedral XY_2Z_2	C_{2v}	Wu (962)
Plane XY_2Z_2	V_h	Kohlrausch (14)
Plane X_5	D_{5h}	Kohlrausch (14), Reitz (735)
Axial XYZ_3	C_{3v}	Kohlrausch (14), Wagner (906)
Plane XYZ_3 (cyclobutanone)	C_{2v}	Trenkler (872)
Plane X_2Y_4	V_h	See the preceding text
HC≡C—C≡CH	$D_{\infty h}$	Glockler and Wall (384)
$X_2(YZ)_2$ (1,3-cyclobutanedione)	V_h	Trenkler (872)
X_6	D_{6h}	Kohlrausch (14), Bossche and Manneback (172) Wilson (930), Manneback (597)
CH_3—C≡CH	C_{3v}	Glockler and Wall (384)
C_2H_6	D_{3h} or D_{3d}	Howard (460)
Trans ClH_2C—CH_2Br	C_s	Mizushima, Morino, and Kozima (630)
X_6Y_2	V_h	Trenkler (871)
$(NH_2)_2CO$	C_{2v}	Kellner (495)
$Y_2XY_2XY_2$ (Al_2Cl_6)	V_h	Kohlrausch and Wagner (532a)
X_6Y_3	D_{3h}	Trenkler (871)
CH_3C≡CCH_3	D_{3h} or D_{3d}	Glockler and Wall (384)
$[(NH_2)_3C]^+$	C_{3h}	Kellner (495)
$C(CH_3)_4$	T_d	Silver (788)

Since there are only six force constants but twelve equations, there are six relations between the frequencies that may serve as a check on the validity of the assumption of valence forces in this case. One of these, which is easily obtained from the above equations, is

$$\lambda_1 + \lambda_2 + \lambda_3 - \lambda_{11} - \lambda_{12} = \left(1 + \frac{2m_Y}{m_X}\right)\frac{\lambda_1\lambda_2\lambda_3}{\lambda_{11}\lambda_{12}} = \frac{2k_1}{m_X}. \tag{II, 241}$$

It serves at the same time to determine k_1. The other force constants can be determined easily from the other equations.

Table 47 gives the observed fundamental frequencies of the molecules C_2H_4, C_2D_4, C_2Cl_4, and N_2O_4, which very probably have the symmetry V_h, as well as the left- and right-hand sides of equation (II, 241). It is seen that the agreement is quite satisfactory. Table 47 gives also some of the force constants obtained from the equations (II, 234–240), assuming an Y—X—Y angle of 120°, and in the case of k_δ'/l_2^2 for C_2H_4 the internuclear distances given in Chapter IV, section 3b. If one and the same force constant is determined from different frequencies, that is, different equations of (II, 234–240), somewhat different values are obtained due to lack of validity of the valence force system. Therefore the frequencies used for each force constant are indicated in the footnotes.

Other molecules. A number of other molecules have been treated on the basis of the valence force field and formulae for the frequencies have been derived. These molecules are collected in Table 48, where references are given to the papers in which they are treated. The table does not include more general treatments (see the following subsection).

(*e*) *Assumption of more general force fields*

While the most general form of the potential energy in most cases contains more constants (quadratic terms) than there are normal frequencies, we have seen in the above that the assumption of central forces or of valence forces leads in general to fewer potential constants than there are frequencies; and therefore one or more of the equations for the frequencies may be used as a check for the validity of the special assumptions made about the forces. Alternatively, one may also use all or part of these additional equations for the *determination of further potential constants*, that is, assume a more general field of force approaching as nearly as possible the most general case. A great deal of work has been done in this direction.

XY_2 molecules. In the case of a *non-linear symmetrical triatomic molecule* XY_2, the assumption of valence or of central forces requires only two potential constants, while there are three frequencies. Therefore, if we start out from the assumption of *central forces* between the three atoms we could introduce an additional force acting on one Y atom when the distance of the other Y atom from X is changed. The potential energy would be, in this case (compare (II, 97)),

$$2V = a_{11}(Q_1^2 + Q_2^2) + a_{33}Q_3^2 + 2a_{12}Q_1Q_2. \tag{II, 242}$$

The resulting formulae for the frequencies are immediately obtained from the general formulae (II, 124–126) by putting $a_{13} = 0$. However, if the observed frequencies (Table 37) are substituted, in general not real but complex values for the force constants are obtained, confirming our previous conclusion that the central force system is not satisfactory for most XY_2 molecules, even if it is modified in the way here described.

If a similar force is introduced in the *valence force system* the potential energy is (compare equation (II, 186)),

$$2V = k_1(Q_1^2 + Q_2^2) + 2k_{12}Q_1Q_2 + k_\delta\delta^2; \tag{II, 243}$$

and in the same way as previously (see p. 168f.), one obtains for the frequencies

$$\lambda_1 + \lambda_2 = \left(1 + \frac{2m_Y}{m_X}\cos^2\alpha\right)\frac{k_1 + k_{12}}{m_Y} + 2\left(1 + \frac{2m_Y}{m_X}\sin^2\alpha\right)\frac{k_\delta}{m_Y l^2},$$

$$\lambda_1\lambda_2 = 2\left(1 + \frac{2m_Y}{m_X}\right)\cdot\frac{k_1 + k_{12}}{m_Y^2}\cdot\frac{k_\delta}{l^2}, \qquad \text{(II, 244)}$$

$$\lambda_3 = \left(1 + \frac{2m_Y}{m_X}\sin^2\alpha\right)\frac{k_1 - k_{12}}{m_Y}.$$

In the case of H_2O these equations lead to the values

$$k_1 = 7.66 \times 10^5, \qquad k_{12} = -0.097 \times 10^5, \qquad \frac{k_\delta}{l^2} = 0.703 \times 10^5 \text{ dynes/cm.}$$

If we apply the equations (II, 244) as previously (p. 172), to the case of a *linear symmetric molecule* ($\alpha = 90°$), we find

$$\lambda_1 = \frac{k_1 + k_{12}}{m_Y},$$

$$\lambda_2 = 2\left(1 + \frac{2m_Y}{m_X}\right)\frac{1}{m_Y}\frac{k_\delta}{l^2}, \qquad \text{(II, 245)}$$

$$\lambda_3 = \left(1 + \frac{2m_Y}{m_X}\right)\frac{(k_1 - k_{12})}{m_Y}.$$

In the case of CO_2 and CS_2 (see Table 41) one obtains $k_1 = 15.5 \times 10^5$ and 7.5×10^5, and $k_{12} = 1.3 \times 10^5$ and 0.6×10^5 dynes/cm respectively, while k_δ/l^2, of course, remains the same as under the assumption of valence forces only.

The smallness of the *interaction constant* k_{12} as compared to k_1 in the cases discussed indicates that the assumption of valence forces is at least a fair approximation [see also Shaffer and Newton (778)]. Therefore also the previous diagrams of the actual form of the oscillations would have to be changed only very slightly.

Instead of introducing as additional force the interaction between Q_1 and Q_2, we may also introduce a central force between the two Y atoms between which no valence force is acting; that is, we have then a *mixture of valence and central forces.* Formulae for this case have been given by Mecke (614).

Pyramidal XY_3 molecules. In the case of four-atomic *pyramidal molecules* XY_3 we have seen that there are only two force constants on the assumption of valence or central forces alone whereas there are four frequencies. We may therefore introduce *two additional force constants.* Naturally there are various possibilities for choosing these force constants, since the most general potential function (II, 153) would have six constants. Starting out from the assumption of valence forces and following Howard and Wilson (462), we choose an additional force between X and one of the Y atoms when the distance of the other Y atoms from X is changed and an additional force tending to change one Y—X—Y angle when the others are changed; that is, we put for the *potential energy* [compare (II, 201) and Fig. 58]

$$2V = k_1(Q_{12}^2 + Q_{13}^2 + Q_{14}^2) + 2k_1'(Q_{12}Q_{13} + Q_{12}Q_{14} + Q_{13}Q_{14})$$
$$+ k_\delta(\delta_{23}^2 + \delta_{24}^2 + \delta_{34}^2) + 2k_\delta'(\delta_{23}\delta_{24} + \delta_{23}\delta_{34} + \delta_{24}\delta_{34}). \qquad \text{(II, 246)}$$

Substituting the δ_{ik} from (II, 202) and the Q_{ik} from (II, 169), we obtain for the potential constants c_{ik} in (II, 153):

$$\begin{aligned}
c_{11} &= 3p^2(k_1 + 2k_1') + \frac{9\cos^2\beta\, s^2}{\cos^2\alpha/2\; l^2}(k_\delta + 2k_\delta'),\\
c_{12} &= -3ps(k_1 + 2k_1') - \frac{9\cos^2\beta\,(2r-1)s}{\cos^2\alpha/2\quad l^2}(k_\delta + 2k_\delta'),\\
c_{22} &= 3s^2(k_1 + 2k_1') + \frac{9\cos^2\beta\,(2r-1)^2}{\cos^2\alpha/2\quad l^2}(k_\delta + 2k_\delta'),\\
c_{33} &= 6r^2(k_1 - k_1') + \frac{81 m_Y^2}{8 m_X^2}\frac{\sin^6\beta}{\cos^2\alpha/2}\frac{k_\delta - k_\delta'}{l^2}\\
c_{34} &= -3r(k_1 - k_1')\sin\beta + \frac{27 m_Y}{8 m_X}\frac{(1+\cos^2\beta)\sin^3\beta}{\cos^2\alpha/2}\frac{(k_\delta - k_\delta')}{l^2},\\
c_{44} &= \tfrac{3}{2}(k_1 - k_1')\sin^2\beta + \frac{9}{8}\frac{(1+\cos^2\beta)^2}{\cos^2\alpha/2}\frac{(k_\delta - k_\delta')}{l^2};
\end{aligned} \qquad \text{(II, 247)}$$

where p, r, and s are given by (II, 158), α is the Y—X—Y angle, and β the angle of XY with the symmetry axis.

Substituting these c_{ik} and the d_{ik} from (II, 157) into (II, 153) gives the following equations for the *frequencies:*

$$\begin{aligned}
\lambda_1 + \lambda_2 &= \left(1 + 3\frac{m_Y}{m_X}\cos^2\beta\right)\frac{k_1 + 2k_1'}{m_Y} + \left(1 + 3\frac{m_Y}{m_X}\sin^2\beta\right)\frac{12\cos^2\beta}{(1+3\cos^2\beta)}\frac{(k_\delta + 2k_\delta')}{m_Y l^2},\\
\lambda_1\lambda_2 &= \frac{12\cos^2\beta}{1+3\cos^2\beta}\left(1 + 3\frac{m_Y}{m_X}\right)\frac{(k_1 + 2k_1')}{m_Y^2}\frac{(k_\delta + 2k_\delta')}{l^2},\\
\lambda_3 + \lambda_4 &= \left(1 + \frac{3}{2}\frac{m_Y}{m_X}\sin^2\beta\right)\frac{(k_1 - k_1')}{m_Y} + \frac{3\left(1 + \cos^2\beta + \dfrac{3}{2}\dfrac{m_Y}{m_X}\sin^4\beta\right)}{(1+3\cos^2\beta)}\frac{(k_\delta - k_\delta')}{m_Y l^2},\\
\lambda_3\lambda_4 &= \frac{3\left(1 + \cos^2\beta + 3\dfrac{m_Y}{m_X}\sin^2\beta\right)}{1+3\cos^2\beta}\frac{(k_1 - k_1')}{m_Y^2}\frac{(k_\delta - k_\delta')}{l^2}.
\end{aligned} \qquad \text{(II, 248)}$$

Comparing these formulae with the previous formulae (II, 204–207), we see that the difference between the k_1 values obtained on the assumption of valence forces from ν_1 and ν_2 on the one hand and from ν_3 and ν_4 on the other is $3k_1'$, and similarly for k_δ. The new values for k_1, k_δ/l^2, k_1', and k_δ'/l^2, as obtained from (II, 248), are given in the last four columns of Table 43. It is seen that in every case k_1' is small compared to k_1, and k_δ' is small compared to k_δ; that is, the simple valence force field is a fairly good approximation in these cases.

Linear X_2Y_2 molecules. In the case of a *linear symmetric four-atomic molecule* X_2Y_2 there are, under the assumption of valence forces, three force constants but five normal frequencies (see p. 180). We may therefore introduce *two further potential constants*, for example, one corresponding to the interaction of adjacent bonds and the other corresponding to the interaction of the two valence angles; that is, we write for the potential energy[66] [compare equation (II, 220)]:

$$2V = k_1Q_{23}^2 + k_2(Q_{12}^2 + Q_{34}^2) + 2k_{12}(Q_{12}Q_{23} + Q_{23}Q_{34}) + k_\delta(\delta_{13}^2 + \delta_{24}^2) + 2k_\delta'\delta_{13}\delta_{24}. \qquad \text{(II, 249)}$$

From this, proceeding in a way entirely similar to that described above for XY_3, we obtain for

[66] A similar expression for the part of the potential energy depending on the Q_{ik} would have been obtained if we had introduced a central force between each Y atom and the non-adjacent X-atom.

the *frequencies* [compare equations (II, 221–225)]:

$$\lambda_1 + \lambda_2 = \frac{2k_1 - 4k_{12}}{m_X} + \left(1 + \frac{m_Y}{m_X}\right)\frac{k_2}{m_Y},$$

$$\lambda_1 \cdot \lambda_2 = \frac{2k_1k_2 - 4k_{12}^2}{m_X m_Y},$$

$$\lambda_3 = \left(1 + \frac{m_Y}{m_X}\right)\frac{k_2}{m_Y}, \qquad \text{(II, 250)}$$

$$\lambda_4 = \frac{k_\delta - k_\delta'}{l_1^2 l_2^2}\left[\frac{l_1^2}{m_Y} + \frac{(l_1 + 2l_2)^2}{m_X}\right],$$

$$\lambda_5 = \left(1 + \frac{m_Y}{m_X}\right)\frac{k_\delta + k_\delta'}{m_Y l_2^2}.$$

Since the most general quadratic potential function contains six independent constants (see p. 148), only one more term would be needed to make (II, 249) the most general expression. This term would be one with $Q_{12}Q_{34}$, that is, the interaction of the two non-adjacent X—Y bonds. Since such an interaction would be expected to be extremely small, the equations (II, 250) should give force constants that lead to a very satisfactory potential energy (II, 249).

In the case of C_2H_2 the following values for the force constants in 10^5 dynes/cm are obtained from the observed frequencies (see Table 45).

$$k_1 = 15.72, \quad k_2 = 5.92, \quad k_{12} = -0.037, \quad \frac{k_\delta}{l_1l_2} = 0.188, \quad \frac{k_\delta'}{l_1l_2} = 0.069.$$

A check is possible by calculating the potential constants for C_2D_2, which should be the same. One obtains for C_2D_2

$$k_1 = 15.85, \quad k_2 = 5.99, \quad k_{12} = -0.026, \quad \frac{k_\delta}{l_1l_2} = 0.188, \quad \frac{k_\delta'}{l_1l_2} = 0.072.$$

The agreement with the potential constants of C_2H_2 is very satisfactory. The remaining differences are almost certainly due to the neglect of anharmonicity (see section 5).

Tetrahedral XY_4 molecules. Much work has been done on *tetrahedral molecules*, which have four fundamental frequencies whereas there are three central force constants and only two valence force constants. Urey and Bradley (882) have first suggested a *combination of the valence and central force fields;* they assume central forces between the corner atoms as well as forces opposing a change of the bond angles. The potential energy is then [compare (II, 182) and (II, 226)]

$$2V = -2a(Q_{12} + Q_{13} + \cdots Q_{34}) + 2\sqrt{6}a(r_1 + r_2 + r_3 + r_4)$$
$$+ k_1(Q_{12}^2 + Q_{13}^2 + \cdots + Q_{34}^2) + k_2(r_1^2 + r_2^2 + r_3^2 + r_4^2) + k_\delta(\delta_{12}^2 + \delta_{13}^2 + \cdots + \delta_{34}^2) \qquad \text{(II, 251)}$$

where Q_{ik}, r_i, and δ_{ik} are explained in Fig. 59 and a is the force constant explained on p. 166. Urey and Bradley have carried through the calculations under the assumption that the force between the corner atoms is a repulsion inversely proportional to the nth power of the distance (like the repulsion between ions in crystals). This leads to a relation between the constants a and k_1 in (II, 251), leaving only three potential constants if a fixed value for n is assumed. Thus there remains one equation between the frequencies as a check for the assumptions about the force field. We refrain from giving the detailed formulae for the frequencies. It will be sufficient to state that Urey and Bradley's assumption about the forces seems to fit the observed frequencies better than either the central or valence force fields. Detailed calculations with the most general force field have been carried out by Rosenthal (747) (748), who has also compared her results with the simpler force fields [see also Dennison (280)].

Plane X_2Y_4 molecules (Sutherland and Dennison's method). Sutherland and Dennison (828) have applied to the X_2Y_4 molecule (assuming point group V_h) a method that seems to be promising also for other cases. They assume the most general potential function for each XY_2 group but then assume an ordinary valence force between the two X atoms.

The potential energy of each XY_2 group may be expressed in terms of its symmetry coordinates, which we call s_1, s_2, and s_3, thus [compare Fig. 56a and equation (II, 109) where they are called S_1, S_2, S_3]:

$$2V_{XY_2} = a_{11}s_1^2 + a_{22}s_2^2 + 2a_{12}s_1s_2 + a_{33}s_3^2. \qquad \text{(II, 252)}$$

If we restrict our considerations to the vibrations in the plane of the molecule (for those perpendicular to the plane, see p. 183) the potential energy of the two XY_2 groups bound by one valence force is given by

$$2V_{X_2Y_4} = a_{11}(s_1'^2 + s_1''^2) + a_{22}(s_2'^2 + s_2''^2) + 2a_{12}(s_1's_2' + s_1''s_2'') + a_{33}(s_3'^2 + s_3''^2) + a_{44}s_0^2 + a_{55}(\sigma_1^2 + \sigma_2^2 + \sigma_5^2 + \sigma_6^2), \quad \text{(II, 253)}$$

where the $'$ and $''$ distinguish the two XY_2 groups, where s_0 is the change of distance of the two X atoms and where σ_1, σ_2, σ_5, and σ_6 are the changes of the angles of the four X—Y bonds with the X—X bond.

We have now to introduce the symmetry coordinates of the X_2Y_4 molecule. As such we use those of Fig. 57. They have been chosen in such a way that wherever possible they are simply a symmetric or antisymmetric combination of the symmetry coordinates of XY_2. Therefore the following relations hold, as can immediately be read from Fig. 57:

$$\begin{aligned} 2s_1' &= S_2 + S_{12}, \\ 2s_1'' &= S_2 - S_{12}, \\ 2s_2' &= S_3 + S_{11}, \\ 2s_2'' &= S_3 - S_{11}, \\ 2s_3' &= S_5 + S_9, \\ 2s_3'' &= S_5 - S_9. \end{aligned} \quad \text{(II, 254)}$$

Furthermore, it is seen from Fig. 57 that

$$s_0 = 2S_1 + \frac{2}{\mu} S_3, \quad \text{(II, 255)}$$

and finally, we have for the changes of angle:

$$\begin{aligned} \sigma_1 &= pS_2 - qS_3 - \left(\frac{1}{a} + p\right) S_5 - rS_6 - pS_9 + tS_{10} - qS_{11} + pS_{12}, \\ \sigma_2 &= pS_2 - qS_3 + \left(\frac{1}{a} + p\right) S_5 + rS_6 + pS_9 - tS_{10} - qS_{11} + pS_{12}, \\ \sigma_5 &= pS_2 - qS_3 + \left(\frac{1}{a} + p\right) S_5 + rS_6 - pS_9 + tS_{10} + qS_{11} - pS_{12}, \\ \sigma_6 &= pS_2 - qS_3 - \left(\frac{1}{a} + p\right) S_5 - rS_6 + pS_9 - tS_{10} + qS_{11} - pS_{12}, \end{aligned} \quad \text{(II, 256)}$$

where

$$\mu = \frac{m_X}{2m_Y}, \; p = \frac{\cos^2 \alpha}{b - a}, \; q = \frac{1 + \dfrac{2m_Y}{m_X}}{b - a} \sin \alpha \cos \alpha,$$
$$r = \frac{1}{a} + \frac{\cos^2 \alpha}{b - a} + \frac{m_X}{2m_Y} \frac{a \cos \alpha}{(b - a)(b - a \sin^2 \alpha)}, \quad t = \frac{\cos^2 \alpha}{b - a} + \frac{m_X}{2m_Y} \frac{1}{(b - a)}. \quad \text{(II, 257)}$$

For explanation of a, b, α, see Fig. 57.

Substituting (II, 254–256) into (II, 253), we obtain

$$\begin{aligned} 2V = {} & 4a_{44}S_1^2 + \left(\frac{a_{11}}{2} + 4a_{55}p^2\right) S_2^2 + \left(\frac{a_{22}}{2} + \frac{4}{\mu^2} a_{44} + 4a_{55}q^2\right) S_3^2 \\ & + \frac{8}{\mu} a_{44}S_1S_3 + (a_{12} + 8a_{55}pq)S_2S_3 \\ & + \left[\frac{a_{33}}{2} + 4a_{55}\left(\frac{1}{a} + p\right)^2\right] S_5^2 + 4a_{55}r^2S_6^2 + 8a_{55}\left(\frac{1}{a} + p\right) rS_5S_6 \\ & + \left(\frac{a_{33}}{2} + 4a_{55}p^2\right) S_9^2 + 4a_{55}t^2S_{10}^2 - 8a_{55}ptS_9S_{10} \\ & + \left(\frac{a_{22}}{2} + 4a_{55}q^2\right) S_{11}^2 + \left(\frac{a_{11}}{2} + 4a_{55}p^2\right) S_{12}^2 + (a_{12} - 8a_{55}pq)S_{11}S_{12}. \end{aligned} \quad \text{(II, 258)}$$

If the coefficients c_{ik} of S_iS_k in this equation and the d_{ik} from (II, 142) are substituted into the equations (II, 133), (II, 135), (II, 138), and (II, 139) the *frequencies* of the nine normal vibrations in the plane of the molecule are obtained in terms of the six *potential constants* a_{11}, a_{22}, a_{12}, a_{33}, a_{44}, and a_{55} of (II, 253), which can thus be determined when the frequencies are known. In addition there would be three checks, one of which might also be used to determine the angle α.

The great advantage of the above described method is that one may now apply the general potential constants a_{11}, a_{22}, a_{12}, and a_{33} of the XY_2 group thus obtained to the discussion of *other molecules containing this group*, for example, the potential constants of CH_2 in C_2H_4 to the molecules H_2CO, $H_2C{=}C{=}CH_2$, and others. Sutherland and Dennison (828) have carried through the calculations only for the "parallel" vibrations of C_2H_4 and H_2CO, that is, the vibrations of symmetry type A_g, B_{3u}, and A_1 respectively. In doing this they neglected the potential constant a_{55} introduced above. Even with this neglect the three equations for the parallel frequencies of H_2CO give three values for the force constant of the C—O bond that agree in a very satisfactory way.

Calculations of the frequencies of X_2Y_4 molecules based on other potential functions have been made by Wilson [quoted by Bonner (163)], Thompson and Linnett (856), Fox and Martin (328), and Manneback and Verleysen (598).

Other molecules. Quadratic potential functions more general than the simple valence force or central force system have been applied to a number of other molecules. Table 49 summarizes these cases and gives references to the original papers.

TABLE 49. MOLECULES TREATED ON THE BASIS OF A MORE GENERAL QUADRATIC POTENTIAL FUNCTION.

Molecule	Point group	References
Non-linear XY_2	C_{2v}	See the preceding text
Linear XY_2	$D_{\infty h}$	See the preceding text
Linear XYZ	$C_{\infty v}$	Engler and Kohlrausch (306), Rosenthal (749)
Pyramidal XY_3	C_{3v}	See the preceding text
Plane XY_3	D_{3h}	Anderson, Lasettre and Yost (52)
Plane XYZ_2	C_{2v}	Silver and Ebers (789)
Linear X_2Y_2	$D_{\infty h}$	See the preceding text
Non-planar non-linear X_2Y_2	C_2	Morino and Mizushima (635), Bailey and Gordon (88)
Tetrahedral XY_4	T_d	See the preceding text
Linear XY_2Z_2 (C_3O_2)	$D_{\infty h}$	Engler and Kohlrausch (306)
Axial XYZ_3	C_{3v}	Voge and Rosenthal (903), Sutherland and Dennison (828), Slawsky and Dennison (796), Linnett (582), Shaffer (777)
Tetrahedral XY_2Z_2	C_{2v}	Wagner (908)
Plane X_2X_4	V_h	See the preceding text
$HC{\equiv}C{-}C{\equiv}CH$	$D_{\infty h}$	Wu and Shen (964), Wu (25)
CH_3CN and CH_3NC	C_{3v}	Linnett (582)
XY_6 (SF_6)	O_h	Eucken and Sauter (314)
$CH_2{=}C{=}CH_2$	V_d	Thompson and Linnett (857), Wu (25)
X_2Y_6 (C_2H_6)	D_{3h} or D_{3d}	Sutherland and Dennison (828), Stitt (810), Linnett (582)
X_8 (S_8)	D_{4d}	Bhagavantam and Venkatarayudu (153)
X_3Y_6 (cyclo-C_3H_6)	D_{3h}	Saksena (754a)
$CH_3{-}CH{=}CH_2$	C_s	Wilson and Wells (945)
X_6Y_6 (C_6H_6)	D_{6h}	Wilson (930), Manneback (597), Bernard, Manneback, and Verleysen (139a)

(*f*) *Intercomparison of force constants in different molecules, characteristic bond frequencies, stretching and bending vibrations, and related matters*

As mentioned before, the calculations described in the preceding subsections have as their ultimate aim the *accurate determination of the field of force*, that is, *of the force constants*, in as many molecules as possible, from the observed fundamental frequencies. The comparison of the different force constants in a given molecule and of the force constants of different molecules with similar groups is very important because this comparison throws light on the *nature of the forces holding the atoms together*, or in other words, on the electronic structure of the molecules.

Invariance of force constants in different molecules. If the restoring force between two atoms is the same in one molecule as in another, one would conclude that the electronic structure of the bond is at least very similar in the two cases. Conversely, if one knows that the electronic structure is the same, one would expect the same force constants. On the basis of an elementary theory of valence one would expect the C—H bond to have essentially the same electronic structure and therefore the same force constant in different molecules, and similarly for the =C=O, —C≡N, and other bonds. This is indeed observed. For example, the C≡N *bond-stretching force constant* in HCN, ClCN, BrCN, ICN, and $(CN)_2$ (see Tables 42 and 45) is approximately 17×10^5 dynes/cm throughout; the C—H *bond-stretching force constant* in HCN, C_2H_2, and C_4H_2 is 5.85×10^5 dynes/cm; *the* C=O *and* C=S *bond-stretching force constants* in SCO (Table 42) are very nearly the same as in CO_2 and CS_2 respectively (Table 41), and similarly in other cases.

However, on closer examination it is found that this *invariance of the force constants holds exactly only if the bond is in similar surroundings;* for example, the C—H stretching force constant in C_2H_4 and H_2CO, where it is adjacent to a double bond, is 5.28; in C_2H_6 and other cases, where it is adjacent to single bonds, it is 4.79; and finally in the free radical it is 4.09, as compared to the above value of 5.85×10^5 dynes/cm when it is adjacent to a triple bond. Similar results are obtained for other bond-stretching force constants and also for the bond-bending force constants. For the latter, the condition of similar surroundings is still more critical, as one would expect. It is not sufficient that the same type of bond (single, double, or triple) be adjacent to the one considered—the atoms at the other end of the adjacent bonds must be the same; for example, the C—H bending constant in CH_2 would be expected to be different from that in CHCl, just as the bending force constant for the C—C—H angle in C_2H_4 is different from that for the H—C—H angle (see p. 183f.). In Table 50 are given the stretching and bending force constants for the more important cases. It is interesting to note that the stretching force constants for the C—C single, double, and triple bonds are approximately in the ratio 1 : 2 : 3.

The above discussion was based on the assumption of valence forces but, as was first pointed out by Sutherland and Dennison (828), it also holds for more general force systems (see p. 189f.). Crawford and Brinkley (240) and others have shown for a number of molecules that by taking over not only the stretching and bending constants *but also the interaction constants* [such as k_{12} in (II, 243) or k_1' and k_δ' in (II, 246)] from other molecules with the same groups some or all of the normal frequencies may be predicted to within 1 or 2 per cent of the observed values. It is clear that such a procedure is of great importance since in this way it is possible

TABLE 50. STRETCHING AND BENDING FORCE CONSTANTS FOR VARIOUS BONDS AND BOND ANGLES.

Bond	Stretching[67] force constant	Bond	Stretching[67] force constant	Bond angle	Bending[67] force constant
≡C—H	5.85*	=C=O	15.5	≡C—H	$0.210^* \cdot r^2_{CH}$
=C—H	5.1	>C=O	12.1	=C<(H)(H)	$0.3_0 \cdot r^2_{CH}$
>C—H	4.79*	C=O($^1\Sigma$)	18.53	—C(H)—H (H)	$0.46^* \cdot r^2_{CH}$
		C=O($^3\Pi$)	11.82		
C—H (radical)	4.09	=C=S	7.5	C=C<(H)(H)	$0.5_1 \cdot r^2_{CH}$
—C≡C—	15.59*	C=S($^1\Sigma$)	8.22	C—C(H)	$0.55^* \cdot r^2_{CH}$
>C=C<	9.6	—O—H	7.66	C—C≡	$0.155^* \cdot r^2_{CC}$
		O—H radical	7.12		
>C—C≡	5.18*			O=C(H)	$1.5 \cdot r^2_{CH}$
		—N—H	6.35		
—C—C—	4.50*	N—H radical	6.03	F—C(H)—	$0.76^* \cdot r^2_{CH}$
C—C (radical)	9.25	>C—F	5.96*	Cl—C(H)—	$0.58^* \cdot r^2_{CH}$
—C≡N	17.73*	>C—Cl	3.64*	Br—C(H)—	$0.52^* \cdot r^2_{CH}$
C≡N (radical)	15.88	C—Cl radical	3.87	I—C(H)—	$0.45^* \cdot r^2_{CH}$
		>C—Br	3.13*		
		>C—I	2.65*		
	$\times 10^5$ dynes/cm		$\times 10^5$ dynes/cm		$\times 10^5$ dynes/cm

[67] The values marked with an asterisk have been calculated by Crawford and Brinkley (240) under the assumption of a more general force field (see text). Values differing from some of these by amounts up to 10 per cent have been given by Linnett (582) (583) and Noether (672) on the basis

to find the correct assignment of the observed frequencies in more complicated molecules or even to obtain approximate values for these frequencies without actual observation.

It must be realized, however, that with increasing accuracy of observations and calculations slight differences in the force constants are to be expected since there will always be some difference in the surroundings of a given group in different molecules. Thus, for, example, while the $>$C—C≡ stretching constant is the same in H_3C—C≡C—H and H_3C—C≡C—CH_3, it seems to be somewhat different in H_3C—C≡N [see Crawford and Brinkley (240) and Linnett (583)]. Relations very similar to those for the force constants hold also for the internuclear distances (see p. 440).

Characteristic bond (group) frequencies. The application of the above considerations to the problem of the correct assignment of the observed vibrational frequencies is greatly simplified by a corollary of the invariance of bond force constants, namely, the *constancy of bond or group frequencies in different molecules*, which was first established by observation and only later explained theoretically on the basis of the constancy of bond force constants. For example, all molecules containing the ≡C—H bond have normal frequencies of about 3300 and 700 cm^{-1} (see, for example, HCN in Table 42 and C_2H_2 in Table 45), all molecules containing the $>$=CH group have normal frequencies of about 3020 cm^{-1}, all molecules containing the $>$C—H group have normal frequencies of about 2960 cm^{-1}, all molecules containing the $>$$CH_2$ or the —CH_3 group have in addition normal frequencies of about 1450 cm^{-1}, and similarly for other cases. Conversely, from the observation of such frequencies it may be concluded that the corresponding group is present, although this should be done with caution on account of the complications mentioned below. Table 51 gives for a number of groups these characteristic vibrational frequencies. The values given usually hold within ± 100 cm^{-1}, in some cases less.

The observation of the characteristic frequencies as indicated in Table 51 has led Mecke (609) (610) (611) to the introduction of the concept of *valence and deformation vibrations* [67a]; that is, the idea that to every bond in a molecule corresponds a vibration in which this bond is stretched and another one of much smaller frequency in which it is bent. This idea seems to be confirmed by the general correspondence between the vibrational frequencies of Table 51 and the force constants of Table 50. We shall in the following refer to these oscillations as *bond-stretching* and *bond-bending vibrations* since the names valence and deformation vibrations appear somewhat ambiguous.

---

of slightly different potential functions.—The values for the diatomic radicals are based on $\Delta G_{\frac{1}{2}}$ rather than ω_e values since the values for the polyatomic molecules are obtained from observed fundamentals rather than zero order frequencies.

[67a] Some authors distinguish these oscillations by the designations $\nu_1, \nu_2, \nu_3 \cdots \delta_1, \delta_2, \delta_3 \cdots$ and in addition the torsion oscillations by $\tau_1, \tau_2, \tau_3 \cdots$. This practice is not adopted here because of the ambiguities to be discussed below (p. 199f.).

At first sight it might appear somewhat puzzling that there should be such characteristic group frequencies, even though there are characteristic force constants, since we know that in a certain normal vibration in general *all* nuclei of a molecule are taking part. But it will be shown below that in many cases the occurrence of characteristic group frequencies is in agreement with theory. On the other hand,

TABLE 51. CHARACTERISTIC FREQUENCIES OF VARIOUS GROUPS.

Group	Bond-stretching vibration	Group	Bond-stretching vibration	Group	Bond-bending vibration
≡C—H	3300	—C≡C—	2050	≡C—H	700
=C—H	3020	C=C	1650	=C(H)H	1100
—C—H	2960	—C—C—	900	H—C—H	1000
—O—H	3680 [67b]	—C—F	1100	C(H)H	1450
—S—H	2570	—C—Cl	650	C(H)(H)H	1450
—N—H	3350	—C—Br	560	C—C≡C	300
C=O	1700	—C—I	500		
—C≡N	2100				
	cm^{-1}		cm^{-1}		cm^{-1}

it must be realized that there are cases in which observed frequencies cannot be classified as belonging to a certain group, or if they belong to a certain group cannot be classified as bond-stretching or bond-bending vibrations.

An *explanation of the occurrence of group frequencies* can easily be given in the cases of the C—H, O—H, N—H vibrations. Since the mass of the hydrogen nucleus is so much smaller than that of the other nuclei in the molecule, the amplitudes of vibrations of the latter will be very much smaller than those of the former. In first approximation we may consider the H nucleus oscillating against an infinitely large mass, and therefore the vibration frequency depends practically only on the

[67b] This figure refers to the gaseous state. In liquids the value is about 3400 cm^{-1} (see p. 334).

force by which the H atom is bound to the rest of the molecule and will be nearly the same for different molecules with the same C—H, or O—H or N—H force constants. According to our previous discussion of a mass suspended by an elastic bar (p. 62), since the H atom is always an end atom, it can only move in the line of the particular bond or perpendicular to it: that is, it can execute only stretching or bending vibrations with frequencies corresponding to the stretching and bending force constants (see above).

If there are two O—H groups as in H_2O, or two ≡C—H groups as in HC≡CH and HC≡C—C≡CH, or two $\diagdown$=C—H groups as in $H_2C{=}O$, or more than two such groups as in C_2H_4, they will of course all vibrate simultaneously and there will be several normal frequencies, but they will all be of nearly the same magnitude as for a single group, differing only by a comparatively small amount, which is an indication of the strength of the interaction of the equivalent groups.

Let us consider as an example the C_2H_2 molecule. Each C—H group, if the rest of the molecule were fixed, would vibrate with the frequency corresponding to the ≡C—H stretching or bending force constant. In the actual molecule, however, the vibration must be symmetric or antisymmetric with respect to the center of symmetry of the molecule; that is, the two C—H groups can oscillate only symmetrically or antisymmetrically with respect to each other. This gives two C—H vibrations in the axis of the molecule and two perpendicular to it (ν_1, ν_3, and ν_4, ν_5 of Fig. 64). Of course in all these four vibrations the C atoms are also moving, but with much smaller amplitude because of their larger mass. In addition there is a C≡C vibration (ν_2 of Fig. 64) in which essentially the C—H groups are moving as if they were rigid, and which therefore has a frequency corresponding to the —C≡C— stretching force constant. The actually observed frequencies (Table 45) are in agreement with these considerations. There are two C—H stretching vibrations of frequency 3374 and 3287 cm^{-1} respectively, two C—H bending vibrations of frequency 612 and 729 cm^{-1} respectively, and one C≡C stretching vibration of frequency 1974 cm^{-1}.

These relations are also borne out by the equations (II, 221–225) for the frequencies, which are based on the assumption of valence forces; the frequency ν_3 depends only on k_2, the C—H stretching force constant [see equation (II, 223)] and if it is considered that $1 + (m_Y/m_X) \approx 1$ it follows from (II, 221) and (II, 222) that

$$\lambda_1 = 4\pi^2\nu_1^2 \approx 2\,\frac{k_1}{m_X}, \qquad \text{(II, 259)}$$

$$\lambda_2 = 4\pi^2\nu_2^2 \approx \frac{k_2}{m_Y}; \qquad \text{(II, 260)}$$

that is, ν_1 corresponds to a C≡C, ν_2 to a C—H stretching vibration. The frequencies are the same as in diatomic molecules C≡C and C—H with the force constants k_1 and k_2 respectively (except that for C—H it is m_H rather than $m_H m_C/(m_H + m_C)$ that is used).

One may also conceive of the two C—H stretching frequencies of C_2H_2 (and similarly the two bending vibrations) as being produced by a *resonance between the two* C—H *oscillators* much the same as the resonance between two coupled pendulums or, in quantum theory, Heisenberg's resonance in helium [see (8), p. 66f.]. The coupling is provided by the C≡C bond. As in the case of the two coupled pendulums, the resultant motion may be considered as a superposition of a symmetric and an antisymmetric vibration of somewhat different frequencies, one somewhat higher and the other somewhat lower than the frequency of one C—H group would be if the rest of the molecule were fixed. The difference in frequency is the larger the larger the coupling. In the present case it is 86 cm^{-1}. In the diacetylene molecule H—C≡C—C≡C—H, the distance of the C—H groups is much larger and therefore the coupling much smaller. One would therefore expect the difference of the two C—H stretching frequencies to be much smaller. Up to now only one vibration has been observed.

If in acetylene one H is replaced by D there is no longer resonance, and therefore there is essentially one C—H and one C—D vibration, as has been shown in more detail by Förster [quoted in (439)] and as is in agreement with the observed frequencies [see Fig. 87, p. 292, and accompanying discussion]. The same considerations apply to HDO, HDS, and other molecules.

In *linear* molecules the bond-stretching and bond-bending vibrations have different symmetry types (they are Σ^+ and Π vibrations[68] respectively). Therefore this distinction can be rigorously made for any mass of the atoms. *For bent molecules this distinction can be made only to a certain approximation and only when the masses of the end atoms of the group are either small or quite different.* Consider, for example, the H_2O molecule, or the CH_2 group in H_2CO. We have seen previously (p. 171) that for these molecules in one of the two totally symmetric oscillations, ν_1, the H atoms move approximately but not exactly along the lines O—H or C—H respectively, in the other, ν_2, perpendicular to these lines, while in the third (antisymmetric) vibration ν_3 they move exactly in these lines (see Fig. 60a). Equations (II, 190) and (II, 191) show indeed that, *if* m_Y *is small compared to* m_X (but only then), ν_1 depends almost entirely on the stretching force constant k_1, ν_2 almost entirely on the bending force constant k_δ. On this basis it is possible to obtain an approximate picture of the form of the oscillations and approximate values for their frequencies without detailed calculations. For H_2O, two O—H stretching frequencies of nearly the same magnitude are expected, one symmetric and the other antisymmetric, and in addition a bending frequency of much smaller magnitude. This is very similar to the situation discussed above for C_2H_2. The frequencies are 3652, 3756, and 1595 cm^{-1}. Three similar frequencies are expected for the CH_2 group, for example in H_2CO. They are found at 2780, 2874, and 1503 cm^{-1} (see Table 51). The form of the vibrations given in the previous Fig. 24 was obtained in the above described way.[69]

Similarly, for NH_3, CH_3, and other XY_3 molecules or groups with $m_Y \ll m_X$ there are three vibrations in which the Y atoms move approximately in the X—Y direction and three in which they move perpendicular to this direction. Of each group of vibrations two are mutually degenerate. In NH_3 the bond-stretching frequencies are 3337 and 3414 cm^{-1}, the bond-bending frequencies 950 and 1628 cm^{-1}; in CH_3Cl the corresponding frequencies are 2966, 3043, and 1355, 1453 cm^{-1}.

If there are *four equivalent bonds* in a molecule there will be four vibrations of similar magnitude in which chiefly these bonds are stretched, and in addition several bending vibrations whose number depends on the number of independent bond angles. For example, in C_2H_4 or $H_2C{=}C{=}CH_2$, in consequence of the coupling, each vibration of the CH_2 group alone (see above) splits into two. In C_2H_4 one is symmetric and the other antisymmetric with respect to the center of symmetry. Thus the three pairs of vibrations $\nu_1 = 3019$ and $\nu_{11} = 2989$, $\nu_5 = 3272$ and $\nu_9 = 3106$, $\nu_3 = 1342$ and $\nu_{12} = 1444$ of C_2H_4 (see Fig. 44) are accounted for, the first four as C—H stretching, the last two as CH_2 bending vibrations. In addition there are, of course, bending vibrations in which the whole CH_2 group is bent against the C—C axis and which can be dealt with in a similar way.

As has been mentioned, the simple considerations given above can only be applied to bonds involving *end atoms* like H *whose mass is small compared to the other*

[68] This should not be confused with the designation π and σ vibrations used by some authors who let π stand for parallel, σ for perpendicular to the axis of symmetry.

[69] It may be noted that in the antisymmetric C—H stretching vibration of H_2CO the H atoms need not move exactly in the C—H direction as they do in free H_2C, since the vibration is not completely determined by symmetry.

atoms. Fig. 60b and c shows clearly that if the latter condition is not fulfilled the two totally symmetric vibrations ν_1 and ν_2 of bent XY_2 can no longer be considered as bond-stretching only and bond-bending only respectively; or, at least, it would be a very poor approximation to do so even though the assumption of valence forces is a good approximation (ν_3 is still strictly a bond-stretching vibration). Cross and Van Vleck (250) have shown that in a bent XYZ molecule, if X and Z are sufficiently different, we may again with more justification distinguish one bending and two stretching vibrations. This is shown for the three-particle system $CH_3 \cdot CH_2 \cdot Cl$ by Fig. 62 taken from Cross and Van Vleck's paper. It is seen that ν_2 is essentially a (CH_3)—(CH_2) stretching vibration, ν_1 a C—Cl stretching vibration and ν_3 a bending vibration.

Apart from the end atoms discussed up to now, experiment shows in many cases that *bonds involving two heavier atoms have characteristic frequencies even if none of them are end atoms* (see Table 51). This fact was first explained by Bartholomé and Teller (124) [see also Bartholomé (120) and Bauermeister and Weizel (129)]. They consider a chain of nearly equally heavy mass points connected by springs of very different stiffness. Suppose ν_1^0, ν_2^0 are the frequencies with which each pair of particles connected by a spring would oscillate if the other particles were not present. Then calculation shows that frequencies in the neighborhood of ν_1^0, ν_2^0 are normal frequencies of this system. This is plausible because if, say, ν_1^0 is excited the particles other than those directly affected will not move appreciably since they are bound by springs of different stiffness. It is as if the propagation of the oscillation in the system were strongly damped. Therefore the actual normal frequency is only slightly changed compared to ν_1^0. Similar considerations hold for the other frequencies. Thus we again obtain *characteristic frequencies for the different bonds.* However, this conclusion is based on the *assumption that the force constants of different bonds are different when the masses are of a similar magnitude.* Therefore in a molecule containing, for example, $\rangle$C—C(—)—O— or $\rangle$C—C(—)—N$\langle$, we would *not* expect a vibration characteristic for each bond since the masses as well as the force constants are of about the same magnitude; whereas in a molecule containing $\rangle$C—C(—)=O, or $\rangle$C—C(—)=N—, or $\rangle$C—C≡N, there would be a characteristic frequency for each of C—C, C=O, C=N, or C≡N.

On the other hand, if the force constants in a chainlike molecule are of similar magnitude but *the masses quite different,* there will again be *characteristic group frequencies,* as we have seen above for the C—H group and similar groups, and as is also true for C—Cl, C—Br, and others (see Table 51), *as long as the characteristic frequencies are far apart from one another.*

How well the characteristic frequencies are defined depends also on the *angle between successive bonds* in the chain of atoms. The above-mentioned authors have shown that if the angle is 90° there will be no transfer of vibrational energy from one bond to the next, and thus the characteristic frequencies are best defined. In the case of the bond-bending frequencies the chance that in a molecule there is another frequency (which need not be a bending frequency) of a similar magnitude is much larger than for the bond-stretching frequencies and thus the *bending frequencies are often not very characteristic.*

In case a chainlike molecule of the above type has *two or more equivalent bonds*, there will again be a resonance which leads to a *splitting of the characteristic frequency*, and again the stronger the coupling between the two equivalent bonds the larger is this splitting. For example, in diacetylene H—C≡C—C≡C—H, the C≡C stretching vibration which has the frequency 1974 cm^{-1} in C_2H_2 splits into two frequencies 2023 and 2183 cm^{-1}; or in $(CN)_2$ the C≡N frequency, which is 2089 cm^{-1} in HCN, splits into the two frequencies 2149 and 2322 cm^{-1}. If there are more than two equivalent bonds, there will be a splitting into correspondingly more different frequencies, and these frequencies will deviate increasingly from the frequency of a single such bond. Thus in a carbon-chain molecule (paraffin) there will be not one characteristic C—C frequency but as many frequencies as there are C—C bonds, and these will be distributed over a rather wide range if there are many C—C bonds. Some of them may be in the neighborhood of other group frequencies. Thus it would be difficult to ascribe particular frequencies in these molecules to the C—C bond, even though in C_2H_6 and similar simple molecules such a C—C frequency does exist (see Table 51).

In the case of long chain molecules, more detailed rough predictions of the normal frequencies than those given by Bartholomé and Teller (124), on the basis of the known force constants, have been made by Thomas and Whitcomb (843), Kirkwood (507), Barriol (118), Whitcomb, Nielsen, and Thomas (922), and Pitzer (694). Such predictions are particularly important for the calculation of the thermodynamic properties of these molecules [see Pitzer (694) and Chapter V, section 1].

Summarizing the preceding discussion of characteristic group frequencies, we may say: *Whenever a particular group, if separated from the rest of the molecule, has a vibrational frequency that differs sufficiently from any vibrational frequency of the rest of the molecule, then this frequency will occur only slightly changed in the whole molecule and will roughly correspond to vibrational motions in that group only, or if there are several equivalent groups to simultaneous motions in those groups.* In symmetrical molecules the condition that the frequency be sufficiently different can be disregarded if the two vibrations, one of the group considered and one of the rest of the molecule, have different species.

Under these conditions, it is possible to obtain rough values for most of the frequencies of a molecule (particularly organic molecules) without actual observation, if the characteristic stretching and bending frequencies of the bonds occurring in the molecule are known [see Mecke (610) (611)]. While such predictions are quite rough they do not require any calculations (such as those described in the preceding subsections, which supply much closer approximations to the actual frequencies if the force constants are known)and are therefore exceedingly helpful in the *analysis of observed vibrational spectra* (see Chapter III).

The *limitations of this procedure* should however always be kept in mind. A good illustration of these limitations is obtained if for a given set of force constants of a molecular model the frequencies are calculated for different values of the masses. The result for the plane vibrations of the X_2CO molecule as calculated by Burkard (185) from the formulae (II, 214–218) for variable m_X is plotted in Fig. 65a (with $2\alpha = 110°$, $k_1 = 10$, $k_2 = 4.45$, $k_\delta/l_2^2 = 0.384$, $k_\delta'/l_2^2 = 0.25 \times 10^5$ dynes/cm). The totally symmetric vibrations ν_1, ν_2, ν_3 (compare Fig. 24) are indicated by solid curves, the antisymmetric vibrations ν_4 and ν_5 by heavy broken curves. As was to be expected, no intersection of the curves representing vibrations of the same species occur while intersection of curves of different species do occur. As thin dotted lines are also indicated the curves for the vibrations ν_1, ν_2, ν_3, of free CX_2 and ν_0 of free C=O assuming the same force constants. Both $\nu_1(CX_2)$ and $\nu_3(CX_2)$ intersect ν_0. But since

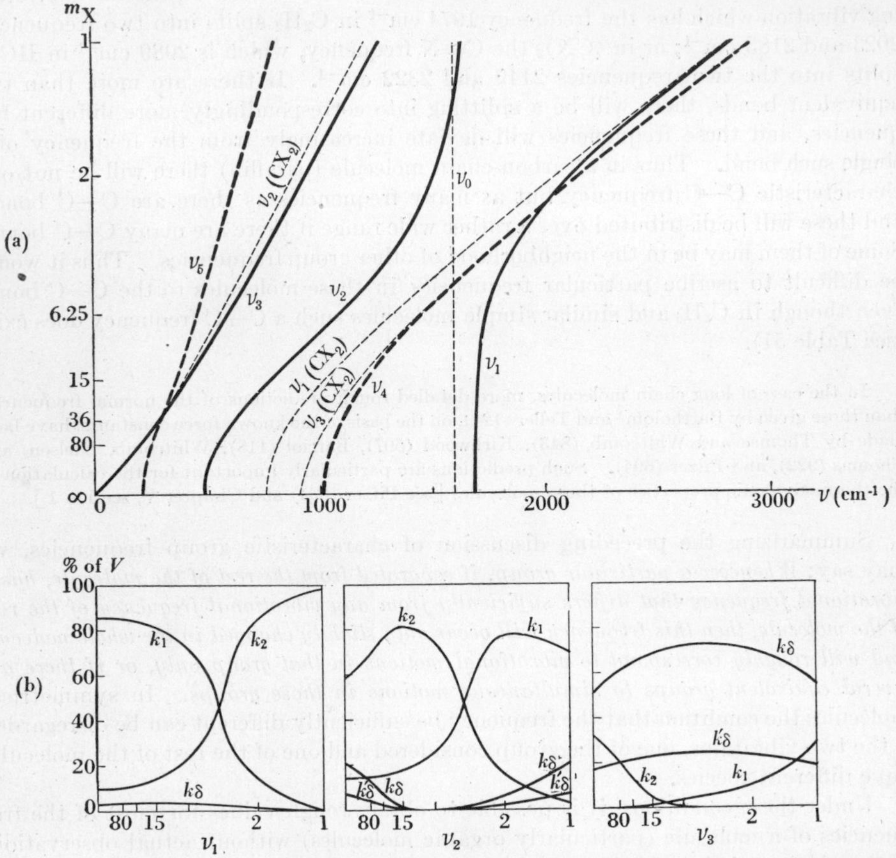

FIG. 65. **Characteristic group frequencies in X_2CO as a function of the mass of X [after Burkard (185)]. (a) Frequencies of the plane vibrations ν_1, ν_2, ν_3, ν_4, ν_5 as a function of m_X. (b) Percentage contribution of terms with k_1, k_2, k_δ, k_δ' to the potential energy in ν_1, ν_2, ν_3.**—The mass scales on the ordinate of (a) and the abscissae of (b) are proportional to $\sqrt{1/m_X}$. Thus the mass increases toward the origin. The contribution of k_δ' to the potential energy in ν_1 is always less than 1.5% and therefore not shown.

the X_2CO vibration ν_4 resulting from $\nu_3(CX_2)$ has different species from the one resulting from ν_0, it has throughout almost the same frequency as $\nu_3(CX_2)$ and may be considered as a characteristic CX_2 frequency for any mass. However, on the basis of the above general rule, when $\nu_1(CX_2)$ and ν_0 have the same value they are no longer characteristic group frequencies in the molecule. Rather a resonance occurs which leads in the complete molecule to quite different frequencies of ν_1 and ν_2. Thus, while for small as well as for large m_X the vibrations ν_1 and ν_2 have characteristic bond-frequency values, this is no longer the case in the intermediate mass region. At the same time it must be realized that while for small m_X the vibration ν_1 (that is, the highest totally symmetric vibration) is a C—X vibration [approaching $\nu_1(CX_2)$, see Fig. 65a], for large m_X it is the C—O vibration (approaching ν_0), while the reverse is the case for ν_2.

A change of mass may not only change a given vibration from one characteristic of one bond to one characteristic of another bond as just discussed, but it may also change a bond-bending into a bond-stretching vibration and vice versa. Such a change would occur in the case of the vibration ν_2 and ν_3 of X_2CO for m_X values smaller than one (which cannot be realized in practice). For such m_X values ν_3 (which for $m_X > 1$ is the CX_2 bending vibration) becomes the CO stretching vibration while ν_2 becomes the CX_2 bending vibration. A very similar change of character does occur for realizable mass values, for example, for H_2CX_2 [see Wagner (908)], where the CH_2 bending vibration with increasing m_X changes into the CX stretching vibration (see also the discussion of the methyl halides in Chapter III, section 3c). These examples may suffice to stress the importance of caution in using the concepts of bond-stretching and -bending vibrations. Such caution is particularly necessary when the frequencies of vibrations of the same species are not very different.

Another way of representing the change of character of vibrations is by plotting the *fractional contribution of the different forces to the total potential energy in a given vibration* [see Burkard (185) and Wagner (908)]. This is done for the three vibrations ν_1, ν_2, ν_3, of X_2CO in Fig. 65b. The four curves in each case give the contribution of the terms with the force constants k_1, k_2, k_δ, and k_δ' respectively in the potential energy function (II, 213). It is seen that ν_1 for small m_X is almost 100 per cent C—X stretching, since essentially only the term with k_2 contributes to the potential energy, while for large m_X it is up to 85 per cent C—O stretching. The reverse holds for ν_2, although the percentages are not as high. For small as well as large m_X there is about 25 per cent bending in ν_2. On the other hand the vibration ν_3, while predominantly bending, has up to 30 per cent stretching character. The vibrations ν_4 and ν_5 (not shown in Fig. 65b) are 100 per cent C—X stretching and whole-molecule-bending respectively for small m_X. For large m_X they mix up to the amount of 18 per cent, since they are no longer so different in frequency.

5. Anharmonicity and Interaction of Vibrations; Limitations of the Concept of Normal Vibrations

As has been stressed before, the concept of normal vibrations rests on the assumption of sufficiently small amplitudes of the oscillations (strictly speaking, infinitesimal amplitudes), when only the quadratic terms in the potential energy need be considered. Actually the amplitudes of the quantized oscillations, though usually small, are by no means infinitesimal and therefore for accurate calculations *cubic, quartic, and higher terms in the potential energy* must be considered; in other words, actually the *oscillations are anharmonic.*

(*a*) *Influence of anharmonicity for non-degenerate vibrations*

A simple potential surface. It is immediately clear that the potential energy always contains higher powers of the displacements than the second, since, just as for diatomic molecules, for very large displacements the potential energy approaches a constant value (corresponding to the dissociation energy). The potential energy of a polyatomic molecule depends on 3N − 6 (or 3N − 5) coordinates and is therefore much more difficult to visualize than it is for a diatomic molecule. Even for a triatomic molecule we would, for a complete representation, have to consider a three-dimensional hypersurface in a four-dimensional space. However, if in a *linear symmetric triatomic molecule* XY_2, for example, we neglect the possibility of bending (or in other words assume an infinitely large bending force constant) we can represent the potential energy as a two-dimensional surface in ordinary three-dimensional space. We may choose the two X—Y distances r_1 and r_2 as the two independent coordinates on which the potential energy depends. Plotting, then, the potential energy for every point in the r_1, r_2 plane we obtain a surface which is most easily visualized by means of a model, made for example of plaster of paris [see Goodeve (387)]. Fig. 66a gives a photograph of such a model for CO_2. Another way of representing such a *potential surface* is by means of *contour lines*, as in Fig. 66b.[70]

The potential surface of the electronic ground state of CO_2, as of any other stable linear XY_2 molecule, has a deep minimum at a point $r_1 = r_2 = r_e$ corresponding to the *equilibrium position*. From this minimum the potential energy increases in all directions, that is, for any change of r_1 or r_2 or both. For large r_1 or large r_2 one has simply a Y atom and an XY molecule; that is, a cross section of the potential energy surface at large r_1 parallel to the r_2 axis gives the potential curve of the XY molecule. Similarly, for large r_2 a cross section parallel to the r_1 axis also gives the potential curve of the XY molecule. If both r_1 and r_2 are large, a plateau is reached corresponding to the energy of the three atoms at great distance from one another. For

[70] In Fig. 66 the fact has been disregarded that the electronic ground state of CO_2, since it is a singlet state, cannot dissociate into CO and O in their ground states [see Herzberg (431)].

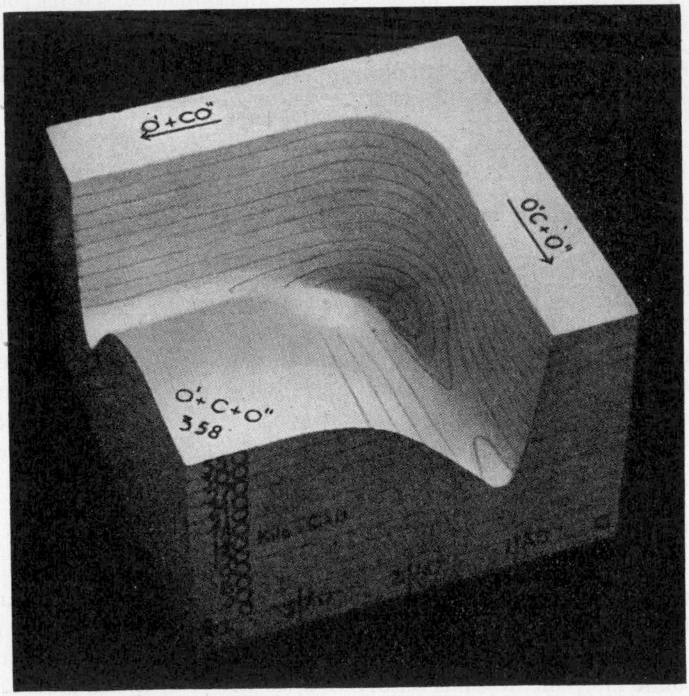

FIG. 66. (a)

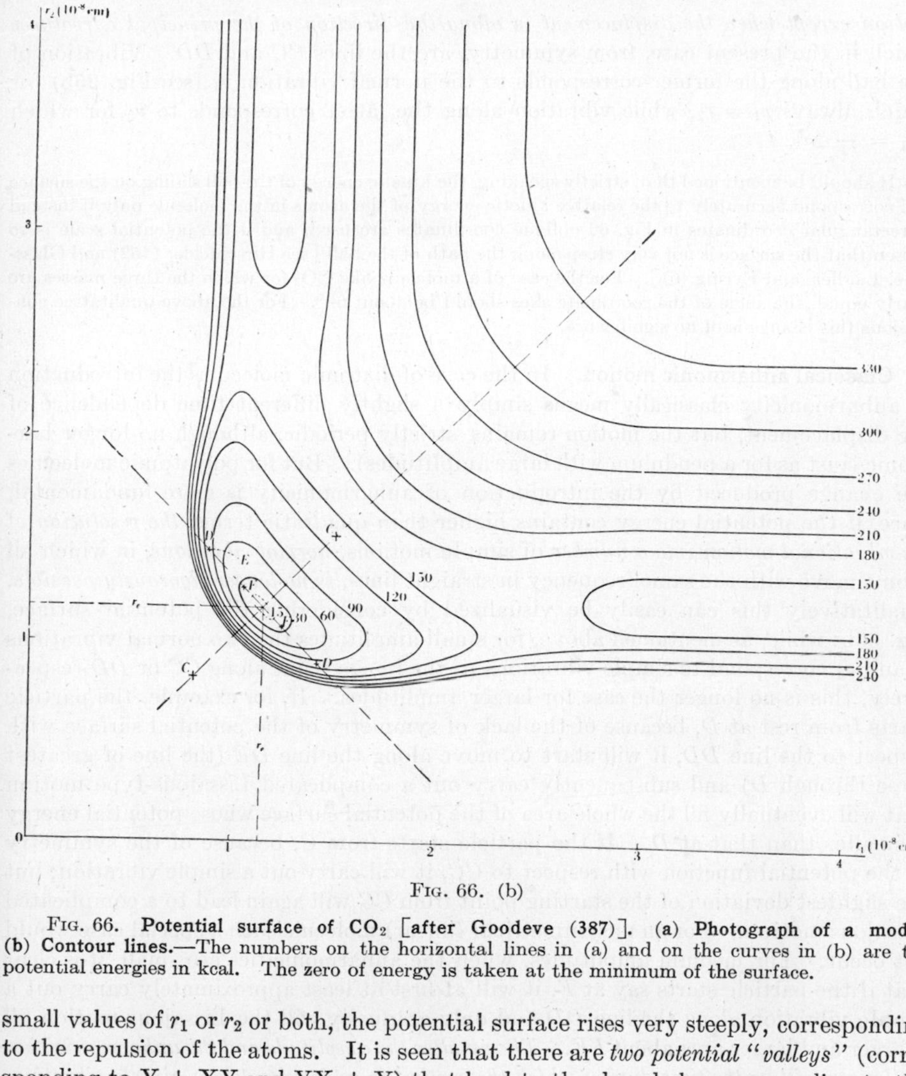

FIG. 66. (b)

FIG. 66. **Potential surface of CO_2 [after Goodeve (387)]. (a) Photograph of a model, (b) Contour lines.**—The numbers on the horizontal lines in (a) and on the curves in (b) are the potential energies in kcal. The zero of energy is taken at the minimum of the surface.

small values of r_1 or r_2 or both, the potential surface rises very steeply, corresponding to the repulsion of the atoms. It is seen that there are *two potential "valleys"* (corresponding to Y + XY and YX + Y) that lead to the deep hole corresponding to the stable XY_2 molecule.

While near the equilibrium positions the potential surface (Fig. 66) can be approximated quite well by a *paraboloid* (which it would be exactly if the potential energy had only quadratic terms), it is obvious from the figure that this is no longer possible for larger amplitudes (compare also Fig. 45 of Molecular Spectra I for the diatomic case).

The *relative motions of the three atoms* (always assuming that they remain on a straight line) *can be represented by the motion of a single mass point* (a small ball) *under the action of gravity on the potential surface* (considered as solid, as in Fig. 66a). It is immediately clear that if the ball is displaced from the minimum it will in general not describe a simple oscillation through the minimum but will carry out a *Lissajous motion except when the displacement is along the direction of the principal curvatures* which in the present case, from symmetry, are the lines CC and DD. Vibration of the ball along the former corresponds to the normal vibration ν_1 (see Fig. 25b) for which always $r_1 = r_2$, while vibration along the latter corresponds to ν_3 for which $\Delta r_1 = -\Delta r_2$.

It should be mentioned that, strictly speaking, the kinetic energy of the ball sliding on the surface will correspond accurately to the relative kinetic energy of the atoms in the molecule only if instead of rectangular coordinates in Fig. 66 oblique coordinates are used, and if the potential scale is so chosen that the surface is not very steep along the path of the ball [see Hirschfelder (452) and Glasstone, Laidler, and Eyring (6)]. For the case of a molecule like CO_2 for which the three masses are nearly equal, the angle of the coordinate axes should be about 60°. For the above qualitative conclusions this change is of no significance.

Classical anharmonic motion. In the case of diatomic molecules the introduction of anharmonicity classically means simply a slightly different time dependence of the displacement, but the motion remains strictly periodic, although no longer harmonic (just as for a pendulum with large amplitudes). But for polyatomic molecules the change produced by the introduction of anharmonicity is more fundamental, since, if the potential energy contains higher than quadratic terms, *the resolution of the vibrational motion into a number* of simple motions, *normal vibrations,* in which all atoms move with the same frequency in straight lines, *is no longer rigorously possible.* Qualitatively this can easily be visualized by considering the potential surface, Fig. 66b: while, as mentioned above, for small amplitudes the two normal vibrations ν_1 and ν_3 correspond to simple vibrations of the image point along CC or DD respectively, this is no longer the case for larger amplitudes. If, for example, the particle starts from rest at D, because of the lack of symmetry of the potential surface with respect to the line DD, it will start to move along the line DE (the line of greatest slope through D) and subsequently carry out a complicated Lissajous-type motion that will eventually fill the whole area of the potential surface whose potential energy is smaller than that at D. If the particle starts from C, because of the symmetry of the potential function with respect to CC, it will carry out a simple vibration; but the slightest deviation of the starting point from CC will again lead to a complicated Lissajous motion. For an unsymmetrical (linear) molecule such a special case would not occur. For medium amplitudes, when the anharmonicities are small, it is clear that if the particle starts say at F, it will at first at least approximately carry out a simple vibration along the line DD and only gradually will the Lissajous motion fill a larger and larger area about FF. *The smaller the amplitude and the anharmonicities, the longer will be the time during which the motion is approximately a simple vibration.*

As previously (p. 73), the classical *kinetic energy* T is given by

$$2T = \sum m_i(\dot{x}_i^2 + \dot{y}_i^2 + \dot{z}_i^2), \tag{II, 261}$$

while the *potential energy* is now given by [compare equation (II, 26)]

$$2V = \sum_i \sum_j k_{ij}q_iq_j + \sum_i \sum_j \sum_k f_{ijk}q_iq_jq_k + \sum_i \sum_j \sum_k \sum_l g_{ijkl}q_iq_jq_kq_l + \cdots, \tag{II, 262}$$

where, as previously, for abbreviation the q_i are the Cartesian displacement coordinates in the order $x_1, y_1, z_1, x_2, y_2, z_2, \cdots$. On transforming to normal coordinates by means of (II, 31) and normalizing to unity (see p. 75), we obtain for the total energy, instead of (II, 35),

$$H = V + T = \tfrac{1}{2}(\dot{\eta}_1^2 + \dot{\eta}_2^2 + \dot{\eta}_3^2 + \cdots) + \tfrac{1}{2}(\lambda_1\eta_1^2 + \lambda_2\eta_2^2 + \lambda_3\eta_3^2 + \cdots)$$
$$+ \sum_i \sum_j \sum_k \alpha_{ijk}\eta_i\eta_j\eta_k + \sum_i \sum_j \sum_k \sum_l \beta_{ijkl}\eta_i\eta_j\eta_k\eta_l + \cdots; \tag{II, 263}$$

that is, while there are no quadratic cross terms there are *cubic and higher cross terms in the normal coordinates* and the total energy is no longer that of a sum of independent (even though anharmonic) oscillators. In a completely unsymmetric molecule all coefficients α_{ijk} and β_{ijkl} would be different from zero, but in a symmetric molecule some of them may be equal to zero. This is because the potential energy must remain unchanged for all symmetry operations permitted by the point group of the molecule. For this reason the *anti-symmetric normal coordinates can only occur with even powers* in (II, 263). For example, for H_2O the coefficients α_{113}, α_{123}, α_{223}, and α_{333} of the cubic terms must be zero since otherwise the potential energy would change by reflection at the plane of symmetry, and similarly for some of the quartic terms. Further simplifications of the anharmonic part of the potential function can only be obtained if assumptions somewhat analogous to the valence force system (used in discussing the harmonic terms) are made [see Redlich (727)].

In the harmonic oscillator approximation, the time dependence of the normal coordinates is given by [see equation (II, 13)]

$$\eta_i = \eta_i^0 \cos(2\pi\nu_i t + \varphi_i). \tag{II, 264}$$

In the case of slightly anharmonic vibrations, the normal coordinates may be expressed as a generalized Fourier series which, for example for the first normal coordinate η_1 of a triatomic molecule, would be

$$\begin{aligned}\eta_1 = {} & \eta_{100}\cos(2\pi\nu_1 t + \varphi_1^{100}) + \eta_{200}\cos(2\pi 2\nu_1 t + \varphi_1^{200}) + \eta_{300}\cos(2\pi 3\nu_1 t + \varphi_1^{300}) \\ & + \cdots + \eta_{110}\cos(2\pi\nu_1 t + \varphi_1^{110})\cos(2\pi\nu_2 t + \varphi_2^{110}) + \eta_{101}\cos(2\pi\nu_1 t + \varphi_1^{101}) \\ & \times \cos(2\pi\nu_3 t + \varphi_3^{101}) + \eta_{111}\cos(2\pi\nu_1 t + \varphi_1^{111})\cos(2\pi\nu_2 t + \varphi_2^{111})\cos(2\pi\nu_3 t + \varphi_3^{111}) \\ & + \eta_{201}\cos(2\pi 2\nu_1 t + \varphi_1^{201})\cos(2\pi\nu_3 t + \varphi_3^{201}) + \cdots. \end{aligned} \tag{II, 265}$$

Here the amplitudes η_{200}, η_{110}, $\cdots$ are small compared to η_{100}. The φ are phase constants. Similar relations hold for η_2 and η_3.

Energy levels. If cubic, quartic, and possibly higher terms are introduced into the wave equation (II, 40) in addition to the quadratic terms, it no longer resolves into a number of independent equations as it did when quadratic terms only were used. In consequence the energy is no longer a sum of independent terms corresponding to the different normal vibrations but contains *cross terms containing the vibrational quantum numbers of two or more normal vibrations*. Considering the well-known formula for the anharmonic vibrations of a diatomic molecule [see equation (III, 56) p. 100 of Molecular Spectra I], it is easy to guess the formula for a *non-linear triatomic molecule* having no degenerate vibrations. Using wave number units (cm^{-1}) for the vibrational energy, we expect

$$\begin{aligned}G(v_1, v_2, v_3) = {} & \omega_1(v_1 + \tfrac{1}{2}) + \omega_2(v_2 + \tfrac{1}{2}) + \omega_3(v_3 + \tfrac{1}{2}) + x_{11}(v_1 + \tfrac{1}{2})^2 \\ & + x_{22}(v_2 + \tfrac{1}{2})^2 + x_{33}(v_3 + \tfrac{1}{2})^2 + x_{12}(v_1 + \tfrac{1}{2})(v_2 + \tfrac{1}{2}) \\ & + x_{13}(v_1 + \tfrac{1}{2})(v_3 + \tfrac{1}{2}) + x_{23}(v_2 + \tfrac{1}{2})(v_3 + \tfrac{1}{2}) + \cdots. \end{aligned} \tag{II, 266}$$

Here ω_1, ω_2, ω_3 are the *frequencies* (in cm^{-1}) *for infinitesimal amplitudes* of the three normal vibrations (corresponding to ω_e for diatomic molecules).[71] They are also called the *zero-order frequencies*. The x_{ik} are the *anharmonicity constants* (corresponding to $\omega_e x_e$ of diatomic molecules). v_1, v_2, v_3 are the vibrational quantum numbers corresponding to the three normal vibrations.

The expression (II, 266) for the vibrational energy levels or its equivalent (II, 268) below has indeed been obtained by rather lengthy calculations on the basis of the old quantum theory by Born and Brody (170) and from the wave equation by Bonner (162), King (502), Shaffer and Nielsen (780), and Darling and Dennison (263). These authors have also found expressions for the vibrational constants ω_i and x_{ik} in terms of the potential constants in (II, 262) [see also Shaffer and Newton

[71] In agreement with common practice the subscript e has been omitted. Some authors use X_i, others ν_i in place of ω_i, and X_{ik} in place of x_{ik}.

(778)]. As was to be expected, they found that the ω_i depend on the force constants k_{ij} only (in just the way previously discussed when anharmonicity was neglected) while the x_{ik} depend on the coefficients of the cubic and quartic terms as well. In the case of non-linear XY_2 there are six anharmonic constants x_{ik} but twelve cubic and quartic potential constants.[72] The latter can therefore not be determined from the x_{ik} unless the x_{ik} have also been obtained for an isotopic molecule, or unless certain of the potential constants are assumed to be zero [see Redlich (727)]. However, the interaction of vibration and rotation (see Chapter IV) supplies additional equations for the cubic constants which can be used for their determination even if no isotopic molecule is observed. In fact, this is the only method that has up to the present time been used, and then only for two cases: CO_2 [Dennison (280)] and H_2O [Darling and Dennison (263)].

Frequently the *vibrational energy*, instead of being referred to the minimum of the potential surface as in (II, 266), is *referred to the lowest vibrational state* ($v_1 = 0$, $v_2 = 0$, $v_3 = 0$) which has an energy

$$G(0, 0, 0) = \tfrac{1}{2}\omega_1 + \tfrac{1}{2}\omega_2 + \tfrac{1}{2}\omega_3 + \tfrac{1}{4}x_{11} + \tfrac{1}{4}x_{22} + \tfrac{1}{4}x_{33} + \tfrac{1}{4}x_{12} + \tfrac{1}{4}x_{13} + \tfrac{1}{4}x_{23} \qquad \text{(II, 267)}$$

above the minimum (*zero-point energy*). We have then

$$\begin{aligned} G_0(v_1, v_2, v_3) = G(v_1, v_2, v_3) - G(0, 0, 0) &= \overset{0}{\omega}_1 v_1 + \overset{0}{\omega}_2 v_2 + \overset{0}{\omega}_3 v_3 \\ &+ \overset{0}{x}_{11} v_1^2 + \overset{0}{x}_{22} v_2^2 + \overset{0}{x}_{33} v_3^2 + \overset{0}{x}_{12} v_1 v_2 + \overset{0}{x}_{13} v_1 v_3 + \overset{0}{x}_{23} v_2 v_3, \qquad \text{(II, 268)} \end{aligned}$$

where $\overset{0}{x}_{ik} = x_{ik}$, as long as no powers higher than the second in the vibrational quantum numbers occur, and where

$$\begin{aligned} \omega_1{}^0 &= \omega_1 + x_{11} + \tfrac{1}{2}x_{12} + \tfrac{1}{2}x_{13}, \\ \omega_2{}^0 &= \omega_2 + x_{22} + \tfrac{1}{2}x_{12} + \tfrac{1}{2}x_{23}, \qquad \text{(II, 269)} \\ \omega_3{}^0 &= \omega_3 + x_{33} + \tfrac{1}{2}x_{13} + \tfrac{1}{2}x_{23}, \end{aligned}$$

Thus the zero-order frequencies ω_i can immediately be obtained if the coefficients $\omega_i{}^0$ and $\overset{0}{x}_{ik}$ in (II, 268) have been obtained from the empirical data.[72a] (Compare the example of H_2O in Chapter III, section 3a).

The observed *fundamentals* ν_i correspond to the transition from a state with $v_i = 1$ to the one with $v_i = 0$ where all other $v_k = 0$. Therefore, from (II, 268) and (II, 269),

$$\begin{aligned} \nu_1 &= \overset{0}{\omega}_1 + \overset{0}{x}_{11} = \omega_1 + 2x_{11} + \tfrac{1}{2}x_{12} + \tfrac{1}{2}x_{13}, \\ \nu_2 &= \overset{0}{\omega}_2 + \overset{0}{x}_{22} = \omega_2 + 2x_{22} + \tfrac{1}{2}x_{12} + \tfrac{1}{2}x_{23}, \qquad \text{(II, 270)} \\ \nu_3 &= \overset{0}{\omega}_3 + \overset{0}{x}_{33} = \omega_3 + 2x_{33} + \tfrac{1}{2}x_{13} + \tfrac{1}{2}x_{23}. \end{aligned}$$

In determining the quadratic potential constants we have in the preceding section always used the observed fundamentals ν_i. Strictly speaking, as is seen from the above, we should have used the frequencies for infinitesimal amplitudes ω_i (zero-order frequencies) which can be obtained from the ν_i and the x_{ik} according to (II, 270). Fortunately, in general, the x_{ik} are small compared to the ω_i and therefore the use of the ν_i in place of the ω_i, which is often necessary because of lack of data, gives

[72] Actually there are fifteen cubic and quartic potential constants but three of these do not influence the anharmonic constants x_{ik}.

[72a] Some authors use X_i in place of $\omega_i{}^0$ used here.

quite a fair approximation. However, one must not be surprised if small inconsistencies appear when the ν_i are used. Usually most of the x_{ik} are negative and therefore, in general, ω_i^0 is slightly larger than ν_i, and ω_i is slightly larger than ω_i^0.

As for diatomic molecules, the anharmonicities x_{ik} and therefore the differences between ω_i, ω_i^0, and ν_i are the largest for vibrations involving the motion of hydrogen atoms. As an example, the values of ω_i, ω_i^0, and ν_i for H_2O may be considered:

$$\omega_1 = 3825.3, \qquad \omega_2 = 1653.9, \qquad \omega_3 = 3935.6;$$

$$\omega_1^0 = 3693.9, \qquad \omega_2^0 = 1614.5, \qquad \omega_3^0 = 3801.8;$$

$$\nu_1 = 3651.7, \qquad \nu_2 = 1595.0, \qquad \nu_3 = 3755.8.$$

In almost all other cases the differences are smaller, for vibrations not involving hydrogen considerably smaller.

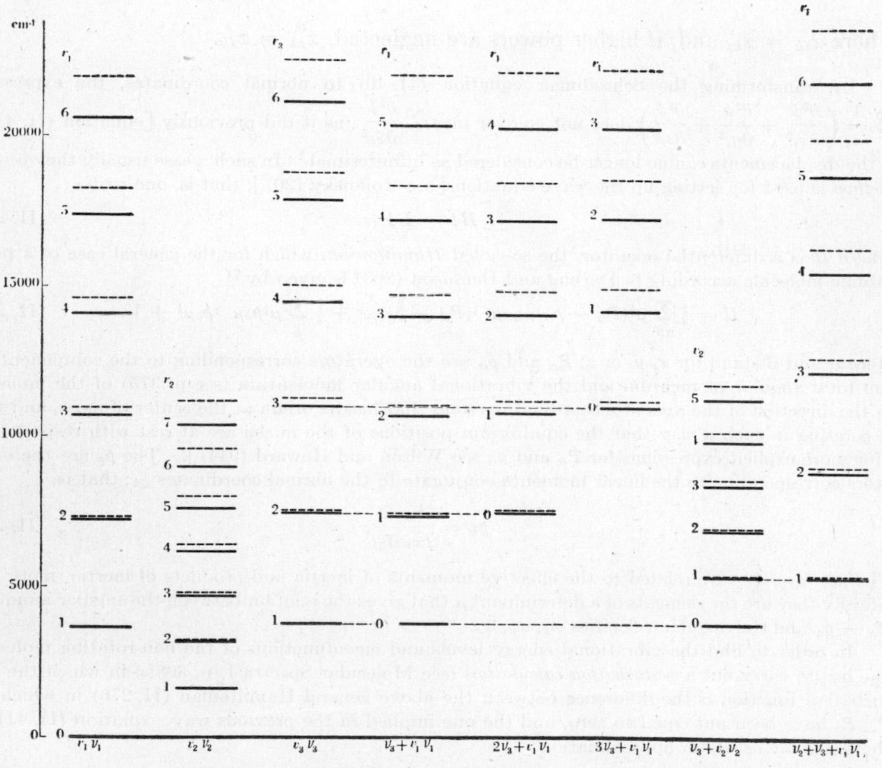

FIG. 67. **Vibrational energy level diagram of H_2O showing the influence of anharmonicity.—The heavy broken lines indicate the positions that the levels would have if anharmonic terms were neglected. Not all levels up to 25000 cm^{-1} are shown. Levels that are repeated in different sets are connected by light broken lines.**

In Fig. 67 the vibrational energy levels of H_2O are plotted (disregarding the resonance phenomenon to be discussed below) and compared to those obtained when anharmonicities are neglected (broken lines). It will be noticed that, for example, since $x_{13} \neq 0$ the series of levels with $v_3 = 1$ and $v_1 = 0, 1, 2, \cdots$ (fourth column) is *not* obtained from the levels with $v_3 = 0$ and $v_1 = 0, 1, 2, \cdots$ (first column) simply by shifting by a constant amount (equal to the first quantum of the vibration ν_3), and similarly in other cases.

The generalization of the above formulae to the case of *more than triatomic molecules* (without degenerate vibrations) is obvious. We have

$$G(v_1, v_2 \cdots) = \sum_i \omega_i(v_i + \tfrac{1}{2}) + \sum_i \sum_{k \geq i} x_{ik}(v_i + \tfrac{1}{2})(v_k + \tfrac{1}{2}) + \cdots, \quad \text{(II, 271)}$$

$$G_0(v_1, v_2 \cdots) = \sum_i \omega_i^0 v_i + \sum_i \sum_{k \geq i} x_{ik}^0 v_i v_k + \cdots, \quad \text{(II, 272)}$$

$$\omega_i^0 = \omega_i + x_{ii} + \tfrac{1}{2} \sum_{k \neq i} x_{ik} + \cdots, \quad \text{(II, 273)}$$

$$\nu_i = \omega_i^0 + x_{ii} = \omega_i + 2x_{ii} + \tfrac{1}{2} \sum_{k \neq i} x_{ik} + \cdots, \quad \text{(II, 274)}$$

where $x_{ik} = x_{ki}$ and, if higher powers are neglected, $x_{ik}^0 = x_{ik}$.

In transforming the Schrödinger equation (II, 40) to normal coordinates, the expression $\sum \frac{1}{m_i}\left(\frac{\partial^2\psi}{\partial x_i^2} + \frac{\partial^2\psi}{\partial y_i^2} + \frac{\partial^2\psi}{\partial z_i^2}\right)$ does not go over into $\sum \frac{\partial^2\psi}{\partial \xi_k^2}$, as it did previously [equation (II, 41)], if the displacements can no longer be considered as infinitesimal. In such a case usually the *operator method* is used for setting up the wave equation [see Rojansky (20)]; that is, one writes

$$H\psi = E\psi \quad \text{(II, 275)}$$

where H is a differential operator, the so-called *Hamiltonian*, which for the general case of a polyatomic molecule according to Darling and Dennison (263) is given by [73]

$$H = \tfrac{1}{2} \sum_{\alpha\beta} \mu^{\frac{1}{4}}(P_\alpha - p_\alpha)\mu_{\alpha\beta}\mu^{-\frac{1}{2}}(P_\beta - p_\beta)\mu^{\frac{1}{4}} + \tfrac{1}{2} \sum_k \mu^{\frac{1}{4}} p_k \mu^{-\frac{1}{2}} p_k \mu^{\frac{1}{4}} + V. \quad \text{(II, 276)}$$

Here α and β stand for x, y, or z; P_α and p_α are the operators corresponding to the components of the total angular momentum and the vibrational angular momentum (see p. 375) of the molecule in the direction of the axes of a coordinate system that has its origin at the center of mass, and that is rotating in such a way that the equilibrium positions of the nuclei are at rest with respect to it [for more explicit expressions for P_α and p_α see Wilson and Howard (944)]. The p_k are the operators corresponding to the linear momenta conjugate to the normal coordinates ξ_k; that is,

$$p_k = \frac{h}{2\pi i}\frac{\partial}{\partial \xi_k}. \quad \text{(II, 277)}$$

The μ_{xx}, μ_{xy}, $\cdots$ are related to the effective moments of inertia and products of inertia, more specifically they are the elements of a determinant μ that gives the relation between the angular momenta $P_\alpha - p_\alpha$ and the angular velocities ω_x, ω_y, ω_z.

In order to find the vibrational energy levels and eigenfunctions of the non-rotating molecule, one has to carry out a *perturbation calculation* (see Molecular Spectra I, p. 308f.) in which the perturbation function is the difference between the above general Hamiltonian (II, 276) in which P_x, P_y, P_z have been put equal to zero, and the one implied in the previous wave equation (II, 41) for the harmonic oscillator approximation:

$$H^0 = \tfrac{1}{2} \sum p_k^2 + \tfrac{1}{2} \sum \lambda_k \xi_k^2. \quad \text{(II, 278)}$$

This is essentially what the above-mentioned authors have done in deriving the energy formula (II, 266).

If in (II, 276) the dependence of the $\mu_{\alpha\beta}$, and therefore of the moments of inertia, on the displacements (normal coordinates) is neglected, and if the coordinate axes are principal axes of inertia, it can be shown fairly easily [see Wilson and Howard (944)] that $\mu_{xx} = \frac{1}{I_x}$, $\mu_{yy} = \frac{1}{I_y}$, $\mu_{zz} = \frac{1}{I_z}$,

[73] Darling and Dennison's Hamiltonian is a slight modification of that originally given by Wilson and Howard (944).

while $\mu_{\alpha\beta} = 0$ for $\alpha \neq \beta$. Here I_x, I_y, I_z are the principal moments of inertia. Therefore, and since then the determinant μ is a constant, (II, 276) simplifies to

$$H = \frac{(P_x - p_x)^2}{2I_x} + \frac{(P_y - p_y)^2}{2I_y} + \frac{(P_z - p_z)^2}{2I_z} + \frac{1}{2} \sum p_k^2 + V. \qquad \text{(II, 279)}$$

Here again, if we disregard rotation, P_x, P_y, P_z must be put equal to zero. It may be noted, however, that other than in the classical theory (see above) we cannot put $P_\alpha - p_\alpha = 0$; that is, if there is a vibrational angular momentum, the molecule cannot be quite non-rotating. In the harmonic oscillator approximation a vibrational angular momentum can occur only for degenerate vibrations; but if anharmonicity, that is, interaction of the vibrations, is taken into account an angular momentum exists in general also in the non-degenerate case, as the terms with p_x, p_y, p_z in the Hamiltonian indicate. Physically this is related to the fact that the nuclei no longer move in straight lines corresponding to one normal coordinate (which has no angular momentum). For a triatomic nonlinear molecule the vibrational angular momentum must obviously be perpendicular to the plane of the molecule (that is, $p_x = p_y = 0$).

Vibrational eigenfunctions. In the harmonic oscillator approximation, the total vibrational eigenfunction is simply a product of oscillator eigenfunctions corresponding to the different normal coordinates [compare equation (II, 42)]. If the anharmonicity is taken into account this is no longer the case, but we can write

$$\psi_v = \psi_1(\xi_1)\psi_2(\xi_2) \cdots \psi_{3N-6}(\xi_{3N-6}) + \chi(\xi_1, \xi_2, \cdots \xi_{3N-6}) \qquad \text{(II, 280)}$$

where χ is the smaller compared to the product $\psi_1\psi_2 \cdots \psi_{3N-6}$ the smaller the anharmonicity.

As has been pointed out previously (p. 104), the vibrational eigenfunction ψ_v in the non-degenerate case can only be symmetric or antisymmetric with respect to any of the symmetry operations permitted by the molecule irrespective of how large the anharmonicity [that is, χ in (II, 280)] is; or, in other words ψ_v must belong to one of the possible symmetry types (see section 3d). It is immediately clear that for a given vibrational level the symmetry type of ψ_v must be the same as that of $\psi_1(\xi_1)\psi_2(\xi_2)\psi_3(\xi_3) \cdots$ in (II, 280). Therefore *the symmetry type (species) of the anharmonic levels is obtained simply by determining in the previously described way the symmetry type of the corresponding harmonic levels.*

It is useful to try to visualize the *form of the eigenfunction* ψ_v. If only one vibration ν_i is excited, that is, if only v_i is different from zero, the eigenfunction depends not only on ξ_i but also on all the other normal coordinates as in the harmonic case (see p. 79), but not in as simple a way. Here, unlike the harmonic case, such a dependence exists even when the zero-point motion of the other normal vibrations is neglected, and is the wave mechanical analogue of the classical fact that a Lissajous motion takes place. However, in order to get a rough picture of the eigenfunction let us for a moment consider the dependence on ξ_i only, assuming all the other ξ to be equal to zero (that is, let us, for example, consider the variation of the eigenfunction along *CC* or *DD* in Fig. 66b). In the case of a *totally symmetric normal coordinate* ξ_i (for example, along *CC* in Fig. 66b), the course of the function ψ_v would be very similar to the vibrational eigenfunction of a diatomic molecule (see Fig. 47, p. 101 of Molecular Spectra I); that is, it is slightly unsymmetrical with respect to the origin, $\xi_i = 0$, the maxima on one side being higher than on the other. On the other hand, for a *normal coordinate that is antisymmetric* with respect to at least one symmetry operation (for example *DD* in Fig. 66b), ψ_v is an even function of ξ_i for even v_i and an odd function of ξ_i for odd v_i, since, when the symmetry operation is carried out, ξ_i goes over into $-\xi_i$ while ψ_v remains unchanged (is symmetric) for even v_i and changes sign (is antisymmetric) for odd v_i (see p. 101). Thus in this case ψ_v is not an unsymmetrically but a symmetrically distorted Hermite function (see Fig. 29); that is, the zero points remain symmetrically placed with respect to $\xi_i = 0$ and the height of corresponding maxima and minima is the same for positive and negative ξ_i. The dependence of ψ_v on the other normal coordinates will again be similar to the harmonic case (see p. 79f.), that is, will be represented mainly by a Gauss-error type of function but somewhat distorted and possibly with small additional humps and maxima.

(*b*) *Influence of anharmonicity for (non-accidentally) degenerate vibrations*

For the degenerate vibrations of molecules with more than two-fold axes, two additional points have to be considered: (1) *For a d_i-foldly degenerate vibration ν_i the energy depends on $v_i + (d_i/2)$* rather than on $v_i + \frac{1}{2}$ as it does for a non-degenerate vibration, for which $d_i = 1$ (see section 2 for the harmonic case). (2). *Certain energy levels that coincide in the harmonic oscillator approximation split into a number of levels when the anharmonicity is taken into account.*

As we have emphasized previously, the symmetry types of the vibrational levels are the same no matter whether the oscillations are harmonic or anharmonic; for example, the state in which a doubly degenerate vibration is excited with $v = 1$ remains doubly degenerate even if the potential is anharmonic. In the harmonic case, if a degenerate vibration is excited by several quanta, or if several degenerate vibrations are excited, a state arises with a degeneracy higher than that of any one of the component vibrations; if on the other hand anharmonicity is taken into account, in general, this high degeneracy does not remain but *a splitting occurs into just those substates that were obtained earlier from group theory* (Tables 32 and 33). This is because, as has been shown in more detail by Tisza (867), an accidental degeneracy occurring in a certain approximation is always removed in a higher approximation and only the necessary degeneracies required by the point group of the molecule remain. This holds rigorously as long as the rotation of the molecule is neglected (for the interaction with the rotation see Chapter IV).

On the basis of the above general theorem it is, for example, immediately concluded that the assertion made in a recent publication cannot be correct, that in SiH_4, assuming tetrahedral symmetry, the state in which the doubly degenerate vibration ν_2 is singly excited is split into two levels of slightly different energy even without rotation, on account of anharmonic terms in the potential energy. Conversely, if the doubling were experimentally established it would prove that SiH_4 is not tetrahedral in its equilibrium position. This example illustrates the importance of taking symmetry considerations into account in dealing with symmetrical polyatomic molecules.

General energy formula for the case of doubly degenerate vibrations. On the basis of the wave equation the following general formula is obtained for the vibrational energy levels of a molecule with doubly degenerate vibrations [see Nielsen (666)]:

$$G(v_1, v_2, v_3, \cdots) = \sum_i \omega_i \left(v_i + \frac{d_i}{2} \right) + \sum_i \sum_{k \geq i} x_{ik} \left(v_i + \frac{d_i}{2} \right) \left(v_k + \frac{d_k}{2} \right) + \sum_i \sum_{k \geq i} g_{ik} l_i l_k + \cdots. \quad \text{(II, 281)}$$

In this equation $d_i = 1$ or 2 depending whether i refers to a non-degenerate or doubly degenerate vibration. The l_i are the previously introduced integral numbers (see p. 81) which assume the values

$$l_i = v_i, v_i - 2, v_i - 4, \cdots 1 \text{ or } 0. \quad \text{(II, 282)}$$

For non-degenerate vibrations $l_i = 0$ and $g_{ik} = 0$. The last term in (II, 281), in which the g_{ik} are small constants of the order of the x_{ik}, gives a number of different sublevels when one or more degenerate vibrations are excited with $v_i > 1$. However, formula (II, 281) gives in general a splitting into fewer levels than the previous group theoretical results (Tables 32 and 33) would indicate. This is due to the fact that, in the approximation in which (II, 281) has been derived, to every l_i correspond two levels of the same energy (see p. 81), whereas according to the previous considerations of symmetry for certain l_i values two non-degenerate states are obtained, and at any rate no more than two-fold degeneracies can occur except for the cubic point groups. For example, if two doubly degenerate vibrations are singly excited, one four-foldly degenerate state is obtained from (II, 281), while according to Table 33 two doubly degenerate states or one doubly degenerate and two non-degenerate states or four non-degenerate states (depending on the point group) result.

Thus in higher approximation many of the levels given by (II, 281) *will split still further*. Explicit formulae for this additional splitting have not been given. However, it appears quite possible that it is of the same order as the l-splitting given by the third and fourth term in (II, 281).

The *zero-point energy* in the present case is, according to (II, 281),

$$G(0, 0, 0, \cdots) = \sum_i \omega_i \frac{d_i}{2} + \sum_i \sum_{k \geq i} x_{ik} \frac{d_i d_k}{4} + \cdots. \qquad \text{(II, 283)}$$

If the *vibrational energy* is *referred to the lowest level* one obtains, similar to (II, 272),

$$G_0(v_1, v_2 \cdots) = \sum_i \omega_i^0 v_i + \sum_i \sum_{k \geq i} x_{ik}^0 v_i v_k + \sum_i \sum_{k \geq i} g_{ik} l_i l_k + \cdots, \qquad \text{(II, 284)}$$

where

$$\omega_i^0 = \omega_i + x_{ii} d_i + \tfrac{1}{2} \sum_{k \neq i} x_{ik} d_k + \cdots. \qquad \text{(II, 285)}$$

and where $x_{ik}^0 = x_{ik}$ if higher powers are neglected. Also we obtain for the fundamentals,

$$\nu_i = \omega_i^0 + x_{ii} + g_{ii} = \omega_i + x_{ii}(1 + d_i) + \tfrac{1}{2} \sum_{k \neq i} x_{ik} d_k + g_{ii}, \qquad \text{(II, 286)}$$

where, as previously, $x_{ik} = x_{ki}$.

Application to linear molecules. In the case of linear molecules, as long as only one degenerate vibration (always of species Π) is excited, l_i is an exactly defined quantum number, $l_i = 0, 1, 2, 3, \cdots$, representing the vibrational angular momentum about the symmetry axis and corresponding to the species Σ, Π, Δ, Φ, $\cdots$. In this case, equation (II, 281) gives all the splittings. The levels with $l_i \neq 0$ are necessarily degenerate (see p. 112). The energy levels are as given in the previous Fig. 52a. Such an energy-level diagram would always apply to *triatomic linear molecules* (XYZ or XY_2), since they have only one degenerate vibration. For them, (II, 281) would simplify to

$$\begin{aligned} G(v_1, v_2, v_3, l_2) = {} & \omega_1(v_1 + \tfrac{1}{2}) + \omega_2(v_2 + 1) + \omega_3(v_3 + \tfrac{1}{2}) + x_{11}(v_1 + \tfrac{1}{2})^2 \\ & + x_{22}(v_2 + 1)^2 + g_{22} l_2^2 + x_{33}(v_3 + \tfrac{1}{2})^2 \\ & + x_{12}(v_1 + \tfrac{1}{2})(v_2 + 1) + x_{13}(v_1 + \tfrac{1}{2})(v_3 + \tfrac{1}{2}) \\ & + x_{23}(v_2 + 1)(v_3 + \tfrac{1}{2}) + \cdots. \end{aligned} \qquad \text{(II, 287)}$$

For the case of linear symmetric XY_2 molecules, Adel and Dennison (37) [see Dennison (280) for a few corrections] have given formulae for the x_{ik} and g_{22} in terms of the cubic and quartic potential constants as well as the ω_i and the moment of inertia.

Similar formulae for the linear XYZ molecule have been given by Adel (33) and Nielsen (654a). Actual values for the anharmonic constants x_{ik} and g_{22} of CO_2 and HCN as obtained from the observed infrared and Raman spectrum will be given in Chapter III.

If *two or more degenerate vibrations* exist as in linear X_2Y_2 and are simultaneously excited, *only the quantum number L of the resultant vibrational angular momentum about the axis is exactly defined* but the individual l_i are still approximately defined as angular momenta (similar to the λ_i of the electrons in diatomic molecules) and formula (II, 281) gives an approximation to the energy. However, the formula does not give a splitting of a level with given l_i values which is actually to be expected on the basis of group theory (see Table 33). For example, if in X_2Y_2 each of the two degenerate vibrations ν_4 and ν_5 (see Fig. 64a) is singly excited, that is, for $v_4 = 1$, $v_5 = 1$, and $l_4 = 1$, $l_5 = 1$, equation (II, 281) gives one energy value only, while on the basis of Table 33 the three states Σ_u^+, Σ_u^-, and Δ_u result. In the approximation assumed in (II, 281), these three states are degenerate with one another; but taking account of the finer interaction of the vibrations will lead to a splitting (see also the next subsection).

It can be shown [see Wu and Kiang (963), and Shaffer and Nielsen (779)] that, for linear X_2Y_2, $g_{45} = 0$, since the two degenerate vibrations have different species. Therefore the part of the energy formula (II, 281) depending on the l_i is simply

$$g_{44}l_4^2 + g_{55}l_5^2 \qquad \text{(II, 288)}$$

Wu and Kiang (963) and Shaffer and Nielsen (779) have given explicit formulae for the g_{44}, g_{55} as well as the x_{ik} in this case in terms of the potential constants.

As an illustration, Fig. 68 gives the lowest energy levels of the form $v_4\nu_4 + v_5\nu_5$ according to (II, 281) with (II, 288). Levels that coincide according to these formulae but split in higher approximation are drawn separately but enclosed in brackets.

Application to some non-linear molecules. The general formula (II, 281) has been shown to hold for plane and pyramidal XY_3 molecules by Silver and Shaffer (790) and Shaffer (776) respectively, and for axial XYZ_3 molecules by Shaffer (777). As in the previous cases, the calculations are based on the wave equation in the form (II, 276). The authors have given detailed formulae for the constants x_{ik} and g_{ik} in terms of the potential constants and the geometric data of the molecule. In these cases the g_{ik} with $i \neq k$ are not zero.

In Fig. 69, which is similar to Fig. 68, the lowest vibrational levels corresponding to an excitation of the two degenerate vibrations of XY_3 are represented. Again, in the approximation in which (II, 281) holds, some of the sublevels coincide that in higher approximation would have different energy.

It should be mentioned that Silver and Shaffer also found an additional small term independent of $v_i + d_i/2$ in the vibrational energy. However, such a term

---

FIG. 68. **Vibrational levels $v_4\nu_4 + v_5\nu_5$ of linear X_2Y_2 taking anharmonicity into account.**—The figure is approximately to scale for C_2H_2. In order to avoid overcrowding of the lettering the subscripts g or u of the species are added in some cases only to one or a few of the levels of a group of given v_4, v_5 for all of which they are the same. The formula used was $G_0 = 603v_4 + 729.5v_5 + 9v_4^2 - 13v_4v_5 - 0.5v_5^2 + 5l_4^2 + 7l_5^2$. The coefficients of l_4^2 and l_5^2 are entirely hypothetical. For the levels 1513 and 1511 the designations Δ, Γ and Σ^+, Σ^-, Δ respectively were inadvertently omitted.

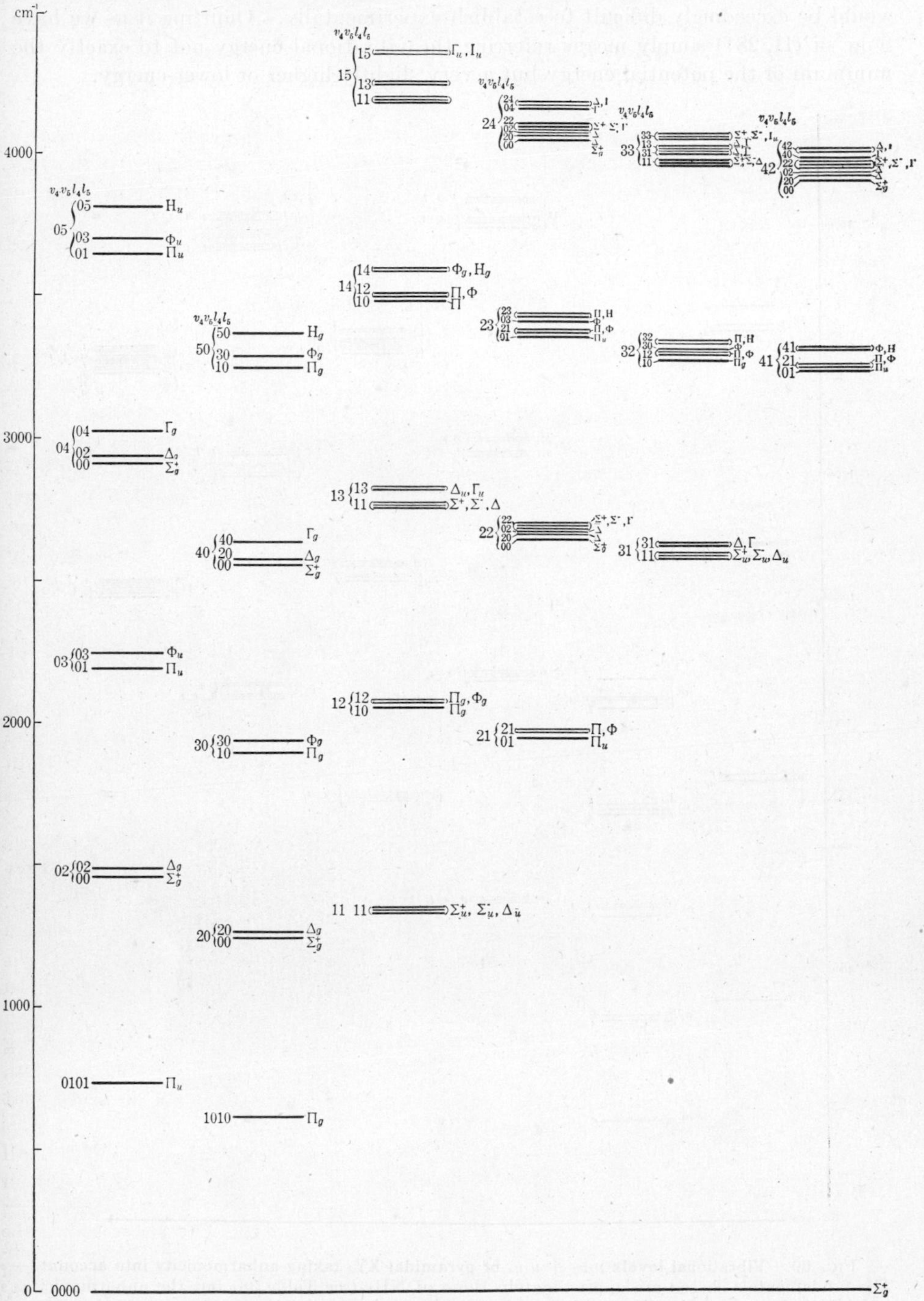

FIG. 68

would be exceedingly difficult to establish experimentally. Omitting it as we have done in (II, 281) simply means referring the vibrational energy not to exactly the minimum of the potential energy but a very slightly higher or lower energy.

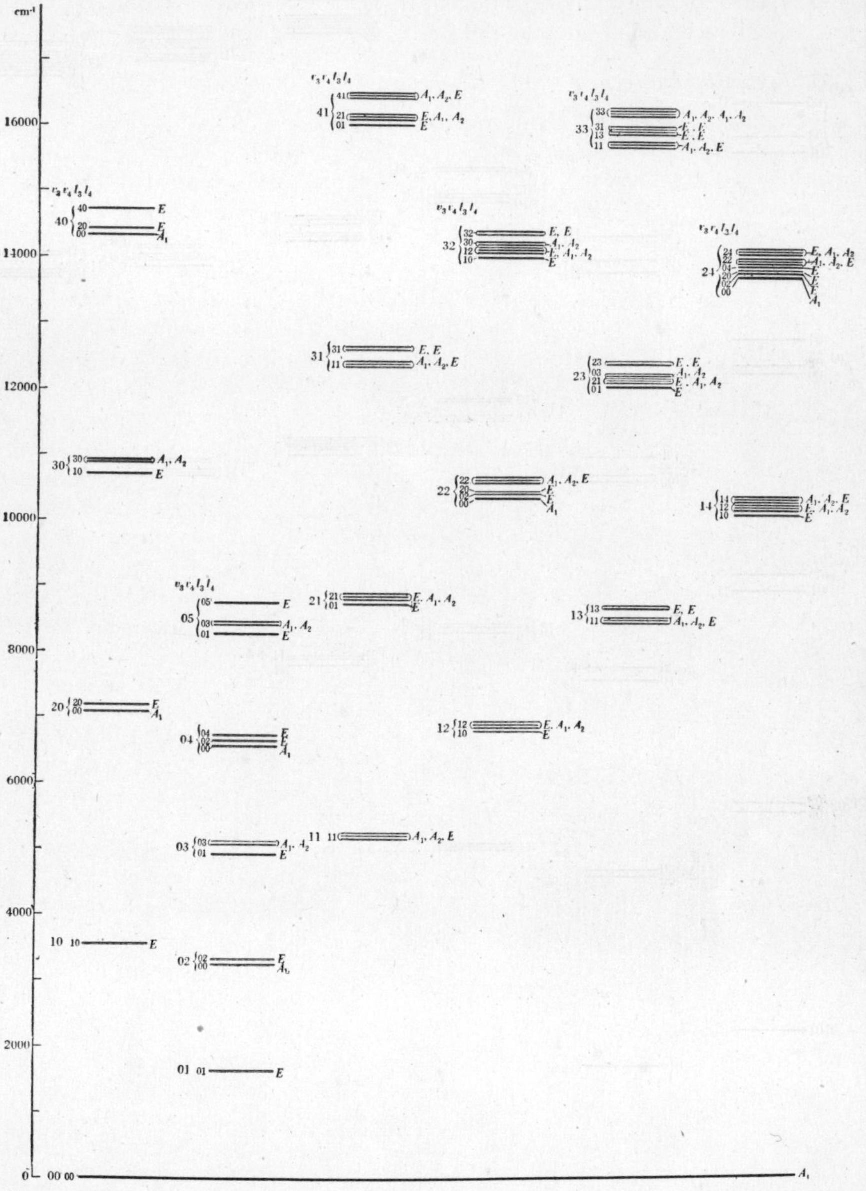

FIG. 69. **Vibrational levels $v_3\nu_3 + v_4\nu_4$ of pyramidal XY_3 taking anharmonicity into account.**—The fundamentals chosen are approximately those of NH_3 (see Table 38) but the anharmonicites are entirely hypothetical. The following formula was used: $G_0 = 3500v_3 + 1600v_4 + 20v_3^2 + 15v_3v_4 + 10v_4^2 + 25l_3^2 + 15l_3l_4 + 20l_4^2$.

It should be realized that in these cases, unlike the case of linear molecules, the l_i defined by (II, 282) do not in general represent the vibrational angular momenta about the symmetry axis. Rather, as we shall discuss in more detail in Chapter IV, section 2a, the vibrational angular momentum of a degenerate vibration ν_i is $\zeta_i(h/2\pi)$, where ζ_i is in general non-integral and has a magnitude smaller than one.

(c) *Accidental degeneracy, Fermi resonance*

Qualitative discussion. In a polyatomic molecule it may happen that two vibrational levels belonging to different vibrations (or combinations of vibrations) may have nearly the same energy, that is, may be *accidentally degenerate.* As was first recognized by Fermi (322) in the case of CO_2, such a "resonance" leads to a *perturbation of the energy levels* which is very similar to the vibrational perturbations of diatomic molecules (see Molecular Spectra I, p. 319f.). The only essential difference is that for diatomic molecules only vibrational levels of different electronic states can have nearly the same energy and thus perturb one another, whereas here two vibrational levels of the same electronic state can have the same energy and perturb each other. For example, in the case of CO_2 the level $v_1 = 0$, $v_2 = 2$, $v_3 = 0$ has almost the same energy as the level $v_1 = 1$, $v_2 = 0$, $v_3 = 0$, since $\nu_1 = 1337$ and $\nu_2 = 667$ cm^{-1} (see Table 41). For diatomic molecules, the perturbation is due to the interaction of vibration and electronic motion whereas here, for polyatomic molecules, the anharmonic terms in the potential energy, that is, the *interactions between different vibrations* are sufficient to produce a perturbation when two levels happen to lie very close together.

As for diatomic molecules, the two vibrational levels that have in zero approximation nearly the same energy *"repel" each other* and the actual levels do not follow accurately a formula of the type (II, 271) or (II, 281). Thus for CO_2 one of the two above-mentioned levels is shifted up and the other down so that the separation of the two levels is much greater than expected. At the same time a *mixing of the eigenfunctions* of the two states occurs. The deviation of the energy values from the formula and the mixing of the eigenfunctions is the stronger the smaller is the zero approximation difference of energy.

Mathematical formulation. In addition to depending, in an inverse way, on the "original" energy difference of the two levels, the magnitude of the perturbation (repulsion) depends, as for diatomic molecules, on the value of the corresponding *matrix element W_{ni} of the perturbation function W:*

$$W_{ni} = \int \psi_n{}^0 W \psi_i{}^{0*} d\tau. \qquad \text{(II, 289)}$$

The perturbation function W is here essentially given by the anharmonic (cubic, quartic, $\cdots$) terms in the potential energy,[74] while $\psi_n{}^0$ and $\psi_i{}^0$ are the zero approximation eigenfunctions of the two vibrational levels that perturb each other. Since W, as we have seen above, has the full symmetry of the molecule (is totally symmetric), $\psi_n{}^0$ must have the same symmetry type as $\psi_i{}^0$ in order to give a non-zero value to W_{ni} and therefore to the magnitude of the perturbation. [If $\psi_n{}^0$ and $\psi_i{}^0$ had different symmetry type, the integrand of (II, 289) would change for at least one

[74] Strictly speaking, it also includes the deviation of the kinetic energy operator [see equation (II, 276)] from the form it has in the zero approximation used. But the corresponding contributions to W_{ni} are usually negligible for the perturbations here under consideration.

symmetry operation and therefore the integral would vanish.] Thus we have the important rule: *Only vibrational levels of the same species can perturb one another*, or, *Fermi resonance can occur only between levels of the same species.* This rule restricts very greatly the occurrence of vibrational perturbations (Fermi resonance) in symmetrical polyatomic molecules.

If the resonance is fairly close the *magnitude of the shift* can be obtained according to first-order perturbation theory (see texts on wave mechanics) from the secular determinant

$$\begin{vmatrix} E_n^0 - E & W_{ni} \\ W_{in} & E_i^0 - E \end{vmatrix} = 0, \tag{II, 290}$$

where E_n^0 and E_i^0 are the unperturbed energies. From this equation one obtains for the perturbed energy E, since according to (II, 289) $W_{in} = W_{ni}^*$,

$$E = \bar{E}_{ni} \pm \tfrac{1}{2}\sqrt{4|W_{ni}|^2 + \delta^2}, \tag{II, 291}$$

where $\bar{E}_{ni} = \frac{1}{2}(E_i^0 + E_n^0)$ is the mean of the unperturbed levels and $\delta = E_n^0 - E_i^0$ is the separation of the unperturbed levels.[75] Formula (II, 291) shows that there is no perturbation if $W_{ni} = 0$ and that, if δ is very small—that is, if the resonance is very close—the shift is $|W_{ni}|$ up for one and down for the other level. If δ is large compared to $2|W_{ni}|$, we can expand (II, 291) into

$$E = \bar{E}_{ni} \pm \left(\frac{\delta}{2} + \frac{|W_{ni}|^2}{\delta}\right), \tag{II, 292}$$

which is essentially the same as one would obtain from second-order perturbation theory applied to each level separately.

In Fig. 70 the position of the perturbed levels for a constant W_{ni} and $\bar{E}_{ni}$ is plotted as a function of δ, the separation of the unperturbed levels. The shift produced by the perturbation is given by

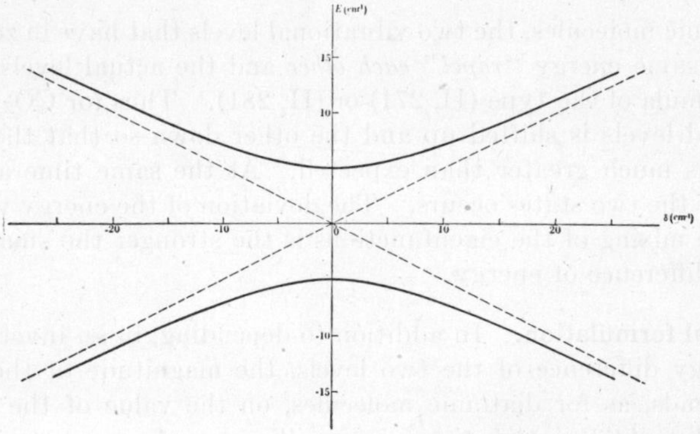

FIG. 70. **Perturbation of two energy levels as a function of the separation of the unperturbed levels.**—The broken lines represent the positions of the unperturbed levels, the solid curves those of the perturbed (actual) levels.

the separation of the heavy curve from the nearest of the broken lines (which represent the unperturbed levels). The shift is largest for $\delta = 0$, that is, for exact resonance.

The *eigenfunctions* of the two resulting states can be shown to be (from standard methods of perturbation theory) the following mixtures of the zero approximation eigenfunctions ψ_n^0 and ψ_i^0:

$$\begin{aligned} \psi_n &= a\psi_n^0 - b\psi_i^0, \\ \psi_i &= b\psi_n^0 + a\psi_i^0, \end{aligned} \tag{II, 293}$$

[75] Formula (II, 291) is seen to be identical with (V, 67) of Molecular Spectra I if it is noted that ϵ in the latter is $E - E_n^0$, that is, is the energy above the lower of the two unperturbed levels. This was not clearly stated in Molecular Spectra I, p. 311.

where

$$a = \left(\frac{\sqrt{4|W_{ni}|^2 + \delta^2} + \delta}{2\sqrt{4|W_{ni}|^2 + \delta^2}} \right)^{\frac{1}{2}}, \quad b = \left(\frac{\sqrt{4|W_{ni}|^2 + \delta^2} - \delta}{2\sqrt{4|W_{ni}|^2 + \delta^2}} \right)^{\frac{1}{2}}. \qquad \text{(II, 294)}$$

If $\delta = 0$ we obtain a fifty-fifty mixture; if δ is very large $\psi_n \rightarrow \psi_n^0$ and $\psi_i \rightarrow \psi_i^0$.

It should be emphasized that the perturbations are due to the same anharmonic terms in the potential energy that cause the terms $x_{ik}v_iv_k$ in the energy formula. The latter are due to the integrated perturbing effect of a large number of vibrational levels of which each contributes a term $|W_{ni}|^2/\delta$ as in (II, 292) whereas the former are due to the effect of one level that is particularly close to the one considered. Also, the terms $x_{ik}v_iv_k$ are always calculated by using as zero approximation the harmonic oscillator levels and eigenfunctions whereas for the perturbations one may instead also use the levels given by (II, 271) or (II, 281) and the corresponding eigenfunctions, to be substituted in (II, 289) and (II, 291).

In a way also the deviations of the characteristic group frequencies from their "normal" values in molecules containing several groups of similar frequencies may be considered as a special case of the vibrational perturbations, that is, if the "normal" values of these group frequencies are considered as the zero approximation. Also in this case such deviations occur only if the two or more vibrations have the same species and are always in such a direction as if they were produced by a "repulsion" (compare the discussion on p. 200).

Application to CO_2 and similar cases. As mentioned before, for CO_2 there is a very close resonance between the levels 1, 0, 0 and 0, 2, 0 since $2\nu_2$ happens to be very nearly equal to ν_1. The level 0, 2, 0 consists of two sublevels 0, 2^0, 0 and 0, 2^2, 0 with $l_2 = 0$ and 2, which have the species $\Sigma_g{}^+$ and Δ_g. According to the preceding discussion only the $\Sigma_g{}^+$ sublevel ($l_2 = 0$) can perturb the 1, 0, 0 level which has the species $\Sigma_g{}^+$ and conversely it alone is perturbed by the 1, 0, 0 level. This is shown in Fig. 71. The separation of the two levels 1, 0, 0 (=1388.3 cm^{-1}) and 0, 2^0, 0 (=1285.5 cm^{-1}) is much larger than would have been expected on the basis of the value for ν_2(=667.3 cm^{-1}). Similarly, of course, the splitting of the two sublevels $l_2 = 0$ and $l_2 = 2$ of 0, 2, 0 is anomalously large (49.9 cm^{-1}). In consequence of the strong perturbation, a strong mixing of the eigenfunctions of the two levels 1, 0, 0 and 0, 2^0, 0 occurs so that the two observed levels can no longer be unambiguously designated as 1, 0, 0 and 0, 2^0, 0. Each actual level is a mixture of the two. Experimentally this mixing is evidenced by the occurrence of two strong Raman lines rather than one (see Chapter III).

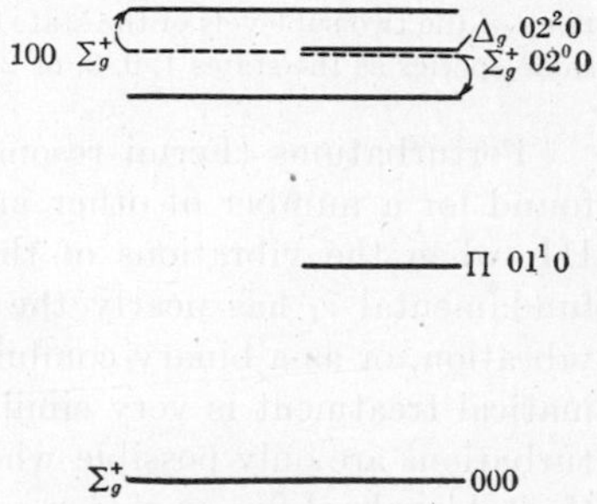

Fig. 71. **Fermi resonance in CO_2.**—The broken lines represent the unperturbed levels which go over, on account of the resonance, into the two levels to which the arrows point.

In consequence of the perturbation (resonance) between the levels 1, 0, 0 and 0, 2^0, 0 of CO_2, there are of course also perturbations between certain higher levels, for instance between 1, 0, 1 and 0, 2^0, 1 or between 1, 1^1, 0 and 0, 3^1, 0 or between the three levels 2, 0^0, 0; 1, 2^0, 0, and 0, 4^0, 0, and so on. A full discussion of these has been given by Adel and Dennison (37) [see also Dennison (280), and Chapter III, p. 275].

If the considerations of p. 205 are applied to linear symmetric XY_2, it is easily seen that the potential energy can only have the cubic terms

$$\alpha_{111}\eta_1^3 + \alpha_{122}\eta_1(\eta_{2a}^2 + \eta_{2b}^2) + \alpha_{133}\eta_1\eta_3^2. \qquad \text{(II, 295)}$$

If this is substituted for W in (II, 289) and if for the eigenfunctions ψ_n^0 and ψ_i^0 products of harmonic oscillator functions (II, 53) and (II, 58) are used, it follows that W_{ni}, if n is a level $v_1 + 1$, v_2, v_3, and i is a level v_1, $v_2 + 2$, v_3, depends only on the potential constant α_{122}. None of the other cubic and none of the quartic terms gives a contribution. For the two levels 1, 0, 0 and 0, 2^0, 0 one finds

$$W_{100,\,02^00} = -\frac{h^{\frac{3}{2}}}{8\sqrt{2}\pi^3c^{\frac{3}{2}}\omega_1^{\frac{1}{2}}\omega_2}\alpha_{122}. \qquad \text{(II, 296)}$$

The value of W_{ni} for all other corresponding pairs of levels differs from $W_{100,\,02^00}$ only by a constant factor. Therefore the separation of all the resonating pairs can be represented by the two constants: δ (the unperturbed separation of the levels 1, 0, 0 and 0, 2^0, 0) and the interaction constant $W_{100,\,02^00}$. The two constants are therefore necessary in addition to the ω_i, x_{ik}, and g_{22} of (II, 287) for a complete representation of the vibrational energy levels of a linear symmetric XY_2 molecule such as CO_2 in which $\nu_1 \approx 2\nu_2$. For CO_2 Dennison (280) obtained the values $\delta = 16.7$ cm^{-1} and $W_{100,\,02^00} = 50.4$ cm^{-1}.[76]

A resonance phenomenon occurs *also in a classical treatment* of the vibrations. For example, for XY_2 during the perpendicular vibration ν_2 there is a slight force in the X—Y direction which has a maximum value twice during one period. This will lead to an excitation of the symmetrical oscillation ν_1 if its period is half as large as that of ν_2, that is, if $2\nu_2 = \nu_1$. Thus if at first only ν_2 is excited, after a while only ν_1 will be excited; and after a further interval only ν_2 will be excited, and so on. The situation is much the same as for two coupled pendulums. The motion may be considered as the superposition of two stationary vibrations of somewhat different frequencies.

If for a linear XY_2 molecule one had $\nu_1 \approx 3\nu_2$ or $2\nu_1 \approx 3\nu_2$, no perturbation would occur since none of the two sublevels of the state 0, 3, 0 with $l_2 = 1$ and 3 (Π_u and Φ_u state respectively) has the same species as the states 1, 0, 0, or 2, 0, 0 which are Σ_g^+ states.

Perturbations (Fermi resonances) similar to those observed for CO_2 have been found for a number of other molecules. They will be discussed briefly in Chapter III, when the vibrations of these molecules are considered. In all these cases a fundamental ν_i has nearly the same frequency as a first overtone, $2\nu_k$ of another vibration, or as a binary combination $\nu_k + \nu_l$ of two other vibrations. The mathematical treatment is very similar to that for XY_2 molecules. As above, such perturbations are only possible when the fundamental ν_i has the same species as one of the sublevels of $2\nu_k$ or $\nu_k + \nu_l$.

Application to H_2O. A somewhat different perturbation has been found to occur for H_2O by Darling and Dennison (263) and is probably of importance also for many other molecules. For H_2O the two vibrations ν_1 and ν_3 have a similar magnitude (3652 and 3756 cm^{-1}) but cannot perturb each other since they have different species. However, the two first overtones $2\nu_1$ and $2\nu_3$ have the same species (A_1) and therefore can perturb each other or, more generally, *any state v_1, v_2, v_3 with $v_1 > 2$ is perturbed by a state $v_1 - 2$, v_2, $v_3 + 2$.* The recognition of this effect by Darling and Dennison (263) has first led to a satisfactory analysis of the vibration spectrum of H_2O (see Chapter III).

The essential difference from the case of CO_2 is that here the matrix element W_{ni} would be zero if only cubic terms of the potential energy were taken into account and harmonic oscillator functions were used for ψ_n^0 and ψ_i^0. However, the quartic terms of the potential with harmonic oscillator functions and the cubic terms with anharmonic oscillator functions give each a contribution. Darling

---

[76] Our $W_{100,\,02^00}$ is Dennison's $\frac{-b}{\sqrt{2}}$. His constant b corresponds to our α_{122} for a certain choice of dimensionless normal coordinates.

and Dennison have shown that

$$W_{ni} = W^{v_1, v_2, v_3}_{v_1-2, v_2, v_3+2} = \tfrac{1}{2}\gamma\sqrt{v_1(v_1 - 1)(v_3 + 1)(v_3 + 2)}, \qquad \text{(II, 297)}$$

where γ depends on the potential constants only.

While the energy levels for which no perturbation occurs are given by (II, 271) those for which a perturbation occurs are determined by (II, 290) or (II, 291) in which E_n^0 and E_i^0 are the unperturbed energies as obtained from (II, 271) and W_{ni} is given by (II, 297). Thus there is one more constant (γ) necessary to represent the observed energy levels than there would be if no resonance occurred (see also Chapter III, p. 282).

A very similar perturbation is to be expected for C_2H_2, since here also ν_1 and ν_3 have a very similar magnitude ($\nu_1 = 3374$ and $\nu_3 = 3287$) but have different species, so that only pairs of levels v_1, v_2, v_3, v_4, v_5 and $v_1 - 2, v_2, v_3 + 2, v_4, v_5$ can perturb each other. The matrix element would be of the same form as for H_2O [equation (II, 297)]. Calculations on this basis have not been carried out, but it seems certain that they would clear up certain discrepancies which have been found in the analysis of the C_2H_2 vibration spectrum [see Herzberg and Spinks (441) and Wu and Kiang (963) (961)]. Similar perturbations will also occur for many other molecules, but for very few are the available data sufficient to carry out a detailed calculation.

In higher approximations perturbations between levels with still greater differences in quantum numbers are possible. But for such pairs higher and higher terms in the potential function would be involved so that the magnitude of the perturbations for the same unperturbed separation would be smaller and smaller. However, it must be realized that the different potential constants, α_{ijk} or β_{ijkl} and so on in (II, 263), may have very different magnitude, as has actually been found for CO_2 [see Dennison (280)], and therefore the effects of certain high-order potential constants may be comparable to those of lower order.

Splitting of the l_i degeneracy. It has been mentioned before (see p. 211) that the *degeneracy of levels with the same l_i values* is not always a necessary degeneracy. In cases where it is not it *will be removed when other levels of the same species are in the neighborhood of the one considered.* For example, in a pyramidal XY_3 molecule the level in which the degenerate vibration ν_3 is excited with $v_3 = 3$, according to (II, 281) consists of two slightly different sublevels with $l_3 = 3$ and $l_3 = 1$. However, while the sublevel $l_3 = 1$ has the species E, the sublevel $l_3 = 3$ has $A_1 + A_2$ (see Fig. 69), that is, is not necessarily degenerate and will split if another level of species A_1 (or A_2) is in the neighborhood. Actually for molecules like NH_3 the vibration ν_1 of type A_1 has a very similar frequency to ν_3 and therefore the level $\nu_1 + 2\nu_3$ is in resonance with $3\nu_3$. The former has sublevels of species A_1 ($l_3 = 0$) and $E(l_3 = 2)$. The first of these can interact with the A_1 sublevel of $3\nu_3$ but not with A_2 and therefore the sublevel $l_3 = 3$ of $3\nu_3$ will split into two fairly widely separated levels. The separation would be expected to be of the same order as, or even greater than, the separation of the sublevels with different l_3.

Since in other cases a suitable perturbing level is bound to be somewhere, even though it may be at a fairly large distance from the one considered, the l_i degeneracy is always split, as has been indicated in Figs. 68 and 69. But the splitting is very slight if the perturbing state is not fairly close.

The preceding considerations show that *the general formulae* (II, 271) *and* (II, 281) *can be expected to give an accurate representation of the vibrational levels of a polyatomic molecule only if no resonances between levels of the same species occur.* It is clear that the more atoms a molecule contains the less likely it is to be free from such resonances, even for the lower vibrational levels. Thus the formulae mentioned must be applied with caution.

(d) Several potential minima

General remarks. It happens frequently that the potential function of a given system of atoms has more than one minimum, that is, that there is more than one equilibrium position. If these minima have different energy and if the potential function in their neighborhood has different shapes, the minima correspond to *different isomers* of the molecule considered, which in general can be separated chemically from one another (for example, CH_3CN and CH_3NC, cis- and trans-$C_2H_2Cl_2$). For most practical purposes and particularly for the consideration of their vibrational levels they can be considered separately as different molecules, except when the barrier separating the minima is very low (as is the case for molecules like $C_2H_4Cl_2$; see Chapter III, section 3f). However, *when the potential minima are identical in height and in shape* the situation is rather different, because exact resonances of the energy levels corresponding to the two minima, and consequently splittings of the energy levels may occur and it may be impossible to separate the different isomers.

We have to distinguish two essentially different cases:

(1) *All non-planar molecules irrespective of whether or not they contain identical atoms have two identical potential minima corresponding to the two equilibrium positions of the nuclei that result from an inversion of all nuclei at the center of mass.* These two configurations cannot be transformed into each other by simple rotations of the whole molecule. The energy levels corresponding to the two configurations are always in exact resonance and therefore, in consequence of the tunnel effect, a splitting into two energy levels, usually very close together, takes place. This is what was previously (p. 27) called *inversion doubling*. For example, for pyramidal XY_3 or for axial ZXY_3, if the Y atoms are numbered $Y^{(1)}$, $Y^{(2)}$, $Y^{(3)}$, by an inversion a configuration is obtained that cannot be obtained by rotations. To be sure, in the cases mentioned the configuration obtained by inversion, because of the identity of the Y atoms, cannot actually be distinguished from the original one. Only if the Y atoms are actually different would that be possible. This is the case for *optical isomers*, the simplest example of which would be a non-planar molecule WXYZ. But in all cases the double degeneracy of all vibrational levels derived for one potential minimum exists and is split in higher approximation. For plane and linear molecules an inversion can always be replaced by a rotation of the whole molecule and therefore the double degeneracy of the vibrational levels and the inversion doubling do not exist (see also p. 27).

(2) *In all cases of molecules with identical atoms, when an exchange of these identical atoms cannot be brought about by a rotation of the whole molecule, or by an inversion, or both, two or more identical potential minima exist.* Examples are molecules like linear

$$X-Y-Y\ (N_2O),\ X_2Y_4\ \text{of point group}\ V_h\ (C_2H_4),\quad \begin{matrix} W \\ \ \\ Z \end{matrix} X-X \begin{matrix} Y \\ \ \\ Y \end{matrix}\ (ClBrC\!\!=\!\!CH_2),$$

X_2Y_6 of point group D_{3h} or D_{3d} (C_2H_6), non-axial $WZXY_3$ (CH_3OH). In all these cases the exchange of the identical nuclei, that is, a transition from one potential minimum to another identical one, can be brought about by an internal rotation, that is, a rotation of one part of the molecule against the other.[77] If the identical atoms have nuclei of the same mass, the vibrational levels in each minimum (if

---

[77] In molecules like symmetrical XY_2 (linear or bent), XY_3 (planar or non-planar), or axial XYZ_3, this type of identical potential minima obviously does not occur.

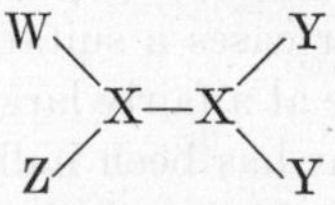

considered separately) will be the same; that is, there will be a degeneracy of each vibrational level of a degree equal to the number of identical minima. This degeneracy will be split in the actual molecule (see below). If the masses of the identical atoms are different the energy levels will be different in the two minima and no resonance will occur. In this case the situation is essentially the same as if the two atoms were different; that is, we have a number of isomers with slightly different properties and particularly a different spectrum, for example $N^{14}N^{15}O$ and

$N^{15}N^{14}O$, or
$$\begin{array}{c}H\\ \diagdown \\ \\ D \diagup\end{array} C\!=\!C \begin{array}{c}H\\ \diagup \\ \\ \diagdown D\end{array} \quad \text{and} \quad \begin{array}{c}D\\ \diagdown \\ \\ H \diagup\end{array} C\!=\!C \begin{array}{c}H\\ \diagup \\ \\ \diagdown D\end{array} \quad , \text{ and others.}$$

It should be noted that the degeneracy introduced by the identical potential minima is not included in the previous treatment of normal vibrations and their symmetry types. This was quite legitimate, since for a strictly quadratic potential energy a transition from one to the other potential minimum would be impossible.

Inversion doubling in NH_3 and similar molecules. The only case of inversion doubling that has been studied in considerable detail both experimentally and theoretically is that of NH_3. Here, assuming a pyramidal structure, if the N atom is moved through the H_3 plane to an equivalent position on the other side an inverted configuration is obtained. If the potential energy is plotted as a function of the distance of the N atom from the H_3 plane, a curve of the form given in Fig. 72a is obtained. Let us for a moment consider the one-dimensional motion of a single particle in a potential field of this form. Suppose the broken horizontal lines are the energy levels that one would obtain if one had two independent (identical) minima not connected by a potential hill (broken curves). Then in consequence of the resonance together with the perturbation, that is, deviation of the potential from the broken curves, a splitting of each degenerate level into two levels indicated by the full horizontal lines in Fig. 72a occurs. *The splitting increases rapidly with increasing v.*

Because of the symmetry of the potential field the corresponding *eigenfunctions* must remain unchanged or can at most change sign for a reflection at the origin $(x \rightarrow -x)$. In a zero approximation, the two functions corresponding to a pair of levels are simply a symmetric and an antisymmetric combination

$$\psi_s = \psi_v(x_0 - x) + \psi_v(x_0 + x),$$
$$\psi_a = \psi_v(x_0 - x) - \psi_v(x_0 + x) \tag{II, 296}$$

of the oscillator wave functions $\psi_v(\xi)$ corresponding to each minimum, where ξ is the displacement from the minimum (counted positive toward the potential hill). These eigenfunctions are given for $v = 0$, 1, and 2 in Fig. 72b. It can be shown [78] that the symmetric function always corresponds to the lower one of a pair of levels. The eigenfunctions tell us that there is an equal probability of finding the particle in the left and in the right bowl. Whereas in classical mechanics, once the particle is at the right with an energy lower than that of the hill it can never go over to the left bowl, in quantum mechanics after a certain length of time the particle will be found in the left bowl, that is, it will *penetrate the potential hill* (tunnel effect). The

[78] See, for example, the somewhat similar problem of the H_2^+ molecule as treated in Pauling and Wilson (18).

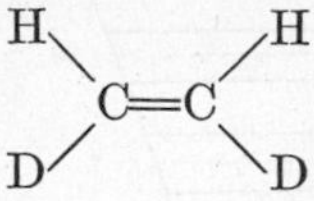

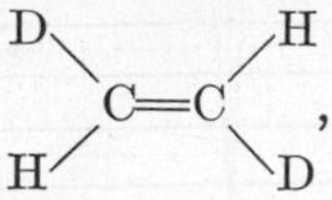

average time it takes the particle to go from one side to the other (to penetrate the hill) is inversely proportional to the energy difference of the two sublevels for a given v. (It is $1/(2\Delta Gc)$ if ΔG is the energy difference in cm^{-1}).

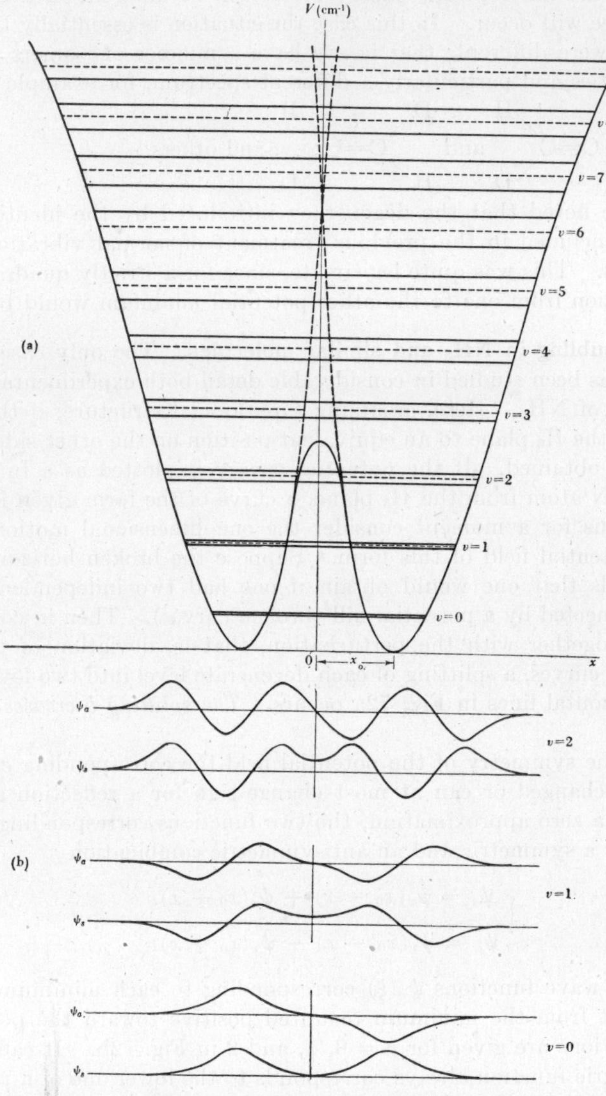

Fig. 72. (a) Potential energy in a pyramidal XY_3 molecule as a function of the distance of the X atom from the Y_3 plane; (b) Eigenfunctions for this case assuming one-dimensional oscillations. —The splitting for $v = 0$ and 1 is much exaggerated.

The mutual interaction of the "right" and "left" energy levels, that is, the magnitude of the doublet splitting, depends on the matrix element W_{ni} defined in (II, 289). It is clear that since the perturbation function is the deviation from the broken-line potential curves in Fig. 72a, W_{ni} will increase very rapidly as the top of the potential hill is approached. [Calculations show that it decreases exponentially as the area of the hill cut off by the energy level increases; see Dennison and Uhlenbeck (284).] Thus the *doublet splitting* and therefore the probability for the particle to go from one side to the other *is exceedingly small for levels considerably below the top of the hill but quite large near the top.* The doublet splitting increases further for the levels above the top until very far above the top it is equal to half the separation of successive unperturbed levels. In other words, far above the top of the hill we have again a simple series of levels with roughly half the spacing that one would have if there were only one minimum (see Fig. 72a). The eigenfunctions of these levels are alternately symmetric and antisymmetric with respect to reflection at the origin (positive and negative levels) similar to the successive levels of the vibration $\nu_2(a_2'')$ of a plane XY_3 molecule (see Fig. 63 and Fig. 42b).

In applying the above considerations to the actual energy levels of NH_3 and similar molecules, it must be realized that none of the normal modes is exactly a one-dimensional oscillation of the N atom against the H_3 plane. Rather in each one of the four normal vibrations (see Fig. 45) the height of the pyramid is changed somewhat, and the splitting of each vibrational level depends on the magnitude of this change in a way that can be roughly obtained from Fig. 72a by finding the potential energy change corresponding to the change of height and interpolating the splitting for this energy. The change of height is in first approximation zero for the two degenerate vibrations ν_3 and ν_4, and therefore one would expect the splitting to be almost independent of v_3 and v_4 (as long as they are small).[79] But the change of height is different from zero, even in a first approximation, for the non-degenerate vibrations ν_1 and ν_2, and therefore the splitting should increase fairly rapidly with increasing v_1 and v_2. The change of height is by far the greatest for ν_2.

These predictions are strikingly confirmed by the observations on NH_3. The splitting of the ground state $v_1 = 0$, $v_2 = 0$, $v_3 = 0$, $v_4 = 0$ is exceedingly small, amounting to only 0.66 cm^{-1}; for the state $v_2 = 1$ the splitting is 35.7, for $v_2 = 2$ it is 312.5 cm^{-1} [see Dennison (280)], while for the state $v_1 = 1$ it is only 1 cm^{-1} even though $\nu_1 = 3337$ against $\nu_2 = 950$ cm^{-1}. For the perpendicular vibrations the observations are not yet accurate enough to detect small splittings under 1 cm^{-1}, but the splitting is certainly not much greater than in the lowest state.

To a fairly good approximation the normal vibration ν_2 of NH_3 is one in which only the distance of the N atom from the H_3 plane is changed; that is, the energy levels of this vibration would be approximately those of a one-dimensional oscillator moving in a potential field of the form of Fig. 72a. Using a certain functional form of this potential, it is possible to determine the splitting of the vibrational levels $v_2\nu_2$ as a function of the potential constants; or, conversely, *from the observed splittings, the constants of this potential, in particular the separation of the two minima and the height of the maximum, can be determined.* Such calculations have been carried out by Morse and Stueckelberg (636), Dennison and Uhlenbeck (284), Rosen and Morse (742), Manning (599), and Wall and Glockler (911). It turns out that the resulting separation of the minima is almost independent of the particular form assumed for the potential. The height of the NH_3 pyramid, which is one-half the separation of

[79] For large v_3 and v_4 the fact has to be taken into account that an inversion can also be brought about by an internal rotation of an NH_2 group by 180°. While the potential hill for this rotation is expected to be much larger than the one discussed in the text, the change of angle for ν_3 and ν_4 will be relatively much larger than the change of height.

the minima, is thus found to be 0.38×10^{-8} cm, which agrees very satisfactorily with the value 0.381×10^{-8} cm obtained from the infrared rotation-vibration spectrum (see Chapter IV, p. 439). The height of the potential hill, according to Manning (599), is 2076 cm^{-1}; that is, the level $2\nu_2$ is just below its top and all higher levels of this vibration are above it.

From the same potential function the splittings for ND_3, ND_2H, and NH_2D may be calculated [see Manning (599), Wall and Glockler (911), and Dennison (280)] and they come out in satisfactory agreement with experiment. For ND_3, for example, the calculated splittings for ν_2 and $2\nu_2$ are 2.5 and 55 cm^{-1} while the observed values are 2.4 and 70 cm^{-1}. The splitting is smaller than for NH_3 because of the smaller energy above the minimum and the smaller tunnel effect due to the larger reduced mass.[80] It has not yet been observed for the ground state of ND_3. A number of levels above the maximum have been observed for ND_3 [see Table 74, p. 297] which agree well with the theoretical results of Manning (599) [see also Dennison (280)]. The splitting for levels other than $v_2\nu_2$ has not as yet been evaluated theoretically.

For PH_3, Sutherland, Lee and Wu (830) have found an inversion doubling of 2.4 cm^{-1} for the state ν_2. From this datum and an approximate value for the height of the pyramid (0.67×10^{-8} cm) the height of the potential maximum is found to be of the same order as for NH_3. The splitting for the ground state obtained on this basis is only 1.51×10^{-4} cm^{-1}, which would be very difficult to observe directly.

The average time for an inversion to take place is, according to the previous formula, 2.5×10^{-11} and 1.1×10^{-7} sec. for NH_3 and PH_3 respectively (in their ground states). These values are respectively 700 and 3.3×10^6 times the periods of oscillation of ν_2.

The only other case of an observable inversion doubling seems to occur for the H_2O_2 molecule. Zumwalt and Giguère (977) have interpreted the observed doublet structure of a photographic infrared band of this molecule as due to an inversion doubling, assuming the model represented by Fig. 2a. In this case the inversion can be brought about by a torsion of the two OH groups about the O—O axis. However, their interpretation of the spectrum is admittedly not final.

Optical isomers. The comparison of NH_3 and PH_3 shows how rapidly the time required for a switching over (an inversion) increases with increasing height of the potential hill and increasing mass of the atoms concerned. For AsD_3 assuming a height of 0.75×10^{-8} cm and a potential maximum of the same height as for NH_3 a value of 1.5×10^{-2} sec. would be obtained. It is therefore clear that with even heavier masses and greater potential hills, as for example in $AsCl_3$ or $BiCl_3$, the time for one switching over will be many powers of ten larger, that is, of the order of hours or even years. As long as there are identical nuclei in the molecule, as in $AsCl_3$, the inverted form is indistinguishable from the non-inverted. But *if all four atoms are different*, as in 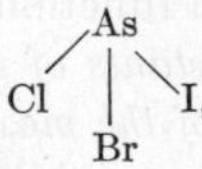

, the two forms would be distinguishable, for example by the fact that one would rotate the plane of polarization of light to the right, the other to the left. They would be optical isomers. While optical isomers of the pyramidal type have not as yet been found, similar isomers of tetrahedral type are well known,

[80] It may be noted that in going through the potential hill the center of mass must be conserved. Therefore the H or D atoms are displaced more than the N atom, the motion corresponding to that of a single particle in the potential field of Fig. 72a with a reduced mass equal to that applying to the vibration ν_2 in a plane XY_3 molecule $\left[\text{that is the factor } m_Y \Big/ \left(1 + \frac{3m_Y}{m_X}\right) \text{ in (II, 210)}\right]$.

namely carbon compounds with four different substituents. Here, in order to transform one molecule into the inverted form, one has to twist one part of the molecule against the other until two substituents are exchanged. The potential hill that has to be surmounted is very large, particularly if the substituents are large groups; the tunnel effect is correspondingly extremely slight. One can estimate [see Hund (465)] that the time that it would take the free molecule to switch over by means of tunneling[81] may be as high as 10^9 years. Thus it is possible to understand that there are bacteria that eat only one of the two forms (the right or left one) of a certain carbon compound. The asymmetric molecules that must be present in these bacteria have not switched over to the opposite configuration since the particular genus came into existence.

Torsional oscillations. The only cases of *identical potential minima, produced by identity of atoms in the molecule*, that have been discussed in greater detail are those involving torsional oscillations.[82] For example, in C_2H_4 the potential energy, plotted as a function of the angle χ of relative rotation of the two CH_2 groups against each other, has two identical minima as shown in Fig. 73a. Similarly, in C_2H_6 and

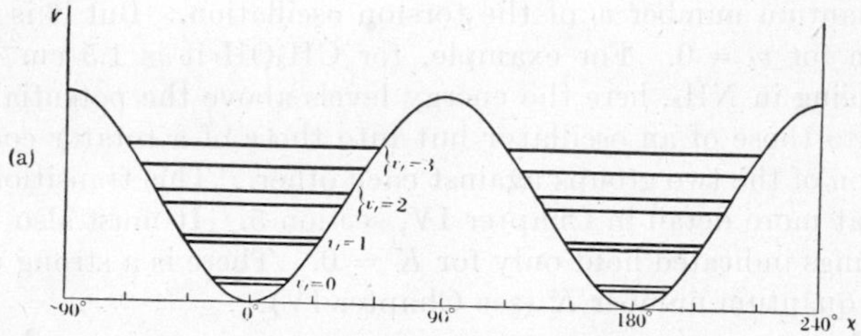

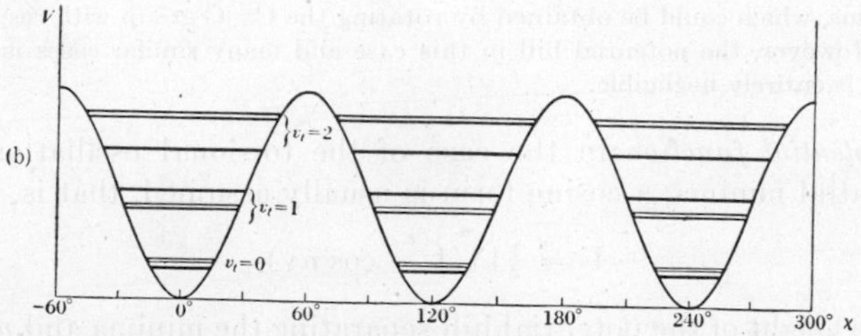

FIG. 73. **Potential energy as a function of the angle of torsion in C_2H_4- and C_2H_6-like molecules and energy levels of the torsion oscillation** (qualitative).—The energy levels are shown only where the kinetic energy is positive. But it must be realized that the eigenfunctions have non-zero (even though small) values also when the potential energy is greater than the total energy (under the potential barriers).

CH_3OH, the potential energy has obviously three identical minima if plotted as a function of the angle of torsion (Fig. 73b). We have, for C_2H_4, for the potential energy V:

$$V(\chi) = V(\chi + \pi) \qquad \text{(II, 299)}$$

[81] The usually observed inversion is due to thermal collisions.

[82] One case that does not involve torsion has recently been discussed by Glockler and Evans (371) on the basis of the data of Buswell, Maycock, and Rodebush (187), namely the linear FHF^- ion in which there appear to be two equilibrium positions for the proton, leading to a doubling of the energy levels. However, the experimental evidence appears to be somewhat ambiguous [see Ketelaar (499)].

and for C_2H_6, CH_3OH, and similar cases,

$$V(\chi) = V(\chi \pm \tfrac{2}{3}\pi). \tag{II, 300}$$

If, as in the examples, the nuclei of the identical atoms have the same mass, the energy levels corresponding to the identical minima are in zero approximation the same; that is, we have in the above cases a two- and three-fold degeneracy respectively. This degeneracy is split, at least partially, if the deviation from a strictly parabolic potential curve, or in other words the passage through the potential barrier, is considered. *Then for C_2H_4 and similar molecules each level of the torsion oscillation $\nu_4(a_u)$ is split into two sublevels. For C_2H_6, CH_3OH, and similar molecules, there is also a splitting into two sublevels but one of these is doubly degenerate* [see Wilson (938) and Koehler and Dennison (517)]. The degenerate sublevel is alternately the upper and lower one. The splittings are shown schematically in Fig. 73. As in the case of inversion doubling, the magnitude of the splitting increases rapidly with increasing vibrational quantum number v_t of the torsion oscillation. But it is in general not negligible even for $v_t = 0$. For example, for CH_3OH it is 1.5 cm^{-1}. Unlike the inversion doubling in NH_3, here the energy levels above the potential maximum do not go over into those of an oscillator but into those of a rotator corresponding to the free rotation of the two groups against each other. This transition will be taken up in somewhat more detail in Chapter IV, section 5. It must also be emphasized that the splittings indicated hold only for $K = 0$. There is a strong dependence on the rotational quantum number K (see Chapter IV).

It may be pointed out that in C_2H_4, for example, in addition to the two identical potential minima mentioned (Fig. 73a) there are also a number of others, for instance that obtained by an exchange of the two C atoms, which could be obtained by rotating the C=C group with respect to the rest of the molecule. However, the potential hill in this case and many similar cases is so high that the ensuing splitting is entirely negligible.

For the *potential function* in the case of the torsional oscillations involving n identical potential minima, a cosine form is usually assumed, that is,

$$V = \tfrac{1}{2}V_0(1 - \cos n\chi), \tag{II, 301}$$

where V_0 is the height of the potential hill separating the minima and n is the number of minima. The *energy levels* corresponding to this potential function cannot be represented by a closed analytical expression except the lower levels in the case of a high potential barrier, that is, when essentially the presence of more than one potential minimum is disregarded. In this case we have a harmonic oscillator with the energy levels

$$G(v_t) = \omega_t(v_t + \tfrac{1}{2}). \tag{II, 302}$$

Here the torsional frequency ω_t, as can easily be seen, is determined by

$$\omega_t = n\sqrt{\frac{V_0 A_1 A_2}{A}}, \tag{II, 303}$$

where A_1 and A_2 are the rotational constants corresponding to the moments of inertia $I_A^{(1)}$ and $I_A^{(2)}$ of the two parts that carry out the torsional motion with respect to each other, where A is the rotational constant corresponding to the moment of inertia I_A $(= I_A^{(1)} + I_A^{(2)})$ of the whole molecule about the axis of torsion, and where

V_0 is in cm^{-1}. For molecules such as C_2H_4 and C_2H_6 with two identical parts, the relation (II, 303) reduces to

$$\omega_t = 2n\sqrt{V_0 A}. \tag{II, 304}$$

For the energy levels in intermediate cases we refer to the work of Nielsen (661), Teller and Weigert (838), Howard (461), Koehler and Dennison (517), and the graphical representations in Chapter IV, Figs. 165 and 166. More complicated cases with several groups carrying out a torsional motion (hindered rotation) have been discussed by Crawford (236) and Pitzer and Gwinn (698).

While in principle it would be possible to obtain the height of the potential hill V_0 (just as in the case of NH_3 above) from the magnitude of the splitting of the levels of given v_t on account of the tunnel effect, in all practical cases thus far this height has been obtained from the position of the levels of different v_t. For example, for C_2H_4 one obtains $V_0 = 8700$ cm^{-1} from (II, 304) with $n = 2$, $\omega_t = 825$ cm^{-1} (see Table 92, p. 326) and $A = 4.87$ cm^{-1} (see Table 132, p. 437). On the basis of the exact relations for the energy levels of a cosine potential (not given here), Kistiakowsky, Lacher, and Stitt (510) obtained $V_0 = 960$ cm^{-1} for C_2H_6 and Koehler and Dennison (517) obtained $V_0 = 470$ cm^{-1} for CH_3OH. In the latter case Crawford (238), on the basis of thermal data (see Chapter V, section 1), obtained $V_0 = 1200$ cm^{-1}. (For further examples see Table 143 p. 520 and the accompanying discussion).

The value of the hindering potential V_0 obtained from the observed energy levels depends considerably on the form of the assumed potential function. Thus Charlesby (196a), assuming a barrier of rectangular shape, obtained for C_2H_6 on the basis of the same energy levels as used by Kistiakowsky, Lacher, and Stitt, the value $V_0 = 600$ cm^{-1}. If it were possible to observe the splitting of levels of a given v_t a valuable check on the shape of the potential function could be obtained.

It should of course be realized that torsional oscillations may also occur when the potential minima are not identical. But few such cases have as yet been studied.

6. Isotope Effect

Introductory remarks. The investigation of the vibrational isotope effect is of even greater importance for polyatomic molecules than for diatomic molecules. Since isotopic molecules have the same electronic structure the *potential function* under the influence of which the nuclei are moving *is the same to a very high order of approximation.*[83] But because of the difference in the masses the vibrational frequencies (levels) are different. Therefore the investigation of the vibrational frequencies of molecules isotopic with the one considered gives *additional equations for the force constants.* It will be remembered that the number of constants in the most general quadratic potential function is usually larger than the number of fundamental frequencies (see p. 159). Thus if only one isotopic species is observed not all the force constants can be evaluated unless simplifying assumptions are made. But with the help of the fundamental frequencies of one or more isotopic molecules a

[83] Strictly speaking, as for diatomic molecules, there may be a slight difference when other electronic states are in the neighborhood of the one considered. But this hardly ever happens for the electronic ground state.

sufficient number of additional equations is in general obtained to determine all constants in the most general (quadratic) potential function.

In addition, the investigation of the isotope effect is invaluable in *correlating the observed vibrational frequencies with the theoretical normal modes of a certain molecule.* It is obvious that the isotope shift of a certain vibrational frequency will be very small when the atom which is replaced by its isotope moves very little in the particular normal mode, whereas the shift will be relatively large when the atom in question has a large amplitude in that normal mode. Thus, for example, for CH_3Cl only one fundamental frequency, 732 cm^{-1}, shows an appreciable isotope splitting corresponding to CH_3Cl^{35} and CH_3Cl^{37}. It must therefore correspond essentially to the C—Cl vibration, a conclusion that is also quantitatively in agreement with the observed shift.

Finally, by means of the (vibrational) isotope effect in certain cases *information about the geometrical structure* may be obtained, since the relative amplitude of vibration of a certain nucleus (to be replaced by an isotope) depends on the geometrical arrangements of the nuclei. For example, the relative amplitude of the X atom in the antisymmetrical vibration of a symmetrical XY_2 molecule is large when the molecule is linear and decreases to zero as the Y—X—Y angle goes to zero. The isotope shift produced by replacing X by an isotope therefore depends critically on the angle, which in turn can be determined from the observed shift.

Up to the present time almost all isotope calculations have been based on the *harmonic oscillator approximation*, that is, on a strictly quadratic potential function. Therefore the *formulae* to be discussed in the following can be expected to *hold rigorously only for the zero-order frequencies* ω_i [see equations (II, 271) and (II, 281)]. Unfortunately there are only very few cases for which these ω_i have been evaluated from a complete vibrational analysis of the infrared and Raman spectra. However, since in general the anharmonic constants x_{ik} are small, the observed fundamentals ν_i represent fair approximations to the ω_i, and the isotope relations should hold for them at least to a certain approximation.

Triatomic molecules. The normal frequencies of a non-linear symmetrical triatomic molecule, assuming the most general quadratic potential function, are given by the previous formulae (II, 124–126). These formulae can be applied only to those isotopic molecules that are also symmetrical; that is, in which the central atom is replaced by an isotope, or in which both "end" atoms are replaced by the same isotope, or in which both these substitutions have been made (for example, H_2O^{16}, H_2O^{18}, D_2O^{16}, D_2O^{18}). In this case we have *for the antisymmetric frequency* ν_3 from (II, 124),

$$\frac{\lambda_3^{(i)}}{\lambda_3} = \left(\frac{\omega_3^{(i)}}{\omega_3}\right)^2 = \frac{m_X m_Y(m_X^{(i)} + 2m_Y^{(i)} \sin^2 \alpha)}{m_X^{(i)} m_Y^{(i)}(m_X + 2m_Y \sin^2 \alpha)}, \tag{II, 305}$$

where the superscript (i) designates quantities referring to the isotopic molecule and where ω_3 has been used instead of ν_3 in (II, 124) in order to emphasize that the formula holds strictly only for the zero-order frequencies. It is seen that the ratio of the frequencies of the antisymmetric vibrations depends only on the masses and on the apex angle 2α of the triangle.

For the *symmetric vibrations* ν_1 and ν_2 of XY_2 one obtains immediately from the general equation (II, 126)

$$\frac{\lambda_1^{(i)}\lambda_2^{(i)}}{\lambda_1\lambda_2} = \left(\frac{\omega_1^{(i)}\omega_2^{(i)}}{\omega_1\omega_2}\right)^2 = \frac{(2m_Y^{(i)} + m_X^{(i)})}{(2m_Y + m_X)}\frac{m_Y{}^2 m_X}{m_Y^{(i)2} m_X^{(i)}}; \qquad \text{(II, 306)}$$

that is, the ratio of the products $\omega_1\omega_2$ formed for the two isotopic molecules is independent of the potential constants and even of the angle α, and depends only on the masses.

In Table 52 the observed values of $\nu_3{}^{(i)}/\nu_3$ and $\nu_1{}^{(i)}\nu_2{}^{(i)}/\nu_1\nu_2$ and the values of $\omega_3{}^{(i)}/\omega_3$ and $\omega_1{}^{(i)}\omega_2{}^{(i)}/\omega_1\omega_2$ as calculated from (II, 305) and (II, 306) are given for the three isotopic pairs H_2O, D_2O; H_2S, D_2S; and H_2Se, D_2Se. The fundamental

TABLE 52. OBSERVED AND CALCULATED ISOTOPE EFFECT IN THE PAIRS $H_2O—D_2O$, $H_2S—D_2S$ AND $H_2Se—D_2Se$.

Isotopic pair	$\frac{\nu_3{}^{(i)}}{\nu_3}$, observed	$\frac{\omega_3{}^{(i)}}{\omega_3}$, calculated	$\frac{\nu_1{}^{(i)}\nu_2{}^{(i)}}{\nu_1\nu_2}$, observed	$\frac{\omega_1{}^{(i)}\omega_2{}^{(i)}}{\omega_1\omega_2}$, calculated
H_2O, D_2O	0.7425	0.7329	0.5390	0.5275_9
H_2S, D_2S	0.7449	0.7184	0.5246	0.5149_9
H_2Se, D_2Se	0.7217	0.7118	0.5003	0.5065_6

frequencies and angles used are those given in the previous Table 37, p. 161. The agreement of observed and calculated values is quite satisfactory, the small differences being due to the use of observed fundamental frequencies rather than zero-order frequencies. It must also be remembered that the influence of anharmonicity for vibrations involving H atoms is particularly large. The rather poor agreement for $\nu_3{}^{(i)}/\nu_3$ of H_2S is probably due to an erroneous measurement of the center of the fundamental ν_3 in either H_2S or D_2S (see Chapter III).

In cases of doubtful assignments of the observed fundamental frequencies of two isotopic XY_2 molecules, equation (II, 306) may serve as a check as to whether the correct identification has been obtained. Also it is possible to calculate one of the zero-order frequencies from (II, 306) if all the others are known.

If one substitutes the observed $\nu_3{}^{(i)}/\nu_3$ in (II, 305) and calculates the angle α, one obtains for H_2O $2\alpha = 137°$ instead of the correct value 105°, while for H_2S and H_2Se one finds $\sin\alpha > 1$, that is, no real solution. This indicates that the angle α obtained from (II, 305) is very sensitive to the use of the correct $\omega_3{}^{(i)}/\omega_3$.

For H_2O the zero-order frequencies are known from a detailed analysis of the vibrational spectrum [see Darling and Dennison (263) and p. 282]. They are

$$\omega_1 = 3825.3, \qquad \omega_2 = 1653.9, \qquad \omega_3 = 3935.6 \text{ cm}^{-1}.$$

For D_2O the available data are not sufficient to determine them directly from the D_2O spectrum. But they can be obtained somewhat indirectly, as follows. The anharmonic coefficients x_{ik} of D_2O can be obtained from those of H_2O if, in analogy to the relation $\omega_e{}^{(i)}x_e{}^{(i)} = \rho^2\omega_e x_e = (\omega_e{}^{(i)2}/\omega_e{}^2)\omega_e x_e$ for diatomic molecules, one assumes [see Bonner (162) and Darling and Dennison (263)] [84]

$$x_{ik}^{(i)} = \frac{\omega_i{}^{(i)}\omega_k{}^{(i)}}{\omega_i\omega_k} x_{ik}. \qquad \text{(II, 307)}$$

[84] This relation has not been rigorously proved but seems to be justified by the good agreement obtained.

With these $x_{ik}^{(i)}$, the zero-order frequencies $\omega_i^{(i)}$ of D_2O are obtained from the observed fundamentals according to (II, 270) as [85]

$$\omega_1^{(i)} = 2757.9, \qquad \omega_2^{(i)} = 1210.2, \qquad \omega_3^{(i)} = 2885.1.$$

Using these "observed" ω values one obtains

$$\left(\frac{\omega_3^{(i)}}{\omega_3}\right)_{\text{observed}} = 0.7331 \qquad \left(\frac{\omega_1^{(i)}\omega_2^{(i)}}{\omega_1\omega_2}\right)_{\text{observed}} = 0.52755,$$

in almost perfect agreement with the calculated values in Table 52. The value of 2α obtained from (II, 305) with the above $(\omega_3^{(i)}/\omega_3)_{\text{observed}}$ is 106° in very good agreement with the value obtained from the rotation vibration spectrum (p. 489).

Because of the influence of anharmonicity, a *calculation of the most general quadratic potential constants* from the equations (II, 124–126) applied to the two isotopes is worth while only when the zero-order frequencies are known. This is the case only for H_2O and D_2O (see above). The resulting values of the a_{ik} in (II, 97) for this case are, in dynes per centimeter,

$$a_{11} = a_{22} = +\,9.568 \times 10^5, \qquad a_{33} = +\,2.037 \times 10^5,$$
$$a_{12} = +\,1.048 \times 10^5, \qquad a_{13} = a_{23} = -\,1.535 \times 10^5.$$

The fact that a_{12} and a_{13} come out to be of a magnitude similar to a_{33} confirms the previous conclusion that the central force system (in which they are neglected) is a poor approximation.

For *linear symmetrical* XY_2 *molecules*, equations (II, 305) and (II, 306) also hold if in the former one puts $\alpha = 90°$. However, since for these molecules ν_1 and ν_2 have different symmetry separate relations may be obtained for them. In this case, unlike that of bent XY_2, (II, 243) is the most general quadratic potential function. The expressions (II, 245) for λ_1 and λ_2 that follow from (II, 243) lead immediately to

$$\frac{\lambda_1^{(i)}}{\lambda_1} = \left(\frac{\omega_1^{(i)}}{\omega_1}\right)^2 = \frac{m_Y}{m_Y^{(i)}},$$

$$\frac{\lambda_2^{(i)}}{\lambda_2} = \left(\frac{\omega_2^{(i)}}{\omega_2}\right)^2 = \left(\frac{\omega_3^{(i)}}{\omega_3}\right)^2 = \frac{m_Y\left(1+\dfrac{2m_Y^{(i)}}{m_X^{(i)}}\right)}{m_Y^{(i)}\left(1+\dfrac{2m_Y}{m_X}\right)} \qquad \text{(II, 308)}$$

For $C^{13}O_2^{16}$ and $C^{12}O_2^{16}$, the observed fundamentals ν_3 [see Nielsen (653) and Chapter III, section 3a] are in the ratio $\nu_3^{(i)}/\nu_3 = 0.9721$, whereas from (II, 308) $\omega_3^{(i)}/\omega_3 = 0.97154$, the difference again being due to the use of fundamentals rather than zero-order frequencies.[86]

All the preceding relations hold only when the two isotopic molecules considered have the same symmetry, but they are *valid for any mass difference of the isotopes.* They do not hold for the case in which in XY_2 only one Y atom is replaced by an isotope. Even then of course the potential energy is the same as in the original molecule; but since the kinetic energy has no longer the same high symmetry the secular equation no longer factors to the same extent as before. Consequently the resultant exact relations between the frequencies and force constants become rather

[85] These are not the final values of Darling and Dennison (263), who have made use of the isotope relations to calculate them (see p. 282).

[86] If the zero-order frequencies are determined as above for H_2O, almost perfect agreement is obtained [see Nielsen (653)].

cumbersome [see Rosenthal (746)]. However, it is obvious that *if the mass difference* $\Delta m_Y = m_Y^{(i)} - m_Y$ of the two isotopes *is small* compared to the mass m_Y, *the isotope shift between* XY_2 *and* $XYY^{(i)}$ *is very nearly half the shift between* XY_2 *and* $XY_2{}^{(i)}$. For the latter shifts we obtain, from (II, 305) and (II, 306) when Δm_Y is small and $m_X^{(i)} = m_X$,

$$\frac{\Delta\lambda_3}{\lambda_3} = \frac{2\Delta\omega_3}{\omega_3} = -\frac{m_X}{(m_X + 2m_Y \sin^2 \alpha)} \frac{\Delta m_Y}{(m_Y + \Delta m_Y)}, \quad \text{(II, 309)}$$

$$\frac{\Delta\lambda_1}{\lambda_1} + \frac{\Delta\lambda_2}{\lambda_2} = \frac{2\Delta\omega_1}{\omega_1} + \frac{2\Delta\omega_2}{\omega_2} = -\frac{2(m_X + m_Y)}{(m_X + 2m_Y)} \frac{\Delta m_Y}{(m_Y + 2\Delta m_Y)}, \quad \text{(II, 310)}$$

and thus for the shifts between XY_2 and $XYY^{(i)}$ just half these values. Here use has been made of the relations

$$\frac{\lambda_3{}^{(i)}}{\lambda_3} = 1 + \frac{\Delta\lambda_3}{\lambda_3} = 1 + \frac{2\Delta\omega_3}{\omega_3},$$

$$\frac{\lambda_1{}^{(i)}\lambda_2{}^{(i)}}{\lambda_1\lambda_2} = 1 + \frac{\Delta\lambda_1}{\lambda_1} + \frac{\Delta\lambda_2}{\lambda_2} = 1 + \frac{2\Delta\omega_1}{\omega_1} + \frac{2\Delta\omega_2}{\omega_2} \quad \text{(II, 311)}$$

which of course hold for small $\Delta\omega$ only. The formulae (II, 309) and (II, 310) have first been derived by Salant and Rosenthal (757).

For linear symmetrical molecules the shifts $\Delta\omega$ between the frequencies of XY_2 and $XY_2{}^{(i)}$ are, according to (II, 308), for small Δm_Y,

$$\frac{\Delta\omega_1}{\omega_1} = -\frac{\Delta m_Y}{2(m_Y + \Delta m_Y)},$$

$$\frac{\Delta\omega_2}{\omega_2} = \frac{\Delta\omega_3}{\omega_3} = -\frac{m_X}{2(m_X + 2m_Y)} \frac{\Delta m_Y}{(m_Y + \Delta m_Y)}, \quad \text{(II, 312)}$$

and again half these values for the shifts between linear XY_2 and $XYY^{(i)}$. For accurate formulae for any mass difference in the latter case, see Rosenthal (746).

Formulae for the isotopic effect in unsymmetrical linear and non-linear triatomic molecules, assuming a small mass difference, have been derived by Adel (32) (33). For the linear molecules he has also discussed in detail the influence of anharmonicity and the interaction of rotation and vibration. The results for the non-linear triatomic molecules may also be applied to molecules such as CH_3CH_2I, CH_3CH_2OH, and others for which there are oscillations in which the three groups CH_3, CH_2, I, or OH oscillate essentially as a whole.

The Teller-Redlich product rule. The same methods that have been applied in the preceding paragraph to triatomic molecules may of course also be applied to molecules having more than three atoms. However, the solution of the most general secular equation becomes increasingly tedious. Such calculations have been carried out for four-atomic molecules XY_3 by Salant and Rosenthal (758) and for five-atomic tetrahedral molecules by Rosenthal (747) (748). Wilson (929) has briefly indicated a perturbation method for the calculation of isotope shifts. Naturally the calculations are somewhat simplified if simplified force systems such as the valence force system are used. Many of the conclusions of the above-mentioned papers can, however be very simply derived from a general theorem given independently by Teller [quoted in (55)] and Redlich (726). This theorem is of fundamental importance for the study of the isotope effect.

Teller and Redlich showed that *for two isotopic molecules the product of the* $\omega^{(i)}/\omega$ *values for all vibrations of a given symmetry type is independent of the potential constants and depends only on the masses of the atoms and the geometrical structure of the molecule.* The previous formulae (II, 305), (II, 306), and (II, 308) are examples of this general theorem. The general formula for any molecule is

$$\frac{\omega_1^{(i)}}{\omega_1}\frac{\omega_2^{(i)}}{\omega_2}\cdots\frac{\omega_f^{(i)}}{\omega_f} = \sqrt{\left(\frac{m_1}{m_1^{(i)}}\right)^{\alpha}\left(\frac{m_2}{m_2^{(i)}}\right)^{\beta}\cdots\left(\frac{M^{(i)}}{M}\right)^{t}\left(\frac{I_x^{(i)}}{I_x}\right)^{\delta_x}\left(\frac{I_y^{(i)}}{I_y}\right)^{\delta_y}\left(\frac{I_z^{(i)}}{I_z}\right)^{\delta_z}}. \quad \text{(II, 313)}$$

Here all quantities referring to the isotopic molecule are marked by the superscript (i) while the quantities referring to the originally considered "ordinary" molecule do not have this superscript; $\omega_1, \omega_2, \cdots \omega_f$ are the (zero-order) frequencies of the f genuine vibrations of the symmetry type considered (see section 4a); $m_1, m_2 \cdots$ are the masses of the representative atoms of the various sets (each set consisting of atoms that are transformed into one another by the symmetry operations permitted by the molecule; see p. 131). The exponents $\alpha, \beta, \cdots$ are the numbers of vibrations (inclusive of non-genuine vibrations) each set contributes to the symmetry type considered (they are the factors of $m, m_h, m_v, \cdots$ in Tables 35 and 36). M is the total mass of the molecule; t is the number of translations of the symmetry type considered (for $\boldsymbol{C}_{2v}$ the column headed T in Table 34; see also the last columns in Tables 12–30). I_x, I_y, I_z are the moments of inertia about the x, y, and z axes through the center of mass (the direction of the axes being chosen in the same way as in the Tables 12–30 of the different symmetry types; δ_x, δ_y, δ_z are 1 or 0 depending on whether or not the rotation about the x, y, z axis is a non-genuine vibration of the symmetry type considered. Both on the left and right hand side (in $\alpha, \beta \cdots t, \delta_x, \delta_y, \delta_z$) a degenerate vibration is counted only once.

For a proof of the theorem (II, 313) the reader must be referred to Redlich's paper (726). Like our previous considerations it is based on the assumption of identical force fields for the isotopic molecules, which is practically always fulfilled. On this assumption *the product rule* (II, 313) *should hold rigorously for the zero-order frequencies* ω_i and at least to a good approximation for the observed fundamentals ν_i (or, in other words, the first vibrational quanta) *for any mass difference.* One can also predict the direction of the deviation in case the ν_i rather than the ω_i are used: If (i) refers to the heavier isotope, since for it the anharmonicity constants $x_{ik}^{(i)}$ are smaller than x_{ik}, and therefore $\omega_i^{(i)} - \nu_i^{(i)}$ is smaller than $\omega_i - \nu_i$, the product $\frac{\nu_1^{(i)}}{\nu_1}\frac{\nu_2^{(i)}}{\nu_2}\cdots\frac{\nu_f^{(i)}}{\nu_f}$ should be slightly greater than the right-hand side of (II, 313).

If there is only one genuine vibration of a certain symmetry type, formula (II, 313) gives directly the frequency of the isotopic molecule in terms of that of the "ordinary" molecule and of the masses and geometrical data; that is, in this case, the relative isotope shift itself (not only the ratio of the product of certain frequencies) is independent of the potential constants.

The reader may verify that (II, 313) for the case of a symmetrical triatomic molecule XY_2 leads to (II, 305), (II, 306), and (II, 308).

As a first example we shall apply (II, 313) to the *plane symmetric* X_2Y_4 *molecule* (point group V_h) *and its isotope* $X_2Y_4^{(i)}$ [see Conn and Sutherland (226) and Gallaway and Barker (345) for C_2H_4 and C_2D_4]. For the totally symmetric species A_g there is no translation and rotation, that is, $t = \delta_x = \delta_y = \delta_z = 0$. There are two sets of nuclei, one (Y_4) on the xy plane ($m_{xy} = 1$ in Table 35) and one (X_2) on the x axis ($m_{2x} = 1$). The first (Y_4) contributes two vibrations to A_g, the second (X_2) one (see Table 35); that is, the mass m_Y appears in (II, 313) to the second, m_X to the first power. Therefore, using the same numbering for the vibrations as on p. 107 and putting $m_X^{(i)} = m_X$, we have

$$\frac{\omega_1^{(i)}\omega_2^{(i)}\omega_3^{(i)}}{\omega_1\omega_2\omega_3} = \frac{m_Y}{m_Y^{(i)}}. \tag{II, 314}$$

For the species A_u again $t = \delta_x = \delta_y = \delta_z = 0$. Now only the set Y_4 contributes one degree of freedom (see Table 35) and therefore we have, for the one (torsional) oscillation of this species,

$$\frac{\omega_4^{(i)}}{\omega_4} = \sqrt{\frac{m_Y}{m_Y^{(i)}}}. \tag{II, 315}$$

There is one non-genuine vibration of species B_{1g}, namely the rotation about the z-axis (perpendicular to the plane of the molecule). Therefore, while $t = \delta_x = \delta_y = 0$, we have for this species $\delta_z = 1$, and we need the moment of inertia about the z axis. In terms of the dimensions a, b, c, introduced on p. 151 (see Fig. 57), the moment of inertia I_z is

$$I_z = 2m_Xa^2 + 4m_Y(b^2 + c^2).$$

The number of degrees of freedom contributed by each set of nuclei is the same as for A_g, and thus (II, 313) becomes

$$\frac{\omega_5^{(i)}\omega_6^{(i)}}{\omega_5\omega_6} = \frac{m_Y}{m_Y^{(i)}}\sqrt{\frac{m_Xa^2 + 2m_Y^{(i)}(b^2 + c^2)}{m_Xa^2 + 2m_Y(b^2 + c^2)}}. \tag{II, 316}$$

The one non-genuine vibration of species B_{1u} is a translation in the z direction. Therefore $t = 1$ but $\delta_x = \delta_y = \delta_z = 0$. The two sets of nuclei contribute one degree of freedom each and we have, according to (II, 313),

$$\frac{\omega_7^{(i)}}{\omega_7} = \sqrt{\frac{m_Y}{m_Y^{(i)}} \cdot \frac{m_X + 2m_Y^{(i)}}{m_X + 2m_Y}}. \tag{II, 317}$$

In a similar way one obtains

$$\frac{\omega_8^{(i)}}{\omega_8} = \sqrt{\frac{m_Y}{m_Y^{(i)}} \cdot \frac{m_Xa^2 + 2m_Y^{(i)}b^2}{m_Xa^2 + 2m_Yb^2}}, \tag{II, 318}$$

$$\frac{\omega_9^{(i)}\omega_{10}^{(i)}}{\omega_9\omega_{10}} = \frac{m_Y}{m_Y^{(i)}}\sqrt{\frac{m_X + 2m_Y^{(i)}}{m_X + 2m_Y}} = \frac{\omega_{11}^{(i)}\omega_{12}^{(i)}}{\omega_{11}\omega_{12}}. \tag{II, 319}$$

The relations (II, 314–319) have proved to be exceedingly helpful in assigning the observed frequencies of C_2H_4 and C_2D_4 to the proper species (see Chapter III, section 3d). From the fundamentals given in Table 47 (p. 184) we obtain, for example, for $\frac{\nu_1^{(i)}\nu_2^{(i)}\nu_3^{(i)}}{\nu_1\nu_2\nu_3}$ the value 0.5085, while according to (II, 314) it should be $m_H/m_D = 0.5004$. Similar agreement is found for the other symmetry types, the observed ratios always being slightly larger than the calculated, as expected (see above).

If the isotopic molecule has lower symmetry than the "ordinary" molecule, the product rule holds rigorously only for the symmetry types of this lower symmetry. For example, if in X_2Y_4 only two Y atoms are substituted the point group is $\boldsymbol{C}_{2v}$ or $\boldsymbol{C}_{2h}$ rather than $\boldsymbol{V}_h$. Let us consider $Y_2XXY_2^{(i)}$. Then the *yz* plane is no longer a plane of symmetry. A_g and B_{3u} of $\boldsymbol{V}_h$ are no longer different. They form A_1 of $\boldsymbol{C}_{2v}$ (see Tables 14 and 13).[87] Similarly A_u and B_{3g} go over into A_2, B_{1g} and B_{2u} into B_1, and B_{1u} and B_{2g} into B_2. We have, therefore, from (II, 313):

$$\frac{\omega_1^{(i)}\omega_2^{(i)}\omega_3^{(i)}\omega_{11}^{(i)}\omega_{12}^{(i)}}{\omega_1\omega_2\omega_3\omega_{11}\omega_{12}} = \frac{m_Y}{m_Y^{(i)}}\sqrt{\frac{(m_X + m_Y + m_Y^{(i)})}{(m_X + 2m_Y)}}, \tag{II, 320}$$

$$\frac{\omega_4^{(i)}}{\omega_4} = \sqrt{\frac{m_Y}{m_Y^{(i)}}\frac{(m_Y + m_Y^{(i)})}{2m_Y}} = \sqrt{\frac{1}{2}\left(1 + \frac{m_Y}{m_Y^{(i)}}\right)}, \tag{II, 321}$$

$$\frac{\omega_5^{(i)}\omega_6^{(i)}\omega_9^{(i)}\omega_{10}^{(i)}}{\omega_5\omega_6\omega_9\omega_{10}} = \frac{m_Y}{m_Y^{(i)}}\sqrt{\frac{M^{(i)}(m_Y + m_Y^{(i)})(b^2 + c^2) + m_X M^{(i)}a^2 - 4(m_Y^{(i)} - m_Y)^2b^2}{M[2m_Y(b^2 + c^2) + m_Xa^2]}}, \tag{II, 322}$$

$$\frac{\omega_7^{(i)}\omega_8^{(i)}}{\omega_7\omega_8} = \sqrt{\frac{m_Y}{m_Y^{(i)}}\frac{[M^{(i)}(m_Y + m_Y^{(i)})b^2 + M^{(i)}m_Xa^2 - 4(m_Y^{(i)} - m_Y)^2b^2]}{M(2m_Yb^2 + m_Xa^2)}}, \tag{II, 323}$$

where M and $M^{(i)}$ are the total masses of the two molecules. It is seen that the results for the less symmetrical molecule are the less detailed. If only one Y atom in X_2Y_4 were replaced by its isotope, the point group would be $\boldsymbol{C}_s$ and there would be only two equations (corresponding to the two symmetry types), one equation for $\frac{\omega_1^{(i)}\omega_2^{(i)}\omega_3^{(i)}\omega_5^{(i)}\omega_6^{(i)}\omega_9^{(i)}\omega_{10}^{(i)}\omega_{11}^{(i)}\omega_{12}^{(i)}}{\omega_1\omega_2\omega_3\omega_5\omega_6\omega_9\omega_{10}\omega_{11}\omega_{12}}$ and the other for $\frac{\omega_4^{(i)}\omega_7^{(i)}\omega_8^{(i)}}{\omega_4\omega_7\omega_8}$. While these equations hold for any mass difference, they are much less useful than (II, 314–319) for the symmetrical case.

If the mass difference is small, equations (II, 314–319) may be brought into a form similar to (II, 309–312) for XY_2. For example, (II, 314) becomes

$$\frac{\Delta\omega_1}{\omega_1} + \frac{\Delta\omega_2}{\omega_2} + \frac{\Delta\omega_3}{\omega_3} = -\frac{\Delta m_Y}{m_Y^{(i)}}, \tag{II, 324}$$

and similarly for the other equations. If only two of the Y atoms are replaced by $Y^{(i)}$ the sum of the relative shifts of ω_1, ω_2, ω_3 is half the amount given by (II, 324); if only one Y is replaced by $Y^{(i)}$ the sum is one quarter of this amount. Thus for small mass differences the more detailed relations of the symmetrical case may also be used for unsymmetrical substitution of isotopes.

As a further example of the application of the product theorem, we consider the *molecule* XYZ_3 *of point group* $\boldsymbol{C}_{3v}$ (examples: CH_3Cl, CCl_3H). There are (see Table 36) three vibrations of species A_1 and three (doubly degenerate) vibrations of species E (see Fig. 91). There is one set of atoms on the planes σ_v (the group Z_3) contributing two vibrations to A_1 and three to E, and two "sets" on

[87] It should be noted that the *z* axis for X_2Y_4 has been assumed to be perpendicular to the plane of the molecule, whereas for $\boldsymbol{C}_{2v}$ it is the two-fold axis (X—X).

the symmetry axis (the atoms X and Y) contributing one vibration each to A_1 and E. There is one translation of type A_1, and one (degenerate) translation and one (degenerate) rotation of type E. Thus we obtain, from (II, 313), for the non-degenerate vibrations,

$$\frac{\omega_1^{(i)}\omega_2^{(i)}\omega_3^{(i)}}{\omega_1\omega_2\omega_3} = \frac{m_Z}{m_Z^{(i)}}\sqrt{\frac{m_X m_Y}{m_X^{(i)} m_Y^{(i)}}\frac{M^{(i)}}{M}}, \tag{II, 325}$$

and for the degenerate vibrations,

$$\frac{\omega_4^{(i)}\omega_5^{(i)}\omega_6^{(i)}}{\omega_4\omega_5\omega_6} = \sqrt{\frac{m_Z^3 m_X m_Y}{m_Z^{(i)3} m_X^{(i)} m_Y^{(i)}}\frac{M^{(i)}}{M}\frac{I^{(i)}}{I}}, \tag{II, 326}$$

where $M = m_X + m_Y + 3m_Z$ is the total mass and $I = m_X a^2 + m_Z(2c^2 + 3h^2) - (1/M)(3m_Z h - m_X a)^2$ [a = XY distance, $c = \frac{1}{2}$ZZ distance, h = height of YZ_3 pyramid] is the moment of inertia about an axis perpendicular to the symmetry axis. It may be noted that on the right in (II, 326), according to the above, the doubly degenerate translation and rotation count each as one only; that is, $M^{(i)}/M$ and $I^{(i)}/I$ occur to the first power only. The reader may apply the relations (II, 325) and (II, 326) to the data on CH_3Cl, CD_3Cl and CH_3Br, CD_3Br given in Table 85 p. 315.

Finally we consider the *tetrahedral* XY_4 *molecule* belonging to the point group $\boldsymbol{T_d}$. For such a molecule we have one genuine vibration of species A_1, one of species E, and two of species F_2 (see p. 140 and Fig. 41). There is one set of atoms lying on the threefold axes (group Y_4, $m_3 = 1$, in Table 36) and contributing one vibration (genuine or non-genuine) to A_1, one to E, one to F_1, and two to F_2. The other set consists of the one atom X at the center ($m_0 = 1$) and contributes one (triply degenerate) vibration to F_2. The three rotations belong to symmetry type F_1, which does not contribute any genuine normal vibrations, and the three translations form one triply degenerate non-genuine vibration of species F_2; that is, $t = 1$ for F_2. Thus we have, for the totally symmetric and for the doubly degenerate vibration,

$$\frac{\omega_1^{(i)}}{\omega_1} = \frac{\omega_2^{(i)}}{\omega_2} = \sqrt{\frac{m_Y}{m_Y^{(i)}}}; \tag{II, 327}$$

and for the two triply degenerate vibrations (F_2),

$$\frac{\omega_3^{(i)}\omega_4^{(i)}}{\omega_3\omega_4} = \frac{m_Y}{m_Y^{(i)}}\sqrt{\frac{m_X}{m_X^{(i)}}\cdot\frac{m_X^{(i)} + 4m_Y^{(i)}}{m_X + 4m_Y}}. \tag{II, 328}$$

These relations agree with those derived in a rather more complicated way (before the product rule was published) by Rosenthal (748). It is noteworthy that according to (II, 327) no isotope shift occurs in ω_1 and ω_2 if only the central X atom is replaced by an isotope. This is in agreement with the previous conclusion (Fig. 41) that in these vibrations the X atom does not move. The relations (II, 327) and (II, 328) will be applied to CH_4 and CD_4 in Chapter III, section 3c.

While the above formulae for XY_4 hold for any mass difference of the isotopes, again somewhat simpler formulae may be obtained if the mass difference is small. In that case also, the isotope shift of ν_1 and ν_2 for $XY_3Y^{(i)}$ is one-quarter of that for $XY_4^{(i)}$ given by (II, 327); for $XY_2Y_2^{(i)}$ it is one-half of that amount, and so on.

Noether (673a) has proposed an empirical rule that is related to the Teller-Redlich product rule. He suggests that the ratio $\nu_k^{(i)}/\nu_k$ is the same for corresponding vibrations of different but similar molecules, for example

$$\frac{\nu_k(CD_3Cl)}{\nu_k(CH_3Cl)} = \frac{\nu_k(CD_3Br)}{\nu_k(CH_3Br)}.$$

Up to now there is no theoretical justification for this rule although the experimental data support it.

Pitzer and Scott (699) have applied with some success the Teller-Redlich product rule to methyl derivatives of benzene considering in a rough approximation the CH_3 group as an isotope of hydrogen. This has proved to be an important aid in the analysis of the vibrational spectra of toluene, the xylenes and mesitylene even though an accurate validity of the product rule cannot be expected in such cases.

Resolution of the symmetry types of a point group into those of a point group of lower symmetry. If the symmetry of an isotopic molecule is lower than that of the ordinary molecule, as in the case of $XY_3Y^{(i)}$, it is necessary for an application of the product theorem to know to which symmetry type of the point group of lower symmetry, say Q, the vibrations of the molecule of higher symmetry, say P, belong. This resolution of the symmetry types of one point group into those of a point group of lower symmetry is of importance also in the discussion of the electronic structure of polyatomic molecules and in the discussion of dissociation processes of polyatomic molecules.

In the case of non-degenerate symmetry types this resolution is very simple. One has only to find out what the characters of each symmetry type of P are with respect to the symmetry operations of Q and see from Tables 12–30 to which symmetry type H such a set of characters belongs. This has already been carried out above for the resolution of V_h into C_{2v} when considering the isotope effect of X_2Y_4.

The resolution is somewhat less obvious for degenerate symmetry types since *in going over to lower symmetry these degeneracies are partly or wholly removed.* In these cases the procedure to be followed is similar to the previously given method of determining the resultant states when two degenerate vibrations are excited (see p. 130). Suppose E is the degenerate symmetry type (representation) of P considered and G, H, $\cdots$ are the symmetry types into which E splits in going to the lower symmetry Q. Let $\chi_E^{(k)}$, $\chi_G^{(k)}$, $\chi_H^{(k)}$ $\cdots$ be the corresponding characters (see p. 108) for a symmetry operation (k) that is common to both point groups P and Q. Then group theory shows [for a proof see Bethe (143) and Mulliken (641)] that, similar to (II, 87) for all such common symmetry elements:

$$\chi_E^{(k)} = \chi_G^{(k)} + \chi_H^{(k)} + \cdots. \quad \text{(II, 329)}$$

Therefore, in order to determine the resolution of E, *one has simply to express the characters* $\chi_E^{(k)}$ *of* E *as a sum of* $\chi_G^{(k)}$, $\chi_H^{(k)}$, $\cdots$ *in such a way that* G, H, $\cdots$ *are the same for all* χ_E *of common symmetry elements* k. This can only be done in one way. In *group theory* language the degenerate representation E of P is no longer irreducible with respect to Q. Its irreducible components obtained from (II, 329) are the representations into which E is resolved in point group Q.

As an example, consider the case of the XY_4 molecule in which one Y is replaced by its isotope. The point groups of the ordinary and the isotopic molecule are T_d and C_{3v} respectively. The symmetry elements common to both are all those of C_{3v}, that is, I, C_3, σ_v. The two non-degenerate species A_1 and A_2 of T_d, of course, go over into A_1 and A_2 of C_{3v}. Similarly, E of T_d goes over into E of C_{3v}, since the characters are the same. The triply degenerate species F_1 of T_d (for which there is no genuine vibration in XY_4) splits, since there are only doubly degenerate species for C_{3v}. The characters of F_1 for the symmetry elements I, C_3, $\sigma_d \equiv \sigma_v$ are $+3$, 0, and -1 (see Table 28). There is only one way in which these three characters can be expressed simultaneously as sums of corresponding characters of C_{3v} (see Table 15): namely, as the sums of the characters of A_2 $(+1, +1, -1)$ and E $(+2, -1, 0)$. Thus F_1 splits into $A_2 + E$. Similarly F_2 is found to split into $A_1 + E$. Thus the two triply degenerate vibrations of a molecule XY_4 each split into a totally symmetric and a doubly degenerate vibration.

If two Y atoms in XY_4 are replaced by their isotopes, we have the point group C_{2v}. Here the operations $\sigma_v(xz)$ and $\sigma_v(yz)$ have in general different characters (see Table 13, p. 106), even though for T_d all reflections σ_d have the same character. A_1 and A_2 of T_d of course go over into A_1 and A_2 respectively of C_{2v}. The characters of E of T_d for I, C_2, $\sigma_v(xz)$, $\sigma_v(yz)$ are $+2$, $+2$, 0, 0 respectively. They are the sums of the characters of A_1 and A_2 of C_{2v}. The characters of F_2 for the same symmetry elements are $+3$, -1, $+1$, $+1$, which are the sums of the characters of A_1, B_1, and B_2. Similarly F_1 resolves into $A_2 + B_1 + B_2$.

The above results for $V_h \rightarrow C_{2v}$, $T_d \rightarrow C_{3v}$, C_{2v} are summarized together with several other cases in Table 53. It must be noted that in some cases the species of one point group can be resolved into those of another point group in a number of different ways; for example, V_h can be resolved into C_{2v} in such a way that the z axis of C_{2v} coincides with the z, or y, or x axis of V_h. The relative orientation of the axes is, if necessary, indicated at the top of the particular subtable of Table 53. Further resolutions can easily be obtained by the reader according to the method outlined above, or by combining the results of Table 53 in a suitable way [see also Mulliken (641)]. For example, the resolution of D_{6h} into C_{3v} is obtained from Table 53 by first looking up the resolution into D_{3h} and then that of D_{3h} into C_{3v}. Also the resolution of D_{ph} into D_p is simply obtained by dropping the g and u or $'$ and $''$. The resolution of C_{2v} into C_s is contained in the subtable for V_h, and similarly in other cases.

Application to the unsymmetrically substituted isotopes of an XY_4 molecule. If in XY_4 only one of the Y atoms is replaced by an isotope $Y^{(i)}$, the symmetry of the resulting molecule $Y^{(i)}XY$ is no longer T_d but C_{3v}. According to the preceding discussion and Table 53 each of the two triply

TABLE 53. RESOLUTION OF SPECIES OF NON-LINEAR POINT GROUPS INTO THOSE OF POINT GROUPS OF LOWER SYMMETRY.

V_h	Resolved into				
	C_{2v}			C_{2h}	C_s
	$x\to x$ $y\to y$ $z\to z$	$x\to z$ $y\to x$ $z\to y$	$x\to x$ $y\to z$ $z\to y$	$z\to z$ $\sigma(xy)\to\sigma(xy)$	$\sigma(xy)\to\sigma(xy)$
A_g	A_1	A_1	A_1	A_g	A'
A_u	A_2	A_2	A_2	A_u	A''
B_{1g}	A_2	B_1	B_1	A_g	A'
B_{1u}	A_1	B_2	B_2	A_u	A''
B_{2g}	B_1	B_2	A_2	B_g	A''
B_{2u}	B_2	B_1	A_1	B_u	A'
B_{3g}	B_2	A_2	B_2	B_g	A''
B_{3u}	B_1	A_1	B_1	B_u	A'

$V_d\equiv D_{2d}$	Resolved into			
	V	C_{2v}	C_2	C_s
	$z\to z$	$z\to z$	$x\to z$	$\sigma_d\to\sigma(xy)$
A_1	A	A_1	A	A'
A_2	B_1	A_2	B	A''
B_1	A	A_2	A	A''
B_2	B_1	A_1	B	A'
E	B_2+B_3	B_1+B_2	$A+B$	$A'+A''$

D_{6h}	Resolved into			
	C_{6v}	D_{3h}	D_{3d}	V_h
	$z\to z$	$z\to z$ $\sigma_v\to\sigma_v$	$z\to z$ $\sigma_v\to\sigma_d$ $C_2'\to C_2$	$z\to z$ $\sigma_h\to\sigma(xy)$ $\sigma_v\to\sigma(yz)$
A_{1g}	A_1	A_1'	A_{1g}	A_g
A_{1u}	A_2	A_1''	A_{1u}	A_u
A_{2g}	A_2	A_2'	A_{2g}	B_{1g}
A_{2u}	A_1	A_2''	A_{2u}	B_{1u}
B_{1g}	B_2	A_1''	A_{2g}	B_{2g}
B_{1u}	B_1	A_1'	A_{2u}	B_{2u}
B_{2g}	B_1	A_2''	A_{1g}	B_{3g}
B_{2u}	B_2	A_2'	A_{1u}	B_{3u}
E_{1g}	E_1	E''	E_g	$B_{2g}+B_{3g}$
E_{1u}	E_1	E'	E_u	$B_{2u}+B_{3u}$
E_{2g}	E_2	E'	E_g	A_g+B_{1g}
E_{2u}	E_2	E''	E_u	A_u+B_{1u}

D_{3h}	Resolved into			
	C_{3v}	C_{2v}		C_s
	$z\to z$	$z\to x$ $C_2\to C_2(z)$ $\sigma_v\to\sigma_v(xz)$	$z\to y$ $C_2\to C_2(z)$ $\sigma_h\to\sigma_v(xz)$	$\sigma_v\to\sigma(xy)$
A_1'	A_1	A_1	A_1	A'
A_1''	A_2	A_2	A_2	A''
A_2'	A_2	B_2	B_1	A''
A_2''	A_1	B_1	B_2	A'
E'	E	A_1+B_2	A_1+B_1	$A'+A''$
E''	E	A_2+B_1	A_2+B_2	$A'+A''$

D_{3d}	Resolved into				
	C_{3v}	C_{2h}	C_2	C_s	C_i
	$z\to z$	$C_2\to C_2(z)$ $\sigma_d\to\sigma_h(xy)$	$C_2\to C_2(z)$	$\sigma_d\to\sigma(xy)$	
A_{1g}	A_1	A_g	A	A'	A_g
A_{1u}	A_2	A_u	A	A''	A_u
A_{2g}	A_2	B_g	B	A''	A_g
A_{2u}	A_1	B_u	B	A'	A_u
E_g	E	A_g+B_g	$A+B$	$A'+A''$	A_g+A_g
E_u	E	A_u+B_u	$A+B$	$A'+A''$	A_u+A_u

T_d	Resolved into			
	C_{3v}	C_{2v}	$V_d\equiv D_{2d}$	C_s
A_1	A_1	A_1	A_1	A'
A_2	A_2	A_2	B_1	A''
E	E	A_1+A_2	A_1+B_1	$A'+A''$
F_1	A_2+E	$A_2+B_1+B_2$	A_2+E	$A'+A''+A''$
F_2	A_1+E	$A_1+B_1+B_2$	B_2+E	$A'+A'+A''$

degenerate vibrations ν_3 and ν_4 of XY_4 (species F_2) splits into a non-degenerate and a doubly degenerate vibration of species A_1 and E respectively, while the species of the vibrations ν_1 and ν_2 remain A_1 and E respectively. The same holds when three Y atoms are replaced by their isotopes, that is, for $YXY_3^{(i)}$. If two Y atoms are replaced by $Y^{(i)}$ the resulting molecule $Y_2XY_2^{(i)}$ belongs to point group C_{2v}; according to Table 53 each of the two triply degenerate vibrations of XY_4 splits into three non-degenerate vibrations of species A_1, B_1, and B_2, while $\nu_2(E)$ splits into two non-degenerate vibrations of species A_1 and A_2, and ν_1 remains totally symmetrical (species A_1).

In Table 54 the correlation of the normal vibrations of the five isotopic molecules of XY_4 is given. On the basis of this table it is now easy to apply the product rule. Thus it is seen that the

TABLE 54. CORRELATION OF THE FUNDAMENTAL VIBRATIONS OF THE ISOTOPIC MODIFICATIONS OF XY_4.

Molecule	Vibrations			
XY_4	$\nu_1(a_1)$	$\nu_2(e)$	$\nu_3(f_2)$	$\nu_4(f_2)$
$Y^{(i)}XY_3$	$\nu_1(a_1)$	$\nu_2(e)$	$\nu_{3a}(a_1)\nu_{3bc}(e)$	$\nu_{4a}(a_1)\nu_{4bc}(e)$
$Y_2^{(i)}XY_2$	$\nu_1(a_1)$	$\nu_{2a}(a_1)\nu_{2b}(a_2)$	$\nu_{3a}(a_1)\nu_{3b}(b_1)\nu_{3c}(b_2)$	$\nu_{4a}(a_1)\nu_{4b}(b_1)\nu_{4c}(b_2)$
$YXY_3^{(i)}$	$\nu_1(a_1)$	$\nu_2(e)$	$\nu_{3a}(a_1)\nu_{3bc}(e)$	$\nu_{4a}(a_1)\nu_{4bc}(e)$
$XY_4^{(i)}$	$\nu_1(a_1)$	$\nu_2(e)$	$\nu_3(f_2)$	$\nu_4(f_2)$

three totally symmetric vibrations of $Y^{(i)}XY_3$ have to be combined with ν_1, ν_3, and ν_4 of XY_4 in (II, 325), while the three doubly degenerate vibrations of $Y^{(i)}XY_3$ have to be combined with ν_2, ν_3, and ν_4 in (II, 326). Thereupon,

$$\frac{\omega_1^{(i)}\omega_{3a}^{(i)}\omega_{4a}^{(i)}}{\omega_1\omega_3\omega_4} = \sqrt{\frac{m_Y}{m_Y^{(i)}}\frac{M^{(i)}}{M}}, \qquad \frac{\omega_2^{(i)}\omega_{3bc}^{(i)}\omega_{4bc}^{(i)}}{\omega_2\omega_3\omega_4} = \sqrt{\frac{m_Y}{m_Y^{(i)}}\frac{M^{(i)}}{M}\frac{I^{(i)}}{I}}, \tag{II, 330}$$

where the different designation of the atoms in XYZ_3 and $Y^{(i)}XY_3$ has been taken into account and where, unlike (II, 326), $I^{(i)}/I$ is independent of the molecular dimensions:

$$\frac{I^{(i)}}{I} = \frac{m_X(3m_Y^{(i)} + 5m_Y) + 4m_Y(5m_Y^{(i)} + 3m_Y)}{8m_YM^{(i)}}.$$

Very similar relations hold between the frequencies of $YXY_3^{(i)}$ and XY_4 as well as $XY_4^{(i)}$. For $Y_2^{(i)}XY_2$ one obtains from (II, 313) four equations corresponding to the four species, which may be easily set up [see also Rosenthal (748)].

The observed fundamentals of the various heavy methanes CH_4, CH_3D, CH_2D_2, CHD_3, CD_4 given in Table 82 represent excellent material for an application of the above isotope formulae for the XY_4 molecule.

It is obvious that for small mass differences the frequencies of the isotopic molecules are close to those of the "ordinary" molecule. However, for molecules with heavy hydrogen such as the heavy methanes, very large shifts may occur. In such cases it is no longer obvious, for example, which of the three totally symmetric vibrations of $Y^{(i)}XY_3$ is ν_1, which ν_{3a}, and which ν_{4a} (see Table 54). However, if it is considered that in going from small to large mass differences *no crossing over of frequencies of the same species can occur* (because of the repulsion in consequence of Fermi resonance see section 5c) it is clear that the largest of the frequencies ν_1, ν_{3a}, ν_{4a} corresponds to the largest of ν_1, ν_3, ν_4 of XY_4, the second largest to the second largest, and so on. Similar considerations apply for other correlations.

It must be emphasized that the correlation lines in Table 54 give a rigorous correlation only between XY_4 (or $XY_4^{(i)}$) and any one of the intermediate molecules, but not between $Y^{(i)}XY_3$ and $Y_2^{(i)}XY_2$ or between $YXY_3^{(i)}$ and $Y_2^{(i)}XY_2$ since in going from $Y^{(i)}XY_3$ (or from $YXY_3^{(i)}$) to $Y_2^{(i)}XY_2$ only one plane of symmetry is conserved and therefore an a_1 vibration of $Y^{(i)}XY_3$ need not necessarily go into an a_1 vibration of $Y_2^{(i)}XY_2$ but may also go into a b_1 or a b_2 vibration (depending on which plane is conserved). For a correlation in which this consideration is taken into account see p. 319f.

CHAPTER III

VIBRATIONAL INFRARED AND RAMAN SPECTRA

1. Classical Treatment

(a) *Infrared vibration spectra*

Active and inactive fundamentals. According to classical electrodynamics any motion of an atomic system that is connected with a *change of its dipole moment* leads to the emission or absorption of radiation.[1] During the vibrational motion of a molecule the charge distribution undergoes a periodic change, and therefore in general (though not always) the dipole moment changes periodically. Since, in the harmonic oscillator approximation, any vibrational motion of the molecule may be resolved into a sum of normal vibrations with appropriate amplitudes, and since the normal vibrations are the only simple periodic motions, *the normal frequencies are the frequencies that are emitted or absorbed by the molecule.* These frequencies lie in the near infrared, as for diatomic molecules. Usually in the infrared it is the *absorption* spectrum that is observed.

Normal vibrations that are connected with a change of dipole moment and, therefore, appear in the infrared are called *infrared active,* while vibrations for which the change of charge distribution (which always occurs) is such that no change of dipole moment arises and which, therefore, do not appear in the infrared, are called *infrared inactive.* In the harmonic oscillator approximation only the fundamentals ν_i are active: the overtones $2\nu_i$, $3\nu_i$, $\cdots$ and combination tones $\nu_i + \nu_k$, and so on, are inactive since the vibrational motion does not contain the frequencies $2\nu_i$, $3\nu_i$, $\cdots$, $\nu_i + \nu_k$, $\cdots$.

The change of the dipole moment may be only a *change of its direction* with respect to a coordinate system fixed in space. (See also the infrared rotation spectrum, Chapter I). For example, in the bending vibration of HCN (Fig. 61), the strong permanent dipole moment which is mainly associated with the C—H bond simply changes its direction, remaining always along the line C—H without an appreciable change of its magnitude. This change of direction is sufficient to cause the bending frequency ν_2 to appear strongly in the spectrum.

In unsymmetrical molecules, every normal vibration is connected with a change of dipole moment; that is, *all normal vibrations are infrared active* (although, of course, there may be great differences in the magnitude of the change and therefore in the intensity of infrared absorption). *Only in symmetrical molecules* may there be *vibrations* during which the change of dipole moment is exactly zero and *which,* therefore, *are infrared inactive.* For example, in the linear symmetric XY_2 molecule (for instance CO_2), during the totally symmetric vibration ν_1 (Fig. 25b), the dipole moment always remains zero as it is in the equilibrium position, and therefore this

[1] A change of the quadrupole moment or of the magnetic dipole moment may also lead to emission or absorption of radiation. But, in the infrared, the intensity of this emission or absorption is entirely negligible [see Teller (836) and Molecular Spectra I, p. 305].

vibration is infrared inactive. On the other hand, for the other two vibrations ν_2 and ν_3, dipole moments perpendicular to and in the molecular axis respectively arise because of the asymmetry of the displaced positions. These dipole moments have the opposite sign after a phase shift of 180°. The two vibrations ν_2 and ν_3 are therefore infrared active. Similarly, it can be seen by inspection of Fig. 64a that only the vibrations ν_3 and ν_5 of the symmetrical linear X_2Y_2 molecule (for instance C_2H_2) are infrared active. (Although in ν_4 the symmetry of the equilibrium position is not conserved, the displaced configuration of the nuclei has still a center of symmetry and therefore no dipole moment.) Furthermore, as can be seen from Fig. 44, only the vibrations ν_7, ν_9, ν_{10}, ν_{11}, ν_{12}, of X_2Y_4 (for instance C_2H_4) are infrared active.

But *not every symmetric molecule has inactive vibrations*. For example, in the non-linear symmetric XY_2 molecule (for instance H_2O) all three normal vibrations (see Fig. 25a) are connected with a change of dipole moment; that is, they are infrared active. On the other hand, a molecule such as X_2YZ_2 of the same point group $\boldsymbol{C_{2v}}$ has one inactive vibration, namely the torsion oscillation of X_2 against Z_2. In this case, the dipole moment is different from zero in the equilibrium position, but, for small amplitudes, does not change its direction or magnitude in the torsion oscillation as it does for all the other oscillations.

Let x, y, z be the axes of a coordinate system fixed in the molecule. Since we are assuming in this chapter, as in the preceding one, that there is no rotation, this coordinate system is also fixed in space. If M_x, M_y, M_z are the three components of the dipole moment $\boldsymbol{M}$ of the molecule in the direction of the coordinate axes in a displaced position of the nuclei, and if M_x^0, M_y^0, M_z^0 are the components of the dipole moment $\boldsymbol{M}^0$ in the equilibrium position, then, for sufficiently small displacements, we can expand M_x thus:

$$M_x = M_x{}^0 + \sum_k \left[\left(\frac{\partial M_x}{\partial x_k} \right)_0 x_k + \left(\frac{\partial M_x}{\partial y_k} \right)_0 y_k + \left(\frac{\partial M_x}{\partial z_k} \right)_0 z_k \right] + \cdots, \quad \text{(III, 1)}$$

where the x_k, y_k, z_k are the displacement coordinates of nucleus k. Similar relations hold for M_y and M_z. If we introduce normal coordinates ξ_1, ξ_2, ξ_3, $\cdots$, we have

$$M_x = M_x{}^0 + \sum_i \left(\frac{\partial M_x}{\partial \xi_i} \right)_0 \xi_i + \cdots \quad \text{(III, 2)}$$

and similarly for M_y and M_z.

Since [see (II, 13)]

$$\xi_i = \xi_i{}^0 \cos (2\pi\nu_i t + \varphi_i), \quad \text{(III, 3)}$$

according to (III, 2), the dipole moment $\boldsymbol{M}$ of the molecule will change with the frequency ν_i of a normal vibration i (that is, this frequency will be active in the infrared) if and only if at least one of the derivatives $\left(\frac{\partial M_x}{\partial \xi_i}\right)_0$, $\left(\frac{\partial M_y}{\partial \xi_i}\right)_0$, $\left(\frac{\partial M_z}{\partial \xi_i}\right)_0$ is different from zero. The intensity of this infrared fundamental band is proportional to the square of the vector representing the change of the dipole moment for the corresponding normal vibration near the equilibrium position; that is, it is proportional to

$$\left(\frac{\partial M_x}{\partial \xi_i} \right)_0^2 + \left(\frac{\partial M_y}{\partial \xi_i} \right)_0^2 + \left(\frac{\partial M_z}{\partial \xi_i} \right)_0^2. \quad \text{(III, 4)}$$

Overtone and combination vibrations. Just as in the case of diatomic molecules (see Molecular Spectra I, p. 99), if the anharmonicity of the vibrations is taken into account, the classical vibrational motion contains also the frequencies $2\nu_i$, $3\nu_i$, $\cdots$, and furthermore (see below) $\nu_i + \nu_k$, $\nu_i - \nu_k$, $2\nu_i + \nu_k$, $\cdots$. Therefore, in the infrared spectrum, *in addition to the fundamentals overtone and combination vibrations may also occur*, if they are connected with a change of dipole moment. However, they will be much weaker than the fundamentals, since the anharmonicities in general are slight, except for very large amplitudes of the nuclei.

If the anharmonicity cannot be neglected, one has to substitute for ξ_i in (III, 2) an expression such as (II, 265) instead of (III, 3). It is seen that then the dipole moment will oscillate with the frequencies $2\nu_i$, $3\nu_i$, $\cdots$, $\nu_i \pm \nu_k$, $2\nu_i \pm \nu_k$, $\cdots$ as well as ν_i; that is, overtone and combination vibrations occur, although with much smaller amplitude than those of the fundamentals.

Apart from this *mechanical anharmonicity*, an *electrical anharmonicity* may also cause overtone and combination vibrations to occur. It is easily seen that an electrical anharmonicity must always be present if it is considered that for a homopolar bond (as in the CH, HCl, $\cdots$ bonds) the dipole moment is zero both for infinitely large and zero internuclear distance. Thus the variation of the dipole moment with internuclear distance is qualitatively as shown in Fig. 74. There is a maximum

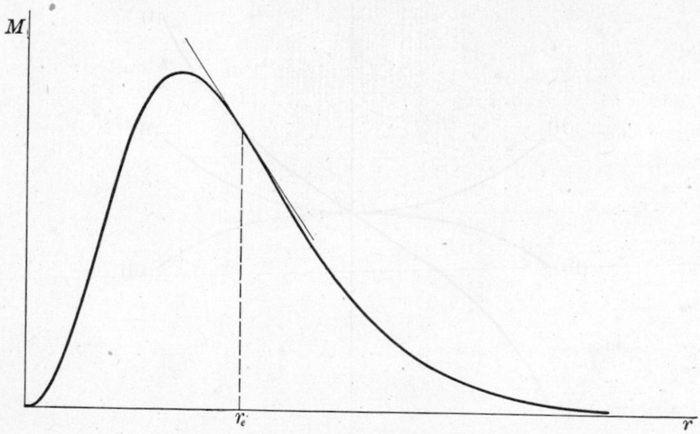

FIG. 74. **Dipole moment as a function of internuclear distance.**—The r value of the maximum may also be larger than r_e.

at an internuclear distance different from the equilibrium distance r_e of the bond. The slope of the curve at the equilibrium position determines the intensity of the fundamental. On account of the deviation from a straight line for larger amplitudes of vibration, even if the potential energy is strictly quadratic, higher terms

$$\sum_{i,k} \left(\frac{\partial^2 M_x}{\partial \xi_i \partial \xi_k} \right)_0 \xi_i \xi_k + \cdots \tag{III, 5}$$

in the development (III, 2) cannot be neglected. Substituting (III, 3) into (III, 5) shows that the dipole moment changes with frequency $2\nu_i$ or $\nu_i \pm \nu_k$, depending on whether $i = k$ or $i \neq k$; that is, overtone and combination vibrations occur. Higher overtone or combination vibrations would occur if still higher terms in (III, 5) were taken into account.

It should be noted that, as can immediately be seen from the above, classically, on the basis of both mechanical and electrical anharmonicity, the amplitude of the change of dipole moment and therefore the intensity of infrared absorption comes out to be the same for $\nu_i + \nu_k$ and $\nu_i - \nu_k$, and similarly in other cases.

(*b*) *Vibrational Raman spectra*

Elementary treatment of fundamentals. As in the case of diatomic molecules (see Molecular Spectra I, p. 88f.), in order that a fundamental frequency shall appear as a shift in the Raman spectrum *the amplitude of the dipole moment induced by the incident radiation must change during the vibration considered.* The magnitude of the induced dipole moment $\boldsymbol{P}$ is given by

$$|\boldsymbol{P}| = \alpha \cdot |\boldsymbol{E}|, \qquad \text{(III, 6)}$$

where $\boldsymbol{E}$ is the electric vector of the incident radiation of frequency ν and α the *polarizability.* If α changes during the vibration i with frequency ν_i, $\boldsymbol{P}$ will change with the frequencies $\nu + \nu_i$ and $\nu - \nu_i$ as well as with the frequency ν; that is, the scattered radiation will contain the frequencies $\nu \pm \nu_i$ in addition to ν. In an unsymmetrical molecule, during all normal vibrations a periodic change of the polarizability takes place and, therefore, all normal frequencies appear in the Raman spectrum, that is, are *Raman active.* However, in symmetrical molecules it may happen that for certain vibrations the polarizability does not change, at least for small amplitudes.

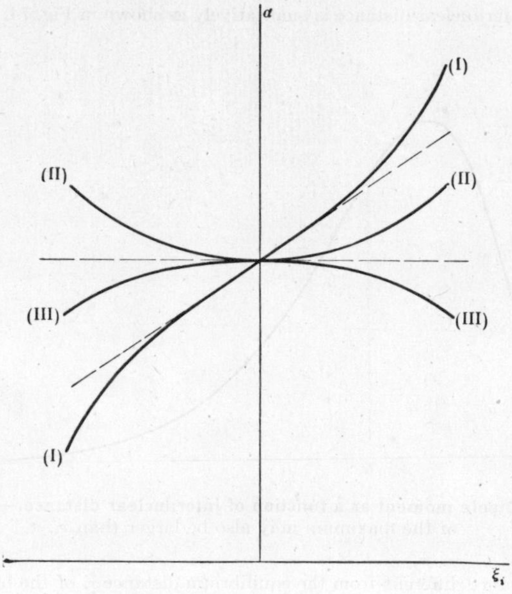

FIG. 75. **Polarizability as a function of normal coordinates (schematic).**

For example, for the linear symmetric XY_2 molecule (for instance CO_2) during the vibration ν_1 (Fig. 25b), the polarizability is larger than the equilibrium value in one half period and smaller in the other. In first approximation, therefore, α changes linearly with ξ_1, as indicated by the curve I in Fig. 75. The vibration ν_1 is, therefore, Raman active. However, during the vibrations ν_2 and ν_3, the polarizability is obviously the same at opposite phases of the motion and therefore its variation with ξ_2 and ξ_3 is of the form given by the curves II or III in Fig. 75, which have a horizontal tangent for $\xi_2 = 0$ or $\xi_3 = 0$. That is, in first approximation, for small amplitudes ξ_i^0, the polarizability does not change: the vibrations ν_2 and ν_3 are Raman inactive. As we shall see later, this holds for the fundamentals even in higher approximation. In a similar manner, Fig. 64 shows that for the linear symmetric X_2Y_2 molecules only the vibrations ν_1, ν_2, and ν_4 are Raman active, whereas ν_3 and ν_5 are inactive. Furthermore, for plane X_2Y_4 (see Fig. 44), the vibrations ν_1, ν_2, ν_3, ν_5, ν_6, ν_8 are Raman active, all others inactive. In these three cases, therefore, none of the infrared-active vibrations (see p. 240) is Raman active. However, according to the above rule, it can be seen from Fig. 25a that all three vibrations of non-linear symmetric (and *a fortiori* unsymmetric) XY_2 are Raman active as well as infrared active.

Mathematical formulation: the polarizability ellipsoid. The relation (III, 6) holds quite generally for any direction of the applied field; but the polarizability is not the same for different directions of $\boldsymbol{E}$, and in general $\boldsymbol{P}$ and $\boldsymbol{E}$ have different directions.[2] Therefore, the x component P_x of $\boldsymbol{P}$ depends in general not only on the x component of $\boldsymbol{E}$ but also on the y and z components; that is, we have

$$P_x = \alpha_{xx}E_x + \alpha_{xy}E_y + \alpha_{xz}E_z, \qquad \text{(III, 7)}$$

and similarly

$$P_y = \alpha_{yx}E_x + \alpha_{yy}E_y + \alpha_{yz}E_z, \qquad \text{(III, 8)}$$

$$P_z = \alpha_{zx}E_x + \alpha_{zy}E_y + \alpha_{zz}E_z. \qquad \text{(III, 9)}$$

Here x, y, z are the axes of a coordinate system fixed in the molecule and, since we are assuming no rotation, fixed in space. The α_{xx}, α_{xy}, $\cdots$ are constants independent of the direction of $\boldsymbol{E}$ and $\boldsymbol{P}$. They are called the *components of the polarizability tensor* $\boldsymbol{\alpha}$. It can be shown that

$$\alpha_{xy} = \alpha_{yx}, \qquad \alpha_{xz} = \alpha_{zx}, \qquad \alpha_{yz} = \alpha_{zy}. \qquad \text{(III, 10)}$$

The polarizability tensor can be visualized most easily by means of the *polarizability ellipsoid*, which is formed in a way very similar to the momental ellipsoid (see p. 13). From (III, 7–9) it follows immediately that the component P_E of $\boldsymbol{P}$ in the direction of $\boldsymbol{E}$ is given by

$$P_E = \alpha_E|\boldsymbol{E}|, \qquad \text{(III, 11)}$$

where α_E may be called the E component of the polarizability. It depends on the α_{xx}, α_{xy}, $\cdots$ and on the direction of $\boldsymbol{E}$.[3] If now $1/\sqrt{\alpha_E}$ is plotted from the origin for each direction it can be shown[4] that an ellipsoid is obtained, the polarizability ellipsoid. In the direction of the principal axes of this ellipsoid, $1/\sqrt{\alpha_E}$ has a maximum or minimum, and, therefore, α_E has a minimum or maximum value. Therefore, in these and only these directions has $\boldsymbol{P}$ the same direction as $\boldsymbol{E}$ (compare again the vibrations of a ball suspended by an elastic bar, p. 62f.) and thus $\alpha_E = \alpha$.

Consequently, if the coordinate system is rotated in such a way that the new coordinate axes $\bar{x}$, $\bar{y}$, $\bar{z}$ coincide with the principal axes of the polarizability ellipsoid,

[2] This is from reasons very similar to those on account of which the restoring force acting on a ball suspended by an elastic bar is in general not directed toward the equilibrium position (see Fig. 22 and accompanying discussion).

[3] The form of this dependence can easily be obtained by forming $P_E|E| = (\boldsymbol{P}\cdot\boldsymbol{E})$ and substituting $E_x = E\cos(E, x)$, and so on. Here $(\boldsymbol{P}\cdot\boldsymbol{E})$ is the scalar product of the two vectors $\boldsymbol{P}$ and $\boldsymbol{E}$.

[4] In a way very similar to that for the momental ellipsoid (see standard texts on mechanics).

the equations (III, 7–9) simplify to

$$P_{\bar{x}} = \alpha_{\bar{x}\bar{x}} E_{\bar{x}}, \qquad P_{\bar{y}} = \alpha_{\bar{y}\bar{y}} E_{\bar{y}}, \qquad P_{\bar{z}} = \alpha_{\bar{z}\bar{z}} E_{\bar{z}}. \tag{III, 12}$$

The constants $\alpha_{\bar{x}\bar{y}}$, $\alpha_{\bar{x}\bar{z}}$, $\alpha_{\bar{y}\bar{z}}$ are zero. For the $\alpha_{\bar{x}\bar{x}}$, $\alpha_{\bar{y}\bar{y}}$, and $\alpha_{\bar{z}\bar{z}}$ we have, as in every orthogonal transformation (see p. 247),

$$\alpha_{xx} + \alpha_{yy} + \alpha_{zz} = \alpha_{\bar{x}\bar{x}} + \alpha_{\bar{y}\bar{y}} + \alpha_{\bar{z}\bar{z}}.$$

From (III, 11) and (III, 12) it follows, with the above definition of the polarizability ellipsoid, that the lengths of its axes are $1/\sqrt{\alpha_{\bar{x}\bar{x}}}$, $1/\sqrt{\alpha_{\bar{y}\bar{y}}}$, $1/\sqrt{\alpha_{\bar{z}\bar{z}}}$.

It is easy to find the *principal axes of the polarizability ellipsoid* for a symmetrical molecule, since, just as for the momental ellipsoid (see p. 13), they must *coincide with the symmetry axes* present and be *perpendicular to any plane of symmetry*. Thus, for a non-linear XY_2 molecule of point group C_{2v} one of the principal axes coincides with C_2 and the other two are perpendicular to it and to (one or the other of) the two planes of symmetry; that is, they coincide with the previous x and y axes (see p. 106). For molecules with one more-than-two-fold axis, for instance CH_3Cl, the polarizability ellipsoid, just as the momental ellipsoid, is a *rotational ellipsoid*, and for molecules of cubic symmetry, for instance CH_4, the polarizability ellipsoid degenerates into a *sphere*. While thus in general the axes of the polarizability ellipsoid coincide with those of the momental ellipsoid, it should be emphasized that in unsymmetrically substituted isotopic molecules (for example $XYY^{(i)}$) the polarizability ellipsoid is the same as in the "ordinary" molecule, while the momental ellipsoid is not. This is because the polarizability depends on the distribution of the charges only, whereas the momental ellipsoid depends on the distribution of the masses.

Every one of the six components of the polarizability in general changes when the nuclei are displaced from their equilibrium positions. For small displacements we can expand the α_{xx}, α_{xy}, $\cdots$ thus [compare (III, 1) for M_x]:

$$\alpha_{xx} = \alpha_{xx}^0 + \sum_k \left[\left(\frac{\partial \alpha_{xx}}{\partial x_k}\right)_0 x_k + \left(\frac{\partial \alpha_{xx}}{\partial y_k}\right)_0 y_k + \left(\frac{\partial \alpha_{xx}}{\partial z_k}\right)_0 z_k\right] + \cdots, \tag{III, 13}$$

and similar relations hold for α_{xy}, α_{yy}, $\cdots$. On introducing normal coordinates, we have [compare (III, 2) for M_x)]

$$\alpha_{xx} = \alpha_{xx}^0 + \sum_i \left(\frac{\partial \alpha_{xx}}{\partial \xi_i}\right)_0 \xi_i + \cdots. \tag{III, 14}$$

If this expression and similar ones for α_{xy} and α_{xz}, together with (III, 3) and

$$E_x = E_x{}^0 \cos 2\pi\nu t, \qquad E_y = E_y{}^0 \cos 2\pi\nu t, \qquad E_z = E_z{}^0 \cos 2\pi\nu t$$

are substituted into (III, 7), we obtain:

$$\begin{aligned} P_x = {} & (\alpha_{xx}^0 E_x{}^0 + \alpha_{xy}^0 E_y{}^0 + \alpha_{xz}^0 E_z{}^0) \cos 2\pi\nu t \\ & + \sum_i \left[\left(\frac{\partial \alpha_{xx}}{\partial \xi_i}\right)_0 E_x{}^0 + \left(\frac{\partial \alpha_{xy}}{\partial \xi_i}\right)_0 E_y{}^0 + \left(\frac{\partial \alpha_{xz}}{\partial \xi_i}\right)_0 E_z{}^0\right] \xi_i{}^0 \\ & \times \tfrac{1}{2}[\cos 2\pi(\nu + \nu_i)t + \cos 2\pi(\nu - \nu_i)t] \end{aligned} \tag{III, 15}$$

and similarly for P_y and P_z. That is, the induced dipole moment oscillates, (a) with the frequency ν of the incident radiation leading to Rayleigh scattering, and (b) with the frequencies $\nu \pm \nu_i$, leading to the Raman scattering.

Thus we see that, in the present approximation, a given *normal vibration* ν_i *will appear in the Raman spectrum if and only if at least one of the six components of the change of polarizability* $\left(\frac{\partial \alpha_{xx}}{\partial \xi_i}\right)_0, \left(\frac{\partial \alpha_{xy}}{\partial \xi_i}\right)_0, \cdots$ *is different from zero.* It is to be noted that, in the classical theory here under discussion, according to (III, 15), the intensity of the Raman line $\nu - \nu_i$ should be the same as that of $\nu + \nu_i$.

Taking as an example again the linear symmetric XY_2 molecule, we see from Fig. 25b that for all three vibrations in the displaced position the polarizability ellipsoid has the same axes as it has in the equilibrium position. If we let the internuclear axis be the z axis, we have, both in the equilibrium and in the displaced position, $\alpha_{xy} = \alpha_{xz} = \alpha_{yz} = 0$, and therefore $\left(\frac{\partial \alpha_{xy}}{\partial \xi_i}\right)_0 = \left(\frac{\partial \alpha_{xz}}{\partial \xi_i}\right)_0 = \left(\frac{\partial \alpha_{yz}}{\partial \xi_i}\right)_0 = 0$ for $i = 1, 2, 3$. In the case of the vibrations ν_2 and ν_3, in addition,

$$\left(\frac{\partial \alpha_{xx}}{\partial \xi_i}\right)_0 = \left(\frac{\partial \alpha_{yy}}{\partial \xi_i}\right)_0 = \left(\frac{\partial \alpha_{zz}}{\partial \xi_i}\right)_0 = 0,$$

since for them the polarizability as a function of the corresponding normal coordinates has a horizontal tangent at the equilibrium position (see Fig. 75). Thus, for ν_2 and ν_3 all six components of the change of polarizability are zero; hence ν_2 and ν_3 cannot occur in the Raman spectrum. On the other hand, for ν_1 while $\left(\frac{\partial \alpha_{xy}}{\partial \xi_1}\right)_0$, $\left(\frac{\partial \alpha_{xz}}{\partial \xi_1}\right)_0, \left(\frac{\partial \alpha_{yz}}{\partial \xi_1}\right)_0$ are also zero, $\left(\frac{\partial \alpha_{xx}}{\partial \xi_1}\right)_0 = \left(\frac{\partial \alpha_{yy}}{\partial \xi_1}\right)_0$ and $\left(\frac{\partial \alpha_{zz}}{\partial \xi_1}\right)_0$ are different from zero and thus ν_1 does occur in the Raman spectrum.

In the case of the non-linear symmetric XY_2 molecule, for the vibrations ν_1 and ν_2 (Fig. 25a) the axes of the polarizability ellipsoid remain the same in the displaced position as in the equilibrium position; that is, $\left(\frac{\partial \alpha_{xy}}{\partial \xi_i}\right)_0 = \left(\frac{\partial \alpha_{xz}}{\partial \xi_i}\right)_0 = \left(\frac{\partial \alpha_{yz}}{\partial \xi_i}\right)_0 = 0$. But $\left(\frac{\partial \alpha_{xx}}{\partial \xi_i}\right)_0, \left(\frac{\partial \alpha_{yy}}{\partial \xi_i}\right)_0$, and $\left(\frac{\partial \alpha_{zz}}{\partial \xi_i}\right)_0$ are not zero for these two vibrations, and thus they are Raman active. For the vibration ν_3, on the other hand, we have $\left(\frac{\partial \alpha_{xx}}{\partial \xi_3}\right)_0 = \left(\frac{\partial \alpha_{yy}}{\partial \xi_3}\right)_0 = \left(\frac{\partial \alpha_{zz}}{\partial \xi_3}\right)_0 = 0$, since in opposite phases α_{xx}, α_{yy}, α_{zz} have the same value (curves II and III in Fig. 75). But the z and x axes (in the plane of the molecule) no longer remain axes of the polarizability ellipsoid during the whole of the vibration; that is, $\left(\frac{\partial \alpha_{xz}}{\partial \xi_3}\right)_0 \neq 0$. Thus also this vibration is Raman active, even though only one of the components of the change of polarizability is different from zero.

Overtone and combination vibrations. Just as for the infrared spectrum, if the anharmonicity is taken into account (since then the classical motion contains the frequencies $2\nu_i$, $3\nu_i$, $\cdots$, $\nu_i \pm \nu_k$, $2\nu_i \pm \nu_k$, $\cdots$), in addition to the fundamentals also overtone and combination vibrations may appear as Raman shifts if they are connected with a change of polarizability. If, as usual, the anharmonicity is small, the intensity of Raman lines corresponding to overtone and combination vibrations will be very small compared to those corresponding to (active) fundamentals.

As for the infrared spectrum, *a mechanical and an electrical anharmonicity* must be considered. In the case of a mechanical anharmonicity, (II, 265) or a similar expression rather than (III, 3) has to be substituted into (III, 14). It is immediately clear that P_x (and similarly P_y and P_z) then oscillates with frequencies $\nu \pm 2\nu_i$, $\nu \pm 3\nu_i$, $\cdots$, $\nu \pm (\nu_i \pm \nu_k)$, $\nu \pm (2\nu_i \pm \nu_k)$, $\cdots$, as well as $\nu \pm \nu_i$. It is also seen that the overtone and combination vibrations involving ν_i occur only when at least one of $\left(\frac{\partial \alpha_{xx}}{\partial \xi_i}\right)_0, \left(\frac{\partial \alpha_{xy}}{\partial \xi_i}\right)_0, \cdots$ is different from zero, that is, when the fundamental also occurs. Thus, the overtones of ν_2 and ν_3 of linear symmetric XY_2 would not occur in consequence of mechanical anharmonicity, but only the overtones $2\nu_1$, $3\nu_1$ of ν_1 and some of its combination frequencies with the other two.

If an electrical anharmonicity is present, we have to take into account higher terms such as

$$\sum_i \sum_k \left(\frac{\partial^2 \alpha_{xx}}{\partial \xi_i \partial \xi_k}\right)_0 \xi_i \xi_k \tag{III, 16}$$

in the development (III, 14). Substituting into (III, 7), we see that the Raman spectrum will show the shifts $2\nu_i$, $\nu_i + \nu_k$, $\cdots$ in addition to the fundamentals. However, here the Raman activity of the overtones and combination tones is independent of the activity of the corresponding fundamentals, since the activity of the former depends on $\left(\frac{\partial^2 \alpha_{xx}}{\partial \xi_i \partial \xi_k}\right)_0 \cdots$ and higher derivatives, whereas the activity of the latter depends on $\left(\frac{\partial \alpha_{xx}}{\partial \xi_i}\right)_0 \cdots$. Thus for the vibrations ν_2 and ν_3 of linear symmetric XY_2, as we have previously seen, $\left(\frac{\partial \alpha_{xx}}{\partial \xi_2}\right)_0 = 0, \left(\frac{\partial \alpha_{xx}}{\partial \xi_3}\right)_0 = 0$ (see curves II or III in Fig. 74), but $\left(\frac{\partial^2 \alpha_{xx}}{\partial \xi_2^2}\right)_0 \neq 0, \left(\frac{\partial^2 \alpha_{xx}}{\partial \xi_3^2}\right)_0 \neq 0$, and similarly for the other polarizability components. Therefore, while the fundamentals ν_2 and ν_3 cannot occur as Raman shifts in any approximation, the first overtones $2\nu_2$, $2\nu_3$ can occur weakly when there is an electrical anharmonicity, that is, when the fact that α is not a constant even during the vibrations ν_2 and ν_3 is taken into account. More simply, since during one oscillation of ν_2 or ν_3 the polarizability α goes twice through a maximum (that is, oscillates with twice the frequency of the mechanical oscillation), it is $2\nu_2$ or $2\nu_3$, not ν_2 or ν_3 that occurs weakly as a Raman shift.

Polarization of Rayleigh and Raman scattering. If the polarizability ellipsoid of a molecule is a sphere (as for instance for CH_4), the direction of the induced dipole moment $\boldsymbol{P}$ coincides for any orientation of the system with the direction of the field $\boldsymbol{E}$ producing it. If, therefore, on irradiation of a gas consisting of such molecules with light of frequency ν (that is, $E = E_0 \cos 2\pi\nu t$), the scattered light of frequency ν (*Rayleigh scattering*) is observed at right angles to the incident beam, it will be completely polarized in the plane at right angles to the incident beam, irrespective of whether the incident light is polarized or not. However, if the polarizability ellipsoid of the scattering system is not a sphere, the direction of $\boldsymbol{P}$ coincides with that of $\boldsymbol{E}$ only if $\boldsymbol{E}$ coincides with one of the axes of the polarizability ellipsoid, but otherwise has a different direction. If a gas (or liquid) containing such molecules with all orientations is irradiated, $\boldsymbol{P}$ is no longer restricted to the plane at right angles to the beam, even though it cannot take all orientations with respect to it with equal probability. Therefore, the *scattered light observed at right angles to the incident beam will no longer be completely polarized.* The *degree of depolarization* will also depend on whether the incident beam is polarized or unpolarized.

In this general case, when the polarizability ellipsoid is not a sphere (anisotropic molecule), we can resolve the polarizability into the sum of a *spherical part* for which the polarizability α^{I} is the average of the three principal polarizabilities

$$\alpha^{\mathrm{I}} = \alpha^{\mathrm{I}}_{xx} = \alpha^{\mathrm{I}}_{yy} = \alpha^{\mathrm{I}}_{zz} = \tfrac{1}{3}(\alpha_{xx} + \alpha_{yy} + \alpha_{zz}) \qquad \text{(III, 17)}$$

and a *completely anisotropic part* α^{II} for which

$$\alpha^{\mathrm{II}}_{xx} + \alpha^{\mathrm{II}}_{yy} + \alpha^{\mathrm{II}}_{zz} = 0, \qquad \text{(III, 18)}$$

and of course:

$$\alpha^{\mathrm{II}}_{xx} = \alpha_{xx} - \alpha^{\mathrm{I}}_{xx}, \qquad \alpha^{\mathrm{II}}_{yy} = \alpha_{yy} - \alpha^{\mathrm{I}}_{yy}, \qquad \alpha^{\mathrm{II}}_{zz} = \alpha_{zz} - \alpha^{\mathrm{I}}_{zz}. \qquad \text{(III, 19)}$$

The *magnitude of the anisotropy* may be indicated by

$$\beta^2 = \tfrac{1}{2}[(\alpha_{xx} - \alpha_{yy})^2 + (\alpha_{yy} - \alpha_{zz})^2 + (\alpha_{zz} - \alpha_{xx})^2 + 6(\alpha^2_{xy} + \alpha^2_{yz} + \alpha^2_{zx})]. \qquad \text{(III, 20)}$$

The sums in (III, 17) and (III, 20) are *invariants of the polarizability tensor*, that is, their value is independent of the orientation of the coordinate axes relative to the polarizability ellipsoid [see (5)]. If the coordinate axes coincide with the axes of the ellipsoid (see the coordinates $\bar{x}$, $\bar{y}$, $\bar{z}$, p. 243), the last term in (III, 20) $[6(\alpha^2_{xy} + \alpha^2_{yz} + \alpha^2_{zx})]$ disappears.

The *degree of depolarization* ρ is defined as the ratio of the intensity of the scattered light polarized[4a] perpendicular to the xy plane, $I_{\perp}$, to that polarized parallel to this plane, $I_{\parallel}$. Here the z axis is taken in the direction of propagation of the incident light, and the direction of observation is perpendicular to the z axis. By averaging over all orientations of the system, it can then be shown [see, for instance, Born (2) and Wolf (949)] that *for natural (unpolarized) incident light*

$$\rho_n = \frac{I_{\perp}}{I_{\parallel}} = \frac{6\beta^2}{45(\alpha^{\mathrm{I}})^2 + 7\beta^2}\cdot \qquad \text{(III, 21)}$$

α^{I} in (III, 21) is always different from zero since the polarizability is always positive ($\alpha^{\mathrm{I}} = 0$ would mean that $\alpha_{\bar{x}\bar{x}} = \alpha_{\bar{y}\bar{y}} = \alpha_{\bar{z}\bar{z}} = 0$, that is, that no scattered radiation would appear). The smallest value of ρ_n is therefore $\rho_n = 0$, which is obtained when $\beta = 0$. The largest value of ρ_n corresponds to the most anisotropic case. This is obtained if all except one, say $\alpha_{\bar{x}\bar{x}}$, of the components of α are zero. Then $\alpha^{\mathrm{I}} = \alpha_{\bar{x}\bar{x}}/3$, $\beta^2 = \alpha_{\bar{x}\bar{x}}$ and therefore $\rho_n^{(\max)} = \frac{1}{2}$, which is thus the maximum degree of depolarization for Rayleigh scattering.

The *degree of depolarization for linearly polarized incident light* (at 90° to the scattered beam) is related to ρ_n by [see Born (2) and Placzek (700)]

$$\rho_l = \frac{\rho_n}{2 - \rho_n}; \qquad \text{(III, 22)}$$

that is, the maximum degree of depolarization in this case is $\rho_l^{(\max)} = \frac{1}{3}$. The Rayleigh scattering of most molecules will have a smaller degree of depolarization than these maximum values.

Sometimes it is also of interest to study the scattering of *circularly polarized incident light*. Obviously, if the polarizability ellipsoid is a sphere, and if the incident light is polarized in a clockwise direction, the light scattered backward (180° to incident beam) should be circularly polarized in a counter-clockwise direction (always looking in the direction of propagation). But if the scattering system is anisotropic,

[4a] Here as in the following text we take the direction of polarization as the direction of the electric vector of the light wave.

the light that is scattered backward may also have a component of clockwise circular polarization. The degree of this reversal ρ_c for circularly polarized incident light may be shown to be related to ρ_n by

$$\rho_c = \frac{I_{\text{clockwise}}}{I_{\text{counter-clockwise}}} = \frac{\rho_n}{1 - \rho_n}, \tag{III, 23}$$

which is zero for $\rho_n = 0$ and has the maximum value 1; that is, for a completely anisotropic system the light scattered backward would be unpolarized.

In order to predict on the basis of classical theory the *state of polarization of a Raman line* corresponding to a normal vibration ν_i, the behavior of the change of polarizability $\left(\frac{\partial \boldsymbol{\alpha}}{\partial \xi_i}\right)_0$ (the derived tensor) must be investigated. For example, for the totally symmetric vibration ν_1 of tetrahedral XY_4 molecules, $\boldsymbol{\alpha}$ retains spherical symmetry during the whole vibration and therefore $\frac{\partial \boldsymbol{\alpha}}{\partial \xi_1}$ is spherical. Consequently, the part of $\boldsymbol{P}$ oscillating with frequency $\nu \pm \nu_1$ will have the same direction as $\boldsymbol{E}$. These Raman lines are, therefore, completely polarized.

More detailed calculations [see Born (2) and Cabannes (3) (189)] show that although the principal axes of the "ellipsoid" corresponding to $\left(\frac{\partial \boldsymbol{\alpha}}{\partial \xi_i}\right)_0$ do not always coincide with those of the polarizability ellipsoid ($\boldsymbol{\alpha}$), the formulae for the polarization are the same as for the Rayleigh scattering except that α^{I} has to be replaced by

$$\alpha_i'^{\mathrm{I}} = \frac{1}{3}\left[\left(\frac{\partial \alpha_{xx}}{\partial \xi_i}\right)_0 + \left(\frac{\partial \alpha_{yy}}{\partial \xi_i}\right)_0 + \left(\frac{\partial \alpha_{zz}}{\partial \xi_i}\right)_0\right] \tag{III, 24}$$

and β^2 has to be replaced by

$$\beta_i'^2 = \frac{1}{2}\left\{\left(\frac{\partial \alpha_{xx}}{\partial \xi_i} - \frac{\partial \alpha_{yy}}{\partial \xi_i}\right)_0^2 + \left(\frac{\partial \alpha_{yy}}{\partial \xi_i} - \frac{\partial \alpha_{zz}}{\partial \xi_i}\right)_0^2 + \left(\frac{\partial \alpha_{zz}}{\partial \xi_i} - \frac{\partial \alpha_{xx}}{\partial \xi_i}\right)_0^2 + 6\left[\left(\frac{\partial \alpha_{xy}}{\partial \xi_i}\right)_0^2 + \left(\frac{\partial \alpha_{yz}}{\partial \xi_i}\right)_0^2 + \left(\frac{\partial \alpha_{zx}}{\partial \xi_i}\right)_0^2\right]\right\}. \tag{III, 25}$$

Thus we have

$$\rho_n = \frac{I_{\perp}}{I_{\parallel}} = \frac{6\beta_i'^2}{45(\alpha_i'^{\mathrm{I}})^2 + 7\beta_i'^2}, \tag{III, 26}$$

and (III, 22) and (III, 23) for ρ_l and ρ_c respectively.[5]

An essential difference from the case of Rayleigh scattering is, however, that the components of $\left(\frac{\partial \boldsymbol{\alpha}}{\partial \xi_i}\right)_0$ in the direction of the principal axes are not necessarily all positive, as are those of $\boldsymbol{\alpha}$. Therefore $\left(\frac{\partial \boldsymbol{\alpha}}{\partial \xi_i}\right)_0$ cannot always be represented by a real ellipsoid. But it is still convenient to visualize the relations by means of such an ellipsoid. In consequence of the possibility of negative values of $\left(\frac{\partial \alpha_{xx}}{\partial \xi_i}\right)_0, \left(\frac{\partial \alpha_{yy}}{\partial \xi_i}\right)_0, \cdots, \alpha_i'^{\mathrm{I}}$ may now be zero without β_i' being zero. This case gives the strongest depolarization, namely:

$$\rho_n^{\max} = \tfrac{6}{7}, \qquad \rho_l^{\max} = \tfrac{3}{4}, \qquad \rho_c^{\max} = 6. \tag{III, 27}$$

For brevity, if a Raman line shows this state of depolarization, it is called *depolarized*, whereas, if the degree of depolarization is smaller, it is said to be (partly or completely) *polarized*.

If a molecule has a plane of symmetry (for example XY_2Z_2), one axis of the polarizability ellipsoid of the equilibrium position is perpendicular to this plane (say in the x direction), and the other two are in the plane. Now, in a displaced position antisymmetric to the plane the polarizability ellipsoid

---

[5] For example, in the above case of the totally symmetric vibration of tetrahedral XY_4, $\beta_i'^2 = 0$ and therefore $\rho_n = 0$, $\rho_l = 0$, $\rho_c = 0$.

has no longer the same axes; but it is easily seen from an examination of Fig. 76 that α_{xx} (the x component of the dipole moment induced by unit field in the x direction) has the same value for this displaced position as it has for the opposite displaced position, that is, it follows a curve like II or III in Fig. 75. Consequently, $\left(\frac{\partial\alpha_{xx}}{\partial\xi_i}\right)_0 = 0$. In a similar manner, it is seen that $\left(\frac{\partial\alpha_{yy}}{\partial\xi_i}\right)_0 = 0$ and $\left(\frac{\partial\alpha_{zz}}{\partial\xi_i}\right) = 0$.

Therefore, according to (III, 24) and (III, 26), $\rho_n = \frac{6}{7}$ for this antisymmetric vibration; that is, the particular Raman line will be depolarized. Similarly, it can be shown that any vibration that is antisymmetric or degenerate with respect to any other symmetry element will give a Raman line with the maximum degree of depolarization if it occurs at all. Therefore, *only Raman lines corresponding to totally symmetric vibrations can have a degree of depolarization smaller than the maximum value* $\frac{6}{7}$, that is, can be polarized.

For molecules belonging to the cubic point groups, we have always $\rho_n = 0$ for the totally symmetric vibrations, since for these

$$\left(\frac{\partial\alpha_{xy}}{\partial\xi_i}\right)_0 = \left(\frac{\partial\alpha_{xz}}{\partial\xi_i}\right)_0 = \left(\frac{\partial\alpha_{yz}}{\partial\xi_i}\right)_0 = 0,$$

$$\left(\frac{\partial\alpha_{xx}}{\partial\xi_i}\right)_0 = \left(\frac{\partial\alpha_{yy}}{\partial\xi_i}\right)_0 = \left(\frac{\partial\alpha_{zz}}{\partial\xi_i}\right)_0,$$

and, therefore, $\beta_i'^2 = 0$ in (III, 26). For the totally symmetric vibrations of all other point groups, ρ_n may have any value between 0 and $\frac{6}{7}$.

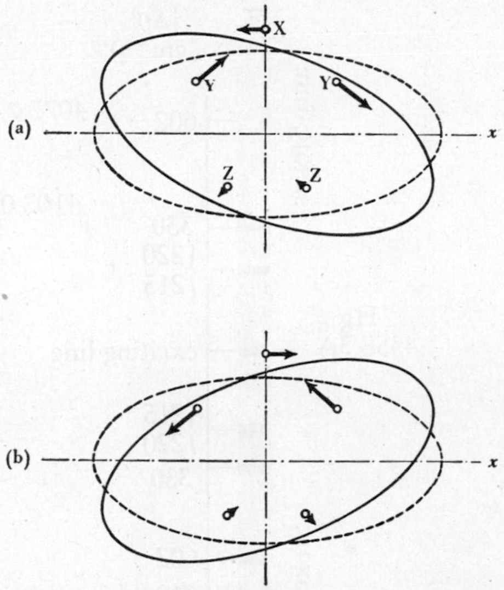

FIG. 76. **Polarizability change for an antisymmetric vibration (schematic).**—The broken-line ellipses indicate the polarizability ellipsoid in the equilibrium position, the solid ellipses in two opposite displaced positions. The intercepts of the latter ellipsoids with the X axis give $1/\sqrt{\alpha_{xx}}$ (see p. 243) which is seen to be the same in the two cases.

2. Quantum-theoretical Treatment

In quantum theory the frequencies of the infrared absorption (or emission) bands and the frequency shifts of the Raman bands are determined by the *energy differences of the vibrational levels* between which the transitions take place. In order to find out which transitions occur and with what intensity, it is necessary to calculate the *transition probabilities*.

(a) *Elementary treatment of fundamentals*

Since for small amplitudes of the nuclei a polyatomic molecule may be considered as a superposition of harmonic oscillators, the results for the harmonic oscillator approximation of diatomic molecules (see Molecular Spectra I, pp. 85 and 93) may be taken over. That is, both in the infrared and Raman spectrum we have the *selection rule*

$$\Delta v_i = \pm 1 \qquad \text{(III, 28)}$$

for each normal vibration ν_i. Since the oscillators, in this approximation, are independent, *no simultaneous jumps of two or more vibrations* can occur. As for diatomic molecules, only those vibrations that are connected with a *change of dipole moment*

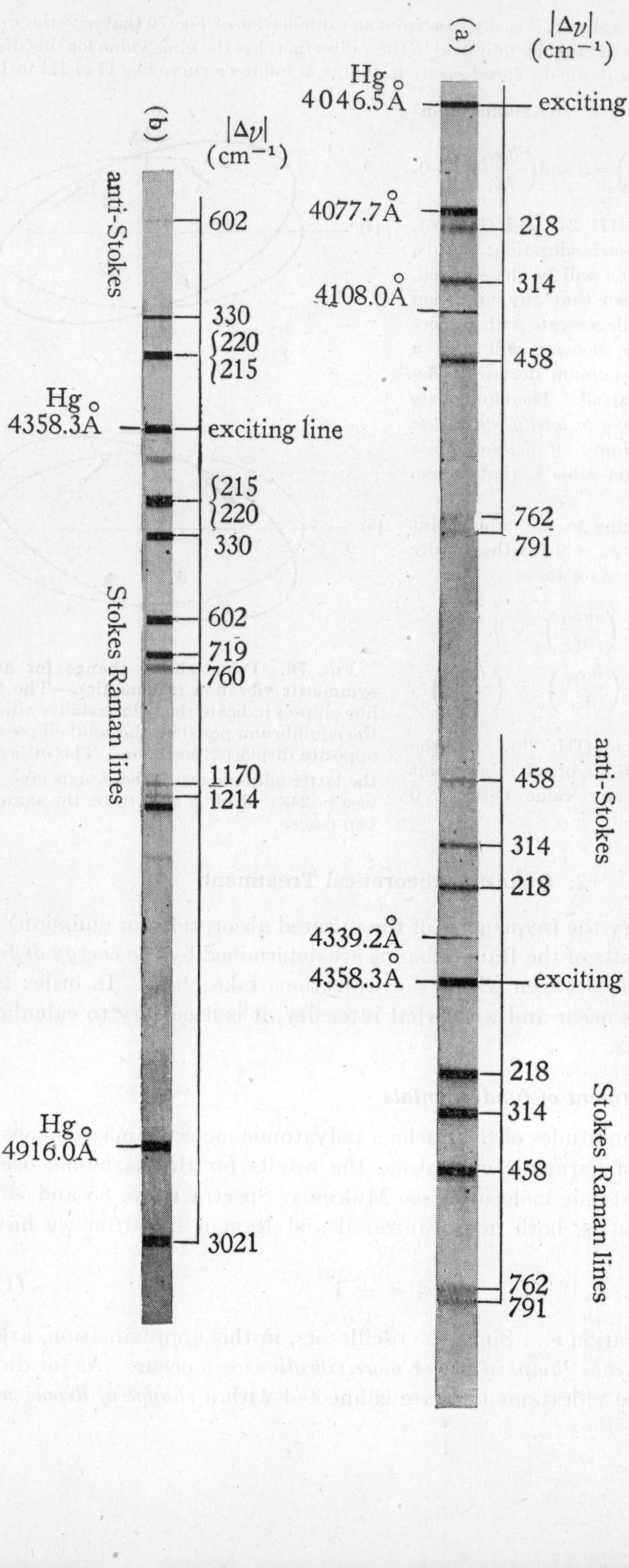

FIG. 77. Raman spectra of CCl_4 and $CHCl_2Br$ showing anti-Stokes lines after Glockler.*—Raman lines are indicated at the top of each spectrogram, Hg lines at the bottom. The exciting line has been reduced in intensity by a screen. In (a) the Stokes Raman lines of CCl_4 occur both through excitation by the line 4358.3 Å and by the line 4046.5 Å.

* The author is greatly indebted to Professor G. Glockler for these two spectrograms.

can have $\Delta v_i = \pm 1$ for an infrared transition, and only those vibrations that are connected with a (linear) *change of polarizability* can have $\Delta v_i = \pm 1$ for a Raman transition. Considering the formula (II, 61) for the vibrational energy levels which applies to this approximation, it is clear that the wave numbers of the infrared bands and the wave-number shifts of the Raman bands are equal to the actual vibrational frequencies (measured in cm^{-1})

$$\nu = \omega_i, \qquad |\Delta\nu| = \omega_i. \qquad \text{(III, 29)}$$

Thus, in this approximation, the quantum-theoretical infrared and Raman vibration spectrum is the same as the classical one, at least as far as the position and occurrence or non-occurrence of the bands is concerned. However, just as for diatomic molecules, there is an important difference as to the intensities of the Raman bands. Whereas according to classical theory the Stokes and anti-Stokes Raman lines $\nu - \omega_i$ and $\nu + \omega_i$ should have the same intensity, according to quantum theory and in agreement with observation the *anti-Stokes lines have a much smaller intensity*, since the number of molecules in the initial state $v_i = 1$ of the anti-Stokes lines is only $e^{-(hc\omega_i/kT)}$ times the number of molecules in the initial state $v_i = 0$ of the Stokes lines (ground state). For diatomic molecules no anti-Stokes vibrational Raman lines have ever been observed. For polyatomic molecules they have been observed, but only for the smaller frequencies, as is shown in Fig. 77 for CCl_4 and $CHCl_2Br$. The intensity ratio to the corresponding Stokes lines is in agreement with the Boltzmann factor.

Since in quantum theory, as in classical theory, the occurrence of a certain fundamental in the infrared or Raman spectrum depends on the presence of a *change of dipole moment or polarizability* respectively, the previous conclusions (section 1) as to the dependence on the symmetry of the molecule remain valid in quantum theory. Thus, for linear symmetrical XY_2, for example, only the totally symmetric vibration ν_1 can change its quantum number v_1 by ± 1 in the Raman effect, whereas only the vibrations ν_2 and ν_3 can change their quantum numbers by ± 1 in the infrared. Conversely, if a triatomic XY_2 molecule exhibits only one (Stokes) Raman line and only two strong infrared bands of a frequency different from that of the Raman line, it can be concluded that the molecule is linear and symmetric, since for a non-linear as well as for a linear unsymmetrical XY_2 molecule all three fundamentals are active both in the infrared and in the Raman effect.

Similar considerations can be applied to other cases (see section 1).

(*b*) *Rigorous vibrational selection rules*

For quantitative calculations of the intensity of infrared and Raman bands and also for the determination of the activity of overtone and combination vibrations [see subsection (d)] it is necessary to develop a more detailed quantum-theoretical treatment.

Infrared spectrum. The dipole moment of the molecule is represented in wave mechanics by the matrix formed from the integrals

$$\int \Psi_n \Psi_m^* M d\tau, \qquad \text{(III, 30)}$$

where M is a vector with components

$$M_x = \sum e_i x_i, \qquad M_y = \sum e_i y_i, \qquad M_z = \sum e_i z_i \qquad \text{(III, 31)}$$

(e_i = charge of particle i having coordinates x_i, y_i, z_i), and where Ψ_n and Ψ_m are the time-dependent eigenfunctions of the system in two states n and m, that is,

$$\Psi_n = \psi_n e^{2\pi i(E_n/h)t}, \quad \Psi_m = \psi_m e^{2\pi i(E_m/h)t}, \quad \Psi_m{}^* = \psi_m{}^* e^{-2\pi i(E_m/h)t}, \tag{III, 32}$$

the asterisk indicating the complex conjugate quantities. The *diagonal elements* of the matrix, that is, the integrals (III, 30) with $n = m$, represent the permanent dipole moments in the states n (since the time dependence cancels out). The *off-diagonal matrix elements* ($n \neq m$) correspond to the transitions from the state n to the state m since they have the time factor $e^{2\pi i(E_n - E_m/h)t}$. The *transition probability* is proportional to the square of the time-independent factor of (III, 30), that is, to the square of

$$[M]^{nm} = \int \psi_n \psi_m{}^* M d\tau. \tag{III, 33}$$

We consider now the transition between two vibrational levels v' and v'' of the molecule produced by dipole radiation. Here v' and v'' stand for the sets of vibrational quantum numbers v_1', v_2', v_3', $\cdots$ and v_1'', v_2'', v_3'', $\cdots$ of the upper and lower state respectively. Strictly speaking, we should substitute the total eigenfunction in (III, 33). But to a usually fairly good approximation, this total eigenfunction is the product of an electronic, a vibrational, and a rotational eigenfunction (see p. 15):

$$\psi = \psi_e \psi_v \psi_r; \tag{III, 34}$$

and therefore, since for the *pure vibration spectrum* the electronic and rotational state remains unchanged, ψ_e and ψ_r simply give a constant factor in (III, 33) so that we conclude that the *vibrational transition probability is proportional to the square of*

$$[M]^{v'v''} = \int \psi_v{}' \psi_v{}''^* M d\tau, \tag{III, 35}$$

where $\psi_v{}'$ and $\psi_v{}''$ are the vibrational eigenfunctions of the upper and lower state respectively. $[M]^{v'v''}$ is also called the *transition moment* of the transition $v' \leftrightarrow v''$.

TABLE 55. SYMMETRY TYPES (SPECIES) [6] OF THE COMPONENTS OF THE DIPOLE MOMENT AND OF THE POLARIZABILITY (INFRARED AND RAMAN SELECTION RULES) FOR THE MORE IMPORTANT POINT GROUPS [SEE PLACZEK (700) AND TISZA (867)].

	C_s	C_i	C_{2h} $[C_2]$[7]	C_{2v}	$V_h \equiv D_{2h}$ $[V \equiv D_2]$[7]	C_{3v}	C_{4v}	C_{5v}, C_{6v}	$C_{\infty v}$	D_3 $[C_3]$[8]	D_4 $[C_4]$[8]	D_5 $[C_5]$[8]	D_6 $[C_6]$[8]	D_{3h} $[C_{3h}]$[8]
M_x	A'	A_u	B_u	B_1	B_{3u}	E	E	E_1	Π	E	E	E_1	E_1	E'
M_y	A'	A_u	B_u	B_2	B_{2u}	E	E	E_1	Π	E	E	E_1	E_1	E'
M_z	A''	A_u	A_u	A_1	B_{1u}	A_1	A_1	A_1	Σ^+	A_2	A_2	A_2	A_2	A_2''
α_{xx}	A'	A_g	A_g	A_1	A_g	A_1, E	A_1, B_1	A_1, E_2	Σ^+, Δ	A_1, E	A_1, B_1	A_1, E_2	A_1, E_2	A_1', E'
α_{yy}	A'	A_g	A_g	A_1	A_g	A_1, E	A_1, B_1	A_1, E_2	Σ^+, Δ	A_1, E	A_1, B_1	A_1, E_2	A_1, E_2	A_1', E'
α_{zz}	A'	A_g	A_g	A_1	A_g	A_1	A_1	A_1	Σ^+	A_1	A_1	A_1	A_1	A_1'
α_{xy}	A'	A_g	A_g	A_2	B_{1g}	E	B_2	E_2	Δ	E	B_2	E_2	E_2	E'
α_{xz}	A''	A_g	B_g	B_1	B_{2g}	E	E	E_1	Π	E	E	E_1	E_1	E''
α_{yz}	A''	A_g	B_g	B_2	B_{3g}	E	E	E_1	Π	E	E	E_1	E_1	E''

[6] For an explanation of the symmetry types and the choice of the coordinate axes, see Tables 12–30 (the symmetry axis of highest symmetry is always the z axis).

[7] For C_2 and $D_2 \equiv V$ omit the subscripts g and u.

[8] For C_3, C_4, C_5, C_6, and C_{3h}, C_{4h}, C_{5h}, C_{6h}, omit the subscripts 1 and 2 of A and B.

[9] For T omit the subscripts 1 and 2 of A and F.

[10] For O omit the subscripts g and u.

TABLE 55.—*Continued*

	D_{4h} $[C_{4h}]$[8]	D_{5h} $[C_{5h}]$[8]	D_{6h} $[C_{6h}]$[8]	$D_{\infty h}$	$D_{2d} \equiv V_d$	D_{3d}	D_{4d}	S_4	S_6	T_d $[T]$[9]	O_h $[O]$[10]
M_x	E_u	E_1'	E_{1u}	Π_u	E	E_u	E_1	E	E_{1u}	F_2	F_{1u}
M_y	E_u	E_1'	E_{1u}	Π_u	E	E_u	E_1	E	E_{1u}	F_2	F_{1u}
M_z	A_{2u}	A_2''	A_{2u}	Σ_u^+	B_2	A_{2u}	B_2	A	B_u	F_2	F_{1u}
α_{xx}	A_{1g}, B_{1g}	A_1', E_2'	A_{1g}, E_{2g}	Σ_g^+, Δ_g	A_1, B_1	A_{1g}, E_g	A_1, E_2	A, B	A_g, E_{2g}	A_1, E	A_{1g}, E_g
α_{yy}	A_{1g}, B_{1g}	A_1', E_2'	A_{1g}, E_{2g}	Σ_g^+, Δ_g	A_1, B_1	A_{1g}, E_g	A_1, E_2	A, B	A_g, E_{2g}	A_1, E	A_{1g}, E_g
α_{zz}	A_{1g}	A_1'	A_{1g}	Σ_g^+	A_1	A_{1g}	A_1	A	A_g	A_1, E	A_{1g}, E_g
α_{xy}	B_{2g}	E_2'	E_{2g}	Δ_g	B_2	E_g	E_2	B	E_{2g}	F_2	F_{2g}
α_{xz}	E_g	E_1''	E_{1g}	Π_g	E	E_g	E_3	E	E_{2g}	F_2	F_{2g}
α_{yz}	E_g	E_1''	E_{1g}	Π_g	E	E_g	E_3	E	E_{2g}	F_2	F_{2g}

Vibrational selection rules exist only when the molecule under consideration has elements of symmetry. In that case it is immediately clear that the integral (III, 35) can be different from zero for a certain transition (that is, that this transition is an allowed one) only when at least one of the components of the integrand $\psi_v'\psi_v''^*M$ remains unchanged for any of the symmetry operations permitted by the symmetry of the molecule in its equilibrium position, or in other words *when at least one of the quantities*

$$\psi_v'\psi_v''^*M_x, \qquad \psi_v'\psi_v''^*M_y, \qquad \psi_v'\psi_v''^*M_z$$

is totally symmetrical. This is the *general vibrational selection rule for the infrared,* which is *rigorous as long as the interaction with rotation and electronic motion is neglected.* In particular it is independent of whether or not the vibrations are harmonic.

From the definition (III, 31) of the dipole moment M it is clear that its components M_x, M_y, M_z have the same behavior with respect to symmetry operations as the translations T_x, T_y, T_z in the direction of the coordinate axes; that is, they belong to one of the species of the point group of the molecule, as indicated in the last column of each of the Tables 12 to 30. For the convenience of the reader, in Table 55 the *species of* M_x, M_y, M_z are collected together for all the more important point groups, as read off from the last columns of Tables 12–30.

With these considerations in mind, the above general selection rule may also be formulated thus: *A vibrational transition* $v' \leftrightarrow v''$ *is allowed only when there is at least one component of the dipole moment* M *that has the same species as the product* $\psi_v'\psi_v''$. The equivalence of this rule and the previous one is immediately obvious for the point groups with non-degenerate species only, since the product of two functions can be symmetric with respect to a symmetry operation only if both factors are symmetric or both antisymmetric with respect to this symmetry operation. However, the above formulation of the general selection rule holds also for point groups with degenerate species. This can easily be verified with the help of Tables 31 and 33, which show that the product of two species is totally symmetric or contains a totally symmetric part only when the two species are the same.

In order to ascertain whether a certain transition $v' \leftrightarrow v''$ is allowed in the infrared, therefore, it is only necessary to see whether the species of $\psi_v'\psi_v''$ (obtained in the same way as outlined in Chapter II, section 3e, for the species of the total vibrational eigenfunction) is the same as that of M_x, M_y, or M_z as given in Table 55.

Raman spectrum. The intensity of scattered light depends on the induced dipole moment $\boldsymbol{P}$ which, similar to (III, 30), is represented by the matrix formed from the integrals,

$$\int \Psi_n \Psi_m^* \boldsymbol{P} d\tau, \tag{III, 36}$$

where $\boldsymbol{P}$ is a vector whose components are given by (III, 7–9). The time-independent part of (III, 36) is

$$[\boldsymbol{P}^0]^{nm} = \int \psi_n \psi_m^* \boldsymbol{P}^0 d\tau, \tag{III, 37}$$

where $\boldsymbol{P}^0$ is the amplitude of $\boldsymbol{P}$. The intensity of a Raman transition $n \leftrightarrow m$ is proportional to the square of $[\boldsymbol{P}^0]^{nm}$. Substituting $\boldsymbol{P}$ from (III, 7–9) we obtain for the components of $[\boldsymbol{P}^0]^{nm}$:

$$\begin{aligned}
[P_x{}^0]^{nm} &= E_x{}^0 \int \alpha_{xx} \psi_n \psi_m^* d\tau + E_y{}^0 \int \alpha_{xy} \psi_n \psi_m^* d\tau + E_z{}^0 \int \alpha_{xz} \psi_n \psi_m^* d\tau, \\
[P_y{}^0]^{nm} &= E_x{}^0 \int \alpha_{xy} \psi_n \psi_m^* d\tau + E_y{}^0 \int \alpha_{yy} \psi_n \psi_m^* d\tau + E_z{}^0 \int \alpha_{yz} \psi_n \psi_m^* d\tau, \\
[P_z{}^0]^{nm} &= E_x{}^0 \int \alpha_{xz} \psi_n \psi_m^* d\tau + E_y{}^0 \int \alpha_{yz} \psi_n \psi_m^* d\tau + E_z{}^0 \int \alpha_{zz} \psi_n \psi_m^* d\tau.
\end{aligned} \tag{III, 38}$$

Here, $E_x{}^0$, $E_y{}^0$, $E_z{}^0$ are the components of the amplitude of the incident light wave, and the integrals

$$[\alpha_{xx}]^{nm} = \int \alpha_{xx} \psi_n \psi_m^* d\tau, \qquad [\alpha_{xy}]^{nm} = \int \alpha_{xy} \psi_n \psi_m^* d\tau, \cdots \tag{III, 39}$$

are the *matrix elements of the six components of the polarizability tensor*. The diagonal matrix elements ($n = m$) of $\boldsymbol{\alpha}$ or $\boldsymbol{P}^0$ correspond to Rayleigh scattering, the off-diagonal elements to Raman scattering, that is, to transitions $n \leftrightarrow m$ induced by the incident light. According to (III, 38) *a Raman transition $n \leftrightarrow m$ is allowed if at least one of the six quantities $[\alpha_{xx}]^{nm}$, $[\alpha_{xy}]^{nm}$, $\cdots$, is different from zero.*

For the *vibrational Raman spectrum* we have again to substitute for ψ_n and ψ_m the vibrational eigenfunctions ψ_v' and ψ_v'' of the upper and lower states. We can then say (compare the selection rule for the infrared spectrum p. 253): *A Raman transition between two vibrational level v' and v'' is allowed if at least one of the six products*

$$\alpha_{xx} \psi_v' \psi_v''^*, \qquad \alpha_{xy} \psi_v' \psi_v''^*, \cdots \tag{III, 40}$$

is totally symmetrical, that is, remains unchanged for all symmetry operations permitted by the symmetry of the molecule.

Similarly to the previous infrared selection rule, the general (and rigorous) Raman selection rule may also be stated in the following somewhat more convenient form: *A Raman transition between two vibrational levels v' and v'' is allowed if the product $\psi_v' \psi_v''$ has the same species as at least one of the six components α_{xx}, α_{xy}, $\cdots$ of the polarizability tensor.*

In order to be able to apply this rule we have to know the species of α_{xx}, α_{xy}, $\cdots$ for the various point groups. It is easy to find the behavior of the α_{xx}, α_{xy}, $\cdots$ with respect to symmetry operations (that is, to find the species) for point groups with non-degenerate species only. Let us suppose that a field is applied in the y direction ($E_y \neq 0$, $E_x = E_z = 0$); then, according to (III, 7), $P_x = \alpha_{xy} E_y$. Since, in the present case, P_x and E_y can only remain unchanged or change sign for a symmetry operation, α_{xy} will remain unchanged or change sign depending on whether P_x and E_y behave in the same or in the opposite way for the symmetry operation considered. P_x and E_y have the same species as the translations T_x and T_y respectively, and therefore α_{xy} has the same species as the product of the species of T_x and T_y. Similarly, α_{xz} and α_{yz} have the same species as the products T_xT_z and T_yT_z respectively. Finally α_{xx}, α_{yy}, α_{zz} have the same species as the products T_xT_x, T_yT_y, T_zT_z respectively, all of which are totally symmetric in the case of point groups with non-degenerate species only. As an example, consider a molecule of point group V_h (for instance plane X_2Y_4). α_{xx}, α_{yy}, α_{zz} have the totally symmetric species A_g, α_{xy} has the species B_{1g} since T_xT_y has species $B_{3u} \times B_{2u} = B_{1g}$, and α_{xz} and α_{yz} have the species B_{2g} and B_{3g}, respectively, for similar reasons. Thus, only those transitions may occur in the Raman effect for which the product $\psi_v'\psi_v''$ has one of the species A_g, B_{1g}, B_{2g}, or B_{3g}.

In Table 55 the *species of the six components of the polarizability tensor* are given *for all the important point groups*, including those with degenerate species. In the latter cases, the species can be obtained in a manner similar to the above but somewhat more complicated (see below). In these cases for α_{xx}, α_{yy} (and for cubic point groups, also for α_{zz}) two species are given. This is meant to indicate that actually only $\alpha_{xx} + \alpha_{yy}$ and $\alpha_{xx} - \alpha_{yy}$ [and for cubic point groups $\alpha_{xx} + \alpha_{yy} + \alpha_{zz}$ and a more complicated linear combination; see Tisza (867)] have a definite species. For most practical cases this amounts to the same thing as assuming that α_{xx}, α_{yy}, (α_{zz}) have the two species indicated.

In the case of the axial point groups (only one more-than-two-fold axis in the z direction), the species of α_{zz}, α_{xz}, and α_{yz} is obtained in a way similar to the one described above for point groups with non-degenerate species only, as follows. If a field is applied in the z direction, we have

$$P_x = \alpha_{xz}E_z, \qquad P_y = \alpha_{yz}E_z, \qquad P_z = \alpha_{zz}E_z. \tag{III, 41}$$

Since T_z for these point groups is non-degenerate, it follows from the last of the above equations that α_{zz} is totally symmetric (see Table 55). The species of T_x and T_y is degenerate in these point groups. Therefore the species of α_{xz} and α_{yz} must also be degenerate and can immediately be obtained from the species of T_x, T_y, and T_z with the aid of the multiplication table 31. In almost all cases it is E or E_1.

For α_{xx}, α_{yy}, and α_{xy} the transformation properties are not as easily obtained since T_x and T_y are both degenerate. Of course, the relation $P_x = \alpha_{xx}E_x$ (field in the x direction) is fulfilled when α_{xx} is totally symmetric, and similarly for α_{yy}. This accounts for one of the species given for α_{xx} and α_{yy} in Table 55. But it is not the only species. Since E_x and E_y are transformed by some of the symmetry operations (for example, a rotation by $360°/p$) into linear combinations of the two, let us assume now that $E_x \neq 0$, $E_y \neq 0$, $E_z = 0$. Then we obtain from (III, 7) and (III, 8):

$$P_x = \alpha_{xx}E_x + \alpha_{xy}E_y, \qquad P_y = \alpha_{xy}E_x + \alpha_{yy}E_y. \tag{III, 42}$$

Carrying out a symmetry operation, we have

$$P_x^t = \alpha_{xx}^tE_x^t + \alpha_{xy}^tE_y^t, \qquad P_y^t = \alpha_{xy}^tE_x^t + \alpha_{yy}^tE_y^t, \tag{III, 43}$$

where the superscripts t indicate the transformed quantities. Since E_x, E_y, P_x, P_y, just as the translations T_x and T_y, have the degenerate species E, they transform for a rotation by $\beta = 360°/p$ in the same way as degenerate normal coordinates [see equations (II, 75], that is,

$$E_x^t = E_x \cos\beta + E_y \sin\beta,$$
$$E_y^t = -E_x \sin\beta + E_y \cos\beta,$$

and similarly for P_x^t and P_y^t. Substituting this and (III, 42) into (III, 43) and equating the coefficients of E_x on the left- and right-hand sides of the two equations obtained, and similarly the coefficients of E_y, we find:

$$\alpha^t_{xx} \cos\beta - \alpha^t_{xy} \sin\beta = \alpha_{xx} \cos\beta + \alpha_{xy} \sin\beta,$$
$$\alpha^t_{xx} \sin\beta + \alpha^t_{xy} \cos\beta = \alpha_{xy} \cos\beta + \alpha_{yy} \sin\beta,$$
$$\alpha^t_{xy} \cos\beta - \alpha^t_{yy} \sin\beta = -\alpha_{xx} \sin\beta + \alpha_{xy} \cos\beta,$$
$$\alpha^t_{xy} \sin\beta + \alpha^t_{yy} \cos\beta = -\alpha_{xy} \sin\beta + \alpha_{yy} \cos\beta.$$

From these equations we obtain for the transformed polarizabilities:

$$\alpha^t_{xx} = \alpha_{xx} \cos^2\beta + 2\alpha_{xy} \sin\beta \cos\beta + \alpha_{yy} \sin^2\beta, \quad \text{(III, 44)}$$
$$\alpha^t_{yy} = \alpha_{xx} \sin^2\beta - 2\alpha_{xy} \sin\beta \cos\beta + \alpha_{yy} \cos^2\beta, \quad \text{(III, 45)}$$
$$2\alpha^t_{xy} = -(\alpha_{xx} - \alpha_{yy}) \sin 2\beta + 2\alpha_{xy} \cos 2\beta. \quad \text{(III, 46)}$$

While it is thus seen that α_{xx} and α_{yy} do not transform in a simple manner (that is, one that agrees with one of the symmetry types), it is immediately seen from (III, 44) and (III, 45) that

$$\alpha^t_{xx} + \alpha^t_{yy} = \alpha_{xx} + \alpha_{yy}. \quad \text{(III, 47)}$$
$$\alpha^t_{xx} - \alpha^t_{yy} = (\alpha_{xx} - \alpha_{yy}) \cos 2\beta + 2\alpha_{xy} \sin 2\beta. \quad \text{(III, 48)}$$

Thus $\alpha_{xx} + \alpha_{yy}$ is symmetric with respect to a rotation by $\beta = 360°/p$ about the p-fold axis. In a similar way it can be shown that $\alpha_{xx} + \alpha_{yy}$ is also symmetric with respect to all other elements of symmetry [using transformation (II, 76) instead of (II, 75)]; that is, $\alpha_{xx} + \alpha_{yy}$ is totally symmetric. On the other hand, as is seen by comparing (III, 48) and (III, 46), $\alpha_{xx} - \alpha_{yy}$ forms a degenerate pair with $2\alpha_{xy}$ having 2β instead of β in (II, 75), that is, this pair has species E_2. For point groups with $p = 3$ (three-fold axis), the species E_2 is identical with E (see p. 89); for those with $p = 4$ (four-fold axis), E_2 splits into two nondegenerate species B; indeed, with $\beta = 90°$ it follows from (III, 46) and (III, 48) that $\alpha^t_{xy} = -\alpha_{xy}$ and $\alpha^t_{xx} - \alpha^t_{yy} = -(\alpha_{xx} - \alpha_{yy})$; that is, α_{xy} and $\alpha_{xx} - \alpha_{yy}$ are antisymmetric with respect to the four-fold axis. The species is B_2 and B_1 respectively in the case of point groups for which this difference exists [see Tisza (867)]. In this way the species for α_{xy}, α_{xx}, α_{yy} given in Table 55 were obtained.

The rule of mutual exclusion. From Table 55 the following rule can immediately be verified: *For molecules with a center of symmetry, transitions that are allowed in the infrared are forbidden in the Raman spectrum;* and conversely, *transitions that are allowed in the Raman spectrum are forbidden in the infrared.* More explicitly, we may also say: In the infrared, only transitions between states of opposite symmetry with respect to a center of symmetry i can take place ($g \leftrightarrow u$); in the Raman effect, only between states of the same symmetry with respect to i ($g \leftrightarrow g$, $u \leftrightarrow u$). The correctness of this latter rule is immediately seen if it is realized that, according to Table 55, for all infrared transitions (all components of M) $\psi_v'\psi_v''$ must be antisymmetric (u) with respect to i, while for all Raman transitions (all components of α), $\psi_v'\psi_v''$ must be symmetric (g). Even without Table 55 it is clear that all components of M change sign for a reflection at i, whereas the components of the polarizability, which behave as the product of two components of M, remain unchanged.

It should be realized that the above *rule of mutual exclusion* does *not* imply that all transitions that are forbidden in the Raman effect occur in the infrared. Some transitions may be forbidden in both, as for example, a transition of a molecule of point group $\boldsymbol{V_h}$ for which $\psi_v'\psi_v''$ has species A_u.

For all molecules without a center of symmetry, with the exception of those belonging to the rare point groups $\boldsymbol{D_{5h}}$, $\boldsymbol{D_{7h}}$, and $\boldsymbol{O}$, there are transitions that can occur both in the infrared and the Raman effect.

Inversion doubling. As we have seen previously (p. 220), all vibrational energy levels of non-planar molecules are doubled on account of the two potential minima corresponding to inversion. The eigenfunction of one sublevel remains unchanged for a reflection at the origin, that of the other changes sign. Because of the above-mentioned behavior of the dipole moment and the polarizability with respect to reflection at the origin, it is immediately clear (since $M\psi_v'\psi_v''$ and $\alpha\psi_v'\psi_v''$ respectively must remain unchanged for such a reflection) that *in the infrared only sublevels of opposite parity can combine with one another* $(+ \leftrightarrow -)$, whereas *in the Raman effect only sublevels of the same parity can combine with one another* $(+ \leftrightarrow +, - \leftrightarrow -)$. Of course, when the inversion doubling of the vibrational levels is unobservably small, this selection rule has no observable influence on the spectrum. We need only consider the previous selection rules for one potential minimum (Table 55). However, when the inversion doubling is noticeable, as for NH_3 (see p. 221), the selection rule for the sublevels is of importance. All Raman and infrared bands are then double as explained in Fig. 78. But, because of the different selection rules in the Raman and infrared spectra, the doublet splitting of the bands is different: it is the *sum* of the splitting of the upper and lower levels in the infrared, while it is the *difference* in the Raman spectrum (see Fig. 78).

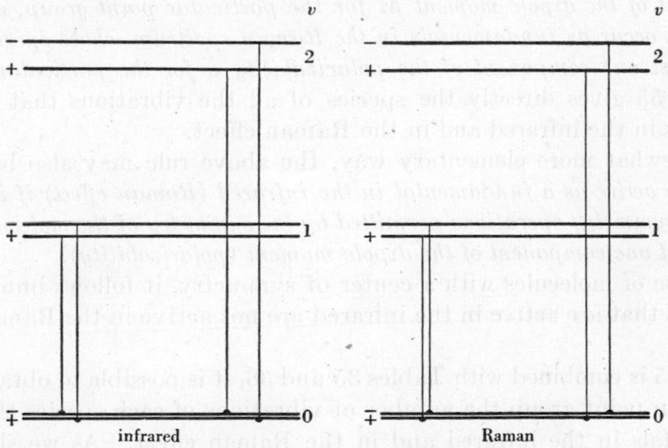

FIG. 78. Influence of inversion doubling on infrared and Raman spectrum.—The small semi-circular dots at the bottom are added in order to make it more apparent at which component level the transition starts.

It is of particular importance that the above selection rule for the infrared (not for the Raman effect) allows also a *transition from one sublevel of a given vibrational level to the other* (see Fig. 78), which, because of the form of the eigenfunctions (see Fig. 72b), has a large intensity. For the ground state of NH_3 such a transition has actually been observed by Cleeton and Williams (215) in the region of extremely short radio waves, at $\lambda = 1.25$ cm (corresponding to 0.8 cm^{-1}), in agreement with expectation from the doubling of the ordinary vibration bands. The observation of this *inversion spectrum* represents one of the most striking confirmations of the predictions of wave mechanics as applied to molecular structure.

In view of the above selection rule for the inversion doublet components, it is seen that the rule of mutual exclusion holds really for all non-planar molecules. This is due to the fact that the potential function of these molecules has a center of symmetry and, therefore, the complete vibrational eigenfunction must remain unchanged or can at most change sign for a reflection at the origin.

Thus, even though at any instant the molecule has no center of symmetry, it behaves as if it had one. It must, however, be understood that when the inversion doubling is unobservably small, as in most cases, infrared and Raman frequencies may coincide except when the molecule has an actual center of symmetry; thus, for unobservably small inversion doubling the rule of mutual exclusion holds only to the restricted extent noted previously.

(c) *More refined treatment of fundamentals*

We apply now the rigorous selection rules derived in the preceding subsection to the fundamentals, that is, the 1–0 transitions occurring in infrared absorption (or emission) and in the Raman spectrum.

General rule. If the lower state of a transition is the vibrationless ground state ($v_1 = 0$, $v_2 = 0$, $\cdots$) the eigenfunction ψ_v'' is totally symmetric (see p. 101). The eigenfunction of a state in which only one vibration is singly excited has the symmetry type of that vibration (see p. 103). Therefore, for a 1–0 transition of a vibration ν_i (fundamental) the product $\psi_v'\psi_v''$ has the symmetry type of the vibration ν_i. Consequently, according to the rigorous selection rules given above, *only those vibrations can occur as fundamentals in the infrared whose species agrees with that of at least one component of the dipole moment* $\boldsymbol{M}$ *for the particular point group*, and *only those vibrations can occur as fundamentals in the Raman spectrum whose species agrees with that of at least one component of the polarizability* $\boldsymbol{\alpha}$ *for the particular point group.* Thus, Table 55 gives directly the species of all the vibrations that are active as fundamentals in the infrared and in the Raman effect.

In a somewhat more elementary way, the above rule may also be formulated: *A vibration is active as a fundamental in the infrared* (*Raman effect*) *if it behaves with respect to all symmetry operations permitted by the symmetry of the molecule in the same way as at least one component of the dipole moment* (*polarizability*).

In the case of molecules with a center of symmetry, it follows immediately that fundamentals that are active in the infrared are not active in the Raman effect, and conversely.

If Table 55 is combined with Tables 35 and 36, it is possible to obtain for a molecule of a given point group the number of vibrations of each species that are active as fundamentals in the infrared and in the Raman effect. As we shall see later, infrared fundamentals of different species have different fine structures which can sometimes be distinguished even without full resolution, and Raman fundamentals have different polarizations and fine structures. Conversely, therefore, if the number of infrared- and Raman-active fundamentals has been found for a certain molecule, particularly when the type of the bands has been ascertained, it is possible to determine the point group to which the molecule belongs; that is, it is possible *to determine its structure from the vibration spectrum only.*

Examples. The three normal vibrations ν_1, ν_2, and ν_3 of non-linear symmetric XY_2 (point group $\boldsymbol{C_{2v}}$) have species A_1, A_1, and B_1 respectively (see Fig. 25a and Table 13). According to Table 55, they are therefore all active as fundamentals in the infrared and in the Raman effect. In particular, ν_1 and ν_2 occur in the infrared with an oscillating dipole moment in the z direction (direction of the two-fold axis) while ν_3 occurs with an oscillating dipole moment in the x direction, in agreement with the classical result (p. 240). In the Raman spectrum, for ν_1 and ν_2 only $[\alpha_{xx}]^{v'v''}$, $[\alpha_{yy}]^{v'v''}$, $[\alpha_{zz}]^{v'v''}$ are different from zero, for ν_3 only $[\alpha_{xz}]^{v'v''}$.

Without reference to Table 55, one may also say: For the 1–0 transition of ν_1 as well as ν_2, the x component $[M_x]^{v'v''} = \int M_x\psi_v'\psi_v''d\tau$ of the amplitude of the dipole moment vanishes because the eigenfunctions ψ_v' and ψ_v'' are symmetric with respect to both planes of symmetry, while M_x changes sign for a reflection at the yz plane, and thus the integrand is not totally symmetric. Similarly, $[M_y]^{v'v''} = \int M_y\psi_v'\psi_v''d\tau$ vanishes for the same transitions since M_y changes sign, and therefore the integrand changes sign for a reflection at the xz plane. But $[M_z]^{v'v''} = \int M_z\psi_v'\psi_v''d\tau$ may be different from zero, since M_z and therefore the integrand remains unchanged for reflections at both planes of symmetry. Therefore, the vibrations ν_1 and ν_2 may occur as fundamentals (1–0 transition) but only with a dipole moment oscillating in the z direction. On the other hand, for the 1–0 transition of ν_3 the eigenfunction ψ_v' and therefore the product $\psi_v'\psi_v''$ is anti-symmetric with respect to the yz plane. Consequently $[M_x]^{v'v''} = \int M_x\psi_v'\psi_v''d\tau$ does not now vanish, since $M_x\psi_v'\psi_v''$ is symmetric with respect to both planes of symmetry, whereas $[M_y]^{v'v''} = \int M_y\psi_v'\psi_v''d\tau$ and $[M_z]^{v'v''} = \int M_z\psi_v'\psi_v''d\tau$ do vanish, since the integrands are antisymmetric with respect to both planes and the yz plane respectively. Thus for this transition the dipole moment changes in the x direction. In a similar way it can be seen, without reference to Table 55, that all three vibrations are Raman active as fundamentals. The polarizability components $\alpha_{xx}, \alpha_{yy}, \alpha_{zz}$ (behaving as the products of the two subscripts) are totally symmetric while α_{xy} is antisymmetric with respect to both planes of symmetry, α_{xz} and α_{yz} are antisymmetric with respect to the yz and xz plane respectively. Therefore, for the 1–0 transition of ν_1 and ν_2, $[\alpha_{xx}]^{v'v''} = \int \alpha_{xx}\psi_v'\psi_v''d\tau$, $[\alpha_{yy}]^{v'v''} = \int \alpha_{yy}\psi_v'\psi_v''d\tau$ and $[\alpha_{zz}]^{v'v''} = \int \alpha_{zz}\psi_v'\psi_v''d\tau$ may be different from zero, while the other integrals vanish; for the 1–0 transition of ν_3 only $[\alpha_{xz}]^{v'v''} = \int \alpha_{xz}\psi_v'\psi_v''d\tau$ may be different from zero.

For a more general molecule of point group $\boldsymbol{C}_{2v}$ (for example X_2YZ_2) for which vibrations of all four species occur, it follows from Table 55 and the above general rule that only the vibrations of species A_1, B_1, and B_2 can occur as fundamentals in the infrared, whereas the vibrations of species A_2 are forbidden (for example the torsion oscillation of X_2YZ_2). On the other hand, all vibrations, including those of species A_2, are active as fundamentals in the Raman effect, but of course with different components of the polarizability.

In the case of a molecule of point group V_h (for example, plane X_2Y_4), according to Table 55 only vibrations of species B_{1u}, B_{2u}, B_{3u} (see Table 14) are active in the infrared as fundamentals. In the case of X_2Y_4 (see Fig. 44) these infrared-active fundamentals are ν_7, ν_9, ν_{10}, ν_{11}, and ν_{12}, in agreement with the previous classical consideration (p. 240). In the Raman effect only vibrations of species A_g, B_{1g}, B_{2g}, and B_{3g} are active, with the polarizability components given in Table 55. For X_2Y_4 the Raman-active fundamentals are ν_1, ν_2, ν_3, ν_5, ν_6, ν_8. It is seen that the infrared-active fundamentals are not active in the Raman effect and the Raman-active fundamentals are not active in the infrared, in agreement with the rule of mutual exclusion, which applies here since there is a center of symmetry. Vibrations of species A_u are inactive as fundamentals both in the infrared and in the Raman effect. For X_2Y_4 only the vibration ν_4, the torsional oscillation, is of this species.

Finally, let us consider, as an example of a molecule with degenerate vibrations, one of point group T_d (for example XY_4). Here the dipole moment has the triply degenerate species F_2 and therefore only the vibrations of species F_2 are active as fundamentals in the infrared. For XY_4 these are the vibrations ν_3 and ν_4 (see Fig. 41). In the Raman effect, since the polarizability components have the species A_1, E, and F_2, only the vibrations of species A_1, E, and F_2 are active as fundamentals. In the case of XY_4 this includes all the fundamentals, ν_1, ν_2, ν_3, ν_4. But in the general case there are also vibrations of species A_2 and F_1 [for example, for $C(CH_3)_4$] which are accordingly forbidden both in the infrared and in the Raman effect.

Alternative treatment; intensities. An alternative way of deriving the selection rules, which at the same time lends itself more easily to the calculation of intensities, is the following. We expand the components of the dipole moment and the polarizability in a power series of the normal coordinates, as in equations (III, 2) and (III, 14) of the classical treatment, and substitute these into $[M_x]^{v'v''}$, $[M_y]^{v'v''}$, $[M_z]^{v'v''}$, and $[\alpha_{xx}]^{v'v''}$, $[\alpha_{xy}]^{v'v''}$, $\cdots$, respectively. In this way we obtain

$$[M_x]^{v'v''} = M_x^0 \int \psi_v{}'\psi_v^{*''}d\tau + \sum_i \left(\frac{\partial M_x}{\partial \xi_i}\right)_0 \int \xi_i \psi_v{}'\psi_v^{*''}d\tau + \cdots, \qquad \text{(III, 49)}$$

and similar equations for $[M_y]^{v'v''}$ and $[M_z]^{v'v''}$; furthermore we obtain

$$[\alpha_{xx}]^{v'v''} = \alpha_{xx}^0 \int \psi_v{}'\psi_v^{*''}d\tau + \sum_i \left(\frac{\partial \alpha_{xx}}{\partial \xi_i}\right)_0 \int \xi_i \psi_v{}'\psi_v^{*''}d\tau + \cdots \qquad \text{(III, 50)}$$

and similar equations for $[\alpha_{xy}]^{v'v''}$, $[\alpha_{xz}]^{v'v''}$, $\cdots$. In these equations v' and v'' stand for the *set* of vibrational quantum numbers in the upper and lower state respectively. The first term on the right in both equations vanishes if $v' \neq v''$ since the vibrational eigenfunctions of different states are orthogonal to one another.

In the harmonic oscillator approximation, according to (II, 42),

$$\int \xi_i \psi_v{}'\psi_v^{*''}d\tau = \int \xi_i \psi_1{}'(\xi_1)\psi_2{}'(\xi_2)\cdots\psi_i{}'(\xi_i)\cdots\psi_1{}''(\xi_1)\psi_2{}''(\xi_2)\cdots \psi_i{}''(\xi_i)\cdots d\xi_1 d\xi_2\cdots d\xi_i\cdots$$

$$= \int \psi_1{}'(\xi_1)\psi_1{}''(\xi_1)d\xi_1 \int \psi_2{}'(\xi_2)\psi_2{}''(\xi_2)d\xi_2 \cdots \int \xi_i\psi_i{}'(\xi_i)\psi_i{}''(\xi_i)d\xi_i \cdots.$$

Because of the orthogonality of the eigenfunctions, the integrals $\int\psi_1{}'(\xi_1)\psi_1{}''(\xi_1)d\xi_1$, $\int\psi_2{}'(\xi_2)\psi_2{}''(\xi_2)d\xi_2 \cdots$ are different from zero only when $v_1{}' = v_1{}''$, $v_2{}' = v_2{}'' \cdots$, while $\int\xi_i\psi_i{}'(\xi_i)\psi_i{}''(\xi_i)d\xi_i$ is different from zero only when $v_i{}' = v_i{}'' \pm 1$. Therefore $[M_x]^{v'v''}$ and $[\alpha_{xx}]^{v'v''}$ and similarly the other components of $\boldsymbol{M}$ and $\boldsymbol{\alpha}$ according to (III, 49) and (III, 50) are different from zero only if only one v_i changes by ± 1.

Thus we see that *in this approximation*, that is, when higher terms in the developments (III, 49) and (III, 50) and the anharmonicity of the vibrations are neglected, *only fundamentals can occur in the infrared and in the Raman effect.* Only when higher terms and the anharmonicity are taken into account does the dipole moment or polarizability associated with overtone or combination vibrations not vanish (see subsection d); that is, these transitions have a much smaller intensity than the fundamentals.

Whether a certain fundamental ν_i, that is, a 1–0 transition, actually occurs depends, just as in the classical treatment, according to (III, 49) and (III, 50), on *whether at least one of the components of* $\left(\frac{\partial \boldsymbol{M}}{\partial \xi_i}\right)_0$ *and* $\left(\frac{\partial \boldsymbol{\alpha}}{\partial \xi_i}\right)_0$ *respectively is different from zero.* This can be decided either in the elementary way indicated in the classical treatment (p. 245), or from the rigorous selection rules given in Table 55. While this table gives the species of the components of $\boldsymbol{M}$ and $\boldsymbol{\alpha}$, it can also be used to determine which components of $\left(\frac{\partial \boldsymbol{M}}{\partial \xi_i}\right)_0$ and $\left(\frac{\partial \boldsymbol{\alpha}}{\partial \xi_i}\right)_0$ are zero, as follows. If according to that table a certain component, say M_x of $\boldsymbol{M}$ or α_{xx} of $\boldsymbol{\alpha}$, is zero for a certain species, it means that $[M_x]^{v'v''}$ or $[\alpha_{xx}]^{v'v''}$, as the case may be, is zero for the 1–0 transition of a vibration of this species, and therefore, since the first term in (III, 49) and (III, 50) is zero, that $\left(\frac{\partial M_x}{\partial \xi_i}\right)_0$ or $\left(\frac{\partial \alpha_{xx}}{\partial \xi_i}\right)_0$, as the case may be, is zero.

To predict the *relative intensities of different infrared fundamentals* of a molecule, one would need actually to calculate the $\left(\frac{\partial M_x}{\partial \xi_i}\right)_0, \left(\frac{\partial M_y}{\partial \xi_i}\right)_0, \left(\frac{\partial M_z}{\partial \xi_i}\right)_0$. The intensity ratio of the fundamentals ν_i in absorption is then given by the ratio of the quantities $\nu_i\{[M]_i^{1,0}\}^2$ where the $[M]_i^{1,0}$ are obtained from the $\left(\frac{\partial M}{\partial \xi_i}\right)_0$ by means of (III, 49). In emission the factor ν_i would have to be replaced by ν_i^4 [see equation (I, 18) of Molecular Spectra I].

The *intensity ratio of the* (Stokes) *Raman lines* is in a similar way given by the ratio of the quantities $(\nu - \nu_i)^4\{[P^0]_i^{1,0}\}^2$, where ν is the wave number of the exciting radiation and $\nu - \nu_i$ is the wave number of the Raman line corresponding to the vibration ν_i. The $[P^0]_i^{1,0}$ are obtained from (III, 38) into which (III, 50) has been substituted.[10a] It must, however, be realized that the polarizability α (and therefore the amplitude of the induced dipole moment $[P^0]_i^{1,0}$) is independent of the incident frequency only when the latter is sufficiently small. If ν is in the neighborhood of an absorption frequency of the molecule, α will increase rapidly with ν, and therefore the intensity of the Raman lines increases more rapidly with ν than is given by the factor $(\nu - \nu_i)^4$. Both effects, the normal dependence on $(\nu - \nu_i)^4$ and the stronger dependence on ν in the ultraviolet when an absorption region is approached, have been observed experimentally for CCl_4 [Ornstein and Rekveld (676) Sirkar (793) and Werth (917)]. For a more detailed discussion, see Placzek (700).

The intensity ratio of the anti-Stokes Raman lines to the corresponding Stokes lines is mainly given by the Boltzmann factor $e^{-(G_v hc/kT)}$. However, here again, for accurate determinations, the frequency factor, which is $(\nu - \nu_i)^4$ for the Stokes and $(\nu + \nu_i)^4$ for the anti-Stokes lines, has to be taken into account.

The absolute intensity of infrared fundamentals in absorption is given by [see for example Mulliken (644)]

$$k_i = \int k_i(\nu)d\nu = \frac{8\pi^3 N\nu_i}{3ch}\{[M]_i^{1,0}\}^2,$$

where k_i is the integrated absorption coefficient and N the number of molecules per cm³ (ν in cm^{-1} or sec.$^{-1}$). From this formula, if k_i is measured, $[M]_i^{1,0}$ and therefore $\left(\frac{\partial M}{\partial \xi_i}\right)_0$ may be obtained, that is, the slope of the curve (Fig. 74) representing the variation of M as a function of ξ_i. Since the amplitude of vibration can easily be obtained, the absolute change of dipole moment connected with the vibration considered may be obtained. However, a direct and accurate determination of the true absorption coefficient is beset with difficulties on account of the small width of the fine structure lines [see for example Wells and Wilson (916)]. An indirect determination from the infrared dispersion has been made for the case of CH_4 by Rollefson and Havens (741). Calculations of the absolute intensity of Raman lines have been carried out by Bell (134a).

(d) *Overtone and combination bands*

General remarks. While transitions for which one $\Delta v_i > 1$ or for which several $\Delta v_i \neq 0$ (overtone and combination bands respectively) are in general much weaker than the fundamentals, they may yet be observed in infrared absorption by using sufficiently thick absorbing layers, and in the Raman effect by using sufficiently long exposure times.

Overtone and combination bands for which $|\Delta v_i| = 2$ or $\sum|\Delta v_i| = 2$, that is, transitions in which either one vibration changes by two quanta or two vibrations by one quantum, are also called *binary combinations;* those for which $|\Delta v_i| = 3$ or

[10a] Such calculations have recently been carried out for some halomethanes by Wolkenstein (949a) and for the dichlorobenzenes by Nordheim and Sponer (673b).

$\sum|\Delta v_i| = 3$ are also called *ternary combinations;* and so on. It is clear from an extension of equations (III, 49) and (III, 50) to include mechanical and electrical anharmonicity (see p. 241) that, in general, ternary combinations are still weaker than binary combinations, and quarternary weaker than ternary, since they involve higher and higher approximations. To illustrate this, Fig. 79 shows schematically the structure of the infrared vibration spectrum of a triatomic molecule. (It might be compared with Fig. 31 of Molecular Spectra I, which holds for a diatomic molecule.) The spectrum consists of a number of *progressions* $\Delta v_i = 0, 1, 2$, starting at every fundamental, binary combination, and so on, and consisting of nearly equidistant bands of very rapidly decreasing intensity. However, this decrease is not always quite regular if the molecule has symmetry, since certain overtone and combination bands may be forbidden by the rigorous selection rules (see subsection b). In fact it may happen in certain cases that a fundamental may be forbidden while certain overtone and combination bands involving the same vibration are allowed.

Overtone bands. For the overtone bands, the lower state is the vibrational ground state (ψ_v'' is totally symmetrical) and, therefore, according to the general rule, p. 253f., *an overtone will be infrared active when at least one component of the dipole moment has the same species as the vibrational eigenfunction* ψ_v' *of the upper state,* and *it will be Raman active if at least one component of the polarizability has the same species as* ψ_v'. The species of the eigenfunction of the upper state is obtained for non-degenerate vibrations from the rule given on p. 101 and for degenerate vibrations from Table 32, while the species of the dipole moment and the polarizability is obtained from Table 55.

For example, while the fundamental $\nu_3(\sigma_u^+)$ of a linear XY_2 molecule is infrared active, its first, third, $\cdots$, overtones ($v_3 = 2, 4, \cdots$) have a ψ_v' of species Σ_g^+ and are therefore infrared inactive (see Table 55); but its second, fourth, $\cdots$ overtones ($v_3 = 3, 5, \cdots$) have a ψ_v' of the same species as the fundamental (Σ_u^+) and are therefore infrared active. On the other hand, according to Table 55, this fundamental ν_3 and its second, fourth, $\cdots$ overtones are inactive in the Raman effect, while the first, third, $\cdots$ overtones are Raman active. Thus we have an *alternation of infrared and Raman activity in the progression* $v_3\nu_3$ ($v_3 = 1, 2, 3, 4, \cdots$), as indicated in Fig. 80. The same holds for all vibrations (degenerate or non-degenerate) that are antisymmetric with respect to a center of symmetry, for example, all infrared-active vibrations of plane X_2Y_4, linear X_2Y_2, and others, as is immediately seen from the g, u rule (p. 124) combined with Table 55. It also holds for the non-degenerate infrared-active vibrations of certain point groups without a center of symmetry: namely, of those for which the totally symmetric vibrations are inactive in the infrared; for example, for the vibration $\nu_2(a_2'')$ of plane XY_3 (see Fig. 63). However, it must be realized that the converse alternation does not hold for the progression of overtones of Raman-active fundamentals. For example, for molecules with a center of symmetry all overtones of a Raman-active fundamental are Raman active, none infrared active.

In the case of a non-linear XY_2 molecule (point group C_{2v}) the totally symmetric fundamentals ν_1 and ν_2, as well as the antisymmetric fundamental ν_3, are both infrared and Raman active. The same holds for all overtones (see Fig. 79). However, the levels with $v_3 = 1, 3, 5, \cdots$ have antisymmetric (B_1) eigenfunctions while the levels with $v_3 = 0, 2, 4, \cdots$ have totally symmetric eigenfunctions. Therefore, in the

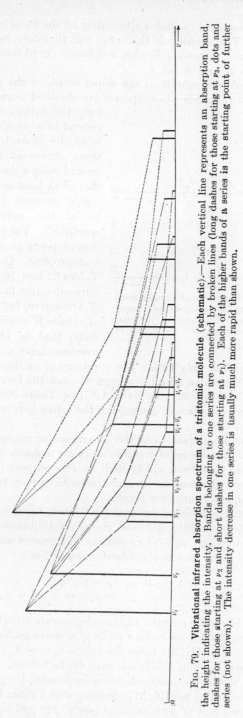

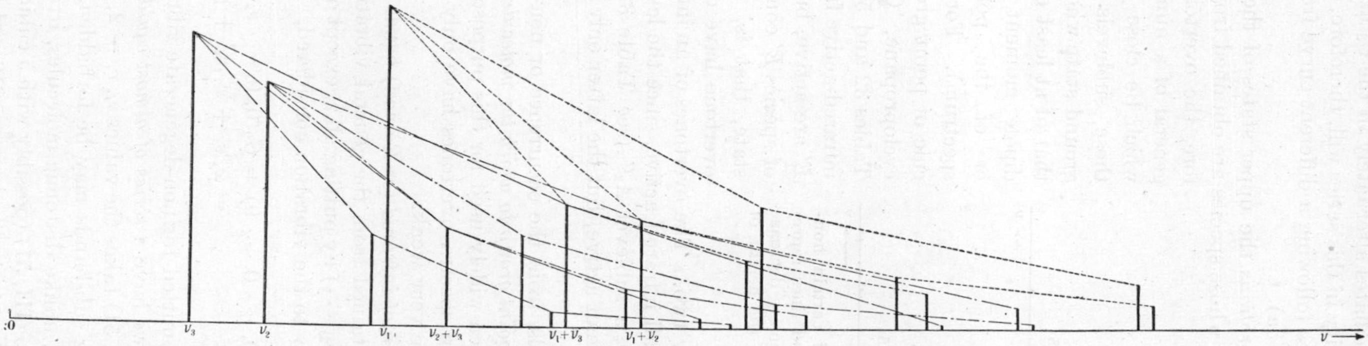

FIG. 79. **Vibrational infrared absorption spectrum of a triatomic molecule (schematic).—Each vertical line represents an absorption band, the height indicating the intensity. Bands belonging to one series are connected by broken lines (long dashes for those starting at ν_3, dots and dashes for those starting at ν_2 and short dashes for those starting at ν_1). Each of the higher bands of a series is the starting point of further series (not shown). The intensity decrease in one series is usually much more rapid than shown.**

series $v_3\nu_3$ the dipole moment oscillates alternately in the C_2 axis and perpendicular to it (see Table 55). The intensity in this series will, therefore, not vary as regularly as in $v_1\nu_1$, $v_2\nu_2$, the even overtones following a different curve from the odd (compare the discussion of H_2O in section 3a).

In the case of *degenerate vibrations* the upper states of the overtone bands are split into a number of sublevels whose species are obtained from Table 32. Therefore, the overtone bands consist in general of a number of "*sub-bands*" which lie close together. But only those sublevels combine with the ground state whose species agree with that of at least one component of the dipole moment (infrared spectrum) or of the polarizability (Raman spectrum). For example, in a molecule of point group $\boldsymbol{D_{3h}}$ (for instance cyclopropane, C_3H_6), according to Tables 32 and 55 all overtones of an infrared-active fundamental of species E' are active, but only the sublevels of species E' combine with the ground state, that is, the first and second overtone have only one "sub-band," the third, fourth, and fifth only two. The overtones of an infrared-inactive fundamental of species E'' are also all infrared active, since the levels with even v_i have a sublevel E', those with odd v_i a sublevel A_2'' (see Table 32). All these fundamentals and overtones are Raman active, but the latter only with the sublevels of species A_1', E', and E''.

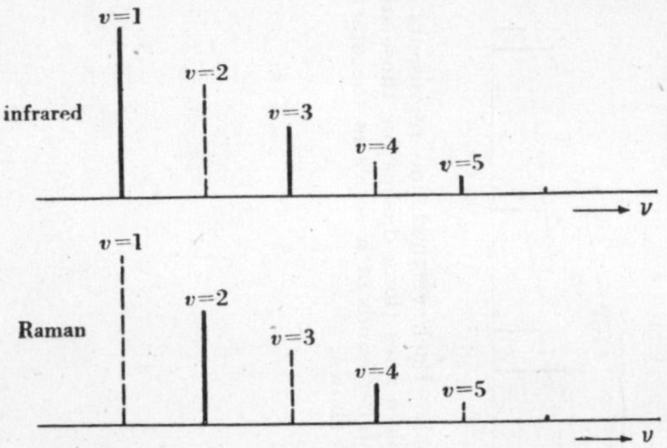

FIG. 80. **Activity of overtones of certain non-totally symmetric vibrations.**—Dotted lines represent forbidden transitions. The intensity decrease is much more rapid than shown by the height of the lines.

Just as for the fundamentals, from the occurrence or non-occurrence of certain overtones *conclusions as to the point group to which the molecule belongs can be drawn.* While infrared spectra have been widely used for this purpose (see section 3 of this chapter), in the case of Raman spectra overtones have only rarely been observed since even the fundamentals are very weak.

The *formula for a progression of infrared* (*or Raman*) *bands* consisting of a fundamental and its overtones is obtained from the general vibrational energy formulae (II, 272) or (II, 284) for $G_0(v_1, v_2, \cdots)$ by putting all v_i except one equal to zero. We obtain, therefore, if i refers now to the vibration considered,

$$\nu = G(0, 0, \cdots v_i \cdots 0) - G(0, \cdots 0 \cdots 0) = G_0(0, 0, \cdots v_i \cdots 0) = \omega_i^0 v_i + x_{ii}^0 v_i^2 + g_{ii} l_i^2 + \cdots, \quad \text{(III, 51)}$$

where the last term has to be omitted for non-degenerate vibrations. Since x_{ii}^0 and g_{ii} are small compared to ω_i^0, we have a *series of almost equidistant bands.* For degenerate vibrations l_i in (III, 51) takes the values $v_i, v_i - 2, \cdots$ 0 or 1 (see p. 210) but some of the corresponding sub-bands may be forbidden. It must be realized that for polyatomic molecules, unlike diatomic molecules, irregular deviations from a smooth variation according to (III, 51) (possibly with a cubic term in v_i) may occur fairly frequently on account of perturbations (see p. 219 and below).

The intensity distribution in a progression $v_i\nu_i$ is determined by the curve that represents the variation of the dipole moment with the normal coordinate (Fig. 74) and the anharmonicity of the vibration. If the latter is known from the observed band positions, the observation of the absolute intensities of a series of overtones can be used to determine the variation of the dipole moment. If certain assumptions are made about the functional form of this variation it is even possible, as was shown by Timm and Mecke (864) and Mecke (615) (616) to obtain the absolute value of the dipole moment of a bond such as the C—H and O—H bonds in a polyatomic molecule from these intensity measurements.

It should also be realized that if the maximum of the curve representing the variation of the dipole moment (Fig. 74) is close to the equilibrium position of the nuclei, it may happen that the intensity distribution in a progression $v_i\nu_i$ may have a maximum at a point different from $v_i = 1$; that is, an overtone may be more intense than a fundamental. However, such a case has not as yet been definitely established.

Summation bands. A combination band for which the lower state is the vibrational ground state of the molecule is also called a summation band, since its wave number in a zero approximation is the sum of the wave numbers of two or more fundamentals or overtones:

$$\nu = v_i\nu_i + v_j\nu_j + v_k\nu_k + \cdots. \qquad \text{(III, 52)}$$

The activity of these summation bands, like that of the overtone bands, is obtained by determining the species of the upper state but now from Tables 31 and 33 together with the rules given on p. 124, and seeing whether it agrees with the species of one of the components of the dipole moment or the polarizability in Table 55. It is particularly important that *inactive fundamentals*, when combined with other fundamentals or overtones, *may give active summation bands*, just as overtones of certain inactive fundamentals may be active in the infrared or Raman spectrum or both. Conversely, certain combination bands may be forbidden even though the fundamentals involved are allowed. For instance it follows immediately from the (g, u) rule (p. 124) that for molecules having a center of symmetry no binary combinations of infrared active fundamentals are infrared active.

Consider as examples some summation bands involving the fundamental $\nu_4(a_u)$ of plane X_2Y_4 (the torsional oscillation), which is inactive both in the infrared and Raman spectrum. The upper state of $\nu_4(a_u) + \nu_7(b_{1u})$ has species $a_u \times b_{1u} = B_{1g}$, and therefore (see Table 55) $\nu_4 + \nu_7$ is allowed in the Raman effect, even though neither ν_4 nor ν_7 is allowed. The upper state of $2\nu_4+\nu_7$ has species $a_u \times a_u \times b_{1u} = B_{1u}$, and therefore $2\nu_4 + \nu_7$ is infrared active but not Raman active. It is thus seen that the frequencies of inactive fundamentals can be obtained from summation bands. On the other hand $\nu_4 + 2\nu_7$, whose upper state has species A_u, is forbidden both in Raman effect and infrared.

The lower the symmetry of a molecule the fewer restrictions there are for the combination bands, as well as for the overtone bands. For axial XYZ_3 molecules (methyl halides), for example, all combination bands (and overtone bands) are allowed both in the infrared and Raman effect, except that the sublevels of species A_2 do not combine with the ground state.

The wave numbers of the summation bands are given more accurately than according to the above zero approximation by $G_0(v_1, v_2, \cdots)$ from (II, 284) or (II, 272) when the appropriate v_i are substituted.

Influence of Fermi resonance. In general, as mentioned before, the intensity of the overtone and combination bands decreases very rapidly with increasing $\sum|\Delta v_i|$

(except when it is exactly zero because of symmetry). However, this state of affairs changes radically if a case of Fermi resonance occurs (see Chapter II, section 5c), for example when a state in which one vibration, say ν_i, is doubly excited has nearly the same energy as the state in which another vibration ν_k is singly excited. As we have seen previously, if the states $2\nu_i$ and ν_k have the same symmetry, a perturbation of the energy levels and, at the same time, a mixing of the eigenfunctions occur. If without resonance $2\nu_i$ and ν_k are infrared active (or Raman active), the fundamental ν_k would, in general, have much greater intensity than the overtone $2\nu_i$. But if the interaction (resonance) is taken into account, *the intensity of the two bands will be more nearly alike*, since in $\int \psi_v' \psi_v^{*\prime\prime} M d\tau$ (or $\int \psi_v' \psi_v^{*\prime\prime} \alpha d\tau$) for ψ_v' the "mixed" functions (II, 293) must now be substituted. In other words, $2\nu_i$ *will "borrow" intensity from* ν_k. Its intensity may then become of the same order as that of a fundamental. If the resonance is complete, the two transitions will have the same intensity, which will be half the "original" intensity of ν_k (since the "original" intensity of $2\nu_i$ was negligible compared to that of ν_k).

Similar considerations apply, of course, if the resonance is between a state in which two different vibrations are singly excited ($\nu_i + \nu_j$) and a state in which only one vibration is singly excited (ν_k). Also, if there is resonance between, say, $3\nu_i$ and $2\nu_k$, then the second overtone of ν_i will have an intensity comparable with that of the first overtone of ν_k, and similarly in other cases. If resonance exists between two combinations of the same order (for example, the two binary combinations $2\nu_1$ and $2\nu_3$ of H_2O; see p. 218) they will tend to have more nearly the same intensity than they would without resonance.

If a state that is in resonance with another corresponds to the excitation of a *degenerate vibration*, a perturbation will usually occur only for one of the sublevels into which the state is resolved when anharmonicity is taken into account (see p. 217) and thus *only one of the sub-bands will have an anomalously large intensity*. For example, for CO_2, where $\nu_1 \approx 2\nu_2$ (see Fig. 71) the upper state of $2\nu_2$ has the two sublevels Σ_g^+ and Δ_g, of which only the first is perturbed by the upper state Σ_g^+ of ν_1. Only the sub-band $2\nu_2(\Sigma_g^+)$ borrows intensity in the Raman effect from ν_1, and, because of the closeness of the resonance, has almost the same intensity as ν_1, whereas the sub-band $2\nu_2(\Delta_g)$, which is also allowed in the Raman effect (see Table 55), has the normal (small) intensity of an overtone and has therefore not been observed.

Difference bands. Finally, we discuss the case of absorption or scattering in which *the initial state is not the vibrationless ground state.* If the vibration ν_i is singly excited in the initial (lower) state and a transition takes place to a state in which the vibration $\nu_k(> \nu_i)$ is singly excited, the frequency of the infrared absorption band (or the frequency shift in the Raman spectrum) is equal to $\nu_k - \nu_i$. Classically (see p. 241) this *difference band* should have the same intensity as the corresponding summation band $\nu_k + \nu_i$. However, in quantum theory the intensity of $\nu_k - \nu_i$ would be expected to be much smaller than that of $\nu_k + \nu_i$ since the number of molecules in the initial state is much smaller, corresponding to the Boltzmann factor $e^{-(hc\nu_i/kT)}$. In addition, the matrix elements $\int \psi_v' \psi_v^{*\prime\prime} M d\tau$ or $\int \psi_v' \psi_v^{*\prime\prime} \alpha d\tau$ for the two transitions will be somewhat different although they will be both of the same order (both zero if the anharmonicity is neglected).

When the rigorous selection rules are applied to a *difference band* $\nu_k - \nu_i$, one finds easily that it *is allowed or forbidden depending on whether the corresponding summation band* $\nu_k + \nu_i$ *is allowed or forbidden:* If the lower state is not the ground state, it is not the species of the eigenfunction of the upper state ψ_v' but the species of the product $\psi_v'\psi_v''$ that must be the same as the species of one of the components of $\boldsymbol{M}$ or $\boldsymbol{\alpha}$, in order that the transition be allowed. But the product $\psi_v'\psi_v''$ for $\nu_k - \nu_i$ has obviously the same species as ψ_v' for $\nu_k + \nu_i$, and thus if $\nu_k + \nu_i$ is allowed (see Table 55) then $\nu_k - \nu_i$ is also allowed. The same applies also in the case of other similar difference bands, such as $2\nu_k - \nu_i$, $3\nu_k - \nu_i$, $\cdots$, $\nu_k + \nu_l - \nu_i$, $\cdots$, $\nu_k - 2\nu_i$, $\nu_k - \nu_i - \nu_j$, and so on.

Since the intensity ratio of a difference band to the corresponding summation band is approximately equal to the Boltzmann factor, such difference bands have been observed in the infrared and Raman effect only in cases where ν_i is small (or, more generally, where the lower state is fairly near the ground state), since then the Boltzmann factor is not too small (compare for example, the difference bands of CO_2 given in Table 56).

A somewhat different type of difference band is *that for which one and the same low-frequency vibration is excited in the upper and lower state* in addition to some other vibrations in the upper state.[10b] In the simplest case, if ν_k is excited by one quantum in the upper, but not in the lower state, whereas ν_i is excited both in the upper and lower state by one quantum, we obtain a band that may be written $\nu_k + \nu_i - \nu_i$. This is represented in the energy-level diagram Fig. 81a. If there were no coupling between the two vibrations, this transition would have the same frequency as the fundamental ν_k and also the same transition probability, since in both cases only $\Delta v_k = 1$. The intensity ratio of $\nu_k + \nu_i - \nu_i$ to ν_k is therefore equal to the Boltzmann factor $e^{-(hc\nu_i/kT)}$ which gives the population of the state $v_i = 1$ relative to the ground state. In consequence of the coupling of the two vibrations, the band $\nu_k + \nu_i - \nu_i$ will not exactly coincide with ν_k but will be displaced by a small amount so that it is actually observable. The intensity is also influenced by the coupling, but, like the frequency, only slightly, so that for small ν_i when the Boltzmann factor is of the order of unity the intensity of $\nu_k + \nu_i - \nu_i$ is of the same order as that of the fundamental ν_k (whereas the intensity of $\nu_k - \nu_i$ was of the same order as $\nu_k + \nu_i$, that is, much smaller than that of ν_k).

If ν_i is sufficiently small, bands $\nu_k + 2\nu_i - 2\nu_i$, $\nu_k + 3\nu_i - 3\nu_i$ may also have a sufficient intensity, to be observed, in spite of the smaller Boltzmann factor. As indicated in Fig. 81a, we then have a series of very nearly equidistant bands with rapidly decreasing intensity, which is the exact analogue of a *sequence* in a diatomic electronic band spectrum (see Molecular Spectra I, p. 171f.). If a sufficiently small ν_i exists, such sequences should join not only onto every allowed fundamental band, but also onto every allowed overtone or combination band, that is, when ν_k in $\nu_k + \nu_i - \nu_i$ is replaced by $2\nu_k$, $\nu_k + \nu_l$, and so on. The case in which $\nu_k = \nu_i$ in $\nu_k + \nu_i - \nu_i$ is of course also possible.

Bands of the type $\nu_k + \nu_i - \nu_i$ have been observed in a number of cases both in the infrared and Raman effect, for example, for CO_2 (see Table 56) and C_2H_2 (see Table 68). Sequences with three members ν_k, $\nu_k + \nu_i - \nu_i$, $\nu_k + 2\nu_i - 2\nu_i$, where ν_k is a second overtone, have been found in the photographic infrared, for methyl acetylene [Herzberg, Patat, and Verleger (440)] and pyrrole [Zumwalt and Badger (976)]. While in the former case the low-frequency ν_i (343 cm^{-1}) is known from

[10b] Dennison (280) calls such bands "upper stage" bands.

Raman data, in the latter case the intensity ratio of the bands in the sequence has been used to obtain from the thus derived Boltzmann factor the value of ν_i ($\sim$650 cm^{-1}).

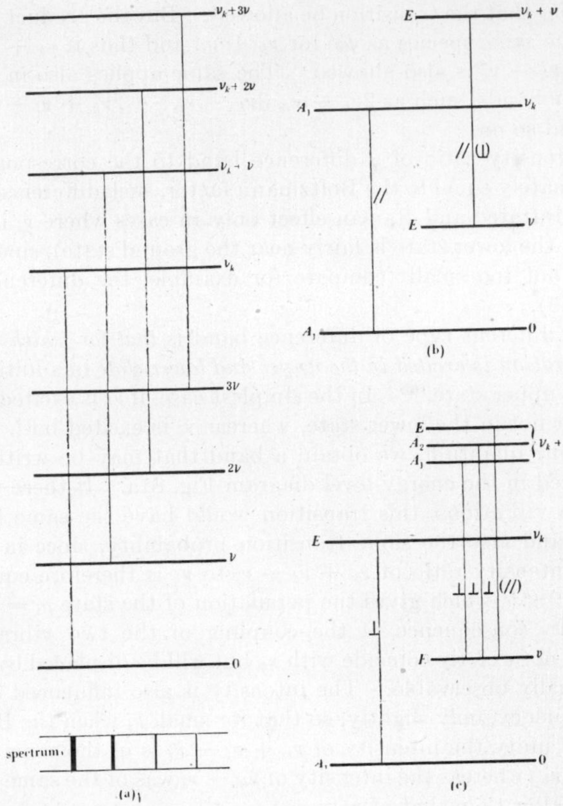

FIG. 81. **Difference bands of the type $\nu_k + \nu_i - \nu_i$; (a) a sequence of difference bands, (b) ν_i degenerate, (c) ν_i and ν_k degenerate.**

If the vibration ν_i involved in the above-described difference bands is *degenerate*, some additional considerations are necessary. First of all, the intensity ratio of the band $\nu_k + \nu_i - \nu_i$ to the band ν_k is twice or three times as great as the Boltzmann factor, depending on whether the degree of degeneracy of ν_i is 2 or 3, because the statistical weight is increased by this factor. For the higher members of a sequence $\nu_k + 2\nu_i - 2\nu_i, \cdots$, the intensity ratio to ν_k is still further increased compared to the non-degenerate case, because the degree of degeneracy of the states $2\nu_i, 3\nu_i, \cdots$ is still higher (see p. 80f.), at least as long as the anharmonicity is neglected. Even if the anharmonicity is not neglected, the splitting of the bands of a sequence into sub-bands will, in general, not be observable since it corresponds to the difference of the splitting of the sublevels in the upper and lower state.

If both ν_k and ν_i in the band $\nu_k + \nu_i - \nu_i$ are non-degenerate, the application of the rigorous selection rules (evaluation of $\int M\psi_v'\psi_v^{*}{}''d\tau$ or $\int \alpha\psi_v'\psi_v^{*}{}''d\tau$) leads to the same result for $\nu_k + \nu_i - \nu_i$ as for ν_k (see above). If, however, ν_i or both ν_i and ν_k are degenerate, for some point groups, additional components of the dipole moment and the polarizability may occur according to the rigorous selection rules. For example, for a molecule of point group C_{3v} (for instance CH_3Cl), if ν_k has species A_1 and ν_i has species E, then both the upper and lower states of the band $\nu_k + \nu_i - \nu_i$ have species E (see Fig. 81b). According to Table 33, the product $\psi_v'\psi_v^{*}{}''$ has the species A_1, A_2, E. Of these, according to Table 55, A_1 gives a parallel band in the infrared ($M_z \neq 0$) and a Raman line with $\alpha_{xx} \neq 0$, $\alpha_{yy} \neq 0$, $\alpha_{zz} \neq 0$ (polarized Raman line; see below), as does ν_k alone, whereas E gives a perpendicular band ($M_z = 0$) in the infrared and a Raman line with $\alpha_{xx} - \alpha_{yy}$, α_{xy}, α_{xz}, α_{yz} all different from zero (depolarized Raman line). Thus one and the same transition ($\nu_k + \nu_i - \nu_i$) can take place according to the rigorous selection rule with two different species of the components of the dipole moment and the polarizability, that is (see Chapter IV), with two different rotational structures (and polarizabilities in the Raman effect). However, only the first of these two species of the transition would have the same transition probability as ν_k. This is easily seen from the series developments (III, 49) and (III, 50) since, if higher terms are neglected, the intensity of a transition in which only v_k changes by one unit (and such is the case for both ν_k and $\nu_k + \nu_i - \nu_i$) depends only on $\left(\frac{\partial \mathbf{M}}{\partial \xi_k}\right)_0$ and $\left(\frac{\partial \boldsymbol{\alpha}}{\partial \xi_k}\right)_0$ respectively, and these in the present case have species A_1. Only in higher approximation, that is, with much smaller intensity, would the E contribution to the transition occur.

If both ν_k and ν_i for a molecule of point group $\boldsymbol{C_{3v}}$ are of species E, the upper state of $\nu_k + \nu_i - \nu_i$ according to Table 33 consists of the three sublevels A_1, A_2, E, whereas the lower state is E (see Fig. 81c). There are *three sub-bands* for $\nu_k + \nu_i - \nu_i$: $A_1 \rightarrow E$, $A_2 \rightarrow E$, and $E \rightarrow E$. For the first two, $\psi_{v'}\psi_{v''}^*$ has species E, and therefore these sub-bands are perpendicular bands ($M_z = 0$) in the infrared (or depolarized in the Raman effect), as is ν_k. The sub-band $E \rightarrow E$, as in the preceding case, can occur both as a parallel and a perpendicular band; but in this case only the perpendicular component (depolarized Raman component) will have an intensity comparable with ν_k since ν_k is a perpendicular band (or depolarized Raman line). Other cases, for other point groups, are similar.

The *formula for* the positions of the *difference bands* in the infrared and for their shifts in the Raman effect is

$$\nu = G(v_1', v_2', \cdots) - G(v_1'', v_2'', \cdots), \tag{III, 53}$$

where now $G(v_1'', v_2'', \cdots) \neq G(0, 0, 0, \cdots)$. It is noteworthy that, as a consequence, the wave number of a difference band $\nu_k - \nu_i$ is *exactly* the difference of the wave numbers of the bands ν_k and ν_i, even if anharmonicity is taken into account, whereas the wave number of the summation band $\nu_k + \nu_i$ is not exactly the sum of the wave numbers of ν_k and ν_i. In other words, the observation of difference bands supplies useful *combination relations* which may serve as checks on the vibrational analysis. Conversely, these difference bands may be used to determine, not only approximately as from summation bands, but accurately, the wave numbers of fundamental bands that cannot be observed directly. For example, from $\nu_k - \nu_i$, if ν_k is known, ν_i follows immediately.

(*e*) *Polarization of Raman lines*

Placzek (700) has shown that the state of polarization of the scattered light can be obtained according to quantum mechanics, if in the classical formula (III, 21) [see also (III, 17) and (III, 20)] the polarizability components α_{xx}, α_{xy}, $\cdots$ are replaced by the corresponding matrix elements $[\alpha_{xx}]^{v'v''}$, $[\alpha_{xy}]^{v'v''}$, $\cdots$. If $v' = v''$, we obtain from (III, 21) the degree of depolarization of the Rayleigh scattering if all molecules are in the state v''; if $v' \neq v''$, we obtain the degree of depolarization of the Raman line corresponding to the transition $v' \leftrightarrow v''$. With the help of Table 55 and the general selection rule that $\psi_{v'}\psi_{v''}^*$ must have the same species as $\boldsymbol{\alpha}$, we can find out which components of $[\boldsymbol{\alpha}]^{v'v''}$ are different from zero for a particular transition and therefore can obtain some information about the state of polarization even without actual calculation of the $[\alpha_{xx}]^{v'v''}$, $[\alpha_{xy}]^{v'v''}$, $\cdots$.

In the case of point groups without degenerate vibrations, Table 55 shows that α_{xx}, α_{yy}, and α_{zz} are totally symmetric. Therefore, if $\psi_{v'}\psi_{v''}^*$ is totally symmetric $[\alpha^{\mathrm{I}}]^{v'v''} = \frac{1}{3}\{[\alpha_{xx}]^{v'v''} + [\alpha_{yy}]^{v'v''} + [\alpha_{zz}]^{v'v''}\}$ is different from zero while for non-totally symmetric $\psi_{v'}\psi_{v''}^*$ it is zero. Consequently, according to (III, 21), *the* *degree of depolarization* ρ_n (for natural incident light) *of totally symmetric Raman lines has a value between* 0 *and* $\frac{6}{7}$, *while for non-totally symmetric Raman lines, it is* $\frac{6}{7}$. By totally symmetric Raman lines we mean lines for which $\psi_v'\psi_v^{*\prime\prime}$ is totally symmetric. Specializing to Raman fundamentals (ψ_v'' totally symmetric), we can therefore say that *Raman lines corresponding to totally symmetric vibrations are polarized, those corresponding to non-totally symmetric vibrations are depolarized.* As will be shown below, this rule holds also for molecules with degenerate vibrations. It must be well understood that for the totally symmetric Raman lines ρ_n may be close to $\frac{6}{7}$, and therefore observation of $\rho_n = \frac{6}{7}$ for a certain Raman line does not definitely exclude the possibility that it is totally symmetric. However, an observation of $\rho_n < \frac{6}{7}$ does prove definitely that the Raman line is totally symmetric.[11]

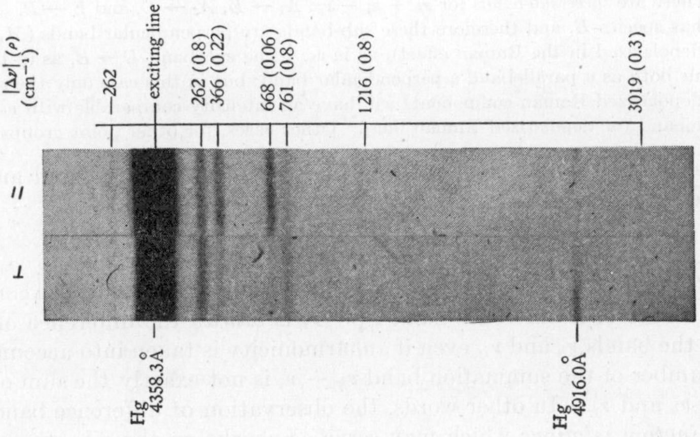

FIG. 82. **Polarization in the Raman spectrum of $CHCl_3$ after Glockler.*—The spectrum at the top was obtained with the analyzer transmitting only light polarized parallel to the *xy* plane (assuming the *z* axis to coincide with the exciting beam); the spectrum at the bottom was obtained with the analyzer transmitting only light polarized perpendicular to the *xy* plane.**

*** The author is greatly indebted to Professor G. Glockler for this spectrogram.**

As in the classical theory, if the incident light is linearly polarized, the maximum degree of depolarization according to (III, 22) is $\rho_l = \frac{3}{4}$, and the maximum degree of reversal for incident circularly polarized light, according to (III, 23), is $\rho_c = 6$.

As an example, Fig. 82 gives the Raman spectrum of $CHCl_3$ taken through an analyzer (a) when only light polarized parallel to the *xy* plane is transmitted, and (b) when only light polarized perpendicular to it is transmitted, assuming that the *z* axis is in the direction of propagation of the incident beam. It is seen that there is a noticeable difference of intensity in the two exposures for $\Delta\nu$ = 366, 668, and 3019 cm^{-1}, whereas for the other Raman lines it is approximately the same. The latter are depolarized and correspond to non-totally symmetric vibrations (see Table 86, p. 316).

As stated previously, for totally symmetric Raman lines α in $\int \alpha\psi_v'\psi_v^{*\prime\prime}d\tau$ must be totally symmetric. In the case of molecules of the cubic point groups (T_d, $O_h \cdots$), this means that α must have cubic symmetry and therefore the polarizability

[11] Even this last statement is not without exception since a strong Coriolis coupling of a non-totally symmetric with a totally symmetric vibration (see Chapter IV) may lead to a lowering of the degree of depolarization of the non-totally symmetric Raman line below the value $\frac{6}{7}$.

ellipsoid must be a sphere, that is, $\alpha_{xx} = \alpha_{yy} = \alpha_{zz}$ while $\alpha_{xy} = \alpha_{xz} = \alpha_{yz} = 0$ (see also Table 55). Consequently the anisotropy β in (III, 20) is zero and thus, according to (III, 21), *for the cubic point groups, the totally symmetric Raman lines are completely polarized* ($\rho_n = 0, \rho_l = 0, \rho_c = 0$), in agreement with the previous classical considerations. The same holds, of course, for the Rayleigh scattering. *For all other molecules*, that is, for all non-cubic point groups, *the degree of depolarization of the totally symmetric Raman lines is intermediate between* 0 *and* $\frac{6}{7}$.

The above-mentioned conclusion that for molecules with degenerate vibrations as well as for those without degenerate vibrations all but the totally symmetric Raman lines are completely depolarized ($\rho_n = \frac{6}{7}$) is obtained in the following way: For those Raman lines for which $\psi_v'\psi_v^{*}{}''$ has a species that occurs only for α_{xy}, α_{xz}, or α_{yz} the previous proof, given for molecules with non-degenerate vibrations only, can be applied without change. If $\psi_v'\psi_v^{*}{}''$ has a species agreeing with one of the species of α_{xx}, α_{yy}, and α_{zz}, we have to remember (see p. 256) that, for axial molecules, of the two species of α_{xx} and α_{yy} (Table 55) the totally symmetric one (A_1, Σ^+, A_1', A_{1g}, $\cdots$) corresponds to $\alpha_{xx} + \alpha_{yy}$ whereas the other (E, B_1, E_2, Δ, E', $\cdots$) corresponds to $\alpha_{xx} - \alpha_{yy}$ (and $2\alpha_{xy}$). Therefore, for these non-totally symmetric species, $\alpha_{xx} + \alpha_{yy} = 0$ and since also $\alpha_{zz} = 0$ (α_{zz} always being totally symmetric for axial molecules), it follows that $\alpha^{\mathrm{I}} = \frac{1}{3}(\alpha_{xx} + \alpha_{yy} + \alpha_{zz}) = 0$; that is, the corresponding Raman lines are depolarized ($\rho_n = \frac{6}{7}$). In the case of the cubic point groups, it can be shown [see Placzek (700) and Tisza (867)] that $\alpha_{xx} + \alpha_{yy} + \alpha_{zz} = 0$ for E and E_g respectively.

Just as in the case of intensity calculations (see p. 261), for actual calculations of the degree of depolarization one would substitute the expansions (III, 50) into (III, 17), (III, 20), and (III, 21). Since $\int \psi_v'\psi_v^{*}{}'' d\tau = 0$ if $v' \neq v''$ and since in first approximation (that is, for the fundamentals) only one term of the sum on the right-hand side of equation (III, 50) is different from zero, it follows immediately that (III, 21) goes over into (III, 26), the classical formula for the degree of depolarization of the Raman lines, which depends only on the $\left(\frac{\partial \alpha_{xx}}{\partial \xi_i}\right)_0$, $\left(\frac{\partial \alpha_{xy}}{\partial \xi_i}\right)_0$, $\cdots$, that is, on the components of the change of polarizability for the normal vibration considered. The above general rule about the polarization of Raman lines could be obtained from (III, 26) in a way perhaps more easily visualized, but the result would be less general since it would hold only for the fundamentals. Cabannes and Rousset (191a) have calculated, under certain simplifying assumptions, for CO_2, CS_2, CO_3^{--}, and C_6H_6 the quantities $\frac{\partial \alpha_{xx}}{\partial \xi_i}$, $\cdots$, and from them the degree of depolarization. In all cases the calculated values are rather larger than the observed. More recently similar calculations have been carried out for some halomethanes by Wolkenstein (949a) and for the dichlorobenzenes by Nordheim and Sponer (673b).

3. Individual Molecules

The vibration spectra of a large number of polyatomic molecules have been investigated, both in the infrared and in the Raman effect. In this section we shall illustrate the theory, as developed in the two preceding sections, by a discussion of the vibration spectra of a number of molecules, containing up to twelve atoms, for which adequate data are available. All the more important molecules of this type have been included and, as far as possible, the data have been brought up to date; also, an attempt has been made to clear up, or at least point out, the numerous inconsistencies that exist in the literature. The tables of observed bands give the wave numbers which, in the opinion of the author, are most reliable. For the molecules not considered in detail, at least references to the original papers are given. A somewhat similar discussion of individual molecules has been given by Wu (26) [see also the tables by Sponer (22) and the collections of Raman data by Kohlrausch (14) and Hibben (10)].

The *numbering of the fundamentals* has been uniformly chosen in accordance with the practice followed in Chapter II. The vibrations are grouped according to their species, which are taken in the order given in Tables 12–30. The largest totally symmetric frequency is called ν_1, the second largest ν_2, and so on. If there are f such vibrations, the largest frequency of the next species is called ν_{f+1}, and so forth. Only one exception to this rule is made: in the case of linear XY_2 and XYZ molecules the perpendicular vibration is always called ν_2 in agreement with a long-established custom. In order to differentiate vibrations of different species the species symbol is usually added in parentheses, for example $\nu_3(a_g)$, $\nu_5(b_{1u})$, $\cdots$, small letters being used for fundamentals, capital letters for overtone and combination bands. The degeneracy of the upper state is also immediately given by the symbol since e, E, π, Π stand for doubly degenerate, f and F for triply degenerate species.[12] For reasons given previously it did not appear advisable to use different symbols (ν, δ, τ as used by many authors) to distinguish stretching, bending and twisting vibrations. If it is desirable to indicate this character and at the same time to indicate in which group the vibration occurs a superscript is added to the symbol thus: $\nu_2{}^{CC}(a_g)$ is a C—C stretching vibration of species A_g; $\nu_{12}^{CH_2}(b_{3u})$ is a CH_2 deformation (bending) vibration of species B_{3u}; and so on. In the tables given in this chapter the wave numbers of fundamentals are printed in heavy type in order to distinguish them clearly from other bands.

Original measurements in the infrared carried out before 1939 have neglected the *vacuum correction* of the wave numbers [see Dennison (280)]. Even though this correction is small (0.82 cm^{-1} at 3000 cm^{-1}), an exact fit of combination differences and an exact agreement with Raman data cannot be expected without introducing it. Therefore this correction has been applied in all the following tables wherever it changes the last significant figure given. Unfortunately, in some papers even after 1939 it is not clearly stated whether ν_{vacuum} or ν_{air} is given.

In some cases, in the following discussions, we shall have to anticipate the results of the investigation of the fine structure of the infrared bands (see Chapter IV) in so far as they tell us the direction of the change of dipole moment.

(a) *Triatomic molecules*

Carbon dioxide, CO_2. One of the most frequently and most thoroughly studied molecules is the carbon dioxide molecule. Fig. 83 gives the more important sections of the observed infrared absorption spectrum under small resolution. Two extremely strong absorption bands at 667.3 and 2349.3 cm^{-1} stand out. In the Raman spectrum, under low resolution, only one strong band at 1340 cm^{-1} is found. This and the two strong infrared bands must be considered as fundamentals. Since any triatomic molecule has only three fundamentals, the above are *the* fundamentals of CO_2. As no one of these occurs both in the Raman and in the infrared spectrum, it follows from the rule of mutual exclusion (see p. 256) that the molecule must have a center of symmetry. For triatomic molecules, this implies that the molecule is *linear and symmetric*.

It should be emphasized that conclusions of this type should be drawn with caution. Even in a completely unsymmetric molecule, the various fundamentals do not all have the same intensity and certain fundamentals may not have been observed in the infrared or in the Raman effect simply

[12] A number of authors indicate the degeneracy by a superscript 2 or 3 in front of ν, δ, τ used for designating the vibrations: ${}^2\nu$, ${}^3\nu$, ${}^2\delta$, ${}^3\delta$, $\cdots$. Since for the electronic states of both molecules and atoms such superscripts are generally used to indicate spin multiplicity, this practice has not been adopted here. Many authors have used π (from parallel) and σ (from the German senkrecht meaning perpendicular) in ν_π, ν_σ, δ_π, δ_σ, ... to indicate whether or not a vibration is symmetric with respect to an axis of symmetry. The use of the complete species symbols makes this superfluous. For this reason and also since the use of π and σ does not seem to lend itself to a consistent scheme applicable to all molecules, it has not been adopted in this book. The π and σ just mentioned have of course nothing to do with the species symbols π and σ used for linear molecules.

because they are too weak, and not because they are forbidden on account of a certain symmetry property of the molecule. In the present case, however, the above conclusion has been confirmed in many different ways, such as the lack of a permanent dipole moment, the rotational Raman spectrum (see p. 21), the fine structure of the infrared bands (see p. 384), and the activity of the overtone and combination bands.

A closer inspection of the spectrum in Fig. 83 shows (see Chapter IV, section 1) that the band at 667.3 is a perpendicular band (species Π_u of the dipole moment, $M_z = 0$, strong central maximum),

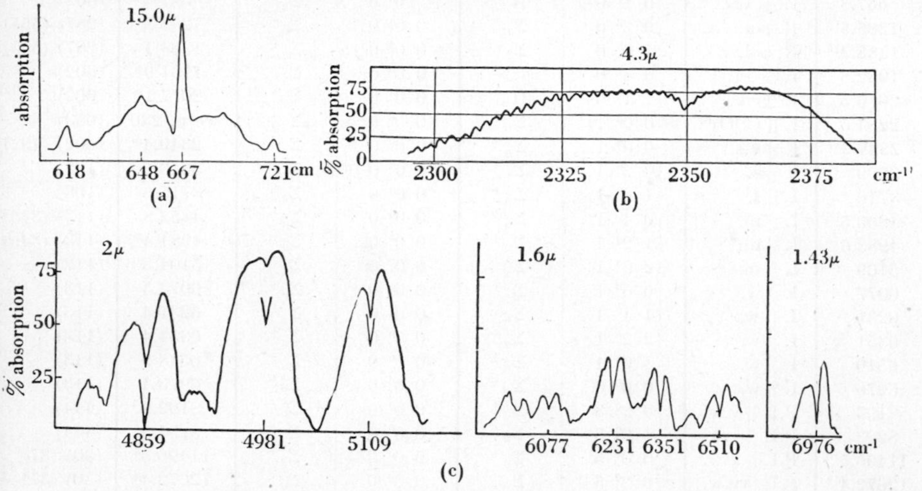

FIG. 83. **Parts of the observed infrared absorption spectrum of CO_2 under low dispersion** [according to Martin and Barker (602) and Barker and Wu (113)].—The equivalent absorbing paths at atmospheric pressure in (a), (b), and (c) are 0.23, 0.10 and 560 cm. respectively.

whereas the band at 2349.3 cm^{-1} is a parallel band (species Σ_u^+ of the dipole moment, central minimum). This identifies the low frequency as ν_2, the high frequency as ν_3 (see Fig. 25b). Investigations with higher dispersion show also that the intense Raman "line" at 1340 cm^{-1} (which should be ν_1) really consists of two lines at 1285.5 and 1388.3 cm^{-1} with an intensity ratio 1 : 0.59 [see Hanson (409) and Langseth and Nielsen (556)]. This observation is less easily explained, since one would expect just one Raman line corresponding to ν_1. However, the average of the two observed Raman lines agrees very closely with double the low frequency ν_2. If $2\nu_2$ is very close to ν_1, a Fermi resonance is to be expected (see p. 217), which will lead to the occurrence of two almost equally intense Raman lines instead of one, as is observed. Then, however, we cannot say that one Raman line is ν_1 the other $2\nu_2^0$, but both are mixtures of ν_1 and $2\nu_2^0$. We write $2\nu_2^0$, since only the one component with $l = 0$ of the state $2\nu_2$ takes part in the resonance. The other component $2\nu_2^2$ ($l = 2$, species Δ_g) would also be allowed in the Raman effect according to the rigorous selection rules (Table 55), but as an overtone it is much weaker than a fundamental and is not observed. $2\nu_2^0$ appears strongly only because of resonance with a fundamental. The degree of depolarization of the two strong Raman lines is small [$\rho \sim 0.18$ and 0.14, respectively, according to Langseth and Nielsen (556)] in agreement with the assumption that both upper states are totally symmetric (Σ_g^+), whereas the line $2\nu_2^2$, if it occurred, would be completely depolarized.

The above interpretation of the strongest Raman lines and infrared bands is confirmed in every way by an investigation of the overtone and combination bands, partly represented in Fig. 83b, [Adel and Dennison (37), Dennison (280)]. In Table 56 we give all the observed infrared and Raman bands together with their interpretation [mostly according to Adel and Dennison (37)]. It should be noted particularly that no overtones $2\nu_3$ and $4\nu_3$ (which would lie approximately at twice and four times the frequency 2349.3) have been observed in the infrared, although $3\nu_3$ and $5\nu_3$ are observed, in agreement with the rigorous selection rules, the states $2\nu_3$, $4\nu_3$ having species Σ_g^+ while $3\nu_3$ and $5\nu_3$ have Σ_u^+. Similarly, no overtones $2\nu_2$, $4\nu_2$ occur in the infrared, but the combination bands $2\nu_2 + \nu_3$, $2\nu_2 - \nu_2$ and others do.

TABLE 56. INFRARED AND RAMAN BANDS OF GASEOUS CO_2.

ν_{vacuum}, observed (cm^{-1})	Band type[13]	Upper state[14]		Lower state[14]		ν, calculated[15] (cm^{-1})	References
		$v_1\, v_2{}^l\, v_3$	Species	$v_1\, v_2{}^l\, v_3$	Species		
667.3	I. $\perp$ v.s.	0 1^1 0	Π_u	0 0^0 0	$\Sigma_g{}^+$	667.3*	(602)
1285.5[16]	R. pol. v.s.	$\Big\{$0 2^0 0	$\Sigma_g{}^+$	0 0^0 0	$\Sigma_g{}^+$	1285.8	(287) (555)
1388.3[16]	R. pol. v.s.	1 0^0 0	$\Sigma_g{}^+$	0 0^0 0	$\Sigma_g{}^+$	1388.1	(287) (555)
1932.5	I. $\perp$ m.	$\Big\{$0 3^1 0	Π_u	0 0^0 0	$\Sigma_g{}^+$	1931.9*	(602)
2076.5	I. $\perp$ m.	1 1^1 0	Π_u	0 0^0 0	$\Sigma_g{}^+$	2077.1*	(602)
2284.5	I. $\parallel$ C^{13}O$_2{}^{16}$	0 0^0 1	$\Sigma_u{}^+$	0 0^0 0	$\Sigma_g{}^+$	(see p. 230)	(653)
2349.3[17]	I. $\parallel$ v.s.	0 0^0 1	$\Sigma_u{}^+$	0 0^0 0	$\Sigma_g{}^+$	2349.4*	(602) (192)
3609	I. $\parallel$ s.	$\Big\{$0 2^0 1	$\Sigma_u{}^+$	0 0^0 0	$\Sigma_g{}^+$	3613.2	(103)
3716	I. $\parallel$ s.	1 0^0 1	$\Sigma_u{}^+$	0 0^0 0	$\Sigma_g{}^+$	3715.6	(103)
4860.5	I. $\parallel$ m.	$\Big\{$0 4^0 1	$\Sigma_u{}^+$	0 0^0 0	$\Sigma_g{}^+$	4852.8	(113) (34a)
4983.5	I. $\parallel$ m.	1 2^0 1	$\Sigma_u{}^+$	0 0^0 0	$\Sigma_g{}^+$	4981.4*	(113) (34a)
5109	I. $\parallel$ m.	2 0^0 1	$\Sigma_u{}^+$	0 0^0 0	$\Sigma_g{}^+$	5104.3	(113)
6077	I. $\parallel$ w.	$\Big\{$0 6^0 1	$\Sigma_u{}^+$	0 0^0 0	$\Sigma_g{}^+$	6074.5	(113)
6231	I. $\parallel$ w.	1 4^0 1	$\Sigma_u{}^+$	0 0^0 0	$\Sigma_g{}^+$	6231.4	(113)
6351	I. $\parallel$ w.	2 2^0 1	$\Sigma_u{}^+$	0 0^0 0	$\Sigma_g{}^+$	6354.4	(113)
6510	I. $\parallel$ w.	3 0^0 1	$\Sigma_u{}^+$	0 0^0 0	$\Sigma_g{}^+$	6518 9	(113)
6976	I. $\parallel$ w.	0 0^0 3	$\Sigma_u{}^+$	0 0^0 0	$\Sigma_g{}^+$	6973.1	(113)
8193	P.I. $\parallel$ v.w.	$\Big\{$0 2^0 3	$\Sigma_u{}^+$	0 0^0 0	$\Sigma_g{}^+$	8192.9	(444)
8293	P.I. $\parallel$ v.w.	1 0^0 3	$\Sigma_u{}^+$	0 0^0 0	$\Sigma_g{}^+$	8295.5	(444)
11496.5[18]	P.I. $\parallel$ v.w.	0 0^0 5	$\Sigma_u{}^+$	0 0^0 0	$\Sigma_g{}^+$	11496.5*	(30) (37)
12672.4[18]	P.I. $\parallel$ v.w.	$\Big\{$0 2^0 5	$\Sigma_u{}^+$	0 0^0 0	$\Sigma_g{}^+$	12672.4*	(30) (37)
12774.7[18]	P.I. $\parallel$ v.w.	1 0^0 5	$\Sigma_u{}^+$	0 0^0 0	$\Sigma_g{}^+$	12774.7*	(30) (37)
618.1	I. $\perp$ m.	0 2^0 0	$\Sigma_g{}^+$	0 1^1 0	Π_u	618.5	(602)
668.3	I. $\perp$	0 2^2 0	Δ_g	0 1^1 0	Π_u	668.1*	(602)
720.5	I. $\perp$ m.	1 0^0 0	$\Sigma_g{}^+$	0 1^1 0	Π_u	720.8*	(602) (34)
1264.8[16]	R. m.	$\Big\{$0 3^1 0	Π_u	0 1^1 0	Π_u	1264.6	(287) (555)
1409.0[16]	R. m.	1 1^1 0	Π_u	0 1^1 0	Π_u	1409.8	(287) (555)
1886	I. $\perp$ w.	0 4^0 0	$\Sigma_g{}^+$	0 1^1 0	Π_u	1880.1	(113)
2094	I. $\perp$ m.	1 2^2 0	Δ_g	0 1^1 0	Π_u	2094.9	(113)
2137	I. $\perp$ m.	2 0^0 0	$\Sigma_g{}^+$	0 1^1 0	Π_u	2131.5	(113)
596.8	I. $\perp$ w.	0 3^1 0	Π_u	0 2^2 0	Δ_g	596.5	(602)
647.6	I. $\perp$ w.	0 3^1 0	Π_u	0 2^0 0	$\Sigma_g{}^+$	646.1	(602)
740.8	I. $\perp$ w.	1 1^1 0	Π_u	0 2^2 0	Δ_g	741.7	(602)
790.8	I. $\perp$ v.w.	1 1^1 0	Π_u	0 2^0 0	$\Sigma_g{}^+$	791.3	(602)
960.8	I. $\parallel$ w.	0 0 1	$\Sigma_u{}^+$	$\Big\{$1 0^0 0	$\Sigma_g{}^+$	961.3	(108) (34)
1063.6	I. $\parallel$ w.	0 0 1	$\Sigma_u{}^+$	0 2^0 0	$\Sigma_g{}^+$	1063.6*	(108) (34)
1242[16]	R. w.	0 4^2 0	Δ_g	0 2^2 0	Δ_g	1248.0	(555) (409)
1305.1[19]	R. v.w.	0 4^2 0	Δ_g	0 2^0 0	$\Sigma_g{}^+$	1297.6	(555)
1325.0	R. v.w.	?		?		?[20]	(555)
1344.1[19]	R. v.w.	1 2^0 0	$\Sigma_g{}^+$	0 2^2 0	Δ_g	1340.4	(555)
1369.4[19]	R. v.w.	1 2^2 0	Δ_g	1 0^0 0	$\Sigma_g{}^+$	1374.1	(555)
1430[16]	R. w.	1 2^2 0	Δ_g	0 2^2 0	Δ_g	1426.8	(555) (409)
1528	R. w.	2 0^0 0	$\Sigma_g{}^+$	0 2^0 0	$\Sigma_g{}^+$	1513.0	(555) (409)

[13] I. = infrared band, R. = Raman line, P.I. = photographic infrared band, pol. = polarized; $\parallel$ = parallel band (species $\Sigma_u{}^+$), $\perp$ = perpendicular band (Π_u), v.s. = very strong, s. = strong, m. = medium, w. = weak, v.w. = very weak. The intensity estimates for infrared bands are only valid for bands in the same region.

[14] Levels connected by braces are in Fermi resonance (see text).

[15] An asterisk indicates the bands that have been used to set up the formula by means of which the others have been calculated. Slight differences between these starred calculated values and the

In Fig. 84 a diagram of all the observed vibrational energy levels of the CO_2 molecule is given. The observed transitions are also indicated. The resonance between ν_1 and $2\nu_2$ is of course repeated for many of the higher levels (see p. 217). Levels that are in resonance are connected by braces in Table 56 and by medium weight solid lines or small arcs in Fig. 84. The quantum numbers given for each level are those of the unperturbed level that preponderates, at least somewhat, in the mixture that the actual level represents.

It is noteworthy that the 02^20 Δ_g sublevel of $2\nu_2$ which does not resonate with ν_1 has been observed in some difference bands and does indeed occur very nearly half-way between the resonating levels $2\nu_2{}^0$ and ν_1.

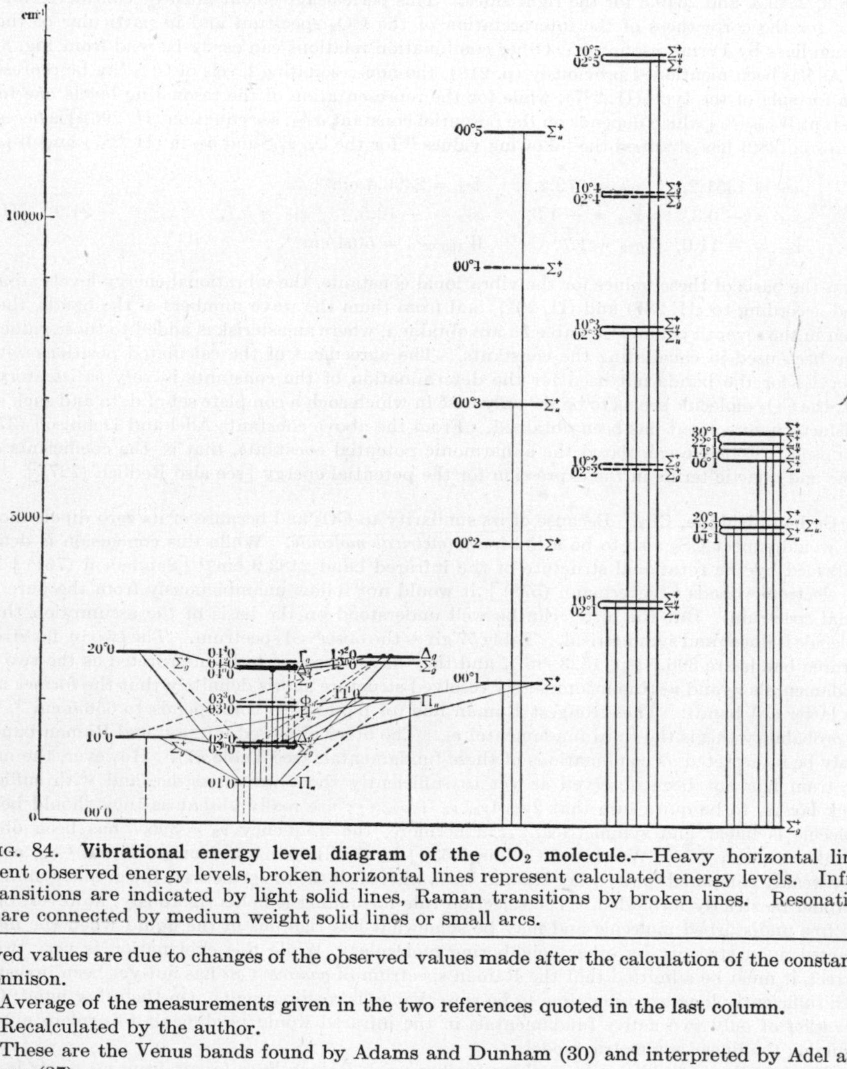

FIG. 84. **Vibrational energy level diagram of the CO_2 molecule.**—Heavy horizontal lines represent observed energy levels, broken horizontal lines represent calculated energy levels. Infrared transitions are indicated by light solid lines, Raman transitions by broken lines. Resonating levels are connected by medium weight solid lines or small arcs.

observed values are due to changes of the observed values made after the calculation of the constants by Dennison.

[16] Average of the measurements given in the two references quoted in the last column.

[17] Recalculated by the author.

[18] These are the Venus bands found by Adams and Dunham (30) and interpreted by Adel and Dennison (37).

[19] Adel and Dennison (37) and Adel (31) considered these lines as spurious. They seemed to believe that a $\Delta_g - \Sigma_g{}^+$ transition cannot occur in the Raman effect. It is, however, allowed according to the rigorous rules (see Table 55).

[20] This is an extremely weak Raman line.

It can be seen immediately from Fig. 84 that, if the interpretation of the observed bands is correct, there should be a number of accurately fulfilled *combination relations*. For example:

$$\nu(01^10 \leftarrow 00^00) + \nu(02^00 \leftarrow 01^10) = \nu(02^00 \leftarrow 00^00);$$
$$\nu(01^10 \leftarrow 00^00) + \nu(10^00 \leftarrow 01^10) = \nu(10^00 \leftarrow 00^00);$$
$$\nu(02^00 \leftarrow 00^00) + \nu(00^01 \leftarrow 02^00) = \nu(00^01 \leftarrow 00^00);$$
$$\nu(10^00 \leftarrow 00^00) + \nu(00^01 \leftarrow 10^00) = \nu(00^01 \leftarrow 00^00).$$

The observed figures for the left sides are 1285.4, 1388.0, 2349.1, and 2349.1 as compared to 1285.5, 1388.3, 2349.3, and 2349.3 for the right sides. This perfect agreement must be considered a strong proof for the correctness of the interpretation of the CO_2 spectrum and in particular of the two Raman lines by Fermi resonance. Other combination relations can easily be read from Fig. 84.

As has been mentioned previously (p. 218), the non-resonating levels of CO_2 can be represented by a formula of the type (II, 287), while for the representation of the resonating levels one further constant $W_{100;02^00}$ [which depends on the potential constant α_{122}; see equation (II, 296)] is necessary. Dennison (280) has obtained the following values [21] for the ω_i, x_{ik}, and g_{22} in (II, 287) and $W_{100;02^00}$:

$$\omega_1 = 1351.2, \quad \omega_2 = 672.2, \quad \omega_3 = 2396.4 \text{ cm}^{-1},$$
$$x_{11} = -0.3, \quad x_{22} = -1.3, \quad x_{33} = -12.5, \quad x_{12} = 5.7, \quad x_{13} = -21.9, \qquad \text{(III, 54)}$$
$$x_{23} = -11.0, \quad g_{22} = 1.7, \quad W_{100;02^00} = 50.4 \text{ cm}^{-1}.$$

If, on the basis of these values for the vibrational constants, the vibrational energy levels are calculated according to (II, 287) and (II, 291), and from them the wave numbers of the bands, the data given in the seventh column of Table 56 are obtained, where an asterisk is added to those values that have been used in calculating the constants. The agreement of the calculated positions with the observed for the bands not used for the determination of the constants is very satisfactory. In fact, the CO_2 molecule seems to be the only case in which such a complete set of data and such a very satisfactory agreement has been obtained. From the above constants Adel and Dennison (37) and Dennison (280) have calculated the anharmonic potential constants, that is, the coefficients of the cubic and quartic terms in the expression for the potential energy [see also Redlich (727)].

Carbon disulfide, CS_2. Because of its similarity to CO_2 and because of its zero dipole moment, one would expect CS_2 also to be a *linear symmetrical molecule.* While this conclusion is definitely confirmed by the rotational structure of the infrared band 2183.9 cm^{-1} [Sanderson (761)] and of the electronic bands [Liebermann (578)], it would not follow unambiguously from the pure vibrational spectrum. But the latter can be well understood on the basis of the assumption that the molecule is linear and symmetrical. Table 57 gives the observed spectrum. The two by far strongest infrared bands are 396.7 and 1523 cm^{-1}, and they have therefore to be interpreted as the two active fundamentals ν_2 and ν_3 (the incompletely resolved structure shows definitely that the former is a $\perp$, the latter a $\|$ band). The strongest Raman line (of liquid CS_2) corresponds to 656.5 cm^{-1}, and in all probability this is the third fundamental ν_1. The other observed infrared and Raman bands can easily be interpreted as combinations of these fundamentals (see Table 57). However, the infrared spectrum has not been observed as yet to sufficiently short wave lengths, and with sufficiently thick layers, to be quite sure that $2\nu_3$, $4\nu_3$, $\nu_3 + \nu_2$, $\cdots$ are really absent as they should be if the molecule is linear and symmetrical. Furthermore, the frequency $\nu_2 = 396.7$ has been observed [Venkateswaran (895), Wood and Collins (953)] in the Raman spectrum of liquid CS_2, although very weakly compared to the other Raman lines, whereas if the molecule were linear and symmetrical, it would be strictly forbidden. The selection rule forbidding it holds rigorously, however, only for the free undisturbed molecule and may be somewhat less rigorous in the liquid when the molecule is acted upon strongly by other neighboring molecules. While this explanation is very probably correct, it must be admitted that the Raman spectrum of *gaseous* CS_2 has not yet been investigated with sufficiently long exposure time to be sure that ν_2 is really absent. On the other hand, the observation of only two active fundamentals in the infrared would be difficult to explain except by assuming the linear symmetric model.

The fact that in addition to ν_1 three further weak Raman lines (apart from ν_2) occur is due to two reasons similar to those responsible for the weak Raman lines of CO_2. First, a Fermi resonance:

---

[21] The constant δ that is also necessary for obtaining the energies of the resonating levels is given by $G(1, 0^0, 0) - G(0, 2^0, 0) = 16.7$ cm^{-1} (see p. 218). $W_{100;02^00}$ is Dennison's $b/\sqrt{2}$.

the Raman line 796.0 is very nearly $2\nu_2$; however, such an overtone would be expected to be very weak if it were not strengthened considerably by a resonance with ν_1. The resonance is much poorer here than in the case of CO_2, and therefore the two lines 656 and 796 have rather different intensities. Second, the lowest frequency ν_2 of CS_2 is so small that at room temperature a considerable fraction

TABLE 57. INFRARED AND RAMAN BANDS OF CS_2.

ν_{vacuum}[22] observed (cm^{-1})	Band type[13]	Upper state		Lower state		References
		$v_1\ v_2^l\ v_3$	Species	$v_1\ v_2^l\ v_3$	Species	
396.7 (gas)	I. (+R.)[23] $\perp$ s.	0 1^1 0	Π_u	0 0^0 0	Σ_g^+	(285)
656.5 (liquid)[24]	R. pol. s.	1 0^0 0	Σ_g^+	0 0^0 0	Σ_g^+	(558)
796.0 (liquid)[24]	R. pol. w.	0 2^0 0	Σ_g^+	0 0^0 0	Σ_g^+	(558)
1523 (gas)	I. $\parallel$ s.	0 0^0 1	Σ_u^+	0 0^0 0	Σ_g^+	(80)
2183.9 (gas)	I. $\parallel$ w.	1 0^0 1	Σ_u^+	0 0^0 0	Σ_g^+	(761)
2329 (gas)	I. ? v.w.	0 2^0 1	Σ_u^+	0 0^0 0	Σ_g^+	(80)
648.3 (liquid)	R. w.	1 1^1 0	Π_u	0 1^1 0	Π_u	(558)
804.9 (liquid)	R. v.w.	0 3^1 0	Π_u	0 1^1 0	Π_u	(558)
878 (gas)	I. ? v.w.	0 0^0 1	Σ_u^+	1 0^0 0	Σ_g^+	(80)

of the molecules is in the $01^10\ \Pi_u$ state and therefore the 1–1 transitions corresponding to the two main Raman lines have appreciable intensity: they are the lines 648.3 and 804.9 cm^{-1}. In agreement with this interpretation they disappear, at low temperatures, in solid CS_2 [Sirkar (794)]. Two further very weak and doubtful Raman lines (not given in Table 57) have been interpreted by Giulotto and Caldirola (363) as due to the isotopic molecule $CS^{32}S^{34}$.

Nitrous oxide, N_2O. The molecule N_2O has the same number of electrons as CO_2 and one might therefore also expect a linear symmetrical structure. However, the vibration and the vibration-rotation spectrum show unambiguously that the molecule, although *linear*, is *not symmetrical*, but has the form N—N—O. The three strongest infrared bands are at 588.8, 1285.0, and 2223.5 cm^{-1}, increasing in intensity in this order [Plyler and Barker (703)]. Because of this intensity relation, and because of the numerical values, none of these bands can be interpreted as an overtone or combination band. Thus there are three infrared-active fundamentals. This fact alone proves that N_2O cannot have a center of symmetry. This is confirmed by the Raman spectrum which (apart from several weak lines) consists of two strong lines with displacements of 1286.5 and 2223.2 cm^{-1} [Langseth and Nielsen (554)], thus agreeing with two of the infrared fundamentals, whereas if N_2O had a center of symmetry only one fundamental should appear in the Raman spectrum and this fundamental should not appear in the infrared. For linear unsymmetric N_2O (point group $\boldsymbol{C_{\infty v}}$), all three fundamentals are Raman (and infrared) active. The absence in the Raman spectrum of the third low fundamental, which as shown by the structure of the infrared band is the perpendicular vibration ν_2, is no contradiction to the assumed model, since quite generally the non-totally symmetric fundamentals are weak in the Raman spectrum compared to the totally symmetric fundamentals (see p. 399). The linearity of the molecule is proved unambiguously by the fine structure of the infrared bands. In this connection it is particularly significant that the band $2\nu_2$ is a $\parallel$ band whereas ν_2 is a $\perp$ band, as it should be for a linear molecule.

The interpretation of the other weaker observed infrared and Raman bands is given in Table 58. The two Raman bands $2\nu_2^0$ at 1167.0 and $2\nu_2^2$ at 1185 cm^{-1} are quite weak compared to the two main Raman bands; yet they seem to be rather too strong for overtones. In the case of the former band, Fermi resonance with ν_1 at 1285.0 cm^{-1} is very probably the reason. This is confirmed by the

[13] On p. 274.

[22] More recently Giulotto and Caldirola (363) have given Raman frequencies that are all approximately 2.2 cm^{-1} smaller than Langseth, Sørensen, and Nielsen's values given in the table.

[23] See text.

[24] Also observed in gaseous CS_2 (though with less accuracy) by Imanishi (466). Liebermann (578), from the ultraviolet spectrum of the gas, obtained $2\nu_2 = 801.89\ cm^{-1}$.

anomalously large intensity of $2\nu_2^0$ in the infrared, where it has as much as half the intensity of the fundamental, whereas $3\nu_2$ for which no resonance occurs is not even observed, although other bands with $\frac{1}{200}$ the intensity of $2\nu_2^0$ have been found in the particular spectral region.

TABLE 58. INFRARED AND RAMAN BANDS OF GASEOUS N_2O, AFTER PLYLER AND BARKER (703) AND LANGSETH AND NIELSEN (554).

ν_{vacuum},[25] observed (cm^{-1})	Band type[13]	Upper state $v_1\ v_2^l\ v_3$	Upper state Species	Lower state $v_1\ v_2^l\ v_3$	Lower state Species	ν_{vacuum}, calculated (cm^{-1})
588.8	I. ⊥ s.	$0\ 1^1\ 0$	Π	$0\ 0^0\ 0$	Σ^+	588.8*
1167.0	I. ‖ m. R. v.w.	$0\ 2^0\ 0$	Σ^+	$0\ 0^0\ 0$	Σ^+	1167.0*
1185	R. v.w.	$0\ 2^2\ 0$?	Δ?[26]	$0\ 0^0\ 0$	Σ^+	1179.1
1285.0	I. ‖ v.s., R. v.s.	$1\ 0\ 0$	Σ^+	$0\ 0^0\ 0$	Σ^+	1285.0*
1867.5	I. ⊥ w.	$1\ 1^1\ 0$	Π	$0\ 0^0\ 0$	Σ^+	1878.5[27]
2223.5	I. ‖ v.s., R. s.	$0\ 0\ 1$	Σ^+	$0\ 0^0\ 0$	Σ^+	2223.5*
2461.5	I. ‖ m.	$1\ 2^0\ 0$	Σ^+	$0\ 0^0\ 0$	Σ^+	2461.5*
2563.5	I. ‖ m.	$2\ 0\ 0$	Σ^+	$0\ 0^0\ 0$	Σ^+	2563.5*
2798.3	I. ⊥ w.	$0\ 1^1\ 1$	Π	$0\ 0^0\ 0$	Σ^+	2799.8
3365.6	I. ‖ w.	$0\ 2^0\ 1$	Σ^+	$0\ 0^0\ 0$	Σ^+	3365.6*
3481.2	I. ‖ m.	$1\ 0^0\ 1$	Σ^+	$0\ 0^0\ 0$	Σ^+	3482.3
4419.5	I. ‖ w.	$0\ 0^0\ 2$	Σ^+	$0\ 0^0\ 0$	Σ^+	4419.5*
4734.7	I. ‖ w.	$2\ 0^0\ 1$	Σ^+	$0\ 0^0\ 0$	Σ^+	4734.7*
579.3	I. ⊥ m.	$0\ 2^0\ 0$	Σ^+	$0\ 1^1\ 0$	Π	578.2
590.3	I. ⊥ m.	$0\ 2^2\ 0$	Δ	$0\ 1^1\ 0$	Π	590.3*
1282	R. v.w.	$1\ 1^1\ 0$	Π	$0\ 1^1\ 0$	Π	1289.7
1828	I. ⊥ v.w.	$(1\ 2^0\ 0$	Σ^+	$0\ 1^1\ 0$	Π	1872.7)[28]
1844	I. ⊥ v.w.	$(1\ 2^2\ 0$	Δ	$0\ 1^1\ 0$	Π	1884.8)[28]
2210	R. v.w.	$0\ 1^1\ 0$	Π	$0\ 1^1\ 0$	Π	2211.0
2776	I. ⊥ v.w.	$0\ 2^0\ 1$	Σ^+	$0\ 1^1\ 0$	Π	2776.8
2785	I. ⊥ v.w.	$0\ 2^2\ 1$	Δ	$0\ 1^1\ 0$	Π	2788.9

From the positions of the observed bands, neglecting resonance terms, the constants in the formula (II, 284) for the vibrational energy levels are found to be (in cm^{-1}):

$$\omega_1^0 = 1288.2_5, \quad \omega_2^0 = 588.0_5, \quad \omega_3^0 = 2237.2_5;$$
$$x_{11} = -3.2_5, \quad x_{22} = -2.2_8, \quad x_{33} = -13.7_5; \qquad \text{(III, 55)}$$
$$x_{12} = +4.7_5, \quad x_{23} = -12.4_5, \quad x_{13} = -26.1_5, \quad g_{22} = +3.0_3.$$

The calculated values for the positions of the Raman and infrared bands on the basis of these constants are given in the last column of Table 58. The wave numbers of bands that have been used in evaluating the constants are indicated by asterisks. The agreement for some of the other bands is not as good as for CO_2, which is to be expected because of the neglect of the influence of resonance. From the above ω_i^0 and x_{ik} the zero-order frequencies ω_i are found, according to (II, 285), to be

$$\omega_1 = 1299.8_3\ cm^{-1}, \quad \omega_2 = 596.4_6\ cm^{-1}, \quad \omega_3 = 2276.5_3\ cm^{-1}. \qquad \text{(III, 56)}$$

These values might have to be changed slightly if resonance is taken into account.

---

[13] On p. 274.

[25] For bands occurring both in the infrared and Raman spectrum, the wave number found in the infrared (converted to vacuum) is given, since it is the more accurate one.

[26] The interpretation of this Raman line is doubtful.

[27] That Wu (26) obtains a much better agreement for this band from his formula is due to an error in sign for his x_{12}. His values are also still based on $1/\lambda_{\text{air}}$ rather than on ν_{vacuum}.

[28] This interpretation given by Plyler and Barker is almost certainly wrong, since the transition $12^00 \leftarrow 01^10$ should be exactly the difference of the observed bands 2461.5 and 588.8 independent of any representation by a formula.

Hydrogen cyanide, HCN. It would be difficult to derive definite conclusions about the structure of the HCN molecule from the vibration spectrum alone. But the fine structure of the infrared bands shows very clearly that HCN is *linear* (see Chapter IV, section 1b). The wave numbers of the observed infrared and Raman bands are given in Table 59. The interpretation given assumes

TABLE 59. INFRARED AND RAMAN BANDS OF HCN VAPOR.

ν_{vacuum}, observed (cm^{-1})	Band type[13]	Upper state $v_1\ v_2^l\ v_3$	Upper state Species	Lower state $v_1\ v_2^l\ v_3$	Lower state Species	ν, calculated (cm^{-1})	References
712.1	I. ⊥ v.s.	$0\ 1^1\ 0$	Π	$0\ 0^0\ 0$	Σ^+	712.1*	(210) (125)
1412.0	I. ∥ s.	$0\ 2^0\ 0$	Σ^+	$0\ 0^0\ 0$	Σ^+	1412.0*	(210)
2062 (liquid)	R. v.w. $HC^{13}N$	$1\ 0\ 0$	Σ^+	$0\ 0^0\ 0$	Σ^+		(252) (432)
2089.0	R. v.s.	$1\ 0^0\ 0$	Σ^+	$0\ 0^0\ 0$	Σ^+	2092.4	(492)
2116.7	I. ⊥ m.	$0\ 3^1\ 0$	Π	$0\ 0^0\ 0$	Σ^+	2112.7	(104)
2800.3	I. ⊥ s.	$1\ 1^1\ 0$	Σ^+	$0\ 0^0\ 0$	Σ^+	2800.3*	(210)
3312.0	I. ∥ s., R. (liquid)[29] m.	$0\ 0^0\ 1$	Σ^+	$0\ 0^0\ 0$	Σ^+	3313.8	(35) (13)
4004.5	I. ⊥ m.	$0\ 1^1\ 1$	Π	$0\ 0^0\ 0$	Σ^+	4006.4	(35)
4992.5	I. ⊥ w.	$2\ 1^1\ 0$	Π	$0\ 0^0\ 0$	Σ^+	4992.5*	(35)
5394[30]	I. ∥ m.	$1\ 0^0\ 1$	Σ^+	$0\ 0^0\ 0$	Σ^+	5391.8	(35)
6521.7	I. ∥ m.	$0\ 0^0\ 2$	Σ^+	$0\ 0^0\ 0$	Σ^+	6521.4	(35)
8585.6	P.I. ∥ w.	$1\ 0^0\ 2$	Σ^+	$0\ 0^0\ 0$	Σ^+	8585.0	(442)
9627.1	P.I. ∥ m.	$0\ 0^0\ 3$	Σ^+	$0\ 0^0\ 0$	Σ^+	9627.1*	(442) (579)
11674.4	P.I. ∥ w.	$1\ 0^0\ 3$	Σ^+	$0\ 0^0\ 0$	Σ^+	11676.3	(71) (579)
12635.9	P.I. ∥ w.	$0\ 0^0\ 4$	Σ^+	$0\ 0^0\ 0$	Σ^+	12635.9*	(71)(579)
14670.7	P.I. ∥ v.w.	$1\ 0^0\ 4$	Σ^+	$0\ 0^0\ 0$	Σ^+	14670.7*	(579)
15552.0	P.I. ∥ v.w.	$0\ 0^0\ 5$	Σ^+	$0\ 0^0\ 0$	Σ^+	15552.0*	(579)
700	I. ⊥ s.	$0\ 2^0\ 0$	Σ^+	$0\ 1^1\ 0$	Π	699.9	(210)
(712)[31]	I. ⊥	$0\ 2^2\ 0$	Δ	$0\ 1^1\ 0$	Π	712.9	(210)
2087.1	I. ⊥ w.	$0\ 4^0\ 0$	Σ^+	$0\ 1^1\ 0$	Π	2089.1	(104) (35)
2102.1	I. ⊥ m.	$0\ 4^2\ 0$	Δ	$0\ 1^1\ 0$	Π	2102.1*	(104) (35)
9568.5	P.I. ∥ w.	$0\ 1^1\ 3$	Π	$0\ 1^1\ 0$	Π	9568.5*	(579)
11613.5	P.I. ∥ v.w.	$1\ 1^1\ 3$	Π	$0\ 1^1\ 0$	Π	11613.5*	(579)
12557.5	P.I. ∥ v.w.	$0\ 1^1\ 4$	Π	$0\ 1^1\ 0$	Π	12557.8	(579)

for ν_1, ν_2, ν_3 the vibrations given in Fig. 61a. ν_1 is essentially the oscillation of the CH group against the N atom; ν_3 is essentially a C—H vibration, but there is no difference in symmetry type between ν_1 and ν_3. Although all three fundamentals are allowed both in the infrared and Raman spectra, ν_1 has not been recorded in the infrared, obviously because the change of dipole moment connected with it is very small, the vibration being similar to that in N_2 and the CH distance remaining practically unchanged. On the other hand, ν_2 has not been observed in the Raman effect, in agreement with the rule that non-totally symmetric vibrations are weak in the Raman spectrum.[32]

In obtaining a formula for the vibrational levels it proves necessary [see Lindholm (579)], since such high overtones of ν_3 are observed, to introduce a cubic term $y_{333} \cdot v_3^3$ in equation (II, 287). The

[13] On page 274.

[29] The Raman line of the liquid occurs at 3213 cm^{-1}; that is, there is a considerable shift between ν_3 (gas) and ν_3 (liquid) [see Chapter V, section 2].

[30] Overlapped by H_2O absorption.

[31] This band is not really observed but only guessed, since it coincides with the band ν_2.

[32] If HCN were not linear the low-frequency fundamental would also be totally symmetrical and would therefore occur strongly in the Raman spectrum.

constants are then, again neglecting resonance effects (in cm^{-1}):

$$\omega_1^0 = 2041.2, \qquad \omega_2^0 = 711.7_0, \qquad \omega_3^0 = 3368.6;$$
$$x_{11} = +52.0, \quad x_{22} = -2.8_5, \quad x_{33} = -55.48, \quad y_{333} = +0.768; \qquad \text{(III, 57)}$$
$$x_{12} = -4.2, \quad x_{13} = -14.4_0, \quad x_{23} = -19.53, \quad g_{22} = +3.2_5.$$

From these, the following values (in cm^{-1}) for the frequencies for infinitesimal amplitudes (zero-order frequencies) are obtained:

$$\omega_1 = 2000.6, \qquad \omega_2 = 729.3, \qquad \omega_3 = 3451.5. \qquad \text{(III, 58)}$$

The weak Raman line, first observed by Dadieu (252) at 2062 cm^{-1} and accompanying the strong line at 2094 cm^{-1}, [33] has been interpreted by Dadieu as belonging to the isomeric form HNC, whose existence in ordinary HCN he believed thus proven. However, according to Herzberg (432), this Raman line must be explained as due to the isotopic molecule $HC^{13}N$, which is certainly there, whose frequency ν_1 would be expected to be about 33 cm^{-1} smaller than that of $HC^{12}N$ and whose concentration in ordinary HCN is just about the same as that derived by Dadieu for HNC. More recently McCrosky, Bergstrom, and Waitkins (607a) have obtained chemical evidence that there is no HNC in pure hydrogen cyanide.

The Raman and infrared spectra of "heavy" hydrocyanic acid, DCN, have also been observed [Dadieu and Kopper (260), Bartunek and Barker (125)], although up to now only the fundamentals are known. They are (corrected for vacuum), in cm^{-1},

$$\nu_1 = 1906,^{33} \qquad \nu_2 = 568.9, \qquad \nu_3 = 2629.3.$$

The latter two agree very well with the values obtained theoretically by using the same potential constants for DCN as for HCN, even though anharmonicity is neglected [see Bartunek and Barker (125)].

Water, H_2O. The observation of a strong far infrared rotation spectrum and the structure of this spectrum (see Chapter I), as well as that of the rotation-vibration spectrum (see Chapter IV), lead unambiguously to the conclusion that H_2O is *not a linear molecule*. This conclusion is also in agreement with the structure of the vibration spectrum. The Raman spectrum of water vapor shows one strong line at 3654.5 cm^{-1} [Johnston and Walker (475), Rank, Larsen, and Bordner (716), Bender (135)] which is obviously the frequency of the symmetrical vibration ν_1 (Fig. 25a), since Raman lines of non-symmetrical vibrations are expected to be weak. Since 3654.5 is very similar to the vibration frequency ($\Delta G_{\frac{1}{2}} = 3568.4$) of the OH radical it cannot correspond to the second symmetrical vibration ν_2 which represents essentially a bending of the OH bond. The observation of further Raman shifts of water vapor is doubtful.

The infrared spectrum [Plyler and Sleator (704), Plyler (702)] gives two very strong bands at 1595.0 and 3755.8 cm^{-1} which are very probably the fundamentals ν_2 and ν_3 respectively (see Fig. 25a). The fundamental ν_1 has only recently [34] been disentangled from the much stronger overlapping ν_3 by Nielsen (667). In agreement with the selection rules for non-linear XY_2 (but not for linear XY_2) the first overtone $2\nu_2$ of ν_2 occurs fairly strongly in the infrared. Further evidence from the vibration spectrum alone that the H_2O molecule is not linear is obtained when the HOH angle is determined from the observed fundamental frequencies according to the valence force system. A value of 120° is obtained (see Table 40). The deviation from 180° can hardly be due to the neglected terms in the valence force system.

A large number of overtone bands of H_2O have been observed in the photographic infrared and the visible region as terrestrial bands in the solar spectrum, particularly in a moist atmosphere, and have been analyzed by Mecke and his co-workers (612) (130) (333). The zero lines (see Chapter IV) of all observed bands are given in Table 60. The assignment of quantum numbers is that given by Mecke (612). The lower state for all observed bands is the 0, 0, 0 state.

It is significant that for all but three of the infrared bands the dipole moment oscillates perpendicular to the axis of symmetry (upper state B_1). In particular, while $\nu_3 + \nu_1$ and even $3\nu_3$ are fairly

[33] These data refer to the liquid state.

[34] Before this observation, on the basis of observed fundamentals only, one might have been tempted to conclude that H_2O is linear since, just as for CO_2 and CS_2, two fundamentals were observed only in the infrared and the third only in the Raman spectrum. This shows clearly how dangerous it is to draw conclusions about the structure of a molecule from the non-observation of certain fundamentals. In the present case, the non-observation of $2\nu_3$ and $4\nu_3$ might have been considered as further evidence that H_2O is linear whereas actually it is not.

strong bands, a band $2\nu_3$ whose upper state is A_1 has not been observed even though it is not forbidden by any rigorous selection rule. The same holds for all other bands with $\Delta v_3 = 2, 4, \cdots$. This difference in the behavior of bands with even and odd v_3 (upper state A_1 and B_1 respectively) is,

TABLE 60. INFRARED AND RAMAN BANDS OF H_2O VAPOR.

ν_{vacuum},[35] observed (cm^{-1})	Band type[36]	Upper state[37] $v_1\ v_2\ v_3$	Species	ν, calculated (cm^{-1})	References
1595.0	I. ||[38] v.s.	0 1 0	A_1	1595.0*	(704) (665)
3151.4	I. || m.	0 2 0	A_1	3151.0*	(704) (665)
3651.7[39]	I. || s., R. s.	1 0 0	A_1	3650.0*	(667) (475) (716) (135)
3755.8	I. ⊥[40] v.s.	0 0 1	B_1	3755.41*	(704) (612) (667)
5332.0	I. ⊥ m.	0 1 1	B_1	5330.6	(704) (702) (612) (667)
6874	I. ⊥ w.	0 2 1	B_1	6866.8	(234a)
7251.6	I. ⊥ m.	1 0 1	B_1	7250.4	(702) (612) (667)
8807.05	P.I. ⊥ s.	1 1 1	B_1	8805.5	(612) (591)
10613.12	P.I. ⊥ s.	2 0 1	B_1	10613.12*	(612) (130)
11032.36	P.I. ⊥ m.	0 0 3	B_1	11032.36*	(612) (130)
12151.22	P.I. ⊥ m.	2 1 1	B_1	12148.5 }*	(612) (130)
12565.01	P.I. ⊥ v.w.	0 1 3	B_1	12567.7 }	(612) (130)
13830.92	P.I. ⊥ w.	3 0 1	B_1	13830.92*	(612) (130)
14318.77	P.I. ⊥ w.	1 0 3	B_1	14318.77*	(612) (333)
15347.91	P.I. ⊥ v.w.	3 1 1	B_1	15346.3 }*	(612) (333)
15832.47	P.I. ⊥ v.w.	1 1 3	B_1	15834.1 }	(612) (333)
16821.61	P.I. ⊥ v.w.	3 2 1	B_1	16822.7	(612) (333)
16899.01	P.I. ⊥ v.w.	4 0 1	B_1	16894.3	(612) (333)
17495.48	P.I. ⊥ v.w.	2 0 3	B_1	17482.6	(612) (333)

conversely, a very definite proof that H_2O really has the symmetry C_{2v}, that is, has a plane of symmetry perpendicular to the plane of the molecule.[41] For an unsymmetrical molecule

such a difference could not occur.

Bonner (162) was the first to derive a fairly satisfactory formula of the form (II, 268) for the observed vibrational levels. However, his formula led to a value for the fundamental ν_1 of 3604 cm^{-1}, in rather serious disagreement with the observed Raman frequency 3654.5 cm^{-1}. Recently this difficulty has been cleared up by Darling and Dennison (263), who have taken into account the perturbation between all pairs of levels of the type v_1, v_2, v_3, and $v_1 - 2$, v_2, $v_3 + 2$ (see p. 218).

[35] The frequencies below 8000 cm^{-1}, with the exception of the band 6874 cm^{-1}, are those of Nielsen (665) (667) reduced to vacuum (assuming that he has not reduced them); the other frequencies are those of Mecke and his co-workers (612) (130) (333) which are stated to be ν_{vacuum}.

[36] || and ⊥ refer here to the direction of the variable part of the dipole moment with respect to the symmetry axis. Nielsen (665) (667) uses the opposite nomenclature, taking || and ⊥ to mean parallel and perpendicular to the axis of least moment of inertia.

[37] The lower state for all observed bands is 0, 0, 0 A_1.

[38] A Raman band has been observed by Johnston and Walker (475) in H_2O vapor at 1648 cm^{-1} but has not been confirmed by Rank, Larsen, and Bordner (716). However, such a Raman line has definitely been established in liquid H_2O [see Rao and Koteswaran (722)].

[39] This is the wave number of the zero line of the infrared band. The center of the Raman band of H_2O vapor has been observed at 3654.5 cm^{-1} by Bender (135). Rank, Larsen, and Bordner (716) give a doublet with frequency shifts 3646.1 and 3653.9 cm^{-1}.

[40] Rank (713) gives a doubtful Raman line at 3804 cm^{-1}.

[41] This might appear trivial but should not be taken for granted without actual proof. It follows of course also from the observed intensity alternation in the rotational structure (see p. 475).

They have obtained the following vibrational constants (all in cm^{-1}):

$$\omega_1^0 = 3693.8_9, \quad \omega_2^0 = 1614.5, \quad \omega_3^0 = 3801.7_8,$$
$$x_{11} = -43.8_9, \quad x_{22} = -19.5, \quad x_{33} = -46.3_7, \quad x_{12} = -20.0_2, \qquad \text{(III, 59)}$$
$$x_{13} = -155.0_6, \quad x_{23} = -19.8_1, \quad |\gamma| = 74.4_6,$$

where γ is the perturbation constant introduced by equation (II, 297). [Very slightly different constants have more recently been given by Nielsen (667).] The wave numbers of the bands calculated from these constants are given in the fourth column of Table 60. The values with an asterisk were used in the determination of the constants. In the case of the two pairs indicated by braces only the average was used. The agreement for all bands is seen to be quite satisfactory. For the zero-order frequencies Darling and Dennison obtained, from the above constants,

$$\omega_1 = 3825.32 \text{ cm}^{-1}, \quad \omega_2 = 1653.91 \text{ cm}^{-1}, \quad \omega_3 = 3935.59 \text{ cm}^{-1}, \qquad \text{(III, 60)}$$

and for the zero-point energy:

$$G(0, 0, 0) = 4631.2_5 \text{ cm}^{-1}. \qquad \text{(III, 61)}$$

Using, in addition to the above vibrational constants, the interaction of vibration and rotation, Darling and Dennison (263) have evaluated the coefficients of the cubic and quartic terms in the potential energy [see also Redlich (727)].

Heavy water, HDO and D_2O. The infrared and Raman spectra of heavy water vapor, HDO and D_2O, have also been investigated, although in not nearly as much detail as those of H_2O. The observed data are given in Table 61.

TABLE 61. OBSERVED INFRARED AND RAMAN SPECTRA OF HDO AND D_2O VAPOR.

Assignment	HDO ν_{vacuum} (cm^{-1})	D_2O ν_{vacuum} (cm^{-1})	References
ν_2	1402 I.	1178.7 I	(112)
ν_1	2719 I., R.	2666 R.	(112) (716)
ν_3	[42]	2789 I.	(112) (121) (263)
$2\nu_2$	2809 I.		(121)
$\nu_1+\nu_2+\nu_3$		6538 I.	(675)
$\nu_2+2\nu_3$	8611.6 P.I.		(446)
$2\nu_1+\nu_3$	9050 P.I.		(443)
$\nu_1+2\nu_3$	10000 P.I.		(443)

It has been shown in Chapter II, section 6, that the observed fundamentals of D_2O are in excellent agreement with those to be expected, on the basis of the isotope relations, from those of H_2O if (and only if) anharmonicity corrections are taken into account. Darling and Dennison's (263) final values for the ω_i, x_{ik}, and the zero-point energy of D_2O obtained from the isotope relations and the observed fundamentals are (in cm^{-1}):

$$\omega_1 = 2758.0_6, \quad \omega_2 = 1210.2_5, \quad \omega_3 = 2883.7_9,$$
$$x_{11} = -22.8_1, \quad x_{22} = -10.44, \quad x_{33} = -24.9_0, \quad x_{12} = -10.5_6, \qquad \text{(III, 62)}$$
$$x_{13} = -81.9_2, \quad x_{23} = -10.6_2, \quad G(0, 0, 0) = 3385.7_4.$$

Hydrogen sulfide, H_2S, HDS, D_2S. The vibrational spectra of H_2S, HDS, D_2S are of course very similar to those of H_2O, HDO, and D_2O respectively. However, they have not as yet been so completely studied and the identifications are not as certain. The available data for H_2S are given in Table 62, those for HDS and D_2S in Table 63. The fine structure of the two photographic infrared bands 9911 and 10100 cm^{-1} only has been analyzed (see Chapter IV, p. 489), and they are found to be perpendicular bands. There is considerable divergence of opinion as to the character of the

[42] Overlapped by $\nu_3(H_2O)$. In the Raman spectrum of the liquid Rao (723a) observed 3363 cm^{-1} for this frequency.

other incompletely resolved infrared bands. Because of the incomplete resolution there is also considerable uncertainty as to the position of the origins of the bands. In Tables 62 and 63 the high dispersion data of Sprague and Nielsen (804) and Nielsen and Nielsen (657) have been given preference wherever they are available. The assignments of the bands given are essentially those of

TABLE 62. INFRARED AND RAMAN BANDS OF GASEOUS H_2S.

ν_{vacuum}, observed[43] (cm^{-1})	Band type[13]	Assignment, Bailey-Thompson-Hale (94)	References
1290	I. v.s.	$\nu_2(a_1)$	(804) (94)
2422	I. m.	$2\nu_2(A_1)$	(94)
2610.8	R.	$\nu_1(a_1)$	(647)
2684	I. s.	$\nu_3(b_1)$	(804) (94)
(2910)[44]	I.	?	(94)
3789	I. s.	$\nu_2+\nu_3(B_1)$	(110) (94)
5154	I.[45] m.	$\nu_1+\nu_3(B_1)$[46]	(803) (94)
9911.05	P.I. ⊥ w.	$3\nu_1+\nu_3(B_1)$	(248) (241)
10100	P.I. ⊥ w.	$\nu_1+3\nu_3(B_1)$	(248) (241)

TABLE 63. INFRARED AND RAMAN BANDS OF GASEOUS HDS AND D_2S.

Assignment, Bailey-Thompson-Hale	HDS		D_2S	
	ν_{vacuum}, observed (cm^{-1})	References	ν_{vacuum}, observed (cm^{-1})	References
?	988	(94)		
ν_2	**1090**	(94) (657)	**934**	(94) (657)
$2\nu_2$	2109	(94)		
ν_1			**1891.6**	(647)
ν_3	(2684)	47	**1999**	(94) (657)
$\nu_1+\nu_2$	2937	(94) (657)	2684[47]	(657)
$\nu_2+\nu_3$	3723	(94)	2797[47]	(94)
$2\nu_1$	3848	(94)		

Bailey, Thompson, and Hale (94), which appear to be preferable to those of Sprague, Nielsen, and Nielsen since they preserve a close analogy to H_2O (absence of transitions with $\Delta v_3 = 0$ and 2 except for ν_2 and $2\nu_2$) and, in the case of the photographic infrared bands, give a species of the upper state in agreement with the fine structure analysis.

---

[13] On p. 274.

[43] It has been assumed that the original infrared data were not corrected for vacuum.

[44] This band has only been observed by Bailey, Thompson, and Hale (94) and not identified.

[45] Sprague and Nielsen do not mention this band in their 1937 paper, although they gave it in an earlier abstract (803).

[46] Bailey, Thompson, and Hale (94) assigned this band to $2\nu_1$, which is an unnecessary contradiction to the analogy to the H_2O spectrum.

[47] It is strange that the two fairly strong bands 2684 and 2797 cm^{-1}, lying in the same spectral region, have not been found by both groups of investigators. It is not unlikely that the band 2684 is really ν_3 of HDS, since it would be expected to have about this magnitude and since it would therefore be observable only in a gas not containing an appreciable amount of H_2S whose ν_3 is at the same place. This change of interpretation would also remove the necessity of assuming $\nu_1 + \nu_2$ to occur for D_2S whereas it does not occur for H_2S or H_2O.

As has been pointed out in Chapter II, section 6, the frequencies ν_1 and ν_2 of H_2S and D_2S give quite a fair agreement with the product rule (see Table 52 and accompanying discussion). However, the frequencies ν_3 of these two molecules give a discrepancy that is rather larger than permitted by the neglect of anharmonicity. It seems likely that the reason for this discrepancy is that the band origin has not been correctly determined (since no rotational analysis has been made).

Nitrogen peroxide, NO_2. Our knowledge of the vibrational structure of the NO_2 molecule is not very satisfactory as yet, particularly because it is practically impossible to observe the Raman spectrum since NO_2 absorbs light throughout the visible and ultraviolet regions. The infrared spectrum shows two very strong absorption bands at 648 and 1621 cm^{-1}, which undoubtedly must be interpreted as the fundamentals ν_2 and ν_3 respectively [Sutherland (825)]. The difficulty is the correct location of the fundamental ν_1. Schaffert (769) found a weak band at 1373 cm^{-1} which he and Sutherland (825) considered for a while as ν_1. However, the study of the ultraviolet spectrum by Harris, Benedict, and King (411) seems to show that $\nu_1 = 1320$ cm^{-1}. Various authors have adopted this latter value,[48] even though more recently in a more detailed paper Harris, King, Benedict,

TABLE 64. INFRARED VIBRATION SPECTRUM OF GASEOUS NO_2.

ν, observed (cm^{-1})	Assignment	References
648 s.	$\nu_2(a_1)$	(84) (769)
1000 v.w.	$\nu_3 - \nu_2$	(825)
1320 from U.V.	$\nu_1(a_1)$?	(411) (414)
1373 w.	$(2\nu_2??)$	(769)
1621 v.s.	$\nu_3(b_1)$	(84) (769) (825)
2220 m.	$\nu_3+\nu_2(B_1)$	(825)
2601 } [49] 2667 }	$2\nu_1(A_1)$	(412)
2910	$\nu_3+\nu_1(B_1)$	(825) (769) (412)
3240	$2\nu_3(A_1)$[50]	(825) (412)
3454 } 3597 }	$\nu_1+\nu_2+\nu_3(B_1)$	(412)
3930	?	(412)
4140	$2\nu_1+\nu_3(B_1)$	(412)
4560	$\nu_1+2\nu_3(A_1)$	(412)
4753	$3\nu_3(B_1)$	(412)

and Pearse (414) try to explain, in a rather artificial way, the ultraviolet spectrum with $\nu_1 = 1373$ cm^{-1}. It appears that a final decision will only be possible after further investigations of the infrared or Raman spectrum. To the author, $\nu_1 = 1320$ cm^{-1} appears somewhat more likely. The infrared band 1373 might possibly be $2\nu_2$ where the large deviation from 2×648 cm^{-1} might be due to a vibration through the linear arrangement of the nuclei for large amplitudes. It may also be that the interpretations of the frequencies 1320 and 1373 cm^{-1} have to be interchanged.

In Table 64 all the observed infrared bands of NO_2 are collected together, the assignment being mostly that of Sutherland and Penney (831). In this as well as any other assignment the predicted structure of the infrared bands does not always seem to agree with the observed envelopes. However, this difficulty does not appear to be serious in view of the very incomplete resolution. The occurrence of the harmonic $2\nu_3$ definitely proves that the molecule *cannot be linear*, as was at one time suggested, on the basis of the infrared spectrum, by Bailey and Cassie (84). The non-linearity

[48] For example, Sutherland and Penney (831) adopt this value, even though in an immediately preceding paper (692) they give $\nu_1 = 1370$. [See also Wu (26)]. In the former paper the infrared band 1373 is not mentioned as an overtone or combination band.

[49] This doublet occurs in the ultraviolet spectrum as the frequency 2624 cm^{-1}. [Harris, King, Benedict, and Pearse (414)].

[50] In his original paper Sutherland gives 3120 cm^{-1} as $2\nu_3$ and a very weak band at 3240, whereas in Sutherland and Penney's summary only 3242 is given.

is confirmed by the fact that the valence force system with the above fundamentals gives a valence angle of 119° (see Table 40) and by the structure of the ultraviolet bands [see Harris and King (413)]. Electron diffraction experiments [Maxwell and Mosley (607)] also prove the non-linear configuration and yield an angle of 130°.

Sulfur dioxide, SO_2. The Raman spectrum of gaseous SO_2 shows one strong line only, at 1150.5 cm^{-1} [Gerding and Nijveld (349)]. The liquid, for which a greater intensity of the spectrum can be obtained, shows in addition two weaker lines at 524.5 and 1336.0 cm^{-1}. These three frequencies must be considered as the fundamentals of SO_2. The fact that all three appear in the Raman spectrum proves that SO_2 is *not a symmetrical linear molecule.*[51] According to Cabannes and Rousset (191), the Raman line 1336.0 cm^{-1} is completely depolarized, whereas the other two are partly polarized, that is, 1336.0 cannot correspond to a totally symmetric vibration (see p. 270). If SO_2 were an unsymmetric linear molecule (O——S—O or O—O—S), the only non-totally symmetric frequency would be the perpendicular vibration, whose frequency would be expected to be the smallest of the three fundamentals. Since the depolarized frequency is the largest, it follows that SO_2 is not linear and since a non-linear triatomic molecule can have a non-totally symmetric vibration only if it has a plane of symmetry perpendicular to the plane of the molecule, it follows that SO_2 is a *non-linear symmetric molecule* (Fig. 25a). Its only non-totally symmetric vibration must accordingly have the frequency $\nu_3 = 1336.0$, while $\nu_1 = 1150.5$ and $\nu_2 = 524.5$ cm^{-1}.

The above conclusion about the structure of SO_2 obtained from the Raman spectrum alone is confirmed by the infrared spectrum, which shows all three fundamentals strongly. Table 65 gives the observed Raman and infrared bands with the assignment of Mecke (611), [see also Badger and Bonner (73)].

TABLE 65. INFRARED AND RAMAN BANDS OF SO_2.

ν_{vacuum}[52] (cm^{-1})	Band type[13]	Assignment, Mecke (611)	References
519[52a](524.5)	I. (gas), R. (liquid) pol.	$\nu_2(a_1)$	(107a) (349)
606	I. (gas)	$\nu_1 - \nu_2$	(86)
1151.2[52c]	I. (gas), R. (gas) pol.	$\nu_1(a_1)$	(86) (107a) (349)
1361[52b](1336.0)	I. (gas), R. (liquid) depol.	$\nu_3(b_1)$	(86) (107a) (349)
1871	I. (gas)	$\nu_2 + \nu_3(B_1)$	(86)
2305	I. (gas)	$2\nu_1(A_1)$	(86)
2499	I. (gas)	$\nu_1 + \nu_3(B_1)$	(86)

From the above fundamentals an apex angle $2\alpha = 120°$ is obtained, assuming a valence force system (Table 40). This agrees rather well with the electron diffraction value 121° ± 5° obtained recently by Schomaker and Stevenson (771). When such a value for the angle is assumed, there are some difficulties as to the contours of the infrared bands [see Badger and Bonner (73)], but since no high dispersion data are available, this does not appear to be serious.

Ozone, O_3. Great difficulties have been experienced in the investigation of the infrared spectrum of ozone, mainly owing to its reactivity and the consequent difficulty in getting entirely rid of impurities. Thus two bands (at 880 and 1355), for a long time considered to be due to O_3 (one even as a fundamental), are actually due to N_2O_5. However, due to the work of Hettner, Pohlman, and Schumacher (449), the main infrared bands are now definitely identified. Their absorption curve

[51] That is, unless one assumes that the weaker lines in the liquid occur only because of the perturbation by the neighboring molecules. This assumption is, however, excluded by the evidence from the infrared spectrum of the gas (see below).

[13] On p. 274.

[52] The data in parentheses refer to observations of liquid SO_2.

[52a] This value was read from the absorption curve given by Barker (107a).

[52b] This is the value given by Bailey, Cassie, and Angus (86). From Barker's curves one reads a value of 1358 cm^{-1}.

[52c] Average of infrared and Raman measurements.

is reproduced in Fig. 85. The wave numbers of the observed bands (mostly from their measurements) are given in Table 66. The Raman spectrum has not yet been obtained [see Sutherland and Gerhard (829)].

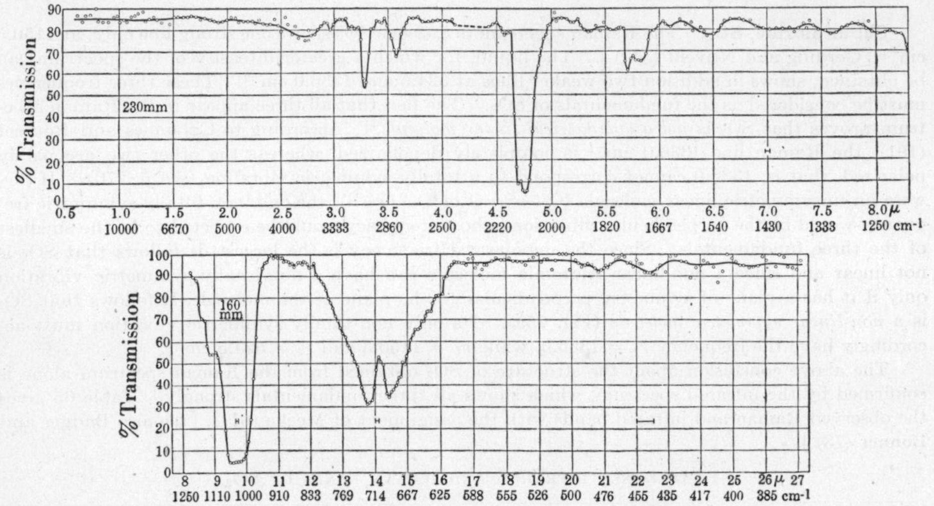

FIG. 85. **Infrared absorption of O_3 [after Hettner, Pohlmann and Schumacher (449)].**—The length of the absorbing path was 30 cm. at the pressures indicated.

TABLE 66. INFRARED SPECTRUM OF GASEOUS OZONE.

ν_{vacuum}, observed (cm⁻¹)	Band type	Assignment, Sutherland-Penney (831)	References
695, 725 } 710	I. doublet s.	$\nu_2(a_1)$	(449)
1043.4[43]	I. doublet? v.s.	$\nu_1(a_1)$	(351) (449) (40)
1724, 1755 } 1740	I. doublet? w.	$\nu_3(b_1)$	(449)
2105	I. ? s.	$2\nu_1(3\nu_2)(A_1)$	(351) (449)
2800	I. ? w.	$2\nu_1+\nu_2(A_1)$	(449)
3050	I. ? w.	$3\nu_1(A_1)$	(449)

If the nuclei in the O_3 molecule were at the corners of an equilateral triangle (point group D_{3h}) there would be only two normal vibrations, a totally symmetric one and a doubly degenerate one (see Fig. 32a), only the latter being infrared active (see Table 55). It is easily seen from the observed frequencies in Table 66 that it is quite impossible to interpret the observed spectrum on the basis of one active and one inactive fundamental only. Thus the equilateral model of O_3 is definitely ruled out.

If O_3 had a linear symmetric structure no binary combinations of active bands could occur (see p. 265), whereas actually a number of such combinations seem to occur. The rotational structure of the 1043.4 band [Adel, Slipher, and Fouts (40)] also definitely rules out the linear symmetrical model as well as the non-symmetrical linear model. Thus only an *isosceles triangle* or a completely unsymmetrical structure remains. The latter structure does not appear to be at all likely for a molecule consisting of three equal atoms. In both cases, there should be three active fundamentals. Choosing the three strongest bands 710, 1043.4, and 2105 cm⁻¹ for these, as was done by Hettner,

[43] On p. 283.

Pohlman, and Schumacher (449), leads to an impossibly large force constant between two of the O atoms, almost double the force constant in the O_2 molecule, as well as to other difficulties. This led Sutherland and Penney (831) to the choice indicated in Table 66, which does give reasonable force constants. However, a difficulty is the weakness of the fundamental band ν_3, which in other non-linear triatomic molecules is strong compared to ν_1. The apex angle on Sutherland and Penney's interpretation would come out from the valence force formulae to be about 127°, which is of the same order as for SO_2. This is satisfactory, since SO_2 and O_3 have the same number of outer electrons. Recently Shand and Spurr (783) have obtained from the electron diffraction pattern of ozone an apex angle of 127° ± 3°. The exact agreement with the above value should not be considered too significant.

The readiness with which O_3 gives off an O atom (in contrast to SO_2) would be much more easily understandable on the basis of an acute-angled model in which one O is at a fairly large distance from an almost unchanged O_2 molecule. It would indeed be possible to interpret the infrared spectrum on this basis. But the above-mentioned electron diffraction work would appear to exclude such a structure.

Further work on the infrared spectrum with high dispersion and longer absorbing columns would be desirable in order to clear up the difficulties mentioned and to obtain more precise values for the angle and the internuclear distances.

Other triatomic molecules. The vibration spectra of a number of other triatomic molecules have been investigated. The fundamentals of some of these have been included in the previous Table 37. In Table 67 references to the more important investigations of these molecules are given and the structures found are indicated. It is particularly noteworthy, and of course very plausible on the basis of the electronic structure, that these investigations, as well as those discussed above,

TABLE 67. FURTHER TRIATOMIC MOLECULES.[53]

Molecule	Structure	Apex angle	References: Raman spectrum	References: Infrared spectrum
OCS	linear, $C_{\infty v}$	180°	(257) (909) (314a)	(81) (125)
N_3^-	linear, $D_{\infty h}$	180°[54]	(557)	(77) (831)
$(BO_2)^-$	linear, $D_{\infty h}$	180°	(669)	
OCN^-, SCN^-, $SeCN^-$	linear, $C_{\infty v}$	180°	(557) (216) (394) (535)	(926) (927)
ClCN, BrCN, ICN	linear, $C_{\infty v}$	180°	(920) (909)	
$HgCl_2$, $HgBr_2$, HgI_2	linear, $D_{\infty h}$	180°	(176) (723)	(913) (801)[55]
HgClBr, HgClI, HgBrI	linear, $C_{\infty v}$	180°	(274a) (330a)	(640a)[55]
$CdCl_2$, $CdBr_2$, CdI_2	linear, $D_{\infty h}$	180°	(889)	
$ZnCl_2$, $ZnBr_2$	linear, $D_{\infty h}$	180°	(889) (760)	
$(HF_2)^-$	linear? $C_{\infty v}$			(187) (499)
H_2Se, HDSe, D_2Se	isosceles triangle, C_{2v}	~90°	(253)	(193)
$(NO_2)^-$	isosceles triangle, C_{2v}	~90°	(559)	(925)
NOCl	triangle, C_s	116°[56]		(85) (133)
ClO_2	isosceles triangle, C_{2v}	137°[57]	(544)	(81) (82) (543)[55]
F_2O	isosceles triangle, C_{2v}	100°[57]		(450) (831)
Cl_2O	isosceles triangle, C_{2v}	115°[57]		(83) (831) (706)
$(UO_2)^{++}$	isosceles triangle, C_{2v}	?	(227) (762a)	(227)

[53] Some of these molecules are discussed more fully by Wu (26).
[54] Confirmed by crystal structure investigations of Hendricks and Pauling (427).
[55] Ultraviolet absorption spectra.
[56] From electron diffraction by Ketelaar and Palmer (500).
[57] From electron diffraction, see Brockway (179).

show that molecules with the same number of outer electrons (that is, in addition to complete inner shells unaffected by the molecule formation) have very similar structures. Thus, CO_2, N_2O, N_3^-, BO_2^-, CS_2, COS, $\cdots$ are linear; O_3, SO_2, SeO_2, $(NO_2)^-$, $\cdots$ are isosceles triangles.

A number of more-than-triatomic molecules may in a certain approximation be considered as three-particle systems. For example, dimethylamine may be considered as consisting of the three particles CH_3, NH, CH_3 which for the so-called *skeletal vibrations* behave like the nuclei in a non-linear symmetric molecule XY_2. Three of the many fundamentals of this molecule, namely 931, 390, and 1073 cm^{-1}, may indeed be correlated to ν_1, ν_2, ν_3, respectively, of XY_2 (Fig. 25a), since they have a smaller magnitude than the internal vibrations of each group. As for XY_2, ν_3 is depolarized in the Raman spectrum. Assuming a valence force system the C—N—C angle can be determined from these frequencies to be 114°. Many similar cases may be found in Kohlrausch's book (14).

(b) *Four-atomic molecules*

Acetylene, C_2H_2. A great deal of work has been done on the Raman spectrum and particularly on the infrared spectrum of the acetylene molecule. Both the vibrational structure and the rotational structure show unambiguously that the molecule is *linear and symmetrical* (point group $D_{\infty h}$).

C_2H_2 in the gaseous state shows only two strong Raman displacements, at 1973.8 and 3373.7 cm^{-1} [Glockler and Morrell (377)], which do not occur as infrared bands even with a fairly long absorbing path. This proves that C_2H_2 must have a *center of symmetry*. Assuming the linear model, there can be no doubt that the two Raman lines correspond to the vibrations $\nu_2(\sigma_g^+)$ and $\nu_1(\sigma_g^+)$ respectively in Fig. 64a, that is, essentially to a C—H and a C≡C vibration.[58] In addition, Bhagavantam and Rao (150) have found a very weak Raman doublet at 589 and 646 cm^{-1} which they interpret as the two branches (see Chapter IV, section 1) of the third Raman active vibration $\nu_4(\pi_g)$ (see Table 55), which appears only weakly because it is not totally symmetric.

In the infrared, C_2H_2 shows two very strong absorption bands at 729.1 and 3287 cm^{-1} which are naturally interpreted as the two infrared-active fundamentals $\nu_5(\pi_u)$ and $\nu_3(\sigma_u^+)$ respectively. This is also in agreement with the type of the band fine structure, the former band being a perpendicular, the latter a parallel band (see Chapter IV). A third fairly strong infrared band of the parallel type occurs at 1328.1 cm^{-1}. It is obviously to be interpreted as the combination $\nu_4 + \nu_5$ of the two perpendicular vibrations. The state $v_4 = 1$, $v_5 = 1$ has three sublevels Σ_u^+, Σ_u^-, Δ_u (see Table 33), the first of which gives rise to the parallel band at 1328.1 cm^{-1}. Many more bands, all of them weak, have been found with longer absorbing paths, both in the ordinary infrared and the photographic infrared (see Table 68). The odd overtones $2\nu_5$, $4\nu_5$, $2\nu_3$, $4\nu_3$ of the active vibrations are definitely absent, and the same holds for binary, quaternary, $\cdots$ combinations of infrared-active fundamentals, such as $\nu_3 + \nu_5$, $3\nu_3 + \nu_5$, $\cdots$, in excellent agreement with expectation on the assumption of the linear symmetric model.

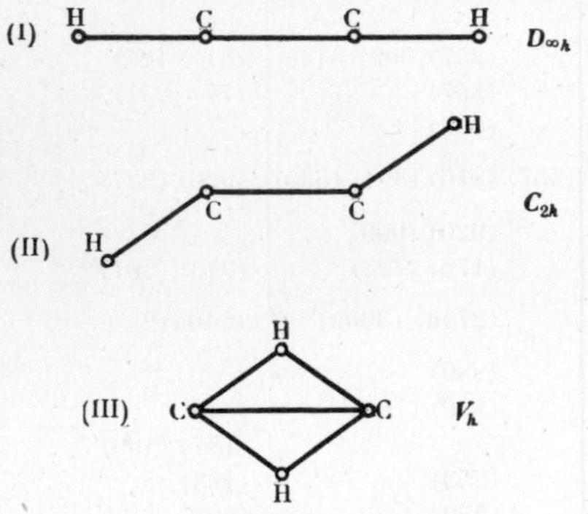

FIG. 86. **Different conceivable models of C_2H_2.**

It ought to be pointed out that the validity of the rule of mutual exclusion of the infrared and Raman spectra and the absence of the odd overtones of the active fundamentals, as well as of other binary and quaternary combinations of infrared-active fundamental bands, would also be compatible with a non-linear model of C_2H_2, as long as it has a center of symmetry. For example, the forms II (point group C_{2h}) and III (point group $D_{2h} = V_h$) in Fig. 86 have a center of symmetry. In form II there would be three vibrations of the totally symmetric species A_g, one of species A_u, and two of species B_u (see Table 35). Thus, as for the linear symmetric model I, there would be three Raman-active vibrations which are infrared inactive. But

[58] It should be noted that for the sake of consistency with our usual designation of fundamentals we have changed the numbering as compared to that used by most authors on C_2H_2. Our ν_1, ν_2, ν_4, ν_5, are their ν_2, ν_1, ν_5, ν_4, respectively.

for this model II there would be three infrared-active fundamentals, whereas only two are observed.[59] For model III there would be two vibrations of type A_g and one vibration each of types B_{1g}, B_{1u}, B_{2u}, B_{3u} (see Table 35). Again there would be three Raman-active and three infrared-active vibrations, whereas only two infrared-active vibrations are observed.[60] Thus, on the basis of the pure vibration spectrum only, both cases II and III are excluded by the fact that only two infrared-active fundamentals are observed. While it may be argued that possibly the third infrared-active fundamental for models II and III is weak, or lies in a region not investigated, or happens to coincide with another band, the rotational fine structure establishes the linear symmetrical structure of C_2H_2 beyond all doubt (see Chapter IV, section 1).

The assignment of the large number of overtone and combination bands has been the subject of much discussion. For every assignment thus far suggested, if the ω_i^0, x_{ik}, and g_{ii} in (II, 284) [compare also (II, 288)] are determined from part of the bands, rather large discrepancies occur between the observed frequencies of at least some bands (not used in the determination of the constants) and the values calculated from the formula. On the basis of the recent work of Darling and Dennison (263) on H_2O (see above), it now seems certain that these discrepancies are due mainly to a mutual perturbation of levels v_1, v_3, and $v_1 + 2$, $v_3 - 2$ (the other quantum numbers being equal), which is expected to be fairly strong since ν_1 and ν_3 have very similar magnitudes. However, new calculations taking account of this have not as yet been completed.

In Table 68, all infrared and Raman bands are given. The assignment that has been adopted in this table is that of Herzberg and Spinks (441) as recently modified by Wu (26). Neglecting resonance effects, this assignment gives a fair representation of all the bands below 10000 cm^{-1}, whereas for three or four bands above 10000 cm^{-1} rather large deviations occur. Another assignment recently proposed by Wu and Kiang (963), which is an extension of the earlier one of Sutherland (824), gives in general about as good an agreement, but it fails rather badly (by 40 cm^{-1}) for the accurately measured band 8512.1 cm^{-1}, whose upper state is not one of a resonating pair. Also, in this assignment, $2\nu_1 + \nu_3$ is considerably more intense than $3\nu_3$, which does not appear to be very plausible, and finally the anharmonic term x_{33} comes out much smaller than one would expect for a C—H vibration. Still another assignment has been given by Mecke and Ziegler (618). It is based on Mecke's treatment of a system containing two identical bonds without the use of normal coordinates. The apparently excellent agreement of this assignment with the observed data has been shown to be due, at least in part, to a mistake in sign [Childs and Jahn (206); see also Wu (26)]. Since all sets of constants ω_i^0, x_{ik}, g_{ii}, and ω_i thus far derived do not take the effect of resonance into account, we refrain from giving such a set, but refer to the papers by Herzberg and Spinks (441) and Wu (961). It seems that for an accurate representation of the higher overtone bands in the visible region the introduction of cubic coefficients is necessary. For an accurate determination of the vibrational constants, new infrared measurements in the region from 3000–7000 cm^{-1} with high dispersion would be very desirable.

Heavy acetylene, C_2HD and C_2D_2. The spectra of the heavy acetylenes have been investigated by a number of investigators, but not in as much detail as that of C_2H_2. Only for C_2HD have overtone and combination bands in the photographic infrared been observed. Tables 69 and 70 give the observed data and the (fairly obvious) assignment of the bands.

Assuming the simple harmonic oscillator approximation, Colby (225) has predicted the fundamental frequencies of C_2HD and C_2D_2 from those of C_2H_2 and has obtained satisfactory agreement. Instead of giving his data, we apply the Teller-Redlich product rule as a check for the assignment of the fundamentals. Comparing C_2H_2 and C_2HD, we obtain from (II, 313), using the point group $C_{\infty v}$ which is common to both,

$$\frac{(\omega_1\omega_2\omega_3)_{C_2HD}}{(\omega_1\omega_2\omega_3)_{C_2H_2}} = \sqrt{\frac{m_H(2m_C + m_H + m_D)}{m_D(2m_C + 2m_H)}},$$
$$\frac{(\omega_4\omega_5)_{C_2HD}}{(\omega_4\omega_5)_{C_2H_2}} = \sqrt{\frac{m_H(2m_C + m_H + m_D)}{m_D(2m_C + 2m_H)}\,\frac{I_{C_2HD}}{I_{C_2H_2}}}. \qquad \text{(III, 63)}$$

[59] Another difference in the spectrum, apart from the rotational fine structure, would be that one Raman line should be depolarized for the linear model, whereas all three should be partly polarized for model II. But the degree of polarization of the weak Raman line of C_2H_2 has not been measured.

[60] As for the linear model, one of the Raman active vibrations would be depolarized.

TABLE 68. INFRARED AND RAMAN SPECTRA OF GASEOUS C_2H_2.

ν_{vacuum}, observed[61] (cm^{-1})	Band type[62]	Upper state[63] $v_1\ v_2\ v_3\ v_4^{l_4}\ v_5^{l_5}$	Upper state[63] Species	Lower state[63] $v_1\ v_2\ v_3\ v_4^{l_4}\ v_5^{l_5}$	Lower state[63] Species	References
611.8[64]	R. v.w.	$0\ 0\ 0\ 1^1\ 0$	Π_g	$0\ 0\ 0\ 0^0\ 0^0$	Σ_g^+	(150) (151) (342)
729.1	I. $\perp$ v.s.	$0\ 0\ 0\ 0\ 1^1$	Π_u	$0\ 0\ 0\ 0^0\ 0^0$	Σ_g^+	(574) (419)
1328.1	I. $\|$ s.	$0\ 0\ 0\ 1^1\ 1^1$	Σ_u^+	$0\ 0\ 0\ 0\ 0^0$	Σ_g^+	(574) (419)
1956	I. $\perp$ w.	$0\ 0\ 0\ 2^0\ 1^1$	Π_u	$0\ 0\ 0\ 0^0\ 0^0$	Σ_g^+	(812)
1973.8	R. v.s.	$0\ 1\ 0\ 0\ 0$	Σ_g^+	$0\ 0\ 0\ 0^0\ 0^0$	Σ_g^+	(377)
2215	I. $\|$? w.	$?0\ 0\ 0\ 0\ 3^1$	Π_u	$0\ 0\ 0\ 0^0\ 0^0$	Σ_g^+	(812)
2701.5	I. $\perp$ m.	$0\ 1\ 0\ 0\ 1^1$	Π_u	$0\ 0\ 0\ 0^0\ 0^0$	Σ_g^+	(574) (611)
3287[65]	I. $\|$ v.s.	$0\ 0\ 1\ 0\ 0$	Σ_u^+	$0\ 0\ 0\ 0^0\ 0^0$	Σ_g^+	(574) (441)
3294?	I. $\|$ w.	$0\ 1\ 0\ 1^1\ 1^1$	Σ_u^+	$0\ 0\ 0\ 0^0\ 0^0$	Σ_g^+	(618)
3373.7	R. s.	$1\ 0\ 0\ 0\ 0$	Σ_g^+	$0\ 0\ 0\ 0^0\ 0^0$	Σ_g^+	(377)
3881	I. $\perp$ w.	$0\ 1\ 0\ 2^0\ 1^1$	Π_u	$0\ 0\ 0\ 0^0\ 0^0$	Σ_g^+	(574)
3897	I. $\perp$ m.	$0\ 0\ 1\ 1^1\ 0$	Π_u	$0\ 0\ 0\ 0^0\ 0^0$	Σ_g^+	(574) (611)
4091	I. $\perp$ m.	$1\ 0\ 0\ 0\ 1^1$	Π_u	$0\ 0\ 0\ 0^0\ 0^0$	Σ_g^+	(574) (611)
(4690)[66]	? I. ?	$1\ 0\ 0\ 1^1\ 1^1$	Σ_u^+	$0\ 0\ 0\ 0^0\ 0^0$	Σ_g^+	(609)
(5250)[66]	? I. ?	$0\ 1\ 1\ 0\ 0$	Σ_u^+	$0\ 0\ 0\ 0^0\ 0^0$	Σ_g^+	(609)
(6500)[66]	? I. ?	$1\ 0\ 1\ 0\ 0$	Σ_u^+	$0\ 0\ 0\ 0^0\ 0^0$	Σ_g^+	(609)
8512.1	P.I. $\|$ s.	$1\ 1\ 1\ 0\ 0$	Σ_u^+	$0\ 0\ 0\ 0^0\ 0^0$	Σ_g^+	(441)
9085	P.I. $\perp$ w.	$1\ 1\ 1\ 1^1\ 0$	Π_u	$0\ 0\ 0\ 0^0\ 0^0$	Σ_g^+	(341) (339) (342)
9151.7	P.I. $\|$ m.	$0\ 3\ 1\ 0\ 0$	Σ_u^+	$0\ 0\ 0\ 0^0\ 0^0$	Σ_g^+	(341) (339)
9177	P.I. $\perp$ v.w.	$0\ 1\ 2\ 0\ 1^1$	Π_u	$0\ 0\ 0\ 0^0\ 0^0$	Σ_g^+	(342)
9366	P.I. $\perp$ v.w.	$?2\ 1\ 0\ 0\ 1^1$	Π_u	$0\ 0\ 0\ 0^0\ 0^0$	Σ_g^+	(341) (339)
9639.8	P.I. $\|$ v.s.	$0\ 0\ 3\ 0\ 0$	Σ_u^+	$0\ 0\ 0\ 0^0\ 0^0$	Σ_g^+	(441)
9667.9	P.I. $\|$ m.	$1\ 1\ 1\ 2^0\ 0$	Σ_u^+	$0\ 0\ 0\ 0^0\ 0^0$	Σ_g^+	(441) (339)
9744.6	P.I. $\|$ w.	$0\ 1\ 2\ 1^1\ 1^1$	Σ_u^+	$0\ 0\ 0\ 0^0\ 0^0$	Σ_g^+	(342)
9835.1	P.I. $\|$ s.	$2\ 0\ 1\ 0\ 0$	Σ_u^+	$0\ 0\ 0\ 0^0\ 0^0$	Σ_g^+	(441)
9905.7	P.I. $\|$ w.	$1\ 1\ 1\ 0\ 2^0$	Σ_u^+	$0\ 0\ 0\ 0^0\ 0^0$	Σ_g^+	(342)
10364.8	P.I. $\perp$ m.	$1\ 0\ 2\ 0\ 1^1$	Π_u	$0\ 0\ 0\ 0^0\ 0^0$	Σ_g^+	(341) (339) (342) (618)
10413.5	P.I. $\perp$ v.w.	$2\ 0\ 1\ 1^1\ 0$	Π_u	$0\ 0\ 0\ 0^0\ 0^0$	Σ_g^+	(342) (618)
11570.7	P.I. $\|$ m.	$?1\ 2\ 1\ 2^0\ 0$	Σ_u^+	$0\ 0\ 0\ 0^0\ 0^0$	Σ_g^+	(340)
11586.4	P.I. $\|$ m.	$?1\ 3\ 0\ 3^1\ 1^1$	Σ_u^+	$0\ 0\ 0\ 0^0\ 0^0$	Σ_g^+	(340)
11600.1	P.I. $\|$ m.	$0\ 1\ 3\ 0\ 0$	Σ_u^+	$0\ 0\ 0\ 0^0\ 0^0$	Σ_g^+	(340)
11663.3	P.I. $\|$ w.	$0\ 2\ 2\ 1^1\ 1^1$	Σ_u^+	$0\ 0\ 0\ 0^0\ 0^0$	Σ_g^+	(342)
11782.9	P.I. $\|$ m.	$2\ 1\ 1\ 0\ 0$	Σ_u^+	$0\ 0\ 0\ 0^0\ 0^0$	Σ_g^+	(418) (342)
12675.7	P.I. $\|$ s.	$1\ 0\ 3\ 0\ 0$	Σ_u^+	$0\ 0\ 0\ 0^0\ 0^0$	Σ_g^+	(419) (340)
12711.0	P.I. $\|$ w.	$0\ 1\ 3\ 2^0\ 0$	Σ_u^+	$0\ 0\ 0\ 0^0\ 0^0$	Σ_g^+	(341) (339) (342)

[61] Wherever available, band origins calculated from the fine structure are given (see Chapter IV).

[62] The intensity estimates are only relative for comparison of neighboring bands. For explanation of symbols see footnote 13, p. 274.

[63] The numbering of the vibrations is that used in Fig. 64a; see also footnote 58, p. 288.

[64] This figure is the difference of the bands at 10413.5 and 9801.7 as well as at 13230.3 and 12618.5. In the Raman spectrum Bhagavantam and Rao (150) (151) have observed a very weak doublet at 589 and 646 cm^{-1} whose center (618 cm^{-1}) is in sufficiently close agreement with the above more accurate value. The two doublet components represent the unresolved *S*, *R* and *P*, *O* branches respectively whereas the *Q* branch, in agreement with the theory [Placzek and Teller (701); see also p. 399], is too weak to be observed. See, however, Glockler and Renfrew (380).

[65] This value is not very accurate because in consequence of overlapping by another band the zero line cannot be accurately determined. From the difference band 2669.0 cm^{-1} one would obtain 3280.8 cm^{-1} for ν_3 [see Mecke and Ziegler (618)].

[66] Measured with very small dispersion.

TABLE 68.—*Continued.*

ν_{vacuum}, observed[61] (cm^{-1})	Band type[62]	Upper state[63]		Lower state[63]		References
		$v_1\ v_2\ v_3\ v_4^{l_4}\ v_5^{l_5}$	Species	$v_1\ v_2\ v_3\ v_4^{l_4}\ v_5^{l_5}$	Species	
12732.7	P.I. ‖ w.	1 3 0 5^1 1^1	Σ_u^+	0 0 0 0^0 0^0	Σ_g^+	(342)
13033.3	P.I. ‖ m.	0 1 3 0 2^0	Σ_u^+	0 0 0 0^0 0^0	Σ_g^+	(341) (339) (342)
13230.3	P.I. ⊥ w.	1 0 3 1^1 0^0	Π_u	0 0 0 0^0 0^0	Σ_g^+	(342) (618)
13532.4	P.I. ‖ w.	0 2 3 0 0	Σ_u^+	0 0 0 0^0 0^0	Σ_g^+	(342)
14597.1	P.I. ‖ w.	0 2 3 2^0 0	Σ_u^+	0 0 0 0^0 0^0	Σ_g^+	(342)
14617.0	P.I. ‖ m.	1 1 3 0 0	Σ_u^+	0 0 0 0^0 0^0	Σ_g^+	(341) (339)
15081	P.I. ⊥ v.w.	?1 1 3 1^1 0	Π_u	0 0 0 0^0 0^0	Σ_g^+	(342)
15600.2	P.I. ‖ m.	0 0 5 0 0	Σ_u^+	0 0 0 0^0 0^0	Σ_g^+	(418) (342)
17518.8	P.I. ‖ w.	0 1 5 0 0	Σ_u^+	0 0 0 0^0 0^0	Σ_g^+	(342)
18088	P.I. ⊥ v.w.	0 1 5 1^1 0	Π_u	0 0 0 0^0 0^0	Σ_g^+	(342)
18430.2	P.I. ‖ w.	1 0 5 0 0	Σ_u^+	0 0 0 0^0 0^0	Σ_g^+	(418)
(716)[67]	I. ⊥ w.	0 0 0 1^1 1^1	$\Sigma_u^+\ \Sigma_u^-\ \Delta_u$	0 0 0 1^1 0	Π_g	(574) (618)
2642.5	I. ⊥ w.	1 0 0 0 0	Σ_g^+	0 0 0 0 1^1	Π_u	(574) (611)
2669.0	I. ⊥ w.	0 0 1 0 0	Σ_u^+	0 0 0 1^1 0	Π_g	(574) (611)
2682.3	I. ⊥ w.	?0 1 0 1^1 1^1	$\Sigma_u^+\ \Sigma_u^-\ \Delta_u$	0 0 0 1^1 0	Π_g	(574) (441)
4076	I. ⊥ v.w.	1 0 0 1^1 1^1	$\Sigma_u^+\ \Sigma_u^-\ \Delta_u$	0 0 0 1^1 0	Π_g	(574) (25)
9602.7	P.I. ‖ m.	0 0 3 1^1 0	Π_u	0 0 0 1^1 0	Π_g	(441) (339) (340)
9801.7	P.I. ‖ w.	2 0 1 1^1 0	Π_u	0 0 0 1^1 0	Π_g	(341) (339)
12618.5	P.I. ‖ w.	1 0 3 1^1 0	Π_u	0 0 0 1^1 0	Π_g	(341) (339) (340)
15521	P.I. ‖ v.w.	0 0 5 1^1 0	Π_u	0 0 0 1^1 0	Π_g	(342)

Substituting for the ω_i the frequencies ν_i of the observed fundamentals (1–0 transitions), we can, of course, not expect exact agreement (see p. 232). We obtain for the left- and right-hand sides of the first equation 0.7288 and 0.7207, and for the second equation substituting the moments of inertia given in Chapter IV (p. 396), 0.7944 and 0.7854. Comparing C_2H_2 and C_2D_2, we obtain from (II, 313), using now point group $D_{\infty h}$,

$$\frac{(\omega_1\omega_2)_{C_2D_2}}{(\omega_1\omega_2)_{C_2H_2}} = \sqrt{\frac{m_H}{m_D}}; \qquad \frac{(\omega_3)_{C_2D_2}}{(\omega_3)_{C_2H_2}} = \sqrt{\frac{m_H}{m_D}\frac{(m_C + m_D)}{(m_C + m_H)}};$$
$$\frac{(\omega_4)_{C_2D_2}}{(\omega_4)_{C_2H_2}} = \sqrt{\frac{m_H}{m_D}\frac{I_{C_2D_2}}{I_{C_2H_2}}}; \qquad \frac{(\omega_5)_{C_2D_2}}{(\omega_5)_{C_2H_2}} = \sqrt{\frac{m_H}{m_D}\frac{m_C + m_D}{m_C + m_H}}. \tag{III, 64}$$

Substituting again the observed fundamentals ν_i, we obtain for the left and right sides of the first equation 0.7147 and 0.7074, of the second equation 0.7384 and 0.7342, of the third equation 0.825 and 0.8340, and of the fourth equation 0.7394 and 0.7342.

It should be noted that, since C_2HD does not have a center of symmetry, all fundamentals are allowed in the infrared as well as in the Raman effect; indeed, all have been observed in the infrared (see Table 69). The selection rules that hold for C_2H_2 and C_2D_2 do not even approximately hold for C_2HD. This is made particularly clear by Fig. 87, in which the amplitudes of the atoms in the vibrations ν_1, ν_2, and ν_3 of C_2H_2 and C_2HD are drawn to scale according to calculations by Förster [see Herzberg, Patat, and Verleger (439)]. It should be noted that in C_2HD, according to Fig. 87, the vibration ν_1 is essentially a C—H vibration whereas ν_3 is essentially a C—D vibration. Correspondingly, as confirmed by more detailed calculations and the observations (Table 69), ν_1 and its overtones are equally intense as, or even more intense than, ν_3 and its overtones, whereas ν_1 is forbidden in C_2H_2.

[67] Very uncertain, derived by Mecke and Ziegler (618) from a disturbance in the galvanometer curve of Levin and Meyer (574) of the 729 band. They did not take into account the fact that this band consists of three subbands.

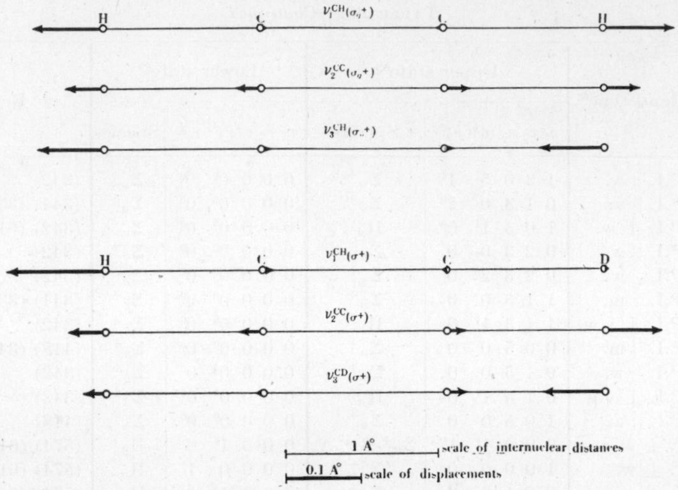

Fig. 87. Parallel vibrations of C_2H_2 and C_2HD drawn to scale.—The scale of displacements in all six diagrams is the same but different from the scale of distances (see bottom of figure). The former apply to the classical motions in the first vibrational state ($v = 1$) of each vibration.

Table 69. infrared and raman spectra of gaseous C_2HD.

ν_{vacuum}, observed[68] (cm^{-1})	Band type[13]	Assignment[69]	References
518.8	I. ⊥ s.	$\nu_4(\pi)$	(711) (225) (812)
683	I. ⊥ s.	$\nu_5(\pi)$	(711) (225) (812)
1202	I. ∥ m.	$\nu_4+\nu_5(\Sigma^+)$[71]	(711) (225) (812) (654)
1330	I. ⊥ w.	$\nu_2-\nu_4$	(812) (654)
1343	I. ∥ m.	$2\nu_5(\Sigma^+)$[71]	(812)
1851.2	I. ∥ m, R. s.	$\nu_2(\sigma^+)$	(377) (812)
2045	I. ? w.	$3\nu_5(\Pi)$, $\nu_3-\nu_4$	(812)
2584	I. ? s.	$\nu_3(\sigma^+)$	(225) (812) (654)
3334.8	I. s. R. s.	$\nu_1(\sigma^+)$	(377) (812) (654)
3950	I. w.	$\nu_1+\nu_4(\Pi)$, $\nu_1+\nu_5(\Pi)$	(812)
5100	I. w.	$2\nu_3(\Sigma^+)$, $\nu_1+\nu_2(\Sigma^+)$	(812)
8409.4	P.I. ∥ m.	$2\nu_1+\nu_2(\Sigma^+)$	(437) (439)
9050.6[70]	P.I. ⊥ v.w.	$2\nu_1+\nu_2+\nu_5(\Pi)$	(439)
9115.5	P.I. ∥ w.	$2\nu_1+\nu_3+\nu_4-\nu_4$	(439)
9138.9	P.I. ∥ m.	$2\nu_1+\nu_3(\Sigma^+)$	(437) (439)
9404.8	P.I. ∥ m.	$\nu_2+3\nu_3(\Sigma^+)$	(439)
9691.9	P.I. ∥ w.	$3\nu_1+\nu_4-\nu_4$	(439)
9706.4	P.I. ∥ s.	$3\nu_1(\Sigma^+)$	(437) (608) (439)
10211[70]	P.I. ⊥ v.w.	$3\nu_1+\nu_4(\Pi)$	(439)
11526	P.I. ∥ w.	$3\nu_1+\nu_2(\Sigma^+)$	(439)
12263.0	P.I. ∥ m.	$3\nu_1+\nu_3(\Sigma^+)$	(439)
12735.0	P.I. ∥ w.	$4\nu_1+\nu_4-\nu_4$	(439)
12746.8	P.I. ∥ m.	$4\nu_1(\Sigma^+)$	(439)

[13] On p. 274.

[68] With the exception of 9050.6 and 10211 cm^{-1} the numbers given for the photographic infrared

TABLE 70. INFRARED AND RAMAN SPECTRA OF GASEOUS C_2D_2.

ν_{vacuum}, observed (cm^{-1})	Band type[13]	Assignment	References
(505)[72]	—	$\nu_4(\Pi_g)$	
539.1	I. ⊥ v.s.	$\nu_5(\Pi_u)$	(711) (812)
1044	I. ∥ m.	$\nu_4+\nu_5(\Sigma_u{}^+)$[71]	(711) (812) (654)
1223	I. ⊥ w.	$\nu_2-\nu_5$	(812)
1610	I. ⊥ v.w.	$3\nu_5(\Pi_u)$[71]	(812)
1762.4	R. s.	$\nu_2(\sigma_g{}^+)$	(377)
1926	I. ⊥ v.w.	$\nu_3-\nu_4$	(812)
2157	I. ⊥? v.w.	$\nu_1-\nu_5$	(812)
2311	I. w.	$\nu_2+\nu_5(\Pi_u)$	(812)
2427	I. s.	$\nu_3(\sigma_u{}^+)$	(812) (654)
2700.5	R. s.	$\nu_1(\sigma_g{}^+)$	(377)
2940	I. v.w.	$\nu_3+\nu_4(\Pi_u)$	(812)
3280	I. w.	$\nu_1+\nu_5(\Pi_u)$	(812)
5120	I. v.w.	$\nu_1+\nu_3(\Pi_u)$	(812)

Cyanogen, C_2N_2. From considerations of electronic structure and valence one would expect C_2N_2 to be a linear symmetrical molecule just as is C_2H_2. The fact that no band has been found that occurs *both* in the Raman and infrared spectrum is in support of this, although it does not prove it. It would be difficult to prove the linear structure in this case by showing that there are only five rather than six fundamentals, since fundamentals may occur in regions that are not well investigated, and since one of the observed bands usually interpreted as a combination might be a fundamental. However, electron diffraction data of Pauling, Springall, and Palmer (687) favor a linear model.[73] Even though it appears that a slightly bent structure, like II in Fig. 86, might also be compatible with these data, we shall for the interpretation of the vibrational spectrum assume that C_2N_2 is *linear*, as has always been done in the literature.[74]

Table 71 gives the vibrational spectrum according to the most recent data of Reitz and Sabathy (737) and Bailey and Carson (78). The assignment of the fundamentals is that due to Woo and Badger (950). Objections to this assignment on the basis of specific heat data [Eucken and Bertram (311)] seem to have been cleared up recently [see Burcik and Yost (184) and Stitt (811)]. A slight difficulty is the great intensity of the Raman line corresponding to ν_4 which is more intense than that

---

bands are band origins calculated from the band fine structure. For the bands observed both in the infrared and Raman spectrum the more accurate Raman frequencies are given.

[69] The designation is similar to that for C_2H_2 (see Fig. 87), which leads to a slight inconsistency in our general nomenclature in that for C_2HD ν_1, ν_2, and ν_3 have the same species and should therefore be designated in the order of their frequencies.

[70] Only the line-like *Q* branch of each of these bands has been measured (in contrast to the other photographic bands). The *Q* branches are very weak and their wave numbers consequently not very accurate.

[71] The state $\nu_4 + \nu_5$, according to Table 33, consists of the substates Σ^+, Σ^-, Δ, of which only the first combines with the ground state. Similarly, for $2\nu_5$ and $3\nu_5$ only the components indicated in the table combine with the ground state.

[72] This is obtained from the infrared band $\nu_4 + \nu_5 = 1044\ cm^{-1}$.

[73] To be sure, they have based their evaluations right from the start on the assumption that it is linear.

[74] An unambiguous proof of the linear structure would be possible if the fine structure of the bands were resolved. Thus far this has not been done.

corresponding to ν_1 whereas usually the totally symmetric vibrations are the more intense ones.[76] Another difficulty is the non-occurrence of $\nu_2 + \nu_5$ in the infrared spectrum, although $\nu_2 - \nu_5$ appears.

TABLE 71. INFRARED SPECTRUM OF GASEOUS AND RAMAN SPECTRUM OF LIQUID C_2N_2, AFTER BAILEY AND CARSON (78), AND REITZ AND SABATHY (737).

ν_{vacuum},[43] observed (cm^{-1})	Type of band	Assignment
226	I. ⊥? s.	$\nu_5(\pi_u)$
506	R. (liquid) s.	$\nu_4(\pi_g)$
618	I. ⊥ m.	$\nu_2 - \nu_5$
732	I. ‖ s.	$\nu_4 + \nu_5(\Sigma_u^+)$
848	R. (liquid) m.	$\nu_2(\sigma_g^+)$
1026	R. (liquid) w.	$2\nu_5(\Sigma_g^+, \Delta_g)$
1102	R. (liquid) v.w.	?
2092	I. ⊥ w.	$\nu_1 - \nu_5$
2149	I. ‖ s.	$\nu_3(\sigma_u^+)$
2322	R. (liquid) v.s.[75]	$\nu_1(\sigma_g^+)$
2562	I. ⊥ s.	$\nu_1 + \nu_5(\Pi_u)$
2662	I. ⊥ s.	$\nu_3 + \nu_4(\Pi_u)$

Ammonia, NH_3 and ND_3. In previous considerations, when the ammonia molecule was used as an example we took the pyramidal structure and the assignment of the fundamental frequencies for granted. We shall now review briefly the spectroscopic evidence that leads to the assumption of this structure, and also leads to the proper selection of the fundamentals. In doing this we shall treat heavy and light ammonia simultaneously as far as possible.

The investigation of the rotational Raman and infrared spectra of ammonia (see Chapter I) has shown that the NH_3 molecule is a symmetrical top with a permanent electric dipole moment. The simplest explanation of this observation is to assume a symmetrical pyramidal structure with the N atom at the top. But it is not the only one. While the observation of the infrared rotation spectrum excludes definitely a plane symmetric structure (point group D_{3h}; see Fig. 1b), since for such a structure no dipole moment could arise, it does not exclude an unsymmetrical structure in which the molecule just happens to have two equal or nearly equal moments of inertia (for example, plane unsymmetrical model of point group C_{2v}, or pyramidal unsymmetrical model of point group C_s). But in this case the molecule would have to have six fundamentals, whereas on the assumption of the symmetrical pyramidal structure (point group C_{3v}) it would only have *four fundamentals*, two totally symmetric (A_1) and two doubly degenerate (E) (see Table 36). The large number of ordinary and photographic infrared bands as well as the Raman bands can, however, be satisfactorily accounted for on the basis of four fundamentals. There is no evidence at all of two more fundamentals. Thus we can take the *symmetrical pyramidal model* as proven.[77]

---

[43] On p. 283.

[75] Daure and Kastler (266) found a Raman line of the gas at 2330 cm^{-1}. Kastler (note added to reprint of Daure-Kastler) found a Raman line of liquid C_2N_2 at the same place within their accuracy (±5 cm^{-1}).

[76] This difficulty may be due to the fact that the Raman spectrum was observed in the liquid state.

[77] If this had been found for NH_3 alone it might conceivably have been due to an approximate coincidence of two pairs of normal frequencies. But the fact that ND_3 also shows only four fundamentals proves the point conclusively.

In Table 72 are given the *fundamentals* of NH_3 and ND_3 as obtained from the infrared and Raman spectra. The numbering of the frequencies is that used in Fig. 58 and in Table 38.[80] According to the selection rules (Table 55) all four fundamentals are both infrared and Raman active. The fact that the degenerate vibrations ν_3 and ν_4 have not been observed in the Raman spectrum is in agreement with the usual weakness of Raman lines corresponding to non-totally symmetric vibrations. At the same time, this consideration together with the fact that the Raman line 3334.2 is polarized confirms the correlation of 3334.2 and 950 cm^{-1} with ν_1 and ν_2 rather than ν_3 and ν_4, quite apart from the structure of the infrared bands. The weakness of ν_3 of NH_3 in the infrared is perhaps

TABLE 72. FUNDAMENTALS OF GASEOUS NH_3 AND ND_3 AS OBSERVED IN THE INFRARED AND RAMAN SPECTRA

	NH_3			ND_3		
	Infrared, ν_{vacuum} (cm^{-1})	Raman, $\Delta\nu_{vacuum}$ (cm^{-1})	References	Infrared, ν_{vacuum} (cm^{-1})	Raman, $\Delta\nu_{vacuum}$ (cm^{-1})	References
$\nu_1(a_1)$	3335.9, 3337.5 [78] ‖ s.	3334.2 v.s. (pol.)	(281) (42) (576) (267)	2419 ‖	2420.0 s.	(624) (385)
$\nu_2(a_1)$	931.58, 968.08 ‖ s.	934.0, 964.3 m.	(785) (42) (576)	748.6, 749.0 ‖	(786) w.	(624) (385)
$\nu_3(e)$	(3414)[79] ⊥	—	(106)	2555 ⊥	—	(624)
$\nu_4(e)$	1627.5 ⊥ v.s.	—	(106)	1191.0 ⊥	—	(624)

less easily understood. It is partly only apparent since ν_3 is overlapped by the strong band ν_1. The value given is obtained from combination bands.[81] A further check on the essential correctness of the fundamentals in Table 72 is obtained when they are substituted into the Teller-Redlich product rule (II, 313).

A very notable feature of the totally symmetric fundamentals ν_1 and ν_2 of NH_3 is that they are double. For ND_3 the splitting occurs for ν_2 but is much smaller than for NH_3 and is apparently too small to be detected for ν_1. As has been discussed earlier (p. 221), this doubling is due to the fact that there are two equilibrium positions for the N atom at the two sides of the H_3 (or D_3) plane (*inversion doubling*). As we have seen previously (p. 222), all vibrational levels are split into two sublevels, a lower positive and an upper negative level, the splitting being relatively the largest for those levels in which the height of the pyramid changes most during the vibration. In the infrared the selection rule is $+ \leftrightarrow -$ (see p. 257), hence each band has two components whose separation is the *sum* of the splittings of the upper and lower levels (see Fig. 78); in the Raman spectrum, where the selection rule is $+ \leftrightarrow +$, $- \leftrightarrow -$, the separation is the *difference* of the splittings of the upper and lower levels. Thus the two components of a Raman vibrational band should not agree exactly with those of the corresponding infrared band. Although such a difference has been observed for ν_2 (see Table 72), the accuracy of the available Raman measurements is not sufficient to obtain from this difference reliable values for the splittings of the upper and lower levels.

However, the individual splittings can be determined in the following way: The selection rule $+ \leftrightarrow -$ holds also, of course, for the pure rotation spectrum, and therefore every line of the rotation spectrum is double (see the observed spectrum in Fig. 12a, p. 33), the line splitting being again the sum of the splittings of the upper and lower states. But here the splitting is the same in the upper and lower states, and therefore the line splitting is just twice the splitting in the vibrational ground

[78] These two values were obtained by the author by interpolating the band origin between the doublets of the *P* and *R* branch given by Dennison and Hardy (281).

[79] Not very certain; see Sutherland (826), who suggests 3450 cm^{-1}, since it fits much better with the isotope relations.

[80] It should be noted that Dennison and his collaborators use a different numbering, exchanging ν_2 and ν_3.

[81] In the earlier literature one or the other of these combination bands at 5053 or 4433 cm^{-1} was taken to be ν_3. See the discussion by Howard (459) who first suggested the lower value.

state, which is thus found to be 0.66 cm^{-1}. Subtracting this from the splitting of the infrared bands ν_1 and ν_2, we get the splittings 0.9 and 35.84 cm^{-1} for their upper states 1, 0, 0, 0 and 0, 1, 0, 0 respectively. The same splitting of the ground state is also obtained by the observation of the transition from the lower to the upper sublevel in the region of short radio waves (0.8 cm^{-1}) by Cleeton and Williams (215) (see p. 257). From these splittings, as has been mentioned previously (p. 224), the height of the NH_3 pyramid has been determined to be 0.38×10^{-8} cm.

In Table 73 are given the wave numbers, band types, and assignments of the considerable number of *overtone and combination bands* that have been observed for NH_3. The assignments are essentially those given by Wu (26). The series of bands ν_1, $2\nu_1$, $3\nu_1$, $4\nu_1$, $5\nu_1$, $6\nu_1$ is most prominent. However, since ν_3 is close to ν_1 the bands $\nu_1 + \nu_3$ and $2\nu_3$ overlap $2\nu_1$; $2\nu_1 + \nu_3$, $\nu_1 + 2\nu_3$, $3\nu_3$ overlap

TABLE 73. OVERTONE AND COMBINATION BANDS OF GASEOUS NH_3.[82]

ν_{vacuum}, observed[83] (cm^{-1})	Band type	Assignment	References
629.3	I. ∥ (w.)	$2\nu_2 - \nu_2$[84]	(785)
1922	R. (w.) I. (v.w.)	$2\nu_2(A_1)$[85]	(42) (770)
2440.1, 2472.6 }*	I. ⊥ (w.)	$\nu_3 - \nu_2$ or $\nu_2 + \nu_4$[86]	(106)
2861	I. ∥ (v.w.)	$3\nu_2(A_1)$[87]	(770)
3219.1	R. (w.) [I ∥ (m.)]	$2\nu_4[A_1(+E)]$	(576) (106)
4176, 4216 }*	I. ∥ (m.)	$\nu_2 + 2\nu_4[A_1(+E)]$?	(106)
4269, 4302 }*	I. ∥ (s.)	$\nu_1 + \nu_2(A_1)$	(106)
4433, 4505 }*	I. ⊥ (s.)	$\nu_2 + \nu_3(E)$	(106)
5053*	I. ⊥ (s.)	$\nu_3 + \nu_4[(A_1) + E]$	(809) (105)
6016	I. ⊥ (m.)	$\nu_2 + \nu_3 + \nu_4[(A_1) + E]$?	(883)
6595, 6624 }	I. ∥? (m.)	$2\nu_1(2\nu_3, \nu_1 + \nu_3)$	(883)
7665	I. ? (w.)	$2\nu_3 + \nu_2(A_1 + E)$? }	(883)
7899	P.I. ? (w.)		(899)
8177, 8202 }	(P.)I. ? (w.)	$2\nu_1 + \nu_4(E)$?	(883) (899)
8460	(P.)I. ? (w.)	$2\nu_3 + \nu_4(A_1 + 2E)$	(883) (899)
9760.4*	P.I. ∥ (m.)	$3\nu_1(A_1)$	(196) (592)
10099.7*	P.I. ∥ (w.)	$\nu_1 + 2\nu_3(A_1 + E)$? }	(196)
10104.9*	P.I. ∥? (w.)		
11364	P.I. ⊥? (w.)	$3\nu_1 + \nu_4(E)$	(75) (592)
12609.2*	P.I. ∥ (w.)	$4\nu_1(A_1)$, $2\nu_1 + 2\nu_3[A_1(+E)]$, ··· }	(75) (196)
12619.8*	P.I. ∥ (w.)		
15440	P.I. ? w.	$5\nu_1(A_1)$, $4\nu_1 + \nu_3(E)$, ···	(65) (477)
18150	P.I. ? (v.w.)	$6\nu_1(A_1)$, $5\nu_1 + \nu_3(E)$, ···	(65) (477)

[82] Amaldi and Placzek (42) give two further very weak Raman bands at 2210 and 2270 cm^{-1}, which are difficult to interpret and of which they themselves say that a confirmation would be desirable.

[83] The wave numbers with an asterisk are zero lines of the bands, obtained from actual fine-structure analysis or from definitely identified Q branches. The other values are band centers which, because of the extent of the bands, are not very accurate.

[84] This is only the transition $2^+ \leftarrow 1^-$.

[85] This is only the transition $2^- \leftarrow 0^+$ (infrared) and $2^- \leftarrow 0^-$ (Raman) respectively.

[86] The latter is the assignment of Sutherland (826), who takes $\nu_3 \approx 3450$ instead of 3414 [Barker (106)].

[87] This is only the transition $3^- \leftarrow 0^+$.

$3\nu_1$; and so on. While ν_1, $2\nu_1$, $3\nu_1$, $\cdots$ are parallel bands, that is, have $M_z \neq 0$ (the upper state has species A_1; see Table 55), the bands overlapping them are perpendicular bands, that is, have $M_z = 0$ (the upper state has species E), or they consist of parallel and perpendicular sub-bands. For example, since ν_1 has species A_1, ν_3 species E, according to Table 31 the upper state of $\nu_1 + \nu_3$ has species E, and therefore the corresponding band is a perpendicular band. The upper state of $2\nu_3$ has two sublevels, of slightly different energy and of species A_1 and E (see Table 32), and therefore the band $2\nu_3$ consists of two sub-bands, a perpendicular and a parallel one. Thus, in the region of 6600 cm^{-1} ($2\nu_1$), we would expect two parallel and two perpendicular bands. Similarly, in the region of 9800 cm^{-1} ($3\nu_1$), we would expect three parallel and three perpendicular bands, and in the region of 12600 cm^{-1} ($4\nu_1$) four parallel and five perpendicular bands. This explains the great complexity and the considerable extent of these "bands." They have been analyzed only partly and the band origins given for some refer to only one or two of the component bands, mostly of the parallel type. Much more work remains to be done before a satisfactory formula for the vibrational levels can be developed, particularly since resonance between certain sublevels of the states $3\nu_1$, $2\nu_1 + \nu_3$, $\nu_1 + 2\nu_3$, $3\nu_3$ and similarly $4\nu_1$, $3\nu_1 + \nu_3$, $\cdots$ will be quite important.

For ND_3 apart from the fundamentals, only the difference bands $2\nu_2 - \nu_2$, $3\nu_2 - \nu_2$, $3\nu_2 - 2\nu_2$, $4\nu_2 - 2\nu_2$ and $4\nu_2 - 3\nu_2$ have been observed [Migeotte and Barker (624)]. Instead of giving these explicitly we give in Table 74 the energy levels 0, v_2, 0, 0 of both NH_3 and ND_3 as far as they have

TABLE 74. ENERGY LEVELS 0, v_2, 0, 0 OF NH_3 AND ND_3 ACCORDING TO DENNISON (280).

Level (v_2)	NH_3 $G(0, v_2, 0, 0)$ cm^{-1}	ND_3 $G(0, v_2, 0, 0)$ cm^{-1}
0^+	0	0
0^-	0.66	$\sim$0
1^+	932.24	745.6
1^-	968.08	749.0
2^+	1597.4	1359
2^-	1910	1429
3^+	2380	1830
3^-	2861[88]	2113
4^+		2495
4^-		2868

been observed, according to Dennison (280) [see also Sheng, Barker, and Dennison (785)]. For a few of the levels, data on the ultraviolet absorption spectrum by Benedict (136) have been used.

Migeotte and Barker (624) have also obtained the fundamentals ν_2 of NH_2D and NHD_2 at 894, 874, and 818, 808 cm^{-1} respectively, showing intermediate values of the inversion splitting.

Trihalides of phosphorus, arsenic, antimony, and bismuth. Thus far only the Raman spectra of the trihalides of phosphorus, arsenic, antimony, and bismuth have been investigated, and these only in the liquid state [PF_3 by Yost and Anderson (968); PCl_3 and PBr_3 by Braune and Engelbrecht (176), Venkateswaran (895), Cabannes and Rousset (191), Niclsen and Ward (670); AsF_3 by Yost and Sherborne (973); $AsCl_3$ and $AsBr_3$ by Braune and Engelbrecht (176), Yost and Anderson (969), Brodskii and Sack (182) and Cabannes and Rousset (191); $SbCl_3$ by Braune and Engelbrecht (176) and Gupta (406); $BiCl_3$ by Bhagavantam (144)]. All these molecules show four fairly strong Raman lines which have been given in the previous Table 38. The occurrence of just four Raman lines is best explained by the assumption that these molecules, like NH_3, form symmetrical pyramids (point group C_{3v}), since in this case we have to expect just four fundamentals which are all Raman active. If the molecules had a plane symmetrical form (point group D_{3h}) there would also be four

[88] This level has not been given by Dennison, but follows from the assignment of the infrared band at 2861 cm^{-1} in Table 73.

fundamentals but only three would be Raman active. Thus, this latter model is excluded.[89] An unsymmetrical model is, of course, extremely unlikely, but such a model is not easily eliminated on purely spectroscopic reasons. It would lead to six Raman active fundamentals, but two might be weak and a few additional very weak Raman lines have indeed been found for some of the compounds. However, the state of polarization of the Raman lines is also in agreement with the symmetrical pyramidal model, two Raman lines being partly polarized, the other two completely depolarized. The fact that reasonable values for the force constants and the valence angles are obtained on the assumption of this model (see Tables 38 and 43) is further proof for it. Finally, it has been confirmed for some of the molecules by electron diffraction experiments [see Brockway (179)].

The degree of depolarization of the Raman lines has, of course, also been used to decide which observed frequencies correspond to the totally symmetric ($\rho_n < \frac{6}{7}$) and which to the degenerate normal vibrations ($\rho_n = \frac{6}{7}$).

Boron trifluoride, BF_3. The two most plausible models for BF_3 are the pyramidal and the plane symmetrical form (point groups $\boldsymbol{C_{3v}}$ and $\boldsymbol{D_{3h}}$ respectively). In both cases, there would be four fundamentals, of species $2A_1 + 2E$ in the first and $A_1' + A_2'' + 2E'$ in the second case (see Table 36); but, according to Table 55, in the first case all four would be active in both the infrared and Raman spectrum, whereas in the second case, the (only) totally symmetric vibration (species A_1') would be inactive in the infrared, the antisymmetric vibration (species A_2'') would be inactive in the Raman effect. Observation shows three infrared-active fundamentals [Gage and Barker (344)] and two strong Raman lines [Yost, DeVault, Anderson, and Lasettre (970)], only one of which coincides with one of the infrared bands. While this result favors the plane model, it might be argued that the fourth fundamental (which gives the strongest line in the Raman spectrum) is weak only and not missing in the infrared, and thus the pyramidal model might still be correct.

However, unambiguous proof for the plane model (excluding at the same time all unsymmetrical models) comes from the isotope effect. Boron has two isotopes B^{10} and B^{11}, of abundance ratio 1 : 4. For the pyramidal model (as well as for any unsymmetrical model), the B atom has a non-zero amplitude for any one of the normal vibrations and therefore every fundamental should be a doublet consisting of bands of relative intensity 1 : 4. In a plane model, as a glance at Fig. 63 shows, the B atom does not move in the totally symmetric vibration $\nu_1(A_1')$, and therefore this vibration, unlike the other three, should not show an isotope splitting. Actually, it is found that while the three infrared-active fundamentals are such isotopic doublets with the correct intensity ratio, the one Raman active fundamental (888 cm^{-1}) that does not appear in the infrared is single, thus proving the plane model and at the same time identifying the Raman line at 888 cm^{-1} as the totally symmetric fundamental ν_1.

The assignment of the other observed fundamentals (Table 75) to the normal vibrations of Fig. 63 is simple on the basis of the previous isotope relations (Chapter II, section 6). From (II, 313) we have:

$$\frac{\omega_2(B^{10}F_3)}{\omega_2(B^{11}F_3)} = \sqrt{\frac{m_{B^{11}}}{m_{B^{10}}} \cdot \frac{m_{B^{10}} + 3m_F}{m_{B^{11}} + 3m_F}} = \frac{(\omega_3\omega_4)_{B^{10}F_3}}{(\omega_3\omega_4)_{B^{11}F_3}}. \quad \text{(III, 65)}$$

The only pair of frequencies (see Table 75) that gives the proper ratio for $\frac{\nu_2^{(i)}}{\nu_2}$ is 719.5 and 691.3 cm^{-1}, giving the ratio 1.0408, whereas the value of the square root is 1.04087. The other two frequencies [90] give $\frac{\nu_3^{(i)}\nu_4^{(i)}}{\nu_3\nu_4} = 1.039$. The overtone and combination bands are easily interpreted on the basis of these fundamentals and the selection rules of Table 55. They are given in Table 75. It is noteworthy that the second overtone $3\nu_2$ of the antisymmetric vibration ν_2 seems to be present, whereas $2\nu_2$ is absent in agreement with the selection rules of the plane model ($\boldsymbol{D_{3h}}$) but of no other model.

As has been pointed out previously (see Table 44 and accompanying discussion), the application of the valence force system leads to a satisfactory representation of the observed fundamentals of BF_3 with reasonable values of the force constants, if the plane model is assumed.

---

[89] While it might be that a molecule with three Raman-active fundamentals has four fairly strong Raman lines because of Fermi resonance (see p. 266), it would appear to be impossible that such a resonance would occur for a whole group of molecules.

[90] The smaller of the two is by definition ν_4, the larger ν_3. ν_3 of $B^{10}F_3$ is less accurately measured.

In a way similar to the above, the molecules BCl_3 and BBr_3 have been found to have the plane symmetrical structure. Their fundamentals are given in the previous Table 44 (for references, see Table 78). For all three boron halides, this structure has been confirmed by electron diffraction experiments [see Pauling (17)] and for BF_3 and BCl_3 also by measurements of the dielectric constant which show that the dipole moment is zero [see Nespital (651) and Linke and Rohrmann (580)].

TABLE 75. INFRARED AND RAMAN SPECTRA OF GASEOUS BF_3.

ν_{vacuum}, observed[43] (cm^{-1}) $B^{11}F_3$	$B^{10}F_3$	Band type	Assignment	References
480.4	**482.0**	I. (s.) R. (m.)	$\nu_4(e')$	(344) (970)
691.3	**719.5**	I. (s.) ‖	$\nu_2(a_2'')$	(344)
[91]	711.3	I. (w.)	$\nu_2+\nu_4-\nu_4$	(344)
888	**888**	R. (s.)	$\nu_1(a_1')$	(970)
1178		I. (w.)	?[92]	(91)
1445.9	**1497**	I. (v.s.)	$\nu_3(e')$	(344)
1831	1928	I. (w.)	$\nu_3+\nu_4(E')$ [or $\nu_1+2\nu_4(E')$]	(91)
2058		I. (w.)	$3\nu_2(A_2'')$	(91)
2250		I. (w.)	$\nu_1+\nu_3(E')$ [or $2\nu_1+\nu_4(E')$]	(91)
2903.2	3008.2	I. (w.)	$2\nu_3(E')$	(344)
3260		I. (w.)	$2\nu_1+\nu_3(E')$?	(91)

Phosphorus, P_4. Up to now, only the Raman spectrum of the P_4 molecule has been investigated, [Venkateswaran (890)]. Three Raman shifts: 363, 465, and 606 cm^{-1} have been found in the liquid.[92a] The lines corresponding to the first two shifts are completely depolarized; that is, they correspond to non-totally symmetric vibrations, whereas the lines with a shift 606 cm^{-1}, which are the most intense, have a degree of depolarization $\rho_n = 0.05$ (which is 0 within the accuracy of the measurements).

One would expect the P atoms in the P_4 molecule to be equivalent. The only models for which this would be the case are the plane square model (point group D_{4h}) and the tetrahedral model (point group T_d). In the first case there would be (see Table 36) five fundamentals, one each of species A_{1g}, B_{1g}, B_{1u}, B_{2g}, E_u, of which the three even ones (g) would be Raman active (Table 55). In the second case there would be only three fundamentals of species A_1, E, and F_2, all of which would be Raman active. Thus, in both cases, three Raman lines are to be expected for each exciting line; two of them should be depolarized, as observed.

An unambiguous decision between the two models on a spectroscopic basis would only be possible if the infrared spectrum were investigated, since with the first model no infrared band should coincide with a Raman band, whereas, in the second case, the only infrared-active fundamental (species F_2) should agree with one of the two depolarized Raman lines. But there are two arguments based on the available Raman data only which strongly favor the tetrahedral model, even though they do not definitely prove it: (1) For this model the degree of depolarization of the Raman line corresponding to the totally symmetric vibration should be zero (see p. 271), as it seems to be, whereas for the plane square model it would have a value between 0 and $\frac{6}{7}$. (2) On the basis of the central force system the ratio of the frequencies for the tetrahedral model should be $1 : \sqrt{2} : 2$, with the totally symmetric vibration highest [see equation (II, 177)], whereas for the square model [on the basis

[43] On p. 283.

[91] Overlapped by CO_2 absorption.

[92] This band would fit $\nu_2 + \nu_4$ but since the upper state would have species $a_2'' \cdot e' = E''$ (see Table 31), this combination is forbidden according to Table 55. It has not been found by Gage and Barker (344).

[92a] Venkateswaran has found these shifts also in solid yellow phosphorus, and the last two in phosphorus vapor.

of valence forces; see Kohlrausch (14)] the totally symmetric vibration should have approximately the same frequency as one of the other Raman-active vibrations. The actual fundamental frequencies do at least approximately fulfill the first condition. Bhagavantam and Venkatarayudu (153) have treated the tetrahedral model on the basis of the valence force system, which gives an even better representation of the frequencies than the central force system since two (rather than one) force constants are used. The investigation of the electron diffraction by phosphorus vapor [Maxwell, Hendricks, and Mosley (606)] has led very definitely to the conclusion that the molecule is tetrahedral.

Formaldehyde, H_2CO and D_2CO. The formaldehyde molecule has always been assumed to have the plane symmetrical Y form (point group C_{2v}; see Fig. 24), although *a priori* (except for considerations of directed valence) a pyramidal form with only one plane of symmetry (point group C_s) would also appear to be possible. However, the latter form is definitely excluded by the observation of an intensity alternation (3 : 1) in the rotational structure of the infrared and ultraviolet bands [see p. 479f. and (288)]. On the basis of the vibrational spectrum of H_2CO alone, it would be difficult to arrive at such a decision, since for both models all six fundamentals (see Fig. 24) are active both in the infrared and Raman spectrum (see Table 55). While there would be some differences between the two models in the infrared activity of the combination vibrations and in the polarization of the fundamentals in the Raman spectrum, the available data [93] do not allow a decision on this basis. The only evidence from the available data on the vibration spectrum that definitely favors the plane model is that the product relations (see Chapter II, section 6) applied to the observed frequencies of H_2CO and D_2CO are well fulfilled on the assumption of the plane model. In what follows we shall assume this model.

Seven fairly strong infrared absorption bands have been observed for H_2CO as well as D_2CO by Nielsen (662) and Ebers and Nielsen (295) (296). The wave numbers of these bands are given in Table 76. One of them must be a combination or overtone band. Since only two C—H (C—D) stretching vibrations are expected (see p. 196), one of the three high-frequency bands must be the combination or overtone band, and it is natural to choose the weakest of the three, which has the shortest wave length. Its frequency is indeed very nearly double that of the band at 1503 cm^{-1} for H_2CO and 1105.7 cm^{-1} for D_2CO.[94] Thus the six remaining strong bands must be the funda-

TABLE 76. INFRARED VIBRATION SPECTRUM OF GASEOUS H_2CO AND D_2CO AFTER EBERS AND NIELSEN (662) (295) (296).

Assignment	Type of band	H_2CO ν_{vacuum}[95]	D_2CO ν_{vacuum}
$\nu_6(b_2)$	I. ⊥ s.	1167[95a]	938[95b]
$\nu_5(b_1$	I. ⊥ s.	1280[95a]	990[95b]
$\nu_3(a_1)$	I. ‖ s.	1503	1105.7
$\nu_2(a_1)$	I. ‖ v.s.	1743.6[95c]	1700
$2\nu_6(A_1)$?	I. ‖ w.	2081[96]	
$\nu_1(a_1)$	I. ‖ s.	2780	2055.8
$\nu_4(b_1)$	I. ⊥ v.s.	2874	2159.7
$2\nu_3(A_1)$	I. ‖ s.	2973	2208

[93] The Raman spectrum of aqueous solutions of H_2CO which has been investigated is of no help here, since H_2CO certainly does not remain unchanged in solution.

[94] It would also fit $\nu_2 + \nu_5$, but this would be a ⊥ band whereas the observed band is a ‖ band.

[95] For H_2CO, Salant and West (759) give in addition the weak bands 4590, 5240, 5430, 5650, 6940, 7140, 8000 cm^{-1} measured under low dispersion. They are easily interpreted as $\nu_1 + \nu_2$ (or $\nu_2 + \nu_4$), $\nu_1 + 2\nu_5$, $2\nu_1$, $2\nu_4$, $2\nu_1 + \nu_2$, $2\nu_4 + \nu_2$, $3\nu_1$ respectively.

[95a] Ebers and Nielsen give 1165 and 1278 cm^{-1} for these bands. However their values do not refer to the zero lines but to the first lines of the *P* branches.

[95b] These values are not zero lines but refer simply to prominent features of the bands.

[95c] This is the value given by Nielsen (662) (corrected for vacuum). In the paper by Ebers and Nielsen (296) the value 1750 is given instead.

mentals, if it is assumed that there are no further strong infrared bands beyond the region investigated. Of the two high-frequency fundamentals, the parallel-type band corresponds obviously to the symmetrical C—H stretching vibration ν_1, the perpendicular-type band to the antisymmetrical C—H vibration ν_4 (see Fig. 24). The two low-frequency fundamentals are naturally assigned to the two bending vibrations ν_5 and ν_6. Of the remaining two parallel vibrations, the higher one, which shows only a small isotope shift, must be considered as the vibration ν_2, which is essentially a C—O vibration, whereas the other, with a large isotope shift, is the C—H bending vibration ν_3. The great intensity of the first overtone $2\nu_3$ of ν_3 is probably due to Fermi resonance with ν_1 (see p. 266).

Using the valence force relations for the frequencies, (II, 214)–(II, 219), Ebers and Nielsen have obtained, from the assumed fundamentals of H_2CO, values for the force constants which appear to be quite reasonable. In addition, using these force constants, they have calculated the fundamentals of D_2CO, with results in fair agreement with the observed. While this is a fairly strong argument in favor of their choice of fundamentals, it must be pointed out that the electronic band spectrum seems to lead to different frequencies for the bending vibrations ν_5 and ν_6 [see Herzberg and Franz (435), Gradstein (397), Sponer and Teller (802), and footnote 96)].

Hydrogen peroxide, H_2O_2. A number of different models have been suggested at one time or another for the H_2O_2 molecule. Up to now, the investigations of its spectrum, because of great experimental difficulties, have not led to an unambiguous result as to its structure. But these investigations do rule out definitely some of the proposed models and make others unlikely. Table 77 gives the observed Raman spectrum of the liquid [Simon and Fehér (791)] and the infrared spectrum of the vapor [Bailey and Gordon (88), Zumwalt and Giguère (977)]. It is seen that the two strongest Raman lines occur also as infrared bands, proving that the H_2O_2 molecule does not have a center of symmetry (see p. 256),[97] This excludes definitely the linear symmetric and the bent model of point group C_{2h} (I and II in Fig. 86 for acetylene). Three other models that have been suggested are given in Fig. 88. Since the degree of depolarization of the Raman lines other than 877 cm^{-1} and the type of the infrared bands is not known, it is not possible to decide between these three models on the basis of the vibration spectrum alone. According to Penney and Sutherland (691), the third model (c), in which the H atoms are in two different planes through O—O approximately at right angles to each other (point group C_2, see Fig. 2a), is strongly favored by the theory of directed valence; however, according to Hellmann [p. 267 in (7)] model (b) is equally possible according to this theory. Bailey and Gordon (88) have interpreted the available data on the basis of model (c). They obtained reasonable values for the force constants, assuming valence forces.

TABLE 77. INFRARED AND RAMAN SPECTRA OF H_2O_2.

ν_{vacuum}, vapor (infrared) (cm^{-1})	$\Delta\nu_{vacuum}$, liquid (Raman) (cm^{-1})	Assignment (Bailey and Gordon)	References
870 (m.)	877 (v.s.) (pol.)	$\nu_4(a)$	(88) (893) (262) (791) (895)
	903 (v.w.)	spurious	(893)
1370 (s.)		$\nu_6(b)$	(88)
	1408 (w.)	$\nu_3(a)$	(791)
	1435 (w.)	$\nu_2(a)$	(791)
2869 (m.)		$\nu_1(a)$	(88)
3417 (s.)	3407 (m.)	$\nu_5(b)$	(88) (791)
10283.7 (v.w.) 10291.1 (v.w.)		$3\nu_5(B)$	(977)

[96] This band, while not mentioned by Ebers and Nielsen (295), was found by Patty and Nielsen (684) and according to a private communication by Nielsen is still considered by him to be a genuine H_2CO band. However, its suggested interpretation as $2\nu_6$ requires a suspiciously large anharmonic term x_{66}.

[97] For the liquid the infrared spectrum gives complete numerical agreement of the two frequencies [see Bailey and Gordon (88)].

Their assignments are given in the third column of Table 77. But these assignments must be considered as decidedly tentative.

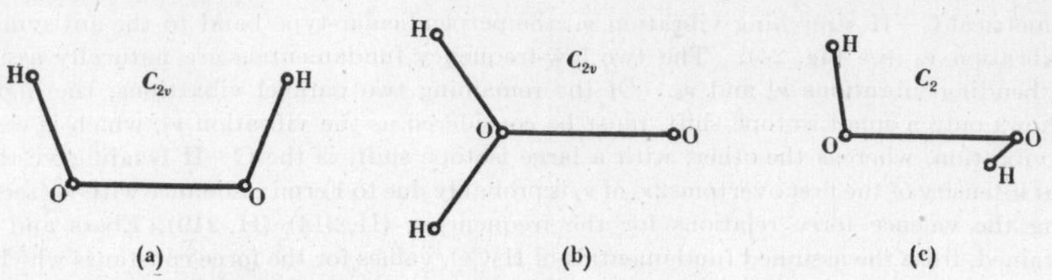

FIG. 88. **Suggested models of H_2O_2.—(c) is non-planar, see Fig. 2(a).**

More recently Zumwalt and Giguère (977) have investigated the fine structure of two photographic bands on the basis of which they consider both model (a) and model (b) excluded. However, their published photometer curve does not appear very convincing. They have interpreted the occurrence of two equally intense bands, close together, as due to an inversion doubling, which would only be possible for model (c).

TABLE 78. FURTHER FOUR-ATOMIC MOLECULES.[53]

Molecule	Structure: Point group	Structure: Description	References: Raman spectrum	References: Infrared spectrum
C_2I_2	C_{2v}[98]	non-linear in solution (?)	(377)	(304)
PH_3, PD_3	C_{3v}	pyramidal, valence angle 99°[99]	(732) (968) (459) (421) (830)	(740) (338) (572)
AsH_3, AsD_3	C_{3v}	pyramidal, valence angle 97.5°	(459) (830) (273) (421)	(674a) (572)
NF_3	C_{3v}	pyramidal, valence angle $\geqq$ 110°		(92)
ClO_3^-, BrO_3^-, IO_3^-	C_{3v}	pyramidal, valence angle 89°	(784) (545) (750a)	(683) (21)
BCl_3, BBr_3	D_{3h}	plane symmetric	(52)	(195)
SO_3	D_{3h}	plane symmetric (?)	(350)	(348)
NO_3^-, $CO_3^=$	D_{3h}	plane symmetric	(767) (49) (720) (637) (199) (545)	(765) (764) (928)
Cl_2CO	C_{2v}	plane symmetric	(48)	(89) (852)
Cl_2CS	C_{2v}	plane symmetric	(844)	(852)
F_2SO, Cl_2SO	C_s ?	pyramidal	(142) (603) (896)	(887)
S_2Cl_2	C_2	(like H_2O_2) ?	(635)	
N_3H	C_s	plane, N_3 group linear	(306)	(438) (268) (318) (319) (188)
HNCO	C_s	(?)	(394)	(445)
$\left(HC\langle{}^{O}_{O}\rangle\right)^-$	C_{2v}	(?)	(14) (10) (299) (325a)	
HNO_2	?	(?)		(476a)

[53] On p. 287.

[98] Very probably this molecule is linear in the gaseous state but a change of structure or a violation of Raman selection rules occurs in solution, the only state investigated.

[99] Stevenson (806) obtained an angle of 93° from the spectroscopic value for one of the moments of inertia and an assumed value for the P—H distance.

The Raman spectrum of D_2O_2 has been investigated by Fehér (320), who found the three shifts 2510, 1009, and 877 cm^{-1}. The agreement of the last of these with the H_2O_2 frequency 877 cm^{-1} proves that this frequency corresponds mainly to an O—O oscillation.

Other four-atomic molecules. In Table 78 are given the structures of several further four-atomic molecules, as well as references to the more recent work on their Raman and infrared spectra on which the conclusions as to the structures are based. It should again be noticed that molecules with the same number of external electrons have the same geometrical structure. For example, the ions NO_3^- and CO_3^{--} as well as the SO_3 molecule have the same plane symmetrical structure (D_{3h}) as the isoelectronic molecules BF_3, BCl_3, BBr_3. ClO_3^-, BrO_3^-, IO_3^- have the same pyramidal structure as the isoelectronic PF_3, AsF_3. A number of more-than-four-atomic molecules may for certain purposes be considered as four-particle systems, for example $N(CH_3)_3$, $P(CH_3)_3$, $As(CH_3)_3$, $HC(CH_3)_3$, $Al(CH_3)_3$, $HCCl_3$, and others [see Kohlrausch (14) and the more recent work of Fehér and Kolb (321), Rosenbaum and Ashford (744), and Rosenbaum, Rubin, and Sandburg (745)].

(c) *Five-atomic molecules*

Carbon suboxide, C_3O_2. Electron diffraction data [Brockway and Pauling (181), Boersch (157)] point to a linear symmetric form for the C_3O_2 molecule, whereas the observation of a small (non-zero) dipole moment favors an asymmetrical structure [Le Fèvre and Le Fèvre (573)]. Valence considerations strongly favor the linear model and therefore, despite the contrary evidence of the dipole measurements, it is usually used as a basis for the interpretation of the spectrum. Unfortunately, the available Raman and infrared data [Engler and Kohlrausch (306), Lord and Wright (590)] are not sufficiently complete to decide the question unambiguously.

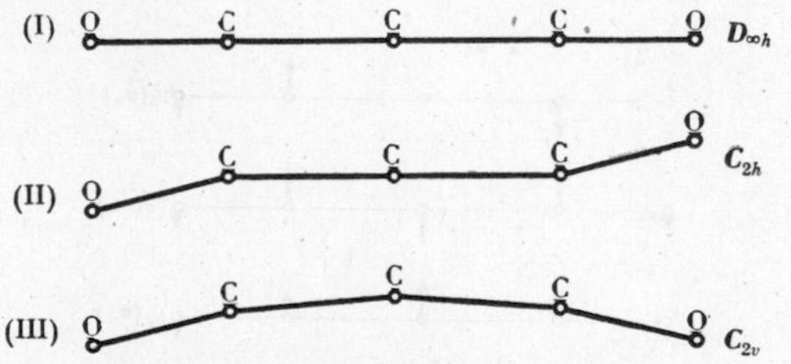

FIG. 89. **Possible models of C_3O_2.**

Since it appears that the electron diffraction data are also compatible with one of the bent forms II (point group C_{2h}) and III (point group C_{2v}) of Fig. 89, we shall include them with the linear model (I in Fig. 89, point group $D_{\infty h}$) in at least part of our discussion. For the three models the number of vibrations of the various species and their activities (I. = infrared active, R. = Raman active) are (see Tables 35, 36, and 55):

$$\text{I} \quad (D_{\infty h}) : 2\,\Sigma_g^+(\text{R.}),\ 2\,\Sigma_u^+(\text{I.}),\ 1\,\Pi_g(\text{R.}),\ 2\,\Pi_u(\text{I.}).$$

$$\text{II} \quad (C_{2h}) : 3\,A_g(\text{R.}),\ 2\,A_u(\text{I.}),\ 4\,B_u(\text{I.}).$$

$$\text{III} \quad (C_{2v}) : 4\,A_1(\text{I., R.}),\ 1\,A_2(\text{R.}),\ 3\,B_1(\text{I., R.}),\ 1\,B_2(\text{I., R.}).$$

Here the numbers in front of the symbols are the numbers of normal modes of the particular species. For models I and II, since there is a center of symmetry the rule of mutual exclusion should hold. Actually there are two fairly close coincidences: the Raman lines 1114 and 2200 and the infrared bands 1126 and 2190 cm^{-1}. It is very doubtful whether the difference is outside the limits of error and therefore, at any rate, it is not possible to exclude model III on the basis of the rule of mutual exclusion.

Five Raman lines have been observed (see Table 79). On the basis of models I and II only three fundamentals are Raman active. Therefore two of the Raman lines must be overtone or combination bands if one of these models is correct. This assumption is not necessary for model III, for which all fundamentals are Raman active. This point favors model III somewhat, since there does not seem to be a Fermi resonance which could make plausible the relatively high intensity of the overtone or combination bands in the Raman spectrum. On the other hand, the observed infrared spectrum can be represented by means of four (infrared-) active fundamentals, in agreement with model I, whereas models II and III would have six and eight active fundamentals respectively. Also, Engler and Kohlrausch (306) carried out calculations for model I on the basis of a generalized valence force system and, using the Raman frequencies, the only frequencies then available, predicted the infrared-active fundamentals, which turned out to be in rough agreement with those later observed by Lord and Wright (590). To be sure, similar calculations have not been carried out as yet for the other two models.

While there is thus no particular point in favor of model II, it seems that the spectroscopic arguments for model III have about as much weight as those for model I. Further investigations, with higher dispersion, and including polarization measurements of the Raman lines, will be necessary to settle this question definitely.

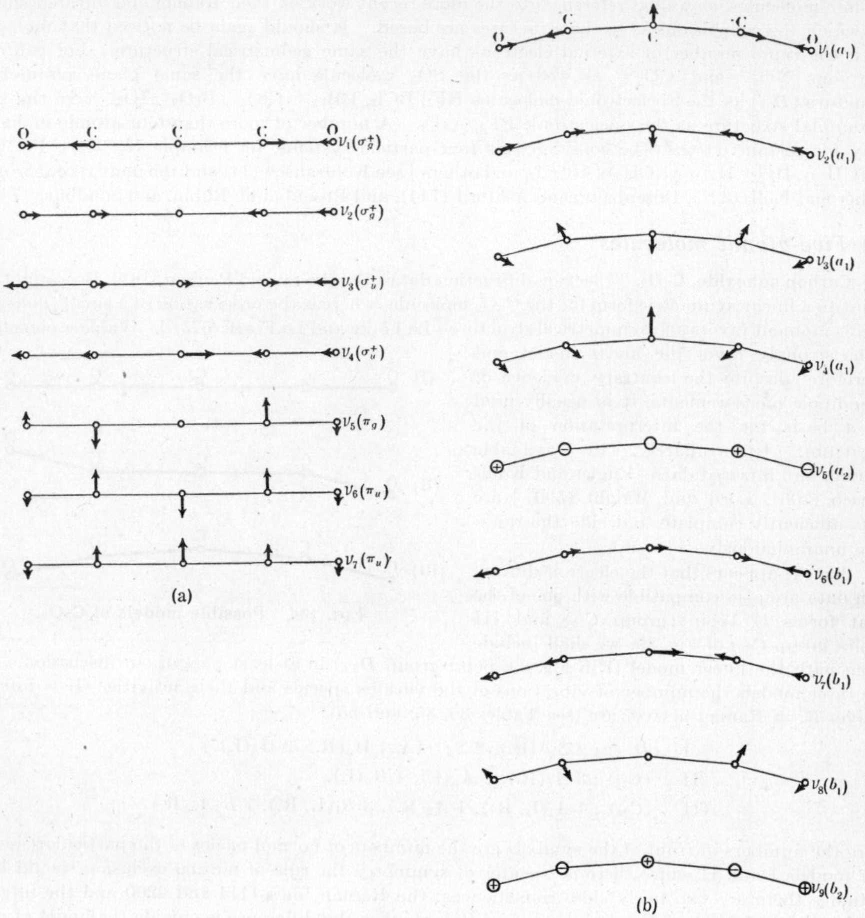

FIG. 90. **Normal vibrations of (a) linear and (b) bent C_3O_2 (schematic).**

In Fig. 90 the expected normal modes for models I and III are given schematically. The three vibrations ν_5, ν_6, and ν_7 of the linear model are doubly degenerate. A final assignment of the observed infrared and Raman bands (Table 79) to the fundamentals of Fig. 90 is not yet possible. In Table 79, three different assignments are given, two based on the linear model (I) and one based on the bent model (III). The assignment of the high-frequency fundamentals is fairly unique: the fairly strong Raman band 2200 cm^{-1} must be the symmetric C═O stretching vibration ν_1 (in either model), since its frequency is very similar to that of ν_3 of CO_2. The strong infrared band 2290 cm^{-1} must be the corresponding antisymmetric C═O vibration ν_3 in model I and ν_6 in III (compare the similar situation in C_2H_2). Similarly, the strong Raman and infrared bands 843 and 1570 cm^{-1} must be the symmetric and antisymmetric C═C vibrations ν_2 and ν_4 of model I and ν_2 and ν_7 of III.[100]

[100] These two frequencies may be compared to the similar pair of C═C vibrations in $H_2C{=}C{=}CH_2$ which are 1071 and 1389 cm^{-1}.

The assignment of the bending vibrations is much less certain. On the linear model, the third strong Raman line 586 cm^{-1} must be interpreted as ν_5. It is difficult to understand why this non-totally symmetric vibration should be so strong in the Raman effect. This difficulty does not appear

TABLE 79. RAMAN SPECTRUM OF LIQUID AND INFRARED SPECTRUM OF GASEOUS C_3O_2 AFTER ENGLER AND KOHLRAUSCH (306) AND LORD AND WRIGHT (590).

ν, observed (cm^{-1})	Band type	Assignments[101] Model I Lord and Wright	Model I Herzberg	Model III Herzberg
544	I. (s.)	$4\nu_7-\nu_7[\Pi_u]$[102]	$2\nu_7-\nu_7[\Pi_u]$[102]	$\nu_9+\nu_5-\nu_5[B_2]$
557	I. (v.s.)	$3\nu_7(\Pi_u)$	$\nu_7(\pi_u)$	$\nu_9(b_2)$
586	R. (s.)	$\nu_5(\pi_g)$	$\nu_5(\pi_g)$	$\nu_4(a_1)$
637	I. (s.)	$\nu_2-\nu_7[\Pi_u]$	$\nu_6(\pi_u)$	$\nu_3(a_1)$
779	I. (m.)	$\nu_5+\nu_7(\Sigma_u^+)$	$\nu_4-\nu_2[\Sigma_u^+]$	$\nu_9+\nu_5(B_1)$
843	R. (s.)	$\nu_2(\sigma_g^+)$	$\nu_2(\sigma_g^+)$	$\nu_2(a_1)$
889	I. (m.)	$\nu_6+\nu_7-\nu_7[\Pi_u]$	$\nu_4+\nu_7-\nu_7-\nu_5[\Pi_u]$	$\nu_8-\nu_5[B_2]$
909	I. (m.)	$\nu_6(\pi_u)$	$\nu_4-\nu_5[\Pi_u]$	
1024	I. (w.)	$\nu_2+\nu_7(\Pi_u)$	$2\nu_5+\nu_6-\nu_2[\Pi_u]$?	
1114	R. (w.)	$\nu_6+\nu_7(\Sigma_g^+, \Delta_g)$	$2\nu_7(\Sigma_g^+)$	
1126	I. (w.)	$3\nu_7+\nu_5(\Sigma_u^+)$	$\nu_5+\nu_7(\Sigma_u^+)$ $2\nu_2-\nu_7[\Pi_u]$	$\nu_8(b_1)$
1176	R. w.	$2\nu_5(\Sigma_g^+, \Delta_g)$	$2\nu_5(\Sigma_g^+, \Delta_g)$	$2\nu_4(A_1)$
1225	I. (v.w.)	$\nu_2+\nu_5-\nu_7[\Sigma_u^+]$	$\nu_6+\nu_5(\Sigma_u^+)$	$\nu_3+\nu_4(A_1)$
1387	I. (m.)	$\nu_7+2\nu_5(\Pi_u)$, $\nu_2+3\nu_7(\Pi_u)$	$\nu_2+\nu_7(\Pi_u)$	$\nu_2+\nu_9(B_2)$
1470	I. (w.)	$\nu_6+\nu_5(\Sigma_u^+)$	$\nu_2+\nu_6(\Pi_u)$	$\nu_2+\nu_3(A_1)$
1570	I. (v.s.)	$\nu_4(\sigma_u^+)$	$\nu_4(\sigma_u^+)$	$\nu_7(b_1)$
1670	I. (m.)	$\nu_2+\nu_7+\nu_5(\Sigma_u^+)$ $3\nu_7+2\nu_5(\Pi_u)$	$3\nu_7(\Pi_u)$	$3\nu_9(B_2)$
1760	I. (m.)	$\nu_2+\nu_6(\Pi_u)$	$2\nu_5+\nu_7(\Pi_u)$	$\nu_5+\nu_7(B_2)$
1850	I. (w.)	$2\nu_2+\nu_7(\Pi_u)$ $\nu_3-2\nu_7[\Sigma_u^+]$	$2\nu_5+\nu_6(\Pi_u)$	
1980	I. (w.)	$\nu_7+2\nu_6(\Pi_u)$	$3\nu_6(\Pi_u)$	$\nu_2+\nu_8(B_1)$
2190	I. (s.)	$\nu_6+2\nu_5(\Pi_u)$	$2\nu_2+\nu_7(\Pi_u)$ $\nu_4+\nu_5(\Pi_u)$	$\nu_1(a_1)$
2200	R. (m.)	$\nu_1(\sigma_g^+)$	$\nu_1(\sigma_g^+)$	
2290	I. (v.s.)	$\nu_3(\sigma_u^+)$	$\nu_3(\sigma_u^+)$	$\nu_6(b_1)$
2410	I. (s.)	$\nu_1+\nu_7(\Pi_u)$ $\nu_2+\nu_4(\Sigma_u^+)$	$\nu_2+\nu_4(\Sigma_u^+)$	$\nu_2+\nu_4(A_1)$
3150	I. (m.)	$\nu_1+\nu_6(\Pi_u)$ $\nu_2+\nu_3(\Sigma_u^+)$	$\nu_2+\nu_3(\Sigma_u^+)$	$\nu_2+\nu_6(B_1)$
3380	I. (w.)	$\nu_4+2\nu_6(\Sigma_u^+)$	$2\nu_2+\nu_4(\Sigma_u^+)$?	$\nu_8+\nu_3(B_1)$
3790	I. (m.)	$\nu_1+\nu_4(\Sigma_u^+)$	$\nu_1+\nu_4(\Sigma_u^+)$	$\nu_1+\nu_4(A_1)$
4590	I. (w.)	$\nu_1+\nu_3(\Sigma_u^+)$[103]	$\nu_1+\nu_3(\Sigma_u^+)$[103]	$2\nu_6(A_1)$

[101] It should be noted that our numbering of the perpendicular vibrations of the linear model is different from that used by Engler and Kohlrausch and by Lord and Wright, but is consistent with our general practice (see Fig. 90). For difference bands the symbol added in square brackets is not the species of the upper state (as for the other bands) but the species of the transition moment.

[102] On the linear model this may also be the maximum of the P branch of $3\nu_7$ and ν_7 respectively whose R branch then would be at 557 cm^{-1}, the separation being of the right order of magnitude [see Lord and Wright (590)].

[103] This band would fit much better as $2 \times 2290 = 2\nu_3$. But such a transition would be forbidden according to the linear model. This is another argument against the linear model.

on the basis of model III. For the other two perpendicular vibrations, ν_6 and ν_7, of the linear model Lord and Wright (590) have given two alternative assignments: $\nu_7 \approx 200$ cm^{-1}, $\nu_6 = 550$, and $\nu_7 \approx 190$, $\nu_6 = 909$. The low value of ν_7 was based on a theoretical estimate of Engler and Kohlrausch (306). For both choices of ν_6 and ν_7, Lord and Wright have been able to assign all other observed infrared bands. The second of these assignments is given in Table 79. However, in both assignments the very strong infrared band at 637 cm^{-1} has to be interpreted as $\nu_2 - \nu_7$, whereas $\nu_2 + \nu_7$ (which should be stronger than $\nu_2 - \nu_7$) is very much weaker (about $\frac{1}{10}$). An assignment (based on the linear model) that is not open to such a serious objection is also given in Table 79. It assumes that the two very strong long-wave-length bands at 557 and 637 are the two fundamentals ν_7 and ν_6 respectively. The conclusion of Engler and Kohlrausch (306) on which Lord and Wright's assignment is based, that ν_7 should be much smaller than ν_5 and ν_6, was obtained under the implicit assumption that there is no appreciable interaction between the C—C—C and C—C—O angular displacements. However, the similar assumption for C_2H_2 would lead to ν_5 being smaller than ν_4, whereas actually it is found to be larger [see Childs and Jahn (206)]. In addition, Engler and Kohlrausch assumed, for want of further information, that the C—C—C bending constant is the same as the C—C—O bending constant, which is certainly an exceedingly rough approximation. There is therefore no need to assume that ν_7 must be as low as 200 cm^{-1}.

The assignments of the deformation vibrations based on the bent model (III) given in Table 79 must be considered as only tentative.

Methane, CH_4 and CD_4. It is usually taken for granted that the four H atoms in CH_4 form a regular tetrahedron whose center is occupied by the C atom. This assumption is usually based on elementary concepts of valence [see, for example, Van Vleck (884)]. However it is very strikingly confirmed by the structure of the infrared and Raman spectra. As we have seen in Chapter I, section 3, CH_4 does not exhibit a rotational Raman spectrum as does, for example, NH_3. This is only compatible with the *tetrahedral model*, since only molecules of cubic symmetry have no rotational Raman spectra, and since the tetrahedral model is the only possible model of cubic symmetry (see p. 42).

This model is further confirmed by the vibration spectrum. Of the three observed Raman bands [Dickinson, Dillon, and Rassetti (287), MacWood and Urey (594)], $\Delta\nu = 2914.2$, 3022, 3071.5, the most intense one (2914.2) is completely polarized [Bhagavantam (146)], which again can in general occur only for a cubic point group.[104] As discussed previously (see p. 140), a five-atomic tetrahedral molecule has only four fundamentals, a totally symmetric (A_1), a doubly degenerate (E), and two triply degenerate (F_2) vibrations (see Fig. 41). According to Table 55 all four are Raman active but only the two triply degenerate vibrations ν_3 and ν_4 are infrared active. It is very significant that the infrared spectrum of CH_4 shows two (and only two) extremely intense bands [see Cooley (229), Dennison (276) (280)] at 1306.2 and 3018.4 cm^{-1} and that all the other much weaker infrared bands can be interpreted on the basis of these two active fundamentals and two inactive fundamentals. If CH_4 were plane and symmetrical (point group $\boldsymbol{D_{4h}}$) or pyramidal[105] (point group $\boldsymbol{C_{4v}}$) there would be three and four active fundamentals respectively. Furthermore, one of the infrared-active fundamentals (3018.4 cm^{-1}) is also Raman active (3022 cm^{-1}), which would be quite incompatible with a plane symmetrical model ($\boldsymbol{D_{4h}}$) for which the rule of mutual exclusion holds. The rotational structure of the bands supplies further confirmation for the tetrahedral model (see Chapter IV).

According to the above the identification of the three fundamentals ν_1, ν_3, ν_4 (Fig. 41) is immediately given.[106] Similar reasons also lead to the identification of these three fundamentals in CD_4. For both molecules the *fundamentals* are given in Table 80. The fourth frequency ν_2 cannot be identified with the third weak Raman band (3071.5 cm^{-1} for CH_4, 2108.1 cm^{-1} for CD_4): first, because this frequency, on account of the form of the vibration involved (Fig. 41), cannot be as high as those of the C—H(C—D) stretching vibrations ν_1 and ν_3; and second, because in the infrared spectrum of CH_4, combination bands of the active fundamentals with an inactive vibration of frequency 1520 cm^{-1}

[104] To be sure, polarization measurements are usually not very accurate and even for a non-cubic point group a vibration may have a very low degree of depolarization. Therefore the above argument alone would not be sufficient to prove the tetrahedral structure of CH_4.

[105] If the C atom were not at the center of the H_4 tetrahedron, that is, if the point group were $\boldsymbol{C_{3v}}$, there would be six active fundamentals.

[106] Whether the higher or the lower of the infrared-active fundamentals is called ν_3 (or ν_4) is, of course, simply a matter of definition. In agreement with the usual practice we assign the lower subscript to the higher frequency.

have been observed. Thus one is led to $\nu_2 \approx 1520$ cm^{-1} for CH_4 and it is necessary to consider the Raman line 3071.5 cm^{-1} as the first overtone of this frequency. The analogous interpretation for CD_4 gives $\nu_2 \approx 1054$ cm^{-1}. The fact that for both CH_4 and CD_4 the first overtone $2\nu_2$ occurs in the Raman spectrum whereas the fundamental (which is Raman active) has not been observed can be easily explained as due to Fermi resonance. It is seen that $2\nu_2$ is close to ν_1. Of the two sublevels

TABLE 80. FUNDAMENTAL FREQUENCIES (1–0 TRANSITIONS) OF GASEOUS CH_4 AND CD_4.

Assignment	CH_4		CD_4	
	ν_{vacuum} (cm^{-1})	References	ν_{vacuum} (cm^{-1})	References
$\nu_1(a_1)$	2914.2 (R.)	(287) (146) (594) (610)	2084.7 (R.)	(594)
$\nu_2(e)$	(1526)[107]	(276)	(1054)[108]	(594)
$\nu_3(f_2)$	3020.3 (I. R.)[109]	(656) (287) (594)	2258.2 (I. R.)[110]	(594) (658)
$\nu_4(f_2)$	1306.2 (I.)	(656)	995.6 (I.)[111]	(658)

of the upper state of $2\nu_2$ (having species A_1 and E; see Table 32) the one of species A_1 can resonate with the upper state of ν_1, and since ν_1 gives the strongest Raman line even a slight interaction will cause $2\nu_2$ to appear much more strongly than it otherwise would.

The correctness of the assignment of the fundamentals can be further checked by the *isotope relations*. Taking the observed values for ν_1 in Table 80 one obtains $\nu_1{}^D/\nu_1{}^H = 0.7154$ (where the superscripts D and H refer to CD_4 and CH_4 respectively), while according to (II, 327) $\omega_1{}^D/\omega_1{}^H = \sqrt{m_H/m_D} = 0.7074$. Similarly $\nu_3{}^D\nu_4{}^D/\nu_3{}^H\nu_4{}^H = 0.5699$, while according to (II, 328) $\omega_3{}^D\omega_4{}^D/\omega_3{}^H\omega_4{}^H = 0.5597$. Considering the neglect of anharmonicity, the agreement is quite satisfactory (compare Table 52). Conversely, by assuming exact agreement for the zero-order frequencies and further assuming that if $\omega_i{}^H = (1 + \alpha_i)\nu_i{}^H$ for CH_4, $\omega_i{}^D = [1 + (\omega_i{}^D/\omega_i{}^H)\alpha_i]\nu_i{}^D$ for CD_4, Dennison (280) calculated the *zero-order frequencies* ω_i from the observed fundamentals ν_i. He obtained for CH_4 (in cm^{-1})

$$\omega_1 = 3029.8, \quad \omega_2 = 1390.2,^{112} \quad \omega_3 = 3156.9, \quad \omega_4 = 1357.6,$$

and for CD_4 (in cm^{-1})

$$\omega_1 = 2143.2, \quad \omega_2 = 983.4,^{112} \quad \omega_3 = 2336.9, \quad \omega_4 = 1026.4.$$

It must be realized, however, that the second of the above assumptions is only roughly fulfilled [compare equation (II, 286)], so that the values obtained cannot claim a high degree of accuracy.

A large number of *overtone and combination bands* of CH_4 have been observed in the ordinary and photographic infrared. They are given in Table 81. As mentioned in Molecular Spectra I (p. 525), some bands occurring in the long-wave-length part of the spectra of the planets Jupiter, Saturn, Uranus, and Neptune have been definitely identified with bands of CH_4 observed with thick absorb-

---

107 This value is not very accurate since it is determined indirectly. The value given here is taken from the band $\nu_2 + \nu_3 = 4546$ cm^{-1}. It is approximately the average of the value 1520 frequently quoted in the literature and obtained from $\nu_2 + \nu_4 = 2823$ cm^{-1} and the value obtained from the Raman band $2\nu_2 = 3071.5$ (neglecting anharmonic terms). Dennison (280) has included the effect of anharmonicity and obtained a value of 1499.4 cm^{-1}.

108 This is half the frequency of the Raman shift 2108.1 cm^{-1} observed by MacWood and Urey (594), assuming that it corresponds to $2\nu_2$ and neglecting anharmonicity. Dennison gives 1036.4 cm^{-1} including anharmonicity.

109 Average of infrared and Raman values.

110 Average of Raman and infrared determination. Dennison (280) gives 2259.4 for the infrared value but the vacuum corrected value of Nielsen and Nielsen (658) is 2258.4 cm^{-1}.

111 Dennison (280) gives 996.5. But this seems to be a mistake since Nielsen and Nielsen (658) give 995.86 and the vacuum correction is −0.27.

112 From the Raman line $2\nu_2$ (see Table 81).

ing layers in the laboratory [Wildt (924), Mecke (613), Dunham (294), Adel and Slipher (39)]. Since the absorbing layers on Uranus and Neptune are much thicker than can be obtained in the laboratory, the spectra of these planets give further CH_4 bands not found in the laboratory, extending down to 4400 Å.

The assignment of the higher overtone and combination bands is rather uncertain for several reasons: (1) Because of the anharmonicity, the overtones of the triply degenerate infrared-active vibrations split into a number of sub-bands whose separations and relative intensities are difficult to

TABLE 81. OVERTONE AND COMBINATION BANDS OF CH_4.

ν_{vacuum}, observed[113] (cm^{-1})	Assignment	Species of upper states	References
1720 I.	$\nu_4-\nu_3$	F_2	(224) (205)
2600 I.	$2\nu_4$	$(A_1+E)+F_2$	(229)
2823 I.	$\nu_2+\nu_4$	$(F_1)+F_2$	(229)
3071.5* R. w.	$2\nu_2$	A_1+E	(287) (886)
4123 I.	$\nu_2+2\nu_4$	$(A_1+A_2+2E+F_1)+F_2$	(229) (633)
4216.3* I.	$\nu_1+\nu_4$	F_2	(674) (656)
4313.2* I.	$\nu_3+\nu_4$	$(A_1+E+F_1)+F_2$	(656)
4546* I.	$\nu_2+\nu_3$	$(F_1)+F_2$	(633)
5585* I.	$\nu_3+2\nu_4$	$(A_1+E+2F_1)+3F_2$	(674) (886)
5775 I.	$\nu_1+\nu_2+\nu_4$	$(F_1)+F_2$	(674) (633)
5861 I.	$\nu_2+\nu_3+\nu_4$	$(A_1+A_2+2E+2F_1)+2F_2$	(633)
6006* I.	$2\nu_3$	$(A_1+E)+F_2$	(633) (674)
7514* I.	$\nu_2+2\nu_3$	$(A_1+A_2+2E+F_1)+F_2$	(674)
8421 I.	$2\nu_1+2\nu_4$	$(A_1+E)+F_2$	(674)
8604 I., P. I. v.s.	$2\nu_3+2\nu_4$	$(4A_1+A_2+3E+3F_1)+5F_2$	(674) (434a)
8807* I., P. I. v.s.	$2\nu_1+\nu_3$	F_2	(674) (434a)
8900 P. I. s.	$\nu_1+2\nu_3$	$(A_1+E)+F_2$	(205)
9047* P. I. s.	$3\nu_3$	$(A_1+F_1)+2F_2$	(205)
10114 P. I. s.	$2\nu_1+\nu_2+\nu_3$	$(F_1)+F_2$	(886)
10300* P. I. s.	$\nu_1+\nu_2+2\nu_3$	$(A_1+A_2+2E+F_1)+F_2$	(886)
	$\nu_2+3\nu_3$	$(E+3F_1)+3F_2$	
11270 P. I. s.	$3\nu_1+\nu_3$	F_2	(886) (282)
11620 P. I. m.	$2\nu_1+2\nu_3$	$(A_1+E)+F_2$	(886)
11885 P. I. w.	$\nu_1+3\nu_3$	$(A_1+F_1)+2F_2$	(886)
	$4\nu_3$	$(2A_1+2E+F_1)+2F_2$	
12755 P. I. v.w.	$3\nu_1+\nu_3+\nu_4$	$(A_1+E+F_1)+F_2$	(886)
	$2\nu_1+\nu_2+2\nu_3$	$(A_1+A_2+2E+F_1)+F_2$	
13790 P. I. m.	$4\nu_1+\nu_3$	F_2	(886)
	$3\nu_1+2\nu_3$	$(A_1+E)+F_2$	

predict. (2) Since ν_1 and $2\nu_2$ have approximately the same frequency as ν_3, the combinations $n\nu_1 + m\nu_3$, $2k\nu_2 + m\nu_3$ are close to $(n + m)\nu_3$ and $(k + m)\nu_3$ respectively. They too are split into sub-bands. (3) Perturbations between the sublevels mentioned under (2) and those under (1) will occur if they have the same species and bring about further deviations from a simple quadratic formula of the type (II, 281). (4) The centers of the bands, because of the overlapping discussed under (1), (2), and (3), are often very ill-defined. For these reasons we have not given in Table 81 the bands observed in the planetary atmospheres only. The assignment of these bands given by Adel and Slipher (39) is certainly not the only possible one,[114] and is not compatible with the assign-

[113] The bands marked with an asterisk have a well-developed central maximum (Q branch); for all others the wave numbers given are not very accurate. Satisfactory intensity estimates are possible at present only for the photographic infrared bands.

[114] Their assignment of a number of bands as $n\nu_4$ with n going up to 16 is hardly tenable.

ment of the laboratory bands given in Table 81, which is an extension of that of Vedder and Mecke (886) and Childs (205). In the third column of Table 81 are given the species of all the sublevels of the respective upper states, including those that cannot combine with the ground state, the latter in brackets. Thus, for example, the upper state of $\nu_3 + 2\nu_4$ consists of seven sublevels, three of which (of species F_2) can combine with the ground state. The observed band at 5585 cm^{-1} is thus presumably a superposition of three sub-bands. Similar considerations apply to the other bands.

The assignment given for the higher overtones is only tentative and it would certainly be premature to give a formula to represent all bands.

For CD_4 up to the present time only two combination bands are known, at 2992.0 and 3102.8 cm^{-1} [Nielsen and Nielsen (658)]. They have been interpreted as $\nu_1 + \nu_4$ and $\nu_3 + \nu_4$ respectively. However, this would require very large anharmonic terms, much larger than for CH_4. It appears very likely that the first of these bands is a fundamental of CD_3H (its frequency agrees exactly with Dennison's (280) predicted value for the C—H stretching frequency of CD_3H),[115] and that the second band is $\nu_1 + \nu_4$ of CD_4 which would imply much smaller anharmonic terms. Further work on CD_4 is desirable and promises to give interesting results.

CH_3D, CH_2D_2, CHD_3. The Raman spectra of the intermediate methanes CH_3D, CH_2D_2, CHD_3 have been investigated by MacWood and Urey (594), the infrared spectra by Ginsburg and Barker (362) and by Benedict, Morikawa, Barnes, and Taylor (137). As has been discussed previously (see p. 236f.), in going from CH_4 to CH_3D or from CD_4 to CHD_3 part of the degeneracies are removed,

TABLE 82. FUNDAMENTALS OF THE VARIOUS DEUTERATED METHANES.[116]

Molecule	$\nu_1(a_1)$	$\nu_2(e)$	$\nu_3(f_2)$	$\nu_4(f_2)$
CH_4	2914.2 R.	1526 I.* R.*	3020.3 I. R.	1306.2 I.
CH_3D	2204.6 I. ‖ (R.)	1476.7 I. ⊥ (R.*)	2982.2 I., 3030.2 I.	1306.4 I. ‖ (R.),[116a] 1156.0 I. ⊥ (R.*)
CH_2D_2	2139.0 R.	1450 I. (R.*),[116b] 1285.6 R.	2974.2 R., 3020 I., 2255 I.	1034.4 I. R., 1235.2 I., 1090.2 I. (R.*)
CHD_3	2141.1 R.	1299.2 (I.) R.	2992.0 I. ‖[117] 2268.6 (I.) R.	982 (I.) R.*, 1046 (I.) R.
CD_4	2084.7 R.	1054 R.*	2258.2 I. R.	995.6 I.

each of the triply degenerate vibrations ν_3 and ν_4 splitting into a non-degenerate and a doubly degenerate vibration. All six resulting vibrations ($3A_1 + 3E$) are infrared active and Raman active. For CH_2D_2 all degeneracies are removed and we have nine different fundamentals, all but one of which are infrared active and all of which are Raman active.

All the fundamentals of the three compounds have been found, most of them both in the infrared and in the Raman spectrum. They are given in Table 82 together with those of CH_4 and CD_4 in an arrangement similar to Table 54 (p. 238), so that the correlation of the frequencies is clearly shown. In principle the correct correlation can be obtained, according to the correlation rules previously given (p. 238), from the band types (whether ‖ or ⊥, and so on). But actually, since the band type has not been established for many of the observed bands, the correlation is based on

---

[115] The observed spacing in the fine structure also fits better for CD_3H than for CD_4.

[116] R.* refers to fundamentals obtained from observed overtones in the Raman spectrum. If a fundamental has been observed both in the Raman and in the infrared spectrum the value given is the more accurate one of the two. Which observation is not used is indicated by putting I. or R. in brackets. The average is taken when neither I. nor R. is in brackets. All values are corrected to vacuum.

[116a] MacWood and Urey (594) give a Raman band at 1330.1 cm^{-1}. The difference from the infrared band is difficult to understand.

[116b] MacWood and Urey (594) give a Raman band at 1332.9 cm^{-1} which may have to replace 1450 cm^{-1} in the table.

[117] This is the band ascribed by Nielsen and Nielsen (658) to CD_4 which above on this page was shown very probably to be due to CD_3H. Benedict, Morikawa, Barnes, and Taylor (137) give a CD_3H band at 3000 cm^{-1} which within their limit of accuracy agrees with the above figure.

calculations by Dennison and Johnston (283) and Dennison (280). They have predicted the frequencies of the intermediate methanes from those of CH_4 and CD_4 by computing the most general potential constants of CH_4 and using these same constants in deriving the frequencies of CH_3D, CH_2D_2, CHD_3 with the help of formulae developed by Rosenthal (747). However, for a few fundamentals these authors have not taken the correlation rules into account. This has been done throughout in Table 82. In place of the potential calculations one could also have used the isotope relations [for example equation (II, 330)] to find the correct correlation. It may be left to the reader to check whether they are actually fulfilled.

Childs and Jahn (207) have observed a photographic infrared band of CH_3D of the $||$ type at 9020.8 cm^{-1}, which is probably one of the sub-bands of the second overtone of 3030.2 cm^{-1}.

Carbon tetrachloride, CCl_4. The Raman spectrum of liquid CCl_4 has been investigated by a large number of workers [see Kohlrausch (13) (14)], but no one seems to have investigated the spectrum of the vapor in any detail. The infrared spectrum has been investigated by only a few authors [most recently by Schaefer and Kern (766) and Barchewitz and Parodi (101)] and also only for the liquid state.[118] It seems fairly safe, however, to assume that the difference between the spectrum of the vapor and that of the liquid is slight in view of the symmetrical, inert character of the molecule.

Considering the similarity to CH_4, it seems natural to assume that CCl_4 also has the *symmetrical tetrahedral structure* (point group T_d). But at some times this has been doubted by various investigators of the spectrum. We shall see, however, that more recent investigations definitely support the tetrahedral model.

The Raman spectrum yields the eight displacements given in Table 83, some of which occur both as Stokes and as anti-Stokes lines (see Fig. 77). Three of the shifts, 145, 434, and 1539 cm^{-1}, correspond to exceedingly weak Raman lines. They are in all probability overtone or combination bands and are indeed very readily interpreted as such (see Table 83). Moreover, a frequency as high as 1539 would not be expected as a fundamental of CCl_4, the vibration frequency of diatomic CCl being 844 cm^{-1}. Thus there remain four or five observed Raman-active fundamentals, depending on whether the two lines 762.0 and 790.5 cm^{-1} are considered as a doublet or as two distinct frequencies. The first assumption is in agreement with the tetrahedral model since for it just four fundamentals, all active in the Raman effect, should occur. Furthermore, as it should be for the tetrahedral model, one of the Raman lines (with $\Delta\nu = 460$ cm^{-1}) is almost completely polarized, whereas the other three are completely depolarized (see p. 270f.). Finally, the infrared spectrum seems to show only two fundamentals, the band 305 cm^{-1} and the doublet band 775 cm^{-1}, in agreement with expectation for the tetrahedral model. While the Raman fundamental 218 cm^{-1} is probably outside the range investigated in the infrared, the Raman band 460 cm^{-1} is certainly within this range and is definitely absent. In consequence of this observation and in view of the complete polarization[119] in the Raman effect, this fundamental must be considered as the totally symmetric vibration $\nu_1(a_1)$. The two infrared-active fundamentals 775 and 305 cm^{-1} can only be $\nu_3(f_2)$ and $\nu_4(f_2)$ respectively (see Fig. 41), and thus the fourth Raman-active fundamental at 218 cm^{-1} must be $\nu_2(e)$. All of the numerous other infrared bands of liquid CCl_4 found by Schaefer and Kern can be interpreted on the basis of these four fundamentals (see Table 83). The doublet structure of ν_3 is explained in this interpretation without difficulty as due to a Fermi resonance [Placzek (700)] between ν_3 and $\nu_1 + \nu_4$. As can be seen from Table 83, the upper state of $\nu_1 + \nu_4$ has nearly the same energy as the average of the two doublet components and also has the same species (F_2) as the upper state of ν_3.

Langseth (547) found a fine structure of the Raman lines ν_1, ν_2, and ν_4 of CCl_4, which he thought was due to a deviation from tetrahedral symmetry. But Wu and Sutherland (957) have shown that this fine structure can be explained as due to a partly resolved isotope effect produced by the presence of the isotopic modifications CCl_4^{35} (31.6 per cent), $CCl_3^{35}Cl^{37}$ (42.2 per cent) and $CCl_2^{35}Cl_2^{37}$ (21.1 per cent).[120] This has been further confirmed by intensity measurements in the fine structure carried out by Menzies (621).

---

[118] Schaefer and Kern (766) state that for the most intense bands they used also the vapor but do not give separate measurements.

[119] If the slight observed depolarization of 5 per cent is real it can easily be explained [see Placzek (700)] as due to the effect of the isotopes $CCl_3^{35}Cl^{37}$ and $CCl_2^{35}Cl_2^{37}$, for which the polarizability change is not exactly spherically symmetric.

[120] The molecules $CCl^{35}Cl_3^{37}$ (4.7 per cent) and CCl_4^{37} (0.4 per cent) are not abundant enough to produce a noticeable contribution.

While thus all observations are in agreement with the symmetrical tetrahedral model, it is necessary to inquire whether they are not also in agreement with a less symmetrical model. As in the case of CH_4 (see above), the plane symmetrical structure (point group D_{4h}) is immediately ruled out by the fact that one of the fundamentals (775 cm^{-1}) occurs in the infrared as well as the Raman spectrum.[121] Two other conceivable models are a pyramid with the C atom at the vertex (point

TABLE 83. RAMAN SPECTRUM AND PART [123] OF THE INFRARED SPECTRUM OF LIQUID CCl_4.

Raman shifts[124] $\Delta\nu_{\text{vacuum}}$, (cm^{-1})	Infrared[125] bands ν_{vacuum}, (cm^{-1})	Assignment[126]	References
145 v.w.		$\nu_1 - \nu_4 [F_2]$	(43)
217.9 s. depol.		$\nu_2(e)$	(547)
314.0 s. depol.	305	$\nu_4(f_2)$	(547) (819) (101)
434 v.w.		$2\nu_2(A_1+E)$	(43)
$\left\{\begin{array}{l}455.1\\458.4\\461.5\end{array}\right.$ v.s. pol.		$\left.\begin{array}{l}CCl_2{}^{35}Cl_2{}^{37}\\CCl_3{}^{35}Cl^{37}\\CCl_4{}^{35}\end{array}\right\}\nu_1(a_1)$	(547) (547) (547)
	635 w.	$2\nu_4(A_1+E+F_2)$	(766)
762.0 m. depol.	768 v.s. $\big\}$	$\nu_3(f_2),\ \nu_1+\nu_4(F_2)$	(547) (766)
790.5 m. depol.	797 v.s.		(547) (766)
	982 m. $\big\}$	$\nu_2+\nu_3(F_2),\ \nu_1+\nu_2+\nu_4(F_2)$	(766)
	1006 m.		(766)
	1068 w. $\big\}$	$\nu_3+\nu_4(A_1+E+F_2),\ \nu_1+2\nu_4(A_1+E+F_2)$	(766)
	1107 w.		(766)
	1218 m. $\big\}$	$\nu_1+\nu_3(F_2),\ 2\nu_1+\nu_4(F_2)$	(766)
	1253 m.		(766)
	1529 m. $\big\}$	$2\nu_3(A_1+E+F_2)$	(766)
1539 v.w. pol.[127]	1546 s.	$2\nu_1+2\nu_4(A_1+E+F_2)$	(710) (766)
	1575 m.	$\nu_1+\nu_3+\nu_4(A_1+E+F_2)$	(766)

group C_{4v}) or a pyramid with an H atom at the vertex and the C atom on the axis (point group C_{3v}, for example, when there is only a slight deviation from tetrahedral symmetry, as has at one time been assumed). These models would have the fundamentals $2A_1 + 2B_1 + B_2 + 2E$ and $3A_1 + 3E$ respectively (see Table 36), all of which would be Raman active (see Table 55). Apart from the fact that the number of observed reasonably strong Raman lines would be less than the number of Raman active fundamentals (even if the doublet at 775 is counted as two fundamentals), the important point is that only one polarized Raman shift (460 cm^{-1}) occurs, whereas for the above two models two or three should occur and they should also be the strongest Raman lines. Furthermore, in the case of the model of symmetry C_{3v}, if the doublet 775 is counted as two lines [122] the number of depolarized Raman lines is four instead of the three required ones. Thus we can consider also the models of symmetry C_{3v} and C_{4v} as ruled out. For still less symmetrical models, the number of

---

[121] The fact that both components (762 and 790 cm^{-1}) occur in both cases excludes the possibility of a chance coincidence.

[122] Otherwise there would be no reason to assume any other but the tetrahedral model.

[123] Only the part below 1600 cm^{-1} and only bands with an absorption coefficient $k_{\text{maximum}} > 4$ are given.

[124] For the degree of depolarization see Rao (717) and the literature quoted by him.

[125] Barchewitz and Parodi (101) give in addition the bands at 217?, 529, 370, 247 cm^{-1} observed in paraffine solution.

[126] Sublevels of species F_1 are omitted, since they combine with the ground state neither in the Raman nor in the infrared spectrum.

[127] Polarization from Rao (718).

polarized Raman lines would *a fortiori* not be correct, so we can consider the symmetrical tetrahedral model as proven.

It should not be overlooked that the following difficulty, which is independent of the CCl_4 model assumed, still remains to be solved: Since the fundamental frequencies of CCl_4 are small, a considerable fraction of the molecules is in the states in which one (or even more than one) vibration is excited by one or two quanta. This gives rise to comparatively strong anti-Stokes lines, as observed for CCl_4. It should also give rise to certain *difference bands* in the Raman spectrum, which however have not been observed. While the ordinary difference bands such as $\nu_1 - \nu_2$, $\nu_3 - \nu_2$, $\cdots$ would be expected to be very weak, just as other binary combinations, and therefore are not readily observed (only one such band, $\nu_1 - \nu_4 = 145$ cm^{-1}, has been found for CCl_4), there are others such as $\nu_1 + \nu_2 - \nu_2$, $2\nu_2 - \nu_2$, $\cdots$ whose intensity, apart from the Boltzmann factor, should be of the same order of magnitude as that of the fundamentals ν_1, ν_2, $\cdots$ respectively. In general, these difference bands coincide very nearly with the corresponding fundamentals and are therefore not resolved in Raman measurements, but in the case of perturbations (Fermi resonance) some of them will be widely separated from the fundamentals. In the present case the transition $\nu_1 + \nu_4 - \nu_4$ whose upper state is split because of resonance, should consist of the two lines at $790.5 - 314.0 = 476.5$ cm^{-1} and $762.0 - 314.0 = 448.0$ cm^{-1}, which would be well separated from ν_1. According to calculations by Horiuti (457), they should have about $\frac{1}{5}$ of the intensity of the strong line 460 cm^{-1}. That they have not been found may be due to the fact that these difference lines are much broader than the main line. But more experimental data, particularly for CCl_4 vapor and with long exposure times, are required to clear up this point.

For a discussion of the force field in CCl_4 see Tables 39 and 46 and the accompanying discussion.

Methyl chloride, CH_3Cl. Since, as has been shown, CH_4 is a symmetrical tetrahedral molecule one would expect CH_3Cl to have a *three-fold axis of symmetry* (the C—Cl axis); that is one would expect it to belong to point group $\boldsymbol{C_{3v}}$. This conclusion is confirmed by the investigation of the vibrational spectrum and particularly of the rotational fine structure (see Chapter IV). The symmetrical model ($\boldsymbol{C_{3v}}$) has six fundamentals, three totally symmetric ones (A_1) and three degenerate ones (E), all of which are both infrared and Raman active, whereas any less symmetric model would give nine active fundamentals. Actually the infrared as well as the Raman spectrum can be analyzed in terms of six fundamentals (see Table 84). Unfortunately, with one exception (see Table 84), the Raman spectrum has been investigated only for *liquid* CH_3Cl [first by Dadieu and Kohlrausch (256)], while the infrared data refer to the gaseous state [first studied by Bennett and Meyer (138)]. Slight inconsistencies are due to this fact.

The form of the six normal vibrations is represented in Fig. 91. The totally symmetric vibrations ν_1, ν_2, ν_3 (species A_1) give rise to $\|$ bands in the infrared (only $M_z \neq 0$; see Table 55), the degenerate vibrations ν_4, ν_5, ν_6 give rise to $\perp$ bands ($M_z = 0$). Actual observation shows that there are indeed three $\|$ and three $\perp$ bands[128] by whose combinations all other bands can be explained. In addition it is found, in agreement with theoretical expectation, that the Raman lines corresponding to the $\|$ bands are polarized, those corresponding to the $\perp$ bands are depolarized. To be sure there are four strong $\|$ bands in the infrared which might be considered as fundamentals, but two of them at 2878.8 and 2966.2 cm^{-1} form a fairly close doublet whose center coincides very nearly with twice the wave number of the $\perp$ band at 1453.2 cm^{-1}. It can therefore be concluded that a Fermi resonance occurs [Adel and Barker (36)] and that the doublet corresponds to one fundamental only (see below).

More detailed calculations show, in agreement with what has been said about group frequencies in Chapter II, section 4f, that the two vibrations ν_1 and ν_4 of frequency about 3000 cm^{-1} are essentially C—H stretching vibrations, that the lowest $\|$ vibration ν_3 is essentially a C—Cl vibration, that the lowest $\perp$ vibration ν_6 is essentially a $H_3\equiv$C—Cl bending vibration, and that the two intermediate vibrations are essentially CH_3 deformation vibrations (see Fig. 91). The observed Cl isotope effect is also in agreement with this assignment, at least for the $\|$ bands. The largest isotope shift, of 6 cm^{-1} occurs for ν_3, the C—Cl vibration [Barker and Plyler (111)]. A much smaller shift, of 0.8 cm^{-1}, has been observed for ν_2 [Nielsen and Nielsen (659)], and the shift for ν_1 is smaller than the resolution used for this band (that is, smaller than 0.8 cm^{-1}). The shift for ν_3 agrees well with the one calculated by considering CH_3Cl as a diatomic molecule CH_3—Cl.

The observed overtone and combination bands of CH_3Cl are given together with the fundamentals in Table 84. All but two very weak Raman shifts 2683 and 1106 cm^{-1} are readily explained

[128] How they are distinguished will be explained in the next chapter.

TABLE 84. INFRARED AND RAMAN SPECTRA OF CH_3Cl.[129]

ν_{vacuum} (cm^{-1}) Infrared (vapor)	$\Delta\nu_{vacuum}$ (cm^{-1}) Raman[130] (liquid)	Assignment[131]	References	ν_{vacuum} (cm^{-1}) Infrared (vapor)	Assignment[131]	References	ν_{vacuum} (cm^{-1}) Infrared (vapor)	Assignment	References
732.1 $\parallel$ (v.s.)	709 (v.s. pol.)	$\nu_3^{CCl}(a_1)$	(138)	3978.9[133] $\perp$	$\nu_1+\nu_6(E)$	(655)	6894 ?	$\nu_1+\nu_4+\nu_6$, $2\nu_1+\nu_6$	(634)
			(111)	4046.0[133] $\parallel$	$\nu_4+\nu_6(A_1+[A_2]+E)$	(655)	8690 P.I. (m.)	$3\nu_1(A_1)$	(846)
1015.0[132] $\perp$ (m.)	1016 (w. depol.)	$\nu_6(e)$	(138)				8890 P.I. (w.)	$2\nu_1+\nu_4(E)$	(846)
	1106 (w.)	?		4088.4[133] $\parallel$	$\nu_2+2\nu_5(A_1+E)$	(655)	8990 $\perp$ P.I. (v.w.)	$\nu_1+2\nu_4(A_1+E)$	(898) (846)
1354.9 $\parallel$ (s.)	1370 (v.w. pol.)	$\nu_2^{CH_3}(a_1)$	(138)	4174.7 $\perp$		(655)	9155 P.I. (v.w.)	$3\nu_4(A_1+[A_2]+E)$	(846)
			(655)	4229.4 $\parallel$	$\nu_1+\nu_2(A_1)$	(655)	9690 P.I. (v.w.)	$3\nu_1+\nu_6(E)$	(846)
1454.6[132] $\perp$ (m.)	1446 (w. depol.)	$\nu_5^{CH_3}(e)$	(138)	4382.6 $\perp$	$\nu_4+\nu_2(E)$	(655)	10050 P.I. (v.w.)	$3\nu_1+\nu_2$, $3\nu_4+\nu_6$, ···	(655) (846)
2461.0 $\perp$ (w.)		$\nu_5+\nu_6(A_1+[A_2]+E)$	(655)	4452.3 $\perp$	$\nu_4+\nu_5(A_1+[A_2]+E)$	(655)	11239 ? P.I. (w.)	$4\nu_1$, $3\nu_1+\nu_4$	(846) (901) (864)
	2683 (v.w.)	$2\nu_2(A_1)$?		5400 $\perp$	$\nu_2+\nu_4+\nu_6$?	(655)	11265 $\parallel$ P.I. (w.)	$2\nu_1+2\nu_4$, $\nu_1+3\nu_4$, $4\nu_4$	(901) (846)
	2815 (w.)	$\nu_2+\nu_5(E)$		5900 $\perp$	$\nu_4+\nu_1$, $\nu_4+2\nu_5$	(655)	13800[133] P.I. (v.w.)	$5\nu_1$, $4\nu_1+\nu_4$, ···	(655)
2878.8 $\parallel$ (s.)	2861 (m.)	$2\nu_5(A_1+E)$	(138)	6013.5[133] $\parallel$	$2\nu_4(A_1+E)$	(655)			
2966.2 $\parallel$ (v.s.)	2955 (v.s. pol.)	$\nu_1^{CH}(a_1)$	(138)	6076[133] $\perp$		(655)			
3041.8[132] $\perp$ (s.)	3036 (m. depol.)	$\nu_4^{CH}(e)$	(138)						

[129] Wherever the bands of CH_3Cl^{35} and CH_3Cl^{37} are resolved the data given refer to the former.

[130] The Raman data given are due to Wagner (906). References to the numerous earlier papers on the Raman spectrum of CH_3Cl may be found in Kohlrausch (13) (14) and Hibben (10). More recently Nielsen and Ward (670) have observed two Raman shifts 726 and 2968 cm^{-1} for the vapor. These agree much better than the corresponding Raman shifts of the liquid with the infrared spectrum of the vapor.

[131] The numbering is that used in Fig. 91, which is different from that used by Barker, Dennison, and their co-workers. The symbols in parentheses give the species of the upper states. Transitions to the substates A_2 are not allowed (see Table 55). This has been overlooked by Nielsen and Barker (655). Consequently their interpretation of the overtones has been slightly changed. In the designation of the overtones, $2\nu_5$ has been taken for the lower, ν_1 for the higher of the two resonating levels.

[132] The position of the zero line in these bands is incorrectly given in all papers on the subject. It is not the frequency of the line $n = 0$ in Bennett and Meyer's (138) nomenclature but approximately half-way between the lines $n = 0$ and $n = -1$. It is exactly half-way between these lines only if $\zeta = 0$ and $A' = A''$, $B' = B''$ (see Chapter IV, section 2b). The deviation because of $\zeta \neq 0$ has been taken into account in the above table only for the fundamentals.

[133] There are a few inconsistencies in the last two decimal places of these figures in Nielsen and Barker's paper which have been corrected as far as possible.

in terms of the above fundamentals. The Raman shift 2683 may be $2\nu_2$, but a reasonable explanation for the shift 1106 cannot be given. Possibly it might be related to the fact that the Raman spectrum refers to liquid rather than gaseous CH_3Cl and that in the liquid double molecules occur.

The resonance between $2\nu_5$ and ν_1 mentioned above influences, of course, only the sublevel A_1 of $2\nu_5$ whereas $2\nu_5(E)$ remains uninfluenced.[134] That is why only two strong || bands appear in this region but no ⊥ band of comparable intensity. It should also be noted that the resonance is not very close as indicated by the unequal intensities of the Raman and infrared bands.

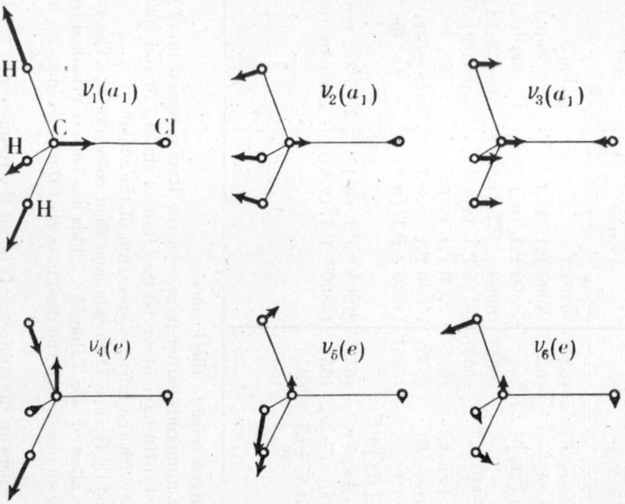

FIG. 91. **Normal vibrations of CH_3Cl (schematic).**—Only side views, and only one component of each degenerate vibration are given.

It is interesting that for some of the overtone and combination bands (for example $2\nu_4$) at least two of the sublevels of the upper state give rise to sub-bands as indicated in Table 84. The complicated structure of the photographic infrared bands is probably due to the overlapping of several such sub-bands as well as to the approximate equality of ν_1 and ν_4, and to the isotope effect.

Further work on CH_3Cl is necessary before a satisfactory formula to represent all the vibrational levels can be developed.

CH_3F, CH_3Br, CH_3I, CD_3Cl, CD_3Br. The other methyl halides have not been investigated quite as fully as CH_3Cl. Naturally their spectra are very similar to that of CH_3Cl. We can therefore omit a full discussion and give in Table 85 only the fundamental frequencies according to Bennett and Meyer (138) and Barker and Plyer (111). Data about overtone and combination bands are given by Moorhead (634), Vierling and Mecke (901), Verleger (898), Naudé and Verleger (650), and Thompson (846). For a discussion of the Raman spectra of the liquids see Wagner (906) and the books of Kohlrausch (13) (14) and Hibben (10). The Raman spectrum of CH_3Br vapor has been investigated by Nielsen and Ward (670). The fundamentals of "heavy" methyl chloride and bromide (CD_3Cl and CD_3Br) recently obtained by Noether (672) are also included in Table 85. The Fermi resonance discussed above for CH_3Cl occurs also for the other molecules. Therefore both resonating levels have been given in Table 85. As for CH_3Cl, the high-frequency component of the doublet is from two to four times as strong as the low-frequency component, indicating that the unperturbed ν_1 is nearer the larger of the two frequencies.

It is interesting to study the way in which the frequencies change in going from the lighter to the heavier halides. For this purpose Fig. 92 gives a graphical representation of the four spectra. It is seen that the frequencies ν_1, ν_4, ν_5 are remarkably constant throughout the series of molecules

[134] It may be noted that $2\nu_3$ very nearly equals ν_5 but no perturbation occurs because the states have different species.

TABLE 85. FUNDAMENTAL FREQUENCIES OF THE METHYL HALIDES,[135] AFTER BENNETT AND MEYER (138), BARKER AND PLYLER (111), AND NOETHER (672).

	ν_{vacuum} (cm^{-1})					
	CH_3F	CH_3Cl^{35}	CH_3Br^{79}	CH_3I	CD_3Cl	CD_3Br
$\nu_3(a_1)$	1048.2	732.1	611	532.8	695	577
$\nu_6(e)$	1195.5	1015.0	952.0	880.1	775	717
$\nu_2(a_1)$	1475.3	1354.9	1305.1	1251.5	1029	987
$\nu_5(e)$	1471.1	1454.6	1445.3	1440.3	1058	1053
$2\nu_5(A_1)$	2861.6	2878.8	2861	2861	2103	2088
$\nu_1(a_1)$	2964.5	2966.2	2972	2969.8	2161	2151
$\nu_4(e)$	2982.2	3041.8	3055.9	3060.3[136]	2286	2293

CH_3X. This is in agreement with their interpretation as C—H vibrations and also follows directly from calculations in which a pure valence force system is assumed [see Wagner (906)] as well as when a somewhat more general force system is assumed [Slawsky and Dennison (796), Linnett (582)]. The frequency ν_3 decreases considerably in going from CH_3F to CH_3I in agreement with the assumption that it is a C—X vibration, and similarly the bending vibration ν_6 decreases although

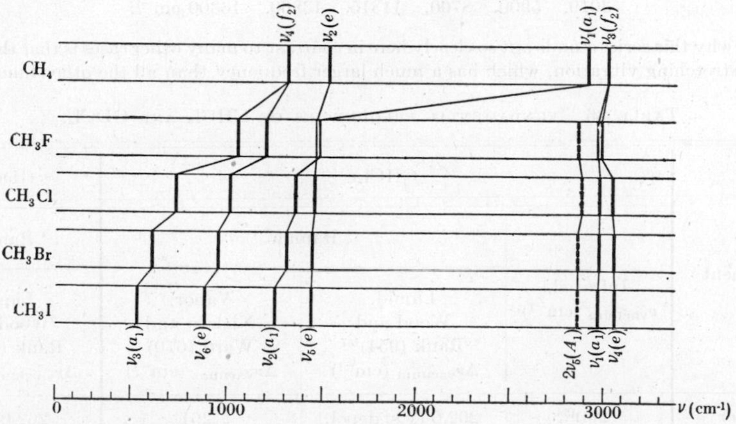

FIG. 92. **Observed fundamental vibration spectra of the methyl halides and methane.**—The broken lines indicate the frequencies $2\nu_5(A_1)$ which are in resonance with $\nu_1(a_1)$. The unperturbed ν_1 would be between ν_1 and $2\nu_5$ (see text).

not quite as strongly. These changes are easily understood on the basis of the change in mass alone (the change of the C—X force constant acts in the same direction). However, the cause of the appreciable change of the symmetrical C—H bending vibration ν_2 is perhaps less easily visualized. The reason for this change is that ν_3, which has the same species, pushes ν_2 upward. If for a moment we consider the CH_3 group as one atom, then ν_3 would increase up to about 3000 cm^{-1} if we were to decrease the atomic weight of X to 1; that is, the curves ν_3 and ν_2 as functions of the mass would intersect each other. Actually, such an intersection does not take place because of the mechanical

[135] The frequencies of the degenerate fundamentals have been corrected as indicated in footnote 132 on p. 313 including the correction due to ζ. All frequencies have been corrected for vacuum. The numbering is that of Fig. 91.

[136] This is the value given by Lagemann and Nielsen (546) (but corrected for vacuum and for the effect of ζ), Bennett and Meyer's numbering of the fine structure lines being incorrect in this case.

coupling of the motions of the different atoms (the cross terms in the kinetic energy). In actual fact the CH_3 deformation frequency ν_2, if we could continuously decrease the mass of X, would eventually go over into the valence stretching vibration $\nu_1 = 2914$ cm^{-1} of methane, whereas the C—X stretching frequency ν_3 (together with ν_6) would go over into the deformation frequency ν_4 of methane [see particularly Wagner (906)]. This is shown in Fig. 92.

The preceding considerations supply a good example for the previous discussion of the limitations of the concept of group frequencies (see p. 199f.). A frequency will remain constant in a series of molecules containing the same group (or groups) and can be assigned to that group only *as long as no other frequency belonging to a normal mode of the same species is in its neighborhood.* This condition is well fulfilled for ν_1, ν_4, ν_5 in all halides (and even in CH_4) but for ν_2 and ν_3 only for the heavy halides. For CH_3I, ν_3 may well be called the C—I vibration frequency and ν_2 the symmetrical CH_3 deformation frequency, but the corresponding statement for CH_3F would have much less significance.

Chloroform, $CHCl_3$. The infrared spectrum of chloroform is only very incompletely known. Most of the work has been limited to the investigation of liquid (or dissolved) $CHCl_3$ with low dispersion particularly in the region of the fundamentals. [Emschwiller and Lecomte (304) and Barchewitz and Parodi (101)]. While a number of overtone and combination bands have been observed in the shorter-wave-length region, including the photographic infrared [Ellis (302), Timm and Mecke (864), Vierling and Mecke (901), Maione (595), Corin (231), Carrelli and Tulipano (194), Rumpf and Mecke (751), Herzberg and McKay (436)], we give in Table 86 only the fundamentals. But it should be mentioned that the series of overtone bands of the C—H stretching vibration has been observed to $v = 6$ far into the visible region [Ellis (302), Rumpf and Mecke (751)]. The series is (in the liquid)

$$3019, \quad 5900, \quad 8700, \quad 11315, \quad 13860, \quad 16300 \text{ cm}^{-1}.$$

The reason why this series stands out so clearly here in contrast to many other cases is that there is only one C—H stretching vibration, which has a much larger frequency than all the other fundamentals.

TABLE 86. FUNDAMENTAL FREQUENCIES OF $CHCl_3$ AND $CDCl_3$.

Assignment	$CHCl_3$			$CDCl_3$,
	Infrared, ν_{vacuum} (cm^{-1})	Raman		Raman
		Liquid, Wood and Rank (954)[137] $\Delta\nu_{vacuum}$ (cm^{-1})	Vapor, Nielsen and Ward (670) $\Delta\nu_{vacuum}$ (cm^{-1})	Liquid, Wood and Rank (954)[137] $\Delta\nu_{vacuum}$ (cm^{-1})
$\nu_6^{CCl_3}(e)$	260[138]	262.0 (v.s. depol.)	261	262.0 (v.s.)
$\nu_3^{CCl_3}(a_1)$	364[138]	365.9 (s. pol.)	363	366.5 (s.)
$\nu_2^{CCl}(a_1)$	667[139]	668.3 (s. pol.)	672	650.8 (s.)
$\nu_5^{CCl}(e)$	760[139]	761.2 (s. depol.)	760	737.6 (s.)
$\nu_4(e)$	1205[139]	1215.6 (m. depol.)	1217	908.3 (m.)
$\nu_1^{CH}(a_1)$	3033[140]	3018.9 (s. pol.)	3030	2256.0 (s.)

The Raman spectrum of liquid $CHCl_3$ has been investigated by many investigators [for detailed references see Kohlrausch (13) (14) and Hibben (10)] and recently also the spectrum of the vapor by Nielsen and Ward (670). There are six strong Raman shifts, as given in Table 86. In addition, a very weak line occurs for the liquid with a shift of 1505 cm^{-1} very probably corresponding to the first overtone of the band 761.2 cm^{-1}. Thus the number of Raman fundamentals is in agreement with that expected on the symmetrical (C_{3v}) model. Furthermore, three of the strong Raman bands

[137] Independent values of similar accuracy have been obtained by Redlich and Pordes (731a).
[138] Measured by Barchewitz and Parodi (101) in $CHCl_3$ dissolved in paraffine.
[139] Measured by Emschwiller and Lecomte (304) in liquid $CHCl_3$.
[140] Measured by Ginsburg, according to a footnote in Voge and Rosenthal (903), in $CHCl_3$ vapor.

are partly polarized, the other three completely depolarized [see (191) and Fig. 82, p. 270], as should be the case. Thus, as in the case of the methyl halides, the first three can be assigned to ν_1, ν_2, ν_3, the other three to ν_4, ν_5, ν_6 in order of decreasing frequency. Further confirmation of this assignment is obtained if the fundamental frequencies of $CHCl_3$ are calculated by assuming the same force constants as in CH_4, CH_3Cl, and CCl_4. This has been done by Voge and Rosenthal (903), who found very satisfactory agreement with the observed frequencies. Finally, the Raman spectrum of "heavy" chloroform $CDCl_3$ has also been investigated and is included in Table 86. The fundamentals of $CDCl_3$ observed in this way also agree with those predicted from the force constants [Voge and Rosenthal (903)].

Most of the fundamental frequencies of $CHCl_3$ may be ascribed fairly definitely to certain bonds or groups (see Fig. 91 and Table 86). ν_1 is without doubt essentially a C—H stretching vibration. ν_2 is essentially the symmetrical C—Cl stretching vibration (corresponding to ν_3 of CH_3Cl) and ν_5 the corresponding degenerate frequency. ν_3 and ν_6 are the symmetrical and degenerate CCl_3 deformation vibrations and finally ν_4 is the bending vibration of the C—H group against the rest of the molecule. In agreement with this interpretation, the isotope shifts for $CDCl_3$ are practically zero for ν_3 and ν_6, are small for ν_2 and ν_5 but very large for ν_1 and ν_4.

Methylene chloride, CH_2Cl_2. There can hardly be any doubt, considering the structures of CH_4, CH_3Cl, $CHCl_3$, CCl_4, that methylene chloride has a tetrahedral structure of point group $\boldsymbol{C_{2v}}$ (similar to CH_2D_2, see above). The infrared and Raman data thus far available do not contradict this assumption but they are hardly sufficient to prove it definitely. Besides, the assignment of

TABLE 87. INFRARED [141] AND RAMAN SPECTRA OF LIQUID CH_2Cl_2 AFTER CORIN AND SUTHERLAND (233) AND WAGNER (908).

Assignment[142]	Infrared ν_{vacuum} (cm^{-1})	Raman $\Delta\nu_{vacuum}$ (cm^{-1})	Assignment[142]	Infrared ν_{vacuum} (cm^{-1})	Raman $\Delta\nu_{vacuum}$ (cm^{-1})
$\nu_4{}^{CCl_2}(a_1)$	[143]	**283** (s. pol.)[144]	$\nu_2+\nu_4(A_1)$	1548 (w.)	—
$\nu_3{}^{CCl}(a_1)$	**704** (v.s.)[145]	**700** (v.s. pol.)[144]	$\nu_3+\nu_7(B_1)$	1613 (w.)	—
$\nu_9{}^{CCl}(b_2)$	**737** (v.s.)[145]	**736** (s. depol.?)	$\nu_5+\nu_7(B_2)$	2057 (w.)	—
$\nu_7(b_1)$ (rocking)	**899** (m.)	**898** (v.w. depol.?)	$\nu_2+\nu_9(B_2)$	2136 (w.)	—
$\nu_3+\nu_4(A_1)$	935 (w.)	—	$\nu_2+\nu_7(B_1)$	2314 (m.)	—
$\nu_4+\nu_9(B_2)$?	1060 (w.)	1057 (v.w. pol.?)[146]	$\nu_5+\nu_8(B_1)$	2414 (w.)	—
$\nu_5(a_2)$?(torsion)	**1155** (w.)	**1148** (w. depol.)	$2\nu_8(A_1)$	2524 (w.)	—
$\nu_4+\nu_7(B_1)$	1192 (w.)	—	$\nu_2+\nu_8(B_2)$	2673 (w.)	
$\nu_3+2\nu_4(A_1)$?	1222 (w.)	—	$2\nu_2(A_1)$	—	2822 (v.w.)
$\nu_8(b_2)$ (rocking)	**1266** (v.s.)	**1255** (v.w.)	$\nu_1{}^{CH}(a_1)$	**2984** (s.)	**2985** (s. pol.)[144]
$\nu_2{}^{CH_2}(a_1)$	**1429** (s.)	**1417** (m. depol.)	$\nu_6{}^{CH}(b_1)$	**3048** (s.)	**3045** (s. depol.)
$2\nu_9(A_1)$	—	1464 (v.w.)	$4\nu_1$, $4\nu_6$?	11309 (v.w.)[147]	

[141] In addition, some very weak bands below 3049 and some weak bands above 3049 cm^{-1} have been found and assigned by Corin and Sutherland (233). Between 525 and 1450 cm^{-1} the absorption spectrum has also been measured, apparently under lower dispersion, by Lecomte (565) and Emschwiller and Lecomte (304).

[142] For somewhat different assignments see Corin and Sutherland (233) and Wu (25).

[143] Outside the region investigated, except in the work of Barchewitz and Parodi (101) on CH_2Cl_2 in paraffine solution. They give the bands 263, 500, 312, 222 cm^{-1} in the order of their intensities. It appears doubtful whether these bands belong to CH_2Cl_2.

[144] In the vapor, according to Nielsen and Ward (670), these lines occur at 280, 712, and 2997 cm^{-1},

[145] These two bands are due to Lecomte (565). Emschwiller and Lecomte (304) give 714 instead of 704.

[146] This band is not given by Wagner (908) but is given by both Trumpy (875) and Kohlrausch and Ypsilanti (534).

[147] This band was found by Timm and Mecke (864) and Vierling and Mecke (901) in the photographic infrared.

the fundamentals is still somewhat controversial. With one exception all infrared and Raman investigations refer to the liquid state.

As for CH_2D_2, there are nine fundamentals distributed over the various species as follows (see Tables 13 and 35):

$$4A_1 + A_2 + 2B_1 + 2B_2.$$

They are given qualitatively in Fig. 93, where it has been assumed that the CH_2 plane is the xz plane (parallel to the plane of the paper) and the two-fold axis the z axis. According to our experience with H_2CO and other molecules, we expect three vibrations characteristic of the CH_2 group (two symmetrical ones, a stretching and a deformation vibration, and one antisymmetrical stretching

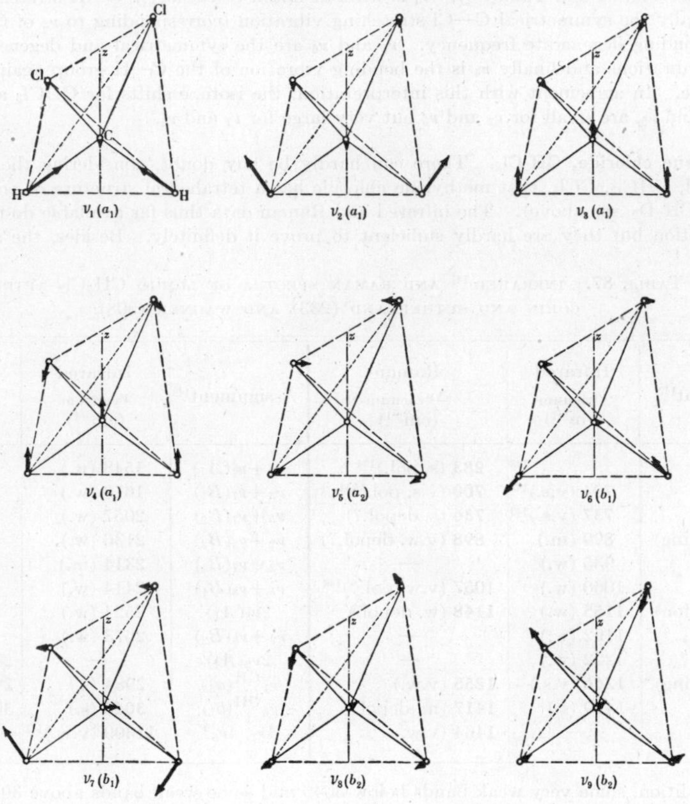

FIG. 93. **Normal vibrations of CH_2Cl_2 (schematic).**—The CH_2 plane is parallel, the CCl_2 plane perpendicular to the plane of the paper (shown in oblique projection). In ν_5, ν_8, ν_9 the H atoms move perpendicularly to the plane of the paper. In ν_5, ν_6, ν_7 the Cl atoms move perpendicularly to the CCl_2 plane, that is, parallel to the plane of the paper; in all others they move in the CCl_2 plane.

vibration) and three similar vibrations characteristic of the CCl_2 group. They are the vibrations $\nu_1(a_1)$, $\nu_2(a_1)$, $\nu_6(b_1)$ and $\nu_3(a_1)$, $\nu_4(a_1)$, $\nu_9(b_2)$ respectively of Fig. 93. This leaves the three vibrations $\nu_5(a_2)$, $\nu_7(b_1)$, and $\nu_8(b_2)$ which cannot be localized in a particular group. $\nu_5(a_2)$ is the torsion oscillation about the z axis mentioned previously (p. 240); ν_7 and ν_8 are the rocking vibrations of the CH_2 and CCl_2 groups respectively against the rest of the molecule.

All nine fundamentals are active in the Raman effect and all but $\nu_5(a_2)$ are active in the infrared. Table 87 gives the observed Raman and infrared bands. Five of the fundamentals are easily identified. The three lowest observed frequencies are obviously the three CCl_2 vibrations, since we know from the discussion of CCl_4, $HCCl_3$, and CH_3Cl that the C—Cl stretching vibration has a frequency of about 700, and since the CCl_2 deformation frequency must be appreciably lower. This assignment also fits with the observed degrees of depolarization, only one of the three vibrations [$\nu_9(b_2)$] being non-totally symmetric. Furthermore, it fits calculations assuming a valence force system in a triatomic system Cl—CH_2—Cl [see Kohlrausch (14)]. The two high-frequency Raman shifts and infrared bands 2985 and 3047 cm^{-1} evidently are the C—H stretching frequencies of the CH_2 group: one, $\nu_1(a_1)$, totally symmetric and therefore polarized, and the other, $\nu_6(b_1)$, anti-symmetric with respect to the CCl_2 plane and therefore depolarized.

By comparison with the vibration spectra of H_2CO and the methyl halides, it would be expected that the fairly strong Raman and infrared band of CH_2Cl_2 at 1423 cm^{-1} would correspond to the CH_2 deformation vibration $\nu_2(a_1)$. Against this assignment the objection may be raised that in the Raman effect this frequency is completely depolarized, whereas $\nu_2(a_1)$ should be partly polarized. However, a very similar frequency has been observed in many other methylene compounds in which it is polarized [see Wagner (908)], and as has been pointed out earlier, polarized Raman lines may in exceptional cases have degrees of depolarization up to $\frac{6}{7}$.

The assignment of the three remaining frequencies $\nu_5(a_2)$ (torsional oscillation), $\nu_7(b_1)$ (rocking of CH_2 in its plane), $\nu_8(b_2)$ (rocking of CH_2 perpendicular to its plane) is much less certain. In Table 87 for ν_5 and ν_8 the assignment by Wagner (908) has been adopted; Wagner has carried out a detailed calculation on the basis of a somewhat generalized valence force system and has also compared CH_2Cl_2 with a number of other methylene compounds. The fact that $\nu_5(a_2)$ seems to occur (though weakly) in the infrared, contrary to the selection rules, must be explained as due to the influence of the neighboring molecules. It should not occur in the infrared spectrum of the vapor.[148] For $\nu_7(b_1)$ Wagner has left the assignment open, suggesting that ν_7 is overlapped by ν_9. In Table 87, following Corin and Sutherland (233), the frequency 898 cm^{-1} has been chosen for $\nu_7(b_1)$ since this is a fairly strong infrared band that cannot be explained as a combination or overtone band.

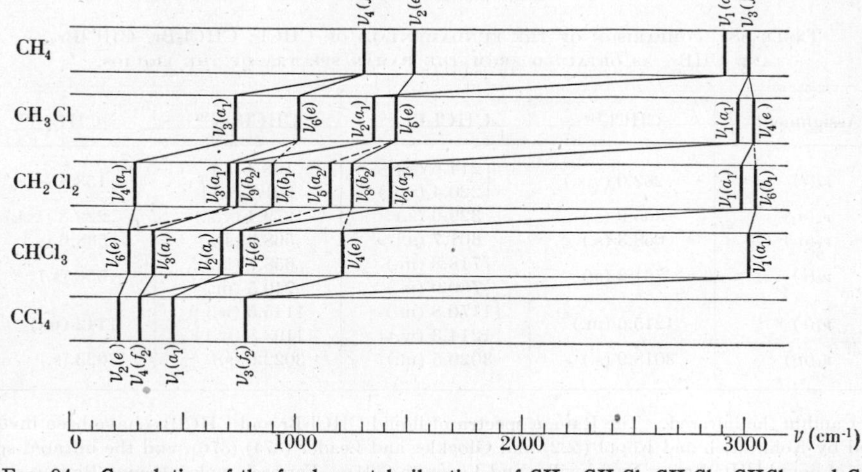

Fig. 94. Correlation of the fundamental vibrations of CH_4, CH_3Cl, CH_2Cl_2, $CHCl_3$, and CCl_4. —The broken correlation lines correspond to vibrations that are antisymmetric with respect to the (only) plane of symmetry that is preserved in going from CH_3Cl or from $CHCl_3$ to CH_2Cl_2.

The other observed infrared and Raman bands are easily assigned as overtone and combination bands, as indicated in Table 87.

In Fig. 94 the fundamental vibration spectra of CH_4, CH_3Cl, CH_2Cl_2, $CHCl_3$, CCl_4 are represented graphically on the basis of the data of Tables 80, 83, 84, 86, 87. Frequently in the literature attempts have been made to correlate the frequencies in these and similar molecules, but quite often such correlations have been made without regard for the non-crossing rule of vibrations of the same species (see p. 200). It must be realized that the correlation between CH_4 and CH_2Cl_2 is different

[148] Wu (962) assigns $\nu_5(a_2)$ to 898 cm^{-1} which seems much less likely since it occurs with medium intensity in the infrared.

depending on whether one imagines two of the H atoms of CH_4 to be simultaneously transformed into Cl atoms or whether one imagines first one H atom transformed into Cl (giving CH_3Cl) and then the other. In the first correlation there are throughout the transition two planes of symmetry, in the second only one. In the first case, therefore, intersections of vibrations of different species of C_{2v} may take place, whereas in the second case only vibrations of different species of the common point group C_s may intersect. This second case is the one represented in Fig. 94. In order to carry out this correlation one has to realize that the e vibrations of CH_3Cl split into one symmetric (a') and one antisymmetric (a'') vibration of C_s (indicated by solid and broken correlation lines respectively), while a_1 vibrations go over into a' vibrations. On the other hand, since it is the CCl_2 plane that remains a plane of symmetry, the a_1 and b_2 vibrations of CH_2Cl_2 go over into a' vibrations of C_s while a_2 and b_1 go over into a''. With this and the non-crossing rule, the correlation in the upper part of Fig. 94 is given. The correlation of CCl_4—$CHCl_3$—CH_2Cl_2 (lower part of Fig. 94) is similar, except that the plane in common is now the CH_2 plane; therefore a_1 and b_1 go over into a', and a_2 and b_2 into a''. Fig. 94 shows that in this correlation none of the vibrations of CH_4 goes over completely into the corresponding vibration of CCl_4. Kohlrausch and Wagner (531), as well as Wu (962), have given a correlation similar to Fig. 94 including as further intermediate steps CH_3D, CH_3F, CH_2FCl, $CHFCl_2$, $CDCl_3$, $CFCl_3$. However, their assignments for CH_2Cl_2 are slightly different from that assumed here.

$CHCl_2Br$ and $CHClBr_2$. A number of halogen derivatives of methane with two different halogens as well as hydrogen have been investigated in recent years, particularly by Glockler and his collaborators. As examples we choose bromodichloromethane and chlorodibromomethane. One would expect these molecules to have one plane of symmetry only (the CHBr and CHCl planes respectively), that is to belong to point group C_s. On this assumption each would be expected (see Table 35) to have six normal vibrations that are symmetric with respect to the plane of symmetry (species A') and three that are antisymmetric to it (species A''). All fundamentals are active both in the Raman

TABLE 88. COMPARISON OF THE FUNDAMENTALS OF $CHCl_3$, $CHCl_2Br$, $CHClBr_2$, AND $CHBr_3$ AS OBTAINED FROM THE RAMAN SPECTRA OF THE LIQUIDS.

Assignment	$CHCl_3$[149]	$CHCl_2Br$[150]	$CHClBr_2$[150]	$CHBr_3$[151]
$\nu_6(e)$	262.0 (v.s.)	214.6 (s.) 220.4 (v.s.)	168.3 (s.) 201.1 (s.)	153.8 (s.)
$\nu_3(a_1)$	365.9 (s.)	329.6 (s.)	279.4 (v.s.)	222.3 (v.s.)
$\nu_2(a_1)$	668.3 (s.)	601.7 (s.)	568.9 (s.)	538.6 (s.)
$\nu_5(e)$	761.2 (s.)	718.8 (m.) 760.0 (m.)	658.7 (m.) 749.5 (m.)	656 (s.)
$\nu_4(e)$	1215.6 (m.)	1170.8 (m.) 1214.3 (w.)	1145.5 (w.) 1193.8 (w.)	1142 (m.)
$\nu_1(a_1)$	3018.9 (s.)	3020.5 (m.)	3022.6 (s.)	3023 (s.)

effect and in the infrared. The Raman spectra of liquid $CHCl_2Br$ and $CHClBr_2$ have been investigated by Kohlrausch and Köppl (522) and Glockler and Leader (374) (376), and the infrared spectrum of liquid $CHClBr_2$ by Emschwiller and Lecomte (304). For each molecule nine Raman shifts, which are presumably all fundamentals, have been found, and for $CHClBr_2$ four infrared bands which coincide within the rather large uncertainty of the measurements with four of the Raman shifts. Since no polarization data are available and the infrared data have been obtained with low dispersion, it is not possible to decide definitely which of the frequencies correspond to the antisymmetric vibrations.

However, a comparison of the spectra of $CHCl_2Br$ and $CHClBr_2$ with those of $CHCl_3$ and $CHBr_3$ is very helpful in finding out to what modes the observed frequencies of the former correspond, since to a certain very rough approximation $CHCl_2Br$ and $CHClBr_2$ may be considered as isotopic molecules of $CHCl_3$ in which one or two Cl atoms are replaced by Br, or of $CHBr_3$ in which

[149] From Table 86.

[150] These data are from Glockler and Leader (376).

[151] These data are from Redlich and Stricks (732).

two or one Br atoms are replaced by Cl atoms [see Wu (26)]. This comparison is given in Table 88. Here it should be noted that every degenerate vibration of $CHCl_3$ or $CHBr_3$ splits into two different frequencies in $CHCl_2Br$ and $CHClBr_2$, which one would expect to be intermediate in frequency between those of $CHCl_3$ and $CHBr_3$. This is seen to be the case from Table 88. One component of each such doublet is an antisymmetric vibration, but as mentioned above it is impossible at present to decide which one. The analogues of the totally symmetric vibrations of $CHCl_3$ and $CHBr_3$ also have, of course, intermediate frequencies.

It is seen from this comparison that while the three highest frequencies of $CHCl_2Br$ and $CHClBr_2$ may definitely be identified as C—H stretching and bending (rocking) vibrations it is not possible to ascribe any one of the others to a definite carbon-halogen band. One can say only that the three lowest are bending frequencies and that the three intermediate are stretching frequencies of the CCl_2Br and $CClBr_2$ groups.

Formic acid, HCOOH and HCOOD. The investigation of the infrared and Raman spectra of formic acid is complicated by the fact that at room temperature the vapor consists mainly of double molecules and only a small fraction of single molecules.[152] In order to ascertain which bands belong to the single molecules (and also to avoid too much overlapping by the bands of the double molecules), it is necessary to investigate the spectrum at an elevated temperature as well as at room temperature. This has been done for the infrared absorption by Bonner and Hofstadter (167) (455), both for light and heavy formic acid, and for the Raman spectrum of the former by Bonner and Kirby-Smith (168). The Raman spectrum of the liquid, investigated by many authors [see Kohlrausch (14)] is probably mostly due to the dimer. Table 89 gives the observed spectra.

TABLE 89. INFRARED AND RAMAN SPECTRA OF MONOMERIC HCOOH AND HCOOD VAPOR AFTER BONNER AND HOFSTADTER (167) (455) AND BONNER AND KIRBY-SMITH (168).

Assignment	HCOOH		HCOOD
	Infrared ν_{vacuum} (cm^{-1})	Raman $\Delta\nu_{vacuum}$ (cm^{-1})	Infrared ν_{vacuum} (cm^{-1})
ν(O—H torsion)?		232	
	658 (m.)		667[153] (w.)
		919	
			953 (w.)
			980 (v.s.)
			{1030 (s.)
ν(C—O)?	1093 (v.s.)		{1163 (v.s.)
	1206[153a]		
		1346	
ν(C=O)	1740 (v.s.)		1764 (v.s.)
			2105[154] (w.)
	2325 (m.)		2346[154] (w.)
ν(CH)	2940 (s.)	2945	3002 (m.)
ν(OH) [ν(OD)]	3570 (m.)	3566	2666 (s.)
3ν(OH)	10202.8[155] (w.)		
4ν(OH)	13285[156] (v.w.)		

[152] For detailed data about the association equilibrium, see Coolidge (230).

[153] This band does not occur in Hofstadter's (455) main table but only in his Table 2.

[153a] This band is due to Herman and Williams (430), who also give slightly different wave numbers for some of the other bands

[154] Possibly due to the dimer.

[155] This photographic infrared band was first observed by Herzberg and Verleger (445) and later studied in more detail by Bauer and Badger (128).

[156] Observed by Thompson (848).

TABLE 90. FURTHER FIVE-ATOMIC MOLECULES.[53]

Molecule	Structure	References: Raman spectrum	References: Infrared spectrum
SiH_4	T_d (tetrahedral)	(815)	(807) (817) (865) (866)
GeH_4	T_d (tetrahedral)		(808) (570) (818)
CF_4, SiF_4	T_d (tetrahedral)	(971)	(93)
$SiCl_4$, $TiCl_4$, $GeCl_4$, $SnCl_4$	T_d (tetrahedral)	(416) (873) (14) (191)	(763)
CBr_4	T_d (tetrahedral)	(547)	(101)
$SiBr_4$, $GeBr_4$, $SnBr_4$	T_d (tetrahedral)	(874) (832) (14)	
SO_4^{--}, SeO_4^{--}	T_d (tetrahedral)	(14)	
SiO_4^{----}	T_d (tetrahedral)		(604)
ClO_4^-, IO_4^-	T_d (tetrahedral)	(325) (58) (541) (727a)	
CrO_4^{--}, MoO_4^{--}, WoO_4^{--}	T_d (tetrahedral)	(671) (264) (356) (262) (842)	
PO_4^{---}, AsO_4^{---}	T_d (tetrahedral)	(792) (321a) (626)	
$CdBr_4^{--}$	T_d ?	(275)	
NH_4^+	T_d (tetrahedral?)	(622) (456) (47) (234) (335)	(705) (99) (335)
HCF_3, $HCBr_3$, $DCBr_3$, HCI_3	C_{3v} (tetrahedral)	(13) (14) (786) (369) (732)	(595) (304) (901) (864) (101)
$HSiCl_3$, $HSiBr_3$	C_{3v} (tetrahedral)	(175) (425) (273) (274)	
$HGeCl_3$, $HGeBr_3$	C_{3v} (tetrahedral)	(904)	
$HSnCl_3$, $HSnBr_3$	C_{3v} (tetrahedral)	(832)	
$ClCBr_3$, $BrCCl_3$	C_{3v} (tetrahedral)	(955) (569)	
$FCCl_3$, $ClCF_3$	C_{3v} (tetrahedral)	(372) (484)	
$BrSiCl_3$	C_{3v} (tetrahedral)	(426)	
$OPCl_3$	C_{3v}	(547) (191)	
$SPCl_3$	C_{3v}	(839)	
CH_2F_2	C_{2v} (tetrahedral)	(373)	
CH_2Br_2, CH_2I_2	C_{2v} (tetrahedral)	(876) (880) (534) (64) (908) (670)	(868) (304) (901) (864) (101) (97)
CF_2Cl_2, CF_2Br_2	C_{2v} (tetrahedral)	(175) (374)	
CCl_2Br_2	C_{2v} (tetrahedral)	(569)	
CH_2ClF, CH_2ClBr, CH_2ClI, CH_2BrI	C_s (tetrahedral)	(367) (64) (908)	(304)
$CHFCl_2$, $CHClF_2$	C_s (tetrahedral)	(366) (376) (370) (175)	
$CHBrF_2$, $CHFBr_2$	C_s (tetrahedral)	(365) (376)	
$CHBrClF$	C_1 (tetrahedral)	(376)	
SO_2Cl_2	C_{2v} ?	(191)	
H_2C—CO	C_{2v}	(536)	(354)
HNO_3, DNO_3	C_{2v} ?	(41) (57) (199) (198) (821) (96) (731) (792a)	(925) (67) (476a)
H_2N—CN	C_s ?	(478b)	

A study of the fine structure of a photographic infrared band seems to show (see Chapter IV) that HCOOH is a plane molecule of the form 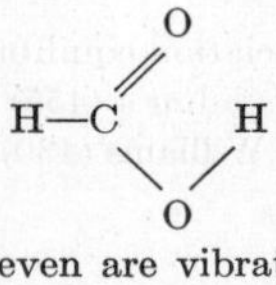 [see Bauer and Badger (128)]. Of the nine fundamentals, according to Table 35, seven are vibrations in the plane of the molecule and two are perpendicular to it. However, the available data are not sufficient to give a complete

[53] On p. 287.

assignment. The most one can do is to ascribe certain observed frequencies to certain bonds by comparison with other molecules and by consideration of the isotope effect. This is indicated in Table 89.

The infrared spectrum of DCOOH and DCOOD has been investigated by Herman and Williams (430).

Other five-atomic molecules. A large number of further five-atomic molecules have been investigated. Most of them have the same structure as one of the examples treated in the foregoing. However in almost all cases the data are less complete. Table 90 gives references to recent work on the Raman and infrared spectra of most of these molecules. The structure of the molecules is also indicated although in many cases it is not unambiguously established. The observed fundamental frequencies of some of the tetrahedral molecules are included in the previous Table 39.

Certain vibrations of a number of more than five-atomic molecules may be considered essentially as due to a five-atomic system. For example, in the neopentane molecule $C(CH_3)_4$ we expect to find four vibrations that can be attributed to the vibrations of the CH_3 groups as a whole against the central carbon atom, just as in, say, CCl_4 [see Rank (714), Kohlrausch and Köppl (521), Rank and Bordner (715), Wall and Eddy (910), Silver (787) (788), Ballaus and Wagner (95)]. These skeletal vibrations have also been discussed for $Si(CH_3)_4$ by Rank and Bordner (715), Wall and Eddy (910), and Silver (787) (788); for $Si(C_2H_5)_4$ by Anderson (50); for $Sn(CH_3)_4$ by Pai (682) and Anderson (50); for $Pb(CH_3)_4$ by Duncan and Murray (293), Pai (682), and Anderson (50); for $Pb(C_2H_5)_4$ by Duncan and Murray (293) and Pai (682); and for $[N(CH_3)_4]^+$ by Edsall (298) and Silver (788). In addition to the skeletal vibrations there are, of course, the "inner" vibrations of the CH_3 groups.[157] Apart from the tetrahedral molecules there are many other less symmetrical molecules that may be considered as five-particle systems, such as CH_3OH, CH_3CN, $(CH_3)CCl_3$, $CH_3 \cdot CH_2 \cdot COH$ and so on, if the OH, CN, CH_3, $\cdots$ groups are considered as one particle [see Kohlrausch (14)].

(*d*) *Six-atomic molecules*

Diacetylene, HC≡C—C≡CH. On the basis of valence considerations one would expect diacetylene to be a *linear molecule*. But a definite spectroscopic proof of this is still lacking, although electron diffraction data [Pauling, Springall, and Palmer (687)] are in favor of this assumption. For the linear model there are the following normal vibrations, which are represented in Fig. 95:

$$3\Sigma_g^+(\text{R.}), \qquad 2\Sigma_u^+(\text{I.}), \qquad 2\Pi_g(\text{R.}), \quad 2\Pi_u(\text{I.});$$

that is, there are five non-degenerate $\|$ vibrations and four doubly degenerate $\perp$ vibrations. Since there is a center of symmetry the rule of mutual exclusion should hold. While this is found to be the case it may be due to the fact that the infrared spectrum has not been investigated below 600 cm^{-1}, where three of the Raman lines lie.

The five Raman lines of liquid C_4H_2 observed by Timm and Mecke (862) (see Table 91) cannot be interpreted as the five Raman-active fundamentals ν_1, ν_2, ν_3, ν_6, ν_7, since ν_1 certainly is the symmetrical CH frequency which must have a frequency of about 3350 cm^{-1} (from analogy to C_2H_2; see Fig. 95) and which is not observed. One may, however, interpret the five observed Raman bands as the fundamentals ν_2, ν_3, ν_6, ν_9, ν_7 (in order of their frequency),[158] if one assumes that the Raman-inactive frequency ν_9 occurs in the liquid because of the interaction with other molecules (see also C_2I_2, p. 302) or that in the liquid state the molecule is not exactly linear as it probably is in the gaseous state. At any rate it seems hardly possible to interpret the frequency 411 cm^{-1} as anything but a fundamental.

In the ordinary and photographic infrared of *gaseous* C_4H_2, a number of absorption bands have been found by Bartholomé (119), Bartholomé and Karweil (122), and Gänswein and Mecke (346). By comparison with C_2H_2 the infrared active $\|$ vibrations ν_4 and ν_5 are easily identified as the bands 3350 (C—H vibration) and 2023 cm^{-1} (C≡C vibration). Assuming that the Raman band 411 cm^{-1} is ν_9, there are three possibilities for ν_8: 690 cm^{-1}, 708 cm^{-1}, and 1235 cm^{-1}, all of which are very intense infrared bands. In favor of ν_8 = 1235 cm^{-1} is the fact that it is the most intense band. But 1235 appears altogether too high for a $\perp$ vibration of a linear molecule. We therefore choose

---

[157] The latter have been included in the detailed theoretical treatment by Silver (788).

[158] Timm and Mecke (862) have interpreted the three Raman lines of lowest frequency as ν_9, $(\nu_3 - \nu_9)$, ν_7 respectively, of which the first *two* are forbidden. In addition, it is hard to understand on this assignment why $\nu_3 + \nu_9$ is not observed.

$\nu_8 = 708$, considering the maximum 690 cm^{-1} as due to $\nu_8 + \nu_7 - \nu_7$ which, because of the smallness of ν_7 and the double degeneracy of the lower state, should have almost the same intensity as ν_8. The interpretation of the other bands on this basis is given in Table 91. The assignments of Timm and Mecke (862) are added in square brackets.

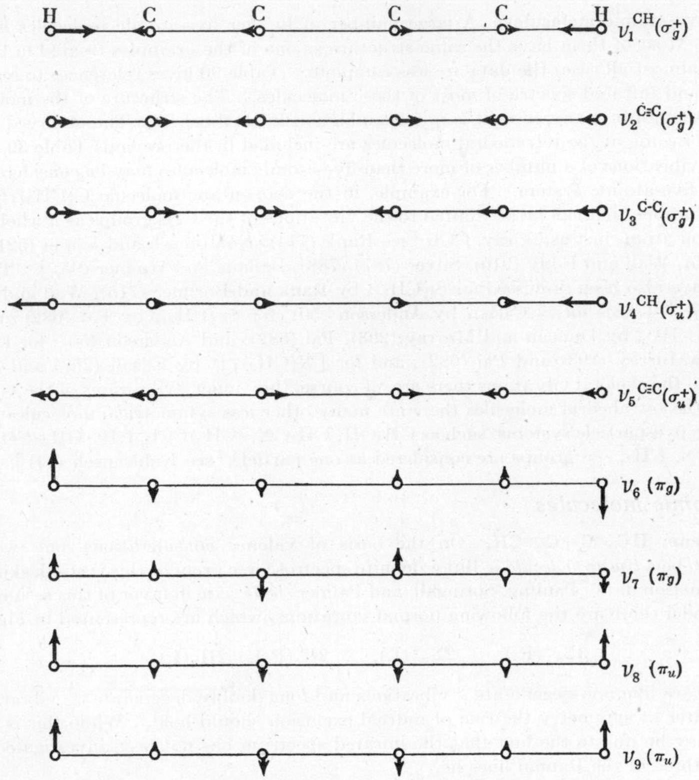

FIG. 95. **Normal vibrations of diacetylene (schematic).**

While ν_1 has not been observed, it is clear that it must have a magnitude very similar to ν_4 (=3350 cm^{-1}), since even in C_2H_2 the frequencies of the two C—H vibrations are close together (3288 and 3374 cm^{-1}) and since here the two C—H groups interact still less with each other (see p. 196).

The frequency of the C—C vibration ν_3 obtained from the Raman spectrum is very low compared to the C—C frequencies of molecules such as $(CN)_2$, C_2H_6, and others. If, on the basis of a generalized valence force system including interaction terms between nonadjacent atoms, the valence stretching-force constants are evaluated from the observed frequencies of the || vibrations [see Bartholomé and Karweil (122) and Wu and Shen (964)[159]], it is found that the C—C force constant, in consequence of the low value of ν_3, comes out more than 20 per cent lower than for C_2H_6 and 45 per cent lower than for the isosteric C_2N_2. On the other hand, according to the electron diffraction data mentioned above, the C—C single bond distance for C_4H_2 is smaller than for C_2H_6 and about the same as for C_2N_2, indicating a tighter bond than for C_2H_6. Further work on the Raman and infrared spectra is required to clear up this difficulty. It does not seem entirely impossible that ν_3 as well as ν_1 has escaped observation and that the observed Raman frequency 644 cm^{-1} is not ν_3. On the other hand, the work of Woo and Chu (951) on the ultraviolet spectrum seems to show that 644 cm^{-1} is a fundamental frequency of the electronic ground state.

---

[159] Wu and Shen have also given formulae for the perpendicular vibrations [see Wu (26)].

TABLE 91. RAMAN SPECTRA OF LIQUID AND INFRARED SPECTRUM OF GASEOUS HC≡C—C≡CH AFTER TIMM AND MECKE (862), BARTHOLOMÉ AND KARWEIL (122), AND GÄNSWEIN AND MECKE (346).

ν_{vacuum} (cm^{-1})	Description of band	Assignment	ν_{vacuum} (cm^{-1})	Description of band	Assignment
231	R. (v.w.)	$\nu_7(\pi_g)[\nu_9]$	3120	I. (m.)	$\nu_3+\nu_5+\nu_6(\Pi_u)$ or $\nu_4-\nu_7(\Pi_u)$
411	R. (w.)	$\nu_9(\pi_u)[\nu_3-\nu_9]$	**3350**	I. (s.)	$\nu_4^{CH}(\sigma_u^+)$
488	R. (m.)	$\nu_6(\pi_g)[\nu_7]$	(3350)	not observed	$\nu_1^{CH}(\sigma_g^+)$
644	R. (m.)	$\nu_3^{C-C}(\sigma_g^+)$	3550	I. (v.w.)	$\nu_4+\nu_7(\Pi_u)$
690	I. (s.)	$\nu_8+\nu_7-\nu_7(\Pi_u)$	3920	I. (m.)	$\nu_3+\nu_4(\Sigma_u^+)$?
708	I. (s.)	$\nu_8(\pi_u)[\nu_7+\nu_9]$	6500	I. (v.w.)	$\nu_1+\nu_4(\Sigma_u^+)$
1235	I. (v.s.)	$\nu_6+\nu_8(\Sigma_u^+)[\nu_7+\nu_8]$	8950?	I. (v.w.)	$\nu_1+\nu_2+\nu_4(\Sigma_u^+)$?
1444	I. (m.)	$\nu_6+\nu_8+\nu_7(\Pi_u)[\nu_6+\nu_8]$	11688	P.I. (w.)	$\nu_2+3\nu_4(\Sigma_u^+)$
2023	I. (s.)	$\nu_5^{C\equiv C}(\sigma_u^+)$	11782	P.I. (v.w.)	$\nu_1+2\nu_4+\nu_5(\Sigma_u^+)$
2183	R. (v.s.)	$\nu_2^{C\equiv C}(\sigma_g^+)$	11809	P.I. (v.w.)	$3\nu_1+\nu_5(\Sigma_u^+)$
2950	I. (w.)	$\nu_2+\nu_8(\Pi_u)$	11865	P.I. (v.w.)	$2\nu_1+\nu_2+\nu_4(\Sigma_u^+)$
2980	I. (w.)	$\nu_5+2\nu_6(\Sigma_u^+)$	12147	P.I. (v.w.)	$\nu_2+3\nu_4+\nu_6(\Pi_u)$?
3030	I. (m.)	$\nu_2+\nu_7+\nu_8(\Sigma_u^+)$	12199	P.I. (v.w.)	$\nu_1+2\nu_4+\nu_5+\nu_6(\Pi_u)$?
			12236	P.I. (v.w.)	$3\nu_1+\nu_5+\nu_6(\Pi_u)$?
			12706	P.I. (m.)	$\nu_1+3\nu_4(\Sigma_u^+)$

Ethylene, C_2H_4 and C_2D_4. Valence considerations suggest the *plane symmetrical form* (point group V_h) for the ethylene molecule. Observation shows that there are no bands that occur in the Raman as well as the infrared spectrum of C_2H_4 gas,[160] proving that the molecule has a center of symmetry (see p. 256). While this does not necessarily imply that C_2H_4 has the symmetrical plane form, any other form with a center of symmetry but the symmetrical plane form seems very unlikely on the basis of valence considerations [see for example Penney (690)]. At any rate the fulfillment of the rule of mutual exclusion fairly definitely rules out the model in which the two CH_2 groups are in planes at right angles to each other, and the model in which the four H atoms are in a plane parallel to but not coinciding with the C—C axis. It does not rule out the model in which the H_4 plane intersects the C—C axis at the mid-point (see model II of C_2H_2, Fig. 86). Although the other Raman and infrared data at present available are not sufficient to decide unambiguously against this model,[161] we shall, in what follows, assume the plane symmetrical model, which has been taken for granted by most investigators.

In Table 92 are given the observed infrared and Raman bands of C_2H_4 and C_2D_4. As we have seen previously (p. 150, Fig. 44), C_2H_4 (C_2D_4) has twelve normal vibrations of the following species and activities:

$$3A_g(\text{R.}),\quad A_u(\text{inactive}),\quad 2B_{1g}(\text{R.}),\quad B_{1u}(\text{I.}),\quad B_{2g}(\text{R.}),\quad 2B_{2u}(\text{I.}),\quad 2B_{3u}(\text{I.})$$

(assuming as previously that the z axis is perpendicular to the plane of the molecule and the x axis is the C═C axis). The numbering of the vibrations in Table 92 is in agreement with our general practice (see p. 272) and is defined more particularly by Fig. 44, except that in the table either CH, CC, or CH_2 is added as a superscript to the symbol whenever the vibration considered is known to be essentially a C—H or C═C stretching or CH_2 deformation vibration.[162]

[160] For liquid C_2H_4 a few coincidences have been found which can be explained as due to a breakdown of the selection rules in consequence of intermolecular forces.

[161] On this model there would be four instead of three totally symmetric fundamentals, that is, four instead of three polarized Raman lines corresponding to fundamentals, while only three are found.

[162] It appears that hardly any two authors use the same designations for the vibrations of C_2H_4. Conn and Sutherland (226) give a comparative table of the various notations. Even though the notation used here is different from any suggested before, it appears to be necessary to introduce it, for the sake of consistency with the designation for the other molecules treated here. It must also be kept in mind that the designation of the species is not unique (see p. 108), since it depends on

TABLE 92. INFRARED AND RAMAN BANDS OF C_2H_4 AND C_2D_4 [163] [AFTER LEVIN AND MEYER (574) AND GALLAWAY AND BARKER (345) (C_2H_4 AND C_2D_4, INFRARED, GAS); DICKINSON, DILLON AND RASETTI (287), AND BHAGAVANTAM (148) (C_2H_4, RAMAN, GAS); GLOCKLER AND RENFREW (378) (C_2H_4, RAMAN, LIQUID); AND DE HEMPTINNE, JUNGERS AND DEFLOSSE (422) (C_2D_4, RAMAN, LIQUID)].

Assignment	C_2H_4		C_2D_4	
	Infrared (gas) ν_{vacuum} (cm^{-1})[164]	Raman[165] $\Delta\nu_{vacuum}$ (cm^{-1})	Infrared (gas) ν_{vacuum} (cm^{-1})[164]	Raman (liquid) $\Delta\nu_{vacuum}$ (cm^{-1})
$\nu_4(a_u)$	**(825)**[166]		**(580)**[167]	
$\nu_8(b_{2g})$		**943** (w.) liquid		**780** (w.)
$\nu_7(b_{1u})$	**949.2** (v.s.) Q[168]		**720.0** (v.s.) Q[168]	
$2\nu_7-\nu_7$			723.4 (m.)	
$\nu_{10}(b_{2u})$	**995** (m.)		**(712)**[167, 169]	
$\nu_6(b_{1g})$		**(1050)**[170]		**(883)**[170]
$\nu_3{}^{CH_2}(a_g)$		**1342.4** (v.s.) gas, pol.		**981** (m.)
$\nu_{12}^{CH_2}(b_{3u})$	**1443.5** (s.) || Q[168]		**1077.9** (s.) || Q[168]	
$\nu_2{}^{CC}(H_2C^{12}{=}C^{13}H_2)$		1602 (v.w.) liquid		
$\nu_2{}^{CC}(a_g)$		**1623.3** (v.s.) gas, pol.		**1515** (v.s.)
$2\nu_4(A_g)$		1656 (w.) gas		
$\nu_7+\nu_8(B_{3u})$	1889.6 (m.) || Q[168]		1495.7 (m.) || Q[168]	
$\nu_6+\nu_{10}(B_{3u})$	2047.0 (w.) || Q[168]		1595.1 Q	
$\nu_3+\nu_{10}(B_{2u})$	2325 (w.)[171]			
$2\nu_{12}^{CH_2}(A_g)$		2880.1 (w.) gas		
$\nu_{11}^{CH}(b_{3u})$	**2989.5** (s.) || Q[168]		**2200.2** (m.) || Q[168]	
$\nu_1{}^{CH}(H_2C^{12}{=}C^{13}H_2)$		2997 (v.w.) liquid		
$\nu_1{}^{CH}(a_g)$		**3019.3** (v.s.) gas, pol.		**2251** (v.s.)
$\nu_9{}^{CH}(b_{2u})$	**3105.5** (s.)	3075 (w.) liquid	**2345** (v.s.)	
$2\nu_2{}^{CC}(A_g)$		3240.3 (v.w.) gas		
$\nu_5{}^{CH}(b_{1g})$		**3272.3** (v.w.) gas		**2304** (w.)
$\nu_1+\nu_7(B_{1u})$			2969 (s.)[172] Q	
$\nu_3+2\nu_7+\nu_{10}(B_{2u})$	4206.7 (w.)			
$\nu_5+\nu_{10}(B_{3u})$			3049.0 Q (w.)	
$\nu_3+3\nu_{10}(B_{2u})$	4323.1 (m.)		3204.3 (w.)	
$\nu_1+\nu_{12}(B_{3u})$			3345.3 Q (m.)	
$\nu_5+\nu_{12}(B_{2u})$	4514.3 (m.)		3387.8 (w.)	
$\nu_2+\nu_9(B_{2u})$	4727.7 (v.w.)		3862.8 (v.w.)	
$\nu_1+3\nu_7(B_{1u})$			4429.5 Q (w.)	
$\nu_1+\nu_{11}(B_{3u})$			4478.6 Q (w.)	
$\nu_5+\nu_9(B_{3u})$			4628.6 Q (w.)	

which are called x, y, and z axes. Our choice has the advantage that the z axis is the axis of largest moment of inertia, the x axis the axis of smallest moment of inertia. This choice of axes agrees with that of Kohlrausch.

[163] Figures in brackets are not directly observed Raman or infrared bands. Photographic infrared bands are not included (see text).

[164] It has been assumed that the data given by Gallaway and Barker (345) were corrected for vacuum, although this is not explicitly stated in their paper and although they took over some of Meyer and Levin's data without correcting them for vacuum. Q indicates a central maximum which is to be expected for species B_{1u} and B_{3u} (see Chapter IV, section 3).

[165] All Raman lines observed for the gas have also been observed in the liquid with very slightly smaller shifts.

(Footnotes continued on following page)

The three strongest (and polarized) Raman lines 3019.3, 1623.3, and 1342.4 cm^{-1} can only be identified with the three totally symmetric vibrations $\nu_1(a_g)$, $\nu_2(a_g)$, $\nu_3(a_g)$. This is also in agreement with the fact that the first is expected to be essentially a C—H vibration (ν_1^{CH}); the second a C═C vibration (ν_2^{CC}), and the third a symmetrical CH_2 deformation vibration ($\nu_3^{CH_2}$), for which the frequencies observed are reasonable values. The correctness of this identification is further confirmed by the fact that the three strongest Raman lines of C_2D_4 also fit this assignment: we have large isotope shifts for the vibrations ν_1^{CH} and $\nu_3^{CH_2}$ but a small isotope shift for ν_2^{CC} and the isotope relation (II, 314) is fulfilled satisfactorily.

The high-resolution work of Gallaway and Barker (345) has shown that the strong infrared bands 2989.5, 1443.5 cm^{-1} of C_2H_4 and 2200.2, 1077.9 cm^{-1} of C_2D_4 have the dipole moment oscillating in the direction of the C═C axis (|| bands, $M_x \neq 0$; see Chapter IV, section 2b). Therefore they must be identified with the two vibrations of species B_{3u} (see Fig. 44), which are essentially C—H (C—D) stretching and CH_2 (CD_2) deformation vibrations (ν_{11}^{CH} and $\nu_{12}^{CH_2}$). The considerable shift between C_2H_4 and C_2D_4 is in agreement with isotope relation (III, 319). Similarly Gallaway and Barker (345) have shown that for the remaining strong infrared bands 3105.5 and 949.2 cm^{-1} of C_2H_4 and 2345 and 720.0 cm^{-1} of C_2D_4 the dipole moment is oscillating perpendicular to the C═C axis, for the higher frequency in the plane, for the lower frequency perpendicular to the plane of the molecule. Therefore they must be interpreted as the C—H (C—D) stretching vibration $\nu_9^{CH}(b_{2u})$ and the whole-molecule-bending vibration $\nu_7(b_{1u})$. The isotope shift of the latter, which is the only one of its species, is in agreement with the isotope relation (II, 317). The only remaining infrared-active vibration $\nu_{10}(b_{2u})$ for C_2H_4 is very probably to be identified with the $\perp$ band at 995 cm^{-1}, which is badly overlapped by the much stronger band 949.2 cm^{-1} and therefore not very accurately measured. For C_2D_4 the corresponding band has not been separated from the overlapping band 720.0 cm^{-1}.

The remaining four fundamentals $\nu_4(a_u)$, $\nu_5(b_{1g})$, $\nu_6(b_{1g})$, and $\nu_8(b_{2g})$ are much less definitely identified. One of them, the Raman-active $\nu_5(b_{1g})$, is a C—H vibration (see Fig. 44) and should have a frequency of about 3000 cm^{-1}. But, curiously, in addition to the very strong Raman line $\nu_1^{CH} = 3019.3$ cm^{-1}, as many as five weaker Raman lines have been found in this region. Two of these (2880.1 and 3240.3 cm^{-1}) are readily explained as first overtones of $\nu_{12}^{CH_2}$ and ν_2^{CC} respectively, which have probably "borrowed" some of their intensity (by Fermi resonance) from ν_1^{CH}, all three upper levels having symmetry type A_g. The line 2997 cm^{-1} is very probably the ν_1 band of the isotopic molecule H_2C^{12}═$C^{13}H_2$, just as the weak line 1602 cm^{-1} corresponds to ν_2 of this molecule [see Glockler and Renfrew (378)]. Thus two possibilities remain for $\nu_5(b_{1g})$: 3075 and 3272.3 cm^{-1}. The latter seems more likely to the writer, since it appears in the gas whereas the former does not, and since otherwise it would have to be explained as a ternary combination, which is extremely unlikely to occur in the Raman spectrum.[173] The line 3075 cm^{-1} could then be interpreted [Glockler and Renfrew (378)] as $\nu_9^{CH}(b_{2u})$ occurring weakly in liquid C_2H_4 because of a breakdown of the selection rules. In C_2D_4 the Raman line 2304 cm^{-1} would have to be taken as $\nu_5(b_{1g})$.

The two remaining Raman-active fundamentals $\nu_6(b_{1g})$ and $\nu_8(b_{2g})$, which represent whole-molecule-bending vibrations (Fig. 44) would be expected to have low frequencies. It seems, therefore, plausible to identify the Raman line 943 cm^{-1} of C_2H_4 as one of these fundamentals, very probably $\nu_8(b_{2g})$, particularly since the corresponding line 780 cm^{-1} of C_2D_4 fulfills the isotope relation (II, 318) fairly well. This identification also leads to a very satisfactory interpretation of the

---

[166] From the interpretation of the Raman band 1656 cm^{-1}. (See also p. 328).

[167] This value is obtained from the corresponding one for C_2H_4 by means of the isotope relations.

[168] The wave numbers given for these bands refer to the band centers (zero lines, see Chapter IV); the wave numbers of the maxima of the Q branches are slightly different.

[169] Conn and Sutherland (226) give a fairly strong maximum at 727 cm^{-1} which has not been confirmed by Gallaway and Barker (345). The latter authors give a value 740 cm^{-1} for this vibration.

[170] From the interpretation of the infrared bands 2047.0 cm^{-1} and 1595.1 cm^{-1} respectively.

[171] This band is due to Coblentz (224).

[172] This band is given by Conn and Sutherland (226) but not by Gallaway and Barker (345).

[173] However Conn and Sutherland (226) as well as Burcik, Eyster and Yost (183), and Gallaway and Barker (345) prefer the assignment $\nu_5^{CH}(b_{1g}) = 3072.3$ cm^{-1}.

infrared band 1889.6 cm^{-1} of C_2H_4 (and 1495.7 of C_2D_4) as $\nu_7 + \nu_8$, since the upper state of this combination has species B_{3u}, leading to a || band in agreement with observation.[174]

No Raman lines are left to be assigned to $\nu_6(b_{1g})$. Its value can be obtained from an interpretation of the || band 2047.0 cm^{-1} of C_2H_4 (and similarly 1595.1 of C_2D_4) as $\nu_6 + \nu_{10}(B_{3u})$. This accounts for the character of the band and gives a frequency ν_6 (1050 cm^{-1} for C_2H_4, 883 cm^{-1} for C_2D_4) in agreement with the expectation that it should be of similar magnitude to ν_{10}.

The last remaining vibration, the twisting vibration $\nu_4(a_u)$, is active neither in the Raman nor in the infrared spectrum. Combination bands containing ν_4 may of course be active, and thus may be used for its determination, although it is difficult to be quite sure about the assignment. The weak Raman line 1656 cm^{-1} observed in the gas can hardly be interpreted other than as a binary combination containing ν_4. Conn and Sutherland (226) have assumed it to be $\nu_4 + \nu_7(B_{1g})$, giving $\nu_4 \sim 700$. However, ν_7 is a weak Raman line and no strong Raman line of species B_{1g} occurs. On the other hand, Wu (26) has suggested $1656 = 2\nu_4$. This seems much more probable because the species of the upper state of $2\nu_4$ is A_g and another level of the same species giving rise to a very strong Raman line (1623.3) is very close to it, so that $2\nu_4$ may be strengthened by Fermi resonance. Thus we obtain $\nu_4 \approx 825$ cm^{-1}. Measurements of the specific heat of C_2H_4 at various temperatures have also led, in a way that will be explained in Chapter V, to a torsional frequency of about 800 rather than 700 cm^{-1} [Eucken and Parts (313), Burcik, Eyster, and Yost (183)].[175]

Very recently Rasmussen (724) has found in C_2H_4 a very weak infrared absorption band at 800 cm^{-1} which cannot be a combination band.[175a] While he interpreted this band as $\nu_{10}(b_{2u})$ (for which Gallaway and Barker found 995 cm^{-1}), it seems very likely that his band represents the torsional oscillation which can be made weakly active for the higher rotational levels on account of Coriolis interaction with the vibration $\nu_{10}(b_{2u})$ (see p. 467).

Some further weak infrared bands and their interpretations as overtone and combination bands, mainly after Gallaway and Barker (345), are given in Table 92. A number of photographic infrared bands have been found by Badger and Binder (72), Bonner (163), Gänswein and Mecke (346) [see also Thompson (849) and Wu (25)]. They are not included in Table 92.

On the whole, it appears that the fundamentals chosen in Table 92 are reliable, with the possible exception of $\nu_5(b_{1g})$. The force constants obtained from them on the basis of a simple valence force potential are given in the previous Table 47. On the basis of older assignments Manneback and his co-workers [Delfosse (272), Manneback and Verleysen (598), Manneback (596), Verleysen and Manneback (900), Bernard and Manneback (139), de Hemptinne and Manneback (423)], as well as Wu (26), have calculated force constants for the C_2H_4 molecule.

The Raman spectra of partly deuterated ethylenes [C_2H_3D, $C_2H_2D_2$ (three forms), and C_2HD_3] have been investigated by de Hemptinne, Jungers, and Delfosse (422). Their results are in general in agreement with the calculations of Manneback and coworkers, quoted above [see also Wu (959)]; however they are not as yet quite complete.

Tetrachloroethylene, C_2Cl_4. It is to be expected that tetrachloroethylene has the same symmetry (V_h) as ethylene. Raman and infrared spectra support this assumption although they do not unambiguously prove it. In Table 93 the observed infrared and Raman bands are given. With one exception (782–784 cm^{-1}), the rule of mutual exclusion is fulfilled. In view of the fairly large number of frequencies and the consequent likelihood of a chance coincidence, and in view of the fact that the Raman data refer to the liquid state only, this exception cannot be considered as evidence that the molecule does not have a center of symmetry. On the contrary, the lack of any other coincidences is fairly strong evidence that a center of symmetry exists. The three strong polarized Raman lines 1571, 447, and 237 cm^{-1} must correspond to the three totally symmetric vibrations ν_1, ν_2, ν_3. Unlike C_2H_4, here ν_1 (that is, the largest of the three) is the C=C vibration. The assignment of the other Raman bands and the infrared bands is much less certain, particularly since the infrared spectrum of the gas has not been investigated below 600 cm^{-1}. A tentative assignment is given in Table 93. For the Raman lines it is essentially that of Wittek (948).

---

[174] If one had chosen 943 cm^{-1} as $\nu_6(b_{1g})$, the infrared band at 1889.6 cm^{-1} could not be interpreted as $\nu_6 + \nu_7$ since the upper state of this combination would have the species A_u and thus could not be reached in infrared absorption.

[175] A further confirmation of the higher value, if such were necessary, could be obtained by measuring the state of depolarization of the Raman line 1656 cm^{-1}.

[175a] Note added in proof: Still more recently this band has also been reported by Thompson and Harris (855a).

Some force constants obtained from the proposed fundamentals are given in the previous Table 47. However, in view of the uncertainties of the assignments, it seems premature to draw far-reaching conclusions from them about the electronic structure of C_2Cl_4, as has been done with older assignments by several authors.

TABLE 93. INFRARED AND RAMAN BANDS OF C_2Cl_4 AFTER WU (958), DUCHESNE AND PARODI (292), AND WITTEK (948).

Assignment	Infrared (gas) ν_{vacuum} (cm^{-1})	Raman (liquid)[176] $\Delta\nu_{vacuum}$ (cm^{-1})	Assignment	Infrared (gas) ν_{vacuum} (cm^{-1})	Raman (liquid)[176] $\Delta\nu_{vacuum}$ (cm^{-1})
$\nu_2-\nu_3$		218 (v.w.)	$\nu_6+\nu_{12}(B_{2u})$?	755 (m.)	
$\nu_3^{\mathrm{CCl_2}}(a_g)$		**237** (s.) pol.	$\nu_{11}(b_{3u})$	**782** (v.s.)	
$\nu_7(b_{1u})$ (or ν_{10}?)	**332** (solution)		$\nu_2+\nu_6(B_{1g})$?		784 (v.w.)
$\nu_6(b_{1g})$		**347** (m.) depol.	$\nu_2+\nu_{12}(B_{3u})$?	802 (s.)	
$\nu_{12}(b_{3u})$?	**387** (solution)		$\nu_9^{\mathrm{CCl}}(b_{2u})$	**913** (v.s.)	
$\nu_2^{\mathrm{CCl}}(a_g)$		**447** (v.s.) pol.	$\nu_5^{\mathrm{CCl}}(b_{1g})$		**1000** (v.w.)
$2\nu_3(A_g)$		464 (w.)	$2\nu_8(A_g)$		1025 (w.) pol.
$\nu_8(b_{2g})$		**512** (m.) depol.	$\nu_2+\nu_5(B_{1g})$?		1441 (w.?)
$\nu_3+\nu_6(B_{1g})$		574 (w.)	$\nu_1^{\mathrm{CC}}(a_g)$		**1571** (s.) pol.
$\nu_{12}+\nu_{10}$?		631 (v.w.)	$\nu_1+\nu_3(A_g)$, $2\nu_9(A_g)$		1819 (w.)
$\nu_7+\nu_{12}(B_{2g})$?		726 (v.w.)	$\nu_1+\nu_2(A_g)$		1998 (w.)

Cis and trans $C_2H_2Cl_2$. As is well known, there are three isomers of dichloroethylene which may be written:

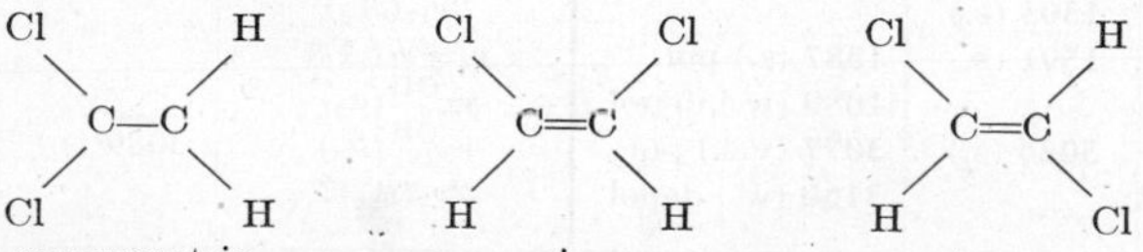

unsymmetric cis trans

The spectrum of the first, unsymmetrical one has as yet been investigated only in the infrared from 525 to 1450 cm^{-1} under low dispersion [Emschwiller and Lecomte (304)]. But numerous investigations of the Raman spectrum of cis and trans $C_2H_2Cl_2$ have been made, as well as two investigations of their infrared spectra. Also the corresponding deutero compounds, cis and trans $C_2D_2Cl_2$, have been investigated [Trumpy (879)].

TABLE 94. NUMBER, SPECIES AND ACTIVITY OF FUNDAMENTALS OF CIS AND TRANS $C_2H_2Cl_2$: CORRELATION TO POINT GROUP $V_h(C_2H_4$ AND $C_2Cl_4)$.

Cis $C_2H_2Cl_2(C_{2v})$				Trans $C_2H_2Cl_2(C_{2h})$			
Number	Species	Activity	Correlation to V_h	Number	Species	Activity	Correlation to V_h
5	A_1	R., I.	$3A_g+2B_{2u}$	5	A_g	R.	$3A_g+2B_{1g}$
2	A_2	R.	A_u+B_{2g}	2	A_u	I.	A_u+B_{1u}
4	B_1	R., I.	$2B_{1g}+2B_{3u}$	1	B_g	R.	B_{2g}
1	B_2	R., I.	B_{1u}	4	B_u	I.	$2B_{2u}+2B_{3u}$

[176] The polarization data are from Heidenreich (420), who used circularly polarized light.

If we assume that the dichloroethylenes have as high a symmetry as possible trans $C_2H_2Cl_2$ would belong to point group $\boldsymbol{C_{2h}}$ whereas cis $C_2H_2Cl_2$ would belong to $\boldsymbol{C_{2v}}$. The number of fundamentals of each species and their activities are indicated in Table 94 (compare Table 35). Table 94 contains also the species of $\boldsymbol{V_h}$ into which the vibrations of these molecules would go over if the masses of the two H atoms or the two Cl atoms were gradually changed, transforming the molecules into C_2Cl_4 and C_2H_4 respectively (see Table 53). For the spectrum of trans $C_2H_2Cl_2$, since it has a center of symmetry, the rule of mutual exclusion (p. 256) should hold, whereas for cis $C_2H_2Cl_2$, it should not. This is indeed observed as shown by Table 95, which gives the infrared and Raman

TABLE 95. INFRARED AND RAMAN SPECTRA OF CIS AND TRANS $C_2H_2Cl_2$ (BELOW 4000 cm^{-1}), AFTER WU (958),[177] TRUMPY (876), AND PAULSEN (688).

Cis $C_2H_2Cl_2$			Trans $C_2H_2Cl_2$		
Assignment	Infrared (gas) ν_{vacuum} cm^{-1}	Raman (liquid) $\Delta\nu_{vacuum}$[178] cm^{-1}	Assignment	Infrared (gas) ν_{vacuum} cm^{-1}	Raman (liquid) $\Delta\nu_{vacuum}$[178] cm^{-1}
$\nu_5(a_1)$		**173** (s.) pol.	$\nu_5(a_g)$		**349** (s.) pol.
$\nu_7(a_2)$		**406** (s.) depol.	$\nu_6(a_u)$	**620** (m.)	
$\nu_{12}(b_2)$	**570** (s.)	**563** (m.) depol.	$\nu_8(b_g)$?		**758** (m.) pol.
$\nu_{11}(b_1)$	**694** (s.)		$\nu_{12}^{CCl}(b_u)$	**820** (s.)	
$\nu_4^{CCl}(a_1)$	(711)?[179]	**711** (s.) pol.	$\nu_4^{CCl}(a_g)$		**844** (s.) pol.
$2\nu_7(A_1)$?		807 (v.w.)	$\nu_{11}(b_u)$	**917** (s.)	
$\nu_{10}^{CCl}(b_1)$	**857** (s.)		$\nu_{10}^{CHCl}(b_u)$	**1200** (s.)	
$\nu_6(a_2)$		**876** (w.) depol.	$\nu_3^{CHCl}(a_g)$		**1270** (s.) pol.
$\nu_3^{CHCl}(a_1)$		**1179** (s.) pol.	$\nu_2^{CC}(a_g)$		**1576** (s.) pol.
$\nu_9^{CHCl}(b_1)$	**1303** (s.)		$2\nu_{12}(A_g)$		1626 (v.w.) pol.
$\nu_2^{CC}(a_1)$	**1591** (s.)	**1587** (s.) pol.	$\nu_3+\nu_5(A_g)$		1692 (w.) pol.
$2\nu_{10}(A_1)$?		1689 (w.) depol.?	$\nu_1^{CH}(a_g)$		**3071** (s.) pol.
$\nu_1^{CH}(a_1)$	**3086** (s.)	**3077** (v.s.) pol.	$\nu_9^{CH}(b_u)$	**3089** (s.)	
$\nu_8^{CH}(b_1)$? $2\nu_2(A_1)$?		**3160** (w.) depol.	$2\nu_2(A_g)$?		3142 (w.) depol.?

spectra of the two molecules. Furthermore, the number of observed Raman lines of trans $C_2H_2Cl_2$ is smaller than that of cis $C_2H_2Cl_2$, as is to be expected from Table 94. More particularly the number of depolarized Raman lines for trans $C_2H_2Cl_2$ should be much smaller than for cis $C_2H_2Cl_2$ there being only one such fundamental for the former. Indeed, only one depolarized Raman line is found for trans $C_2H_2Cl_2$, whereas five are found for cis $C_2H_2Cl_2$.

The five totally symmetric vibrations of the two molecules are easily identified as corresponding to the strong polarized Raman lines. They are numbered $\nu_1, \cdots, \nu_5$ in order of decreasing frequencies. The only remaining Raman-active fundamental $\nu_8(b_g)$ of trans $C_2H_2Cl_2$ should be depolarized. Since it is a (low-frequency) bending vibration perpendicular to the plane of the molecule (see Fig. 96), it cannot be identified with the only observed depolarized (?) Raman line 3142 cm^{-1}. Wu (959) has used 758 for it, apparently assuming that the measured degree of depolarization 0.70 (given by both Trumpy and Paulsen) is actually $\frac{6}{7} = 0.86$. It may just as well be that ν_8 is not observed and that all the Raman lines of trans $C_2H_2Cl_2$ other than $\nu_1, \cdots, \nu_5$ are combination and overtone bands. It seems probable that of the strong infrared bands of trans $C_2H_2Cl_2$ the four highest ones are those of species B_u ($\nu_9, \cdots, \nu_{12}$), since they are vibrations in the plane of the molecule (see Fig. 96), whereas the two vibrations ν_6 and ν_7 of species A_u are vibrations perpendicular to the

[177] The infrared spectrum of the liquid between 525 and 1450 cm^{-1} has been studied by Lecomte (565) and Emschwiller and Lecomte (304). The agreement with the spectrum of the vapor obtained by Wu is not very good.

[178] The values given are averages of Trumpy's and Paulsen's figures with slightly higher weight given to the latter because of the higher dispersion used.

[179] No band is given here in Wu's original paper, but one is given in his book (26).

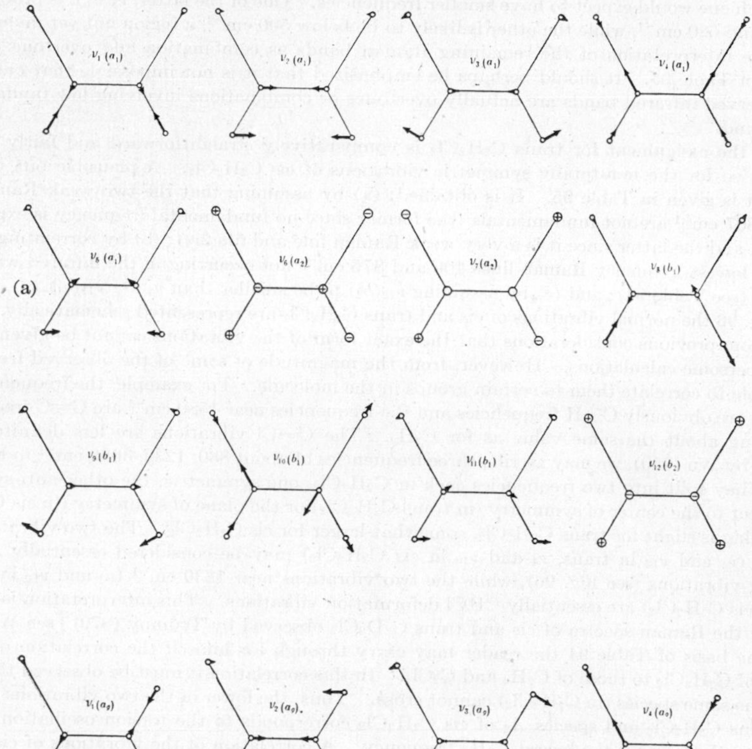

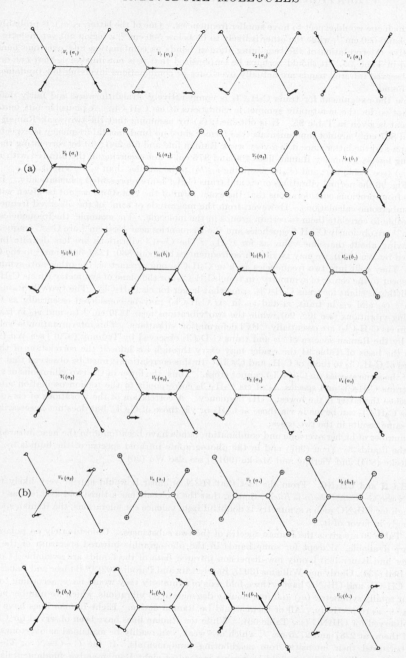

FIG. 96. **Normal vibrations of (a) cis- and (b) trans-$C_2H_2Cl_2$ (schematic).—For cis-$C_2H_2Cl_2$ the species designations are based on the assumption that the plane of the molecule is the xz plane (see Table 13).**

plane, which one would expect to have smaller frequencies. One of the latter, $\nu_6(a_u)$, is probably the infrared band 620 cm^{-1}, while the other is likely to be below 500 cm^{-1}, a region not yet investigated. A tentative interpretation of the remaining Raman bands as combination and overtone bands is indicated in Table 95. It should perhaps be emphasized that it is not impossible that one or two of the observed infrared bands are actually overtones or combinations involving low fundamentals not yet found.

While the assignment for trans $C_2H_2Cl_2$ is comparatively straightforward and fairly definite, this is not so for the non-totally symmetric vibrations of cis $C_2H_2Cl_2$. A plausible but tentative assignment is given in Table 95. It is obtained: (a) by assuming that the two weak Raman lines 1689 and 807 cm^{-1} are not fundamentals (the former since no fundamental frequency is expected in this region and the latter since it is a very weak Raman line and fits $2\nu_7$); (b) by correlating the two remaining lowest-frequency Raman lines 406 and 876 cm^{-1} not occurring in the infrared with $\nu_7(a_2)$ and $\nu_6(a_2)$ (see Table 94); and (c) by assuming $\nu_{12}(b_2)$ to be smaller than $\nu_8 \cdots \nu_{11}(b_1)$.

In Fig. 96 the normal vibrations of cis and trans $C_2H_2Cl_2$ are represented schematically. It will be clear from previous considerations that the exact form of the vibrations cannot be given without very cumbersome calculations. However, from the magnitude of some of the observed frequencies it is possible to correlate them to certain groups in the molecule. For example, the frequencies near 3100 cm^{-1} are obviously C—H frequencies and the frequencies near 1580 cm^{-1} are C═C frequencies, both having about the same value as for C_2H_4. The C—Cl vibrations are less definite but, as suggested by Wu (959), we may ascribe three frequencies of about 830, 1230, 3089 cm^{-1} to the CHCl group. They split into two frequencies each in $C_2H_2Cl_2$, one symmetric, the other anti-symmetric with respect to the center of symmetry (in trans $C_2H_2Cl_2$) or the plane of symmetry (in cis $C_2H_2Cl_2$). The splitting is slight for trans $C_2H_2Cl_2$, somewhat larger for cis $C_2H_2Cl_2$. The two vibrations near 830 cm^{-1} (ν_4 and ν_{12} in trans, ν_4 and ν_{10} in cis $C_2H_2Cl_2$) may be considered essentially as C—Cl stretching vibrations (see Fig. 96), while the two vibrations near 1230 cm^{-1} (ν_3 and ν_{10} in trans, ν_3 and ν_9 in cis $C_2H_2Cl_2$) are essentially CHCl deformation vibrations. This interpretation is well confirmed by the Raman spectra of cis and trans $C_2D_2Cl_2$ observed by Trumpy (879) [see Wu (959)].

On the basis of Table 94 the reader may carry through for himself the correlation of the frequencies of $C_2H_2Cl_2$ to those of C_2H_4 and C_2Cl_4. In this correlation it must be observed that vibrations of the same species (in $C_2H_2Cl_2$) cannot cross. Thus, the lower of the two vibrations of species A_u of trans $C_2H_2Cl_2$ and species A_2 of cis $C_2H_2Cl_2$ corresponds to the torsion oscillation $\nu_4(a_u)$ of C_2H_4, since the latter is the lowest C_2H_4 frequency. A correlation of the vibrations of cis to those of trans $C_2H_2Cl_2$ can be made via those of C_2H_4 or via those of C_2Cl_4, but does not necessarily lead to the same results in the two cases.

A number of higher overtone and combination bands have been found in the near infrared spectra of the liquids by Yeou (967) and in the photographic infrared spectra of the liquids by Timm and Mecke (864) and Vierling and Mecke (901) [see also Wu (26)].

CH_3CN and CH_3NC. From the fact that HCN is linear it would appear very likely that in CH_3CN the C—C≡N *chain is linear*, that is, that the molecule has a three-fold axis just as CH_3Cl. Although for CH_3NC such a symmetry is doubtful from valence considerations, the Raman spectrum is strongly in favor of it.

In Table 96 are given the Raman spectra of the two substances. Unfortunately no polarization data are available. Except for some bands in the photographic infrared spectrum of the vapor [Badger and Bauer (68)], only low-dispersion infrared data of the liquids are available [Coblentz (224), Bell (134), Gordy and Williams (392), Barchewitz and Parodi (102),[180] Badger and Bauer (68)].

If CH_3CN and CH_3NC have a three-fold axis of symmetry they would have, according to Table 36, four totally symmetric (a_1) and four doubly degenerate (e) vibrations, which we number $\nu_1 \cdots \nu_4$ and $\nu_5 \cdots \nu_8$ respectively. All of them would be Raman active. Eight Raman lines have indeed been observed for CH_3NC (see Table 96). While ten Raman lines have been observed for CH_3CN, two of them, at 2287 and 2725 cm^{-1}, which are weak, can readily be explained as overtones which have borrowed their intensity from neighboring fundamentals. If the C—C≡N or C—N═C chains were not linear, there would be twelve instead of eight Raman-active fundamentals. The agreement of the number of observed Raman-active fundamentals with that expected on the sym-

[180] These authors give a considerable number of bands below 390 cm^{-1}, most of which they also find in CH_3I. It appears very unlikely that these bands are due to CH_3CN or CH_3I. No intensity estimates and no absorption curves are given.

metrical model (C_{3v}) is of course by no means a very strong argument for this model; but the following considerations support it further.

As for CH_3Cl, one would expect for CH_3CN and CH_3NC a totally symmetric and a degenerate C—H stretching vibration and a similar pair of CH_3 deformation vibrations. These are indeed

TABLE 96. RAMAN SPECTRA OF CH_3CN AND CH_3NC AFTER REITZ AND SKRABAL (738) AND DADIEU (252).

Assignment	CH_3CN[181] $\Delta\nu_{vacuum}$, cm^{-1}	CH_3NC $\Delta\nu_{vacuum}$, cm^{-1}
$\nu_8(e)$	380[182] (s.)	290 (s.)
$\nu_4{}^{CC}(a_1)$, $\nu_4{}^{C-N}(a_1)$	918[182] (s.)	928 (m.)
$\nu_7(e)$ CH_3 rocking	1124[183] (v.w.)?	1041 (w.)
$\nu_3{}^{CH_3}(a_1)$	1376 (m.)	1414 (m.)
$\nu_6{}^{CH_3}(e)$	1440[182, 184] (m.) (b.)	1456 (w.)
$\nu_2{}^{C\equiv N}(a_1)$	2249[185] (s.)	2161[182] (s.)
$2\nu_7$?	2287 (w.)	
$2\nu_3{}^{CH_3}(A_1)$	2725 (w.)	
$\nu_1{}^{CH}(a_1)$	2942 (v.s.)	2951 (s.)
$\nu_5{}^{CH}(e)$	2999[185] (s.) (b.)	3002 (w.)

observed (Table 96),[186] with nearly the same frequencies as for CH_3Cl (see Table 84). For CH_3CN one would further expect a C≡N vibration similar to that of HCN, for which it is at 2089 cm^{-1}, and a C—C vibration similar to that of C_2H_6, for which it is at 993 cm^{-1}. These are indeed observed for CH_3CN at 2249 and 918 cm^{-1}.[187] The two remaining frequencies 1124 and 380 cm^{-1} are apparently those of the two remaining degenerate vibrations ν_7 and ν_8, which correspond essentially to the two bending vibrations of the linear system (H_3)—C—C≡N). The first of these is mainly an H_3≡C—C bending vibration, or in other words a CH_3 rocking vibration (analogous to the bending vibration $\nu_6 = 1016$ cm^{-1} of CH_3Cl); the second is mainly a C—C≡N bending vibration (analogous to the bending vibrations $\nu_5 = 226$, $\nu_4 = 506$ cm^{-1} of C_2N_2).[188]

If CH_3CN and CH_3NC did not have a three-fold axis there should be two adjacent Raman lines in place of every one of the lines ν_5, ν_6, ν_7, ν_8. The fact that this is not observed might be compatible with a non-linear model in the case of ν_5 and ν_6, which are CH_3 vibrations and for which the splitting would be expected to be slight. But the fact that no such splitting occurs for the whole-molecule vibrations ν_7 and ν_8 is strongly in favor of the symmetrical model.

---

[181] b. refers to broad Raman lines.

[182] These bands have also been observed in the infrared.

[183] In place of this an infrared band has been found at 1040 cm^{-1}.

[184] In their Table 1 Reitz and Skrabal (738) give the value 1445 cm^{-1} for this band.

[185] In place of these two bands Gordy and Williams (392) have found in the infrared two strong bands at 2283 and 3077 cm^{-1}. The shift in wave length in both cases is about the same and it appears very likely that the infrared and Raman bands are really identical, particularly since Bell (134) gives 2980 cm^{-1} for one of the infrared bands. Linnett (582) makes this assumption only for the band 2249–2283.

[186] Linnett (582) considered the strong Raman line 2999 cm^{-1} as a combination band which has borrowed intensity from the fundamental 2942, and he took the infrared band 3077 cm^{-1} as $\nu_5{}^{CH}(e)$ (see footnote 185). However, there is no binary combination in this region and it would be difficult to understand a ternary combination having such a great intensity. Also the Raman line 2999 cm^{-1} is distinctly broad, indicating a degenerate upper state (see Chapter IV), which is not compatible with Linnett's explanation. We assume therefore that the frequency found for ν_5 in the low-dispersion infrared work is too high, just as that found for ν_2.

[187] It is of course also possible that 1124 cm^{-1} is the C—C vibration. A decision would be possible from polarization measurements.

[188] It is significant that the frequency ν_8 of CH_3CN is just about half-way between ν_4 and ν_5 of C_2N_2, the difference between ν_4 and ν_5 in C_2N_2 being due to resonance.

Crawford and Brinkley (240) have predicted the fundamentals of CH_3CN from force constants taken from other molecules. Their predicted values agree within 2.5 per cent with the fundamentals of Table 96, except for the very weak and doubtful Raman line 1124 cm^{-1}.[189]

It is particularly noteworthy that the C≡N frequency undergoes only a slight change in going from CH_3CN to CH_3NC, indicating that the CN groups in the two molecules are very similar to each other. This fact, which cannot be understood on the basis of elementary valence theory, shows that the free valency of the CN radical is essentially not localized at one atom (C) but is a property of the radical as a whole. It can occur at either end without changing the character of the CN group.

Methyl alcohol, CH_3OH and CH_3OD. The fact that the H_2O molecule is bent suggests strongly that the C—O—H chain in CH_3OH is bent as well. This conclusion is confirmed by the spectrum of the molecule, although more definitely by the rotational fine structure (see Chapter IV) than by the vibrational structure. One would expect the H atom of OH to be located in one of the planes of symmetry of the CH_3 group, so that the molecule would have just one plane of symmetry, the C—O—H plane (point group C_s). According to Table 35 there should be eight vibrations symmetric with respect to this plane (a') and four antisymmetric (a''). Four of the symmetric vibrations should be close to the four antisymmetric vibrations, since if C—O—H were linear they would coincide and form four doubly degenerate vibrations. The splitting would be expected to be slight if detectable at all for the internal vibrations of the CH_3 group, but it should be large for the deformation vibrations of the (H_3)—C—O—H chain. This is in fact observed, as shown by Table 97. In order to bring out the relations to the symmetrical model we use the designation of the vibrations appropriate to the latter model (see CH_3CN), distinguishing the two components into which a doubly degenerate vibration of the symmetrical form splits by ′ and ″ if they are resolved.

Four strong Raman lines are observed in the region of the C—H vibration (2900 cm^{-1}). Two of them are easily explained as a combination and an overtone respectively made strong by Fermi resonance with the two fundamentals ν_2^{CH} and ν_5^{CH} [see Wu (26)], although it is not impossible that one is the second component of ν_5. The two infrared bands at 1455 and 1477 cm^{-1} are not well resolved, but are readily explained as corresponding to $\nu_3^{CH_3}$ and $\nu_6^{CH_3}$ of CH_3CN. There is also an indication of the expected doubling for $\nu_6^{CH_3}(e)$. The strong Raman line and strong || infrared band at 1034 cm^{-1} obviously is the C—O vibration corresponding to ν_4^{CC} of CH_3CN. The vibration ν_7 corresponding to a swinging of the whole H_3 group perpendicular to the C—O axis (for which a larger splitting is to be expected) may be assigned to any two of the three Raman lines 1056, 1112, and 1171 cm^{-1}.

There remains the bending vibration of the OH group against C—O. Clearly here the splitting should be largest: The bending in the C—O—H plane should give a frequency of the same order as the bending vibration in H_2O (of course modified by the different distribution of masses). It may be identified either with the polarized Raman line 1056 cm^{-1} or the infrared band 1340 cm^{-1}. The latter alternative, suggested by Noether (673), is perhaps slightly more probable on account of the existence of a corresponding band in CH_3OD with approximately the correct isotope shift. The structure of this band is apparently of the hybrid type (see Chapter IV, section 2b), as it should be since the dipole moment changes neither exactly || nor exactly ⊥ to the axis. The bending vibration perpendicular to the C—O—H plane, which may also be considered as a torsion oscillation of OH about the C—O axis, would be expected to have a much smaller frequency. According to Koehler and Dennison (517) it corresponds to the strong region of absorption at 270 cm^{-1} found by Lawson and Randall (560). The potential energy as a function of the angle of the C—O—H plane with a fixed plane of symmetry of the CH_3 group would be expected to have three identical maxima and minima, and if the maxima are not very high, as seems probable, even for comparatively small energies this torsional oscillation will go over into a rotation. This seems to be indicated by the wide absorption region 380–860 cm^{-1}, as suggested by Borden and Barker (see also p. 494 and p. 498).

The O—H stretching vibration corresponds obviously to the infrared band 3682 cm^{-1}, which has a frequency very similar to that of the symmetrical OH stretching vibration ν_1 of H_2O. The very considerable difference between the frequency measured in the Raman spectrum and that measured in the infrared spectrum is due to the fact that the Raman frequency refers to the liquid. The infrared spectrum of the liquid shows the OH band at 3400 cm^{-1}, just as does the Raman spectrum [Errera and Mollet (307), Buswell, Deitz, and Rodebush (186), and others]. The great difference between the liquid and gaseous OH frequency is due to association of CH_3OH molecules in consequence of

[189] The agreement is better than with Linnett's choice of fundamentals (see footnote 186).

hydrogen bonding (see the references just quoted and Chapter V, section 2). Since the hydrogen bonding is due to the OH groups, the frequencies not involving the OH group are affected by the transition from vapor to liquid by not more than the usual amount.

The second overtone of the OH vibration $3\nu_1^{OH}$ has been observed by Herzberg and Verleger (445) in the photographic infrared at 10530 cm^{-1}. The third overtone of the OH and the second overtone of the CH vibration have been found in the photographic infrared at 13700 and 8230 cm^{-1} respectively by Badger and Bauer (66), who have also reinvestigated $3\nu_1^{OH}$ and found several other weaker bands.

TABLE 97. RAMAN AND INFRARED SPECTRA OF CH_3OH AND CH_3OD BELOW 4000 cm^{-1}, AFTER HALFORD, ANDERSON, AND KISSIN (407), BORDEN AND BARKER (169), BARKER AND BOSSCHIETER (109), AND NOETHER (673).

Assignment	CH_3OH		CH_3OD	
	Raman[190] $\Delta\nu_{vacuum}$ (liquid)[191] cm^{-1}	Infrared ν_{vacuum} (gas) cm^{-1}	Raman $\Delta\nu_{vacuum}$ (liquid) cm^{-1}	Infrared ν_{vacuum} (gas) cm^{-1}
$\nu_8''(e)$OH twisting		270 (s.)[192]		
$2\nu_8, \cdots$		380–860 (m.)		
$\nu_4^{CO}(a')$	1029 (s.)	1033.9 (v.s.) ∥	1033 (s.)	1040.3 (v.s.)
$\nu_7'(e)CH_3$ rocking	1056 (m.) pol.		942 (m.)	
	1112 (m.)		1071 (m.)	
	1153 (w.)		1154 (w.)	
$\nu_7''(e)CH_3$ rocking	1171 (m.)		1179 (m.)	
$\nu_4+\nu_8''$?		1209 (v.w.)[193]		1207 (w.)
		1260 (v.w.)[193]	1226 (w.)	1232 (v.w.)[193]
$\nu_8'(e)$OH bending	1370? pol.	1340 (m.) (⊥)		863 (s.)
$\nu_6'^{CH_3}(e)$		(1430)? (w.)[193]		1427 (m.)
$\nu_6''^{CH_3}(e)$, $\nu_3^{CH_3}(a')$ }	1458 (s.) depol.	{1455 (m.)	1465 (s.)	1459 } (s.)
		1477 (m.)		1480 } (s.)
$2\nu_4^{CO}$		2053 (w.) ∥		2064.8 (m.) ∥
$\nu_6+\nu_7$?	2588 (w.)		2591 (w.)	
$\nu_2^{CH}(a')$	2837 (v.s.) pol.	2844 (s.) ∥	2836 (v.s.)	2849 (s.) ∥
$\nu_3+\nu_6''$ (or ν_5''?)	2914 (s.)	overlapped	2914 (s.)	
$2\nu_6''(A')$	2942 (v.s.) pol.	overlapped	2947 (v.s.)	
$\nu_5^{CH}(e)$	2987 (s.)	2977 (s.) ⊥	2988 (s.)	2964 (s.) ⊥
$\nu_1^{OH}(a')$	3400 (s.)	3682 (m.)	2494 (s.)	2719.7 (s.)

The spectrum of CH_3OD also given in Table 97 confirms to some extent the above conclusions and assignments. Only the frequencies associated with the OH group are appreciably altered. The Raman and infrared spectra of CH_2DOD have been investigated by Halford, Anderson, and Kissin (407) and Barker and Bosschieter (109), the infrared spectrum of CD_3OH and the infrared and Raman spectra of CD_3OD by Noether (673).

Other six-atomic molecules. In Table 98 are given references for the infrared and Raman spectra of further six-atomic molecules that have been investigated, as well as the structures, if any, that have been suggested. Again one may deal with the skeletal vibrations of molecules such as $C_2(CH_3)_4$ by considering them as six-particle systems of the ethylene type.

---

[190] The polarization data are from Trumpy (877).

[191] Nielsen and Ward (670) give for the vapor the three shifts 1032, 2845, and 2955 cm^{-1}.

[192] Obtained by Lawson and Randall (560).

[193] These "bands," given only by Noether (673), are only very slight humps on his transmission curves.

TABLE 98. FURTHER SIX-ATOMIC MOLECULES.[53]

Molecule	Structure	References	
		Raman spectrum	Infrared spectrum
C_2Br_4	V_h (plane)	(709)	
Cis, trans $C_2H_2Br_2$	C_{2v}, C_{2h} (plane)	(258)	(304)
Cis, trans $C_2H_2I_2$	C_{2v}, C_{2h} (plane)		(304)
CCl_2CF_2	C_{2v} (plane)	(415)	
C_2H_3Cl, C_2H_3Br, C_2H_3I	C_s (plane)	(530)	(304) (958) (864) (901)
C_2HCl_3	C_s (plane)	(13) (408)	(958) (864) (901)
$C_2H_2O_2$ (glyoxal)	C_{2h} (plane)?		(850)
$C_2O_2Cl_2$ (oxalylchloride)	C_{2h} (plane)?	(756) (525) (841) (533a)	
$(C_2O_4)^{--}$	V_h (plane)?	(405)	
N_2O_4[194]	V_h (plane)?	(823) (290)	(822) (769) (412)
N_2H_4	no symmetry	(479) (395)	(332) (334)
NH_2COH (formamide)	C_s	(755) (524) (840)	
$ClH_2C—CN$, $Cl_2HC—CN$	C_s	(201)	(102)
$Cl_3C—CN$	C_{3v}	(201)	
CH_3SH	C_s	(906)	(846) (860)
PCl_5	D_{3h} (bipyramid with P at center)	(638)	
$SbCl_5$	C_{4v} (quadratic pyramid?)	(728) (730)	
$HClO_4$	C_s (like CH_3OH)	(325) (727a)	
$AgClO_4$		(649a)	
H_2SeO_3	two forms?	(891)	
H_2CNOH	C_s (plane)		(975)
$SO_2 \cdot OH \cdot Cl$	no symmetry	(€03) (671)	(888)

(e) Seven-atomic molecules

Sulfur hexafluoride, SF_6. Electron diffraction measurements [Braune and Knoke (177), Brockway and Pauling (180)] first strongly suggested that in sulfur hexafluoride the sulfur atom occupies the center of a *regular octahedron* whose corners are occupied by the fluorine atoms (see Fig. 3d). This highly symmetrical structure has been fully confirmed by Raman and infrared investigations [Yost, Steffens, and Gross (974), Eucken and Ahrens (310)].

None of the bands observed in the Raman spectrum is observed in the infrared. This indicates that the molecule has a center of symmetry (see p. 256)[195] as has the octahedral model. If we did not accept the electron diffraction evidence that this is the correct model we might also consider a plane symmetrical model of point group D_{6h}. For the octahedral model there are (see Table 36) six normal vibrations of the following species and activities:

$$1A_{1g}(\text{R.}),\quad 1E_g(\text{R.}),\quad 2F_{1u}(\text{I.}),\quad 1F_{2g}(\text{R.}),\quad 1F_{2u}\ (\text{inactive}).$$

For the plane hexagonal model there are ten normal vibrations of the species and activities

$$1A_{1g}(\text{R.}),\quad 1A_{2u}(\text{I.}),\quad 1B_{1g}\ (\text{inactive}),\quad 1B_{1u}\ (\text{inactive}),\quad 1B_{2u}\ (\text{inactive}),$$
$$2E_{1u}(\text{I.}),\quad 2E_{2g}(\text{R.}),\quad 1E_{2u}\ (\text{inactive}).$$

The observed infrared and Raman bands are given in Table 99. The observation of three Raman lines is compatible with either model. But the fact that there are only two infrared bands of out-

[53] On p. 287.

[194] See Table 47, p. 184.

[195] The objection that the infrared bands corresponding to the Raman bands may be weak (and vice versa) can hardly be raised in the case of a molecule which would have 15 fundamentals if it had not a high symmetry.

standing intensity, that is, two infrared-active fundamentals, and the fact that all other infrared bands can be interpreted as combinations of these fundamentals with the three Raman-active fundamentals and just one inactive fundamental is strongly in favor of the octahedral model.

The form and designation of the fundamentals of an octahedral XY_6 molecule have been given in the previous Fig. 51. In view of the rule that totally symmetric vibrations give rise to the strongest Raman lines, it appears certain that the very strong Raman line 775 cm^{-1} corresponds to the only

TABLE 99. INFRARED AND RAMAN SPECTRA OF SF_6 AFTER EUCKEN AND AHRENS (310) AND YOST AND STEFFENS AND GROSS (914).

ν_{vacuum} (cm^{-1})	Assignment	ν_{vacuum} (cm^{-1})	Assignment
(363)[196]	$\nu_6(f_{2u})$	**965** I. gas (v.s.)	$\nu_3(f_{1u})$
524 R. liquid (w.)	$\nu_5(f_{2g})$	1163 I. gas (v.w.)	$\nu_4+\nu_5(F_{1u})$
545 I. gas (w.)	$\nu_2+\nu_5-\nu_4$?	1205 I. gas (v.w.)	$\nu_3+\nu_4-\nu_6$?
617 I. gas (v.s.)	$\nu_4(f_{1u})$	1262 I. gas (s.)	$\nu_2+\nu_4+\nu_6-\nu_6$
644 R. liquid (w.)	$\nu_2(e_g)$	1282 I. gas (s.)	$\nu_2+\nu_4(F_{1u})$
730 I. gas (w.)	$\nu_3+\nu_6-\nu_4$?	1380 I. gas (v.w.)	$\nu_1+\nu_4(F_{1u})$
775 R. gas (v.s.)	$\nu_1(a_{1g})$	1578 I. gas (m.)	$\nu_2+\nu_3(F_{1u})$
830 I. gas (v.w.)	$\nu_2+\nu_5-\nu_6$?	1703 I. gas (m.)	$\nu_1+\nu_3(F_{1u})$
885 I. gas (v.w.)	$\nu_5+\nu_6(F_{1u})$		

totally symmetric fundamental $\nu_1(a_{1g})$. The two weak Raman lines, 644 and 524 cm^{-1} are then $\nu_2(e_g)$ and $\nu_5(f_{2g})$, respectively (since from Fig. 51 one would expect $\nu_2 > \nu_5$, the latter being a sort of deformation vibration). The two strong infrared bands at 965 and 617 cm^{-1} are $\nu_3(f_{1u})$ and $\nu_4(f_{1u})$, and the remaining weaker infrared bands can be assigned as indicated in Table 99, according to Eucken and Ahrens (310) (with a slight modification). It is significant that in agreement with the selection rules (see p. 262) the first overtones of the infrared-active fundamentals, $2\nu_3$ and $2\nu_4$, do not occur in the infrared spectrum. The assignment of the four weak infrared bands at 545, 730, 830, 1205 cm^{-1} as difference bands is not satisfactory, since the corresponding summation bands do not occur. The frequency of the inactive vibration ν_6 is taken from specific heat measurements [see Eucken and Ahrens (310)]. Its value is not very certain since it is confirmed only by rather weak combination bands. Further work on the infrared spectrum, particularly at longer and shorter wave lengths than investigated by Eucken and Ahrens and with higher dispersion, would be very desirable.

Methyl acetylene, CH_3—C≡CH. In view of the linearity of acetylene, it is to be expected that in methyl acetylene the four atoms >C—C≡CH are arranged in a straight line. This is confirmed by the vibrational structure of the infrared and Raman spectra as well as by the fine structure of the infrared bands (see Chapter IV). The observed Raman and infrared bands are given in Table 100.

If the symmetrical model is correct we have a case very similar to CH_3CN (see p. 332), except that there is one more atom in the linear chain (the two molecules have the same number of electrons). Instead of four totally symmetric and four doubly degenerate vibrations, there are now five of each species. The additional totally symmetric vibration is essentially the C—H stretching vibration of the C≡C—H group and is readily identified as the infrared band 3429 cm^{-1},[197] whereas the additional degenerate vibration is essentially the C—H bending vibration of the C≡C—H group, which by comparison with C_2H_2 (see p. 290) is easily identified as the very strong infrared band (and strong depolarized Raman line) 642 cm^{-1}. The other fundamentals have magnitudes very similar to those of CH_3CN and are consequently easily identified as given in Table 100. Since we number the vibrations of a given species in the order of their magnitudes, the numbering here is somewhat different from that for CH_3CN. The character of the oscillations (C≡C stretching, and the like) is indicated in Table 100. The remaining infrared and Raman bands (except the photographic infrared band and the Raman bands at 2128 and 2134 cm^{-1}) have been interpreted by Crawford (237) in the way

[196] From specific heat data [Eucken and Ahrens (310)].

[197] The large shift in going to the liquid (Raman line 3305 cm^{-1}) is indicative of a large effect of the intermolecular field.

TABLE 100. RAMAN AND INFRARED SPECTRA OF $CH_3C{\equiv}CH$, AFTER GLOCKLER AND DAVIS (368), GLOCKLER AND WALL (386), AND CRAWFORD (237).

Assignment[198]	Raman[199] $\Delta\nu_{vacuum}$ (cm^{-1})	Infrared[200] (gas) ν_{vacuum} (cm^{-1})	Assignment[201]	Raman[199] $\Delta\nu_{vacuum}$ (cm^{-1})	Infrared[200] (gas) ν_{vacuum} (cm^{-1})
$\nu_{10}(e)$ (C—C≡C bending)	**336** (v.s.) depol. (liquid)		$\nu_3+2\nu_{10}-2\nu_{10}[A]$ }	2128 (w.) (gas)	
$\nu_4-\nu_5[A_1]$		454 (v.w.)	$\nu_3+\nu_9-\nu_9[A_1]$ }		
$\nu_9(e)$ (C≡C—H bending)	**643** (s.) depol. (liquid)	**642** (v.s.)	$\nu_3+\nu_{10}-\nu_{10}[A_1]$	2134 (w.) (gas)	
$\nu_8-\nu_{10}[A_1, E]$, $\nu_4-\nu_9[E]$		720 (w.)	$\nu_3^{C\equiv C}(a_1)$	**2142** (v.s.) pol. (gas)	2150 (s.)
$\nu_5^{C-C}(a_1)$	**930** (s.) pol. (gas)	**926** (s.) ‖ doublet	$\nu_6-\nu_{10}[A_1, E]$		2341 (m.)
$\nu_9+\nu_{10}(A_1, E)$		971 (w.)	$\nu_3+\nu_{10}$, $\nu_1-\nu_9$, $\nu_7+\nu_8$		2514 (m.)
$\nu_8(e)$ (CH_3 rocking)	**1035** (v.w.) (liquid)	**1041** (w.)	$2\nu_4(A_1)$	2736 (w.) (liquid)	
$\nu_7-\nu_{10}[A_1, E]$		1110 (m.) ‖ ?	$\nu_4+\nu_7(E)$		2808 (w.)
$\nu_3-(\nu_9+\nu_{10})?[A_1, E]$		1187 (s.) ‖	$2\nu_7(A_1, E)$	2867 (s.) (liquid)	
$2\nu_9(A_1, E)$		1260 (v.s.) ‖	$\nu_2^{CH}(a_1)$	**2941** (v.s.) pol. (gas)	
$\nu_4^{CH_3}(a_1)$	**1382** (s.) depol. (liquid) }	1340 − 1550 (s.)[202]	$\nu_6^{CH}(e)$	**2971** (m.) (liquid)	**2994** (v.s.)
$\nu_7^{CH_3}(e)$	**1448** (m.) (liquid) }		$\nu_1^{CH'}(a_1)$	3305 (m.) (liquid)	**3429** (s.)
$\nu_3-\nu_9[E]$		1515 (m.)	$\nu_1+\nu_{10}(E)$		3750 (w.)
$\nu_4+\nu_{10}(E)$		1721 (v.w.)	$\nu_1+\nu_9(E)$		4050 (m.)
$\nu_3-\nu_{10}[E]$		1815 (w.)	$3\nu_1+2\nu_{10}-2\nu_{10}[A_1]$ }		9691.5 (v.w.) ‖[204]
$3\nu_9(A_1, E)$		1894 (w.)	$3\nu_1+\nu_9-\nu_9[A_1]$ }		
$2\nu_8(A_1, E)$, $\nu_7+\nu_9(A_1, E)$[203]		2100 (w.)	$3\nu_1+\nu_{10}-\nu_{10}[A_1]$		9697.1 (w.) ‖[204]
			$3\nu_1(A_1)$		9702.6 (m.) ‖[204]

[198] For the difference bands the species of the dipole moment is given in square brackets, not the species of the upper state.

[199] Except for one weak line the Raman data are those of Glockler and his co-workers since the dispersion used by them is much larger than that used by Crawford. The polarization data are from Crawford (237).

[200] Infrared bands considered as doubtful by Crawford (237) have been omitted.

[201] H′ distinguishes the acetylenic hydrogen from the methyl hydrogen.

[202] A broad band with incompletely resolved fine structure.

[203] This band may also be the isotopic band of $CH_3C^{12}{\equiv}C^{13}$—H corresponding to $\nu_3^{C\equiv C}(a_1)$ [see Cleveland and Murray (218)].

[204] Observed in the photographic infrared by Herzberg, Patat, and Verleger (440). Some further extremely weak bands at still lower wave lengths are given by Badger and Bauer (69).

given in Table 100. It should be noted that, as for the other methyl compounds, the overtones $2\nu_4$ and $2\nu_7$ of the CH_3 deformation vibrations occur with a fair intensity in the Raman spectrum, due to Fermi resonance with $\nu_2^{CH}(a_1)$. The latter is, therefore, actually higher than it would be without this resonance.

A full discussion of the force constants in CH_3—C≡CH has been given by Crawford (237) and Crawford and Brinkley (240). These authors have also been able to predict the vibration frequencies by assuming force constants as observed in C_2H_2, CH_3X, C_2H_6, obtaining a very satisfactory agreement with the observed figures.

Allene, CH_2=C=CH_2. Allene is an isomer of methyl acetylene. From valence theory one would expect that the C=C=C chain is linear but that the two CH_2 groups lie in planes at right angles to each other and passing through the C=C=C axis, that is, that the molecule belongs to point group V_d (see Fig. 2m). The infrared and Raman data thus far available can be interpreted on this basis but are hardly sufficient to prove the V_d structure. However this structure is also supported by the observation that the dipole moment is zero [Watson, Kane, and Ramaswany (912a)].

The numbers, species, and activities of the fundamentals on the assumption of point group V_d are (see Tables 36 and 55)

$$3A_1(\text{R.}), \quad 1B_1(\text{R.}), \quad 3B_2(\text{R., I.}), \quad 4E(\text{R., I.});$$

that is, all eleven fundamentals are Raman active and seven of them are infrared active. The observed spectrum is given in Table 101.

TABLE 101. RAMAN AND INFRARED SPECTRA OF CH_2=C=CH_2, AFTER LINNETT AND AVERY (584) AND EYSTER (317).[204a]

Assignment	Raman (liquid) $\Delta\nu_{vacuum}$ (cm^{-1})	Infrared (gas) ν_{vacuum} (cm^{-1})	Assignment	Raman (liquid) $\Delta\nu_{vacuum}$ (cm^{-1})	Infrared (gas) ν_{vacuum} (cm^{-1})
$\nu_{11}(e)$ (C=C=C bending)	353 (m.)		$\nu_1^{CH}(a_1)$	2993 (v.s.)[206]	
$2\nu_{11}(A_1, B_1, B_2)$	705 (w.)		$\nu_8^{CH}(e)$	3061 (s.)[206]	overlapped
$\nu_4(b_1)$ (twisting)	(820)[205]		$\nu_3+\nu_8(E)$[208]?		4200 (v.w.)
$\nu_{10}(e)$ (C=C=C bending)	838 (m.)	852 (v.s.)	$3\nu_5(B_2)$		8739.0 (w.)
$\nu_9(e)$ (CH_2 rocking)		1031 (m.)	$2\nu_1+\nu_5(B_2)$		8776.6 (w.)
$\nu_3^{C=C}(a_1)$	1071 (v.s.)[206]		?		8922[209] (v.w.)
$\nu_4+\nu_{11}(E)$		1165 (v.w.)?	$2\nu_8+\nu_5(B_2)$		8978 (w.)
$\nu_7^{CH_2}(b_2)$		1389 (s.)	$2\nu_8+\nu_1(B_2)$		9012 (w.)
$\nu_2^{CH_2}(a_1)$	1432 (s.)[206]		$3\nu_8(E, E)$		9076.7 (w.)
$2\nu_{10}(A_1, B_1, B_2)$	1684 (w.)	1700 (m.)	?		9718 (v.w.)
$\nu_6^{C=C}(b_2)$	1956 (v.w.)	1980 (s.)	?		10420 (v.w.)
$\nu_7+\nu_9(E)$		2420 (w.)	$2\nu_5+\nu_1+\nu_6(B_2)$		10710[209] (v.w.)
$2\nu_2(A_1)$	2858 (w.)		$3\nu_5+2\nu_7(B_2)$ / $3\nu_8+\nu_7+\nu_9(B_2, B_2)$		11418 (w.)
$\nu_5^{CH}(b_2)$		2960 (m.)[207]	$5\nu_5(B_2)$		13904 (v.w.)

[204a] Note added in proof: In a recent paper Thompson and Harris (855a) have reinvestigated the infrared spectrum under somewhat higher dispersion. From the observed contours they have confirmed that the bands 1389 and 1700 cm^{-1} are $\|$ bands and the bands 852 and 1031 cm^{-1} are $\perp$ bands as required by the assignment given in Table 101. Thompson and Harris' wave numbers differ slightly from those of Linnett and Avery given here.

[205] From specific-heat data (see text).

[206] These figures are the averages of the values given by Linnett and Avery (584) and Kopper and Pongratz (537).

[207] Broad band.

[208] A number of ternary combinations would give better agreement.

[209] These are misprinted in Eyster's paper.

The three totally symmetric vibrations, ν_1, ν_2, ν_3, are essentially the two symmetrical CH_2 vibrations (C—H stretching and bending) taking place in phase, and the analogue of the vibration ν_1 of CO_2 considering the CH_2 groups as one particle each. They are readily identified with the strong Raman lines 2993, 1432, and 1071 cm^{-1}. The first two are similar to the corresponding C_2H_4 vibrations (Table 92) and the third agrees well with the value one obtains from the equations for a linear triatomic system using the same force constant for the C=C bond as found for C_2H_4 [see Kopper and Pongratz (537)].

Two of the b_2 vibrations are similar to $\nu_1^{CH}(a_1)$ and $\nu_2^{CH_2}(a_1)$ except that the atoms in the two CH_2 groups move with a phase shift of 180°. The third b_2 vibration corresponds to ν_3 of CO_2. These three vibrations are to be identified with the infrared bands 2960, 1389, and 1980 cm^{-1}, respectively. The last is again in almost exact agreement with calculations on the basis of the C=C force constant of C_2H_4. Of the four degenerate fundamentals one is essentially a vibration in which in both CH_2 groups the antisymmetric C—H stretching vibration (ν_3 in Fig. 25a) takes place. It can only be identified with the strong Raman line 3061 cm^{-1}. The three remaining degenerate fundamentals are essentially the three bending vibrations of a linear five-particle system (H_2=C=C=C=H_2). They are most probably to be assigned as indicated in Table 101. The last remaining fundamental $\nu_4(b_1)$ is the twisting vibration of the two CH_2 groups about the C=C=C axis. From specific heat data, assuming all the other fundamentals as known, Linnett and Avery (584) obtained for it 820 cm^{-1}. They assumed that the Raman line 838 cm^{-1} is a superposition of ν_4 (which is Raman active only) and $\nu_{10}(e) = 852$ cm^{-1} (which is known from the infrared spectrum). A plausible interpretation of the other infrared and Raman bands as combinations and overtone bands is given in Table 101. On the whole, the assignment appears to be very satisfactory.

If the agreement between the two Raman lines 1684 and 1956 cm^{-1} with the infrared bands 1700 and 1980 is not a chance coincidence it would exclude a plane model of the allene molecule, since that would have a center of symmetry and therefore the rule of mutual exclusion should hold. However, it has to be considered that the Raman lines are weak, that the agreement is not very good, and that the Raman spectrum refers to the liquid rather than to the gaseous state. Further work, particularly an investigation of the Raman spectrum of the gas, is necessary to decide this question unambiguously. Also an investigation of the degree of depolarization of the Raman lines would be desirable.

Ethylene oxide, C_2H_4O. Valence considerations lead to the expectation that in the ethylene oxide molecule the three heavier atoms form an isosceles triangle and the two CH_2 groups form planes at right angles to the C_2O plane. Again the Raman and infrared spectra can be interpreted on this assumption, but they do not supply an unambiguous proof for it. Under the above assumption the point group is $\boldsymbol{C}_{2v}$ and the distribution of the fifteen fundamentals over the four species is

$$5A_1(\text{I., R.}),\qquad 3A_2(\text{R.}),\qquad 4B_1(\text{I., R.}),\qquad 3B_2(\text{I., R.}),$$

where (I.) and (R.) indicate infrared and Raman activity.[210]

For each symmetry type there is one vibration that is essentially a C—H stretching vibration which must be of the order 3000 cm^{-1}: The two CH_2 groups may oscillate in phase or with a 180° phase difference, performing either the symmetrical or the antisymmetrical CH_2 vibration (ν_1 and ν_3 of Fig. 25a). While four Raman lines are observed in the 3000 cm^{-1} region (Table 102), only one of them is depolarized, whereas of the four fundamentals mentioned all but one should be depolarized. It is therefore necessary to assume that the four fundamentals coincide in pairs as indicated in Table 102, and that the two remaining Raman frequencies 2915 and 2958 cm^{-1} are overtones strengthened by Fermi resonance.

In a similar way the CH_2 deformation vibration may take place in phase or with 180° phase shift in the two CH_2 groups, giving rise to two frequencies, $\nu_2(a_1)$ and $\nu_{10}(b_1)$, in the 1450 cm^{-1} range. That one of these frequencies is rather higher than in C_2H_4 (see Table 92) is due to the fact that here it is pushed up by the lower fundamental, $\nu_3^{C_2O}(a_1)$, whereas in C_2H_4 it is pushed down by the higher fundamental $\nu_2^{CC}(a_g)$ of the same species (see p. 200).

Three further fundamentals would be expected to correspond essentially to the deformations of the C_2O ring [or in other words to the vibrations of a three-particle system $(CH_2)_2O$]. They should give rise to strong Raman lines and it is plausible to identify them with the three strongest Raman lines below 2000 cm^{-1}: 1267, 863, and 806 cm^{-1}, as has always been done in the literature. This

[210] If the H atoms were in the C_2O plane, the point group would also be $\boldsymbol{C}_{2v}$, but the distribution over the species would be $6A_1$, $2A_2$, $5B_1$, $2B_2$ (see Table 34)

leads indeed to very reasonable values for the C—C and C—O force constants (similar to those in C_2H_6 and CH_3OH) but, as has been emphasized by Kohlrausch and Reitz (527), it appears to be in disagreement with the degree of depolarization of one of the lines (806 or 863 cm^{-1}), since only one of the three should be depolarized. To be sure, for totally symmetric vibrations the degree of de-

TABLE 102. RAMAN AND INFRARED SPECTRA OF ETHYLENE OXIDE (C_2H_4O), AFTER KOHLRAUSCH AND REITZ (527)[211], BONNER (165), AND LINNETT (581).[212]

Assignment	Raman (liquid) $\Delta\nu_{vacuum}$ (cm^{-1})	Infrared (gas) ν_{vacuum} (cm^{-1})[213]	Assignment	Raman (liquid) $\Delta\nu_{vacuum}$ (cm^{-1})	Infrared (gas) ν_{vacuum} (cm^{-1})[213]
$\nu_{11}-\nu_{15}$?	509 (v.w.)?		$\nu_3+\nu_{15}(B_2)$		1948 (w.)
$\nu_{15}(b_2)$ CH_2 twisting	**704** (v.w.)?	**685** (w.)[214]	$\nu_{11}+\nu_{12}(A_1)$		2021 (m.)
$\nu_5^{C_2O}(a_1)$	**806** (s.) depol.	**808** (w.)	$\nu_3+\nu_{12}(B_1)$		2118 (w.)
$\nu_{12}^{C_2O}(b_1)$	**863** (s.) depol.	**865** (v.s.)[215]	$\nu_4+\nu_{11}(B_1)$		2290 (m.)[214]
$\nu_8(a_2)$ CH_2 twisting	**1023** (v.w.)		$\nu_{10}+\nu_{11}(B_1)$		2635 (w.)
$\nu_4(a_1)$ CH_2 bending	**1120** (m.) pol.		$\nu_2+\nu_3(A_1)$		2785 (v.w.)
$\nu_{14}(b_2)$ CH_2 rocking, $\nu_{11}(b_1)$ CH_2 bending	**1153** (w.) depol.	**1151** (s.)[214]	$2\nu_{10}(A_1)$	2915 (s.) pol.	
$\nu_3^{C_2O}(a_1)$	**1267** (v.s.) pol.	**1263** (s.)[215]	$2\nu_2(A_1)$	2958 (s.) pol.	
$\nu_7(a_2)$ CH_2 rocking, $2\nu_{15}(A_1)$	**1379** (v.w.)?		$\nu_1^{CH}(a_1)$, $\nu_9^{CH}(b_1)$	**3007** (v.s.) pol	3029 (v.s.)
$\nu_{10}^{CH_2}(b_1)$	**1469** (v.w.) depol.	**1453** (w.)	$\nu_6^{CH}(a_2)$, $\nu_{13}^{CH}(b_2)$	**3061** (m.) depol.	
$\nu_2^{CH_2}(a_1)$	**1487** (w.) pol.	**1495** (m.)	$\nu_1+\nu_{12}$, $\nu_9+\nu_{12}$, ⋯		3875 (m.)
$2\nu_5(A_1)$		1616 (w.)	$\nu_1+\nu_3$, $\nu_9+\nu_3$		4291 (m.)
$\nu_5+\nu_{12}(B_1)$		1638 (m.)	$\nu_2+\nu_{13}(B_2)$		4586 (w.)
$2\nu_{12}(A_1)$		1727 (w.)	$2\nu_6(A_1)$, $2\nu_{13}(A_1)$, ⋯		6171 (w.)
$\nu_8+\nu_{12}(B_2)$, $\nu_4+\nu_{15}(A_1)$		1844 (v.w.)			

polarization may have any value between 0 and $\frac{6}{7}$. It may occasionally be so close to $\frac{6}{7}$ that a polarized line cannot be distinguished from a depolarized one. Another explanation would be that the CH_2 twisting vibration $\nu_8(a_2)$ (see below) has the same frequency as $\nu_5{}^{C_2O}(a_1)$ and predominates in the Raman effect [see Linnett (581)].[216] However, heat-capacity measurements by Kistiakowsky and Rice (513) do not seem to be compatible with two fundamentals at 806 cm^{-1} (if all the other fundamentals are correctly identified).

In addition to the CH_2 and C_2O frequencies there are the frequencies in which the whole molecule is bent or twisted. First, we expect two vibrations in which the CH_2 groups remain perpendicular and symmetrical to the C_2O plane but are bent symmetrically or antisymmetrically with respect to the yz plane [CH_2 bending, $\nu_4(a_1)$, and $\nu_{11}(b_1)$]. Then there are two vibrations in which the CH_2 groups move in their planes [CH_2 rocking, $\nu_7(a_2)$ and $\nu_{14}(b_2)$], and finally two vibrations in which

211 Other Raman data have been obtained, previous to Kohlrausch and Reitz, by Timm and Mecke (863), Ananthakrishnan (46) and Bonner (165), but are not as complete. The values given are not averages but those of Kohlrausch and Reitz only.

212 Some further bands in the photographic infrared have been found by Mecke and Vierling (617) and Eyster (316).

213 Average of Bonner's and Linnett's data wherever both are available, except above 3000 cm^{-1}, where Linnett's data are given since his dispersion seems to have been much greater.

214 For these bands the data of Bonner and Linnett diverge rather more than corresponds to their stated accuracy.

215 These bands have three maxima corresponding apparently to P, Q, and R branches.

216 Kohlrausch and Reitz (527) have suggested that the two lower C_2O vibrations coincide at 863 cm^{-1}. But then the infrared band of the gas at 808 cm^{-1} would be difficult to explain.

the CH_2 groups are twisted with respect to their symmetry axis [CH_2 twisting, $\nu_8(a_2)$ and $\nu_{15}(b_2)$]. A tentative assignment of these vibrations is given in Table 102.[217]

The interpretation of the other infrared and Raman bands as overtone and combination bands [mostly due to Linnett (581)] is given in Table 102.

Some authors have used the frequencies of the C_2O ring deformation vibrations to calculate, on the assumption of a central (or valence) force system, the angle at the vertex of the C_2O triangle. But apart from the uncertainty in the assignment of these frequencies, and apart from the limitations of these force systems, such calculations neglect the fact that those CH_2 vibrations that have the same species as the C_2O vibrations push the zero-approximation C_2O frequencies up or down as the case may be.

Other seven-atomic molecules. In Table 103 are given references for the Raman and infrared spectra of some further seven-atomic molecules. The suggested structures, which are indicated in the table as obtained from the spectra, are in most cases less certain than for the molecules considered in detail above, and also the fundamentals are less definitely identified.

TABLE 103. FURTHER SEVEN-ATOMIC MOLECULES.

Molecule	Structure	References	
		Raman spectrum	Infrared spectrum
SeF_6, TeF_6	$\boldsymbol{O_h}$ regular octahedron	(974)	(754)
$SiF_6^=$, $SnCl_6^=$, $PbCl_6^=$	$\boldsymbol{O_h}$ regular octahedron (?)	(820) (728) (729) (402)	
$SnBr_6^=$	$\boldsymbol{D_{4h}}$ octahedron (?)	(730)	
$SbCl_6^-$	$\boldsymbol{O_h}$ regular octahedron (?)	(728) (729)	
$B(OH)_3$	$\boldsymbol{C_{3h}}$ plane ?	(478) (625)	(775)
N_2O_5	?	(197) (200) (821)	
H_2SO_4, H_2SeO_4	?	(14)	
$CH_3C{\equiv}CCl$, Br, I	$\boldsymbol{C_{3v}}$	(218a)	
CH_3NH_2, CH_3ND_2	$\boldsymbol{C_s}$	(518) (506) (299a)	(213) (680) (861) (847) (79) (214)
CH_3N_3	$\boldsymbol{C_s}$ (N_3 linear)	(482)	(156) (319)
CH_3NCO	$\boldsymbol{C_s}$ (NCO linear?)	(537) (394)	(319)
CH_3NO_2, CD_3NO_2	$\boldsymbol{C_s}$	(689) (915) (619) (945a)	(915) (945a)
CCl_3NO_2, CBr_3NO_2	$\boldsymbol{C_s}$	(948a)	
$CCl_3CO_2^-$	$\boldsymbol{C_s}$	(948a)	
CH_3COH, CD_3COD	$\boldsymbol{C_s}$	(520) (952) (350a) (350b)	(347) (855) (635a)
CH_3COCl, CH_3COBr	$\boldsymbol{C_s}$	(305) (524)	
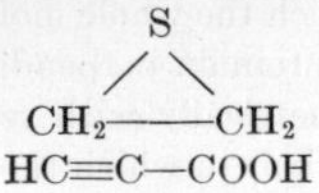	$\boldsymbol{C_{2v}}$ (like C_2H_4O)	(854)	(316) (854)
$HC{\equiv}C{-}COOH$	$\boldsymbol{C_s}$ ($HC{\equiv}C{-}C$ linear)	(753)	
C_2H_3CN	$\boldsymbol{C_s}$ (plane ?)	(863) (738)	
$^+H_3N{-}NH_2$, $^+D_3N{-}ND_2$	$\boldsymbol{C_s}$	(298) (299a)	

(*f*) *Eight-atomic molecules*

Ethane, C_2H_6 and C_2D_6. The most important eight-atomic molecule is the ethane molecule. Considering the previous results about the structure of the methyl group, there can be no doubt that in C_2H_6 the C—C axis is a *three-fold axis of symmetry*. The only question as to the structure is whether there is in addition a plane of symmetry perpendicular to the axis (*eclipsed model*, point group $\boldsymbol{D_{3h}}$) or

[217] This is essentially the assignment of Linnett (581) but takes into account the new weak Raman lines found by Kohlrausch and Reitz (527) and the results of the heat-capacity measurements of Kistiakowsky and Rice (513).

a center of symmetry (*staggered model*, point group D_{3d}), or whether there is *free rotation* of the two CH_3 groups relative to each other about the C—C axis.

The assumption that there is free rotation (or nearly free rotation) is definitely ruled out by the work of Kemp and Pitzer (496), Kistiakowsky, Lacher and Stitt (509) (510) and Wilson (938) on the entropy and the low-temperature heat capacity of C_2H_6 and C_2D_6 [see also Schaefer (768)].[218] At the temperature of liquid air the contributions to the specific heat of all vibrations except the torsional oscillation are negligible. The excess over the ordinary translational and rotational specific heat ($3R$) is thus entirely due to the torsion of the two CH_3 groups with respect to each other. If there were free rotation (or nearly free rotation) this contribution should be $\frac{1}{2}R$ (or somewhat larger for nearly free rotation), whereas for larger frequencies of the torsional oscillation (that is, for strongly hindered rotation) the contribution would go to zero (see Chapter V, section 1, in particular Fig. 170a). The fact that it is found to be about $0.3R$ shows definitely that the rotation is not free. From the heat capacity data, the first two levels of the torsional oscillation in C_2H_6 were found to be at 275 cm^{-1} and 520 cm^{-1}. A third level at 725 cm^{-1} is rather uncertain. These energy levels can be represented by a cosinelike potential curve for the torsion with a barrier height of 965 cm^{-1} (2750 cal/mole). The barrier height, however, depends considerably on the assumed form of the potential (see Chapter II, section 5d).

Taking now a non-zero torsional vibration as proven, we need only discuss the eclipsed (D_{3h}) and the staggered (D_{3d}) forms of C_2H_6. Table 104 gives for both forms the designation, species, activity, and description of the fundamental vibrations. The form of the vibrations is easily visualized if it is realized that the four fundamentals of CH_3 (see Fig. 45) occur once with both CH_3 groups vibrating in phase and once with these groups vibrating in opposite phase and that in addition there are the C—C stretching vibration, two bending vibrations of $H_3{\equiv}C{-}C{\equiv}H_3$ similar to those of C_2H_2 (see p. 181), and the torsional oscillation. These vibrations have been illustrated in the previous Fig. 49. It should be noted that in order to avoid having different numberings for the two models the order of the symmetry types for D_{3d} in Table 104 is not the usual one, as it is for D_{3h}.

TABLE 104. DESIGNATION, SPECIES, ACTIVITY, AND DESCRIPTION OF THE NORMAL VIBRATIONS OF THE TWO MODELS OF C_2H_6.

Designation	Description	D_{3h} (eclipsed model)		D_{3d} (staggered model)	
		Species	Activity	Species	Activity
ν_1	CH stretching	A_1'	R. pol.	A_{1g}	R. pol.
ν_2	CH_3 deformation	A_1'	R. pol.	A_{1g}	R. pol.
ν_3	C—C stretching	A_1'	R. pol.	A_{1g}	R. pol.
ν_4	Torsion	A_1''	inactive	A_{1u}	inactive
ν_5	CH stretching	A_2''	I. $\parallel$	A_{2u}	I. $\parallel$
ν_6	CH_3 deformation	A_2''	I. $\parallel$	A_{2u}	I. $\parallel$
ν_7	CH stretching	E'	I. $\perp$, R. depol.	E_u	I. $\perp$
ν_8	CH_3 deformation	E'	I. $\perp$, R. depol.	E_u	I. $\perp$
ν_9	Bending	E'	I. $\perp$, R. depol.	E_u	I. $\perp$
ν_{10}	CH stretching	E''	R. depol.	E_g	R. depol.
ν_{11}	CH_3 deformation	E''	R. depol.	E_g	R. depol.
ν_{12}	Bending	E''	R. depol.	E_g	R. depol.

It is seen from Table 104 that of the twelve fundamentals in the staggered model (D_{3d}) only six are Raman active, whereas for the eclipsed model (D_{3h}) nine are Raman active. While eight or nine Raman-lines have been observed for the gas,[219] this is not sufficient evidence to exclude the staggered model, since some of them may correspond to overtone or combination bands.

The observed infrared and Raman bands for both C_2H_6 and C_2D_6 are given in Table 105. In the last two columns an assignment is given for both assumptions about the structure of the molecule [see Crawford, Avery, and Linnett (239) and Stitt (810)].

[218] The fine structure of the infrared bands leads to the same conclusion [see Howard (461) and Chapter IV].

[219] The additional lines observed for the liquid may be due to a violation of the selection rules.

TABLE 105. VIBRATION SPECTRA OF C_2H_6 AND C_2D_6.

C_2H_6 ν_{vacuum} (cm^{-1})	C_2D_6 ν_{vacuum} (cm^{-1})	Assignments[220]	
		D_{3h}	D_{3d}
Infrared:[221]			
820.82 ⊥ (s.)	**∼601** (s.)	$\nu_9(e')$	$\nu_9(e_u)$
1379.0 ‖ (w.)	**∼1072** (s.)	$\nu_6{}^{CH_3}(a_2'')$	$\nu_6{}^{CH_3}(a_{2u})$
1414 ⊥ (w.)		$\nu_4+\nu_{12}(E')$	$\nu_4+\nu_{12}(E_u)$
1486.0 ⊥ (s.)	**∼1102** (s.)	$\nu_8{}^{CH_3}(e')$	$\nu_8{}^{CH_3}(e_u)$
1740 ? (w.)		$\nu_4+\nu_{11}(E')$	$\nu_4+\nu_{11}(E_u)$
2230 ⊥ ? (w.)		$2\nu_{12}(E')$, $\nu_2+\nu_9(E')$	$\nu_2+\nu_9(E_u)$
2302 ‖ ? (w.)[222]	1654 (w.)	$\nu_9+\nu_{11}(A_2'')$	$\nu_9+\nu_{11}(A_{2u}, E_u)$
2368.7 ‖ ? (w.)	1907 (w.)	$\nu_3+\nu_6(A_2'')$	$\nu_3+\nu_6(A_{2u})$
2660 ⊥ (w.)		$\nu_3+2\nu_9(E')$[223]	$\nu_8+\nu_{12}(A_{2u}, E_u)$
2753.3 ‖ (m.)		$\nu_2+\nu_6(A_2'')$	$\nu_2+\nu_6(A_{2u})$
2894 ‖ (m.)	**2087** (s.) }	$\nu_8+\nu_{11}(A_2'')$	$\nu_8+\nu_{11}(A_{2u})$
2954 ‖ (s.)	**2111** (s.) }	$\nu_5{}^{CH}(a_2'')$	$\nu_5{}^{CH}(a_{2u})$
2994.3 ⊥ (m.)	**2236** (s.)	$\nu_7{}^{CH}(e')$	$\nu_7{}^{CH}(e_u)$
3006 ‖⊥ (m.)		{ $\nu_6+2\nu_9(A_2'')$ $\nu_3+\nu_9+\nu_{12}(A_2'')$	$\nu_3+\nu_9+\nu_{12}(A_{2u}, E_u)$
3100–3150 ⊥ (w.)		$\nu_8+2\nu_9(E', E')$	$\nu_8+2\nu_9(A_{2u}, E_u, E_u)$
3185 ‖ (w.)		{ $2\nu_4+\nu_8+\nu_{12}(A_2'')$ $\nu_2+\nu_3+\nu_9(E')$	$2\nu_4+\nu_8+\nu_{12}(A_{2u}, E_u)$ $\nu_2+\nu_3+\nu_9(E_u)$
	2414 (w.)	$\nu_4+\nu_{10}(E')$	$\nu_4+\nu_{10}(E_u)$
3222 ‖ (w.)		$\nu_3+\nu_9+\nu_{11}(A_2'')$	$\nu_3+\nu_9+\nu_{11}(A_{2u}, E_u)$
	2710 ? (w.)	$\nu_1+\nu_9(E')$	$\nu_1+\nu_9(E_u)$
Inactive (from specific heat):[224]			
275	**200**	$\nu_4(a_1'')$	$\nu_4(a_{1u})$
Raman:[225]			
617 liquid (v.w.)[226]		$2\nu_4(A_1')$	$2\nu_4(A_{1g})$
813 liquid (v.w.)		$\nu_9(e')$	$[\nu_9(e_u)]$
975 (w.)		$\nu_3(C^{12}—C^{13})$	$\nu_3(C^{12}—C^{13})$
993.0 (s.) pol.	**852** (s.) pol.	$\nu_3{}^{C-C}(a_1')$	$\nu_3{}^{C-C}(a_{1g})$
(1155) indirect	∼970 (w.) depol.	$\nu_{12}(e'')$	$\nu_{12}(e_g)$
1344?		impurity?	

[220] Only the symmetry types of the active components are given.

[221] For C_2H_6, with several exceptions, this is the list given by Crawford, Avery, and Linnett (239), who have incorporated in addition to their own data those of Levin and Meyer (574) and Bartholomé and Karweil (123). The five bands 1379.0, 1486.1, 2368.7, 2753.3, and 2994.3 cm^{-1} are from the more accurate measurements of Smith and Woodward [(798) and private communication for which the author is greatly indebted to Dr. Smith], the band 820.82 cm^{-1} from Owens and Barker (681). A few further very weak bands given by Avery and Ellis (63) are not included. For C_2D_6 the data are due to Stitt (810).

[222] This band may be identical with 2368.7 cm^{-1} since it is given only by Bartholomé and Karweil (123) who do not mention the 2368.7-cm^{-1} band.

[223] This is the assignment by Bartholomé and Karweil (123). That by Crawford, Avery, and Linnett (239) as $\nu_{10} - \nu_4$ is not compatible with the fact that this band does *not* disappear at liquid-air temperatures [see Avery and Ellis (63)].

[224] See Kistiakowsky, Lacher, and Stitt (510).

[225] Except where otherwise stated, the data for C_2H_6 refer to the gaseous state and are taken from Bhagavantam (145) and Lewis and Houston (576). The liquid data for C_2H_6 are from Glockler and Renfrew (379) and Goubeau and Karweil (396). The data for C_2D_6 are those of Stitt (810), which refer to the liquid state.

[226] Specific-heat data give $2\nu_4 = 520$ cm^{-1}.

TABLE 105.—*Continued.*

C_2H_6 ν_{vacuum} (cm^{-1})	C_2D_6 ν_{vacuum} (cm^{-1})	Assignments[220]	
		D_{3h}	D_{3d}
(1375) calculated	1158 (m.) pol.	$\nu_2^{CH_3}(a_1')$	$\nu_2^{CH_3}(a_{1g})$
1460 (m.) depol.	1055 (m.) depol.	$\nu_{11}^{CH_3}(e'')$	$\nu_{11}^{(CH_3)}(e_g)$
1491 liquid (w.)		$\nu_8^{CH_3}(e')$	$[\nu_8^{CH_3}(e_u)]$
	1930 (v.w.)	$2\nu_{12}(A_1', E')$	$2\nu_{12}(A_{1g}, E_g)$
2744 (w.)	~2300 (w.) pol.	$2\nu_2(A_1')$	$2\nu_2(A_{1g})$
2778 (v.w.)		$2\nu_6(A_1')$	$2\nu_6(A_{1g})$
2899.2 (s.) pol.	2083 (s.) pol. }	{ $\nu_1^{CH}(a_1')$	$\nu_1^{CH}(a_{1g})$
2955.1 (s.) pol.	2147 (s.) pol. }	$2\nu_{11}(A_1')$	$2\nu_{11}(A_{1g})$
2939.5 (w.)		{ $\nu_8+\nu_{11}(E'')$, $2\nu_8(A_1', E')$ }	$2\nu_8(A_{1g}, E_g)$
2963 liquid (w.)	2225 (s.) depol.	$\nu_{10}^{CH}(e'')$	$\nu_{10}^{CH}(e_g)$

There can be no doubt that the strong polarized Raman doublet of C_2H_6 at 2899–2955 cm^{-1} (for C_2D_6 at 2083–2147 cm^{-1}) is due to resonance between the totally symmetric C—H vibration ν_1 and the overtone of one of the CH_3 deformation vibrations, most probably 1460 cm^{-1}, since only one totally symmetric fundamental is expected in this region for both models. The other strong polarized Raman line, 993.0 cm^{-1} (852 for C_2D_6), is obviously ν_3, that is, essentially the C—C stretching vibration. The third totally symmetric vibration, $\nu_2^{CH_3}$, does not seem to appear in the Raman spectrum of C_2H_6, unless one wants to correlate it with the questionable line 1344 cm^{-1}, which according to Crawford, Avery, and Linnett is due to C_2H_4. However ν_2 does appear for C_2D_6 as the medium-intense polarized Raman line 1158 cm^{-1}. For C_2H_6, $\nu_2^{CH_3} = 1375$ cm^{-1} is obtained from the very plausible and apparently only reasonable assignment of the infrared || band at 2753.3 cm^{-1} as $\nu_2 + \nu_6$.

The two fundamentals ν_5 and ν_6 should give rise to parallel infrared bands (see Table 104). They are most probably to be identified with the infrared doublet[227] at 2954–2894 cm^{-1} (2111–2087 cm^{-1} for C_2D_6) and the band at 1379 cm^{-1} (1072 cm^{-1} for C_2D_6) respectively. The appearance of a doublet rather than a single band must again be due to resonance (see Table 105).

The three fundamentals ν_7, ν_8, ν_9 should give rise to ⊥ infrared bands. Since there are no other such fundamentals they have to be identified with the only three medium or strong ⊥ infrared bands 2994.3, 1486.0, and 820.82 cm^{-1}, respectively (2236, 1102, and 601 cm^{-1} for C_2D_6) which also have a very reasonable magnitude.

Of the three Raman-active fundamentals ν_{10}, ν_{11}, ν_{12}, only $\nu_{11}^{CH_3}$ is readily identified as the medium-strength depolarized Raman line 1460 cm^{-1} (1055 cm^{-1} for C_2D_6). The C—H stretching vibration ν_{10} may be any one of the four weak Raman lines 2744, 2778, 2939.5, and 2963 cm^{-1}. Karweil and Schaefer (485) used 2778, whereas Crawford, Avery, and Linnett (239) as well as Stitt (810) chose 2963. In Table 105 the latter choice is adopted since it seems to lead to a somewhat simpler assignment of the combination and overtone bands.[228] The corresponding C_2D_6 band is the strong depolarized Raman line 2225 cm^{-1}. The frequency ν_{12} has been called the "uncertain frequency" of C_2H_6. Specific-heat data exclude the possibility that it has a value below 1000 cm^{-1} [see, for example, Schaefer (768)]. No Raman line above 1000 cm^{-1} that might be identified with this frequency has been observed. Crawford, Avery, and Linnett (239) have assigned the infrared band 1414 cm^{-1} as $\nu_4 + \nu_{12}(E', E_u)$ where ν_4 is taken as 275 cm^{-1} from specific-heat data (see above). This gives a value of about 1140 for ν_{12}. A similar value, 1170 cm^{-1}, is obtained if the Raman band 970 cm^{-1} of C_2D_6 is interpreted as ν_{12} and the product rule applied to determine ν_{12} of C_2H_6 [see Stitt (810)]. We shall therefore use an average value 1155 cm^{-1} for ν_{12} of C_2H_6.

It is seen from the above that we can use the same fundamentals on the assumption of either model (D_{3h} or D_{3d}) of C_2H_6. One way to decide which model is correct would be to establish whether

[227] The close coincidence of this doublet with the Raman doublet 2955–2899 cm^{-1} is very strange, since they cannot possibly correspond to the same transition.

[228] A more definite decision would be possible by a measurement of the state of polarization of the four Raman lines in question.

or not the fundamentals ν_7, ν_8, and ν_9 observed in the infrared occur in the Raman spectrum as well, since, according to Table 104, for D_{3h} they should but for D_{3d} they should not occur. According to the assignment in Table 105 the infrared band ν_7 seems to be definitely absent in the Raman spectrum. But there are two weak Raman bands at 1491 and 813 cm^{-1} that agree fairly well with the infrared fundamentals $\nu_8 = 1486$ and $\nu_9 = 821$ cm^{-1}. However, it would be premature to conclude that this proves the correctness of the eclipsed model (D_{3h}), since the Raman lines in question have been observed in liquid C_2H_6 only, and it may be that they are due to a violation of the selection rules for D_{3d} caused by intermolecular forces. It is necessary to observe these Raman lines in the gas in order to obtain an unambiguous decision.[229] A decision would also be possible through the observation of overtone or combination bands which are forbidden for the one but allowed for the other model. For example, for D_{3h}, one component of $2\nu_{12}$ is infrared active, whereas for D_{3d} both components are inactive (see Tables 32 and 55). The infrared band 2230 cm^{-1} does indeed fit the interpretation $2\nu_{12}$ but is not suitable for a proof of the model D_{3h} as was thought by Karweil and Schaefer (485), since for D_{3d} it can also be interpreted as $\nu_2 + \nu_9(E_u)$, although perhaps not as well.

The interpretation of the other infrared and Raman bands on either model is indicated in Table 105. It is essentially that of Crawford, Avery, and Linnett (239) and Stitt (810). Since there is as yet no band that can definitely not be interpreted on the basis of either model, the question as to which is the correct model is still open. The agreement of the calculated and observed values for the overtone and combination bands is perhaps somewhat better for D_{3h} than for D_{3d}.[230] Some higher overtone and combination bands have been observed by Levin and Meyer (574), Adel and Slipher (39), and Gänswein and Mecke (346).

Theoretical calculations of the potential hill opposing free rotation, by Gorin, Walter, and Eyring (393), point to D_{3h} as the more probable model, whereas the heats of hydrogenation of unsaturated cyclic hydrocarbons according to Conn, Kistiakowsky, and Smith (228) favor D_{3d}. The latter model is also suggested by the fact that in disubstituted ethanes the trans form is the stable one (see below) and that apparently in cyclopentane the C atoms are not all in one plane [see Aston, Schumann, Fink, and Doty (62) (61) and Pitzer and Gwinn (697)]. Some more direct evidence from the spectrum of C_2H_6 would of course be very desirable.

Detailed formulae for the relation between force constants and fundamental vibrations in C_2H_6 have been discussed by Howard (460), Glockler and Wall (384), Stitt (810), and Crawford and Brinkley (240).

$C_2H_4Cl_2$, $C_2H_2Cl_4$, and the question of rotational isomerism. The fact that there is no free rotation in ethane suggests that in substituted ethanes likewise free rotation does not exist. If this is true one would expect two (*rotational*) *isomers* of, for example, 1,2-dichloroethane (CH_2Cl—CH_2Cl) or 1,2-tetrachloroethane ($CHCl_2$—$CHCl_2$), with different relative orientations of the two halves of the molecule with respect to each other, somewhat similar to the cis and trans isomers of dichloroethylene (CHCl═CHCl). That such rotational isomers have never been separated chemically is easily explained as due to the smallness of the potential hill separating the modifications. If it is of the same order (3000 calories) as for C_2H_6, unlike the case of the dichloroethylenes, a transformation can readily take place at ordinary temperatures.

Raman and infrared spectra do indeed give rather definite indications of the presence of at least two rotational isomers in these substances [see Kohlrausch (14)]. There are many more strong Raman lines than would be expected if only one modification (with[231] or without free rotation) were present, and at low temperatures a considerable number of these disappear, indicating that at low temperatures the thermodynamically stabler isomer predominates [for $C_2H_4Cl_2$ and other dihaloethanes, see Mizushima, Morino, and co-workers (629) (630) (631) (538); for $C_2H_2Cl_4$, see Langseth and Bernstein (549)].

Up until 1940 only the cis and trans isomers, illustrated in Fig. 97a and d for the case of $C_2H_4Cl_2$, were usually considered. These two forms belong to point groups C_{2v} and C_{2h}, respectively. However, if the forces opposing torsion of the two CH_2Cl (or $CHCl_2$) groups with respect to each other are at all similar to those in ethane, one would not expect that both the cis (eclipsed) and the trans

---

[229] Coriolis interaction with active fundamentals (see p. 458) cannot cause a violation of the selection rules in this case.

[230] A somewhat different assignment for the fundamentals has been suggested by Barker (107), but a number of strong arguments against it have been given by Stitt (811).

[231] The statement by Kohlrausch (14) that if there were completely free rotation the Raman lines would not be sharp appears to be incorrect.

(staggered) forms could be stable, but rather that there are *either three stable eclipsed* (Fig. 97a, b, d) *or three stable staggered* (Fig. 97d, e, f,) *configurations*, that is, that there are three potential minima

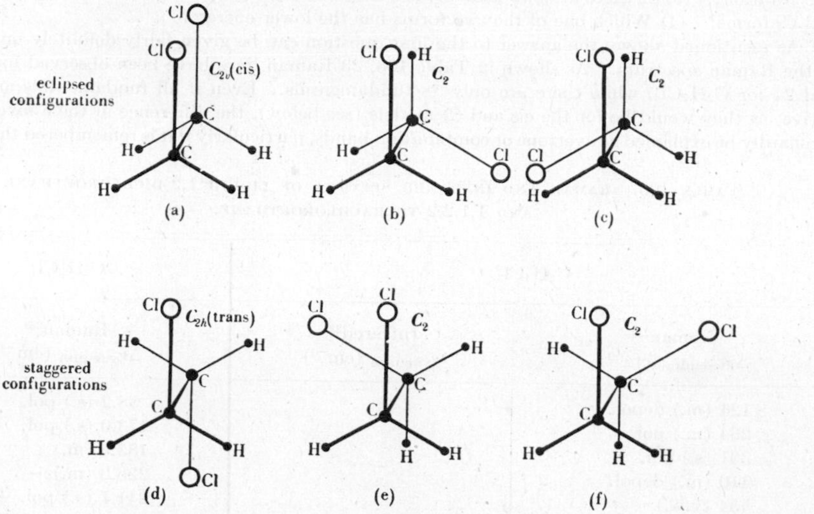

FIG. 97. **Conceivable rotational isomers of $C_2H_4Cl_2$.**

separated by approximately 120°, as indicated in Fig. 98. Unlike the case of C_2H_6, there are no longer any reasons of symmetry requiring the angle to be exactly 120°, but the angle should be the same between a and b as between a and c (and similarly between d and e as between d and f).[232] In the unsymmetrical forms there is still a two-fold axis of symmetry perpendicular to the C—C axis and therefore they belong to point group C_2. For brevity we shall henceforth call them the C_2 forms (eclipsed or staggered). The two eclipsed C_2 forms, and similarly the two staggered C_2 forms, can be transformed into each other by an inversion followed by a rotation about the C—C axis. They are optical isomers (see p. 224) and, since they have identical potential minima, they have the same energy levels and therefore the same spectrum and the same chemical properties. For all practical purposes they may be considered as one modification. Of course, on account of the finite height of the potential barrier separating the C_2 forms, each level is split into two sublevels (each of which corresponds to both configurations; see p. 221), but this splitting will be noticeable only for the higher levels of the torsional oscillation.

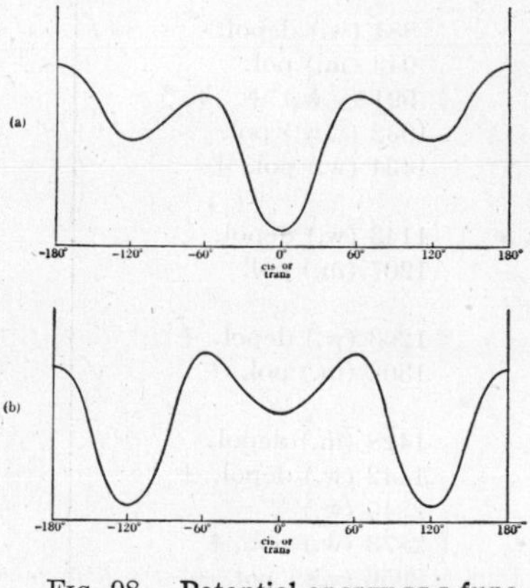

FIG. 98. **Potential energy as a function of the angle of twist about the C—C axis in $C_2H_4Cl_2$ (a) when the symmetrical cis or transform is the more stable one, (b) when it is less stable than the C_2 forms.**

In both cases when the cis or when the trans form has a minimum, two possibilities have to be distinguished: of the three minima, either the minimum corresponding to the cis (or trans) form is the lowest (Fig. 98a) or the (identical) minima corresponding to the C_2 forms are lowest (Fig. 98b).

[232] There is, of course, a remote possibility that the deviation from 120° is so great that the two minima corresponding to the C_2 forms coalesce. This would correspond to the old idea that the cis and trans forms are both stable.

Thus the following questions present themselves: (1) Are there one or two isomers in the di- and tetrahaloethanes? (2) If there is only one, is it the cis or trans form, or the one with completely free rotation? (3) If there are two isomers, are they the cis and trans, the cis and C_2, or the trans and C_2 forms? (4) Which one of the two forms has the lower energy?

As mentioned above, the answer to the first question can be given fairly definitely on the basis of the Raman spectrum. As shown in Table 106, 23 Raman lines have been observed for $C_2H_4Cl_2$ and 24 for $C_2H_2Cl_4$, while there are only 18 fundamentals. Even if all fundamentals are Raman active, as they would be for the cis and C_2 models (see below), the difference in the above numbers can hardly be explained as overtone or combination bands, particularly if it is remembered that usually

TABLE 106. RAMAN AND INFRARED SPECTRA OF LIQUID 1,2-DICHLOROETHANE AND 1,1,2,2-TETRACHLOROETHANE.

$C_2H_4Cl_2$		$C_2H_2Cl_4$
Raman[233] $\Delta\nu_{vacuum}$ (cm^{-1})	Infrared[234] ν_{vacuum} (cm^{-1})	Raman[235] $\Delta\nu_{vacuum}$ (cm^{-1})
124 (m.) depol.		88.2 (s.) pol.
264 (m.) pol.		173.0 (s.) pol.
301 (s.) pol. +		183.6 (m.)
410 (m.) depol.		226.0 (m.) −
454 (v.w.)		241.7 (s.) pol. +
653 (v.s.) pol.	656 (s.)	288.8 (s.) depol. +
676 (m.) depol.	676 (s.)	294.6 (m.) −
	707 (s.)	352.9 (v.s.) pol. +
753 (v.s.) pol. +		366.7 (m.) −
	759 (w.)	546.4 (m.) depol. +
	818 (v.w.)	648.1 (s.) pol. +
881 (w.) depol.	878 (m.)	765.1 (m.) pol. −
943 (m.) pol.	937 (m.)	801.6 (v.s.) pol.
991 (v.w.) +		812.1 (m.)
1032 (v.w.) pol.	1015 (w.)	1018.1 (m.)
1054 (w.) pol. +		1028.1 (w.)
	1092 (v.w.)	1118 (v.w.)
1143 (w.) depol.		1171 (v.w.)
1207 (m.) pol.		1203.1 (w.)
	1243 (m.)	1217.0 (m.) depol.
1263 (w.) depol. +		1245.2 (m.)
1302 (m.) pol. +		1278.8 (w.)
	∼1400 (m.)	1307.3 (w.)
1428 (m.) depol.		2989.1 (m.) pol.
1442 (w.) depol. +		
2845 (w.)		
2873 (w.) pol. +		
2956 (v.s.) pol. +		
3002 (m.) depol. +		
	∼3000 (s.)	

[233] Raman shifts as given by Kohlrausch and Wittek (533) on the basis of their own and earlier measurements. A + indicates that the line does not disappear in the solid at −40° C according to Mizushima and Morino (628).

[234] According to Cheng and Lecomte (204).

[235] The frequencies are those of Langseth and Bernstein (549). The polarization data are from Trumpy (878) (879). He gives an additional polarized Raman line at 395 cm^{-1}. A + indicates that the intensity of a line increases at lower temperature, a − that it decreases according to Langseth and Bernstein. The above authors have also investigated the corresponding deutero compound.

not all allowed fundamentals do occur. In addition it is found that a considerable number of lines of both substances disappear or become decidedly weaker at lower temperatures (see Table 106). This can hardly be explained except by assuming two modifications, one of which is thermodynamically less stable (higher potential minimum in Fig. 98) than the other. Indeed, in the case of $C_2H_2Cl_4$, by careful intensity measurements at different temperatures, Langseth and Bernstein (549) have determined the energy difference between the two isomers as 1100 cal/mole ($=385$ cm^{-1}). This is also in agreement with the fact that at room temperature the Raman lines of the less stable form are all weaker than those of the more stable form.

It does not appear to be possible to assume, as has been done at one time by several investigators, that the appearance of additional Raman lines at higher temperatures is due not to a second isomer but to the fact that with increasing amplitude of the torsional oscillation the selection rules break down. The selection rules are rigorous for any amplitude of vibration as long as the equilibrium position is the one for which they are derived (see p. 253f.) and as long as the interaction of vibration and rotation is slight. However, in agreement with the selection rules new lines may appear, whose lower states are higher vibrational levels; but the stronger of these lines should all lie close to the low temperature lines unless large anharmonicities or strong resonances occur (see p. 312). Since this is not the case we consider the existence of two isomers as established.

In order to be able to answer the third and fourth questions on the basis of the spectrum, it is necessary to know the species and activities of the fundamentals of the various forms. On the basis of Tables 35 and 55 one finds for the cis form of either $C_2H_2Cl_4$ or $C_2H_4Cl_2$

(point group $\boldsymbol{C_{2v}}$): $6A_1$(R., I.), $4A_2$(R.), $5B_1$(R., I.), $3B_2$(R., I.);

for the trans form

(point group $\boldsymbol{C_{2h}}$): $6A_g$(R.), $4A_u$(I.), $3B_g$(R.), $5B_u$(I.);

and for the C_2 form, either staggered or eclipsed,

(point group $\boldsymbol{C_2}$): $10A$(R., I.), $8B$(R., I.).

It is very significant that for $C_2H_4Cl_2$ ten Raman lines persist at low temperatures. Six of these are polarized, three are depolarized, and for one the degree of depolarization has not been measured. This is in almost perfect agreement with the predictions for the trans form, which requires nine Raman-active fundamentals (six giving polarized, three depolarized lines), whereas it would be difficult to reconcile with either of the other two models, for both of which all eighteen fundamentals are Raman active. Assuming the trans form, one Raman line has to be explained as an overtone, and this is very readily possible for the line 2873 cm^{-1}, which is almost certainly the first overtone of 1442 cm^{-1}, and is strengthened by resonance with 2956 cm^{-1} (similar to C_2H_6; see above). In support of the conclusion that *the trans form is the one that is stable at low temperatures* is also the fact that none of the Raman lines persisting at low temperature has a corresponding infrared band (the rule of mutual exclusion must hold for $\boldsymbol{C_{2h}}$).

The above conclusion, based entirely on spectroscopic arguments, is in agreement with the results of dipole measurements, which give a value zero at low temperatures, and with electron diffraction data [see Beach and Palmer (131) and Beach and Turkevich (132)[236]].

The question whether the less stable form of $C_2H_4Cl_2$ is the cis or C_2 form cannot be so definitely answered at the present time. In principle a decision would be possible from the number of polarized and depolarized Raman lines (6 and 12 for $\boldsymbol{C_{2v}}$, 10 and 8 for $\boldsymbol{C_2}$). But thus far, not all Raman fundamentals of this model have been found, probably due to the comparatively small concentration of the second isomer at room temperature. Four of the Raman lines disappearing at low temperatures are polarized, eight are depolarized, which is in agreement with either model. As mentioned above, theoretically the staggered C_2 structure seems much more plausible for the second less stable form of $C_2H_4Cl_2$ [see also Langseth, Bernstein, and Bak (550) and Edgell and Glockler (297)].

Assuming the cis structure for the less stable isomer, Wu (960) and Kohlrausch and Wittek (533) have given detailed assignments of the observed Raman and infrared bands. Even though it would be easy to change the assignment to fit the C_2 model we shall not give it here, because of the large uncertainties involved.

---

[236] The assertion of these authors that there is no other potential minimum than that corresponding to the trans configuration cannot be considered as definitely proven since the concentration of the other isomer at room temperature may be small.

In the case of tetrachloroethane Langseth and Bernstein (549) have concluded, on the basis of the number of Raman-active vibrations below 400 cm^{-1}, that neither of the two forms that exist (as shown by the temperature dependence) can be the trans form and that the modification of lowest energy is the cis form. However, without detailed calculations (which they have not carried out), it would not appear to be possible in molecules as complicated as these to determine the number of fundamentals with a frequency below a certain value.[237] Electron diffraction work by Schomaker

TABLE 107. FURTHER EIGHT-ATOMIC MOLECULES.

Molecule	Structure	References: Raman spectrum	References: Infrared spectrum
S_8	D_{4d} (puckered octagon)	(149) (153) (891a) (897) (673c)	(114)
C_2Cl_6, C_2Br_6	D_{3h} or D_{3d} (like C_2H_6)	(13) (420) (627) (48) (407a) (532a)	
Si_2H_6, Si_2Cl_6	D_{3h} or D_{3d} (or free rotation)	(816)	
B_2H_6	D_{3h} or D_{3d} (like C_2H_6)?	(51) (813)	(814)
Al_2Cl_6, Al_2Br_6	V_h	(743) (350c) (532a)	
$CH_3NH_3^+$, $CH_3ND_3^+$	D_{3h} or D_{3d}?	(298) (299a)	
$^+H_3N—NH_3^+$, $^+D_3N—ND_3^+$	D_{3h} or D_{3d}?	(298) (299a)	
$CH_2Br—CH_2Br$, $CH_2I—CH_2I$	C_{2h} and C_2	(14) (424) (190) (628) (481)	(204)
CH_3CHCl_2, CH_3CHBr_2, CH_3CHI_2	C_s	(483)	(304)
$C_2H_2Br_4$	C_{2h} and C_2	(14)	
$C_2F_4Cl_2$, $C_2F_4Br_2$	C_{2h} and C_2	(383)	
$CFCl_2CFCl_2$	C_{2h} and C_2?	(382)	
C_2F_4HCl	C_s, and C_1	(383)	
C_2H_4ClBr, C_2H_4ClI	C_s, C_1	(14) (190) (905)	(204) (905)
$CH_3—CCl_3$, $CH_3—CF_3$	C_{3v}	(463) (415) (522) (752a)	
$CCl_3—CF_3$	C_{3v}	(383)	
$CHCl_2—CH_2Cl$, $CHBr_2—CH_2Br$	C_s?	(522) (463) (483)	
$CF_2Cl—CFCl_2$	C_s and C_1	(383)	
C_2HCl_5	C_s	(13) (14)	(966)
C_2H_5Cl, C_2H_5Br, C_2H_5I	C_s	(14) (907) (424) (908)	(249)
CH_2DCH_2Br	C_s and C_1	(551)	
CCl_3CHFCl	C_1	(383)	
C_2H_3FClBr	C_1	(375)	
$CF_2ClCHFCl$, $CF_2BrCFClBr$	C_1	(383)	
$(COOH)_2$	C_{2v}? or V_h?	(59) (451) (721) (756)	
$CNCH{=}CHCN$	C_{2h}		(737)
CH_3COOH	C_s?	(14) (786) (755a) (300)	(361) (70) (429)
CCl_3COOH	C_s?	300	(269)
$CHCl_2COOH$	C_1	(300)	
$(NH_2)_2CO$	C_{2v}	(14) (679) (739)	(494) (495)
$(NH_2)_2CS$	C_{2v}?	(532)	
$C_2H_3—C{\equiv}C—H$	C_s plane?	(368) (862)	(122) (346)
C_2H_4NH	C_s		(316)
CH_3NO_3	C_s	(199) (948b) (568a)	(568a)

[237] Also they have not proven the assignment of all Raman lines to the different isomers but only those indicated by + or − in Table 106.

and Stevenson (771a) seems to show very definitely that in $C_2H_2Cl_4$ the staggered (trans and C_2) configurations occur and not the cis form. This work does not allow a decision as to which of the two staggered forms is the more stable one, but a proper modification of Langseth and Bernstein's arguments makes it appear probable that it is the trans form.

Summarizing, we can say that it appears certain that there are *two rotational isomers* of $C_2H_4Cl_2$ as well as of $C_2H_2Cl_4$, and that, at least for $C_2H_4Cl_2$, the *trans form is the more stable one.* The latter conclusion very probably applies also to $C_2H_2Cl_4$, and in both cases it is very likely that the second isomer has the C_2 form.

Rotational isomerism occurs, of course, in many other similar cases, not only when Cl is replaced by another halogen but also in cases such as C_2H_4XY where X is any halogen and Y is CH_3, C_2H_5, and so on. In fact, it was the observation that in C_3H_7X, C_4H_9X, $\cdots$ the C—X stretching vibration is split into two whereas it is single in CH_3X and C_2H_5X that first led Kohlrausch and his co-workers to the assumption of rotational isomerism.

TABLE 108. DESIGNATION, DESCRIPTION, SPECIES, AND ACTIVITY OF THE NORMAL VIBRATIONS OF CYCLOPROPANE; AND COMPARISON WITH THOSE OF C_2H_4O.

Designation	Description	D_{3h}		C_{2v}[239]		Corresponding vibration of C_2H_4O
		Species	Activity	Species	Activity	
ν_1	CH stretching	A_1'	R. pol.	A_1	I. R. pol.	$\nu_1^{CH}(a_1)$
ν_2	CH_2 deformation	A_1'	R. pol.	A_1	I. R. pol.	$\nu_2^{CH_2}(a_1)$
ν_3	ring deformation	A_1'	R. pol.	A_1	I. R. pol.	$\nu_3^{C_2O}(a_1)$
ν_4	CH_2 twisting	A_1''	inactive	A_2	R. depol.	$\nu_8(a_2)$
ν_5	CH_2 bending	A_2'	inactive	B_1	I. R. depol.	$\nu_{11}(b_1)$
ν_6	CH stretching	A_2''	I. $\|\|$	B_2	I. R. depol.	$\nu_{13}^{CH}(b_2)$
ν_7	CH_2 rocking	A_2''	I. $\|\|$	B_2	I. R. depol.	$\nu_{14}(b_2)$
ν_8	CH stretching	E'	I. $\perp$ R. depol.	A_1+B_1	I. R. pol. +depol.	$\nu_1^{CH}(a_1)$ $+\nu_9^{CH}(b_1)$
ν_9	CH_2 deformation	E'	I. $\perp$ R. depol.	A_1+B_1	I. R. pol. +depol.	$\nu_2^{CH_2}(a_1)$ $+\nu_{10}^{CH_2}(b_1)$
ν_{10}	CH_2 bending	E'	I. $\perp$ R. depol.	A_1+B_1	I. R. pol. +depol.	$\nu_4(a_1)+\nu_{11}(b_1)$
ν_{11}	ring deformation	E'	I. $\perp$ R. depol.	A_1+B_1	I. R. pol. +depol.	$\nu_5^{C_2O}(a_1)$ $+\nu_{12}^{C_2O}(b_1)$
ν_{12}	CH stretching	E''	R. depol.	A_2+B_2	I. R. depol.	$\nu_6^{CH}(a_2)$ $+\nu_{13}^{CH}(b_2)$
ν_{13}	CH_2 rocking	E''	R. depol.	A_2+B_2	I. R. depol.	$\nu_7(a_2)+\nu_{14}(b_2)$
ν_{14}	CH_2 twisting	E''	R. depol.	A_2+B_2	I. R. depol.	$\nu_8(a_2)+\nu_{15}(b_2)$

Other eight-atomic molecules. A large number of further eight-atomic molecules have been investigated. Table 107 gives references to the more recent work on a number of these molecules as well as the most probable structures. The information about most of these molecules is less complete than for those discussed above and the conclusions as to the structure are necessarily less certain.

The molecules C_2Cl_6, Si_2H_6, Si_2Cl_6, and B_2H_6 would be expected to have a structure similar to C_2H_6. The Raman spectra do not contradict this conclusion but are not sufficient to establish it beyond doubt. The infrared spectra, except for B_2H_6, have not as yet been investigated and consequently not all fundamentals are as yet known. The question as to the potential hill preventing free rotation in these molecules has also not been definitely settled.[238]

Unlike the case of the di- and tetrahaloethanes discussed above, for $CF_2ClCHFCl$ and similar haloethanes one would expect three rotational isomers [see Glockler and Sage (383)] but the spectroscopic evidence is not yet sufficient to establish this.

[238] For B_2H_6 Stitt (814) obtains a potential hill of between 4000 and 5000 or 15,000 cal/mole depending on the interpretation of the spectrum.

[239] The plane of the three C atoms is assumed to be the *xz* plane.

(g) *Nine-atomic molecules*

Cyclopropane, C_3H_6. The cyclopropane molecule is related to the ethylene oxide (C_2H_4O) molecule treated earlier, differing from it only in that the oxygen atom is replaced by the isoelectronic CH_2 group. Since the three C atoms in cyclopropane are equivalent one would expect them to form an *equilateral triangle* and as in C_2H_4O the CH_2 groups would be expected to be in planes at right angles to that of the triangle. Thus the point group would be $\boldsymbol{D_{3h}}$. This is confirmed by the investigation of the Raman and infrared spectra.

In Table 108 are given the designations, descriptions, species and activities of the fundamental vibrations of cyclopropane, assuming point group $\boldsymbol{D_{3h}}$ (see Tables 22, 36, 55 and Fig. 36). Also the corresponding species and activities of point group $\boldsymbol{C_{2v}}$ are given to provide for the possibility that this represents the symmetry of the molecule, and at the same time to facilitate the correlation to C_2H_4O. The last column gives the corresponding vibrations of C_2H_4O (see Table 102). The assignment of the observed frequencies to the theoretically expected fundamentals can be carried out in much the same way as for C_2H_4O. It is therefore not necessary to give a detailed discussion. Table 109 gives the observed infrared and Raman spectra with essentially the assignments given by Linnett (581), as modified by Kistiakowsky and Rice (512) and Smith (797). The two Raman- and infrared-inactive fundamentals ν_4 and ν_5 and one of the e'' fundamentals, which has not been directly observed,

TABLE 109. RAMAN AND INFRARED SPECTRA OF CYCLOPROPANE (C_3H_6).

Assignment	Raman, liquid[240] $\Delta\nu_{vacuum}$,	Infrared, gas[241] ν_{vacuum},	Assignment	Raman, liquid[240] $\Delta\nu_{vacuum}$,	Infrared, gas[241] ν_{vacuum},
$\nu_{13}-\nu_{14}(A_1')$	382 (v.w.)		$\nu_2+\nu_{10}(E')$, $\nu_5+\nu_9(E')$		2493 (w.)
$\nu_2-\nu_{11}$?		654[242] (w.)			
$\nu_9-\nu_{14}$?		694[242] (w.)	$\nu_3+\nu_9(E')$		2631 (w.)
$\nu_{14}(e'')$ twisting	740 (w.) depol.		$2\nu_9(A_1'+E')$	2854 (w.) pol.	
$\nu_{11}^{C_3}(e')$	866 (v.s.) depol.[243]	868 (v.s.)	$4\nu_{14}(A_1'+2E')$?	2952 (w.) pol.	
$\nu_7(a_2'')$ rocking		872 (v.s.)	$2\nu_2(A_1')$	3011 (s.) pol.[243]	
$\nu_4(a_1'')$ twisting	(1000)		$\nu_8^{CH}(e')$		3024.4 (v.s.)⊥
$\nu_{10}(e')$ bending	1022 (w.)	1027.6 (s.)⊥	$\nu_1^{CH}(a_1')$	3029 (s.) pol.[243]	
$\nu_5(a_2')$ bending	(1070)		$\nu_{12}^{CH}(e'')$	3080 (m.) dep.	
$\nu_{13}(e'')$ rocking	(1120)		$\nu_6^{CH}(a_2'')$		3103.0 (v.s.)||
$\nu_3^{C_3}(a_1')$	1189 (v.s.) pol.[243]		$\nu_2+\nu_5+\nu_{10}(E')$?		3580 (w.)
$\nu_9^{CH_2}(e')$	1435 (m.) depol.	1432[244] (s.)	$\nu_1+\nu_{11}(E')$, $\nu_8+\nu_{11}(E')$		3845 (w.)
$2\nu_{14}(A_1'+E')$	1454 (m.)				
$\nu_2^{CH_2}(a_1')$	1504 (w.) pol.		$\nu_3+\nu_6(A_2'')$, $\nu_3+\nu_8(E')$		4200[245] (w.)
$\nu_4+\nu_{14}(E')$		1739 (m.)			
$\nu_7+\nu_{14}(E')$		1779 (w.)	$\nu_1+\nu_9(E')$, $\nu_8+\nu_9(E')$		4450[245] (m.)
$\nu_{10}+\nu_{11}(A_1'+E')$	1873 (v.w.)	1888 (s.)			
$\nu_5+\nu_{10}(E')$		2084 (s.)	$\nu_6+2\nu_5(A_2'')$, etc.		5130 (v.w.)
$\nu_9+\nu_{14}(A_2'')$		2178 (w.)	$2\nu_8(E')$, etc.		6020 (w.)
$\nu_2+\nu_{11}(E')$		2330 (w.)			

240 Average of the data of Ananthakrishnan (46) and Harris, Ashdown, and Armstrong (410), wherever both are available. The polarization data are from Ananthakrishnan (46) and Linnett (581).

241 Average of the data of Bonner (165) and Linnett (581) except for the bands 868, 872, 1027.6, 3024.4, 3103.0 cm^{-1} taken from the higher dispersion work of Smith (797) and 3580 cm^{-1} observed only by King, Armstrong, and Harris (503).

242 These two absorption maxima which possibly form one band have been observed by Bonner (165) only, although Linnett (581) looked for them. They are therefore possibly due to an impurity.

243 Also observed in the gas.

244 This band has two maxima (PR?).

245 These two bands were resolved by King, Armstrong, and Harris (503) only, whereas Bonner (165) and Linnett (581) measured them as one band.

are based on rather doubtful identifications of overtone bands. Otherwise the assignments are quite satisfactory. The selection rules are well observed, a fact which strongly supports the assumed model (D_{3h}). The observation of both a strong || and a strong ⊥ infrared band in the 3000 cm^{-1} region would not be compatible with a completely plane model if the two bands are both fundamentals. This supports the assumption that the CH_2 planes are at right angles to the plane of the three C atoms. An evaluation of the force constants from the observed fundamentals on the basis of a generalized valence force system has been carried out by Saksena (754a).

A number of photographic infrared bands have been observed by Eyster (316). Their assignment is not very certain.

Dimethyl ether, $(CH_3)_2O$. The dimethyl ether molecule $(CH_3)_2O$ is also in some respects similar to C_2H_4O, except that there is no bond between the two C atoms and that there are CH_3 instead of CH_2 groups. One would expect the point group to be C_{2v} unless, as is unlikely, there is completely free rotation of the CH_3 groups about the C—O bonds. The designation, species, character and activity of the 21 fundamentals are given in Table 110. In this table the designations symmetrical

TABLE 110. DESIGNATION, CHARACTER, SPECIES,[246] AND ACTIVITY OF THE NORMAL VIBRATIONS OF $(CH_3)_2O$.

Description	A_1	A_2	B_1	B_2
C—H non-symmetric stretching	$\nu_1{}^{CH}(a_1)$	$\nu_8{}^{CH}(a_2)$	$\nu_{12}^{CH}(b_1)$	$\nu_{18}^{CH}(b_2)$
C—H symmetric stretching	$\nu_2{}^{CH}(a_1)$	—	$\nu_{13}^{CH}(b_1)$	—
CH_3 non-symmetric deformation	$\nu_3{}^{CH_3}(a_1)$	$\nu_9{}^{CH_3}(a_2)$	$\nu_{14}^{CH_3}(b_1)$	$\nu_{19}^{CH_3}(b_2)$
CH_3 symmetric deformation	$\nu_4{}^{CH_3}(a_1)$	—	$\nu_{15}^{CH_3}(b_1)$	—
CH_3 ⊥ rocking[247]	—	$\nu_{10}(a_2)$	—	$\nu_{20}(b_2)$
CH_3 || rocking[247]	$\nu_5(a_1)$		$\nu_{16}(b_2)$	
C—O stretching	$\nu_6{}^{CO}(a_1)$	—	$\nu_{17}^{CO}(b_1)$	—
C—O—C bending	$\nu_7{}^{C_2O}(a_1)$	—	—	—
CH_3 twisting	—	$\nu_{11}(a_2)$	—	$\nu_{21}(b_2)$
Activity	R. pol. I.	R. depol.	R. depol. I.	R. depol. I.

and non-symmetrical stretching and deformation indicate that the displacements within the CH_3 groups are essentially the totally symmetric and degenerate vibrations respectively of free CH_3 (ν_1, ν_2 and ν_3, ν_4 respectively in Fig. 45). Of course, if both CH_3 groups in $(CH_3)_2O$ carry out, for example, a symmetrical stretching vibration, then unlike the case of free CH_3 the amplitudes of the six H atoms will not be exactly the same, but those of the two H atoms in the C—O—C plane will be somewhat different from those of the other H atoms. As emphasized before, an exact distinction between vibrations of the same species, particularly if they have similar frequencies, does not exist.

Table 111 gives the observed Raman and infrared spectra of $(CH_3)_2O$. The three C—O—C vibrations are easily identified as indicated in this table by comparing with the C—O vibration of CH_3OH and by considering the state of polarization of the Raman lines. The weaker bands accompanying the strong C—O bands 1122 and 940 cm^{-1} in the infrared are apparently transitions from excited vibrational levels. That these secondary bands agree better with the Raman bands than the main infrared bands is probably due to the shift between liquid and vapor. If the C—O—C frequencies are substituted into the valence force equations of a triatomic system, very reasonable values for the C—O stretching- and bending-force constants as well as for the C—O—C angle (116°) are obtained [see Kohlrausch (14)]. However, it would be wrong to place too much reliance in the accuracy of the values obtained, since the neighboring CH_3 vibrations of the same symmetry type will shift the C—O—C frequencies somewhat from the values they would have if the CH_3 groups were single particles.

The other fundamentals cannot as definitely be identified. A tentative assignment is given in the table. All the six CH_3 deformation frequencies seem to be superimposed at about 1450 cm^{-1},

---

[246] The C—O—C plane is assumed to be the *xz* plane.

[247] ⊥ and || respectively to the C—O—C plane.

all infrared-active C—H stretching frequencies in the broad band at 2914 cm^{-1}. The assignment of the two CH_3 twisting oscillations ν_{11} and ν_{21} is particularly uncertain. It may be the pair of Raman lines 160 and 300 cm^{-1} [as has been suggested by Pitzer (696)], or the pair 702 and 583 cm^{-1}, or neither one of them. Indirect information about these twisting vibrations may be derived from

TABLE 111. RAMAN AND INFRARED SPECTRA OF $(CH_3)_2O$.

Assignment	Raman $\Delta\nu_{vacuum}$[248] (liquid)	Infrared ν_{vacuum} (gas)[249]	Assignment	Raman $\Delta\nu_{vacuum}$[248] (liquid)	Infrared ν_{vacuum} (gas)[249]
?$\nu_{11}(a_2)$?[250]	160? (w.) pol.		$\nu_3+\nu_7$, $\nu_4+\nu_7$, ···		1861 (m.)
?$\nu_{21}(b_2)$?[250]	300? (v.w.) pol.		$\nu_3+\nu_{21}$, $\nu_4+\nu_{21}$, ···		2032 (m.)
$\nu_7^{C_2O}(a_1)$	**414** (w.) pol.	(below 440)	$\nu_6+\nu_{17}(B_1)$		2100 (s.)
?$\nu_{21}(b_2)$?	583? (w.)	610 (w.)	$\nu_3+\nu_6$, $\nu_4+\nu_6$, ···		2324 (w.)
?$\nu_{11}(a_2)$?	702? (v.w.)		$\nu_{14}+\nu_6$, $\nu_{15}+\nu_6$,		2399 (m.)
$\nu_6^{CO}(a_1)$	**918** (s.) pol.	920 (m.)	$\nu_{19}+\nu_6$		
		940 (s.)	$\nu_{10}+\nu_{15}$, ···?		2652 (w.)
$\nu_{17}^{CO}(b_1)$	**1100** (w.)	1102 (w.)	$\nu_2^{CH}(a_1)$	**2812** (v.s.) pol.	
		1122 (m.)	$2\nu_4$?	2863 (s.) pol.	
ν_5, ν_{10}, ν_{16}, ν_{20}	1155? (w.) depol.	1180 (s.)	$2\nu_{15}$? ν_{12}^{CH}, ν_{13}^{CH}, ν_{18}^{CH}	**2916** (s.) pol.	**2914** (v.s.)
ν_3, ν_4, ν_9, ν_{14}, ν_{15}, ν_{19}	**1448** (s.) depol.	**1466** (v.s.)	$2\nu_3$?	2950 (s.) pol.	
			$\nu_1^{CH}(a_1)$	**2986** (s.) pol.	
$\nu_7+\nu_{16}$, $\nu_7+\nu_{20}$		1605 (w.)	$\nu_2+\nu_3$?		4273 (m.)

thermodynamic data. By assuming the assignment of Table 111 with the exception of ν_{11} and ν_{21}, Kistiakowsky and Rice (513) obtained from the heat capacity of $(CH_3)_2O$ a potential barrier opposing free rotation of 2500 cal/mole, while Kennedy, Sagenkahn, and Aston (498) obtained from the entropy 3100 cal/mole (see Chapter V, section 1). The frequency of the twisting vibrations is therefore by no means zero but of the same order as in ethane (see Table 105).

Propylene, CH_3—CH=CH_2. The propylene molecule, even in its most symmetrical configuration, has only one plane of symmetry (point group C_s). It would be difficult to establish the presence of this plane of symmetry from a study of the vibration spectrum. Even when this is assumed, on account of the lack of symmetry, the determination of the fundamental frequencies from the observed Raman and infrared spectra is by no means an easy task. In order to accomplish it, Wilson and Wells (945) have calculated the fundamentals below 2000 cm^{-1} from assumed values of the force constants taken from other molecules (including many interaction terms in the valence force treatment) and have used the calculated values as an aid in assigning the observed bands. In this case a knowledge of only the order of magnitude of the stretching and bending vibrations is not sufficient on account of the large number of vibrations and the fact that their interaction is very little restricted by symmetry.

In Table 112 are given the observed Raman and infrared spectra of propylene. There are fourteen fundamentals symmetric with respect to the plane of symmetry (species A') and seven antisymmetric ones (species A''), all of which are infrared and Raman active. The assignment of the fundamentals below 2000 cm^{-1} in Table 112 is essentially that of Wilson and Wells. Their calculated values are also given. The assignment of the fundamentals above 2000 cm^{-1} is essentially that of Fox and Martin (330), who have obtained them by comparison with similar molecules containing CH, CH_2, and CH_3 groups. The descriptions of the vibrations given in the table are similar to those used previously. The assignments for the rocking and twisting vibrations are rather uncertain, as are the interpretations of the overtone and combination bands.

[248] Average of Kohlrausch's (518) and Ananthakrishnan's (48) data. Values with a question mark have been observed by one observer only.

[249] After Crawford and Joyce (245).

[250] The tentative assignment of these bands is due to Pitzer (696), who realized that it contradicts the observed polarization of the bands.

If the fundamentals of Table 112 (except the CH_3 twisting frequency) are assumed to be correct the height of the potential barrier opposing free rotation of the CH_3 can be calculated from heat-capacity data (see Chapter V, section 1). The value obtained by Wilson and Wells (945) and confirmed by Telfair and Pielemeier (835) (834) is 2100 cal/mole (740 cm^{-1}).

TABLE 112. INFRARED AND RAMAN SPECTRA OF PROPYLENE ($CH_3—CH═CH_2$).

Assignment	Infrared[251] ν_{vacuum} (cm^{-1})	Raman (liquid)[252] $\Delta\nu_{vacuum}$ (cm^{-1})	Calculated (cm^{-1})
$\nu_{21}(a'')$ C—CH_3 twisting?	(177)[253]		
$\nu_{14}(a')$ C═C—C bending	417 (m.)	432 (m.) pol.	428
$\nu_{20}(a'')$ C═CH_2 twisting	578 (s.)	580 (w.b.) depol.	574
$\nu_{20}+\nu_{21}(A')$?	755 (w.)		
$\nu_{13}(a')$ C—C stretching	919 (s.)	920 (s.) pol.	917
$\nu_{19}(a'')$ C—H bending	936 (v.w.)*		940
$\nu_{14}+\nu_{20}(A'')$?	963 (v.w.)*		
$\nu_{18}(a'')$ CH_3 rocking ?	996 (w.)		1000
$\nu_{12}(a')$ CH_3 rocking ?	1043 (v.w.)*		1022
$\nu_{17}(a'')$ CH_2 rocking ?	1166 (w.)		1100
$\nu_{11}(a')$ CH_2 rocking ?	1224 (m.)		1111
	1244 (w.)		
$\nu_{10}(a')$ CH bending	1287, 1317 (w.)	1297 (v.s.) pol.	1297
$\nu_9(a')$ CH_3 symmetric deformation	1399 (w.)		1386
$\nu_8(a')$ CH_2 deformation	1416 (w.)	1415 (m.) pol.	1429
$\nu_7(a')$ CH_3 non-symmetric deformation	1448 (s.)	1448 (w.)	1471
$\nu_{16}(a'')$ CH_3 non-symmetric deformation	1472 (m.)		1474
$\nu_{19}+\nu_{20}(A')$?	1489, 1508, 1520 (v.w.)		
$\nu_6(a')$ C═C stretching	1647 (s.)	1648 (v.s.) pol.	1681
$\nu_{17}+\nu_{20}(A')$?	1718 (w.)		
$2\nu_{13}(A')$	1830 (m.)		
$2\nu_{18}(A')$?	1976 (v.w.)		
$\nu_7+\nu_{20}(A')$?	2035 (w.)		
$2\nu_{17}(A')$?	2320 (w.)		
$2\nu_{10}(A')$?	2574 (w.)		
$\nu_8+\nu_{10}(A')$?		2732 (w.) pol.	
$\nu_7+\nu_{10}(A')$?		2763 (v.w.)	
$2\nu_9(A')$?		2795 (v.w.)	
$2\nu_8(A')$?		2823 (w.) pol.	
$\nu_5(a')$ CH_3 symmetric stretching	2852 (m.)*	2857 (w.) pol.	
$2\nu_7(A')$	2884 (m.)*	2890 (m.) pol.	
$\nu_4(a')$ CH_3 non-symmetric stretching	2916 (s.)*	2924 (s.) pol.	
$2\nu_{16}(A')$	2942 (s.)*		
$\nu_{15}(a'')$ CH_3 non-symmetric stretching	2960 (m.)*	2956 (v.w.)	
$\nu_3(a')$ CH_2 symmetric stretching	2979 (s.)*	2990 (w.) pol.	
$\nu_2(a')$ CH stretching	3012 (m.)*	3010 (s.b.) pol.	
$\nu_6+\nu_8(A')$	3067 (w.)*		
$\nu_1(a')$ CH_2 non-symmetric stretching	3081 (m.)*	3087 (w.)	

[251] Except for the bands marked by * these are the data for the gas given by Wilson and Wells (945). The three bands at 936, 963, and 1043 cm^{-1} were measured by Avery and Ellis (63) in the solid at −195° and the bands above 2800 cm^{-1} by Fox and Martin (330) in solution in CCl_4.

[252] Data of Ananthakrishnan (45), except for the weak line 1448 cm^{-1} which was only found by Bourguel and Piaux (173).

[253] From the interpretation of the band 755 cm^{-1}, very uncertain.

Other nine-atomic molecules. In Table 113, references to Raman and infrared investigations of other nine-atomic molecules are given, as well as the structures derived. However it must be understood that in no case can these structures be considered as established beyond doubt.

TABLE 113. OTHER NINE-ATOMIC MOLECULES.

Molecule	Structure	References	
		Raman spectrum	Infrared spectrum
$Ni(CO)_4$	T_d? (tetrahedral)[254]	(293) (932)	(87) (243)
$(CH_3)_2S$	C_{2v} [like $(CH_3)_2O$]	(518) (620) (326)	(326) (851)
$(CH_3)_2Se$	C_{2v} [like $(CH_3)_2O$]	(289)	
C_4H_4O (furan)	C_{2v} (plane)	(14) (736)	(600) (601) (693)
C_4H_4S (thiophene)	C_{2v} (plane)	(14) (736) (894)	(600) (601) (100) (117)
C_2H_5OH	C_s	(14) (907) (773)	(67) (308) (773) (974a)
C_2H_5SH	C_s	(907)	
C_2H_5SeH	C_s	(254)	
C_2H_5CN, C_2H_5NC	C_s	(14)	
CH_3—CO—NH_2	C_s	(14) (762) (478a)	
CH_3—CS—NH_2	C_s	(532)	
CCl_3—CO—NH_2	C_s	(948a)	
$(CH_3)_2Zn$	D_{3h}(?)	(894) (682)	(859)
$(CH_3)_2Hg$	D_{3h}(?)	(682) (858)	
CH_2Cl—CH_2OH	C_s ?	(632)	
$CNCH_2$—COOH	C_s?	(252)	

(*h*) *Ten-atomic molecules*

Dimethyl acetylene, CH_3—C≡C—CH_3. From our previous discussion of methyl acetylene it is practically certain that in dimethyl acetylene (CH_3—C≡C—CH_3) the four C atoms lie on a straight line, that is, that *the molecule has a three-fold axis of symmetry.* As for C_2H_6, we have the three possibilities for the structure of CH_3—C≡C—CH_3: point group D_{3h}, point group D_{3d}, or free rotation of the two CH_3 groups about the C—C≡C—C axis. In fact, if the potential barrier opposing free rotation in C_2H_6 is due to the interaction between the H atoms one would expect the potential barrier to be very slight in CH_3—C≡C—CH_3 because of the much larger distance of the two CH_3 groups from each other. Specific heat and entropy determinations [Crawford and Rice (246), Osborne, Garner, and Yost (678), Kistiakowsky and Rice (513)] do indeed point strongly to free rotation.

If there is free rotation, the symmetry of the molecule is D_3 in a general position of the two CH_3 groups (see Fig. 2j). However, since in this event an internal rotation by an arbitrary angle does not change the potential energy, the symmetry types of the normal vibrations are the same as for a special position of the CH_3 groups, for example those corresponding to D_{3h}, as has been shown in more detail by Howard (460), who called this symmetry D'_{3h}. For instance, whereas for a molecule of symmetry D_3 (without free rotation) there is only one species of degenerate vibrations (see Table 15), in the present case there are two just as for D_{3h} (or D_{3d}). In one of them the atoms on the axis move symmetrically, in the other anti-symmetrically, with respect to the plane perpendicular to the axis through the mid-point, even though this plane is not an element of symmetry of the molecule in its most general position. We can therefore use the same nomenclature as for D_{3h}. Also the selection rules are found to be the same as for D_{3h} [see Howard (460)]; they are not as stringent as those for D_{3d}.

Table 114 gives the designation and description of the normal modes of dimethyl acetylene for both D_{3h} and D_{3d}, the former including the case of free rotation. The torsional oscillation for the

[254] It must be mentioned that there is a serious discrepancy between this structure and the observed Raman spectrum in that the three strongest Raman lines are polarized whereas on the tetrahedral model there are only two totally symmetric vibrations.

TABLE 114. DESIGNATION, CHARACTER, SPECIES, AND ACTIVITY OF THE NORMAL VIBRATIONS OF $CH_3—C≡C—CH_3$.

Description \ D_{3h}: / D_{3d}:	A_1' / A_{1g}	A_1'' / A_{1u}	A_2'' / A_{2u}	E' / E_u	E'' / E_g
C—H stretching	ν_1^{CH}	—	ν_6^{CH}	ν_9^{CH}	ν_{13}^{CH}
C≡C stretching	$\nu_2^{C≡C}$	—	—	—	—
CH_3 deformation	$\nu_3^{CH_3}$	—	$\nu_7^{CH_3}$	$\nu_{10}^{CH_3}$	$\nu_{14}^{CH_3}$
CH_3 rocking	—	—	—	ν_{11}	ν_{15}
C—C stretching	ν_4^{C-C}	—	ν_8^{C-C}	—	—
C—C≡C—C bending	—	—	—	ν_{12}	ν_{16}
CH_3 twisting (torsion)	—	ν_5	—	—	—
Activity, D_{3h} and free rotation	R. pol.	inactive	I. ‖	I. ⊥, R. depol.	R. depol.
Activity, D_{3d}	R. pol.	inactive	I. ‖	I. ⊥	R. depol.

case of free rotation has, of course, zero frequency. There is one more vibration of each species than there is for C_2H_6 except of species $A_1''(A_{1u})$ which corresponds to the internal rotation or torsional oscillation.

In Table 115 are given the observed infrared and Raman spectra of the $CH_3—C≡C—CH_3$ molecule. Its interpretation is greatly facilitated by our knowledge of the spectra of ethane and of acetylene. Of the six observed polarized Raman lines only four can be fundamentals. Two can indeed easily be explained as first overtones (see Table 115) which occur strongly because of Fermi resonance. The remaining four have a plausible magnitude for ν_1, ν_2, ν_3, and ν_4. The strong infrared band at 2975 cm^{-1} is obviously the superposition of the two infrared-active C—H stretching vibrations $\nu_6(a_2'')$ and $\nu_9(e')$. Because of the small interaction between the two CH_3 groups one would expect the two CH_3 deformation vibrations $\nu_3(a_1')$ and $\nu_7(a_1'')$ (whose only difference is that in one the two CH_3 groups vibrate in phase, in the other in opposite phase) to have very nearly the same frequency. Similar conclusions hold for $\nu_{10}(e')$ and $\nu_{14}(e'')$ as well as for $\nu_{11}(e')$ and $\nu_{15}(e'')$. With these considerations in mind, the identification of the C—H stretching and bending vibrations is easily made as given in Table 115. Because of the larger interaction between the C—C bonds the pairs ν_4, ν_8 and ν_{12}, ν_{16} would be expected to have a considerably larger separation. The assignment given seems plausible, although that for ν_8 and ν_{12} does not appear to be unique.

By using the force constants of the CH_3 groups as determined from C_2H_6, and by using five of the fundamentals to determine the remaining force constants of a valence-type potential (with certain interactions), Crawford (235) has calculated the other frequencies and found very satisfactory agreement with the observed frequencies [see also Crawford and Brinkley (240)]. In particular, nearly the same frequencies are found for each of the two members of the pairs mentioned above. This agreement is good evidence that the assumed model is correct. However it does not provide an answer to the question whether or not there is free rotation of the CH_3 groups (see above). It is interesting to note that the force constant for the C—C bond in $CH_3—C≡C—CH_3$, just as in $CH_3—C≡C—H$, is appreciably higher than in C_2H_6. This corresponds to the fact that the internuclear distance is smaller (see Chapter IV).

The remaining infrared and Raman bands have been assigned by Crawford (235) as combination and overtone bands, as indicated in Table 115.

Azomethane, $(CH_3)_2N_2$. Because of the similarity of the electronic structure of N_2 and CO, one might expect azomethane to have a Y structure similar to that of acetone $(CH_3)_2CO$. However, the absence of a dipole moment [West and Killingsworth (921)] definitely excludes this possibility. Instead, one has to assume that the two CH_3 groups are symmetrically placed on either end of the N_2 group. This is confirmed by the Raman and infrared spectra, since there are only two (probably chance) coincidences of infrared and Raman bands, whereas many more would be expected for a Y structure. Two possibilities remain: The linear structure $CH_3—N=N—CH_3$ or the bent trans structure 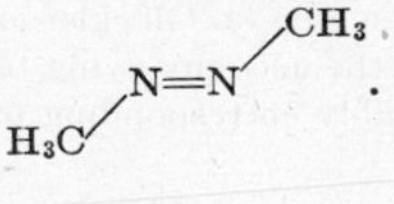. Valence theory and electron diffraction data [Boersch (158)] favor the second structure but the question cannot yet be regarded as settled. In either case one would expect free rotation of the CH_3 groups about the N—C axes. In the first model we have exactly the same symmetry as for CH_3—C≡C—CH_3, that is, the same species and the same number of fundamentals of each species. In the second model in the general position of the CH_3 groups there

TABLE 115. RAMAN AND INFRARED SPECTRA OF CH_3—C≡C—CH_3 AFTER GLOCKLER AND DAVIS (368)[255] AND CRAWFORD (235).

Assignment	Raman[256] (liquid) $\Delta\nu_{vacuum}$ (cm^{-1})	Infrared (gas) ν_{vacuum} (cm^{-1})	Assignment	Raman[256] (liquid) $\Delta\nu_{vacuum}$ (cm^{-1})	Infrared (gas) ν_{vacuum} (cm^{-1})
$\nu_{16}-\nu_{12}$?	144 (w.)[257]		$\nu_{12}+\nu_{14}(A_2'')$		1664?
$\nu_{12}(e')$	213 (w.)		$\nu_7+\nu_{16}(E')$		1763 (w.)
$\nu_4-\nu_{16}$?	351 (w.)[258]		$\nu_4+\nu_8(A_2'')$		1841 (m.)
$\nu_{16}(e'')$	**374** (v.s.) depol.[259]		$\nu_7+\nu_{12}+\nu_{16}(E')$ }		1919 (w.)
$2\nu_{12}(A_1', E')$	400 (v.w.)		$\nu_3+\nu_{12}+\nu_{16}(A_2'')$ }		
$\nu_4-\nu_{12}$?	508 (v.w.)[257]	507 w.	$2\nu_{15}(E')$, }		2065 (s.)
$\nu_{12}+\nu_{16}(A_2'')$		568 (s.)[260]	$\nu_{11}+\nu_{15}(A_2'')$ }		
$\nu_{11}-\nu_{16}$, $\nu_4+\nu_{12}-\nu_{12}$ }	687 (m.)	671 (v.w.)	$\nu_8+\nu_{15}(E')$		2139 (w.)
			$2\nu_8$ { (C^{13} isotope	2201.4 (w.)	
$\nu_4^{C-C}(a_1')$	**697.4** (s.) pol.		ν_2 bands) }	2280.2 (w.)	
$2\nu_{16}(A_1'+E')$	**773.5** (s.) pol.		$2\nu_8(A_1')$	**2234.6** (v.s.) pol.	
?	788.5 (m.)		$\nu_2^{C\equiv C}(a_1')$	**2312,7** (v.s.) pol.	
$\nu_{11}-\nu_{12}(E')$	834 (v.w.)[257]	835 (w.)	$\nu_3+\nu_8(A_2'')$		2500 (s.)
$\nu_{12}+\nu_{15}-\nu_{16}$		877 (w.)	$2\nu_3(A_1')$, $2\nu_7(A_1')$	2736.8 (m.)	
?	971 (w.)[258]		$\nu_6-\nu_{12}$, $\nu_9-\nu_{12}$		2785 (w.)
$\nu_{15}(e'')$	**1029** (m.) depol.		$2\nu_{14}(A_1', E')$	2861.8 (s.)	
$\nu_{11}(e')$		**1050** (m.)	$\nu_1^{CH}(a_1')$	**2920** (v.s.) pol.	
$\nu_8^{C-C}(a_2'')$		**1126** (w.)	$\nu_{13}^{CH}(e'')$	**2961** (s.)	
$\nu_{12}+\nu_{15}(A_2'', E'')$	1243 (v.w.)	1240 (m.)	$\nu_6^{CH}(a_2'')$, $\nu_9^{CH}(e')$		**2975** (v.s.)
$\nu_{12}+3\nu_{16}$?		1340 (w.)	$\nu_2+\nu_4(A_1')$	2996 (v.w.)[258]	
$\nu_3^{CH_3}(a_1')$	**1379** (v.s.) pol.[259]		$\nu_2+\nu_8(A_2'')$		3450 (w.)
$\nu_7^{CH_3}(a_2'')$		**1380** (m.)	$3\nu_7(A_2'')$		4140 (w.)
$\nu_{11}+\nu_{16}(A_2'')$		1425 (w.)	$\nu_6+\nu_{14}(E')$, }		
$\nu_{14}^{CH_3}(e'')$	**1447** (s.) depol.		$\nu_9+\nu_{14}(A_2'')$, }		4425 (m.)
$\nu_{10}^{CH_3}(e')$		**1468** (s.)	$\nu_9+\nu_{10}(E')$ }		
			$2\nu_9(E')$?		6100 (w.)

is no element of symmetry. However, for special positions of the CH_3 groups there is a center of symmetry (C_i) or a plane of symmetry (C_s) or both (C_{2h}). The latter symmetry also applies to the C—N═N—C group. If the potential energy does not depend on the angle of rotation of the CH_3 groups, the normal vibrations will fall into species exactly analogous to those of point group C_{2h}. These are uniquely related to those of D_{3d} and therefore also D_{3h}, in the following way (see p. 237):

D_{3h}:	A_1',	A_1'',	A_2',	A_2'',	E',	E'';
D_{3d}:	A_{1g},	A_{1u},	A_{2g},	A_{2u},	E_u,	E_g;
C_{2h}:	A_g,	A_u,	B_g,	B_u,	$A_u + B_u$,	$A_g + B_g$.

[255] As corrected by Glockler and Renfrew (381).

[256] Most of the frequencies are from Glockler and Davis (368), the polarization data from Crawford (235).

[257] These lines are reported only by Kohlrausch, Pongratz, and Seka (526).

[258] These lines are reported only by Gredy (400).

[259] These lines are accompanied by sequences (see p. 267), as can be seen on Glockler and Davis' spectrogram. However, these authors do not give the frequencies of the accompanying bands.

[260] This band has two maxima with a separation of 14 cm^{-1}, probably corresponding to a P and an R branch.

Because of this relation we may use the same designation of the fundamentals for the two models, namely that given in Table 114 for dimethyl acetylene. Of course, for every degenerate vibration of the linear model there are two non-degenerate ones of the bent model. The activities of the vibrations for the models $\boldsymbol{D_{3h}}$ and $\boldsymbol{D_{3d}}$ are given in Table 114. For the $\boldsymbol{C_{2h}}$ structure, A_g and B_g would be Raman active (the former polarized) and A_u and B_u would be infrared active. For free rotation all transitions would be allowed.

Unfortunately, the Raman and infrared spectra of azomethane are not nearly as completely known as those of dimethyl acetylene. A tentative assignment is given in Table 116. It is significant that the C—H stretching, CH_3 deformation, and rocking vibrations can be assigned satisfactorily to frequencies that are nearly the same as those in $CH_3—C \equiv C—CH_3$. While one would not expect the degenerate C—H or CH_3 frequencies of the linear model to split appreciably in the non-linear model, the two C—N=N—C bending frequencies should split very considerably. Bending in the C—N=N—C plane should be opposed by a greater restoring force than bending perpendicular to the plane. It would indeed be difficult to interpret the four observed infrared and Raman bands below 800 cm^{-1} by means of only two bending frequencies, whereas with the assumption of four such frequencies a very reasonable assignment is possible, as shown.[261] This is a fairly strong argument in favor of the bent model. The four bending frequencies are, according to this assignment,

TABLE 116. RAMAN AND INFRARED SPECTRA OF AZOMETHANE AFTER WEST AND KILLINGSWORTH (921).

Assignment	Raman (liquid, −60°) $\Delta\nu_{vacuum}$ (cm^{-1})	Infrared (gas, +20°) ν_{vacuum} (cm^{-1})	Assignment	Raman (liquid, −60°) $\Delta\nu_{vacuum}$ (cm^{-1})	Infrared (gas, +20°) ν_{vacuum} (cm^{-1})
$2[\nu_{16}(b_g)](A_g)$	548 (w.)		$\nu_4+\nu_{11}(A_u+B_u)$		1922 (w.)
$\nu_{16}(a_g)$	596 (s.) pol.		$\nu_8+\nu_{15}(A_u+B_u)$		2189 (m.)
$\nu_{12}(b_u)+\nu_{16}(b_g)$		700 (w.)	$\nu_3+\nu_{11}(A_u+B_u)$		2391 (m.)
$\nu_{12}(a_u)+\nu_{16}(a_g)$		730 (w.)	$\nu_8+\nu_{14}(A_u+B_u)$		2596 (m.)
$\nu_4{}^{C-N}(a_g)$	922 (m.)		$2\nu_3{}^{CH_3}(A_g)$, $2\nu_7{}^{CH_3}(A_g)$	2733 (w.)	
$\nu_{15}(a_g+b_g)$, $\nu_{11}(a_u+b_u)$	1023 (w.)	1013 (s.)	$2\nu_{10}^{CH_3}(A_g)$, $2\nu_{14}^{CH_3}(A_g)$	2854 (w.)	
$\nu_8{}^{C-N}(b_u)$		1110 (s.)	$\nu_1{}^{CH}(a_g)$	2914 (v.s.) pol.	
$2[\nu_{16}(a_g)](A_g)$	1182 (m.)		$\nu_{13}^{CH}(a_g+b_g)$	2985 (s.) depol.	
$\nu_3{}^{CH_3}(a_g)$, $\nu_7{}^{CH_3}(b_u)$	1376 (s.)	1430 (v.s.)	$\nu_6{}^{CH}(b_u)$, $\nu_9{}^{CH}(a_u+b_u)$		3030 (v.s.)
$\nu_{14}^{CH}(a_g+b_g)$, $\nu_{10}(a_u+b_u)$	1442 (v.s.) pol.		$\nu_1+\nu_{10}$, $\nu_6+\nu_3$		4360 (m.)
$\nu_2{}^{N=N}(a_g)$	1576 (m.)		$\nu_1+\nu_6(B_u)$?		5650 (w.)

$$\nu_{12}(a_u) = 134, \quad \nu_{16}(b_g) = 274, \quad \nu_{12}(b_u) = 426, \quad \nu_{16}(a_g) = 596 \text{ cm}^{-1}.$$

Further investigation of azomethane would be very desirable in order to settle this point definitely.

Other ten-atomic molecules. Table 117 gives references to work on the Raman and infrared spectra of further ten-atomic molecules. While the structures given are probably correct, the spectroscopic data in almost all cases are not sufficient to establish them unambiguously.

(*i*) *Eleven-atomic molecules*

The number of eleven-atomic molecules for which both infrared and Raman data are available is very limited. We consider only the propane molecule.

Propane, $CH_3—CH_2—CH_3$. From the fact that there is no free rotation in ethane one would conclude that there is also none in propane. Therefore one would expect propane to belong to point

[261] It must be realized that the infrared spectrum has not been observed below 500 cm^{-1}. Therefore $\nu_{12}(a_u)$ and $\nu_{12}(b_u)$ have not been directly observed, but only their combinations with ν_{16}.

TABLE 117. FURTHER TEN-ATOMIC MOLECULES.

Molecule	Structure	References: Raman spectrum	References: Infrared spectrum
$(CH_3)_2CO$	C_{2v}	(14) (300) (774) (480) (219)	(232) (707) (774)
C_4H_6 (butadiene)	C_{2h} or C_{2v}	(255) (174)	(122) (346) (724b)
C_4H_4NH (pyrrole)	C_{2v} planar	(14) (894) (159) (160) (736) (588)	(976) (601) (588)
$C_2H_5NH_2$	C_s?	(14) (10) (907)	
$C_2H_5NO_2$	C_s	(948a)	
C_2H_5COH	C_s?	(520)	(493)
C_2H_5COCl, C_2H_5COBr	C_s?	(523) (524)	
C_2H_5NCO	C_s?	(537)	
C_2H_5SCN, C_2H_5NCS	C_s?	(252)	
$C_2H_4(OH)_2$	C_{2v}?	(14) (10)	(67)
$C_2H_4(CN)_2$	C_{2v}?	(534)	
$C_2H_4(OH)(CN)$	C_s	(534)	(845)
$C_2H_4(OH)(SH)$	?		(845)
C_3H_5CN (cyclopropyl-cyanide)	C_{2v}?	(738)	
$(CH_3)_2NH$	C_s	(518) (298)	
$H_2C=CH—CH_2OH$		(14) (10)	(67)
$C^+(NH_2)_3$, $C^+(ND_2)_3$	C_{3h}?	(679) (495)	
$Cl_3C—C(OH)_2$ | H			(269)
$C_2H_5C≡CH$	C_s	(218b)	
$C_2H_5C≡CBr$, $C_2H_5C≡CI$	C_s	(649b)	
$CH_2=C(CH_3)(CN)$, $CNCH=CHCH_3$	C_s	(737)	

TABLE 118. DESIGNATION, CHARACTER, SPECIES AND ACTIVITY OF THE NORMAL VIBRATIONS OF PROPANE C_3H_8.

Description	A_1	A_2	B_1	B_2
CH_2 stretching	$\nu_1^{CH}(a_1)$			$\nu_{22}^{CH}(b_2)$
CH_3 non-symm. stretching	$\nu_2^{CH}(a_1)$	$\nu_{10}^{CH}(a_2)$	$\nu_{15}^{CH}(b_1)$	$\nu_{23}^{CH}(b_2)$
CH_3 symm. stretching	$\nu_3^{CH}(a_1)$		$\nu_{16}^{CH}(b_1)$	
CH_2 deformation	$\nu_4^{CH_2}(a_1)$			
CH_3 non-symm. deformation	$\nu_5^{CH_3}(a_1)$	$\nu_{11}^{CH_3}(a_2)$	$\nu_{17}^{CH_3}(b_1)$	$\nu_{24}^{CH_3}(b_2)$
CH_3 symm. deformation	$\nu_6^{CH_3}(a_1)$		$\nu_{18}^{CH_3}(b_1)$	
CH_2 rocking			$\nu_{19}(b_1)$	$\nu_{25}(b_2)$
CH_3 rocking	$\nu_7(a_1)$	$\nu_{12}(a_2)$	$\nu_{20}(b_1)$	$\nu_{26}(b_2)$
C—C stretching	$\nu_8^{C-C}(a_1)$		$\nu_{21}^{C-C}(b_1)$	
C—C—C bending	$\nu_9^{C_3}(a_1)$			
CH_2 twisting		$\nu_{13}(a_2)$		
CH_3 twisting		$\nu_{14}(a_2)$		$\nu_{27}(b_2)$
Activity	R. pol. I.	R. depol.	R. depol. I.	R. depol. I.

group C_{2v}, the three C atoms forming an isosceles triangle, one H atom of each of the CH_3 groups lying in the plane of this triangle and the other H atoms being arranged in pairs symmetrical with respect to this plane. The available spectroscopic data are not sufficient to prove this structure. Rather we shall interpret these data on the assumption that the symmetry is C_{2v}.

TABLE 119. RAMAN AND INFRARED SPECTRA OF C_3H_8 (PROPANE).

Assignment[261a]	Raman $\Delta\nu_{vacuum}$[262] (cm^{-1}) (liquid)	Infrared ν_{vacuum}[263] (cm^{-1}) (gas)	Assignment	Raman $\Delta\nu_{vacuum}$[262] (cm^{-1}) (liquid)	Infrared ν_{vacuum}[263] (cm^{-1}) (gas)
$\nu_{14}(a_2)$?		(333)[264]	$\nu_8+\nu_{21}(B_1)$		1730 (w.)
$\nu_9^{C_3}(a_1)$	375 (w.)		$\nu_{20}+\nu_{21}(A_1)$		1936 (w.)
$\nu_{20}-\nu_{14}[B_2]$		720 (m.) Q	$\nu_6+\nu_{18}(B_1)$		(2640)
$\nu_{26}(b_2)$		748 (s.) Q	$2\nu_{18}(A_1)$	2725 (m.)	
		864 (w.) Q	$\nu_6+\nu_{24}(B_2)$	2761 (v.w.)	
$\nu_8^{C-C}(a_1)$	867 (s.)	870 (w.)	$2\nu_5(A_1)$	2872 (v.w.)	
$\nu_{21}^{C-C}(b_1)$	(940)[265]	922 (m.) Q	$\nu_{16}^{CH}(b_1)$		2885 (m.) Q
		925 (m.) Q	$\nu_3^{CH}(a_1)$?	2903 (s.)[266]	
$\nu_{20}(b_1)$	1054 (m.)	1053 (m.) Q	$2\nu_4$, $2\nu_5$ ··· ?	2920 (s.)[266]	
$\nu_7(a_1)$, $\nu_{19}(b_1)$	1152 (w.)	1152 (m.)	$\nu_2^{CH}(a_1)$	2946 (s.)	2960 (m.)
$\nu_{25}(b_2)$		1179 (m.) Q	$\nu_1^{CH}(a_1)$, $\nu_{10}^{CH}(a_2)$	2967 (m.)	
$\nu_{12}(a_2)$	1278 (w.)		$\nu_{15}^{CH}(b_1)$, $\nu_{23}^{CH}(b_2)$		2968 (s.) Q
$\nu_{20}+\nu_{14}(B_2)$		1338 (m.) Q	$\nu_{22}^{CH}(b_2)$		2980 (m.)
$\nu_6^{CH_3}(a_1)$		1370 (m.)			3190 (w.)
$\nu_{18}^{CH_3}(b_1)$		1375 (s.)			3350 (v.w.)
$\nu_{11}^{CH_3}(a_2)$, $\nu_5^{CH_3}(a_1)$	1451 (s.)				
$\nu_{17}^{CH_3}(b_1)$		{ 1465 (s.) Q			
$\nu_4^{CH_2}(a_1)$		{ 1468 (s.)			
$\nu_{24}^{CH_3}(b_2)$		{ 1470 (v.s.) Q			

The number, species and activity of the fundamentals of C_3H_8 on the assumption of the C_{2v} structure is, according to Tables 34 and 55:

$$9A_1\text{(I., R. pol.)}, \quad 5A_2\text{(R. depol.)}, \quad 7B_1\text{(I., R. depol.)}, \quad 6B_2\text{(I., R. depol.)}.$$

The vibrations will be similar to those of $(CH_3)_2O$ (see Table 110) except that the O atom is replaced by a CH_2 group, and therefore we expect in addition to the stretching, deformation, rocking, bending, and twisting oscillations of the CH_3 groups also those of the CH_2 group while the vibrations of the

[261a] Note added in proof: In a very recent paper Pitzer (696a) has reviewed the assignments for propane. The main differences of his assignment from that of Table 119 are: (1) that he considers the doubtful Raman line 940 cm^{-1} as $\nu_{12}(a_2)$ and as distinct from the infrared band 922 cm^{-1} while he uses 1278 cm^{-1} [which is $\nu_{12}(a_2)$ in the table] as $\nu_{13}(a_2)$ for which the above table does not give an assignment, (2) that he considers the infrared band 1338 cm^{-1} as a fundamental, namely $\nu_{19}(b_1)$, and (3) that he considers the infrared band 720 cm^{-1} as $\nu_{21} - \nu_{14}$ and therefore obtains $\nu_{14} = 202$ cm^{-1} while he derives from thermodynamic data $\nu_{27}(b_2) = 283$ cm^{-1}.

[262] Average of the data of Kohlrausch and Köppl (521) and Bhagavantam (145).

[263] Wave numbers given by Wu and Barker (965), except for the bands 1730, 3190, and 3350 cm^{-1} given by Bartholomé (119). For the last two bands, as well as for the band 2640 cm^{-1}, Wu and Barker only give contours but do not list the wave numbers. Q means that a sharp central maximum is observed.

[264] From the difference band 720 cm^{-1}.

[265] This Raman band has been given as a strong one by Daure (265) only, whereas Bhagavantam and Kohlrausch and Köppl, who find more lines otherwise, do not report it. It is probably due to an impurity.

[266] These two lines were measured as one at 2914 cm^{-1} by Kohlrausch and Köppl (521).

C—C—C chain replace those of the C—O—C chain. Table 118 gives the resulting designations and descriptions of the various vibrations. It should be realized, however, that for a molecule with so many atoms the division of the vibrations of each species indicated in this table may be a very poor approximation. In particular, vibrations of the same species that are expected to have a similar magnitude, such as $\nu_7(a_1)$ and $\nu_8(a_1)$ (corresponding to CH_3 rocking and C—C stretching), may be considerably mixed; that is, for example, ν_7 is actually a combination of a CH_3 rocking and a C—C stretching motion, and ν_8 is a different combination of the same motions.

Table 119 gives the observed Raman and infrared spectra up to 3400 cm^{-1}. Further infrared bands at higher frequencies are given by Bartholomé (119) and Wu and Barker (965) and two photographic infrared bands by Gänswein and Mecke (346). The eight C—H stretching vibrations are expected in the frequency range 2700–3100 cm^{-1}, the CH_3 and CH_2 rocking vibrations in the range 1200–900 cm^{-1}, the two C—C stretching vibrations near 900 cm^{-1}, and the C—C—C bending vibration and the twisting vibrations below 500 cm^{-1}. A further guide in the assignment may be found in the rule (to which numerous exceptions occur) that the totally symmetric vibrations (A_1) give the strongest Raman lines, and the consideration that a sharp central maximum in an (unresolved) infrared band (indicated by Q in Table 119) indicates that the dipole moment is perpendicular to the axis of smallest moment of inertia (see Chapter IV, section 4), which in the present case means that the dipole moment has species B_1 or B_2. The assignment given in Table 119 takes all these considerations into account. It is essentially that of Wu and Barker (965). In view of the lack of polarization data for the Raman spectrum and the large number of fundamentals of each species, the assignment can only be considered as tentative.

Other eleven-atomic molecules. References to work about other eleven-atomic molecules are given in Table 120.

TABLE 120. FURTHER ELEVEN-ATOMIC MOLECULES.

Molecule	Structure	References: Raman spectrum	References: Infrared spectrum
C_5H_5N (pyridine)	C_{2v}, planar	(14) (10) (44) (141) (428) (515a) (430a)	(224) (567a) (880a) (515a)
C_5H_6 (cyclopentadiene)	C_{2v}, planar	(735) (736)	
$H_3C\cdot HC{=}CH\cdot COH$ (crotonaldehyde)	C_s?	(524)	
C_2H_5COOH	no symmetry	(14) (10)	(429)
$HCOOC_2H_5$	?	(14) (10)	
CH_3—$COOCH_3$	?	(14) (10)	
C_4H_6O (cyclobutanone)	C_{2v}?	(529)	
$CH_3CH_2CH_2Cl$, Br, I	?	(14) (632)	
$CH_3CBr_2CH_3$, $CH_3CI_2CH_3$ and other halo-propanes		(14) (483)	
$OH\cdot CH_2\cdot CH_2\cdot NH_2$ (ethanol amine)			(845)
HOOC—CH_2—COOH (malonic acid)		(14)	(322a)
$C_2H_5NO_3$		(948b)	

(*j*) *Twelve-atomic molecules*

For twelve-atomic molecules higher symmetries are possible than for ten- and eleven-atomic molecules. In such cases fewer fundamentals occur, and since there are more different symmetry types the vibrations are better determined by symmetry alone. The best example is the benzene (C_6H_6) molecule which will be the only one to be considered here in detail.

Benzene, C_6H_6 and C_6D_6. The question of the structure of benzene is one that has occupied chemists for many decades. Kékulé's model explains most of the chemical evidence. On this model C_6H_6 would belong to the point group D_{3h}. However, modern valence theories suggest a still more symmetrical model in which all C—C bonds are equivalent and consequently there is a six-fold axis of symmetry. Assuming the H atoms to be in the plane of the C atoms and in their most symmetrical position the point group of this model is D_{6h} (see Fig. 1i). Other models that have been considered at one time or another are the puckered ring of symmetry D_{3d} and the model in which the H atoms are in a plane different from that of the C atoms, that is, in which the point group is C_{6v}.

Since the point group D_{6h} has all the symmetry elements of D_{3h}, D_{3d}, and C_{6v}, we can obtain the normal vibrations for these lower symmetries from those of the D_{6h} model (see Table 53). According to Table 23 there are twelve different species for D_{6h}. The numbers of vibrations of each species [267] are for $C_6H_6(C_6D_6)$:

$$2A_{1g},\quad 0A_{1u},\quad 1A_{2g},\quad 1A_{2u},\quad 0B_{1g},\quad 2B_{1u},\quad 2B_{2g},\quad 2B_{2u},\quad 1E_{1g},\quad 3E_{1u},\quad 4E_{2g},\quad 2E_{2u}.$$

TABLE 121. DESIGNATION,[268] DESCRIPTION, SPECIES, AND ACTIVITY [269] OF THE NORMAL VIBRATIONS OF BENZENE (C_6H_6).

Description[270]	D_{6h}				
	A_{1g} (R. pol.)	$A_{2g}(-)$	A_{2u}(I.)	$B_{1u}(-)$	$B_{2g}(-)$
C—H stretching	ν_1^{CH}, [2]	—	—	ν_5^{CH}, [13]	—
C—C stretching	ν_2^{CC}, [1]	—	—	—	—
C—H ‖ bending	—	$\nu_3^{H\parallel}$, [3]	—	—	—
C—H ⊥ bending	—	—	$\nu_4^{H\perp}$, [11]	—	$\nu_7^{H\perp}$, [5]
C—C—C ‖ bending	—	—	—	$\nu_6^{C\parallel}$, [12]	—
C—C—C ⊥ bending	—	—	—	—	$\nu_8^{C\perp}$, [4]
D_{3h}	A_1' (R. pol.)	$A_2'(-)$	A_2''(I.)	A_1' (R. pol.)	A_2''(I.)
D_{3d}	A_{1g} (R. pol.)	$A_{2g}(-)$	A_{2u}(I.)	A_{2u}(I.)	A_{1g} (R. pol.)
C_{6v}	A_1 (R. pol., I.)	$A_2(-)$	A_1 (R. pol., I.)	$B_1(-)$	$B_1(-)$

TABLE 121.—*Continued.*

Description[270]	D_{6h}				
	$B_{2u}(-)$	E_{1g} (R. dep.)	E_{1u}(I.)	E_{2g} (R. dep.)	$E_{2u}(-)$
C—H stretching	—	—	ν_{12}^{CH}, [18]	ν_{15}^{CH}, [7]	—
C—C stretching	ν_9^{CC}, [14]	—	ν_{13}^{CC}, [19]	ν_{16}^{CC}, [8]	—
C—H ‖ bending	$\nu_{10}^{H\parallel}$, [15]	—	$\nu_{14}^{H\parallel}$, [20]	$\nu_{17}^{H\parallel}$, [9]	—
C—H ⊥ bending	—	$\nu_{11}^{H\perp}$, [10]	—	—	$\nu_{19}^{H\perp}$, [17]
C—C—C ‖ bending	—	—	—	$\nu_{18}^{C\parallel}$, [6]	—
C—C—C ⊥ bending	—	—	—	—	$\nu_{20}^{C\perp}$, [16]
D_{3h}	$A_2'(-)$	E'' (R. dep.)	E' (R. dep. I.)	E' (R. dep., I.)	E'' (R. dep.)
D_{3d}	$A_{1u}(-)$	E_g (R. dep.)	E_u(I.)	E_g (R. dep.)	E_u(I.)
C_{6v}	$B_2(-)$	E_1 (R. dep., I.)	E_1 (R. dep., I.)	E_2 (R. dep.)	E_2 (R. dep.)

267 Here we assume that the C_2 axis (not the C_2' axis) goes through a C—H group (see Table 23).

268 The numbering is in accordance with our usual practice (see p. 272). However for the convenience of the reader the numbering used by Wilson (930) and adopted by several others has been added in square brackets.

269 (−) means inactive both in the Raman and infrared spectrum.

270 ‖ and ⊥ refer to the plane of the molecule.

There are only twenty fundamentals, ten non-degenerate and ten doubly degenerate. In order to obtain a rough idea of the frequencies and forms of these oscillations we may again distinguish C—H and C—C stretching and bending vibrations. The bending may occur in the plane or perpendicular to the plane of the molecule. In this way we obtain the characterization of the fundamental vibrations given in Table 121. A graphical representation is given in the previous Fig. 50. It should be remembered that the diagrams give the exact normal modes only for those species for which there

TABLE 122. RAMAN SPECTRUM OF LIQUID C_6H_6 AND C_6D_6 AFTER ANGUS, INGOLD, AND LECKIE (56).

Assignment[271]	C_6H_6			C_6D_6		
	$\Delta\nu_{vacuum}$[272] (cm^{-1})	Intensity[273]	Polarization	$\Delta\nu_{vacuum}$[274] (cm^{-1})	Intensity[273]	Polarization
$[\nu_{20}^{C\perp}(e_{2u})]$	404	w.	?	337	w.	?
$\nu_{18}^{C\parallel}(e_{2g})$	**605.6**	2.1	depol.	**576.7**	1.2	depol.
	685	w.				
$\nu_{11}^{H\perp}(e_{1g})$	**848.9**	0.9	depol.	**661.2**	1.4	depol.
$\nu_2(C_5{}^{12}C^{13}H_6)$	984	0.6[275]		939.5[276]	1.0	
$\nu_2^{CC}(a_{1g})$	**991.6**	10.0[275a]	s. pol.	**944.7**	10.0	s. pol.
$\nu_2+\nu_{18}-\nu_{18}$	999	w.	?			
$[\nu_{14}^{H\parallel}(e_{1u})]$	1030	w.				
$\nu_{17}^{H\parallel}(e_{2g})$	**1178.0**	2.2	depol.	**867.2**	2.3	depol.
$\nu_6+\nu_{20}(E_{1g})$	1404	w.				
$[\nu_{13}^{CC}(e_{1u})]$	1478	w.		1327	w.	
$\nu_{16}^{CC}(e_{2g})$ }	**1584.8**	1.9	depol.	**1558.6**	2.0	depol.
$\nu_2+\nu_{18}(E_{2g})$ }	**1606.4**	1.6	depol.			
$2\nu_{11}(A_{1g}, E_{2g})$	1693	w.		1327	w.	
$2\nu_{10}(A_{1g})$?	2293	w.				
$2\nu_3(A_{1g})$	2454	w.		1931	w.	
$\nu_{13}+\nu_{14}(A_{1g}, E_{2g})$	2543	w.		2145		
$\nu_{10}+\nu_{13}(E_{2g})$	2618	w.				
$\nu_9+\nu_{14}(E_{2g})$?	2925	w.				
$2\nu_{13}(A_{1g}, E_{2g})$	2948	w.				
$\nu_{15}^{CH}(e_{2g})$	**3046.8**	4.8	depol.	**2263.9**	6.1	depol.
$\nu_1^{CH}(a_{1g})$	**3061.9**[277]	10.6	pol.	**2292.3**	10.6	pol.
$\nu_2+\nu_{16}(E_{2g})$				2461	w.	
$2\nu_{16}(A_{1g}, E_{2g})$ }	3164	w.		3110	w.	
$2\nu_2+2\nu_{18}(A_{1g}, E_{2g})$ }	3187	w.	pol.			
?				2739	w.	

[271] Assignments corresponding to forbidden transitions assumed to take place because of intermolecular forces in the liquid are put in square brackets.

[272] These are averages of the values of various investigators as given by Angus, Ingold, and Leckie (56). A number of further very weak Raman bands have been observed by Grassmann and Weiler (399) and Ananthakrishnan (44), [see also Angus, Ingold, and Leckie (56)].

[273] For the stronger lines the integrated intensities as measured by Angus, Ingold, and Leckie (56) are given.

[274] A number of further weak bands corresponding to the very weak C_6H_6 bands not given [272] are omitted.

[275] This datum is from Cheng, Hsueh, and Wu (203).

[275a] Paulsen (688a) gives 26.8 as the intensity of this line.

[276] From Langseth and Lord (552).

[277] Observed in the vapor at 3069 cm^{-1} by Bhagavantam and Rao (152).

is only one vibration. Table 121 also shows the correlation to the less symmetrical models as well as the activities in all cases.

The number and degeneracy as well as the approximate frequency of the fundamentals is the same for all four models. In order to decide which model is correct one has to see with which of the four sets of selection rules the Raman and infrared spectra agree. Since the selection rules for

TABLE 123. INFRARED SPECTRUM OF C_6H_6 AND C_6D_6 VAPOR AFTER BAILEY, HALE, INGOLD, AND THOMPSON (90).

Assignment[278]	C_6H_6 ν_{vacuum}[279] (cm^{-1})	C_6H_6 Intensity	C_6H_6 Type[280]	C_6D_6 ν_{vacuum} (cm^{-1})	C_6D_6 Intensity	C_6D_6 Type[280]
$\nu_{16}-\nu_{13}[E_{1u}]$	87[281]	v.w.				
$\nu_2+\nu_{18}-\nu_{13}[E_{1u}]$	117[281]	v.w.				
$\nu_{17}-\nu_{14}[E_{1u}]$	140[281]	v.w.				
$\nu_{11}-\nu_4[E_{1u}]$	168[281]	v.w.				
$\nu_4^{H\perp}(a_{2u})$	**671**	s.	*PQR*	**503**	s.	*PQR*
$\nu_{17}-\nu_{20}[A_{2u}]$	793	v.w.	*PR*			
?	962	v.w.		826	v.w.	
$\nu_{14}^{H\parallel}(e_{1u})$	**1037**	s.	*PQR*	**813**	s.	*PQR*
?				926	v.w.	
$\nu_7-\nu_{20}[E_{1u}]$	1143	v.w.				
?				1009		*PR*
$\nu_{11}+\nu_{20}[E_{1u}]$	1240	v.w.				
?	1377	w.				
$\nu_{17}+\nu_{20}(A_{2u})$				1212	v.w.	
$\nu_{13}^{CC}(e_{1u})$	**1485**	s.	*PQR*	**1333**	m.	*PQR*
$\nu_4+\nu_{11}(E_{1u})$	(1529)			1154	w.	*PQR*
$\nu_{14}+\nu_{18}(E_{1u})$	1617	w.		1385	v.w.	
$\nu_2+\nu_4(A_{2u})$	(1669)			1450	m.	
$\nu_{18}+\nu_{19}(A_{2u})$	1807	m.		1616	m.	
?				1630	v.w.	
$\nu_{11}+\nu_{14}(A_{2u})$	1906	v.w.				
$\nu_{11}+\nu_{19}(E_{1u})$	1964	m.		1692	w.	
$\nu_{10}+\nu_{17}(E_{1u})$	2288	w.				
$\nu_9+\nu_{17}(E_{1u})$	2856	v.w.				
$\nu_{13}+\nu_{16}(E_{1u})$ }	3045[282]	s.	*PQR*			
$\nu_2+\nu_{16}+\nu_{18}(E_{1u})$ }	3073[282]	w.				
$\nu_{12}^{CH}(e_{1u})$ }	**3099**[282]	s.	*PQR*	**2293**	s.	
$\nu_1+\nu_{12}(E_{1u})$				4544	w.	

[278] For the difference bands the species given (in the square bracket) is that of the dipole moment of the transition, not of the upper state; for all other bands it is the species of the dipole moment as well as that of the upper state (see p. 267).

[279] Values in parentheses have been observed in liquid C_6H_6 only.

[280] Some bands have been resolved into two or three maxima, apparently representing the maxima of rotational branches. This is indicated by *PR* and *PQR*. The frequency given refers in these cases to the central maximum or the mean frequency of the two branches.

[281] From Barnes, Benedict, and Lewis (116). The gap between this group and the other bands is probably not real but due to lack of observations.

[282] These three bands are from Leberknight (561), who apparently used higher dispersion, the weak band at 3073 cm^{-1} being taken from his absorption curve since he does not give it in his table for the vapor. The same three bands with different relative intensities and without the *P* and *R* maxima occur also in the liquid and solid [see Leberknight (561)] as well as in solution in CCl_4 [see Fox and Martin (328)].

the D_{6h} model are the most stringent [a vibration that is Raman (infrared) active for D_{6h} is also active for the other models; some vibrations that are inactive for D_{6h} become active for these other models], it is best to start out by trying to fit the observed spectrum to the D_{6h} model and then see whether contradictions to the selection rules occur that can be explained by assuming a lower symmetry.

The Raman and infrared spectra of C_6H_6 have been investigated by a large number of investigators [for the former see Kohlrausch (14), Hibben (10)]. Probably the most complete data are due to Ingold and his co-workers who have also investigated the spectrum of C_6D_6. Their results are given in Tables 122 and 123.

Since the plane regular hexagon model of C_6H_6 (D_{6h}) has a center of symmetry, the rule of mutual exclusion should hold for the Raman and infrared spectra. The fact that a number of coincidences actually occur (compare Tables 122 and 123) was at one time taken as evidence that the D_{6h} (as well as the D_{3d}) model was incorrect.[283] However, the Raman data given refer to the liquid state, whereas the infrared data refer to the gaseous state. As we have seen in some of the previous examples, the molecular fields in the liquid may bring about a slight breakdown of the selection rules. Indeed, Ingold and his co-workers found that if they compared the infrared spectrum of the liquid with its Raman spectrum the number of coincidences was greatly increased. By a detailed consideration of every individual coincidence they showed that there is no coincidence that is sufficiently significant to rule out the D_{6h} model. In particular, in the only case of an apparent coincidence of a strong Raman line of the liquid with a strong infrared band of the vapor (3045 cm^{-1}), measurements of the infrared spectrum of the liquid yield 3032 cm^{-1} [see Leberknight (561)], which does not coincide with the Raman shift of the liquid.[284]

The preceding considerations, of course, do not by any means prove the D_{6h} model. They show only that the observed coincidences may not be genuine and do not rule out the D_{6h} model. It will be remembered that the D_{3d} model also requires the absence of such coincidences. We have now to see whether all the infrared and Raman bands can be understood on the D_{6h} model.

The only two strong polarized *Raman lines*, 992 and 3062 cm^{-1}, correspond obviously to the only two totally symmetric vibrations ν_2^{CC} and ν_1^{CH} of D_{6h}. The corresponding frequencies for C_6D_6 are 945 and 2292 cm^{-1}. For any other model there are more totally symmetric vibrations and therefore there should be further polarized Raman lines. (Of course, they might be some of the weak lines whose states of polarization have not been measured.) The five remaining non-totally symmetric Raman-active fundamentals are readily identified with five of the six fairly strong depolarized Raman lines if it is assumed [Wilson (931)] that the pair 1585–1606 cm^{-1} is a resonance doublet of the fundamental $\nu_{16}^{CC} \approx 1596$ cm^{-1} and the combination $\nu_{18}^{C||} + \nu_2^{CC} = 1597$ cm^{-1} (both have the same species E_{2g}). This assumption is confirmed by the fact that for C_6D_6 only one line instead of the pair is observed, since here $\nu_{18}^{C||} + \nu_2^{CC}$ is considerably smaller than ν_{16}^{CC} and no resonance occurs. The great difference between the frequencies of the two carbon stretching vibrations, ν_2 and ν_{16}, is due to the fact that in the former (the C_6 "breathing" vibration) the C atoms do not move in the direction of the C—C bonds but toward or away from the center. A more detailed calculation gives indeed very nearly the observed frequency ratio.

There are four *infrared-active fundamentals* for the D_{6h} model: $\nu_4, \nu_{14}, \nu_{13}, \nu_{12}$. These are readily identified with the five infrared bands of outstanding intensity 671, 1037, 1485, 3045, and 3099 cm^{-1}, the last two forming a resonance doublet of the fundamental $\nu_{12}^{CH}(e_{1u})$ and the combination $\nu_{16}^{CC}(e_{2g}) + \nu_{13}^{CC}(e_{1u})$, one of whose sublevels has the right species. Since ν_{16} itself forms a resonance doublet with $\nu_2 + \nu_{18}$ (see above), we expect a third infrared band $\nu_2 + \nu_{18} + \nu_{13}(E_{1u})$ of comparable intensity which is observed at 3073 cm^{-1}.[285] Only one band is observed in this region for C_6D_6, since

---

[283] More coincidences are found if some of the very weak lines omitted in the tables are included in the comparison.

[284] It may be mentioned that Bailey, Hale, Ingold, and Thompson measured the two infrared bands of the vapor at 3045 and 3073 cm^{-1} as one band at 3060 cm^{-1}, which coincides closely with the strongest Raman line 3062 cm^{-1} of the liquid. But the Raman spectrum of the vapor [Bhagavantam and Rao (152)] gives 3069 cm^{-1} for this line.

[285] Rumpf and Mecke (751) consider the fact that three infrared bands are observed near 3000 cm^{-1} when only one fundamental is expected in this region for the D_{6h} model as "unambiguous proof" that the D_{6h} model is incorrect. They do not consider the possibility of Fermi resonance, which has been amply proven in many instances. In the present case the resonances assumed must

neither resonance occurs for it. This represents a strong argument in favor of the interpretation for C_6H_6. It is very satisfactory that the infrared frequencies $\nu_4^{H\perp}$, $\nu_{14}^{H\parallel}$, ν_{13}^{CC}, and ν_{12}^{CH} have similar magnitudes to the Raman frequencies $\nu_{11}^{H\perp}$, $\nu_{17}^{H\parallel}$, ν_{16}^{CC}, and ν_1^{CH}, which have similar characters (see Table 121).

It is very significant that none of the Raman frequencies ν_{15}, ν_{16}, ν_{17}, $\nu_{18}(e_{2g})$ of C_6H_6 and C_6D_6 have been observed in the infrared spectrum of the vapor, whereas for the $\boldsymbol{D_{3h}}$ model they would be infrared active (see Table 121). Further strong evidence for the $\boldsymbol{D_{6h}}$ model (and at the same time for the correctness of the assignment of the active fundamentals) is supplied by the application of the Redlich-Teller product relations to the active C_6H_6 and C_6D_6 vibrations (see Chapter II, section 6). It is found that for all five active species the observed product ratios agree within 1 or 2 per cent with the theoretical values[286] [see Ingold and co-workers (55)]. A final argument in favor of the

TABLE 124. CALCULATED INACTIVE FUNDAMENTAL FREQUENCIES OF C_6H_6 AND C_6D_6 AFTER LORD AND ANDREWS (587).

Designation	C_6H_6 $\nu_{calculated}$ (cm^{-1})	C_6D_6 $\nu_{calculated}$ (cm^{-1})	Designation	C_6H_6 $\nu_{calculated}$ (cm^{-1})	C_6D_6 $\nu_{calculated}$ (cm^{-1})
$\nu_3^{H\parallel}(a_{2g})$	1190	930	$\nu_9^{CC}(b_{2u})$	1854	1844
$\nu_5^{CH}(b_{1u})$	3063	2294	$\nu_{10}^{H\parallel}(b_{2u})$	1145	816
$\nu_6^{C\parallel}(b_{1u})$	1008[288]	960[288]	$\nu_{19}^{H\perp}(e_{2u})$	1160	1070
$\nu_7^{H\perp}(b_{2g})$	1520	1480	$\nu_{20}^{C\perp}(e_{2u})$	404 observed	314
$\nu_8^{C\perp}(b_{2g})$	538	395			

$\boldsymbol{D_{6h}}$ model may be derived from a study of the Raman spectra of benzene derivatives [Kohlrausch (519)]. It is found that of the disubstitution products the 1, 4 substituted benzenes are decidedly less rich in lines than the 1, 2 and 1, 3 benzenes, and that of the tetrasubstituted products the 1, 2, 4, 5 benzenes are less rich in lines than the 1, 2, 3, 4, and the 1, 2, 3, 5 substituted benzenes. This is obviously because the 1, 4 and 1, 2, 4, 5 substituted benzenes are more symmetrical (point group $\boldsymbol{V_h}$) than the others, and therefore a number of Raman lines are forbidden for them, whereas for the others all Raman lines are allowed. However, such a difference in symmetry is possible only if C_6H_6 has the *plane regular hexagonal structure.*[287] From all these arguments we may take the $\boldsymbol{D_{6h}}$ model as proven.

The observed Raman and infrared spectra give us directly only eleven of the twenty fundamentals of benzene. Because of the large number of inactive fundamentals it would be almost hopeless to determine them from observed combination bands. However, Wilson (930), on the assumption of a valence force system, has derived formulae for all the frequencies in terms of six force constants (corresponding to the six types of motion in Table 121. Kohlrausch (517a) (519) and Lord and Andrews (587), have determined five of these force constants from five of the observed fundamentals. The sixth constant (the C—C—C⊥ bending constant) cannot be determined immediately, since the

necessarily occur if ν_2, ν_{18}, and ν_{13} have been identified correctly; there is no *ad hoc* assumption. Also Rumpf and Mecke do not seem to realize that even on their proposed $\boldsymbol{D_{3d}}$ model (H atoms alternately above and below the C_6 plane) only two C—H stretching vibrations are infrared active (as is also the case for the $\boldsymbol{D_{3h}}$ and $\boldsymbol{C_{6v}}$ models; see Table 121) so that in any case at least one of the three observed bands would have to be considered as a binary combination strengthened by resonance. Finally, Rumpf and Mecke seem to have overlooked the fact that for C_6D_6 only one band is observed in the region of the C—D stretching vibrations, a fact that lends strong support to the $\boldsymbol{D_{6h}}$ model.

[286] In each case the observed ratios of the products of the C_6D_6 frequencies of a given species to the products of the corresponding C_6H_6 frequencies are larger than the theoretical ones, as may be expected because of the influence of anharmonicity (see p. 232). The intensity ratio of corresponding lines of C_6H_6 and C_6D_6 is also in agreement with theoretical expectation, as has been shown by Lord and Teller (589).

[287] For the trisubstitution products the symmetrical one (1, 3, 5) is distinguished from the others also for the other benzene models ($\boldsymbol{D_{3h}}$, $\boldsymbol{D_{3d}}$, $\boldsymbol{C_{6v}}$).

[288] Ingold and co-workers give 766 and 730 cm^{-1}, respectively, for this vibration.

two C—C—C⊥ bending vibrations, ν_8 and ν_{20}, are both inactive. Lord and Andrews (587) have given various reasons for assigning the weak Raman line 404 cm^{-1} of liquid C_6H_6 to one of these vibrations, namely $\nu_{20}^{C\perp}(e_{2u})$; particularly because it cannot be assigned as any allowed difference band. This assignment has been confirmed by the agreement obtained with it between the calculated and observed heat capacity of solid C_6H_6 at low temperatures [Lord, Ahlberg, and Andrews (586)], as well as by work on the electronic spectrum [Sponer, Nordheim, Sklar, and Teller (800)]. It is then possible to determine the sixth potential constant and to predict all the other frequencies. Table 124 gives the values for the inactive frequencies thus obtained, according to Lord and Andrews (587). From the agreement of calculated and observed values (not shown here) for those active frequencies that are not used for the determination of the force constants, it may be concluded that the inactive frequencies are correct within ±50 cm^{-1}.

TABLE 125. FURTHER TWELVE-ATOMIC MOLECULES.

Molecule	Structure	References	
		Raman spectrum	Infrared spectrum
$B_3N_3H_6$	D_{3h}	(244)	(244)
$(PNCl_2)_3$	D_{3h} ?	(323)	
C_6H_5F, Cl, Br, I	C_{2v}	(14) (10) (519) (799)	(566) (101) (496a)
$C_6H_4Cl_2$, Br_2, FCl, etc.	C_{2v} or V_h	(14) (519) (688a) (522a) (799) (673b)	(568)
$C_6H_3Cl_3$, Br_3, Cl_2Br, etc.	D_{3h} or C_{2v} or C_s	(14) (10) (519)	(567)
$C_6H_2Cl_4$	V_h or C_{2v}	(259) (648)	
C_6HCl_5	C_{2v}	(259) (648)	
C_6Cl_6	D_{6h}	(259) (648)	
C_4H_8 (cyclobutane)	D_{4h}	(945b)	(945b)
$C_2H_4(NH_2)_2$	C_{2v} ?	(14) (10)	
$(CH_3CO)_2$ (diacetyl)	C_{2v} ?	(417) (525)	
$C_2H_5C_2H_3$ (butene 1)	C_s ?	(530) (173)	
cis, trans $CH_3C_2H_2CH_3$ (butene 2)	C_{2v} and C_{2h} ?	(355)	(355)
C_3H_5COOH (cyclopropane carbonic acid)	C_s ?	(528)	
$(CH_3)_2C{=}N{-}OH$ acetoxime	C_{2v} ?	(140)	
$CH_3(CH_2)_2OH$, $(CH_3)_2CH\cdot OH$	?	(14) (261) (774)	(67) (974a)
$CH_3(CH_2)_2CN$		(14)	

Recently Pitzer and Scott (699) have tried to identify some of the weaker bands in the Raman and infrared spectra of liquid C_6H_6 and C_6D_6 with the inactive fundamentals. Many of the values adopted by them differ by more than 100 cm^{-1} from the calculated values in Table 124. Therefore and since the same bands can also be interpreted as overtone and combination bands (see Table 12[illegible] and 123), the question of the exact values of the inactive fundamentals cannot be considered a[illegible] definitely settled.

A tentative assignment of some of the other infrared and Raman bands as overtone and combina[illegible] tion bands of the active and inactive fundamentals is indicated in Tables 122 and 123, essentiall[illegible] according to Ingold and co-workers. The calculated values of the inactive fundamentals ν_6, ν_{10}, $\nu_{1[illegible]}$ ν_3, ν_7, and ν_9 are confirmed by these observed overtone and combination bands. Further infrare[illegible] bands at shorter wave lengths have been given by Leberknight (561), Barnes and Brattain (117[illegible] Barchewitz (98), Battista (126), and in the photographic infrared by Rumpf and Mecke (751).

The weak Raman line 984 cm^{-1} accompanying the strong line 991.6 cm^{-1} was first interprete[illegible] by Gerlach (353) as due to $C_5{}^{12}C^{13}H_6$. This has been confirmed by Cheng, Hsueh, and Wu (20[illegible] and Rao (719), who found the intensity of the satellite to be 6 per cent of that of the main line, [illegible] is to be expected for the isotope line. Langseth and Lord (552) found the corresponding line f[illegible] C_6D_6 and gave a satisfactory explanation of the fact that the observed isotope shift is in both cas[illegible] somewhat larger than the calculated: In $C_5^{12}C^{13}H_6$ the vibration ν_6^{CH} has the same symmetry as ν_2^{CC} (both are totally symmetric), and therefore presses ν_2 downward, whereas in $C_6^{12}H_6$ they have different symmetry and do not influence each other. Langseth and Lord have also given an explanation of several further very weak satellites of 992 cm^{-1}.

The partly deuterated benzenes have been investigated by Redlich and Stricks (731b), Klit and Langseth (516), Langseth and Lord (553), and Ingold (467). As far as the data go at present they support the above interpretation of the C_6H_6, C_6D_6 spectra.

Other twelve-atomic molecules. The number of further twelve-atomic molecules whose infrared and Raman spectra have been investigated is very large. Table 125 gives references for some of the more important ones. The molecule $B_3N_3H_6$ (triborine-triamine) is particularly interesting since it is isoelectronic with benzene, and seems to have a very similar structure. The spectrum has been analyzed fairly completely. For few of the other molecules are very complete data and analyses available.

While we conclude our discussion of the vibration spectra of individual molecules with the twelve-atomic molecules it should be stressed that a very considerable amount of work has been done on more-than-twelve-atomic molecules. In many instances very interesting and definite conclusions about their structure have been derived even though frequently the situation is more complex than for the molecules treated here.

As one particularly striking example of the help given by the study of vibration spectra in the analysis of molecular structure the recent investigations of cyclohexane, C_6H_{12}, may be mentioned [Kohlrausch and Wittek (533a), Rasmussen (724a)]. It appears to follow rather definitely both from the Raman and the infrared spectrum that cyclohexane has the "chair" form of point group $\boldsymbol{D_{3d}}$ (puckered $\boldsymbol{C_6}$ ring) rather than a planar ($\boldsymbol{D_{6h}}$) or a "cradle" ($\boldsymbol{C_{2v}}$) form. This result confirms incidentally the greater stability of the staggered arrangement of the substituents about a C—C single bond indicated above for the di- and tetra-haloethanes (p. 351).

There is no question that further study of vibration spectra will greatly increase our knowledge of molecular structure.

CHAPTER IV

INTERACTION OF ROTATION AND VIBRATION, ROTATION-VIBRATION SPECTRA

Thus far we have considered rotation and vibration as independent motions of the molecule. Naturally, rotation can take place simultaneously with vibration and, as for diatomic molecules, this gives rise to the fine structure of infrared and Raman vibration bands. The investigation of this fine structure, whenever it is well resolved, leads to very accurate and reliable information about the structure of the particular molecules. From this fine structure can be determined moments of inertia, internuclear distances, and valence angles, in many cases with greater accuracy than by any other method.

1. Linear Molecules

(*a*) *Energy levels*

Elementary treatment. In a first rough approximation the energy of a linear molecule carrying out simultaneous rotation and vibration is simply the sum of the rotational and vibrational energy, as discussed in Chapters I and II:

$$T = G(v_1, v_2, \cdots) + F(J)$$

where G and F are given by the relations (II, 281) and (I, 1) respectively. Actually, as for diatomic molecules, the interaction of vibration and rotation manifests itself in that the *rotational constant* $B = h/8\pi^2cI$ occurring in $F(J)$ (see Chapter I, section 1) *has a slightly different value for the different vibrational levels.* This is because the moment of inertia I changes during a vibration in such a way that the average value of $1/I$ is not exactly the same as its value in the equilibrium position. In first approximation one would expect a formula to hold similar to that for diatomic molecules,

$$B_{v_1v_2v_3\cdots} = B_e - \alpha_1(v_1 + \tfrac{1}{2}) - \alpha_2(v_2 + \tfrac{1}{2}) - \cdots, \qquad \text{(IV, 1)}$$

where the α_i are small compared to B_e, which is the rotational constant for the equilibrium position, and where in higher approximation higher powers of $(v_i + \frac{1}{2})$ would occur.

Since the α_i for the two components of a degenerate vibration are the same, the corresponding terms in (IV, 1) can be taken together and we can write

$$B_{[v]} = B_e - \sum_i \alpha_i \left(v_i + \frac{d_i}{2} \right), \qquad \text{(IV, 2)}$$

where d_i is the degree of degeneracy of the vibration ν_i ($d_i = 1$ for non-degenerate, $d_i = 2$ for doubly degenerate vibrations; see p. 82) and $[v]$ stands for all the vibrational quantum numbers. The summation in (IV, 2) extends over all vibrations (the degenerate ones counted only once). The constants α_i are usually positive for the non-degenerate vibrations, as for diatomic molecules, but may be positive or negative for the degenerate vibrations since the average moment of inertia does not necessarily increase during a perpendicular vibration.

The interaction of vibration and rotation is also responsible for the introduction of the *rotational constant D* (see Chapter I, section 1), which represents the *influence of centrifugal stretching.* This influence, just as for diatomic molecules, will be slightly different for the different vibrational levels; that is, we have to use D_v instead of D in (I, 1) and expect a formula similar to (IV, 2) to hold for D_v. However, since D itself is always only a very small correction term, its dependence on v can usually be neglected, at least in all practical cases known thus far. In fact, only in very few cases of linear polyatomic molecules has even the influence of D been established.

According to the above, the *rotational energy of a linear molecule in a non-degenerate vibrational state* is given by

$$F_{[v]}(J) = B_{[v]}J(J+1) - D_{[v]}J^2(J+1)^2, \qquad \text{(IV, 3)}$$

where in future we shall usually neglect the term $D_{[v]}J^2(J+1)^2$.

If the molecule is in a *degenerate vibrational state* ($\Pi, \Delta, \cdots$) there is a vibrational angular momentum $l(h/2\pi)(l = 1, 2, \cdots)$ about the internuclear axis and, just as for diatomic molecules (see Molecular Spectra I, p. 122f.), the symmetric-top energy formulae have to be applied; that is, apart from an additive constant,

$$F_{[v]}(J) = B_{[v]}[J(J+1) - l^2] - D_{[v]}[J(J+1) - l^2]^2, \qquad \text{(IV, 4)}$$

where the second term is usually negligible, and where $B_{[v]}$ is given by (IV, 2). The additive constant is usually taken into the term $\sum g_{ii}l_i^2$ of the vibrational formula for this case. One may also take the term $-B_{[v]}l^2$ in (IV, 4), which is constant for a given vibrational level, into the vibrational formula and may therefore use (IV, 3) instead of (IV, 4) also for the degenerate vibrational levels. However, it must be realized that for the degenerate vibrational levels (just as for diatomic molecules) *J must be larger than or at least equal to l:*

$$J = l, l+1, l+2, \cdots; \qquad \text{(IV, 5)}$$

that is the levels $J = 0, 1, \cdots, l-1$ do not occur.

According to the preceding discussion, the *total energy of vibration and rotation of a linear molecule* is thus given by

$$\begin{aligned} T &= G(v_1, v_2, \cdots) + F_{[v]}(J) \\ &= \sum \omega_i \left(v_i + \frac{d_i}{2}\right) + \sum\sum x_{ik}\left(v_i + \frac{d_i}{2}\right)\left(v_k + \frac{d_k}{2}\right) + \sum_i g_{ii}l_i^2 \\ &\qquad + B_{[v]}J(J+1) - D_{[v]}J^2(J+1)^2. \qquad \text{(IV, 6)} \end{aligned}$$

For every vibrational state we have a set of rotational levels as in Fig. 4, p. 15, but with slightly different spacings in the different vibrational levels, and for degenerate levels with the first rotational levels (up to $l - 1$) missing.

It should be emphasized that in consequence of (IV, 2) the B value of the lowest vibrational level $B_{[0]}$ is different from B_e, the B value for the equilibrium position. For an accurate determination of the *equilibrium moment of inertia* and the *equilibrium internuclear distances,* B_e has to be used. Any moments of inertia or internuclear distances determined from $B_{[0]}$ are average values averaged in a certain way over the various positions assumed by the nuclei during the zero-point vibrations. The internuclear distances determined from electron or X-ray diffraction are, of course, also averages for the $v_1 = 0$, $v_2 = 0$, $\cdots$ state. However, the difference between the r_0 and r_e values is always very small, much smaller than the limit of error of electron diffraction measurements. While in principle the difference can easily be detected spectroscopically, only in a very few cases are all data (that is all α values) known for a determination of B_e from $B_{[0]}$.

Symmetry properties of rotational levels. As we have seen in Chapter I, section 1, the rotational levels of linear molecules are *positive* or *negative* depending on whether the total eigenfunction remains unchanged or changes sign for an *inversion.* In the lowest vibrational level (as in Chapter I) and in all totally symmetric excited vibrational levels (species Σ^+) of the electronic ground state, the even rotational levels are positive, the odd negative (see Fig. 4). This holds assuming that the electronic ground state is also totally symmetric. In Σ^- vibrational levels (just as for Σ^- electronic states of diatomic molecules), the even rotational levels are negative, the odd positive. In Π, Δ, $\cdots$ vibrational levels (as for Π, Δ, $\cdots$ electronic states of diatomic molecules), for each value of J there is a positive and a negative level of very slightly different energy (see below) whose order alternates thus: $+\ -$, $-\ +$, $+\ -$, $\cdots$ or $-\ +$, $+\ -$, $-\ +$, $\cdots$. This is shown in Fig. 99a.

In the case of linear molecules with a center of symmetry (point group $D_{\infty h}$, for example CO_2 and C_2H_2), the positive rotational levels are *symmetric*, the negative *antisymmetric* with respect to a *simultaneous exchange of all pairs of identical nuclei* for all vibrational levels that are symmetric with respect to an inversion (species Σ_g^+, Σ_g^-, Π_g, Δ_g, $\cdots$), while the reverse is the case for all vibrational levels that are antisymmetric with respect to an inversion (species Σ_u^+, Σ_u^-, Π_u, Δ_u, $\cdots$). Fig. 99b gives a few examples. All these relations are the same as for the various electronic states of diatomic molecules and the proof is quite analogous to that given in Molecular Spectra I (p. 260f.) if the electronic eigenfunctions considered there are replaced by the vibrational eigenfunctions. For diatomic molecules the vibrational eigenfunctions are always totally symmetric; here the electronic eigenfunction is assumed to be totally symmetric. While the latter is practically always true for the electronic ground state, it is not generally the case for excited electronic states, for which, therefore, different rules apply.

The *ratio of the statistical weights of the symmetric and antisymmetric rotational levels* is determined by the spins of the pairs of identical nuclei in the way discussed in Chapter I, section 1 (see Table 2, p. 18). As an important example it may be noted that, for Bose statistics of the identical nuclei, in a Σ_g^+ state the even and in a Σ_u^+ state the odd rotational levels have the greater statistical weight (and conversely for Fermi statistics).

More detailed theory of the rotational constants α_i; Coriolis interaction. In the case of diatomic molecules the variation α of the rotational constant B with the vibrational quantum numbers is, in a first approximation, due to two reasons (see Molecular Spectra I, p. 116): (1) the fact that even for *harmonic oscillations* the mean value of $1/I$ (which is, apart from a constant, the mean B) is not equal to $1/I_e$ although the mean value of r is r_e; and (2) the fact that in consequence of the *anharmonicity* of the vibrations (cubic and higher terms in the potential energy) the mean value of r during a vibration is larger than r_e. These same influences act also in the (linear) polyatomic molecule. But there is in addition a third reason, the presence of the *Coriolis interaction.*

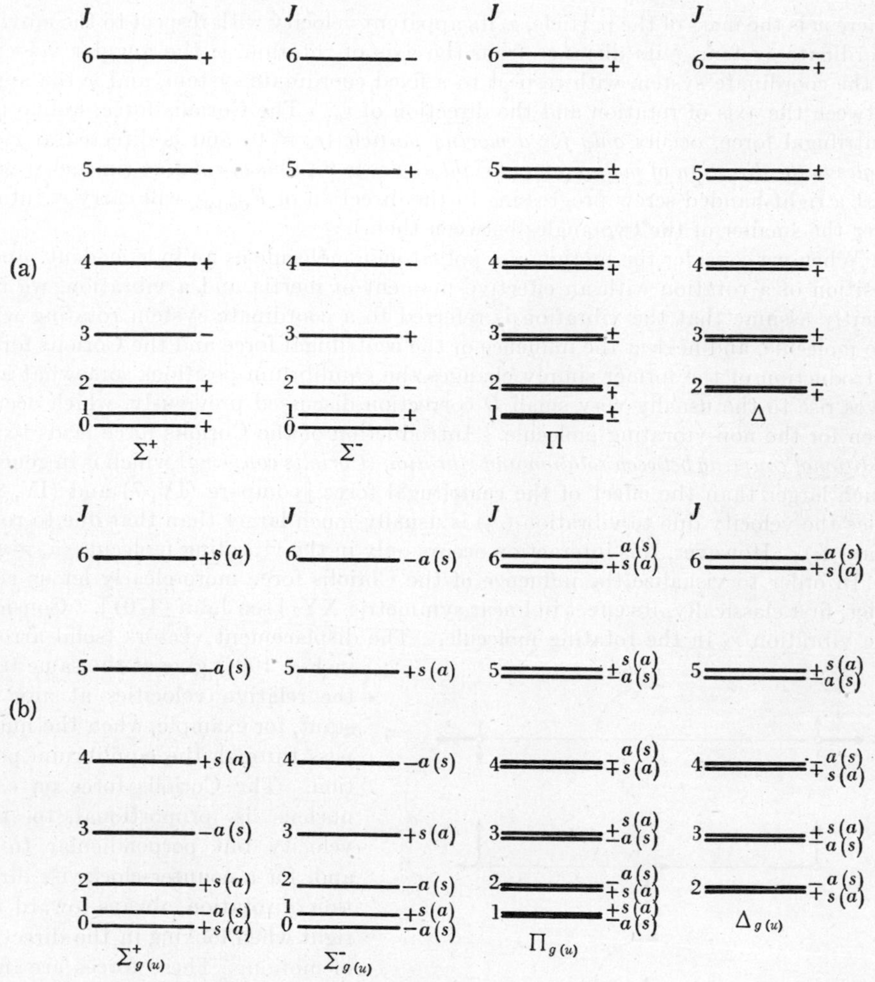

FIG. 99. **Symmetry properties of the rotational levels in various species of vibrational levels of linear molecules (a) of point group $C_{\infty v}$, (b) of point group $D_{\infty h}$.**—In (b) the symmetry properties for vibrational levels that are antisymmetric with respect to the center of symmetry (u) are added in brackets.

As is well known, if the motion of a particle is referred to a *uniformly rotating coordinate system,* apart from the acceleration produced by the acting forces, two additional accelerations appear, the centrifugal acceleration and the Coriolis acceleration, which may be thought of as due to two apparent forces, the *centrifugal force* and the *Coriolis force* respectively. The magnitude of the former force is given by

$$F_{\text{centrifugal}} = mr\omega^2, \tag{IV, 7}$$

the magnitude of the latter by

$$F_{\text{Coriolis}} = 2mv_a\omega \sin \varphi, \tag{IV, 8}$$

where m is the mass of the particle, v_a its apparent velocity with respect to the moving coordinate system, r its distance from the axis of rotation, ω the angular velocity of the coordinate system with respect to a fixed coordinate system, and φ the angle between the axis of rotation and the direction of v_a. The Coriolis force, unlike the centrifugal force, occurs *only for a moving particle* ($v_a \neq 0$) and is directed at *right angles to the direction of motion and at right angles to the axis of rotation* (in such a way that a right-handed screw progressing in the direction of F_{Coriolis} will carry v_a into ω over the smaller of the two angles between them).

When we consider the motion of a polyatomic molecule as an independent superposition of a rotation with an effective moment of inertia and a vibration, we implicitly assume that the vibration is referred to a coordinate system rotating with the molecule, and neglect the influence of the centrifugal force and the Coriolis force. Introduction of the former simply changes the equilibrium positions somewhat and gives rise to the usually very small D correction discussed previously, which occurs even for the non-vibrating molecule. Introduction of the Coriolis force leads to an *additional coupling between rotation and vibration* (*Coriolis coupling*) which is in general much larger than the effect of the centrifugal force [compare (IV, 7) and (IV, 8)] since the velocity due to vibration (v_a) is usually much larger than that due to rotation ($r\omega$). However, this interaction occurs only in the vibrating molecule ($v_a \neq 0$).

In order to visualize the influence of the Coriolis force more clearly let us consider, first classically, its effect in linear symmetric XY_2 [see Jahn (470)]. Consider the vibration ν_3 in the rotating molecule. The displacement vectors (solid arrows in Fig. 100c) give at the same time the relative velocities at any instant, for example, when the nuclei pass through the equilibrium position. The Coriolis force on each nucleus is proportional to this velocity but perpendicular to it, and, for a counter-clockwise direction of rotation, always toward the right when looking in the direction of motion. These forces are indicated by the broken-line arrows in Fig. 100. It is seen from this figure that during the vibration ν_3 these forces tend to excite the perpendicular vibration ν_2, but with the frequency of ν_3. Conversely, when the vibration ν_2 is excited in the rotating molecule the Coriolis forces are as given in Fig. 100b and tend to excite the parallel vibration ν_3, but with the frequency ν_2.

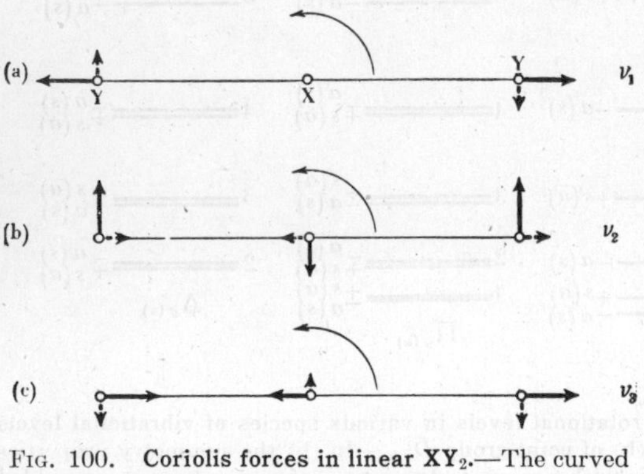

FIG. 100. **Coriolis forces in linear XY_2.**—The curved arrow indicates the direction of rotation.

If the frequencies of ν_2 and ν_3 were nearly the same, a strong excitation of one of the two vibrations would take place if the other were first excited, in consequence of this Coriolis coupling. However, this excitation will be very weak when, as is usually the case, ν_2 and ν_3 have widely different frequencies. In any case it will lead to the effect that, in the rotating coordinate system, the *nuclei will not move in straight lines* when, say, the vibration ν_3 is excited, *but in ellipses* which are the flatter the smaller the coupling, that is, the smaller the velocity of rotation or the larger the difference in frequency between ν_2 and ν_3. This motion is indicated in Fig. 101 for the three fundamental vibrations of linear symmetric XY_2. Since the ellipses for one particular vibration are traced out by each nucleus in the same sense, it is clear that an additional, *vibrational, angular momentum* arises which involves a change of energy. In the case of the vibration ν_1 the Coriolis force does not produce any coupling with another vibration; it produces only a coupling with the rotation (see Fig. 100a). It is the same effect that occurs also for diatomic molecules and produces a deviation of $[1/I]_{\text{average}}$ from $1/I_e$ even for strictly harmonic oscillations [(1), p. 372]. It is usually not considered as a Coriolis coupling.

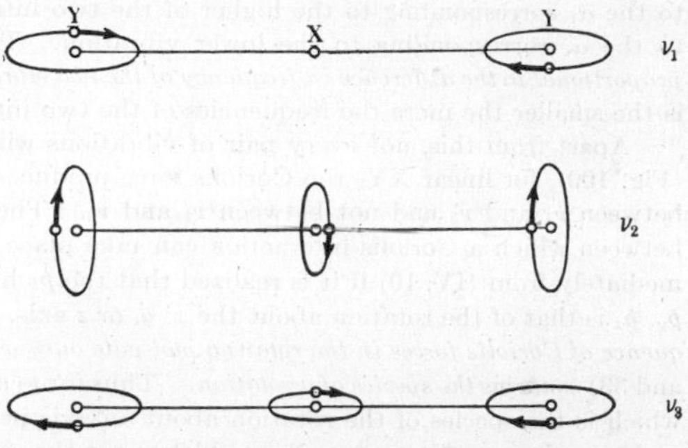

Fig. 101. Classical motion of nuclei in linear XY_2 on account of Coriolis interaction.—This is not an oblique, but a perpendicular projection. The heavy arrows indicate the instantaneous velocities of the nuclei referred to a coordinate system rotating with the molecule. The molecule is assumed to be rotating in a counter-clockwise direction about an axis perpendicular to the plane of the paper. The width of the ellipses representing the path of the nuclei is greatly exaggerated. With decreasing speed of rotation this width decreases to zero.

In order to take account of the interaction of vibration and rotation in the *wave mechanical treatment* of a polyatomic molecule, we have to use in the wave equation (II, 275) the *general Hamiltonian* (II, 276). The energy levels are obtained by carrying out a perturbation calculation treating the difference between the general Hamiltonian (II, 276) and the harmonic oscillator–rigid rotator Hamiltonian,

$$H = \frac{P_x^2}{2I_x} + \frac{P_y^2}{2I_y} + \frac{P_z^2}{2I_z} + \tfrac{1}{2}\sum p_k^2 + \tfrac{1}{2}\sum \lambda_k \xi_k^2, \tag{IV, 9}$$

as the perturbation function (see p. 208f.).

If in this perturbation calculation, the anharmonic terms in the potential energy and the dependence of the $\mu_{\alpha\beta}$ (that is, essentially of the moments of inertia) on the normal coordinates are neglected, the *influence of the Coriolis interaction* is obtained. The Hamiltonian is then (II, 279) with $V = \frac{1}{2}\sum \lambda_k \xi_k^2$. Its only difference from (IV, 9) consists in the replacement of P_x, P_y, P_z by $P_x - p_x$, $P_y - p_y$, $P_z - p_z$, where p_x, p_y, p_z are the components of the vibrational angular momentum operator. The latter quantities are defined by

$$p_x = \sum m_i(y_i\dot{z}_i - z_i\dot{y}_i), \quad p_y = \sum m_i(z_i\dot{x}_i - x_i\dot{z}_i), \quad p_z = \sum m_i(x_i\dot{y}_i - y_i\dot{x}_i) \tag{IV, 10}$$

where x_i, y_i, z_i are the displacement coordinates and $\dot{x}_i$, $\dot{y}_i$, $\dot{z}_i$ are the velocity components of nucleus i with respect to a coordinate system rotating with the molecule. In transforming to normal coordinates one obtains [see Wilson and Howard (944) and Jahn (470)]

$$p_x = \sum_{i<k} p_x^{(i,k)} = \sum_{i<k}{}^{(x)} \zeta_{ik}^{(x)}(\xi_i p_k - \xi_k p_i), \tag{IV, 11}$$

and similarly for p_y and p_z, where the $\zeta_{ik}^{(x)}$, $\zeta_{ik}^{(y)}$, $\zeta_{ik}^{(z)}$ are constants depending on the masses, the equilibrium internuclear distances and the force constants,[1] and where the p_k and p_i are linear momentum operators corresponding to the normal coordinates ξ_k and ξ_i [see equation (II, 277)].

[1] For explicit formulae for the $\zeta_{ik}^{(x)}, \cdots$ see Jahn (468) who calls them $C_j^{(rs)}/\mu_s$.

In agreement with the preceding classical treatment, a vibrational angular momentum arises from the Coriolis interaction of two normal vibrations. The total vibrational angular momentum is the sum of contributions from each pair of interacting vibrations, as shown by equation (IV, 11). As always, introduction of this perturbation into the wave equation leads to a mutual *repulsion* of the two "original" vibrational levels which increases, in the present case quadratically, with increasing rotational quantum number J. In other words *the higher of the two vibrational levels has a larger, the lower a smaller B value, than without this interaction*; that is, there is a negative contribution to the α_i corresponding to the higher of the two interacting vibrations and a positive contribution to the α_i corresponding to the lower vibration. The magnitude of this contribution is *inversely proportional to the difference in frequency of the two vibrations*, since the vibrational angular momentum is the smaller the more the frequencies of the two interacting vibrations differ (see above).

Apart from this, not every pair of vibrations will give a contribution. As we have seen above (Fig. 100), for linear XY_2 the Coriolis force produces an interaction only between ν_2 and ν_3 but not between ν_1 and ν_3 and not between ν_1 and ν_2. The general rule specifying the vibrational states between which a Coriolis interaction can take place has been given by Jahn (470). It follows immediately from (IV, 10) if it is realized that the p_k have the species of ξ_k and that the species of p_x, p_y, p_z is that of the rotation about the x, y, or z axis. Therefore, *two vibrations will interact in consequence of Coriolis forces in the rotating molecule only when the "product" of their species* (see Tables 31 and 33) *contains the species of a rotation.* Thus for ν_2 and ν_3 of linear symmetric XY_2, $\pi_u \times \sigma_u^+ = \Pi_g$, which is the species of the rotation about an axis perpendicular to the internuclear axis; but for ν_1 and ν_3 we have $\sigma_g^+ \times \sigma_u^+ = \Sigma_u^+$, which is not the species of any rotation (see Table 24). Similarly one finds that for linear symmetric X_2Y_2 only the pairs of vibrations ν_1 and ν_4, ν_2 and ν_4, ν_3 and ν_5, interact (see Fig. 64), but none of the other pairs.

If in the general Hamiltonian (II, 276) the vibrational angular momenta p_x, p_y, p_z, as well as the anharmonic terms of the potential energy, are neglected, but not the dependence of the moments of inertia on the normal coordinates, one obtains the "harmonic" contribution to the α_i mentioned under (1), p. 372. If the anharmonicity is then introduced but the vibrational angular momenta are still neglected one obtains an additional contribution to the α_i which in general, as for diatomic molecules, is the largest of the three contributions.

Summarizing, we can thus write, for the rotational constants α_i which represent the main influence of the coupling of rotation and vibration [see Herzberg (434)],

$$\alpha_i = \alpha_i^{(\text{harm.})} + \alpha_i^{(\text{anh.})} + \alpha_i^{(\text{Cor.})} \tag{IV, 12}$$

where the superscripts refer to the three contributions discussed above. By actually carrying out the perturbation calculation the α_i can be expressed in terms of the potential constants and the internuclear distances. For linear symmetric XY_2 molecules such formulae have been given by Dennison (280)[1a] [correcting earlier results of Adel and Dennison (38)], for linear XYZ molecules by Nielsen (654a), and for linear symmetric X_2Y_2 molecules by Shaffer and Nielsen (779) [correcting earlier results of Wu (961)]. For the general case of linear molecules (as well as non-linear molecules) detailed formulae have been given by Nielsen (666) [see also Sayvetz (762b)]. We give here only the formulae for the linear XY_2 molecule:

$$\begin{aligned}
\alpha_1^{(\text{harm.})} &= -\frac{6B_e^2}{\omega_1}, & \alpha_1^{(\text{anh.})} &= -\frac{24B_eI_e^{\frac{3}{2}}}{\omega_1^3}\frac{\alpha_{111}}{hc}, & \alpha_1^{(\text{Cor.})} &= 0;\\
\alpha_2^{(\text{harm.})} &= +\frac{B_e^2}{\omega_2}, & \alpha_2^{(\text{anh.})} &= -\frac{8B_e^3I_e^{\frac{3}{2}}}{\omega_1^2\omega_2}\frac{\alpha_{122}}{hc}, & \alpha_2^{(\text{Cor.})} &= \frac{4B_e^3\omega_2}{(\omega_3^2-\omega_2^2)}\\
\alpha_3^{(\text{harm.})} &= +\frac{2B_e^2}{\omega_3}, & \alpha_3^{(\text{anh.})} &= -\frac{8B_e^3I_e^{\frac{3}{2}}}{\omega_1^2\omega_3}\frac{\alpha_{133}}{hc}, & \alpha_3^{(\text{Cor.})} &= -\frac{8B_e^2\omega_3}{(\omega_3^2-\omega_2^2)}.
\end{aligned} \tag{IV, 13}$$

[1a] It should be noted that Dennison's α_i are our α_i/B_e.

In these equations I_e is the moment of inertia in the equilibrium position and α_{111}, α_{122}, and α_{133} are the anharmonic potential constants in (II, 263).[2] As explained above, the contribution of the Coriolis interaction to α_1 is zero. The other two contributions to α_1 agree exactly with those for diatomic molecules (see Molecular Spectra I, p. 116). As expected, the Coriolis contributions to α_2 and α_3 are opposite in sign, and depend on the difference of the frequencies ω_2 and ω_3. By using the above formulae, if B_e and the ω_i are known, the anharmonic potential constants can be determined from the observed α values (see also below).

It should be mentioned that according to Nielsen and Shaffer [(668) and private communication] the terms here called $\alpha_2^{(\text{harm.})}$ and $\alpha_3^{(\text{harm.})}$ are also due to Coriolis interaction; that is, for linear XY_2 the harmonic contributions to the α_2 and α_3 are zero. However, the terms here called $\alpha_2^{(\text{harm.})}$ and $\alpha_3^{(\text{harm.})}$ are at least formally of a different type from $\alpha_2^{(\text{Cor.})}$ and $\alpha_3^{(\text{Cor.})}$ in that they do not depend on the separation of ω_2 and ω_3. It appears that there is room for divergence of opinion as to the correct interpretation of the various terms even though there is no disagreement about the correct formulae for the α_i.

l-type doubling. As we have seen previously, the Π, Δ, $\cdots$ vibrational levels of linear polyatomic molecules are doubly degenerate on account of the equivalence of the two directions of the angular momentum $\boldsymbol{l}$. Just as for the Π, Δ, $\cdots$ electronic states of diatomic molecules, this double degeneracy is removed with increasing rotation; that is, *a splitting into two components for each J occurs whose separation increases with increasing J.* This is indicated, greatly exaggerated, in Fig. 99.

Qualitatively there are two reasons for this l-type doubling [see Herzberg (434)].

(1) As we have seen previously, when a $\perp$ vibration (species Π) is singly excited in a linear molecule we may take as the two component motions either (a) the vibrations in two planes at right angles to each other or (b) the rotational vibrations about the symmetry axis in the clockwise and counter-clockwise directions (see Fig. 27a) with angular momentum $l = \pm 1$. If in the former scheme the molecule is rotating, for a vibration in a plane ($\sigma_v^{||}$) parallel to the axis of rotation no change of moment of inertia of the molecule will take place as long as the vibrations are harmonic, since the nuclei move parallel to the axis of rotation. But for a vibration in a plane ($\sigma_v^{\perp}$) perpendicular to the axis of rotation the moment of inertia about the axis will change, since it is the sum of the original moment of inertia and that about the symmetry axis of the molecule (which is not zero in a displaced position). Thus we expect slightly different effective B values for the two component vibrations. If we use scheme (b) we have, when the nuclei are swinging around the symmetry axis, essentially the same situation as in a molecule with slightly bent equilibrium position; that is, we have a slightly asymmetric top for which the double degeneracy of levels with $K \neq 0$ of the corresponding symmetric top is removed, the splitting increasing with J (see Fig. 18). In the present case K is identical with l. Thus on either scheme (a) or (b), a doubling is expected on account of the fact that *in the displaced position the molecule is a slightly asymmetric top.*

(2) Since the *Coriolis force*, according to (IV, 8), depends on the sine of the angle between the axis of rotation and the velocity v_a of the particle (in the rotating coordinate system), it is clear that the component vibration in the plane $\sigma_v^{||}$ is unaffected by it, and only the component vibration in $\sigma_v^{\perp}$ is affected (see also Fig. 100). Since the Coriolis force is proportional to the speed of rotation we expect a splitting that increases with increasing J.

On the basis of the above considerations and the close analogy to the Λ-type doubling of diatomic molecules, we conclude that *the l-type splitting* in a vibrational

[2] Dennison gives his formulae in terms of potential constants which are referred to certain dimensionless normal coordinates σ_i which are related to the η_i in (II, 263) by $\sigma_i = 2\pi\sqrt{\omega_i c/h}\,\eta_i$. Here it must be remembered that the η_i in (II, 263) are normalized to unity.

Π state is given by

$$\Delta\nu = qJ(J + 1), \qquad \text{(IV, 14)}$$

where q is of the order $\alpha_i^{(\text{harm.})} + \alpha_i^{(\text{Cor.})}$ the subscripts i referring to the ⊥ vibration that is excited in the Π state.

In a more formal way the l-type doubling may also be explained as a special case of the perturbations to be discussed below. Since only levels of equal parity (+, −) and equal J influence one another, a Σ^+ vibrational state will affect only one set (Π^+) of component levels of a Π state, namely that for which the even levels are + and the odd −, while the other set (Π^-) is uninfluenced (see Fig. 99). If, as usual, the Σ^+ state is above the Π state ($\omega_3 > \omega_2$ for XY_2), the Π^+ component is shifted down (increasingly so with increasing J) while the Π^- component is unchanged.

A more detailed theory of the l-type doubling has been given by Nielsen and Shaffer (668). They consider the whole effect as due to Coriolis interaction. However, their resultant formula for q shows no dependence on the frequency difference between the two interacting vibrations, as one would expect for Coriolis interaction. Also, in the few cases in which experimental data are available [see Herzberg (434)], Nielsen and Shaffer's formulae gives q values about twice the observed values. It therefore appears not definitely decided whether there is also a contribution due to the asymmetry of the displaced position.

Perturbations. The interaction of rotation and vibration, which causes the above systematic changes of the energy levels, may also cause somewhat more *irregular changes—perturbations*—similar to the perturbations found in diatomic molecules, where, however, they can only be produced by the interaction of rotation and electronic motion. Just as for diatomic molecules, these perturbations are always produced by the *mutual interaction of two states of nearly the same energy, of the same J value, and of the same parity* (+, −) *and symmetry with respect to an exchange of identical nuclei* [see Molecular Spectra I, p. 313, and Kronig (542)]. But while for diatomic molecules these two states always belong to two different electronic states, here they may belong to the same electronic state (ground state) but different *vibrational* states. We may subdivide the perturbations according to their *appearance* into *vibrational and rotational perturbations* (just as for diatomic molecules), and according to their *cause* into *Fermi perturbations* and *Coriolis perturbations* [or *homogeneous* and *heterogeneous* perturbations; Mulliken (642)].

A *vibrational perturbation* consists of a shift of a vibrational level from its normal position accompanied by a *change of the effective B value* from its normal value; or the latter effect alone may occur. The shift of a vibrational level even without rotation, as we have seen previously, is always due to Fermi resonance, that is, to an interaction with another vibrational level which lies close to the one considered, and has the *same species*. If the molecule is rotating this Fermi interaction will in general also produce a change of the effective B values of the two interacting levels. According to Adel and Dennison (38) we have, for the *actual B values of two resonating levels n* and *i*,

$$B_n = a^2B_n^0 + b^2B_i^0,$$
$$B_i = b^2B_n^0 + a^2B_i^0, \qquad \text{(IV, 15)}$$

where B_n^0 and B_i^0 are the unperturbed B values [as obtained from (IV, 2)] and where a and b are the fractional contributions of the unperturbed to the perturbed vibrational eigenfunctions according to the previous formula (II, 293). Since $a^2 + b^2 = 1$, it follows that

$$B_n + B_i = B_n^0 + B_i^0. \qquad \text{(IV, 16)}$$

The sum of the B values is unchanged by the perturbation.

A second type of vibrational perturbation can be produced by Coriolis interaction. If two vibrational levels of different species lie close together they cannot interact in the rotationless state; but there will be an increasing repulsion of the rotational levels of the same J with increasing J if the two vibrational levels can have Coriolis interaction (see p. 376); that is, the effective B values will be changed as compared to the unperturbed values even though there is no shift of the pure vibrational levels. For example, such a perturbation could occur between a Σ_u^+ and a Π_u vibrational level, or between Σ_g^+ and Π_g, and so on. However, it must be realized that this interaction will affect only one component of a Π state if the interaction is with a Σ state; that is, the Π state will show an anomalously large (or small) l-type doubling. Such an anomalously large l-type doubling has been found (but not explained) by Funke and Lindholm (342) for the upper state $2\nu_1 + \nu_3 + \nu_4$ of the C_2H_2 band at 10,413 cm^{-1}, for which $q = 0.0084$ cm^{-1}, whereas for all other observed states in which the vibration $\nu_4(\Pi_g)$ is singly excited q is between 0.0059 and 0.0067 cm^{-1}. The perturbing state in this case is probably $\nu_1 + 2\nu_2 + \nu_3(\Sigma_u^+)$. Certain observed irregularities in the B values of C_2H_2 are probably also due to Coriolis interaction.

Both causes, Fermi interaction and Coriolis interaction, may also lead to typically *rotational perturbations.* If the Fermi interaction between two states of the same species is very slight and yet the two states are very close together (which may happen for the higher vibrational levels), and if at the same time the B values are such that the unperturbed term curves plotted as functions of J intersect each other (see Fig. 124, p. 312 of Molecular Spectra I), then only the levels in the neighborhood of this point of intersection will be perturbed, and we have a typical rotational perturbation.

Similarly, when the two vibrational states have different species but can have Coriolis interaction ($\Delta l = \pm 1$), near the point of intersection of the term curves this Coriolis interaction will produce a shift of the energy levels from the normal positions, that is, a rotational perturbation. The difference from the preceding case is, however, that since the Coriolis interaction increases with increasing J the shift from the normal position does not go back to zero for large J even though the difference in energy between levels of the same J is then fairly large.

Both types of rotational perturbations have been observed by Funke (340) in the rotation-vibration spectrum of C_2H_2. Just as for diatomic molecules, extra lines appear in the perturbation region because of the mixing of the eigenfunctions (see Molecular Spectra I, p. 318).

(*b*) *Infrared spectrum*

Selection rules. As in the diatomic case, in a good approximation, the selection rules for the pure vibration spectrum and for the pure rotation spectrum are not changed by the interaction of vibration and rotation (for a proof see section 2b). Thus also for the rotation-vibration spectrum in the infrared only those *vibrational* transitions occur (see Table 55) for which M_z has the species Σ_u^+ or M_x and M_y have the species Π_u (where the subscript u is to be dropped for point group $C_{\infty v}$), that is,

$$\Delta l = 0, \pm 1, \qquad \Sigma^+ \nleftrightarrow \Sigma^-, \qquad g \nleftrightarrow g, \qquad u \nleftrightarrow u, \qquad \text{(IV, 17)}$$

and only those *rotational* transitions for which

$$\Delta J = 0, \pm 1 \qquad (J = 0 \nleftrightarrow J = 0); \qquad + \leftrightarrow -; \qquad s \nleftrightarrow a. \qquad \text{(IV, 18)}$$

However, while in the rotation spectrum only $\Delta J = +1$ is significant, here $\Delta J = 0$, and -1 are also significant. But, as for diatomic molecules, $\Delta J = 0$ is forbidden when $l = 0$ for both upper and lower state (Σ—Σ transitions).

Since for all known linear polyatomic molecules the *electronic* ground state is $^1\Sigma$ we need not discuss the influence of a possible electronic angular momentum $\mathbf{\Lambda}$ on the rotation-vibration spectrum. Its place is taken by the vibrational angular momentum $\boldsymbol{l}$, and therefore the *structure of the infrared bands of linear polyatomic molecules is similar in all respects to that of the corresponding electronic bands of diatomic molecules.*

It may be noted that the rotational selection rules (IV, 18) do not restrict the vibrational transitions any further than the vibrational selection rules (IV, 17). On the other hand, transitions that are forbidden according to the vibrational selection rules $\Sigma^+ \nleftrightarrow \Sigma^-$, $g \nleftrightarrow g$, $u \nleftrightarrow u$ also lead to conflicts with the rotational selection rules. Even when the interaction of vibration and rotation is very large the rotational selection rules hold rigorously for dipole radiation, and since in the infrared the intensity of quadrupole radiation is 10^{-9} of the intensity of dipole radiation, we cannot expect to observe any forbidden transitions of the type $g \leftrightarrow g$ or $u \leftrightarrow u$ in absorption in the gaseous state. However, for extremely large interaction of vibration and rotation, transitions violating the rule $\Delta l = 0, \pm 1$ (for example, $\Delta \leftrightarrow \Sigma$) or the rule $\Sigma^+ \nleftrightarrow \Sigma^-$ may occur weakly. But no such case has as yet been observed.

Types of infrared bands. According to the above selection rules the following three types of infrared bands of linear molecules can occur.

(1) *Transitions for which $l = 0$ in both the upper and lower state* (|| bands, Σ—Σ transitions). For these bands only $\Delta J = \pm 1$ can occur; that is, they have only one P and one R branch but no Q branch. They are of exactly the same type as the rotation-vibration bands of diatomic molecules with $^1\Sigma$ ground state, or as $^1\Sigma$—$^1\Sigma$ electronic bands of diatomic molecules (compare Molecular Spectra I).

(2) *Transitions for which $\Delta l = \pm 1$* ($\perp$ bands, Π—Σ, Δ—Π, $\cdots$ transitions). For these bands $\Delta J = 0$ as well as $\Delta J = \pm 1$ is possible, and therefore in addition to a P and an R branch a Q branch appears which is, in fact, stronger than either the P or R branch. There is no analogue of these bands in the rotation-vibration spectra of diatomic molecules, since for the latter the angular momentum about the internuclear axis cannot change. However, these $\perp$ bands are of exactly the same type as $^1\Pi$—$^1\Sigma$, $^1\Delta$—$^1\Sigma$, $\cdots$ electronic bands of diatomic molecules.

(3) *Transitions for which $\Delta l = 0$ but $l \neq 0$.* (|| bands, Π—Π, Δ—Δ, $\cdots$ transitions). For these bands, also, $\Delta J = 0, \pm 1$ is possible and therefore they also have a Q as well as a P and an R branch. Here, however, since only $M_z \neq 0$, the *Q branch is weak*, the intensity decreasing from the first line instead of reaching first a maximum as for the P and R branches. These bands are exactly analogous to the rotation-vibration bands of diatomic molecules with a Π (or Δ) ground state and to $^1\Pi$—$^1\Pi$, $^1\Delta$—$^1\Delta$, $\cdots$ electronic bands.

For any linear polyatomic molecule all three types of bands occur.

Σ—Σ bands [type (1)]. In the case of unsymmetrical linear molecules (point group $C_{\infty v}$), vibrational bands Σ^+—Σ^+ and Σ^-—Σ^- will have a structure of type (1); in the case of symmetrical linear molecules (point group $D_{\infty h}$) the vibrational bands

$\Sigma_u^+ - \Sigma_g^+$, $\Sigma_g^+ - \Sigma_u^+$, $\Sigma_u^- - \Sigma_g^-$, $\Sigma_g^- - \Sigma_u^-$ (where the upper state is written first) will have this structure. Of these the most important ones and the only ones thus far observed are the $\Sigma^+ - \Sigma^+$ and $\Sigma_u^+ - \Sigma_g^+$ bands respectively. For $C_{\infty v}$ molecules the strong 1—0, 2—0, · · · transitions of the || vibrations, for $D_{\infty h}$ molecules the 1—0, 3—0, · · · transitions of the antisymmetric || vibrations are of the $\Sigma^+ - \Sigma^+$ type. In addition, certain combination bands will have this type. Fig. 102 shows in an energy-level diagram the *possible rotational transitions* for $\Sigma_u^+ - \Sigma_g^+$ and the resulting spectrum. This diagram also represents $\Sigma^+ - \Sigma^+$ transitions for $C_{\infty v}$ molecules if the difference between *a* and *s* and between heavy and light lines is disregarded.

As for diatomic molecules, the *formulae for the two branches R* and *P* corresponding to $\Delta J = +1$ and -1 respectively are

$$R(J) = \nu_0 + 2B' + (3B' - B'')J + (B' - B'')J^2, \quad \text{(IV, 19)}$$

$$P(J) = \nu_0 - (B' + B'')J + (B' - B'')J^2, \quad \text{(IV, 20)}$$

where, as usual, J is the rotational quantum number of the *lower* state ($\equiv J''$) and ν_0 the band origin. Higher terms in J^3 and J^4 have to be added if the D correction is not negligibly small (see Molecular Spectra I, p. 191). The two branches can also be represented by the single formula

$$\nu = \nu_0 + (B' + B'')m + (B' - B'')m^2, \quad \text{(IV, 21)}$$

where $m = J + 1$ for the R branch and $m = -J$ for the P branch. Thus we may also say that we have *a single series of lines in which one line is missing* (*zero gap*), this missing line separating the two branches. For the fundamentals, $B' - B''$ is very small and therefore the lines are almost equidistant.

As examples, observed spectrograms of N_2O, HCN, CO_2, and C_2H_2 are reproduced in Figs. 103, 104, 105, and 106. It is seen that in each branch the intensity first in-

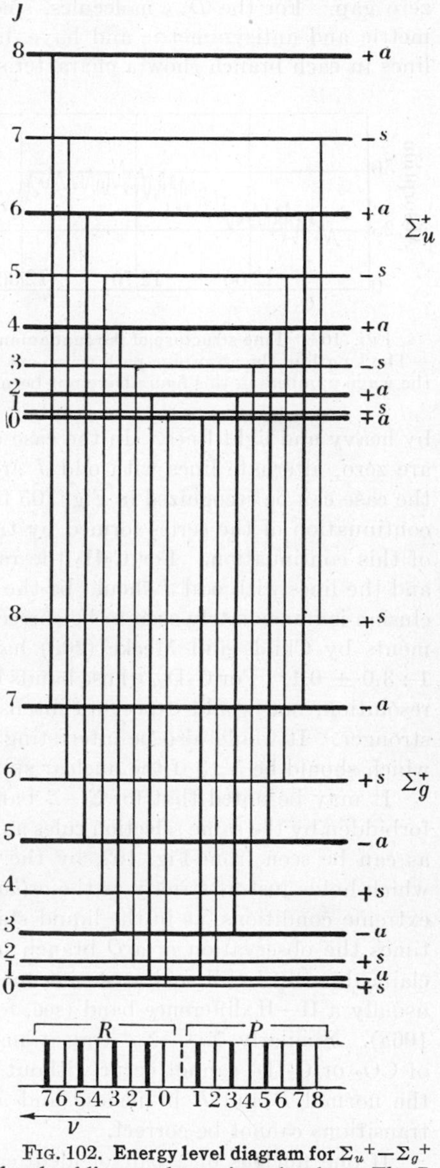

Fig. 102. Energy level diagram for $\Sigma_u^+ - \Sigma_g^+$ bands of linear polyatomic molecules.—The B values in the upper and lower state have been assumed to be the same. At the bottom the resulting spectrum is shown schematically. The intensity alternation indicated refers to molecules like C_2H_2 with Fermi statistics as "resultant" statistics of the nuclei (see p. 17).

creases to a maximum and then decreases, in agreement with the thermal distribution of the rotational levels discussed in Chapter I. Also, Fig. 106b shows that at lower temperature the intensity maxima in the *P* and *R* branches are closer to the zero gap. For the $\boldsymbol{D_{\infty h}}$ molecules, since the rotational levels are alternately symmetric and antisymmetric and have different statistical weights (see Table 2), the lines in each branch show a characteristic *intensity alternation* indicated in Fig. 102

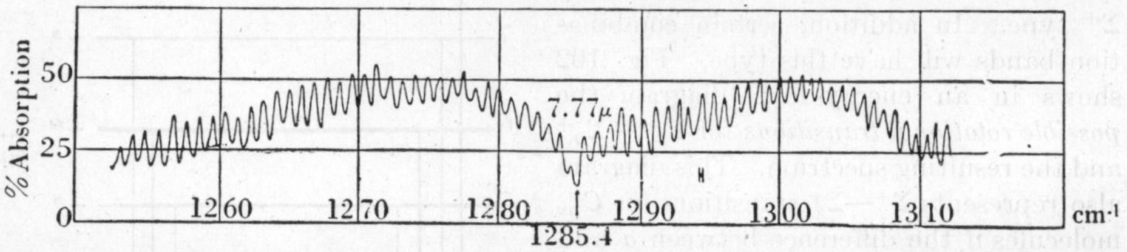

FIG. 103. **Fine structure of the fundamental ν_1 of N_2O at 7.77μ** [after Plyler and Barker (703)]. —The length of the absorbing path was 6 cm. at a pressure of 10 cm. Unlike the data in Table 58 the wave numbers in this figure have not been converted to vacuum.

by heavy and light lines. In the case of CO_2, since the spins of the identical nuclei are zero, alternate lines with odd *J* are completely missing. That this is actually the case can be recognized in Fig. 105 from the fact that the *R* lines do not form the continuation of the series formed by the *P* lines but lie half way between the lines of this continuation. For C_2H_2 the ratio of the statistical weights should be 1 : 3 and the lines with odd *J* should be the more intense (see Table 2). The latter conclusion is immediately seen to be correct from Fig. 106. Careful intensity measurements by Childs and Mecke (209) have indeed shown that the intensity ratio is 1 : 3.0 ± 0.1. For C_2D_2, whose bands have not as yet been observed under sufficient resolution, one would expect an intensity alternation 2 : 1 with the even lines the stronger. It would also be interesting to study the intensity alternation in $C_2{}^{13}H_2$, which should be 5 : 3 if the nuclear spin of C^{13} is $\frac{1}{2}$ (see Table 2).

It may be noted that for Σ—Σ bands of $\boldsymbol{D_{\infty h}}$ molecules a *Q* branch is not only forbidden by the same selection rules as for $\boldsymbol{C_{\infty v}}$ molecules (see above) but in addition, as can be seen from Fig. 102, by the very strong rule *symmetric* ↔ *antisymmetric*, which holds just as strictly as the *ortho* ↔ *para* rule for H_2. Therefore, even under extreme conditions, as in the liquid state, this *Q* branch cannot occur. At various times the observation of a *Q* branch in a Σ—Σ band of a $\boldsymbol{D_{\infty h}}$ molecule has been claimed. But in all such cases it can be explained as due to an overlapping band, usually a Π—Π difference band (see, for example, the line near the zero gap in Fig. 106a). Similarly, $\Sigma_g{}^+$—$\Sigma_g{}^+$ transitions (for example the first or third overtone of ν_3 of CO_2 or C_2H_2) cannot occur without violation of the $s \leftrightarrow a$ rule, at least not with the normal *P* and *R* branches,[3] and all interpretations of infrared bands as such transitions cannot be correct.

If one nucleus of a pair of identical nuclei in a linear molecule of point group $\boldsymbol{D_{\infty h}}$ is replaced by an *isotope*, that is, if the two nuclei are no longer exactly identical, the reason for the distinction between symmetric and antisymmetric rotational levels disappears, and *no intensity alternation* should occur. This has indeed been found

[3] The *Q* branch could occur as quadrupole radiation or enforced dipole radiation.

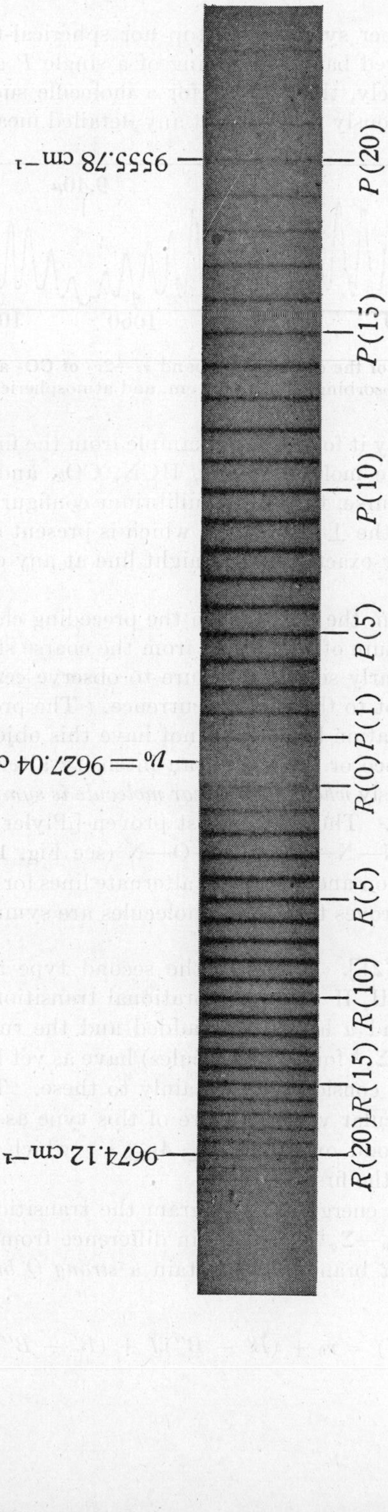

FIG. 104. **Fine structure of the second overtone $3\nu_3$ of HCN in the photographic infrared at 1.038μ [after Herzberg and Spinks (442)].**—The length of the absorbing path was 600 cm at a pressure of 190 mm.

for C_2HD [Herzberg, Patat, and Spinks (437)], as shown by the spectrogram in Fig. 107.

As we shall see, neither symmetrical-top nor spherical-top nor asymmetric-top molecules can give infrared bands consisting of a single P and R branch separated by a zero gap. Conversely, therefore, if for a molecule such a simple structure is found it shows unambiguously and without any detailed measurements *that the molecule is linear.* In this way it follows, for example from the fine structures reproduced in Fig. 103–106, that the molecules N_2O, HCN, CO_2, and C_2H_2 are linear. (By "linear" we mean, of course, that the equilibrium configuration is linear. Due to the zero-point energy of the $\perp$ vibrations, which is present even in the lowest state, the nuclei are not usually exactly on a straight line at any one time but are so only on the average).

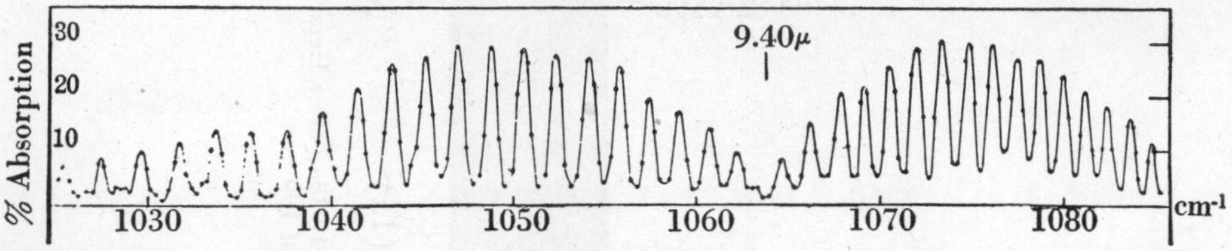

FIG. 105. **Fine structure of the combination band ν_3—$2\nu_2$ of CO_2 at 9.40μ** [after Barker and Adel (108)].—An absorbing path of 700 cm. and atmospheric pressure was used.

It will be recalled from the examples in the preceding chapter that it is not easy to prove the linear structure of a molecule from the coarse structure of the vibration spectrum alone, particularly since the failure to observe certain bands may be due to their weakness and not to their non-occurrence. The proof of linearity from the fine structure of the vibration bands does not have this objection.

Moreover, the presence or absence of an intensity alternation in such a simple band proves unambiguously *whether the linear molecule is symmetric* (point group $D_{\infty h}$) *or not* (point group $C_{\infty v}$). Thus it was first proven [Plyler and Barker (703)] that N_2O has the structure N—N—O not N—O—N (see Fig. 103). Similarly, the observed intensity alternation and missing of alternate lines for C_2H_2 (Fig. 106) and CO_2 (Fig. 105) respectively proves that these molecules are symmetrical.

Π—Σ bands [type (2)]. Bands of the second type are the Π—Σ^+, Π—Σ^-, Δ—Π $\cdots$, Σ^+—Π, Σ^-—Π, Π—Δ, $\cdots$ vibrational transitions, where for $D_{\infty h}$ molecules the subscripts g and u have to be added and the rule $g \leftrightarrow u$ applies. Only Π—Σ^+ transitions (Π_u—Σ_g^+ for $D_{\infty h}$ molecules) have as yet been completely resolved and we shall restrict our considerations mainly to these. The fundamentals of the infrared-active perpendicular vibrations are of this type as well as their even overtones $3\nu_\perp$, $5\nu_\perp$, $\cdots$ (the odd overtones $2\nu_\perp$, $4\nu_\perp$, $\cdots$ which are active only for $C_{\infty v}$ molecules give bands of the first type).

Fig. 108 shows in an energy-level diagram the transitions allowed according to the selection rules for Π_u—Σ_g^+. The main difference from the first type is that in addition to the P and R branches we obtain a *strong Q branch.* The formula for the Q branch is

$$Q(J) = \nu_0 + (B' - B'')J + (B' - B'')J^2. \qquad \text{(IV, 22)}$$

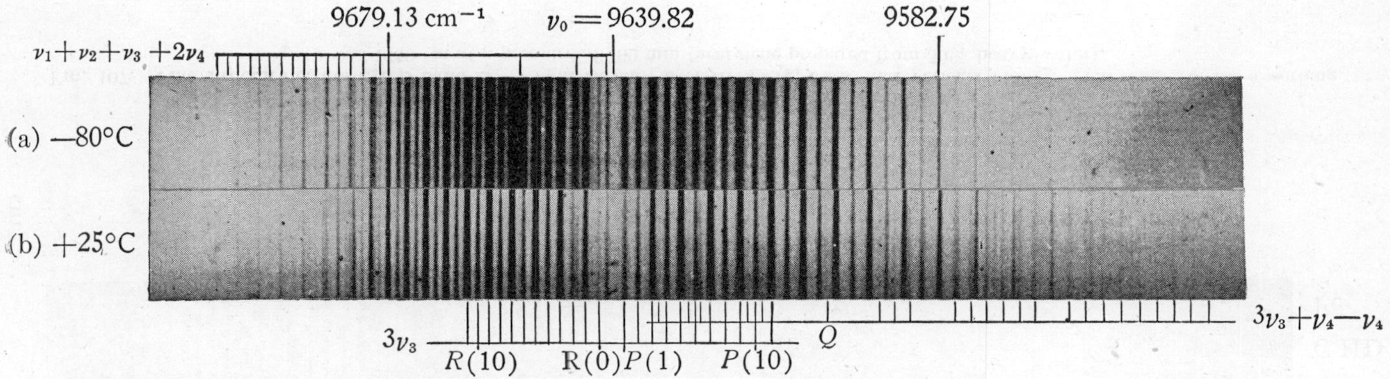

FIG. 106. **Fine structure of the second overtone $3\nu_3$ of C_2H_2 at 1.037μ as well as of the accompanying bands $3\nu_3 + \nu_4 - \nu_4$ (1.041μ) and $\nu_1 + \nu_2 + \nu_3 + 2\nu_4$ (1.034μ) at room temperature and at −80° C.**—Only the first lines of P and R branch of the main band are indicated by long leading lines. The short leading lines at the bottom refer to the band $3\nu_3 + \nu_4 - \nu_4$ (see p. 390), those at the top to the band $\nu_1 + \nu_2 + \nu_3 + 2\nu_4$. The length of the absorbing path was 400 cm at a pressure of 760 mm.

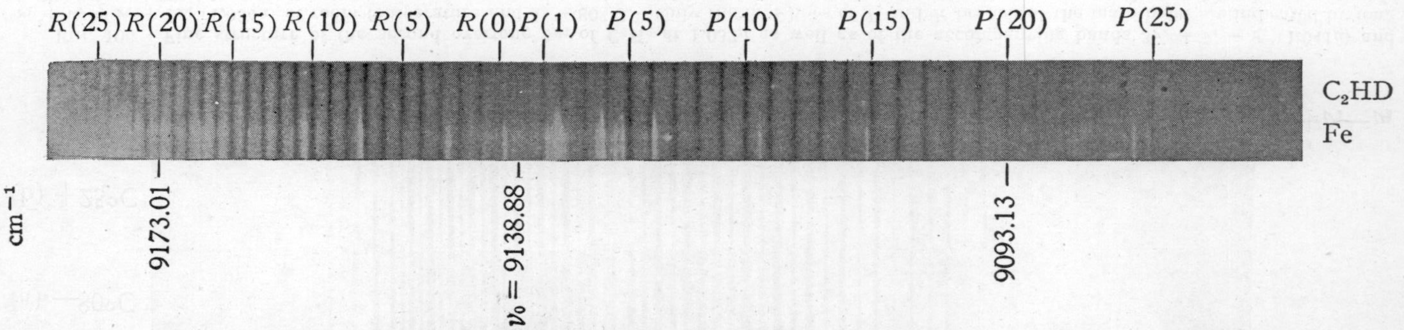

FIG. 107. **Fine structure of the C_2HD band $2\nu_1 + \nu_2$ at 1.095μ** [after Herzberg, Patat and Verleger (439)].—The length of the absorbing path was 1650 cm at a pressure of 760 mm (acetylene prepared from 50% heavy water).

For the fundamentals and low overtone and combination bands, $B' - B''$ is very small and therefore all lines of the Q branch fall practically together, giving rise to a *very strong "line" in the center of the band.* Fig. 109 and 110 give as examples the observed fine structures of the fundamentals ν_2 of HCN and ν_5 of C_2H_2. For higher combination bands the Q lines no longer coincide but form a separate branch. Such bands have been found in the photographic infrared spectrum of C_2H_2 [Funke and Herzberg (341), Funke (339) (340), Mecke and Ziegler (618), Funke and Lindholm (342)].

According to Fig. 108, in addition to the line $P(0)$ *the line* $P(1)$ *should likewise be missing.* This is, however, difficult to ascertain when the Q branch is not resolved, since it usually tails off toward longer wave lengths and overlaps the place where $P(1)$ would be if it occurred (see Fig. 109 and Fig. 110). For Σ—Π transitions, instead of $P(1)$ the line $R(0)$ will be missing near the band origin (compare the completely analogous cases of electronic bands of diatomic molecules, Molecular Spectra I, Chapter V, 3b).

Fig. 108 shows clearly that on account of the l-type doubling and the selection rule $+ \leftrightarrow -$ the *upper states of the Q lines are slightly different from those of the P and R lines.* For most purposes this difference can be neglected since the l-type splitting is very small, but for very precise measurements it must be taken into account.

It is also seen from Fig. 108 that for $D_{\infty h}$ molecules in every branch of Π—Σ (and similarly Σ—Π) bands, there is an *intensity alternation,* just as for Σ—Σ bands. This is seen clearly in the C_2H_2 fundamental ν_5 of Fig. 110. It must, however, be stressed that for Δ—Π and Π—Δ bands, which otherwise would be quite similar to the Π—Σ and Σ—Π bands, such an intensity alternation would not occur, since for them

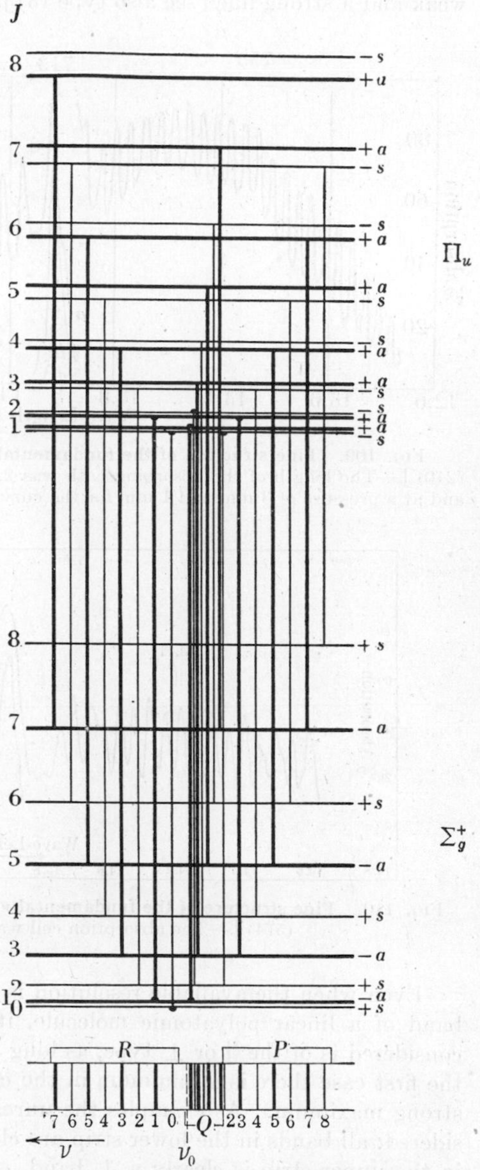

Fig. 108. Energy level diagram for Π_u—Σ_g^+ bands of linear polyatomic molecules.—See caption of Fig. 102. In the schematic spectrum below, the lines of the Q branch have been spread out even though actually, since $B' = B''$ is assumed, they should all coincide. The magnitude of the l-type splitting in the upper state has been greatly exaggerated.

each rotational level of both upper and lower state is a close doublet of a symmetric and antisymmetric level and therefore each "line" is a close doublet of a weak and a strong line [see also type (3)].

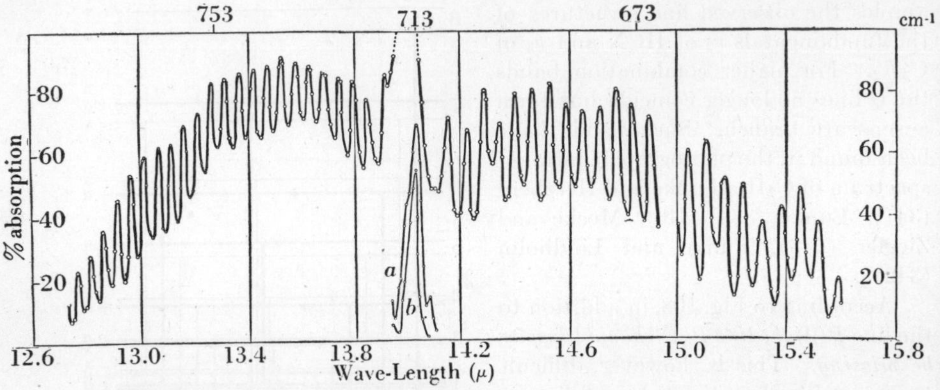

FIG. 109. **Fine structure of the fundamental $\nu_2(\pi)$ of HCN at 14.04μ** [after Choi and Barker (210)].—The length of the absorbing path was 2.5 cm at a pressure of 350 cm for the main curve and at a pressure of 3 mm and 1 mm for the curves a and b respectively.

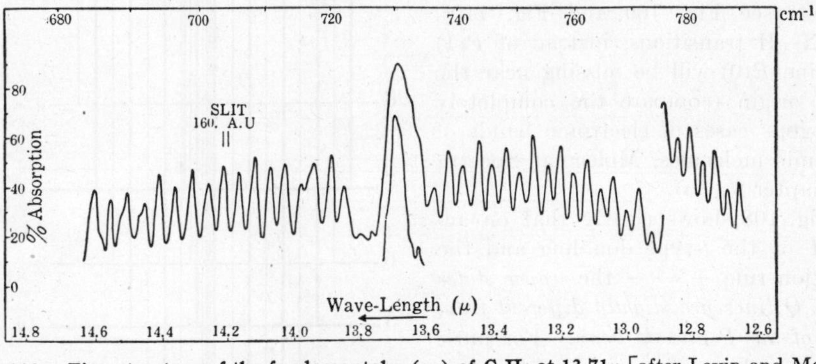

FIG. 110. **Fine structure of the fundamental $\nu_5(\pi_u)$ of C_2H_2 at 13.71μ** [after Levin and Meyer (574)].—The absorption cell was 4 cm long and "partly filled."

Even when the available resolution is not sufficient to resolve the branches of a band of a linear polyatomic molecule, it is possible to decide whether the band considered is of the || or ⊥ type, as long as the dispersion is not too small, since in the first case there is a minimum in the center of the *envelope*, in the second case a strong maximum. As examples the unresolved CO_2 bands in Fig. 83 may be considered: all bands in the lower strip are clearly || bands whereas the fundamental ν_2 in the upper strip is clearly a ⊥ band. In this way, in many cases, the types of the bands of the linear molecules discussed in Chapter III have been determined.

It should be realized that the || bands of symmetric top molecules also consist of a P, a Q, and an R branch (see section 2 of this chapter). The observation of such a band with three branches, in itself, is therefore not sufficient proof for the linearity of a molecule.

Π—Π bands [type (3)]. Bands of the third type are the Π—Π, Δ—Δ, · · · vibrational transitions, of which, however, up to now only Π—Π transitions have been observed. No fundamentals are of this type, but only *difference bands* belonging to a sequence that starts out with a Σ—Σ band (see p. 267). For example, when

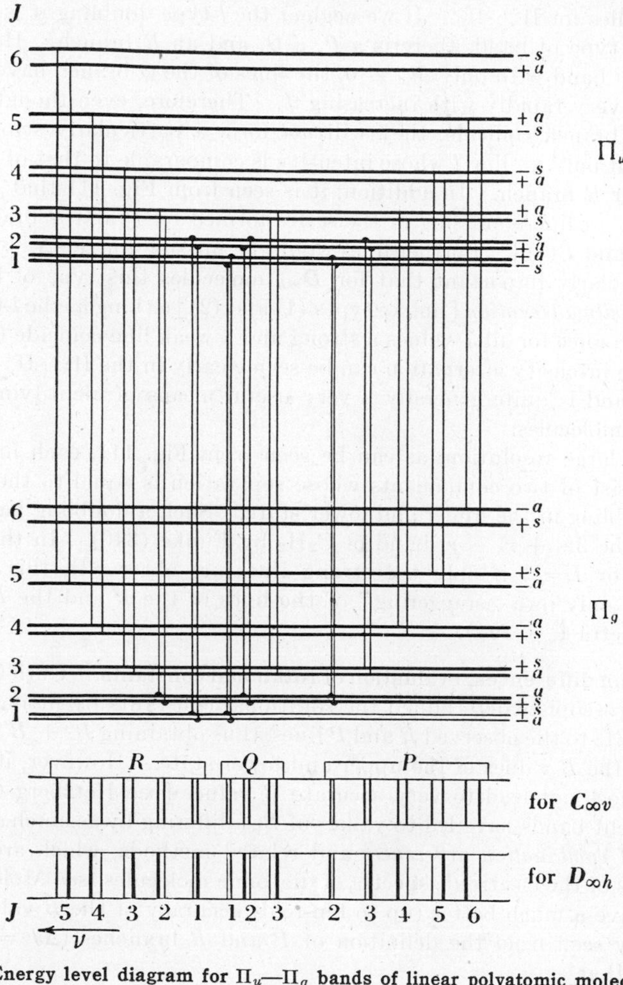

FIG. 111. **Energy level diagram for Π_u—Π_g bands of linear polyatomic molecules.**—See captions of Figures 102 and 108. The two strips at the bottom give schematically the spectrum for the case of $C_{\infty v}$ and $D_{\infty h}$ molecules respectively; in the latter case under the assumption that the antisymmetric levels have the greater weight (as in C_2H_2).

in the upper and lower state of the 1–0 transition of a || vibration a ⊥ vibration is singly excited, we obtain a Π—Π band whose wave number is of course very close to that of the 1–0 transition (Σ—Σ band). The Π—Π bands can be observed in absorption only when an appreciable fraction of the molecules is in the first excited state of a ⊥ vibration, and even then they are strongly overlapped by the much stronger Σ—Σ band to which they belong. The first case of this type was observed by Herzberg and Spinks (441) for C_2H_2 and can be seen in Fig. 106a (short leading lines at bottom); Fig. 106b shows that at low temperature the Π—Π band disappears, since then the fraction of molecules in the first excited state of the ⊥ vibration is too small.

Fig. 111 shows in an energy-level diagram the transitions allowed according to the selection rules for Π_u—Π_g. If we neglect the l-type doubling it is seen that, as for the second type of band, there is a P, a Q, and an R branch. However, here, since this is a || band with only $M_z \neq 0$, the lines of the Q branch have an intensity that decreases very rapidly with increasing J. Therefore, even though most of the lines of the Q branch coincide, they will *not* form a particularly strong "line" as for type (2), but only a "line" whose intensity is comparable to that of an individual line of the P or R branch. In addition, it is seen from Fig. 111 that on both sides of this Q "line" a line is missing in the series formed by P and R branch [namely, the lines $R(0)$ and $P(1)$]. Further lines would be missing for Δ—Δ, ··· transitions.

It is particularly important that for $D_{\infty h}$ molecules this type of band *will not exhibit an intensity alternation* [unlike types (1) and (2)] as long as the l-type doubling is not resolved, since for all J values a strong and a weak line coincide (see Fig. 111). This lack of an intensity alternation can be seen clearly in the Π_u—Π_g band of C_2H_2 in Fig. 106a, and is quite generally a very useful means of identifying such Π—Π bands in $D_{\infty h}$ molecules.

With very large resolution, as can be seen from Fig. 111, each line of a Π—Π band will consist of two components whose separation is equal to the difference of the l-type doubling in the upper and lower state. Such a doubling has indeed been observed for the $3\nu_3 + \nu_4 - \nu_4$ band of C_2H_2 by Funke (340). In the case of zero nuclear spin (or $I = \frac{1}{2}$ if only the strong lines are observed), this doubling will manifest itself only in a "staggering" of the lines of the P and the R branch (see Molecular Spectra I, p. 294).

Combination differences, evaluation of rotational constants. Until fairly recently, infrared spectroscopists determined the rotational constants by fitting a formula of the type (IV, 21) to the observed R and P lines, thus obtaining $B' + B''$ and $B' - B''$ and therefore the B values of the upper and lower state. However, it appears that this method does not lead to very accurate B values [see Herzberg (433)]. Thus for CO_2 different bands have led to values of $B_{[0]}$ differing by as much as 0.6 per cent. The *method of combination differences* and related methods, which are always used in the analysis of the electronic spectra of diatomic molecules (see Molecular Spectra I, p. 191f.), give a much better (up to ten-fold) accuracy of the B values.

It is easily seen from the definition of P and R branches ($\Delta J = -1$ and $+1$ respectively) that

$$R(J) - P(J) = F'(J + 1) - F'(J - 1) = \Delta_2 F'(J), \quad \text{(IV, 23)}$$

$$R(J - 1) - P(J + 1) = F''(J + 1) - F''(J - 1) = \Delta_2 F''(J); \quad \text{(IV, 24)}$$

that is, by forming these differences it is possible to *separate the upper and lower state.* If now (IV, 3) is substituted for $F(J)$,

$$\Delta_2 F(J) = 4B(J + \tfrac{1}{2}), \quad \text{(IV, 25)}$$

where the D correction has been neglected, as is usually although not always possible for linear polyatomic molecules (for its inclusion see Molecular Spectra I, p. 198).

According to (IV, 25), the observed $\Delta_2F(J)$ for the upper and lower state should fall on a straight line if plotted against J, and $4B'$ or $4B''$ is immediately obtained as the slope of this line, or as the average of the $[\Delta_2F(J)/(J+\frac{1}{2})]$ values. The formation of the combination differences also gives an extremely valuable and *critical check on the correctness of the analysis and on the consistency of the data* if two or more bands with the same lower or the same upper state have been measured, since the $\Delta_2F(J)$ values for the common state formed from the two or more bands must agree for every J value within the accuracy of the measurements. The B value of the common state is then, of course, determined from the *average* of the respective $\Delta_2F(J)$ values.

The *difference of the B values* $B' - B''$ for a given band can be obtained (from the same data) with still greater accuracy from $R(J-1) + P(J)$ or, if a resolved Q branch is present, from $Q(J)$. According to (IV, 19) and (IV, 20),

$$R(J-1) + P(J) = 2\nu_0 + 2(B' - B'')J^2; \qquad \text{(IV, 26)}$$

in other words, when $R(J-1) + P(J)$ is plotted against J^2 a straight line is obtained whose slope is $2(B' - B'')$ and whose intercept on the ordinate axis gives an accurate value of ν_0, that is, of the zero line. Similarly, according to (IV, 22), when $Q(J)$ is plotted against $J(J+1)$ a straight line of slope $(B' - B'')$ and intercept ν_0 is obtained.

It is usually advisable to determine only one B value from the combination differences, and all others by finding first the $B' - B''$ value from $R(J-1) + P(J)$ or $Q(J)$. This gives a greater relative than absolute accuracy of the B values for the various vibrational states and therefore the rotational constants α, which involve the differences of B values, can be obtained more reliably than if each B were determined independently (see the examples below).

In the case of accurate measurements of Π—Σ bands it must be realized that from $R(J) - P(J)$ or $R(J-1) + P(J)$ one obtains only the B value of one *l-doubling component* of the Π state, whereas the Q branch, if resolved, gives that of the other component. The difference of these two B values is the constant q of the l-type doubling [see equation (IV, 14)]. It can also be obtained from the combination differences $R(J) - Q(J)$ and $Q(J+1) - P(J+1)$, which have been discussed in more detail in Molecular Spectra I, p. 201 and p. 278.

If a sufficient number of B values has been determined, the B_e and α_i in (IV, 2) follow immediately.

In the case of unresolved bands a rough B value may nevertheless be obtained if the maxima of P and R branch can be measured (see Fig. 83). The formulae for the intensity distribution [see equation (I, 10)] give, for the *separation of the maxima*,

$$\Delta\nu = \sqrt{\frac{8kTB}{hc}} = 2.358\sqrt{T\cdot B}\ (\text{cm}^{-1}), \qquad \text{(IV, 27)}$$

where T is the absolute temperature.

Examples. As an example for the agreement of combination differences, Table 126 gives the $\Delta_2F(J)$ values for the lowest state of HCN as determined from the bands ν_2 at 14 μ and $3\nu_3$ at 1.04 μ (see Table 59), the first measured by the ordinary infrared methods and the second measured in the photographic infrared. It is seen that the agreement is very satisfactory indeed. In Fig. 112 the values

TABLE 126. COMBINATION DIFFERENCES $\Delta_2F''(J) = R(J-1) - P(J+1)$ FOR THE BANDS ν_2 AND $3\nu_3$ OF HCN, AS DETERMINED FROM THE DATA OF BARTUNEK AND BARKER (125) AND HERZBERG AND SPINKS (442).

J	$\Delta_2F''(J) = R(J-1) - P(J+1)$		J	$\Delta_2F''(J) = R(J-1) - P(J+1)$	
	ν_2 band	$3\nu_3$ band		ν_2 band	$3\nu_3$ band
1	—	8.86	12	74.06	73.89
2	14.63	14.80	13	79.75	79.78
3	21.21	20.75	14	85.69	85.72
4	26.53	26.64	15	91.63	91.58
5	32.53	32.53	16	97.56	97.46
6	38.42	38.46	17	103.48	103.36
7	44.43	44.30	18	109.16	109.24
8	50.45	50.26	19	115.09	115.16
9	56.15	56.15	20	120.66	120.92
10	62.16	62.10	21	127.05	126.92
11	68.12	67.95	22		132.73

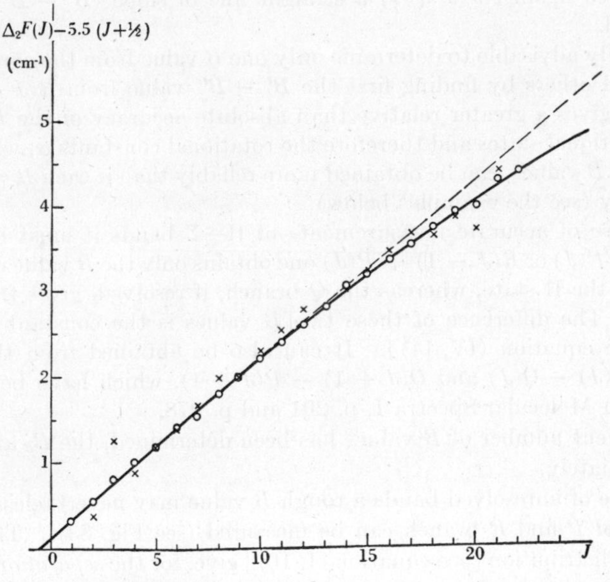

FIG. 112. **Δ_2F curve of HCN in its lowest state.**—Circles represent photographic infrared measurements of $3\nu_3$, crosses represent ordinary infrared measurements of ν_2.

$\Delta_2F(J) - 5.5(J + \frac{1}{2})$ are plotted[4] against J. It is seen that the photographic infrared measurements of $3\nu_3$ (circles) are somewhat more consistent than the far infrared measurements of ν_2 (crosses), since they give a smoother curve. From the slight deviation of the curve from a straight line the rotational constant D can be determined (see Molecular Spectra I, p. 199). The slope of the curve (for $J = 0$) gives

[4] 5.5 is an approximate value of $4B$. If Δ_2F itself had been plotted, the accuracy of plotting in a graph of this size would not allow detection of small deviations from the straight line.

$4B_{[0]} - 5.5$. One obtains $B_{[0]} = 1.4784\ cm^{-1}$ and[5] $D = 3.3 \times 10^{-6}\ cm^{-1}$. Using this D value also for all the other bands whose fine structure has been measured, the B values for the various vibrational levels as given in Table 127 have been obtained.

TABLE 127. B VALUES FOR THE VARIOUS VIBRATIONAL LEVELS OF HCN, FROM THE DATA OF HERZBERG AND SPINKS (442), AND LINDHOLM (579).[6]

v_1	v_2^l	v_3	$B_{[v]}$, observed	$B_{[v]}$, calculated	v_1	v_2^l	v_3	$B_{[v]}$, observed	$B_{[v]}$, calculated
0	0^0	0	1.4784	1.4784	1	1^1	3	1.436_5	1.4374
0	1^1	0	1.4789	1.4791	0	0^0	4	1.4343	1.4352
1	0^0	2	1.4477	1.4475	0	1^1	4	1.435_5	1.4359
0	0^0	3	1.4463	1.4460	1	0^0	4	1.423	1.4259
0	1^1	3	1.4472	1.4467	0	0	5	1.423	1.4244
1	0^0	3	1.4366	1.4367					

They can be represented by the formula

$$B_{[v]} = 1.4878 - 0.0093(v_1 + \tfrac{1}{2}) + 0.0007(v_2 + 1) - 0.0108(v_3 + \tfrac{1}{2}). \quad \text{(IV, 28)}$$

The values calculated from this formula are also given in Table 127. The agreement is very satisfactory. It could be somewhat further improved by introduction of a small term with $(v_3 + \frac{1}{2})^2$ [see Lindholm (579)].

From (IV, 28), the rotational constant for the equilibrium position is

$$B_e(\text{HCN}) = 1.4878\ \text{cm}^{-1},$$

and from it the moment of inertia in the equilibrium position is found to be [see equation (I, 2)]

$$I_e = 18.816 \times 10^{-40}\ \text{gm cm}^2,$$

which may be compared to the average moment of inertia in the lowest vibrational state as obtained from B_{000} in Table 127:

$$I_{[0]} = 18.935 \times 10^{-40}\ \text{gm cm}^2.$$

It is interesting to note that from the measurements of the ν_2 band (Fig. 109) by Bartunek and Barker (125) one obtains from $P(J) + R(J - 1)$, in the manner described above, $B' - B'' = 0$; a result that does not agree well with the B_{010} and B_{000} values obtained from the photographic infrared bands (Table 127). The reason for this apparent discrepancy is not a lack of accuracy of the infrared measurements [as Lindholm (579) assumed], but lies in the l-type doubling. The B_{010} value in Table 127 is obtained from a Π—Π band, and therefore represents an average of the B values of the two l-doubling components, while the B_{010} value from ν_2 refers to the upper level of the P and R branches only (see Fig. 108). The B_{010} value of the upper level of the Q lines of ν_2 would therefore be greater than B_{010} of Table 127 (approximately 1.4794). It follows that the Q branch should be slightly shaded toward shorter wave length even though P and R lines are perfectly equidistant. Such an asymmetry of the Q branch can indeed be seen in Fig. 109. The fact that $B_{010}^{Q} > B_{010}^{PR}$ (and not the reverse) is also in agreement with expectation.

---

[5] This D value is the average of the values given by Herzberg and Spinks (442) and Lindholm (579). The latter author unfortunately does not give the wave numbers of the band lines measured by him, so that it is impossible to take the average of all combination differences.

[6] In order to have a consistent set, Lindholm's B values for some of the levels have been increased by 0.0005 cm^{-1}, since he used for them a smaller D value, which changes the resulting B value by about this amount.

As a second example we consider the infrared spectrum of the CO_2 molecule. The B_{000} value can be obtained from the combination differences $\Delta_2F(J)$ of the bands ν_2 and ν_3. However, it appears that the resonance pair of difference bands $\nu_3 - 2\nu_2$, $\nu_3 - \nu_1$ has been measured with much greater accuracy [Barker and Adel (108)]. Therefore it is better to start out from the state 0, 0, 1. In Fig. 113 the values $\Delta_2F'(J) - 1.4800(J + \frac{1}{2})$ are plotted for this state as obtained from the three bands ν_3, $\nu_3 - 2\nu_2$, $\nu_3 - \nu_1$. It is seen that the consistency of the $\Delta_2F'(J)$ from $\nu_3 - 2\nu_2$ and $\nu_3 - \nu_1$

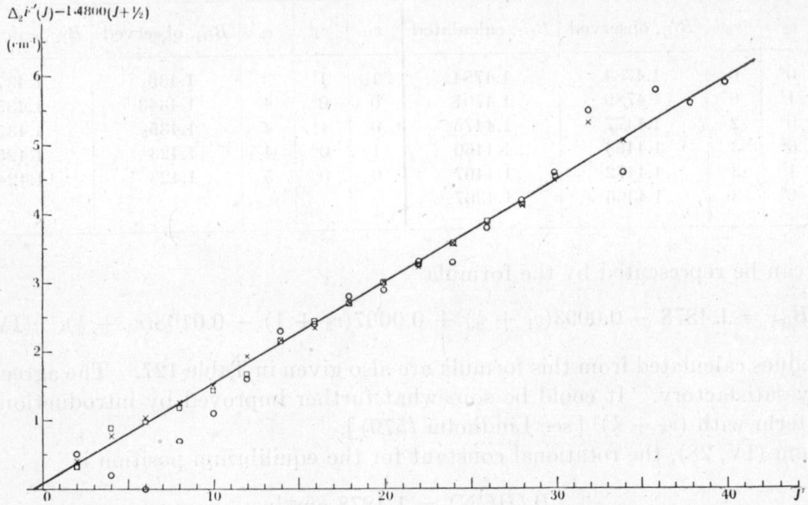

FIG. 113. **Δ_2F curve of CO_2 in the state 0 0 1, obtained from the bands ν_3, $\nu_3 - 2\nu_2$, $\nu_3 - \nu_1$.** Circles represent values obtained from ν_3, squares those from $\nu_3 - 2\nu_2$ (9.40μ), crosses those from $\nu_3 - \nu_1$ (10.41μ).

is better than the consistency of those from ν_3. From the slope of the straight line, using only the data from $\nu_3 - 2\nu_2$ and $\nu_3 - \nu_1$, the correction to be applied to 1.4800 in order to get $4B_{001}$ is obtained, yielding $B_{001} = 0.3866$ cm^{-1}.

The value of B_{000} can now be obtained more accurately than from the Δ_2F'' of the bands ν_3 and ν_2 by determining $B' - B''$ for ν_3 in the manner described above. Since here alternate lines are missing we have to form $R(J) + P(J)$ rather than $R(J-1) + P(J)$. According to (IV, 19) and (IV, 20), we have

$$R(J) + P(J) = 2\nu_0 + 2B' + 2(B' - B'')J(J+1). \qquad \text{(IV, 29)}$$

In Fig. 114 $R(J) + P(J) + 0.005\,J(J+1)$ for the band ν_3 is plotted against $J(J+1)$. The slope of the straight line obtained gives $2(B' - B'') + 0.005 = -0.0008$ from which, with the above B_{001} value it follows that $B_{000} = 0.3895$ cm^{-1}.[7] This value, as can be seen from the graphs in Fig. 113 and Fig. 114 can at the worst have an error of ± 0.0002 cm^{-1}, corresponding to 0.05 per cent, which is 10 times the accuracy of the values obtained from (IV, 21) using the same measurements (see p. 390). Applying the same method to obtain the B values of the lower states of $\nu_3 - 2\nu_2$, $\nu_3 - \nu_1$ and of the upper state of ν_2 the values given in Table 128 are found [see Herzberg (434)]; the relative accuracy of these values is again greater than their absolute accuracy.

The rotational constant α_3 of CO_2 is immediately obtained as the difference $B_{000} - B_{001} = 0.0029$ cm^{-1}. In determining the rotational constants α_1 and α_2, account must be taken of two facts: (1) that the two levels $0, 2^0, 0$ and $1, 0^0, 0$ are in resonance with each other and (2) that the observed value for B_{010} refers to only one l-doubling component (the upper states of P and R lines). This B_{010}^{PR} cannot therefore be used to obtain α_2. However, α_1 and α_2 can be determined from the B

---

[7] This may be compared with the less accurate value 0.394 cm^{-1} which follows from the rotational Raman spectrum (see p. 21).

values of the two resonating levels ν_1 and $2\nu_2$ if the constants a^2 and b^2 in (IV, 15) or (II, 293) are known. The values $a^2 = 0.57$ and $b^2 = 0.43$ given by Adel and Dennison (37) give $\alpha_1 = 0.00056$, $\alpha_2 = -0.00062$ cm^{-1}.[8] In view of the uncertainty of the values of a^2 and b^2, these α values are uncertain within ± 0.00010 cm^{-1}.

Assuming the above α values we obtain, for the equilibrium position of CO_2,

$$B_e = 0.3906 \text{ cm}^{-1}.$$

The moments of inertia corresponding to B_e and B_0 are respectively

$$I_e = 71.67 \times 10^{-40} \text{ gm cm}^2 \quad \text{and} \quad I_{[0]} = 71.87 \times 10^{-40} \text{ gm cm}^2.$$

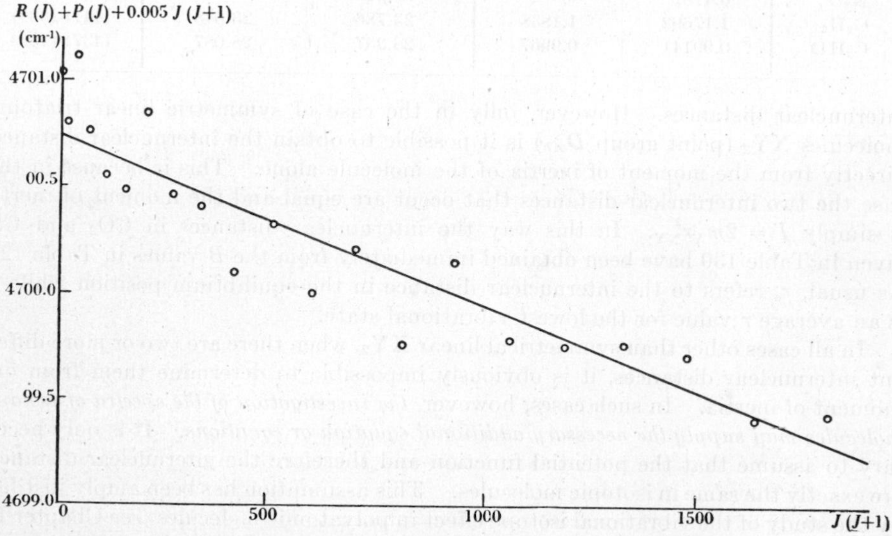

FIG. 114. **Determination of $B' - B''$ from $R(J) + P(J)$ for the band ν_3 of CO_2.**

TABLE 128. OBSERVED B VALUES OF THE VARIOUS VIBRATIONAL LEVELS OF THE CO_2 MOLECULE.

ν_1	$\nu_2{}^l$	ν_3	$B_{[v]}$ (cm^{-1})	Obtained from
0	0^0	0	0.3895_0	B_{001} and $B' - B''$ of ν_3
0	1^1 (*PR*)	0	0.3899_5	B_{000} and $B' - B''$ of ν_2
0	2^0	0 }	0.3899_6	B_{001} and $B' - B''$ of $\nu_3 - 2\nu_2$
1	0^0	0 }	0.3897_1	B_{001} and $B' - B''$ of ν_3 ν_1
0	0^0	1	0.3866_0	$\Delta_2 F'(J)$ of $\nu_3 - 2\nu_2$ and $\nu_3 - \nu_1$

The only other linear polyatomic molecules for which fairly extensive data are available are C_2H_2 and C_2HD. In order to save space we do not give all the $B_{[v]}$ values here; instead, we collect in Table 129 the $B_{[0]}$, B_e, $I_{[0]}$, and I_e values for these as well as for all the other linear molecules investigated.

Determination of internuclear distances: isotope effect. The data of greatest interest in connection with the geometrical structure of linear molecules are the

[8] From the α_2 value the value $B_{010} = 0.39012$ cm^{-1} is obtained whose difference from B^{PR}_{010} is half the constant q of the l type doubling [see equation (IV, 14)]. Thus we find $q = 0.00034$ cm^{-1}.

TABLE 129. ROTATIONAL CONSTANTS $B_{[0]}$ AND B_e AND MOMENTS OF INERTIA $I_{[0]}$ AND I_e OF LINEAR POLYATOMIC MOLECULES.[9]

Molecule	$B_{[0]}$ (cm^{-1})	B_e (cm^{-1})	$I_{[0]}$ (10^{-40} g cm^2)	I_e (10^{-40} g cm^2)	References
HCN	1.4784	1.4878	18.935	18.816	See text above
DCN	1.2088		23.159		(125) (439)
CO_2	0.3895	0.3906	71.87	71.67	See text above
CO_2^+	0.3806[10]		73.55		(639)
CS_2	0.1092[11]		256.4		(578)
N_2O	0.4182[11a]		66.94		(703)
C_2H_2	1.17692	1.1838	23.786	23.648	(441) (339)
C_2HD	0.99141	0.9967	28.237	28.087	(437) (439)

internuclear distances. However, only in the case of symmetric linear triatomic molecules XY_2 (point group $D_{\infty h}$) is it possible to obtain the internuclear distances directly from the moment of inertia of the molecule alone. This is because in this case the two internuclear distances that occur are equal and the moment of inertia is simply $I = 2m_Y r_{XY}^2$. In this way the internuclear distances in CO_2 and CS_2 given in Table 130 have been obtained immediately from the B values in Table 129. As usual, r_e refers to the internuclear distance in the equilibrium position while r_0 is an average r value for the lowest vibrational state.

In all cases other than symmetrical linear XY_2, when there are two or more different internuclear distances, it is obviously impossible to determine them from one moment of inertia. In such cases, however, *the investigation of the spectra of isotopic molecules may supply the necessary additional equation or equations.* It is only necessary to assume that the potential function and therefore the internuclear distances are exactly the same in isotopic molecules. This assumption has been amply justified by the study of the vibrational isotope effect in polyatomic molecules (see Chapter II, section 6), and particularly by the study of the rotational and vibrational isotope effect in diatomic molecules. Except for diatomic molecules with low-lying excited electronic states (for which theoretically a slight difference of the order of 0.001×10^{-8} cm is to be expected) it has always been found that the internuclear distance in isotopic diatomic molecules is the same within the accuracy of the measurements ($\pm 0.0002 \times 10^{-8}$ cm), in agreement with expectation.[12] Thus, since the linear polyatomic molecules here to be considered have no low-lying electronic states, we can be certain that the *internuclear distances in isotopic molecules are the same* within much less than 0.001×10^{-8} cm. It must be realized that this equality would be expected to hold *exactly only for the equilibrium internuclear distances r_e* but not as accurately for the average (effective) internuclear distances r_0 in the lowest vibrational level, since the amplitude of the zero-point vibration is different for different

[9] Some of the moments of inertia given here are slightly different from those in the original papers because of the use of new conversion factors.

[10] From the ultraviolet emission spectrum [Mrozowski (639)].

[11] This value is from the ultraviolet absorption spectrum [Liebermann (578)]. It seems to be more accurate than the infrared value 0.112 cm^{-1} given by Sanderson (761).

[11a] Calculated from Plyler and Barker's (703) data by the author according to the method outlined in the text.

[12] This is due to the fact that the potential functions are almost entirely determined by the electrons and the nuclear charges but are independent of the nuclear masses.

isotopic species. However, even for the r_0 the equality will hold within less than 0.002×10^{-8} cm.[13]

As an example we consider the acetylene molecule (C_2H_2) to which this method was first applied [Herzberg, Patat, and Spinks (437)]. The moments of inertia of C_2H_2 and C_2HD obtained from the spectra are given in Table 129.

The distances being designated as in Fig. 115, the moment of inertia of C_2H_2 in terms of these distances is

$$I = 2m_C a^2 + 2m_H b^2. \tag{IV, 30}$$

The center of mass of C_2HD is no longer in the geometrical center of the molecule but shifted toward the D atom (Fig. 115) by

$$d = \frac{m_D - m_H}{M^i} b, \tag{IV, 31}$$

where M^i is the total mass of the C_2HD molecule. Using the theorem of parallel axes, we obtain for the moment of inertia I^i of C_2HD about its center of mass:

$$I^i = 2m_C a^2 + (m_H + m_D)b^2 - M^i d^2. \tag{IV, 32}$$

If d is substituted from (IV, 31), we have two equations [(IV, 30) and (IV, 32)] for the two unknowns a and b, from which we obtain

$$\begin{aligned} b &= \tfrac{1}{2}r(\text{C}{\equiv}\text{C}) + r(\text{C—H}) = \sqrt{\frac{M^i(I^i - I)}{M(m_D - m_H)}}, \\ a &= \tfrac{1}{2}r(\text{C}{\equiv}\text{C}) = \sqrt{\frac{I - 2m_H b^2}{2m_C}}, \end{aligned} \tag{IV, 33}$$

where M^i is the total mass of the C_2H_2 molecule. If the I_e or I_0 values of Table 129 are substituted into (IV, 33) the internuclear distances r_e(C≡C), r_e(C—H), and r_0(C≡C), r_0(C—H) respectively, given in Table 130, are obtained. According to the above the r_0 values involve a (very slight) systematic error, but the smallness of the difference between r_0 and r_e obtained shows that this error cannot be larger than a few thousandths of an Ångstrom unit, particularly if it is remembered that the r_0 values should be slightly larger than the r_e.[14] This shows that even in cases in which I_e cannot be determined the error introduced by using I_0 will not be more than the amount given.

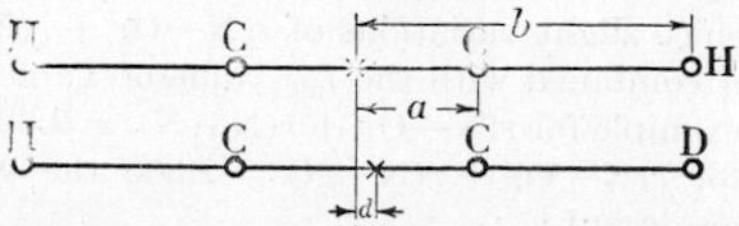

Fig. 115. Dimensions of C_2H_2 and C_2HD.—The × indicates the position of the center of mass.

It would be very interesting to obtain the moment of inertia I_e (or I_0) of C_2D_2, since it would supply an additional equation for the two internuclear distances

$$I(C_2D_2) = 2m_C a^2 + 2m_D b^2, \tag{IV, 34}$$

which would give a check of the values determined from C_2H_2 and C_2HD.

[13] Even in such an unfavorable case as the pair HCl and DCl the r_0 values are as close together as 1.2839 and 1.2816 respectively, whereas r_e for both is 1.2747×10^{-8} cm.

[14] The fact that the difference between r_e(C≡C) and r_0(C≡C) comes out larger than the difference between r_e(C—H) and r_0(C—H), whereas the opposite would be expected, is apparently due to the non-exact validity of the assumption of equal r_0 values in the two isotopes.

In a way similar to the above the internuclear distances $r(C{\equiv}N)$ and $r(C{-}H)$ in HCN given in Table 130 have been obtained from the moments of inertia of HCN and DCN of Table 129 [see Herzberg, Patat, and Verleger (439)]. In this case, since B_e of DCN has not yet been obtained, only the r_0 values can be evaluated. The close agreement of the CH distance in C_2H_2 and HCN is very remarkable.[16]

TABLE 130. INTERNUCLEAR DISTANCES IN LINEAR POLYATOMIC MOLECULES OBTAINED FROM THEIR ROTATION-VIBRATION SPECTRA [15] AND COMPARED WITH THOSE IN THE CORRESPONDING DIATOMIC MOLECULES.

Molecule	Bond	$r_e(10^{-8}$ cm)	$r_0(10^{-8}$ cm)	Corresponding diatomic molecule	
				$r_e(10^{-8}$ cm)	$r_0(10^{-8}$ cm)
CO_2	$r(C{=}O)$	1.1615	1.1632	1.1284	1.1310
CO_2^+	$r(C{=}O)$	—	1.1767		
CS_2	$r(C{=}S)$	—	1.554	1.536	1.538
$C_2H_2(C_2HD)$	$r(C{\equiv}C)$	1.202_8	1.207_4	1.3121	1.3155
	$r(C{-}H)$	1.059_7	1.059_7	1.1201	1.1305
HCN(DCN)	$r(C{\equiv}N)$	—	1.157_4	1.1721	1.1747
	$r(C{-}H)$	—	1.058_7	1.1201	1.1305

It is interesting to compare the internuclear distances of the linear polyatomic molecules with those of the corresponding diatomic molecules or radicals. For this purpose the last two columns in Table 130 are added. It is seen that while for CO_2 and CS_2 the internuclear distances are larger, in C_2H_2 and HCN they are smaller than in the diatomic molecules.

For N_2O, since no isotopic molecule has been investigated, the individual internuclear distances cannot be determined. It is only possible to say that $r(N{-}O) + r(N{-}N) \leq 2.324 \times 10^{-8}$ cm. Electron diffraction data by Schomaker and Spurr (770a) give $r(N{-}O) + r(N{-}N) = 2.32 \pm 0.02 \times 10^{-8}$ cm; but this additional datum is not sufficiently accurate to determine the individual distances since slight variations of $r(N{-}O) + r(N{-}N)$ within the accuracy of the electron diffraction data, if combined with the $I_{[0]}$ value of Table 129 lead to large variations of $r(N{-}O)$ and $r(N{-}N)$ [for example for $r(N{-}O) + r(N{-}N) = 2.320$ one obtains $r(N{-}O) = 1.198$ and $r(N{-}N) = 1.123$ while for $r(N{-}O) + r(N{-}N) = 2.300$ the values $r(N{-}O) = 1.366$ and $r(N{-}N) = 0.934 \times 10^{-8}$ cm are found].

(c) *Raman spectrum*

Selection rules. Just as for the infrared spectrum the vibrational selection rules for the Raman spectrum are in a good approximation unchanged by the interaction of vibration and rotation (see Table 55), and the rotational selection rules are the same as for diatomic molecules, that is, when l is zero in both upper and lower vibrational state,

$$\Delta J = 0, \pm 2, \tag{IV, 35}$$

and when l is different from zero in at least one of the states,

$$\Delta J = 0, \pm 1, \pm 2. \tag{IV, 36}$$

[15] All values are based on the same conversion factors (see appendix p. 538).

[16] If it is assumed that r_e(CH) in HCN is the same as in C_2H_2, I_e(HCN) (of Table 129) gives $r_e(CN) = 1.153 \times 10^{-8}$ cm.

In addition, we have the symmetry rules

$$+ \leftrightarrow -, \qquad s \leftrightarrow a \tag{IV, 37}$$

and the restriction $J' + J'' \geq 0$ [see Placzek and Teller (701)]

Types of Raman bands. From the lowest vibrational level (Σ^+) only the vibrational Raman transitions $\Sigma^+—\Sigma^+$, $\Pi—\Sigma^+$, and $\Delta—\Sigma^+$ can take place (for $D_{\infty h}$ molecules the subscript g should be added to each species symbol). Of these transitions $\Sigma^+—\Sigma^+$ is the most important since the totally symmetric (||) vibrations will give strong Raman bands of this type. According to (IV, 35), such $\Sigma^+—\Sigma^+$ *Raman transitions consist of an S, a Q and an O branch*, each of which, in the case of $D_{\infty h}$ molecules, would have the usual intensity alternation. As in the case of the infrared ⊥ bands, all the lines of the Q branch fall almost together, since $B' - B''$ is very nearly zero. The resulting Raman "line" is much stronger than the individual lines of the O and S branches.

The ⊥ vibrations will give Raman bands of the type $\Pi—\Sigma^+$ (only Π_g being Raman active for $D_{\infty h}$ molecules). For these transitions, in addition to S, Q, and O, *a P and an R branch* ($\Delta J = \pm 1$) *also occur* (each with intensity alternation for $D_{\infty h}$ molecules). However, Placzek and Teller (701) have shown that for these transitions the *Q branch is extremely weak*, only its first line being comparable in intensity with the first lines of the other branches (which are not their strongest lines). Thus these $\Pi—\Sigma^+$ Raman bands would *not* have the very strong and sharp central "line," and even under fairly low dispersion they would appear as broad bands, possibly with two maxima.

Raman transitions of the types $\Delta—\Sigma^+$ (for example, the first overtone of a ⊥ vibration) and $\Pi—\Pi$ (for example, the second band in a sequence that starts out with a $\Sigma—\Sigma$ band) have also the *five branches O, P, Q, R, S*, but for them the Q branch is again very strong, as for $\Sigma—\Sigma$ bands, and forms a characteristic strong central "line." For $D_{\infty h}$ molecules the $\Pi—\Pi$ bands, like the $\Pi—\Pi$ infrared bands, do not show an intensity alternation in the branches, at least as long as the l-doubling is not resolved.

Placzek and Teller (701) have given explicit expressions for the intensity distribution in the branches of all the various types of Raman bands.

Observed Raman bands. Unfortunately, up to the present time the rotational structure of no vibrational Raman band of a linear polyatomic molecule has been resolved. Usually only the *line-like Q branches of the || bands* are observed, the individual lines of the S and O branches being too weak to be recorded even as unresolved maxima. This explains, at the same time, why the Raman bands are usually so nearly like lines. However, Bhagavantam and Rao (150) (151) have found in gaseous C_2H_2 a weak broad doublet band which fits very well the Raman-active ⊥ fundamental ν_4, and which would confirm the above theory, according to which the Q branch in such a band should be practically missing. The absence of the (line-like) Q branch accounts also for the apparent weakness of this band. The same reason explains also the fact that the perpendicular vibrations of HCN and N_2O have not as yet been observed in the Raman spectrum.

2. Symmetric Top Molecules

(*a*) *Energy levels*

Non-degenerate vibrational states. While in a zero approximation the energy of a vibrating and rotating symmetric top is simply the sum of the vibrational energy and the rotational energy (I, 20) of a rigid symmetric top, in higher approximation we have to take into account that both moments of inertia I_B and I_A change periodically during the vibration. In a first approximation (just as for linear molecules) we can apply the formulae for the rigid symmetric top if we use as rotational constants B and A the *average values* $B_{[v]}$ *and* $A_{[v]}$ *during a vibration*, which differ in general from the equilibrium values $B_e = h/8\pi^2 cI_B^e$ and $A_e = h/8\pi^2 cI_A^e$. As in the case of linear molecules, we expect the following relations to hold:

$$B_{[v]} = B_e - \sum \alpha_i^B \left(v_i + \frac{d_i}{2}\right) + \cdots, \quad \text{(IV, 38)}$$

$$A_{[v]} = A_e - \sum \alpha_i^A \left(v_i + \frac{d_i}{2}\right) + \cdots, \quad \text{(IV, 39)}$$

where d_i is the degree of degeneracy of the vibration ν_i ($d_i = 1$ or 2). For the present we consider only non-degenerate vibrational levels, since for the degenerate levels another fairly large coupling effect comes in. To be sure, in such a non-degenerate level, one or more degenerate vibrations may be excited.

According to the above, then, the *total energy of vibration and rotation of a symmetric top molecule* (as long as the vibrational state is *non-degenerate*) is given by

$$T = G(v_1, v_2, v_3, \cdots) + F_{[v]}(J, K), \quad \text{(IV, 40)}$$

where $G(v_1, v_2, v_3, \cdots)$ is given by the previous equation (II, 281) and

$$F_{[v]}(J, K) = B_{[v]}J(J + 1) + (A_{[v]} - B_{[v]})K^2 \quad \text{(IV, 41)}$$

According to (IV, 41) we have for each (non-degenerate) vibrational level a set of rotational levels as in Fig. 8, p. 25, but the spacing is slightly different in the different vibrational levels. Also the restrictions on the J and K values are the same as in Fig. 8.

According to (IV, 38–39), just as for linear molecules, even the rotational constants $B_{[0]}$ and $A_{[0]}$ for the lowest vibrational state *differ slightly from the values* B_e *and* A_e *which correspond to the equilibrium position.* The moments of inertia and internuclear distances obtained from $B_{[0]}$ and $A_{[0]}$ are therefore not exactly, even though fairly closely, the equilibrium values.

In addition to the above effect of the interaction of rotation and vibration there is also *centrifugal distortion.* As discussed in Chapter I [see equation (I, 27)], in the vibrationless state the terms $- D_J J^2(J + 1)^2 - D_{JK} J(J + 1)K^2 - D_K K^4$ have to be added to (I, 20) [see Slawsky and Dennison (795)]. Similar terms with slightly altered coefficients would be expected here in addition to the terms in (IV, 41). The effect of these terms on the rotational levels of the higher vibrational states would of course be quite similar to the effect on the vibrationless state discussed in Chapter I. Up to the present time no data accurate enough to detect the influence of these terms in rotation-vibration spectra are available; we shall therefore in future always neglect them.

The above considerations apply equally to the case of a molecule that is a symmetric top because of symmetry (such as NH_3 and the methyl halides) and to the case of a non-symmetrical molecule for which two of the principal moments of inertia happen to be equal. The formulae (IV, 38) and (IV, 39) have been proved more rigorously on the basis of the wave equation for planar and pyramidal XY_3 molecules by Silver and Shaffer (790) and Shaffer (776) respectively, for axially symmetric XYZ_3 molecules by Shaffer (777) and for the general case by Nielsen (666). These authors have also given explicit formulae for the α_i^A and α_i^B in terms of the geometrical and potential constants of the molecule. As for linear molecules, the α_i are sums of harmonic, anharmonic, and Coriolis contributions [see equation (IV, 12)]. Silver, Shaffer and Nielsen have also found that a constant term $-\alpha_0^A$ and $-\alpha_0^B$ should be added at the right of (IV, 38–39). However, it is of the order of magnitude of the rotational constants D_J and can therefore practically always be neglected.[17]

Degenerate vibrational states. We shall consider here only the case of necessary degeneracy, and not the case of accidental degeneracy. Degenerate vibrational states occur for all molecules that are symmetric tops because of their symmetry (see Chapter II, section 3). For such degenerate states, as was first recognized by Teller and Tisza (837) (836), the *influence of the Coriolis force* is in general much larger than it is for the non-degenerate states or for the degenerate states of linear molecules.

As we have seen above, for linear molecules the Coriolis force produces an interaction of two vibrations of different species, which increases with increasing rotation and results in a contribution to the rotational constant α. This contribution is small as long as the interacting vibrations have rather different frequencies. For symmetric top molecules this same effect also occurs and is responsible for part of the rotational constants α_i. But in addition, since now a rotation about the top axis may occur, the Coriolis force may produce an *interaction between the two components of a degenerate pair of vibrations.*

For example, consider the component ν_{2a} of the degenerate vibration of the triatomic molecule X_3 of point group D_{3h} (equilateral triangle) shown in Fig. 116a. For a counter-clockwise rotation of the molecule the Coriolis forces are as shown by the broken arrows. It is seen from the figure that they tend to excite the other component ν_{2b} of the degenerate vibration (see Fig. 32), and, since the two components have the same frequency, the transition from the one to the other mode of motion (and back) will occur very quickly.[18] Therefore, the influence on the energy levels will be large and cannot be considered as a second-order effect.

The result of this strong Coriolis interaction is a (first-order) *splitting of the degenerate vibrational levels* into two levels whose separation increases with increasing rotation (K) about the top axis and is zero for $K = 0$. As usual in wave mechanics, the two component levels cannot be described by the two modes ν_{2a} and ν_{2b} given in Fig. 32 (or rather the corresponding eigenfunctions), but by a linear combination of the two, such that they no longer influence each other. Such modes are, in the

[17] It may be noted that Silver and Shaffer's (790) α_i and β_i correspond to our α_i^B and $\alpha_i^B - \alpha_i^A$ respectively, whereas in the papers by Shaffer (776) (777) α_n and β_n correspond to our $-\alpha_i^B/B_e$ and $-\alpha_i^A/A_e$.

[18] In fact, it is easily seen that after a rotation by 90°, if the displacement vectors retain their direction in a fixed coordinate system, the first mode goes over into the second.

above example, the clockwise and counter-clockwise circular oscillations (see Fig. 116b and c and p. 75f.). While without rotation about the top axis these two oscillations have the same energy (frequency), with increasing rotation they have increasingly different energies, because in the one (circular oscillation opposite to rotation) the Coriolis force acts as an additional restoring force and increases the frequency, whereas in the other (same sense for circular oscillation and rotation) the Coriolis force is opposite to the restoring force and decreases the frequency.

Fig. 116b and c show that there is a *vibrational angular momentum* about the symmetry axis in each component level of the degenerate vibration (independent of

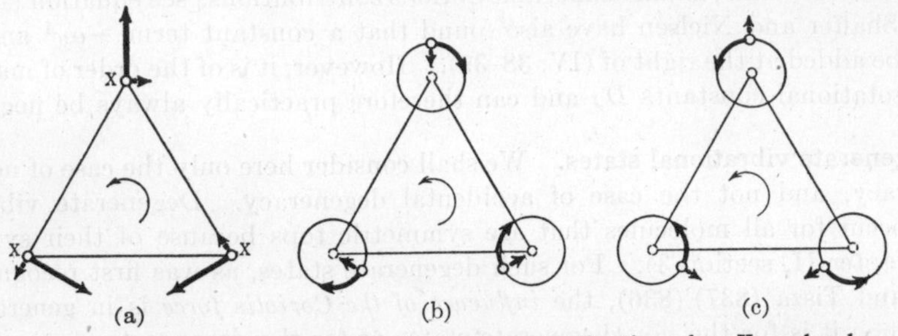

FIG. 116. **Coriolis forces during the degenerate vibration of an X_3 molecule.**

the rotation of the molecule), and we may also consider the splitting that occurs with increasing K as a consequence of the *interaction of the angular momentum due to vibration with that which is due to ordinary rotation about the top axis.*

It must be realized that in the present case the vibrational angular momentum is much larger than the vibrational angular momentum arising from Coriolis interaction with other vibrations as discussed previously for linear molecules (the ellipses in Fig. 101 are very narrow). In the previous case it decreases to zero as the speed of rotation goes to zero, whereas in the present case it persists even for no rotation since the two circular oscillations are solutions to the pure vibrational problem (see Fig. 27b).

It can be shown (see below) that the *magnitude of the vibrational angular momentum* in a degenerate vibrational state in which only one degenerate vibration ν_i is singly excited is $\zeta_i(h/2\pi)$ where $0 \leqq |\zeta_i| \leqq 1$. In the above example (Fig. 116) it can be shown [see Teller (836)] that $|\zeta_i| = 1$, since the nuclei move in circles whose planes are perpendicular to the axis of symmetry. However, in other cases when the molecule has more than one degenerate vibration, values of the constant ζ_i of the vibrational angular momentum intermediate between -1 and $+1$ or even equal to zero may occur. For example, consider the two degenerate vibrations ν_3 and ν_4 of BF_3 (point group D_{3h}) given in Fig. 63. By superimposing ν_{3a} and ν_{3b} (and similarly ν_{4a} and ν_{4b}) with a phase shift of 90°, an elliptical motion of each F nucleus is obtained (compare also S_{3a} and S_{3b} of pyramidal XY_3 in Fig. 58). For different masses of the nuclei and different potential constants different eccentricities are obtained, that is, different values of ζ_i. The maximum value of ζ_i is obtained when a circular motion of each nucleus results, as for example when the masses and potential constants are such that the Y_3 triangle moves as a whole against the X atom (compare ν_{7a} and ν_{7b} of X_3Y_3 in Fig. 36). On the other hand, when in both degenerate components of a vibration the nuclei move in the same line, no superposition will give a vibrational angular momentum; for example, for ν_5 of X_5 (Fig. 38) and ν_8 of X_6 (Fig. 40). In these cases $\zeta_i = 0$, even independent of the masses and potential constants, since there is only one vibration of the particular species. When there is more than one degenerate vibration of a certain species, the exact value of ζ_i for each depends in a complicated manner on the masses as well as the potential constants and the geometrical dimensions (see below).

According to quantum theory the component of the *total* angular momentum along the axis of any symmetric top must be an integral (or, for an odd number of electrons, half integral) multiple of $h/2\pi$. The fact that in general the vibrational angular momentum ζ_i is not integral means, therefore, that the purely rotational angular momentum about the top axis is likewise not integral; but still the sum of the two is integral ($=Kh/2\pi$).

According to Teller (836) and Johnston and Dennison (476), the *formula for the rotational energy levels* in a vibrational state of a symmetric top molecule in which one degenerate vibration ν_i is singly excited, in consequence of the above described Coriolis coupling, is not (IV, 41) but

$$F_{[v]}(J, K) = B_{[v]}J(J+1) + (A_{[v]} - B_{[v]})K^2 \mp 2A_{[v]}\zeta_i K, \qquad \text{(IV, 42)}$$

which differs from (IV, 41) only by the term $\mp 2A_{[v]}\zeta_i K$.[18a] The $-$ sign in this term applies if the vibrational angular momentum p has the same direction as the rotational angular momentum, whereas the $+$ sign applies if they are opposite in direction. This additional term gives a *splitting that increases linearly with increasing K*. One of the component levels has a factor $e^{+il\varphi}$ of its eigenfunction, the other the factor $e^{-il\varphi}$ where l is the previously introduced quantum number of degenerate vibrations (p. 81), which here has the value $l = 1$ since $v_i = 1$. For convenience we shall distinguish the two levels as $+l$ and $-l$ levels. In Fig. 117 is given a schematic energy-level diagram for such a case, which should be compared to Fig. 8. Each level shown in Fig. 117 is still doubly degenerate, the levels with $K \neq 0$ because of the two possible directions of K, those with $K = 0$ because of the vibrational degeneracy which, according to (IV, 42), is not removed for $K = 0$.[19] The splitting is the same for all levels with a given K.

The energy formula (IV, 42) can be easily proven [see Johnston and Dennison (476] on the basis of the rotational part of the total rotational-vibrational energy (II, 279) of the molecule (in which only the dependence of the moments of inertia on the normal coordinates is neglected):

$$H_{\text{rot}} = \frac{(P_x - p_x)^2}{2I_x} + \frac{(P_y - p_y)^2}{2I_y} + \frac{(P_z - p_z)^2}{2I_z}. \qquad \text{(IV, 43)}$$

Here P_x, P_y, P_z are the components of the total angular momentum, p_x, p_y, p_z those of the vibrational angular momentum. In the present case $I_x = I_y = I_B$, $I_z = I_A$, $p_x = p_y = 0$. In addition we put $p_z = +p$ or $-p$, depending on whether the vibrational angular momentum p is parallel or antiparallel to the z axis. We may then write, instead of (IV, 43):

$$H_{\text{rot}} = \frac{(P_x^2 + P_y^2 + P_z^2)}{2I_B} + \frac{P_z^2}{2I_A} - \frac{P_z^2}{2I_B} \mp \frac{pP_z}{I_A} + \frac{p^2}{2I_A}. \qquad \text{(IV, 44)}$$

[18a] While (IV, 42) appears to be a reasonable extension of the formulae given by Teller and Johnston and Dennison who neglected the dependence of B and A on v_i, Shaffer and Nielsen in their papers use A_e in the term $2A\zeta_i K$, not $A_{[v]}$ as has been assumed here.

[19] It should be noted that the levels with $K = 0$ are not missing even though $\zeta \neq 0$. This is because K is the sum of the angular momenta due to pure rotation and to vibration. If the former is opposite and equal to the latter we have $K = 0$.

Using this classical expression for the energy as the Hamiltonian in the wave equation, the energy formula (IV, 42) is immediately obtained by putting

$$P_x^2 + P_y^2 + P_z^2 = J(J+1)\frac{h^2}{4\pi^2}, \qquad P_z = K\frac{h}{2\pi}, \qquad p = \zeta_i \frac{h}{2\pi}, \tag{IV, 45}$$

if the last term $p^2/2I_A$ in (IV, 44) is omitted, which is legitimate since it does not depend on the rotational quantum numbers and can therefore be combined with the vibrational part of the energy.

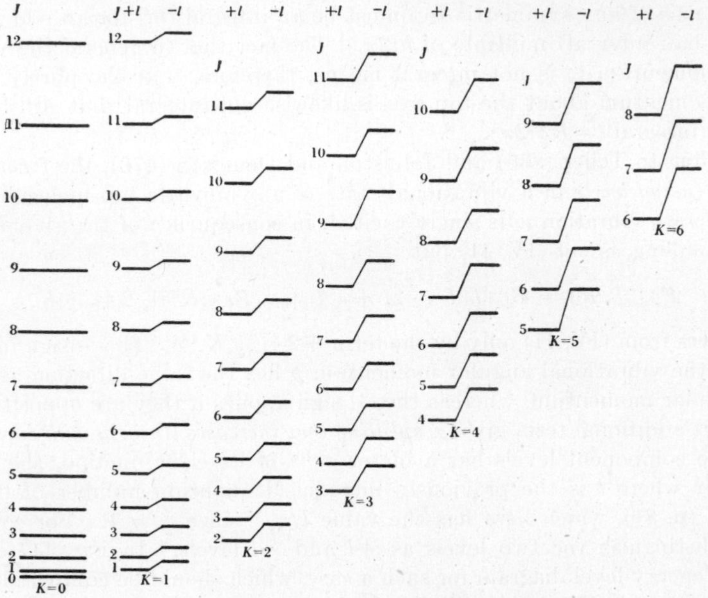

FIG. 117. **Rotational energy levels of a symmetric top molecule in a doubly degenerate vibrational state with $\zeta_i > 0$.**

As for linear molecules, the components p_x, p_y, p_z of the vibrational angular momentum are given by equations of the form of (IV, 11), where the ζ_{ik} are constants depending on the equilibrium internuclear distances, force constants, and masses. However, here $\zeta_{ik}^{(z)}$ may be different from zero even when i and k refer to two components of a degenerate vibration. These are the ζ_i used above, which result in a first-order energy change, whereas all other ζ_{ik} give only a second-order change, that is, a contribution to the rotational constants α_i. Silver and Shaffer (790) and Shaffer (776) (777) have given explicit (rather complicated) formulae for the ζ_i in terms of the masses, force constants, and internuclear distances for the case of planar and pyramidal XY_3 molecules and axial XYZ_3 molecules [see also Jahn (468)].

While the prediction of the ζ_i values usually requires the knowledge of more molecular constants than are actually known, *the sum of the ζ_i values for all vibrations of the same species*, as was first shown by Teller (836), *is independent of the potential constants*, and can be expressed as a very simple function of the moments of inertia. According to Johnston and Dennison (476) [see also Silver and Shaffer (790) (776) (777)], one obtains (using an extreme and simplified force field, since the ζ sum is independent of it) for axial XY_3 molecules (pyramidal or planar)

$$\zeta_3 + \zeta_4 = \frac{I_A}{2I_B} - 1 = \frac{B}{2A} - 1, \tag{IV, 46}$$

and for axial XYZ_3 molecules

$$\zeta_4 + \zeta_5 + \zeta_6 = \frac{I_A}{2I_B} = \frac{B}{2A}. \qquad \text{(IV, 47)}$$

For planar XY_3 molecules, since $I_A = 2I_B$, (IV, 46) simplifies to

$$\zeta_3 + \zeta_4 = 0 \qquad \text{or} \qquad \zeta_3 = -\zeta_4. \qquad \text{(IV, 48)}$$

For X_2Y_6 molecules of point group D_{3h} or D_{3d} (such as ethane) according to Howard (461) the relations

$$\zeta_7 + \zeta_8 + \zeta_9 = 0 \qquad \text{and} \qquad \zeta_{10} + \zeta_{11} + \zeta_{12} = 0 \qquad \text{(IV, 49)}$$

hold. For the numbering of the vibrations in the four cases see Figs. 45, 91, 63, and 49 respectively.

It must be realized that the ζ_i may be *positive or negative* even though the energy according to (IV, 42) depends only on the magnitude of ζ_i. But for positive ζ_i the lower component levels are $+l$ levels, the higher the $-l$ levels, as in Fig. 117, while for negative ζ_i the reverse is the case. It can be shown [see Teller (836)] that for positive ζ_i the *direction of rotation of the dipole moment* during the vibration coincides with the direction of the vibrational angular momentum, while for negative ζ_i they have opposite direction. The direction of rotation of the dipole moment is the same as the direction in which the whole vector diagram of one linear component vibration has to be rotated in order to obtain a second one which, superimposed on the first with an appropriate positive phase difference, gives the rotational oscillation.

If there is only one vibration of a certain (degenerate) species its ζ_i value is independent of the force constants. For example, for the X_3 molecule, the only degenerate vibration has $\zeta_2 = -1$ (see above). It can be seen from Fig. 33a that the superposition of $\nu_{2a}^{(0)}$ and $\nu_{2a}^{(240)}$ of Fig. 32 with a phase difference of 120° gives the clockwise rotational oscillation in Fig. 116b, while $\nu_{2a}^{(240)}$ arises from $\nu_{2a}^{(0)}$ by a counter-clockwise rotation through 120°.

It must be emphasized that the above *sum rules for* ζ, like the product rule for the isotope effect, *hold rigorously only as long as anharmonicity can be neglected and no resonances occur.*

In all the above considerations we have assumed that only one degenerate vibration is singly excited. If *several* (doubly) *degenerate vibrations are multiply excited,* $\mp 2A_{[v]}\zeta_i K$ in (IV, 42) has to be replaced by

$$-2A_{[v]} \sum_i (\pm \zeta_i l_i) K, \qquad \text{(IV, 50)}$$

as has been shown in detail for XY_3 molecules by Silver and Shaffer (790) and Shaffer (776) and for XYZ_3 molecules by Shaffer (777). Here l_i is the quantum number introduced previously (p. 81), which assumes the values v_i, $v_i - 2$, $v_i - 4$, $\cdots$, 1, or 0. Since there are always two values of $\sum (\pm \zeta_i l_i)$ equal in magnitude but opposite in sign, we have again a splitting of each degenerate vibrational level into two ($+l$ and $-l$ level) for $K \neq 0$, as in Fig. 117, the splitting increasing with increasing K.

When a degenerate vibration ν_i is doubly excited we have $l_i = 2$, or 0, and therefore $\sum (\pm \zeta_i l_i) = \pm 2\zeta_i$ or 0 respectively. Thus the substate $2\nu_i(E)$ has twice the splitting of $\nu_i(e)$ while of course the substate $2\nu_i(A_1)$ does not split. For the state $3\nu_i$ we have $l_i = 3$ and 1, corresponding to $\sum (\pm \zeta_i l_i) = \pm 3\zeta_i$ and $\pm\zeta_i$. Thus the substate $3\nu_i(E)$ gives the same splitting as ν_i while the substates $3\nu_i(A_1)$ and $3\nu_i(A_2)$ which together form the state $l_i = 3$ will split by three times this amount. But it must be remembered (see p. 219) that the two states $3\nu_i(A_1)$ and $3\nu_i(A_2)$ may have different energy even without rotation. In this case, therefore, the l_i degeneracy can be removed both by Coriolis interaction of rotation and vibration and by Fermi interaction of different vibrations.

If two degenerate vibrations are singly excited, say in the state $\nu_i + \nu_k$, $\sum (\pm \zeta_i l_i)$ has the values $\pm (\zeta_i + \zeta_k)$ and $\pm (\zeta_i - \zeta_k)$, the first corresponding to the substate $\nu_i + \nu_k(E)$, the second to the two substates $\nu_i + \nu_k(A_1 + A_2)$. The additional excitation of non-degenerate vibrations does not influence the ζ values.

It should be mentioned that for the higher vibrational levels the ζ_i cannot be expected to be exactly the same as for the fundamentals. Rather we expect a dependence of the ζ_i on the v_i similar to that for B_v and D_v:

$$\zeta_i^{[v]} = \zeta_i^e - \sum \alpha_i^\zeta \left(v_i + \frac{d_i}{2}\right).$$

The sum rule holds rigorously only for the ζ_i^e. However, this effect has not as yet been discussed theoretically and the experimental data are not sufficient to establish it.

Symmetry properties of the rotational levels. For a molecule that is accidentally a symmetric top the only symmetry property of the rotational levels is the property "*positive*" or "*negative*"; that is, the *total eigenfunction remains unchanged or changes sign for a reflection at the origin.* If the molecule is *non-planar*, each of the levels considered in the preceding discussion has a positive and a negative sublevel which may be considered as coincident for most practical purposes (however, see below); if the molecule is *planar*, the levels are partly positive and partly negative, as indicated in Fig. 8b for the case in which the electronic and vibrational state is totally symmetric. In a non-totally symmetric vibrational (electronic) level, the property positive-negative will be reversed compared to Fig. 8b if the vibrational (electronic) eigenfunction changes sign for an inversion (obtained by one of the symmetry operations followed by a rotation about the top axis).

If the molecule is a symmetric top because of having a more-than-two-fold axis of symmetry, additional symmetry properties of the rotational eigenfunctions have to be considered, since certain rotations are symmetry operations, depending on the point group to which the molecule belongs. *All the symmetry operations of a point group that are equivalent to rotations form the rotational subgroup.* For example, for the point group $\boldsymbol{C_{3v}}$ the rotations about the three-fold axis belong to the rotational subgroup, but not the reflections at the three planes of symmetry. The rotational subgroup is therefore $\boldsymbol{C_3}$. Similarly, in other cases the rotational subgroup has all the p-fold axes of the point group considered but no other elements of symmetry. Thus the rotational subgroup of $\boldsymbol{D_{3h}}$ is $\boldsymbol{D_3}$; of $\boldsymbol{D_{6h}}$, it is $\boldsymbol{D_6}$; of $\boldsymbol{T_d}$, it is $\boldsymbol{T}$; and so on.

Similar to the case of the vibrational eigenfunctions, the *rotational eigenfunctions may belong to any one of the symmetry types (species) of the rotational subgroup.* For example, for the rotational subgroup $\boldsymbol{C_3}$ of $\boldsymbol{C_{3v}}$ we have the species A and E (see Table 25). Thus the rotational eigenfunctions of molecules such as NH_3 and CH_3F are either of species A or of species E. The rotational eigenfunctions of molecules such as C_3H_6 (cyclopropane) and C_2H_6 (ethane) can have the species A_1, A_2, and E, and similarly in other cases.

The eigenfunctions of the symmetric top may be written [see equation (I, 26)]

$$\psi_r = \Theta_{JKM}(\vartheta) \cdot e^{iM\chi}\, e^{\pm iK\varphi}, \tag{IV, 51}$$

where φ is the angle of rotation about the top axis. It is immediately clear that if φ increases by $2\pi/3$ the rotational eigenfunction will remain unchanged if K is a multiple of 3.[20] In the case of a molecule of point group $\boldsymbol{C_{3v}}$ (or $\boldsymbol{C_{3h}}$, or $\boldsymbol{C_3}$) the rota-

[20] This holds irrespective of the symmetry of the molecule. But only for molecules with a three-fold axis will it lead to any consequence (see further below). Similar statements can of course be made for rotations by $2\pi/p$ for any p if K is a multiple of p.

tional eigenfunctions for $K = 3q$ ($q = 0, 1, 2, \cdots$) have therefore the species A of the rotational subgroup. This holds for both components of the doubly degenerate rotational levels when $K \neq 0$. If K is not a multiple of 3 ($K = 3q \pm 1$) the rotational eigenfunction does not remain unchanged, that is, it is of species E. These symmetry properties are indicated in the energy-level diagram, Fig. 118a where for the present case the difference between A_1 and A_2 should be disregarded. Similar considerations apply to other point groups [see Wilson (933) and below].

What matters for the selection rules is not the species of the rotational eigenfunction alone but the *species of the total eigenfunction* (*over-all species*). Correspondingly a rotational level of a molecule of point group $\boldsymbol{C}_{3v}$ is said to be of species A or E depending on whether the total eigenfunction ψ_{evr} (exclusive of nuclear spin; see below) is of species A or E with respect to the rotational subgroup $\boldsymbol{C}_3$ of $\boldsymbol{C}_{3v}$, and similarly for other point groups.[21]

In order to find the over-all species we have to remember that

$$\psi_{evr} = \psi_e \psi_v \psi_r + \psi'_{evr}, \tag{IV, 52}$$

where ψ_e, ψ_v, ψ_r are the electronic, vibrational, and rotational eigenfunctions and ψ'_{evr} is a small correction term corresponding to the mutual interaction of the three motions. The species of ψ_{evr} is therefore that of the species of the product $\psi_e\psi_v\psi_r$, which is obtained by "multiplying" the species of ψ_e, ψ_v, and ψ_r in the same way as was explained previously for the derivation of the species of the higher vibrational levels (see Chapter II, section 3e). In the case of the rotation-vibration spectra, ψ_e is practically always totally symmetric, and therefore we need only determine the species of $\psi_v\psi_r$. The species of ψ_v with respect to the rotational subgroup is immediately obtained from the species with respect to the complete point group simply by dropping the indices that distinguish species which have the same characters for all p-fold rotations. Thus both species A_1 and species A_2 of $\boldsymbol{C}_{3v}$ (and similarly A' and A'' of $\boldsymbol{C}_{3h}$) belong to species A of the rotational subgroup $\boldsymbol{C}_3$, and of course E of $\boldsymbol{C}_{3v}$ (and $\boldsymbol{C}_{3h}$) remains E of $\boldsymbol{C}_3$. Similarly A_1' and A_1'' of $\boldsymbol{D}_{3h}$ belong to A_1 of the rotational subgroup $\boldsymbol{D}_3$, A_2' and A_2'' to A_2, and E' and E'' to E.

If for a *molecule of point group* $\boldsymbol{C}_{3v}$ the vibrational state has species A_1 or A_2 (either of which is totally symmetric with respect to the rotational subgroup), the rotational levels are of species A or E depending on whether the rotational eigenfunction is of species A or E, that is, the species are those given in Fig. 118a, ignoring the subscripts 1 and 2 of A. However, if the vibrational state is of species E the situation is different. For the rotational levels whose ψ_r is of species A (that is, for the levels $K = 3q$), the product $\psi_v\psi_r$ (and therefore $\psi_e\psi_v\psi_r$) is of species $A \times E = E$ (see Table 31). For the rotational levels whose ψ_r is of species E (that is, for the levels $K = 3q \pm 1$), the product $\psi_v\psi_r$ (and therefore $\psi_e\psi_v\psi_r$) has species $E \times E = A + A + E$ (see Table 33). In the first case ($K = 3q$) there are, except for $K = 0$, two levels of species E for each J, whereas in the second case ($K = 3q \pm 1$) there are three levels, two of species A and one of species E (doubly degenerate). Without the influence of the Coriolis force the sublevels with a given J and $K \neq 0$ have the same energy. But if the Coriolis interaction is taken into account they split into two levels: each of these, for $K = 3q$, is doubly degenerate (species E); for $K = 3q \pm 1$, only one of the levels has species E (for $K = 3q - 1$ the $+l$ level,

[21] It will be recalled that also for linear molecules the symmetry + or − and s or a of the rotational levels depends on the symmetry of the total eigenfunction (exclusive of nuclear spin).

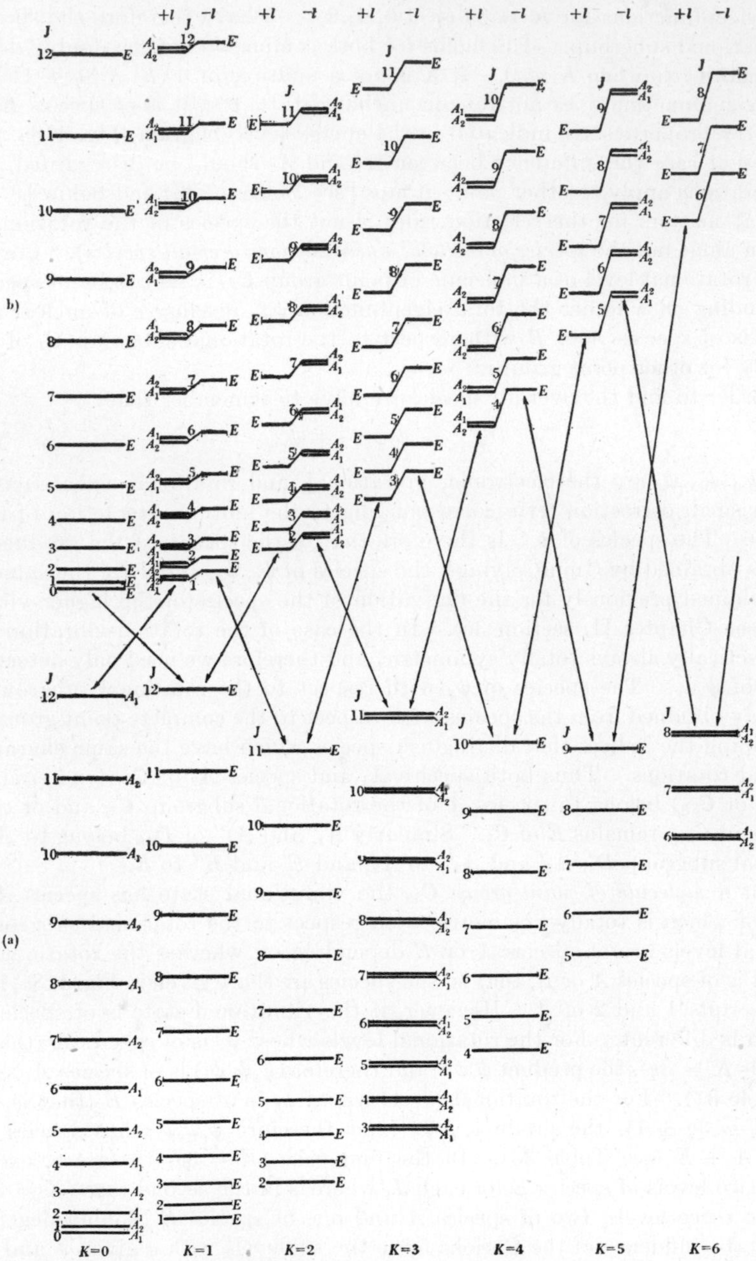

Fig. 118. Symmetry properties of the rotational levels of molecules with a three-fold axis (a) in a totally symmetric vibrational state (b) in a degenerate vibrational state.—The species indicated refer to molecules of point groups D_{3h}, D_{3d} and D_3. If the subscript 1 or 2 of A is dropped they apply to molecules of point groups C_{3v}, C_{3h} and C_3. For point groups D_{3h}, D_{3d}, C_{3v} and C_{3h} part (a) of the figure applies also to A_1'', A_{1u}, A_2 and A'' vibrational states respectively. For A_2', A_2'' of D_{3h} and A_{2g}, A_{2u} of D_{3d} the figure applies if A_1 and A_2 are interchanged. Only those levels

for $K = 3q + 1$ the $-l$ level), while the other consists of two coinciding levels of species A. This is shown in Fig. 118b, again ignoring the subscripts 1 and 2. The degeneracy of the rotational levels of species E persists even if all interactions are taken into account, whereas the pairs of coinciding levels of species A may split [see Wilson (934) and below]. We call the latter splitting K-type doubling. It is indicated in Fig. 118, but has not as yet been observed.

For a *molecule of point group* D_{3h} (and similarly D_{3d}) we have the rotational species A_1, A_2, and E (see above). The rotational eigenfunctions for $K = 0$ are of species A_1 for even J and of species A_2 for odd J, since ψ_r changes sign for a rotation by 180° about an axis perpendicular to the symmetry axis when J is odd but remains unchanged when J is even. For $K = 3q \neq 0$ there is a function ψ_r of species A_1 and one of species A_2 for each J, whereas for $K = 3q \pm 1$ as before the ψ_r is of species E. From this the over-all species of $\psi_e\psi_v\psi_r$ can be determined in the same way as above. The result for the vibrational species A_1', A_1'', E', E'' is shown in Fig. 118, now taking account of the subscripts 1 and 2. For A_2' and A_2'' vibrational levels the A_1 and A_2 in Fig. 118a would have to be interchanged. Other point groups have been considered by Wilson (933).

We have now to consider the *influence of nuclear spin and statistics.* Let us first consider the case of *zero nuclear spin* of the Y nuclei in a non-planar molecule XY_3 of point group C_{3v} (the same considerations would also apply to any C_{3v} molecule if all identical nuclei have zero nuclear spin). The rotation of the molecule by 120° about the top axis is equivalent to two successive exchanges of two pairs of identical nuclei. Therefore, for Bose or Fermi statistics of the identical nuclei the total eigenfunction must remain unchanged; that is, all those energy levels in Fig. 118 whose eigenfunctions do not remain unchanged for such a rotation cannot occur. *Only the levels with over-all species A occur for zero nuclear spin of the identical atoms;* that is, for non-degenerate vibrational states only those with $K = 3q$, and for degenerate vibrational states only half of those with $K = 3q \pm 1$. For a plane XY_3 molecule, in addition, a rotation about one of the two-fold axes is equivalent to the exchange of two identical nuclei. Therefore, for Bose statistics of the identical nuclei of spin zero only the A_1 levels in Fig. 118 can occur, since only for them do the eigenfunctions remain unchanged for such a rotation, that is, such an exchange of nuclei. For Fermi statistics only the A_2 levels in Fig. 118 would occur, since the eigenfunction must be antisymmetric with respect to an exchange of identical nuclei. However, actually, nuclei with zero spin and Fermi statistics do not exist, so that only the first case is realized. Thus in molecules like SO_3, $CO_3^=$, if they have point group D_{3h} (as is very likely), in the non-degenerate vibrational states only the rotational levels with $K = 0, 3, 6, 9, \cdots$ occur (and for $K = 0$ only those with even J) while for degenerate vibrational states only the rotational levels with $K = 1, 2, 4, 5, 7, 8, \cdots$ occur, and of these only one sublevel for every J (see Fig. 118).

Similar considerations apply to other molecules with identical nuclei of spin zero (and Bose statistics). Always *only the rotational levels of totally symmetric over-all species* occur.

If the *spin of the identical nuclei* is *different from zero,* the function ψ_{evr} in (IV, 52) is no longer the total eigenfunction, but we have to add a factor ψ_s, the *nuclear spin*

---

are drawn separately that at least in a sufficiently high approximation form separate levels. The E levels are doubly degenerate, but do not split in any approximation. The K values given at the bottom refer to both (a) and (b). The oblique arrows indicate the possible transitions (see p. 429).

function (compare the analogous situation for diatomic molecules, Molecular Spectra I, p. 144f.). The important point is that by the inclusion of this factor the total eigenfunction can be made to have the proper symmetry with respect to an exchange of any two identical nuclei (symmetric for Bose, antisymmetric for Fermi statistics), even though $\psi_e\psi_v\psi_r$, and therefore ψ_{evr}, does not have the proper symmetry. Therefore, *in general all rotational levels* (for example, in Fig. 118) *can occur although with different statistical weights.* Here the statistical weight is the number of independent eigenfunctions for the level considered.

We shall consider in somewhat more detail only the case of the molecule XY_3 with the spin $I = \frac{1}{2}$ of the identical Y nuclei, both for the planar and non-planar case [see Wilson (933)]. The same considerations apply also to axial XYZ_3 molecules. There are eight possible orientations of the three spins, as shown in Fig. 119; that is, there are eight different spin functions ψ_s. The first and last of these are totally symmetric with respect to all rotations permitted by the symmetry of the molecule; that is, they have species A or A_1 for C_{3v} or D_{3h} respectively. Any permutation of the nuclei that is equivalent to a rotation leaves these functions unchanged. Although this is not the case for the remaining six functions, there are two linear combinations of them, namely

$$\psi_s^{\text{II}} + \psi_s^{\text{III}} + \psi_s^{\text{IV}} \quad \text{and} \quad \psi_s^{\text{V}} + \psi_s^{\text{VI}} + \psi_s^{\text{VII}}$$

that are totally symmetric. The four remaining spin functions that are linearly independent of the four just considered (for example ψ_s^{II}, ψ_s^{III}, ψ_s^{V}, ψ_s^{VI}) are degenerate, since they change by more than just the sign for at least one of the permutations that are equivalent to rotations. Thus we have four totally symmetric and two doubly degenerate spin functions ($4A + 2E$ and $4A_1 + 2E$ respectively).

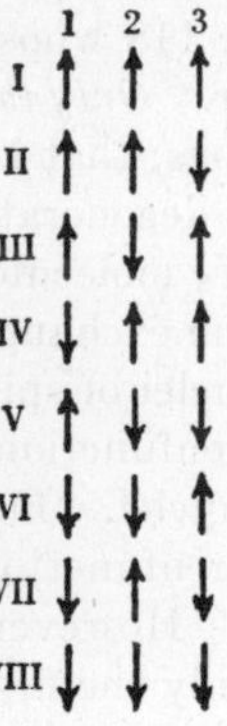

FIG. 119. **Possible spin orientations of three identical nuclei with $I = \frac{1}{2}$.**

The species of the total eigenfunction is obtained from the species of ψ_s and of ψ_{evr} (that is, here, of $\psi_v\psi_r$ since ψ_e is assumed to be totally symmetric) in the same way as the species of $\psi_v\psi_r$ is obtained from those of ψ_v and ψ_r.

In the case of a *non-planar* XY_3 *molecule* (and similarly for any C_{3v} molecule with only three identical atoms of nuclear spin $\frac{1}{2}$ outside the axis of symmetry), the total eigenfunction (inclusive of nuclear spin) will have species A or E for all A rotational levels depending on whether the spin function is A or E, while for E rotational levels the total eigenfunction has species E or $E + 2A$, for spin functions A and E respectively. Even though only levels with species A of the total eigenfunction can occur for either statistics of the nuclei (see above), now, unlike the case of zero nuclear spin, both the A and E rotational levels (species A and E of ψ_{evr}) can occur if combined with appropriate spin functions. But the E rotational levels, in spite of their double degeneracy, have only the same statistical weight (apart from the factor $2J + 1$) as the A levels, because there are only two doubly degenerate spin functions (see above) and because only half the number of spin sublevels are totally symmetric ($E \times E = 2A + E$). Thus, if as usual the K doubling is not resolved, for a *non-degenerate vibrational state* (Fig. 118a) the *rotational levels with* $K = 3q$ ($\neq 0$) *have double the statistical weight of the levels* $K = 3q \pm 1$; that is, we have an alternation 2, 1, 1, 2, 1, 1, 2, $\cdots$. For a *degenerate vibrational level* (Fig.

118b) a *similar alternation* results for both $+l$ and $-l$ sublevels, but for the $+l$ sublevels the rotational levels with $K = 3q + 1$, and for the $-l$ sublevels the rotational levels with $K = 3q - 1$ have the higher statistical weight (since they are the A levels).

For a *planar* XY_3 *molecule* (symmetry $\boldsymbol{D_{3h}}$) and Bose statistics of the nuclei the species of the total eigenfunction must be A_1, for Fermi statistics it must be A_2 (see above). If $I = \frac{1}{2}$ the nuclei always follow Fermi statistics. Since the spin functions have species A_1 and E only, the total eigenfunction can have species A_2 only for the rotational levels of species A_2 and E. Therefore the A_1 rotational levels (Fig. 118) do not occur, since $A_1 \times A_1 = A_1$ and $A_1 \times E = E$. Hence for $K = 0$ in an A_1 vibrational state only the levels with odd J values occur; in an A_2 vibrational state only the levels with even J would occur. Since $E \times E = A_1 + A_2 + E$, and since there are four A_1 and two E spin functions, the E rotational levels have only the weight 2, whereas the A_2 rotational levels have the weight 4. Thus for $K \neq 0$ we have again (as for $\boldsymbol{C_{3v}}$ molecules) the alternation 1, 1, 2, 1, 1, 2, $\cdots$ of the statistical weights as a function of K. It should be noted that the absence of the A_1 rotational levels follows only for $I = \frac{1}{2}$. For higher I values of the identical nuclei, spin functions of species A_2 also occur and therefore all rotational levels can occur. The ratios of the statistical weights for such cases have been given earlier (see Chapter I, p. 28).

In the case of both planar and non-planar XY_3 molecules the *alternation of statistical weights as a function of* K occurs also for larger spin values, but the ratio is reduced. It is 8, 8, 11, 8, 8, 11 $\cdots$ for $I = 1$, the ratio becoming closer to 1 for still larger I values (see Chapter I, p. 28). This holds in the same way for axial molecules like $CHCl_3$.

For molecules of point groups $\boldsymbol{D_{3h}}$ and $\boldsymbol{D_{3d}}$ with six (or more) identical atoms such as ethane (C_2H_6) or cyclopropane (C_3H_6), the spin functions and therefore the statistical weights of the rotational levels are of course different from the above. They have been given by Wilson (933) (938) [for C_2H_6, see also Schäfer (768)]. Wilson has also discussed the species and the statistical weights of the rotational levels of C_6H_6 (point group $\boldsymbol{D_{6h}}$). For pyramidal XY_4 molecules Placzek and Teller (701) have given the statistical weights.

Inversion doubling. For all *non-planar molecules* each single one of the energy levels thus far considered will actually be double on account of the possibility of inversion. In most cases this doubling can be disregarded since it is immeasurably small; but in a few cases, of which NH_3 is a well-known example, it must be considered.

The dependence of the inversion doubling on the vibrational quantum numbers has been considered in Chapter II, section 5d. It is to be expected that on account of the interaction of rotation and vibration the *doublet splitting will also depend on the rotational quantum numbers.* This dependence, as seems plausible and as has been shown in detail by Sheng, Barker, and Dennison (785), can be taken into account by using effective rotational constants $B_{[v]}^{+}$, $B_{[v]}^{-}$ and $A_{[v]}^{+}$, $A_{[v]}^{-}$ for each inversion sublevel. The previous formulae (IV, 38) and (IV, 39) then hold for the average $B_{[v]}$ and $A_{[v]}$, while for the individual $B_{[v]}^{+}$, $B_{[v]}^{-}$, $A_{[v]}^{+}$, $A_{[v]}^{-}$ similar formulae hold with different α_i^B and α_i^A. The difference between α_i^{B+} and α_i^{B-}, and between α_i^{A+} and α_i^{A-} is fairly large for those vibrations ν_i which tend to produce an inversion (ν_2 for NH_3). In other words, the difference of the effective B and A values is large when the in-

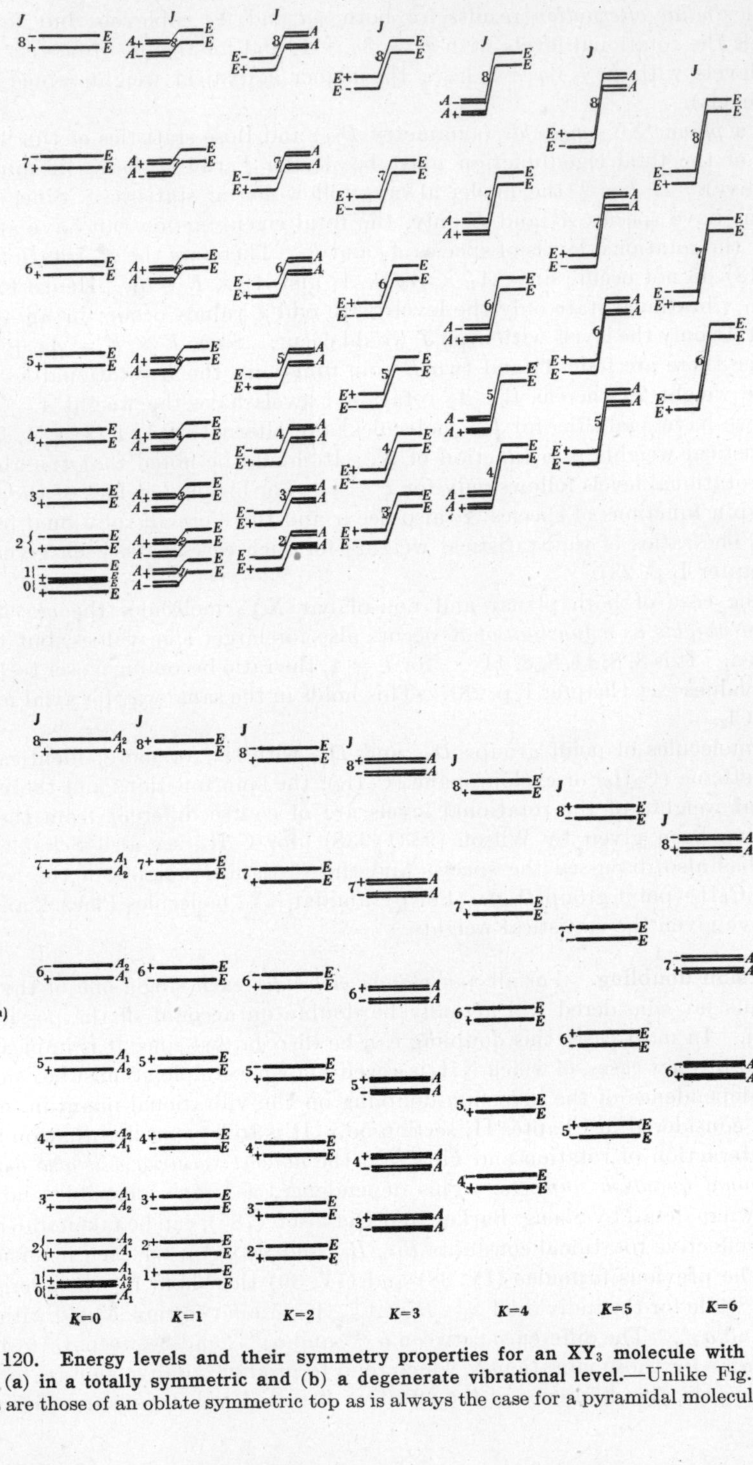

FIG. 120. **Energy levels and their symmetry properties for an XY_3 molecule with inversion doubling (a) in a totally symmetric and (b) a degenerate vibrational level.**—Unlike Fig. 118 here the levels are those of an oblate symmetric top as is always the case for a pyramidal molecule of small

version splitting itself is large. For the lowest vibrational level the difference is usually negligibly small.

If the inversion doubling is not negligible, some special considerations of the *symmetry properties* are necessary. We consider again only the case of an XY_3 molecule of point group $\boldsymbol{C}_{3v}$ (such as NH_3). As we have seen previously (p. 221f.), the vibrational eigenfunction of the lower component of an inversion doublet remains unchanged while that of the upper changes sign for an inversion. Combining this with the +, − character of the rotational levels of an oblate symmetric top (Fig. 8b), we obtain the parities of the rotational levels for a totally symmetric and for a degenerate vibrational level, as indicated on the left of each energy level in Fig. 120. Now it must be realized that the vibrational levels, since each vibrational eigenfunction is the sum or difference of the eigenfunctions of the left and right forms, may be classified according to the species of point group $\boldsymbol{D}_3$. (The potential field has point group $\boldsymbol{D}_{3h}$). It can easily be seen that the "positive" vibrational sublevels of a non-degenerate vibrational state have the (vibrational) species A_1, the "negative" the species A_2. Combining this with the species of the rotational levels in a totally symmetric vibrational level (Fig. 118a), we obtain the over-all species (apart from nuclear spin) given in Fig. 120a at the right of each level. In a similar way the over-all species for a degenerate vibrational level in Fig. 120b are obtained. If now the spin of the identical nuclei is zero, only the A_1 rotational levels can occur. For a totally symmetric vibrational level this means, as before, that only the levels with $K = 0, 3, 6, \cdots$ occur; but if $K = 0$ for even J only the upper, for odd J only the lower doublet component occurs. If the spin of the identical nuclei is $\frac{1}{2}$ (and if they follow Fermi statistics), just as for planar XY_3, only the A_2 and E levels can occur; that is, now all K values occur, but *for* $K = 0$ again *alternately only the upper and lower doublet components occur.* This is the case for NH_3. For $I > \frac{1}{2}$ the spin function may also have species A_2 and therefore all rotational levels occur. This would apply, for example, to ND_3. The statistical weights are the same as for the case of plane XY_3 discussed above.

From the previous formulae for the statistical weights (see above and Chapter I, p. 28), it can easily be checked that *the total statistical weight of each rotational level when the inversion doubling is neglected is the same as the sum of the statistical weights of the inversion sublevels.* Therefore, whenever the inversion doubling is not resolved *it is always possible to disregard its existence entirely* and to consider only one equilibrium position. This would apply, for example, to molecules such as CH_3Cl, CH_3CN, and others.

Perturbations. In symmetric top molecules as in linear molecules, the interaction of rotation and vibration may also lead to somewhat more irregular changes of the energy levels—perturbations.

Once again we have Fermi and Coriolis perturbations, each of which may give rise to *vibrational or rotational perturbations.* Only levels of the same over-all species, the same J, and with $\Delta K = 0, \pm 1$ can perturb one another. Except for the changed species the considerations are perfectly analogous to those given previously for linear molecules. However, it must be realized that the E rotational levels cannot be further split by any interaction of rotation and vibration [see Wilson (934)].

---

height. In order to avoid crowding of lettering A has been used whenever an A_1 and A_2 level are very close together (usually not resolved). Thus in part (a) for $K = 3$, $J = 3$ there are the levels A_2, A_1, A_1A_2 in this or the opposite order. The K values indicated at the bottom refer to both part (a) and part (b).

In contrast to the effect of the Coriolis force discussed above which produces a splitting of degenerate vibrational levels with increasing K and is a first-order effect, the Coriolis perturbations here considered are second- or higher-order effects, being due to the interaction of two different vibrations on account of Coriolis forces. As in the case of linear molecules, this effect is usually quite small. In the case of C_{3v} molecules it is seen immediately from Jahn's rule, given previously (p. 376), that Coriolis perturbations between A_1 and E, A_2 and E, A_1 and A_2, E and E vibrational levels are possible. For the first two pairs, the perturbation would increase with increasing J, for the last two with increasing K. No such cases have as yet been studied in detail. A special case of such perturbations is the K-type doubling mentioned previously, that is, the splitting of a level with a given J and $K \neq 0$ if the over-all species of the two component levels is non-degenerate. But also this splitting has not as yet been observed.

(b) *Infrared spectrum*

Selection rules. It can be shown (see below) that to a good approximation the selection rules for the infrared vibration-rotation spectra of symmetric top molecules are the same as those for the rotation spectra and the vibration spectra separately, except that for the rotational transitions it is now the direction of the *change of dipole moment* (or in other words of the *transition moment*) rather than the direction of the permanent dipole moment that matters.

Thus, if the *transition moment* of the vibrational transition (see Table 55) is *parallel to the top axis* (|| *band*) we have for the rotational quantum numbers

$$\Delta K = 0, \quad \Delta J = 0, \pm 1, \quad \text{if} \quad K \neq 0, \tag{IV, 53a}$$

$$\Delta K = 0, \quad \Delta J = \pm 1, \quad \text{if} \quad K = 0. \tag{IV, 53b}$$

And if the transition moment is *perpendicular to the top axis* (⊥ *band*), we have

$$\Delta K = \pm 1, \quad \Delta J = 0, \pm 1. \tag{IV, 54}$$

If the transition moment has a component both in the direction of the top axis and perpendicular to it, as is usual for an accidentally symmetric top, both the transitions allowed according to (IV, 53) and those allowed according to (IV, 54) may occur (*hybrid band;* see further below).

The proof of the above selection rules proceeds in a way similar to the one indicated for the rotation spectrum in Chapter I, section 2. As there, we start out from the matrix elements (I, 35) of the electric dipole moment; now, however, in a first approximation,

$$\psi = \psi_v\psi_r. \tag{IV, 55}$$

Substituting this into (I, 35) and expressing the M_{xf}, M_{yf}, M_{zf} in terms of the components M_x, M_y, M_z of the dipole moment with respect to a coordinate system fixed in the molecule according to (I, 36), we obtain, instead of (I, 37),

$$R_{xf} = \int M_x\psi_v'\psi_v''^* d\tau_v \int \cos\alpha_x\psi_r'\psi_r''^* d\tau_r + \int M_y\psi_v'\psi_v''^* d\tau_v \int \cos\alpha_y\psi_r'\psi_r''^* d\tau_r + \int M_z\psi_v'\psi_v''^* d\tau_v \int \cos\alpha_z\psi_r'\psi_r''^* d\tau_r \tag{IV, 56}$$

and similar expressions for R_{yf} and R_{zf}. Here M_x, M_y, M_z are no longer constant. The integrals $\int M_x\psi_v'\psi_v''^* d\tau_v, \cdots$ are those occurring in the discussion of the vibration spectrum; they are different from zero only when the vibrational selection rules are fulfilled (see Table 55). The integrals $\int \cos\alpha_x\psi_r'\psi_r''^* d\tau_r, \cdots \int \cos\beta_x\psi_r'\psi_r''^* d\tau_r, \cdots, \int \cos\gamma_x\psi_r'\psi_r'' d\tau_r, \cdots$ are those occurring in the discussion of the rotation spectrum [in equation (I, 37)]; they are different from zero only when the rotational selection rules of the symmetric top are fulfilled (see p. 32). For a transition to take place, both the first and the second integral must be different from zero for at least one of the terms in (IV, 56) or in the similar equations for R_{yf} and R_{zf}. Thus the above selection rules for K and J follow in the same way as for the rotation spectrum.

It will be noted from the above derivation that the validity of the vibrational and rotational selection rules for the rotation-vibration spectrum is dependent upon the possibility of writing ψ as the product $\psi_v\psi_r$. This is no longer possible when the interaction of rotation and vibration is strong and therefore, when this is the case, forbidden transitions may be expected that are not in agreement with the vibrational and rotational selection rules (see also p. 456).

In addition to the above selection rules for the rotational quantum numbers, there are also *selection rules which are concerned with the symmetry properties of the rotational levels.* For all symmetric top molecules we have (as for the pure rotation spectrum),

$$+ \leftrightarrow -, \quad + \not\leftrightarrow +, \quad - \not\leftrightarrow -, \tag{IV, 57}$$

where + and − refer now to the over-all symmetry with respect to inversion. For non-planar molecules this rule is of no consequence as long as the inversion doubling is not resolved, since always a positive and a negative level coincide. But it must be taken into account when the separate inversion doublet components are considered (for example in NH_3; see Fig. 120) and for planar molecules.

If the molecule is a symmetric top on account of its symmetry, there is, in addition, the rule that *only rotational levels of the same over-all species* (apart from nuclear spin) *can combine with one another.* The reason for this rule is the same as for the rule that for homonuclear diatomic molecules symmetric rotational levels do not combine with antisymmetric (see Molecular Spectra I, p. 139). The former rule is just as rigorous as the latter. It holds for all kinds of transitions, even those produced by collisions.[22] Thus (as was pointed out in Chapter I, section 2) NH_3, CH_3Cl, and similar gases (point group $\boldsymbol{C_{3v}}$) consist of *two modifications A and E* which are transformed into each other only extremely slowly, just as are ortho- and para-hydrogen. For molecules of point group $\boldsymbol{D_{3h}}$ there are in general three such modifications (A_1, A_2, E) and similarly in other cases.[23] For NH_3, CH_3Cl, $\cdots$ in the vibrational ground state one modification has only the levels $K = 0, 3, 6, 9, \cdots$, the other only the levels $K = 1, 2, 4, 5, \cdots$.

In a degenerate vibrational state we have to distinguish between $+l$ and $-l$ levels, depending on whether vibrational and rotational angular momentum have the same or opposite sign (see Fig. 117). Teller (836) has shown that *for a transition between an upper degenerate and a lower non-degenerate vibrational state only the $+l$ levels combine with the rotational levels of the non-degenerate state for $\Delta K = +1$, whereas only the $-l$ levels combine with them for $\Delta K = -1$.* The reverse is true if the degenerate state is the lower one (and if we define $\Delta K = K' - K''$ as usual). It is easily seen from Fig. 118 that this rule is in agreement with the rule that only rotational levels of the same species combine with one another. For a transition between two degenerate states we have in general (see p. 268) a || and a ⊥ component ($\Delta K = 0$ and $\Delta K = \pm 1$ respectively). For the former we have $+l \leftrightarrow +l$, $-l \leftrightarrow -l$, whereas for the latter $-l \leftrightarrow +l$ for $\Delta K = +1$ and $+l \leftrightarrow -l$ for $\Delta K = -1$ where the first l refers to the upper, the second to the lower state.[24]

[22] It will be realized that this rule is actually responsible for the fact that for molecules with zero nuclear spin of the identical atoms only levels of one species occur (see p. 409 and Molecular Spectra I, p. 144).

[23] It may be noted that ND_3 has the three modifications A_1, A_2, and E if the inversion doubling is not neglected (see Fig. 120).

[24] These rules are not given by Teller but form a natural extension of Teller's selection rule, and are in agreement with the other symmetry rules.

Rotation and inversion spectrum. Before we discuss the various types of rotation-vibration bands it is appropriate here to reconsider the pure rotation spectrum and the inversion spectrum for cases of XY_3 molecules in which the inversion doubling is resolved and in which the nuclear spin $I(Y) = \frac{1}{2}$ (for example NH_3). It is immediately seen from Fig. 120 that the *rotation lines* ($\Delta J = +1, \Delta K = 0, + \leftrightarrow -$) *are double* with the exception of those with $K = 0$ (since the A_1 levels do not occur). The doublet splitting of the lines is twice the separation of the inversion doublet levels. This splitting is clearly seen in the NH_3 spectrum of Fig. 12a, p. 33. If only lines with $K = 0$ occurred they would be single but alternately shifted in the one or the other direction. Since actually each rotation "line" is a superposition of $J + 1$ lines with $K = 0, 1, \cdots J$, only the line $J = 0$ is single, but the other "lines" have doublet components of slightly different intensity. Alternately the short- and long-wave-length component is the stronger, because the $K = 0$ contribution is missing alternately in the long- and short-wave-length component. This is indeed seen to be the case for NH_3 in Fig. 12a. (The line $J = 0$ has not been observed.) The doublet splitting, according to the above, should depend slightly on J. However, for larger J values, for which this change of doublet width would become noticeable, the splitting of each line into $J + 1$ components likewise becomes appreciable. The resultant somewhat complicated structure has as yet only partially been resolved [see Foley and Randall (324)].

It is seen from Fig. 120 that the transition from one inversion sublevel to the other without change of rotational quantum numbers (*inversion spectrum*) can take place only if $K \neq 0$, since for $K = 0$ only one component level exists for each J. However, for $K \neq 0$ this transition is entirely in conformity with the selection rules and, as mentioned previously (p. 257), has been observed for NH_3 in the region of short radio waves.

Transitions between non-degenerate vibrational levels: parallel bands. In the case of molecules with a more-than-two-fold axis, the change of dipole moment for all allowed transitions between non-degenerate states (see Table 55) is in the direction of the symmetry axis (which is the axis of the top) and therefore only transitions with $\Delta K = 0$ (that is, $\|$ bands) occur. For less symmetric molecules the change of dipole moment for the vibrational transition may also be perpendicular to the top axis and in that case $\Delta K = \pm 1$ (that is, $\perp$ bands) can also occur. In fact, it may happen (for sufficiently low symmetry) that the change of dipole moment makes an intermediate angle with the top axis. In this case both $\Delta K = 0$ and $\Delta K = \pm 1$ may occur, and we have what is called a hybrid band.

Let us first consider a $\|$ band. For such a band only levels of the same K value, that is, levels in the same vertical column in Fig. 121, combine with one another. Considering a particular column, that is, a particular value of K, we obtain, since $\Delta J = 0, \pm 1$, *a sub-band with three simple branches P, Q, and R.* The complete $\|$ band is obtained by *superposition of a number of such sub-bands*, corresponding to the various K values that occur at the temperature of observation. The sub-bands up to $K = 5$ and their superposition are shown schematically in Fig. 122a and b. The $K = 0$ sub-band has no Q branch because of the restriction (IV, 53b).

If, for the moment, we neglect the interaction between vibration and rotation, that is, take $B' = B''$, $A' = A''$, all the sub-bands coincide exactly, since then the spacing of the levels in the upper and lower states is exactly the same (the different columns of levels in Fig. 121 which are responsible for the different sub-bands are identical, apart from a constant shift which is the same for the upper and lower state). In this case, furthermore, all the lines of the Q branch in each sub-band coincide. Thus we obtain *a band with a strong line-like Q branch and a P and an R*

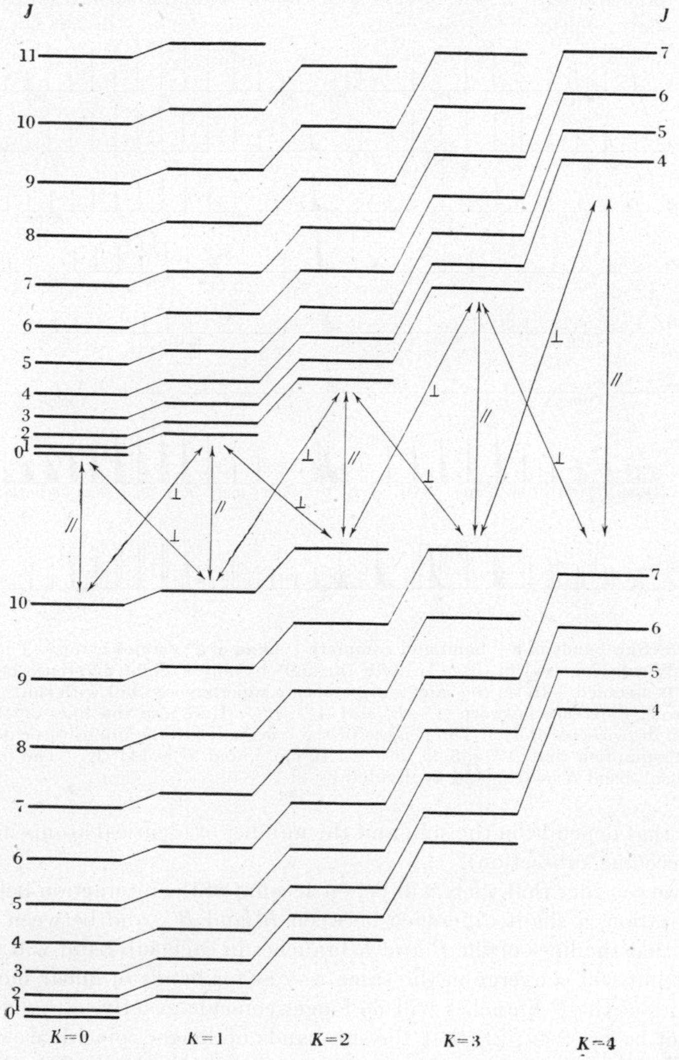

FIG. 121. **Combination of two non-degenerate vibrational levels of a symmetric top molecule.**

branch just like a ⊥ band of a linear molecule. The spacing of the lines in the P and R branches is $2B$, that is, is determined by the moment of inertia about an axis perpendicular to the top axis.

These conclusions are not altered if the symmetry selection rules discussed above are taken into account. They change only the *intensities:* For molecules with a threefold axis the sub-bands with $K = 1, 2, 4, 5, 7, 8, \cdots$ are either missing (nuclear spin $I = 0$) or have lower intensity than the sub-bands with $K = 3, 6, 9, \cdots$ (ratio $1:2$ for $I = \frac{1}{2}$). Furthermore, for molecules of point groups $\boldsymbol{D}_3$, $\boldsymbol{D}_{3h}$, $\boldsymbol{D}_{3d}$, there is an intensity alternation for the sub-band $K = 0$: alternate lines are missing for a plane XY_3 molecule with $I = 0$ or $I = \frac{1}{2}$ of the Y atoms; alternate lines are weaker

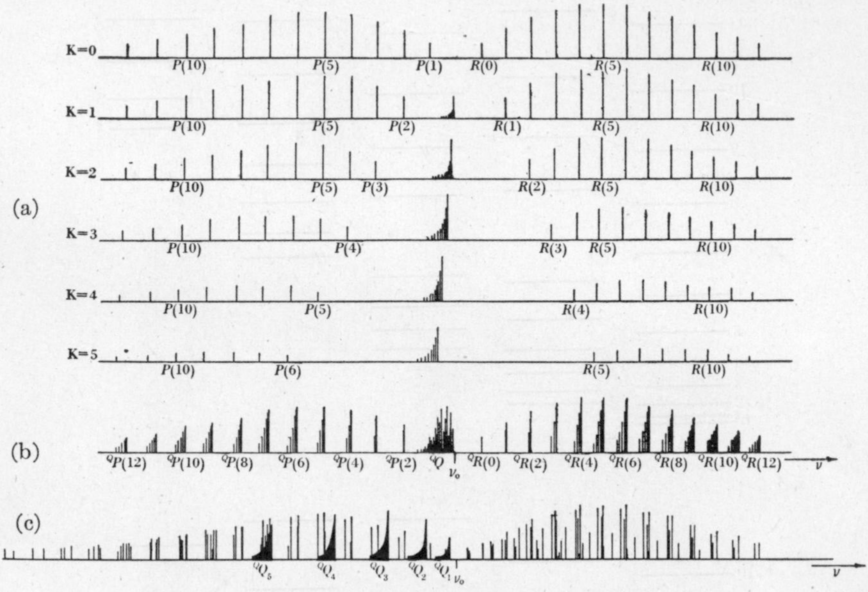

FIG. 122. **Sub-bands of a || band and complete || band of a symmetric top.**—The sub-bands in (a) are directly superimposed in (b). In both (a) and (b) only a slight difference between A'—B' and A''—B'' is assumed. In (c) the same sub-bands are superimposed but with shifts corresponding to a much larger difference between A'—B' and A''—B''. Here also the lines of the Q branches have not been drawn separately. The heights of the lines indicate the intensities calculated on the basis of the assumption that $A'' = 5.25$, $B'' = 1.70$ cm^{-1}, and $T = 144°$ K. The intensities indicated for the sub-band $K = 0$ should be divided by 2.

by a factor that depends on the spin and the number of identical atoms in other cases (see the preceding subsection).

If now we consider that there will be, on account of the interaction between vibration and rotation, a slight difference between B' and B'', and between A' and A'', we see: (1) that the lines of the P and R branches in each sub-band will no longer be equidistant but will converge in the same way as for bands of linear molecules, and that the lines of the Q branches will no longer coincide exactly, although in general they will not be resolved; (2) that the sub-bands no longer coincide exactly.

It is seen immediately from the energy formula (IV, 41) that the lines in each sub-band (of fixed K) are given by the same formulae as for diatomic or linear polyatomic molecules [that is (IV, 19, 20, 22)] except that ν_0 is replaced by ν_0^{sub}, the origin of the sub-band. *Apart from the change of* ν_0^{sub} *the different sub-bands follow exactly the same formula.*[25] However on account of the restriction $J \geq K$, *more and more lines are missing* in the sub-bands at the beginning of the branches (see Fig.

[25] This holds as long as the effect of centrifugal distortion is neglected.

122a). For the *origins of the sub-bands* ($J = 0$) we obtain from (IV, 41)

$$\nu_0^{\text{sub}} = \nu_0 + [(A'_{[v]} - A''_{[v]}) - (B'_{[v]} - B''_{[v]})]K^2. \qquad \text{(IV, 58)}$$

As long as the second term in this equation is small compared to the separation of successive lines ($2B$) in a sub-band, the whole || band will appear to consist, under medium dispersion, of one P, one R, and one line-like Q branch. This is shown in Fig. 122b. It is, however, also clear from this figure as well as from the preceding discussion that (unlike the case of a ⊥ band of a linear molecule) each line consists of a number of components ($J + 1$ in the R branch, J in the P branch).

The simple PQR branch structure has been observed for a number of symmetric top molecules; and conversely, when such a structure is observed, one can conclude definitely *that the molecule investigated is* (at least to a good approximation) *a symmetric top molecule,* if it is known from other evidence that it is not linear.

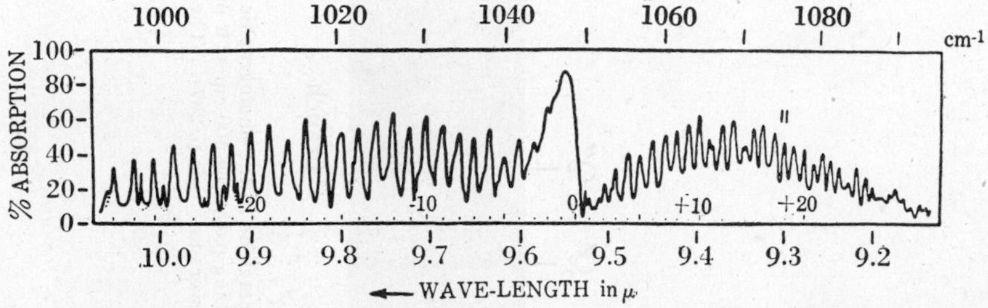

FIG. 123. **Fine structure of the fundamental ν_3 of methyl fluoride at 9.55μ** [after Bennett and Meyer (138)]. The absorbing path was 6 cm. long at a pressure of 4 cm. The numbers above the bottom scale are m values (see p. 381).

As examples we give in Fig. 123 a || band of CH_3F in the ordinary infrared (fundamental ν_3) and in Fig. 124 a || band of CH_3—C≡C—H in the photographic infrared (overtone band $3\nu_1$). The convergence of the lines in the second example is clearly seen. From the average line distance in the bands a rough value of $2B$ is obtained. For a more accurate determination of $B'_{[v]}$ and $B''_{[v]}$ the same methods as for linear molecules have to be applied (see also below).

The second term in (IV, 58) may be small because $(A'_{[v]} - A''_{[v]}) - (B'_{[v]} - B''_{[v]})$ is small, or because only levels with small K values occur, or because of both these reasons. $(A'_{[v]} - A''_{[v]}) - (B'_{[v]} - B''_{[v]})$ will be small for all fundamental bands, but may not be small for overtone bands. Only levels with small K will occur when A − B is large, since then the Boltzmann factor for the levels with higher K is very small. Thus, for CH_3F and CH_3—C≡C—H, for which $A - B$ is approximately 5 cm^{-1}, the first six K values at room temperature contribute more than 75 per cent of the intensity (compare Fig. 10c).

If $(A'_{[v]} - A''_{[v]}) - (B'_{[v]} - B''_{[v]})$ is large, as may happen for overtone bands, but at the same time $B'_{[v]} - B''_{[v]}$ is still small, each sub-band will still have a line-like Q branch; however, the Q branches of the different sub-bands will no longer even approximately coincide, but according to (IV, 58) form a resolved Q branch of "lines," the first of which has $K = 1$. This is shown in Fig. 122c. The P and R branches in this case form a rather irregular background of weaker lines. In such cases it is necessary to have a designation of the lines that will distinguish different sub-bands:

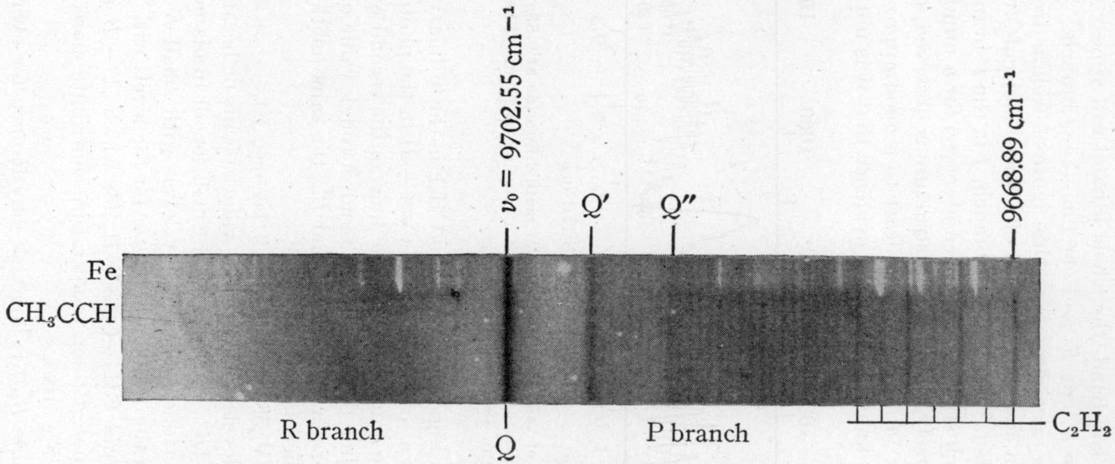

Fig. 124. **Fine structure of the band $3\nu_1$ of methyl acetylene in the photographic infrared at 1.030μ** [after Herzberg, Patat and Verleger (440)].—The length of the absorbing path was 400 cm at a pressure of 760 mm. Q' and Q'' are the Q branches of corresponding bands with excited lower state (see p. 267). The C_2H_2 band at 1.04μ appears as an impurity.

we indicate the ΔJ value by the usual P, Q, R, the ΔK value by a left superscript P, Q, R (for $\|$ bands Q) and the K value by a subscript. Thus ${}^QP_3(3)$ would be the line with $J = 3$ of the P branch of the sub-band with $K = 3$ of a $\|$ band.[26]

If both $(A'_{[v]} - A''_{[v]})$ and $(B'_{[v]} - B''_{[v]})$ are large so that the ${}^QQ$ branches no longer form "lines," the appearance of a $\|$ band becomes rather complicated. We shall not discuss it in detail.

Returning now to the simple type of $\|$ bands, when $A'_{[v]} - A''_{[v]}$ and $B'_{[v]} - B''_{[v]}$ are very small, let us consider the *intensity distribution.* Detailed formulae for the intensities of the various lines in any $\|$ band are given below. The intensity distribution in the P and R branches is essentially determined by the population of the lower levels irrespective of K; that is, it will be represented closely by the previous (upper) curves in Fig. 10, which give the number of molecules with a given J value (summed over all K values). Because of the smaller number of components for small J, the intensity of the R and P lines with small J is relatively smaller than in a band of a linear molecule. The intensity of the individual lines in the Q branch of a sub-band does not follow the curve for the population of the rotational levels (Fig. 10), but decreases rapidly with increasing J, just as for bands of diatomic molecules with $\Delta\Lambda = 0$. The intensity of the Q branch relative to the P and R branches is different in the different sub-bands (see Fig. 122). It is zero for the sub-band $K = 0$ (as in a Σ—Σ band of a linear molecule) and increases rapidly with increasing K. Thus, if $I_A \ll I_B$, that is, if only relatively few K values are of importance, the intensity of the Q branch is small compared to the total intensity of the P and R branches; but with increasing I_A/I_B the Q branch gains in relative intensity, since higher and higher K values are populated. In Fig. 125 is given a graphical representation of the contributions of P, Q, and R branches in a $\|$ band as a function of I_A/I_B, after Teller (836) [see also Gerhard and Dennison (352)]. It may be noted that in agreement with these considerations the Q branch is much less intense for CH_3—C≡C—H (Fig. 124) than for CH_3F (Fig. 123), since in the former I_A/I_B is smaller.

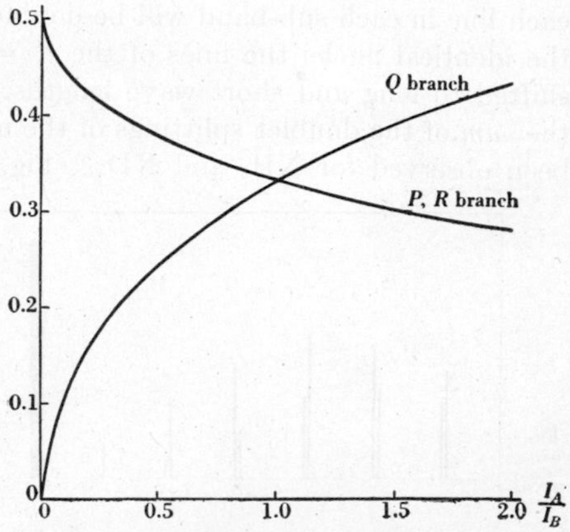

FIG. 125. **Relative intensity of P, Q and R branch as a function of I_A/I_B in a $\|$ band of a symmetric top [after Teller (836)].—This figure holds rigorously only for sufficiently large moments of inertia. For smaller moments of inertia there is a difference in the intensity of P and R branch. But even then the general trend is as indicated.**

The rigorous formulae for the line intensities in the bands of symmetric top molecules were first given, on the basis of the old quantum theory, by Hönl and London (456a) and were later derived on the basis of wave mechanics by Dennison (278), Reiche and Rademaker (734) and others. As mentioned previously (p. 32) the intensity of a given transition in absorption is proportional to the product

$$CA_{KJ}\,\nu\,g_{KJ}\,e^{-F(K,\,J)hc/kT}$$

[26] This nomenclature has of course nothing to do with the superficially similar nomenclature used for multiplet electronic bands of diatomic molecules (see Molecular Spectra I, p. 273).

where C is a constant independent of K and J but depending on the vibrational transition, where g_{KJ} and $F(K, J)$ are statistical weight and term value of the lower state and where A_{KJ} is proportional to the square of the transition moment $R_{xf}^2 + R_{yf}^2 + R_{zf}^2$ summed over all orientations of J. The quantities A_{KJ} in the present case ($\Delta K = 0$) are [see Dennison (279)]

$$\text{for } \Delta J = +1: \quad A_{KJ} = \frac{(J+1)^2 - K^2}{(J+1)(2J+1)}$$

$$\text{for } \Delta J = 0: \quad A_{KJ} = \frac{K^2}{J(J+1)}$$

$$\text{for } \Delta J = -1: \quad A_{KJ} = \frac{J^2 - K^2}{J(2J+1)}$$

where, as always, K and J refer to the lower state. The g_{KJ} in the above expression for the intensity are $2J + 1$ for $K = 0$ and $2(2J + 1)$ for $K \neq 0$ (see p. 27). The intensities indicated in Fig. 122 are based on these formulae.

In the case of C_{3v} molecules for which the *inversion doubling* is not negligible, it is immediately seen from Fig. 120, on the basis of the selection rule $+ \leftrightarrow -$, that each line in each sub-band will be double, with the exception that for $I = 0$ or $\frac{1}{2}$ of the identical nuclei the lines of the $K = 0$ sub-band will be single, but alternately shifted to long and short wave lengths. The doublet splitting of the lines equals the *sum* of the doublet splittings of the upper and lower levels. Such || bands have been observed for NH_3 and ND_3. Fig. 126 gives the fine structure of the funda-

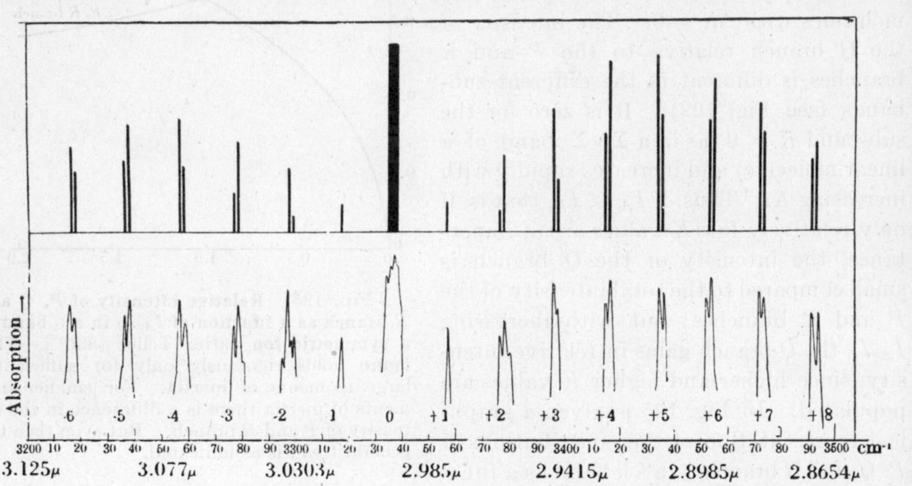

FIG. 126. **Fine structure of the fundamental ν_1 of NH_3 at 3.00μ** [after Dennison and Hardy (281)].—The length of the absorbing path was 60 mm at atmospheric pressure. The numbers above the bottom scale are m values. The wave number scale has not been corrected for vacuum.

mental ν_1 of NH_3, as observed by Dennison and Hardy (281). The theoretical structure and intensity distribution is given at the top. It agrees closely with the observed. As in the case of the rotation spectrum, the unequal intensities of the doublet components are due to the fact that for $K = 0$ the upper and lower levels are alternately missing (see Fig. 120). While this makes very little difference for large J values, where the lines of many sub-bands are contributing to one "line," it does have a considerable effect for small J values. In particular for the first line of P and R branch one component is entirely missing, since only $K = 0$ contributes.

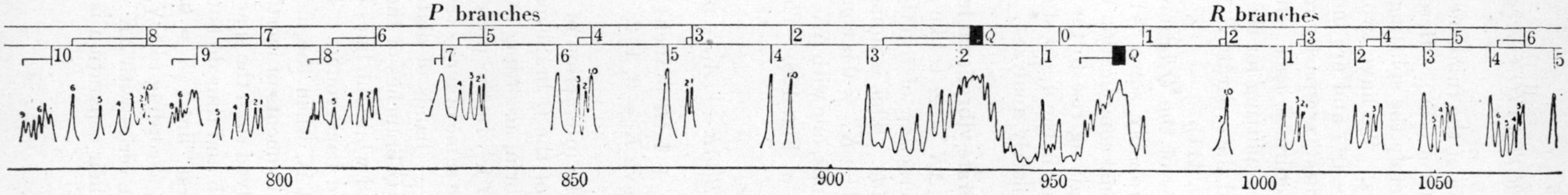

FIG. 127. **Fine structure of the fundamental ν_2 of NH_3 at 10.5μ** [after Sheng, Barker and Dennison (785)].—At the top the two component bands $\nu_2(1^+—0^-)$ and $\nu_2(1^-—0^+)$ are indicated. The numbers on the leading lines are the J values. The numbers directly above some of the maxima are K values. It should be noted that the lines $P(1)$ and $R(0)$ are missing in the $\nu_2(1^+—0^-)$ and $\nu_2(1^-—0^+)$ bands respectively (see p. 422).

In Fig. 127 also the fine structure of the fundamental ν_2 of NH_3 is given, as observed by Sheng, Barker, and Dennison (785). Here the inversion doubling in the upper state is rather large, and therefore the splitting of the "lines" is greater than the separation $2B$ of successive lines. We have two component bands which may be designated $\nu_2(1^+—0^-)$ and $\nu_2(1^-—0^+)$ and are indicated at the top of Fig. 127. At the same time the individual "lines" are partly resolved into their components. The agreement of this k fine structure of each "line" with expectation is very striking. From the difference of this splitting for the two component bands, Sheng, Barker, and Dennison have obtained the separate B_v^+, B_v^-, A_v^+, A_v^- values for the inversion doublet components (see p. 411).

Bands in which neither the lines of the Q branches of each sub-band nor the different sub-bands coincide are probably some NH_3 bands in the photographic infrared. Here we have the additional complication that the upper states consist of a number of vibrational sublevels, giving rise to the overlapping of || bands by $\perp$ bands, and that the inversion doubling in the upper states is large. That is why none of these bands has as yet been completely analyzed.

Transitions between non-degenerate vibrational levels: perpendicular and hybrid bands. A perpendicular band with $\Delta K = \pm 1$, like a || band, consists of a number of sub-bands. Even when the interaction of rotation and vibration is neglected, however, the *sub-bands do not coincide.* This is immediately obvious from Fig. 121 if, for example, the transition $K' = 1 \rightarrow K'' = 0$ is compared with $K' = 0 \rightarrow K'' = 1$. Also there are now two sets of sub-bands, one with $\Delta K = +1$ and one with $\Delta K = -1$. These are shown in the upper part of Fig. 128. The zero lines ν_0^{sub}, according to (IV, 41), are represented by

$$\nu_0^{\text{sub}} = \nu_0 + (A'_{[v]} - B'_{[v]}) \pm 2(A'_{[v]} - B'_{[v]})K + [(A'_{[v]} - B'_{[v]}) - (A''_{[v]} - B''_{[v]})]K^2, \quad \text{(IV, 59)}$$

where the + sign applies to $\Delta K = +1$ (R branch) and the − sign to $\Delta K = -1$ (P branch). In the first case we have $K = 0, 1, 2, \cdots$, in the second case $K = 1, 2, \cdots$ (see Fig. 121).

If, for the moment, we neglect the interaction of vibration and rotation, that is, take $A' = A''$, $B' = B''$, the lines of the Q branches in each sub-band coincide at ν_0^{sub}, and the different Q branches form, according to (IV, 59), a *series of equidistant "lines" with a spacing of* $2(A' - B')$. In the present case the intensity of the Q branch in a sub-band is always of the same order as the intensity of P and R branches together (as for the $\Pi—\Sigma$, $\Delta—\Pi$, $\cdots$ bands of diatomic molecules), and therefore the series of "lines" formed by the Q branches of the different sub-bands forms the most prominent feature of a $\perp$ band while the lines of the P and R branches of the sub-bands form a usually unresolved background. The series of Q branches stands out particularly when $A \gg B$, that is, when the moment of inertia about the top axis is small compared to the other moments of inertia, since then the spacing of the Q branches is much more easily resolved than the spacing in the sub-bands; the latter, in this case, do not overlap too much and thus do not form too strong a background.

It is important to note that the first "line" ($K = 0$) in the series of Q branches with $\Delta K = +1$ ("positive" sub-bands), according to (IV, 59), occurs at $\nu_0 + (A' - B')$, whereas the first line ($K = 1$) in the series with $\Delta K = -1$ ("negative" sub-bands) occurs at $\nu_0 - (A' - B')$. Thus their separation is the same as that of the other

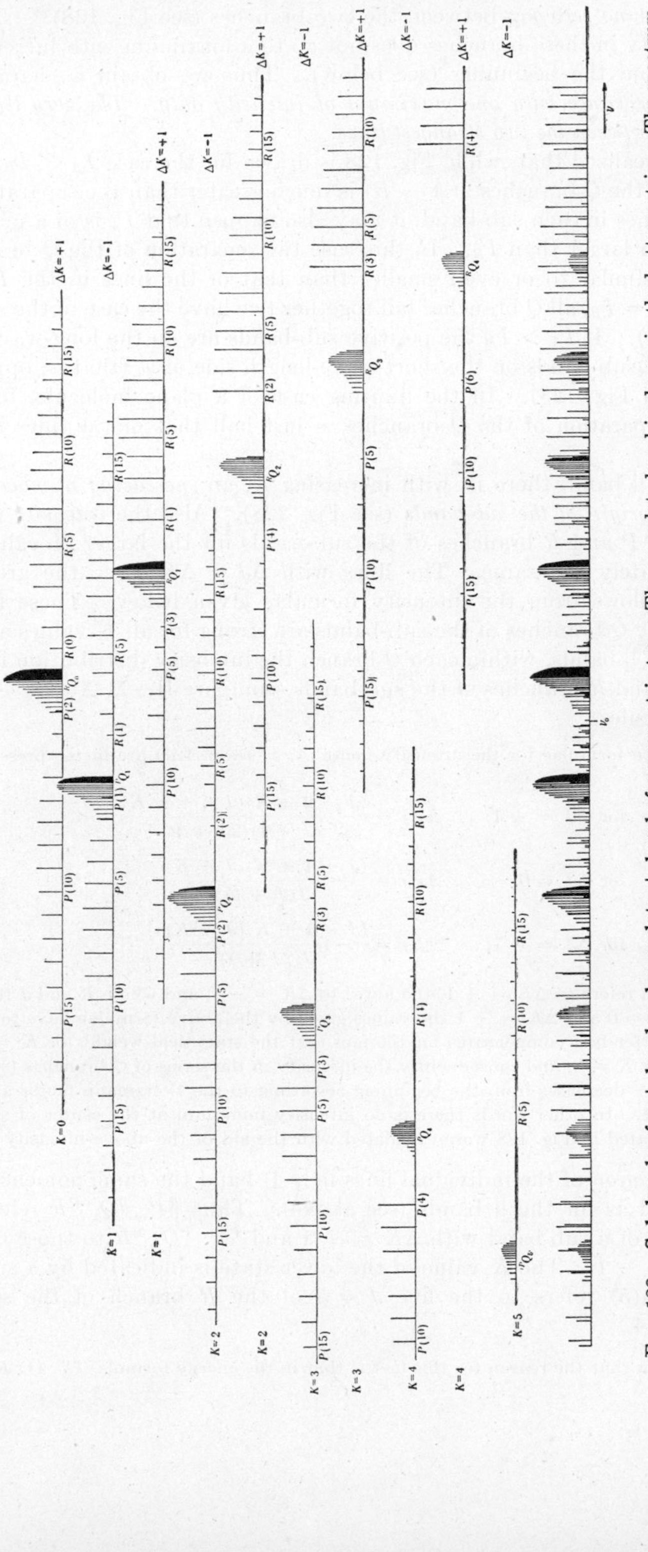

FIG. 128. **Sub-bands of a ⊥ band and complete ⊥ band of a symmetric top.**—The complete band is shown in the bottom strip. The spectrum is drawn under the assumption that $A' = 5.18$, $A'' = 5.25$, $B' = 0.84$, $B'' = 0.85$ cm^{-1} and $\zeta_i = 0$. The intensities were calculated for a temperature of 144° K. It should be realized that if the lines of an individual Q branch are not resolved the resulting "line" would stand out much more prominently than might appear from the spectrum given.

"lines." There is *no zero gap* between the two branches (see Fig. 128).[27] Furthermore, the intensity in these branches does not go to a maximum with increasing K, but decreases from the beginning (see below). Thus we obtain a characteristic *single-branch appearance with one maximum of intensity only. The zero line of the band is half-way between the two strongest lines.*

It should be realized that, while Fig. 128 is drawn for the case $I_A \ll I_B$, so that the separation of the Q branches $2(A - B)$ is much greater than the separation, $2B$, of the P and R lines in each sub-band, it may also happen that I_A is of a magnitude similar to or even larger than I_B. In this case the separation of the Q branches is of a magnitude similar to or even smaller than that of the lines in the P and R branches. If $I_A = I_B$, all Q branches fall together (we have the case of the spherical top; see section 3). If $I_A > I_B$ the positive sub-bands are on the long-wave-length side, the negative sub-bands on the short-wave-length side of ν_0 (that is, opposite to what they are in Fig. 128). In the limiting case of a plane molecule, for which $I_A = 2I_B$, the separation of the Q branches is just half that of the lines in the P and R branches.

Just as for a || band, there is, with increasing K, an *increasing number of lines missing near the origin of the sub-bands* (see Fig. 128). Also the *intensity* of corresponding lines in P and R branches of the sub-bands for the larger K values is no longer approximately the same. The lines with $\Delta J = \Delta K$ have the greater intensity. This follows from the intensity formulae given below. These formulae also show that the Q branches of the sub-bands are strong for all K values and that, unlike the case of || bands, within each Q branch the intensity distribution is similar to that in the P and R branches of the sub-bands (compare Π—Σ, Δ—Π, $\cdots$ bands of diatomic molecules).

The Hönl-London formulae for the intensity factor A_{KJ} (see p. 421) are in the present case of a $\perp$ band

$$\text{for } \Delta J = +1: \qquad A_{KJ} = \frac{(J + 2 \pm K)(J + 1 \pm K)}{(J + 1)(2J + 1)}$$

$$\text{for } \Delta J = 0: \qquad A_{KJ} = \frac{(J + 1 \pm K)(J \mp K)}{J(J + 1)}$$

$$\text{for } \Delta J = -1: \qquad A_{KJ} = \frac{(J - 1 \mp K)(J \mp K)}{J(2J + 1)}$$

where the upper sign refers to $\Delta K = +1$, the lower to $\Delta K = -1$, and where K and J refer to the lower state. For $K = 0$ and $\Delta K = +1$ the values given by the above formulae have to be multiplied by 2. This latter fact compensates for the fact that the statistical weight for $K = 0$ is only half of the weight for $K \neq 0$, and consequently the intensity in the series of Q branches (sub-bands) with $K = 0, 1, 2, \cdots$ decreases from the beginning according to the Boltzmann factor and has no maximum for $K = 1$. In other words there is no intensity minimum at the center of a $\perp$ band. The intensities indicated in Fig. 128 were calculated with the aid of the above intensity formulae.

For the *designation* of the individual lines in a $\perp$ band the same nomenclature is conveniently used as for the || bands (see above). Thus ${}^RP$, ${}^RQ$, ${}^RR$ refer to the P, Q, R branches of a sub-band with $\Delta K = +1$ and ${}^PP$, ${}^PQ$, ${}^PR$ to those of a sub-band with $\Delta K = -1$. The K value of the lower state is indicated by a subscript. For example ${}^PR_4(5)$ refers to the line $J = 5$ of the R branch of the sub-band $K' = 3 \leftarrow K'' = 4$.

[27] It is easily seen that the reason for this fact is that in the energy formula (IV, 41) K^2 and not $K(K + 1)$ occurs.

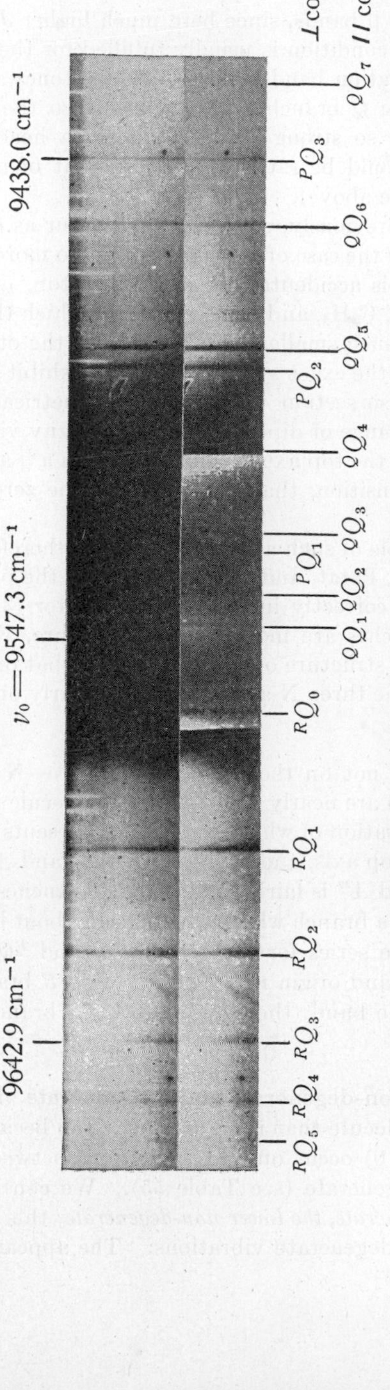

FIG. 129. **Photographic infrared band of N_3H at 1.047μ.**—The length of the absorbing path was 400 cm at a pressure of about 400 mm. Since the central part of the band is so much more intense, in the lower part of the spectrogram the exposure time of the central part has been greatly reduced in making the enlargement. The upper part shows the correct intensity relations. The long leading lines refer to the ∥ component the short leading lines to the ⊥ component of the band.

If the *interaction between rotation and vibration*, that is, the difference between A', B' and A'', B'' is taken into account, the individual lines of a Q branch of a sub-band no longer exactly coincide, and the series formed by the Q branches, according to (IV, 59), *converges slightly*, usually toward shorter wave lengths. In order that the Q lines of a sub-band shall fall at least approximately together, $B' - B''$ must be smaller than is necessary for $\|$ bands, since here much higher J values occur in the Q branches. However, this condition is usually fulfilled for the fundamental bands and low overtone or combination bands. For high overtones, it may happen that the line-like structure of the Q branches is lost, and also the convergence of the series of sub-bands may be so strong that they form a head. In this case the structure of the $\perp$ band would be rather similar to that of a $\|$ band with large $(A' - B') - (A'' - B'')$ (see above).

As has been mentioned previously, a $\perp$ band can occur as a transition between non-degenerate states only in the case of a molecule with no more-than-two-fold axes, that is, for a molecule that is accidentally a symmetric top. Molecules that may be considered here are H_2CO, C_2H_4, and similar ones for which the moment of inertia about one principal axis is much smaller than that about the other two. However, they are not close enough to the exact symmetric top to exhibit all the characteristic features of $\perp$ bands (see also section 4). In less symmetrical molecules that are *nearly symmetric tops*, the change of dipole moment for many vibrational transitions does not lie exactly at 90° to the top axis. Therefore both a $\|$ and a $\perp$ band appear for the same vibrational transition, that is, with the same zero line. One obtains a so-called *hybrid band*.

A very instructive example of such a hybrid band and therefore also of a $\perp$ band has been found by Herzberg, Patat, and Verleger (438) in the photographic infrared spectrum of N_3H, and first correctly interpreted by Eyster (318). It is shown in Fig. 129, where the Q branches are indicated. To be sure, N_3H is not exactly a symmetric top, but the wide structure of the band shows that one moment of inertia is very small, that is, that the three N atoms are very nearly on a straight line with the H atom at one end but not on the axis, thus: $\mathrm{N{-}N{-}N}^{\diagup \mathrm{H}}$. Therefore the other two moments of inertia are nearly equal and the molecule is very nearly a symmetric top. The N—H vibration of which the band represents the second overtone is at an angle $\neq 90°$ to the top axis, and therefore both $\|$ and $\perp$ components occur. Since the difference of A' and A'' is fairly large the ${}^QQ$ branches of the $\|$ component do not all coincide, but form a branch which has its head about half way between the two strongest "lines" of the series formed by the ${}^PQ$ and ${}^RQ$ branches of the $\perp$ component, that is, at the band origin ν_0. The ${}^QP$ and ${}^QR$ branches form the finer structure in the center of the band (the ${}^PP$, ${}^PR$, ${}^RP$, ${}^RR$ branches are too weak to be observed).

Transitions between a non-degenerate and a degenerate vibrational level: perpendicular bands. In a molecule that is a symmetric top because of its symmetry, perpendicular bands ($M_z = 0$) occur only as transitions between vibrational states at least one of which is degenerate (see Table 55). We consider first the case in which *the upper state is degenerate, the lower non-degenerate;* this applies, for example, to the fundamentals of the degenerate vibrations. The appearance of such a band is, of course, very similar to that of the ⊥ band previously discussed (see Fig. 128). The Coriolis splitting of the degenerate vibrational state (Fig. 118) does not lead to a splitting of the band lines (sub-bands) since for $\Delta K = +1$ only the $+l$ levels, for $\Delta K = -1$ only the $-l$ levels of the degenerate state combine with the lower non-degenerate state (according to the rule that only rotational levels of the same over-all species combine with one another, as well as the selection rule for the $+l$ and $-l$ levels).

The *formula for the lines of each sub-band* (fixed K' and K'') is exactly the same as before, that is, the same as for bands of linear molecules. However, the *formula for the zero lines* (Q branches) *of the sub-bands* is different, since in the formula for the energy levels of the upper state (IV, 42) we now have the additional term $\mp 2A_{[v]}\zeta_i K$, where ζ_i is, apart from the factor $h/2\pi$, the vibrational angular momentum of the upper vibrational state. We obtain therefore for the ν_0^{sub}, in place of (IV, 59),[28]

$$\nu_0^{\text{sub}} = \nu_0 + [A'_{[v]}(1 - 2\zeta_i) - B'_{[v]}] \pm 2[A'_{[v]}(1 - \zeta_i) - B'_{[v]}]K + [(A'_{[v]} - B'_{[v]}) - (A''_{[v]} - B''_{[v]})]K^2, \quad \text{(IV, 60)}$$

where the + sign in the third term at the right holds for the R branch ($\Delta K = +1$), the − sign for the P branch ($\Delta K = -1$). In (IV, 60), account is taken of the selection rule for the $+l$ and $-l$ levels (see above). The transitions are indicated in Fig. 118. It will be seen that (IV, 60) goes over into (IV, 59) if $\zeta_i \rightarrow 0$. If the difference between A', B' and A'', B'' is small (as is usually the case) we have again a *series of (almost) equidistant sub-bands of which the line-like Q branches will form the most prominent feature* (see Fig. 128). However, the separation of successive "lines" (Q branches) is no longer $2(A - B)$ but $2[A(1 - \zeta_i) - B]$ which may be smaller or larger than $2(A - B)$ depending on whether ζ_i is positive or negative, that is, depending on whether the direction of rotation of the electric dipole moment in the upper state coincides with, or is opposite to the direction of the angular momentum p of the vibration (see p. 405). As previously, the distance of the first line ($K = 0$) of the R branch from the first line ($K = 1$) of the P branch is the same as the separation of successive lines, that is, there is *no zero gap*. However, the zero line ν_0 is no longer exactly half-way between these two lines. Its distance from the first line of the P branch is $(A - B)$.

Since ζ_i has different values (between +1 and −1) for the different degenerate vibrations of one and the same molecule, the *separations of successive lines in different ⊥ bands of a given symmetric top molecule may vary considerably*, whereas without the influence of the Coriolis force it would be the same in all of them. The observation of such widely different spacings in different ⊥ bands of the same molecule had long been a puzzle before the theory of the Coriolis coupling was developed by Teller and Tisza (837), Teller (836), and Johnston and Dennison (476).

As examples, we give in Fig. 130 and Fig. 131 the fine structure of three ⊥ bands of CH_3Br, one a fundamental in the ordinary infrared, the other two second overtones (or ternary combinations) in the photographic infrared. In both cases only the line-like Q branches are resolved from one another, since $A \gg B$. The typical difference of the appearance of these bands from that of || bands is clearly exhibited. The

[28] The formula given by Johnston and Dennison (476) is somewhat different because they use $K' = K$ instead of, as here, $K'' = K$.

average separation of successive lines in the first band is 7.42 cm^{-1}, in the second 9.3 cm^{-1}. For the other two degenerate fundamentals, the separations 11.9 and 9.0 cm^{-1} respectively have been found. While for the fundamentals the separation of successive "lines" is constant throughout the band (within the accuracy of the

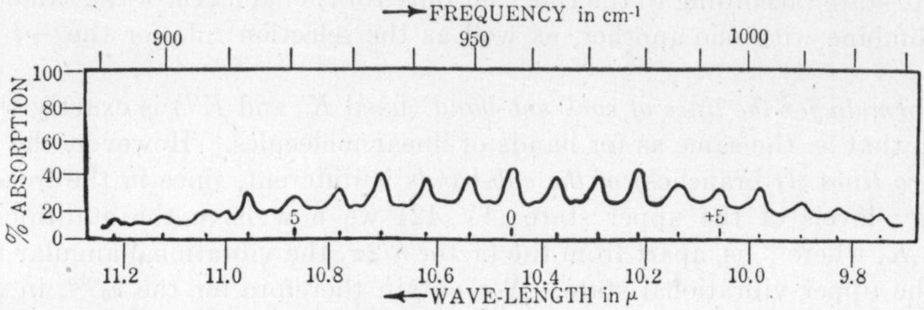

FIG. 130. **Fine structure of the fundamental band ν_6 of CH_3Br at 10.49μ** [after Bennett and Meyer (138)].—The length of the absorbing path was 6 cm at a pressure of 21.8 mm.

measurements) there is a definite convergence in the photographic infrared band, as shown by Table 131, corresponding to the term $[(A'_{[v]} - B'_{[v]}) - (A''_{[v]} - B''_{[v]})]K^2$ in (IV, 60). More recently, for two of the ⊥ bands of CH_3I, Lagemann and Nielsen (546) have been able to resolve partially the structure of the individual Q branches (compare Fig. 128) indicating the effect of slightly unequal B in the upper and lower state.

TABLE 131. WAVE NUMBERS OF THE LINES (SUB-BANDS) IN THE PHOTOGRAPHIC INFRARED BAND AT 1.1 μ OF CH_3Br, AFTER VERLEGER (898).

Assignment (a)	Assignment (b)	ν_{vacuum}[29] (cm^{-1})	$\Delta\nu$	Assignment (a)	Assignment (b)	ν_{vacuum}[29] (cm^{-1})	$\Delta\nu$
$^RQ_9$	$^RQ_{12}$	9124.99*		$^PQ_1$	$^RQ_2$	9036.34	
							9.31
$^RQ_8$	$^RQ_{11}$			$^PQ_2$	$^RQ_1$	9027.03	
							9.51
$^RQ_7$	$^RQ_{10}$	9106.60		$^PQ_3$	$^RQ_0$	9017.52*	
			8.33				9.64
$^RQ_6$	$^RQ_9$	9098.27*		$^PQ_4$	$^PQ_1$	9007.88	
			8.80				9.77
$^RQ_5$	$^RQ_8$	9089.47		$^PQ_5$	$^PQ_2$	8998.11	
			8.97				9.68
$^RQ_4$	$^RQ_7$	9080.50		$^PQ_6$	$^PQ_3$	8988.43*	
			8.41				9.86
$^RQ_3$	$^RQ_6$	9072.09*		$^PQ_7$	$^PQ_4$	8978.57	
			8.72				10.20
$^RQ_2$	$^RQ_5$	9063.37		$^RQ_8$	$^PQ_5$	8968.37	
			8.82				10.18
$^RQ_1$	$^RQ_4$	9054.55		$^PQ_9$	$^PQ_6$	8958.19*	
			9.07				10.05
$^RQ_0$	$^RQ_3$	9045.48*		$^PQ_{10}$	$^PQ_7$	8948.14	
			9.14				

The *numbering of the Q branches* in an observed ⊥ band is not obvious, since there is no zero gap. If there is no intensity alternation (see below), all that can be said is that the zero line must be between the two strongest "lines" in the center of the band. Because of the uncertainty as to which are the strongest lines there are two numberings of the lines of the 1.1μ CH_3Br band compatible with the intensity alternation as given in Table 131. Depending on the assignment chosen, the "lines" can be represented by

$$\nu_0^{sub} = 9045.42 \pm 9.10K - 0.065K^2 \tag{IV, 61a}$$

or

$$\nu_0^{sub} = 9017.55 \pm 9.49K - 0.065K^2. \tag{IV, 61b}$$

[29] The intense lines are marked with an asterisk.

FIG. 131. **Fine structure of the photographic infrared bands of CH_3Br at 1.1μ** [after Verleger (898)].—The length of the absorbing path was 1100 cm at a pressure of 760 mm. The short and long leading lines at the top refer to two different bands. Absorption lines that are not marked are due to the H_2O band at 1.13μ.

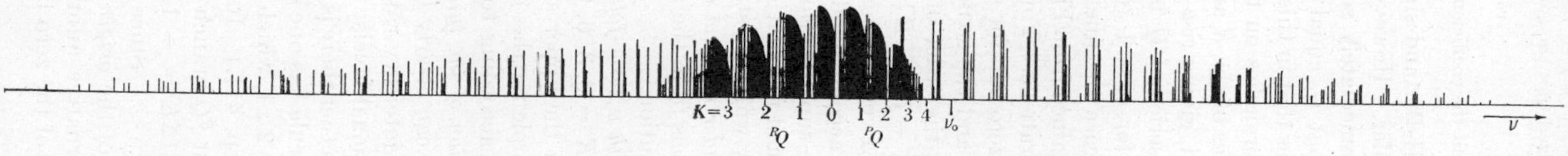

FIG. 132. **Structure of a $\perp$ band with ζ_i close to 1.**—The sub-bands are the same as in Fig. 128 but shifted so as to correspond to $\zeta_i = 0.9826$. It should be noted that the order of the sub-bands (whose zero lines are indicated at the bottom) is the reverse of that in Fig. 128. The origin of the whole band is off to one side. The K values given refer to the lower state.

Here it must be emphasized that the constant term according to (IV, 60) is not ν_0 but

$$\nu_0 + A'(1 - 2\zeta_i) - B';$$

that is, approximately half the coefficient of the linear term has to be subtracted to get ν_0.

In the case of CH_3Br (and similarly for the other methyl halides), ζ_i is of the order of 0.25 or smaller. However, in special cases ζ_i may be as large as +1. In such a case, as is immediately seen from equation (IV, 60), the spacing of the Q branches is $-2B$, that is, equal to the spacing of the lines in a sub-band. The negative sign indicates that in this case the ${}^PQ$ branches would be on the short-wave-length side, the ${}^RQ$ branches on the long-wave-length side of ν_0. Also, if $B' = B''$ and $A' = A''$ the lines of the R and P branches of the different sub-bands would be exactly superimposed, giving rise to one strong P and one strong R branch, in addition to the central series of Q branches. If ζ_i is slightly smaller than +1 or if $B' \neq B''$, $A' \neq A''$ for $\zeta_i = 1$ (or both), this exact coincidence would no longer occur. For such a case the fine structure of a $\perp$ band is shown schematically in Fig. 132, which is obtained from Fig. 128 by superimposing the sub-bands with smaller (and negative) separations. For $\zeta_i = 1 - B/A$ the spacing of the sub-bands (Q branches) would be zero and for slightly smaller ζ_i values the spacing would have a small positive value and the structure would be very similar to that represented in Fig. 132 except that the order of the Q branches is reversed. In all these cases of small spacing $2[A(1 - \zeta_i) - B]$ of the Q branches, if the band is not completely resolved it will have a very strong (though not very sharp) central maximum formed by the Q branches, accompanied by a weaker maximum or shoulder on each side corresponding to the superposition of all the P and all the R branches respectively. Also, in spite of a small moment of inertia I_A, and in spite of the fact that the band is a $\perp$ band, no very wide fine structure will occur. Thus, *in these cases the appearance of a $\perp$ band under medium dispersion is very similar to that of a || band.* This shows that conclusions from the appearance of the bands under low dispersion have to be drawn with caution.

For *molecules with a three-fold axis*, if the identical nuclei have zero spin, only the sub-bands with $K = 0, 3, 6, 9, \cdots$ occur, since in the lower state only the levels $K = 0, 3, 6, \cdots$, in the upper state only the levels $K = 1, 2, 4, 5, \cdots$ occur (see Fig. 118). No example for this case is yet known, however. If the identical nuclei have non-zero spin, according to the previous discussion of the statistical weights, an *intensity alternation of the type strong, weak, weak, strong* $\cdots$ is to be expected. Such an alternation can clearly be seen in the CH_3Br bands reproduced in Fig. 130 and Fig. 131. The intensity ratio is, of course, equal to the ratio of the statistical weights of the rotational levels; this ratio, as we have seen, depends on the nuclear spin as well as on whether there is one set of identical atoms or more. For the simplest case, when the molecule has one set of identical atoms of spin $\frac{1}{2}$ only, as has CH_3Br, the intensity ratio is 2 : 1, which is in good agreement with observation. (For other cases see p. 28 and p. 411.) It should be noted that irrespective of the statistics of the nuclei the first ${}^RQ$ branch ($\Delta K = +1, K = 0$) always forms a strong "line," the first ${}^PQ$ branch ($\Delta K = -1, K = 1$) forms a weak "line" in the series strong, weak, weak, strong, $\cdots$. Since there is no zero gap, this is a welcome criterion for the correct placing of the *origin of the band*, and has been applied in Table 131 to obtain the two alternative numberings as well as in Tables 84 and 85 in order to obtain the positions of the zero lines.

If the molecule has a plane of symmetry perpendicular to the three-fold axis (point group $\boldsymbol{D}_{3h}$), the spectrum exhibits in addition to the above intensity alternation an intensity alternation of the type strong, weak, strong, weak, $\cdots$ within the first positive and the first negative sub-band corresponding to the alternation of statistical weights in the $K = 0$ levels (see above). The same applies to each component band if the inversion doubling is resolved for molecules of symmetry $\boldsymbol{C}_{3v}$. No such case has as yet been resolved.

For *molecules with four-, five-, six-fold axes* a four-, five-, six-fold intensity alternation of the Q branches results. For example, for C_6H_6 the intensities of successive "lines" in a E_{1u}—A_{1g} band would be proportional (apart from the Boltzmann factor) to 10, 11, 9, 14, 9, 11, 10, 11, 9, $\cdots$ [see Wilson (933)].

If the upper and lower states are reversed, that is, if we have a difference band with a degenerate lower state, we obtain, taking into account also the fact that the $+l$, $-l$ selection rule is reversed (see above), instead of (IV, 60)

$$\nu_0^{\text{sub}} = \nu_0 + (A'_{[v]} - B'_{[v]}) \pm 2(A'_{[v]} - A''_{[v]}\zeta_i - B'_{[v]})K + [(A'_{[v]} - B'_{[v]}) - (A''_{[v]} - B''_{[v]})]K^2. \quad \text{(IV, 62)}$$

From this it is seen that, apart from the difference introduced by the slight dependence of A on v_i, a $\perp$ difference band has the same spacing of its Q branches as the corresponding fundamental band whose upper state forms the lower state of the difference band.

Transitions between two degenerate vibrational levels. As we have seen previously, a transition between two doubly degenerate vibrational states of a symmetric top molecule (E—E transition) has a || as well as a $\perp$ component of the oscillating dipole moment; the two have, however, in general very different magnitudes. For the || *component,* since $+l \leftrightarrow +l$, $-l \leftrightarrow -l$, we have a structure very similar to that of an ordinary || band (Fig. 122); but each sub-band with the exception of the one with $K = 0$ is split into two components (compare Fig. 118), which according to (IV, 42) will be given by

$$\nu_0^{\text{sub}} = \nu_0 + [(A'_{[v]} - A''_{[v]}) - (B'_{[v]} - B''_{[v]})]K^2 \mp 2(A'_{[v]}\zeta_i' - A''_{[v]}\zeta_i'')K. \quad \text{(IV, 63)}$$

The splitting would be very small if in the upper and lower state one and the same degenerate vibration were excited by the same amount (1—1 band of a sequence starting with a || band) since then $\zeta_i' = \zeta_i''$. If, however, two different degenerate vibrations are excited in the upper and lower state ($\zeta_i' \neq \zeta_i''$), the splitting will be large, and we will obtain a structure very similar to that of a $\perp$ band with a separation of successive lines approximately equal to $2A(\zeta_i' - \zeta_i'')$. The difference from a $\perp$ band will lie in the fact that there will be an intensity zero for the Q branch with $K = 0$ in the center of the band. As can be seen from Fig. 118b, in the case of molecules with three-fold axes, there will again be the intensity alternation strong, weak, weak, strong, $\cdots$ in the series of Q branches.

For the $\perp$ *component* of an $E - E$ vibrational transition, since $-l \leftrightarrow +l$ for $\Delta K = +1$ and $+l \leftrightarrow -l$ for $\Delta K = -1$ we obtain again a single series of sub-bands as for ordinary $\perp$ bands. The origins of the sub-bands according to (IV, 42) are given by

$$\nu_0^{\text{sub}} = \nu_0 + [A'_{[v]}(1 + 2\zeta_i') - B'_{[v]}] \pm 2[(A'_{[v]} - B'_{[v]}) + (A'_{[v]}\zeta_i' - A''_{[v]}\zeta_i'')]K + [(A'_{[v]} - B'_{[v]}) - (A''_{[v]} - B''_{[v]})]K^2, \quad \text{(IV, 64)}$$

where the upper sign refers to $\Delta K = +1$, the lower one to $\Delta K = -1$. If the difference between A' and A'' and between B' and B'' is small we obtain, as before, a series of almost equidistant "lines" —Q branches—whose spacing is $2(A - B)$ if $\zeta_i' = \zeta_i''$ and otherwise is $2[A(1 + \zeta_i' - \zeta_i'') - B]$. As can be seen from Fig. 118, for a molecule with a three-fold axis this series has again an intensity alternation strong, weak, weak, strong, $\cdots$; but unlike the ordinary $\perp$ bands the zero line is between two weak lines.

No actual case of a transition between two degenerate states has as yet been studied in detail.

Analysis of infrared bands, moments of inertia, and internuclear distances of symmetric top molecules. If in *a* || *band* the *K fine structure* is *not resolved* (that is, if all sub-bands coincide), its structure is essentially the same as a ⊥ band of a linear molecule, and we can obtain the *rotational constants* B' and B'' in the same way as there *from the combination differences* $\Delta_2F'(J) = R(J) - P(J)$ and $\Delta_2F''(J) = R(J-1) - P(J+1)$ respectively (see p. 390). This procedure, if applied to the || bands reproduced in Fig. 123 and Fig. 124, gives the B_0'' values summarized, together with others, in the later Table 132 (p. 437). It is hardly necessary to point out that the $\Delta_2F''(J)$ obtained from different || bands of one and the same molecule must agree for every J value if the lower state is in common. Moreover, the sum of the wave numbers of two succeeding lines in the pure rotation spectrum must also be exactly equal to the appropriate $\Delta_2F''(J)$ value of a rotation-vibration band.[30]

In the case of molecules that are only approximately symmetric tops, the value of B' and B'' obtained in the way indicated above gives the *average* of the two almost identical rotational constants, that is, $\frac{1}{2}(B + C)$ (see Chapter I, section 4).

The rotational constant A cannot be obtained from a || band.

It should be realized that a slight systematic error occurs in the analysis of || bands whose K fine structure is not resolved, due to the fact that with increasing J an increasing number of lines (which do not exactly coincide) contribute to the measured "lines." However, if $A \gg B$ this has a negligible effect, except possibly for the smallest J values; and even if A and B are of similar magnitude the effect is slight, since the intensity of the sub-bands decreases from $K = 1$ on.

If the *K fine structure of a* || *band* is *resolved*, each sub-band can be dealt with in the above-described way. Apart from the very slight influence of centrifugal distortion [in particular the rotational constants D_{JK} in (I, 27)], the $\Delta_2F(J)$ for all sub-bands must agree for every J. Each group of lines of a given J (see Fig. 122 and Fig. 127) is represented by a formula exactly like (IV, 58). If the lines of such a group are plotted against K^2, a straight line should be obtained whose slope gives $(A' - A'') - (B' - B'')$. The slope should be the same for each J value as long as centrifugal stretching terms are neglected. This is shown in Fig. 133 for ${}^QP_K(5)$ and ${}^QP_K(6)$ of the NH_3 band ν_2, according to the data of Sheng, Barker, and Dennison (785) (compare Fig. 127). In both cases the slope gives $(A' - A'') - (B' - B'') = 0.279$ cm^{-1}.[31] Thus, since B' and B'' can be obtained from the sub-bands (see above), $A' - A''$ can be determined, which in the case of a fundamental ν_i yields α_i^A, in the case of an overtone or combination band a multiple of α_i^A or a combination of α_i^A values. But A' and A'' separately cannot be determined.

The only case other than NH_3 for which a fairly complete resolution of a || band has been obtained is that of the N_3H photographic infrared bands (see Fig. 129). Eyster (318) determined B_0'' from the combination differences for each sub-band while $B'_{[v]}$ was obtained by determining $B' - B''$ from $R(J-1) + P(J)$ (see p. 391). The very small difference $B' - B''$ that was obtained explains why the Q branches are such exceedingly sharp "lines."

The rotational constants A can only be obtained from ⊥ *bands.* (In principle also the B values can be obtained from them if the P and R branches of the sub-bands are resolved, but no such case has as yet been investigated.) If the molecule is accidentally (or approximately) a symmetric top, $(A' - B')$ can be obtained from the linear term of the formula (IV, 59) representing the Q heads of the sub-bands (that is, ν_0^{sub}), and $(A' - B') - (A'' - B'')$ from the quadratic term. Then if B' and

[30] For the few J values that the near and far infrared measurements of NH_3 have in common the agreement is very satisfactory [see (956) (281) (115) and (785)].

[31] Sheng, Barker, and Dennison have determined $(A' - A'') - (B' - B'')$ by taking the difference of successive lines, dividing by $2K + 1$ and then averaging. Such a procedure is less accurate than the above.

B'' are known A' and A'' are immediately obtained. Here again, a more accurate way of determining $A' - B'$ and $A'' - B''$ is by means of the *combination differences*, which are immediately obtained from (IV, 41) or (IV, 59):

$$^RQ_{K-1} - {}^PQ_{K+1} = \Delta_2{}^K F''(J, K) = F''(J, K+1) - F''(J, K-1) = 4(A'' - B'')K, \quad \text{(IV, 65)}$$

$$^RQ_K - {}^PQ_K = \Delta_2{}^K F'(J, K) = F'(J, K+1) - F'(J, K-1) = 4(A' - B')K. \quad \text{(IV, 66)}$$

Since these relations hold for any J they apply also to the unresolved Q branches.

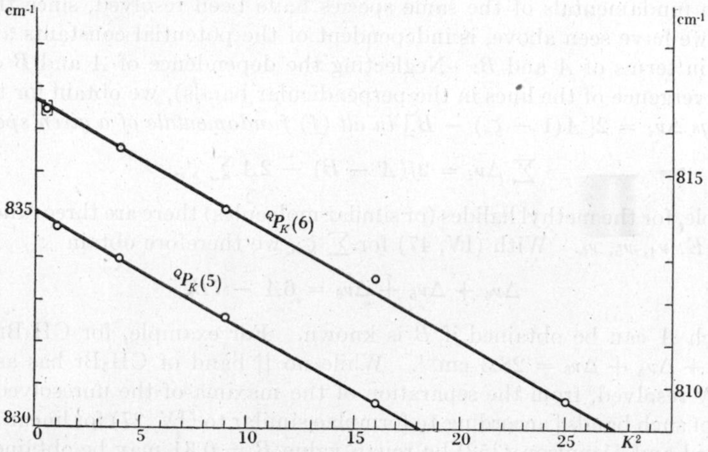

FIG. 133. $^QP_K(5)$ and $^QP_K(6)$ of the NH_3 band 931.58 cm^{-1} as a function of K^2 [after the data of Sheng, Barker and Dennison (785)].—The left-hand scale applies to the lower, the right-hand scale to the upper curve. The point in square brackets represents the average of the two lines with $K = 0$ and $K = 1$ which are not resolved.

In this way the $A - B$ values have been obtained for N_3H by Eyster (318) and for C_2H_4 by Gallaway and Barker (345). For these slightly asymmetric top molecules the constant B of the symmetric top has to be replaced by the average $\tilde{B}$ of the two rotational constants B and C (see p. 488). From the $A - \tilde{B}$ values thus derived from a $\perp$ band, A can be obtained if $\tilde{B}$ is known from $||$ bands (as is the case for C_2H_4) or from the $||$ component of the same hybrid band whose $\perp$ component supplied $A - B$ (as has been done for N_3H).

In the case of $\perp$ *bands of molecules with a more-than-two-fold axis*, when the upper or lower state (or both) is a degenerate vibrational state the constant ζ_i of the vibrational angular momentum enters the formula for the series of Q branches [compare (IV, 60], and therefore $A - B$ *cannot immediately be determined.* While the coefficient of K^2 in the formula for the branches still gives $(A' - B') - (A'' - B'')$, the coefficient of the linear term gives $2(A' - A'\zeta_i - B')$. In order to obtain A' and A'' it is necessary not only to know B' and B'' but also ζ_i. Nor are the combination differences of any help since corresponding ^PQ and ^RQ "lines" no longer have the same upper state (see Fig. 118) and since therefore the combination differences do not permit a complete separation of upper and lower rotational levels. Instead of

(IV, 65) and (IV, 66), one obtains from (IV, 60) (upper state degenerate):

$$^RQ_{K-1} - {}^PQ_{K+1} = 4(A'' - A'\zeta_i - B'')K, \tag{IV, 67}$$

$$^RQ_K - {}^PQ_K = 4(A' - A'\zeta_i - B')K. \tag{IV, 68}$$

In principle ζ_i can be obtained theoretically from the force constants and approximate values for the internuclear distances [compare the formulae of Silver and Shaffer (790) and Shaffer (776) for planar and non-planar XY_3 and of Shaffer (777) for XYZ_3]. But this method is not very practical for a determination of the rotational constant A and has not been attempted in any case.

However, a determination of the constant A is nevertheless possible when all degenerate fundamentals of the same species have been resolved, since the sum of the ζ_i, as we have seen above, is independent of the potential constants and can be expressed in terms of A and B. Neglecting the dependence of A and B on v (that is, the convergence of the lines in the perpendicular bands), we obtain for the *sum of the spacings* $\Delta\nu_i = 2[A(1 - \zeta_i) - B]$ *in all (f) fundamentals of a given species:*

$$\sum \Delta\nu_i = 2f(A - B) - 2A \sum \zeta_i. \tag{IV, 69}$$

For example, for the methyl halides (or similar molecules) there are three fundamentals of species E: ν_4, ν_5, ν_6. With (IV, 47) for $\sum \zeta_i$, we therefore obtain

$$\Delta\nu_4 + \Delta\nu_5 + \Delta\nu_6 = 6A - 7B, \tag{IV, 70}$$

from which A can be obtained if B is known. For example, for CH_3Br (see Fig. 130), $\Delta\nu_4 + \Delta\nu_5 + \Delta\nu_6 = 28.3$ cm^{-1}. While no $||$ band of CH_3Br has as yet been completely resolved, from the separation of the maxima of the unresolved P and R branches of such bands [according to formulae similar to (IV, 27) for linear molecules; see Gerhard and Dennison (352)] a rough value $B = 0.31$ may be obtained. Since $A \gg B$, this is sufficiently accurate for a determination of A from (IV, 70). One obtains $A = 5.08$ cm^{-1}. In a similar manner, the A values of the other methyl halides given in Table 132 have been obtained [see Dennison (280)].

It should be noted that the *accuracy* of the A values determined in this way is not very high, since the dependence of A and B on v has been neglected, and also since the sum rule for ζ_i holds exactly only under the assumption of strictly harmonic oscillations [that is, for the ζ_i^e in (IV, 50)]. If sufficiently accurate measurements were available, some improvement could be obtained if instead of the average $\Delta\nu_i$ the linear terms $2(A' - A'\zeta_i - B')$ of (IV, 60) were used in (IV, 69). Assuming that in the ζ_i sum rule A' and B' can be used, we would obtain first A', from which A'' could be obtained with the help of the coefficient of the quadratic term.

It may be mentioned that the best practical way to obtain the coefficient of K^2 in (IV, 60) is to plot $^RQ_K + {}^PQ_K$ against K^2, which gives a straight line whose slope is $2[(A' - B') - (A'' - B'')]$. The best way to obtain the linear term is to plot $^RQ_K - [(A' - B') - (A'' - B'')]K^2$ and $^PQ_K - [(A' - B') - (A'' - B'')]K^2$ against K, which gives straight lines both of which have the slope $2(A' - A'\zeta_i - B')$. By this procedure the coefficients in the equations (IV, 61a and b) were obtained.

In Table 132 the *rotational constants* $A_{[0]}$ *and* $B_{[0]}$ *of all symmetric top molecules so far investigated* are collected together. The corresponding *moments of inertia* $I_A{}^0$ and $I_B{}^0$ are also given. While in most cases some other $B_{[v]}$ or $A_{[v]}$ values have also been determined (compare the references quoted), in no case are all the $\alpha_i{}^B$ or $\alpha_i{}^A$ values known, so that the B_e and A_e values cannot be determined. Fortunately the α are small, and therefore the moments of inertia and internuclear distances

TABLE 132. ROTATIONAL CONSTANTS AND MOMENTS OF INERTIA OF SYMMETRIC TOP MOLECULES IN THEIR GROUND STATES.

Molecules	$B_{[0]}$ (cm^{-1})[32]	$A_{[0]}$ (cm^{-1})[33]	I_B^0 (10^{-40} gm cm^2)	I_A^0 (10^{-40} gm cm^2)	I_C^0 (10^{-40} gm cm^2)	References
NH_3	9.941[34]	(6.30_9)	2.816	(4.43_7)	$=I_B^0$	(785)
ND_3	5.138[35]	(3.15_7)	5.448	(8.86_8)	$=I_B^0$	(624) (280)
CH_3D	3.878_5	(5.245_1)	7.217_7	(5.337_2)	$=I_B^0$	(362) (207)
CH_3F	0.8496	5.10_0	32.95	5.48_9	$=I_B^0$	(138)
CH_3Cl	0.49	5.09_7	$57._1$	5.49_2	$=I_B^0$	(827)
CH_3Br	0.31[35a]	5.08_2	90	5.50_8	$=I_B^0$	(138)
CH_3I	0.28[35a]	5.07_7	100	5.51_4	$=I_{B0}$	(138)
C_2H_6	0.6621[36]	2.538[36]	42.28	11.03	$=I_B^0$	(798)
$CH_3C\equiv CH$	0.2848		98.29		$=I_B^0$	(440) (69)
BF_3	0.35_5	(0.17_8)	78.9	(158)	$=I_B^0$	(344)
N_3H	[0.3996]	20.346	69.38	1.3759	70.75	(318)
H_2O_2	$[0.82_5]$	10.056	[37]	2.783_8	[37]	(977)
C_2H_4	$[0.911_6]$	4.867	28.08	5.752	33.85	(345)
C_2D_4	$[0.652_2]$	2.437	37.93	11.48_7	49.42	(345)
H_2CO	[1.215][38]	9.404[38]	21.65	2.976_8	24.62	(288)
CH_3NH_2	[0.7385]	3.50	[37]	8.00	[37]	(847) (680)
HCOOH	[0.348]	2.55_4	75.3	10.96	86.3	(128) (848)
cyclo C_3H_6	0.6680[36]		41.91		$=I_B$	(797)
CH_3OH	[0.8032]					(169)

determined from $A_{[0]}$ and $B_{[0]}$ are fairly good approximations to the equilibrium values (compare Tables 129 and 130).

In the case of the nearly symmetric top molecules, as shown on p. 48, the B value obtained from the || bands is really $\tilde{B} = \frac{1}{2}(B + C)$. For plane molecules such as N_3H and C_2H_4 we have the relation

$$I_C = I_A + I_B, \tag{IV, 71}$$

which holds strictly only for the equilibrium moments of inertia, but with very good approximation also for the I^0. Therefore if A is known, B and C can be obtained separately from $\tilde{B}$. The following relations are easily found:

$$B = -(A - \tilde{B}) + \sqrt{A^2 + \tilde{B}^2}, \qquad C = (A + \tilde{B}) - \sqrt{A^2 + \tilde{B}^2}.$$

The values of I_B^0 and I_C^0 obtained from these relations are also given in Table 132. For non-planar molecules B and C can of course not be obtained from $\tilde{B}$.

[32] Values in square brackets are $\tilde{B}$ values.

[33] Values in parentheses are obtained indirectly.

[34] This is the average of the B values for the two inversion doubling components. The pure rotation spectrum (see Chapter I) gives 9.945 cm^{-1}.

[35] From the pure rotation spectrum, Barnes (115) obtained the (less accurate) value 5.13 cm^{-1}.

[35a] From the separation of the P and R maxima of the infrared bands as given by Dennison (280). Sutherland (825a) gives slightly different values.

[36] These values were kindly supplied by Dr. L. G. Smith in private communications and are very slightly changed compared to those in the abstracts by Smith and Woodward (798) and by Smith (797).

[37] Since these molecules are not planar the individual I_B and I_C values cannot be obtained from the observed $\tilde{B}$.

[38] From the ultraviolet spectrum.

Up to now the ζ_i sum rule has been used for the determination of A only in the case of the methyl halides. In the case of NH_3 and ND_3 a much better value of A is obtained from the isotope effect, assuming that internuclear distances and angles are the same in NH_3 and ND_3. Similarly, for CH_3D the value of A can be determined from the B value, assuming tetrahedral structure and $r(\text{C—H}) = r(\text{C—D})$. These indirectly determined A values are put in parentheses in Table 132.

With the knowledge of the A values thus obtained, it is now possible to obtain the following *ζ_i values* from the observed spacings in the perpendicular bands:

NH_3	$\zeta_3 = 0.06$[39]	$\zeta_4 = -0.26$	
ND_3	$\zeta_3 = 0.20$	$\zeta_4 = -0.36$	
CH_3D	$\zeta_4 = 0.24$	$\zeta_5 = -0.27$	$\zeta_6 = 0.66$
CH_3F	$\zeta_4 = 0.099$	$\zeta_5 = -0.294$	$\zeta_6 = 0.280$
CH_3Cl	$\zeta_4 = 0.100$	$\zeta_5 = -0.273$	$\zeta_6 = 0.222$
CH_3Br	$\zeta_4 = 0.053$	$\zeta_5 = -0.232$	$\zeta_6 = 0.208$
CH_3I	$\zeta_4 = 0.058$	$\zeta_5 = -0.216$	$\zeta_6 = 0.187$

The values for ND_3 fit fairly well the sum rule (IV, 46), not used in their derivation.[40] The sum rule (IV, 47) is not fulfilled for CH_3D, probably on account of insufficient resolution of the $\perp$ bands which did not allow an unambiguous assignment of the fine structure lines.

A very striking confirmation of the assumption of *equal internuclear distances in isotopic molecules* is supplied by the I_A values of C_2H_4 and C_2D_4, which should be in the ratio $m_H/m_D = 0.50037$; the observed ratio is 0.5007.

From the moments of inertia in Table 132 some or all of the *internuclear distances and angles* can accurately be determined, although not always without some simplifying assumption. We consider first those molecules for which an unambiguous determination of the geometrical structure is possible:

The simplest case is that of BF_3, which is known to be plane and symmetrical (see p. 298), and for which therefore the observed B value is sufficient to determine the geometrical structure. Since for plane XY_3

$$I_B = \tfrac{1}{2}I_A = \tfrac{3}{2}m_Y r_{XY}^2,$$

we obtain:

$$BF_3: \quad r_0(B - F) = 1.29_1 \times 10^{-8} \text{ cm}.$$

This value is not very accurate, since the B value is obtained only from a partially resolved band. It agrees well with the electron diffraction value $1.30 \pm 0.02 \times 10^{-8}$ cm given by Lévy and Brockway (574a).

For non-planar XY_3 there are two quantities that determine the geometrical structure, for example, the X—Y distance $r(\text{X—Y})$ and the angle β of X—Y with the three-fold axis. These quantities are given by the two moments of inertia I_B and I_A according to the relations:

$$I_A = 3m_Y r^2(\text{XY}) \sin^2\beta;$$

$$I_B = \frac{3m_Y r^2(\text{XY})}{2\left(1 + \dfrac{3m_Y}{m_X}\right)}\left[2 - \left(1 - \frac{3m_Y}{m_X}\right)\sin^2\beta\right]. \qquad \text{(IV, 72)}$$

---

[39] From ζ_4 and the ζ_i sum rule.

[40] Dennison (280) adjusted the ζ_i values slightly so that they would fulfill the ζ_i sum rule exactly. He gives $\zeta_3 = 0.185$, $\zeta_4 = -0.368$.

However, for NH_3 only I_B is reliably determined and therefore it is necessary to use in addition I_B of ND_3 to determine $r(N—H)$ and β. We obtain, by applying the second equation (IV, 72) to both NH_3 and ND_3,

$$NH_3(ND_3): \quad r_0(N—H) = 1.014 \times 10^{-8} \text{ cm}, \quad \beta = 67° 58'.$$

The height of the pyramid is $h_0 = r_0 \cos \beta = 0.381 \times 10^{-8}$ cm, a value that agrees well with that obtained from the magnitude of the inversion doubling (see p. 224), but is more accurate. The H—N—H angle following from the above is $\alpha = 106° 47'$, and the H—H distance $r_0(H—H) = 1.628 \times 10^{-8}$ cm.

For tetrahedral molecules XY_4, and therefore for CH_3D, the structure is completely determined by the one $r(X—Y)$ distance. One finds easily

$$I_B(CH_3D) = \left(\tfrac{8}{3}m_H + m_D - \frac{(m_D - m_H)^2}{m_C + 3m_H + m_D}\right) r_0^2(CH), \qquad (IV, 73)$$

from which, together with the I_B value in Table 132, it follows that

$$CH_4(CH_3D): \quad r_0(C—H) = 1.0936 \times 10^{-8} \text{ cm}.$$

The moments of inertia of C_2H_4 alone are not sufficient to determine the geometrical dimensions, but with the moments of inertia of C_2D_4 they are uniquely determined and in addition the check mentioned above is provided. Gallaway and Barker (345) obtained

$$C_2H_4(C_2D_4): \quad r_0(C—H) = 1.071 \times 10^{-8} \text{ cm},$$
$$r_0(C{=}C) = 1.353 \times 10^{-8} \text{ cm},$$
$$\sphericalangle HCH = 119° 55'.$$

For the remaining molecules of Table 132 additional assumptions have to be made in order to evaluate at least some of the internuclear distances.

In the case of H_2O_2, assuming the asymmetrical C_2 structure given in Chapter III, p. 301, with certain assumed values of the angles and the OH distance equal to that in H_2O, Zumwalt and Giguère (977) obtain

$$H_2O_2: \quad r_0(O—O) = 1.48 \times 10^{-8} \text{ cm}.$$

In the case of N_3H, assuming a linear N—N—N chain with a ratio of the two N—N distances of 1.10 (the end N atoms having the smaller distance), and an N—H distance of 1.012×10^{-8} cm (somewhat smaller than in NH_3), Eyster (318) obtained

$$N_3H: \quad r_0(HN—N) = 1.241 \times 10^{-8} \text{ cm},$$
$$r_0(N—N) = 1.128 \times 10^{-8} \text{ cm},$$
$$\sphericalangle HNN = 110° 52'.$$

In the case of the methyl halides, if the C—H distance is assumed to be the same as in CH_4 (CH_3D), the carbon-halogen distance and the H—C—H angle can be determined. One obtains:

CH_3F:	$r_0(C—F) = 1.39_8 \times 10^{-8}$ cm	$\sphericalangle HCH = 111° 48'$
CH_3Cl:	$r_0(C—Cl) = 1.7_1 \times 10^{-8}$ cm	$\sphericalangle HCH = 111° 52'$
CH_3Br:	$r_0(C—Br) = 1.9_5 \times 10^{-8}$ cm	$\sphericalangle HCH = 112° 6'$
CH_3I:	$r_0(C—I) = 2.0_0 \times 10^{-8}$ cm	$\sphericalangle HCH = 112° 12'$

While the carbon-halogen distances, except r_0(C—F), are not very accurate since the || bands have not yet been completely resolved, the *deviation of the* HCH *angle from the tetrahedral angle* (109° 28′) and its increase in the series of halogens is believed to be genuine, and appears to be rather significant.

If, in a similar way, we assume for C_2H_6 the same C—H distance as in CH_4, we obtain, from the moments of inertia of Table 132:

$$C_2H_6: \quad r_0(\text{C—C}) = 1.573 \times 10^{-8} \text{ cm}, \quad \sphericalangle \text{HCH} = 112° 12'.$$

If in methyl acetylene the C≡C, ≡C—H, distances are taken over from C_2H_2 (see Table 130) and the CH_3 group is assumed to have the same structure as in CH_4, the C—C single-bond distance is found to be

$$CH_3\text{—C≡C—H}: \quad r_0(\text{C—C}) = 1.469 \times 10^{-8} \text{ cm}.$$

If on the other hand the CH_3 group is assumed to have the same structure as in ethane (see above) a value r_0(C—C) = 1.477×10^{-8} cm is obtained.

If in H_2CO (formaldehyde) we assume the same C—H distance as in ethylene we obtain for the C═O distance and the HCH angle from the moments of inertia in Table 132

$$H_2CO: \quad r_0(\text{C═O}) = 1.225 \times 10^{-8} \text{ cm} \qquad \sphericalangle \text{HCH} = 123° 26'.$$

In cyclopropane (C_3H_6), since only one moment of inertia is known, two geometrical data have to be assumed in order to calculate the third one. Assuming a D_{3h} structure (see p. 352f.) and a C—H distance as in ethylene [r_0(C—H) = 1.071 $\times 10^{-8}$ cm] and an HCH angle of 120° one obtains

$$\text{cyclo } C_3H_6: \qquad r_0(\text{C—C}) = 1.296 \times 10^{-8} \text{ cm}.$$

With an HCH angle of 180° and the same C—H distance one obtains

$$\text{cyclo } C_3H_6: \qquad r_0(\text{C—C}) = 1.556 \times 10^{-8} \text{ cm}.$$

If, finally, in CH_3NH_2 the internuclear distances and angles in the CH_3 and NH_2 groups are assumed to be the same as in CH_4 and NH_3 respectively, one obtains for the C—N distance [Owens and Barker (680)]:

$$CH_3NH_2: \quad r_0(\text{C—N}) = 1.48 \times 10^{-8} \text{ cm}.$$

The assumptions made above for the second group of molecules could be tested (or corrected) if the corresponding "heavy" molecules were investigated.

It is very interesting to compare the C—C single-, double-, and triple-bond distances as obtained above for C_2H_6, C_2H_4, and in Table 130 for C_2H_2. They are

$$r_0(\text{C—C}) = 1.573, \qquad r_0(\text{C═C}) = 1.353, \qquad r_0(\text{C≡C}) = 1.207 \times 10^{-8} \text{ cm}.$$

It is very significant that the C—C *single-bond distance in* CH_3—C≡C—H ($1.469.10^{-8}$ cm), where it is adjacent to a triple bond, is *appreciably smaller than in* C_2H_6. Finally, the small but definite *differences between the* C—H *distances* in CH_4, C_2H_4, and C_2H_2 (that is, when they are adjacent to single, double, and triple bonds respectively) are noteworthy, and correspond to the similar differences between the C—H force constants (see Table 50) and C—H bond frequencies (see Table 51). We have

$$r_0(\text{—C—H}) = 1.093_6, \qquad r_0(\text{═C—H}) = 1.071, \qquad r_0(\text{≡C—H}) = 1.059 \times 10^{-8} \text{ cm}.$$

(c) *Raman spectrum*

Selection rules. As with the infrared spectrum, so with the rotation-vibration Raman spectrum, the vibrational and rotational selection rules are to a very good approximation the same as for the pure vibration spectrum (see Table 55) and the pure rotation spectrum respectively. In the most general case, when the top axis is not an axis of symmetry, we have

$$\Delta K = 0, \pm 1, \pm 2 \quad \text{and} \quad \Delta J = 0, \pm 1, \pm 2, \qquad \text{(IV, 74)}$$

with the restriction [41]

$$J' + J'' \geq 2.$$

But if the molecule has symmetry, certain values of ΔK do not occur, depending on the type of the vibrational transition. Placzek and Teller (701) have shown that *transitions with* $\Delta K = 0$ *occur only for vibrational transitions for which* $[\alpha_{zz}]^{nm}$ or $[\alpha_{xx} + \alpha_{yy}]^{nm}$ *or both are different from zero* (z axis = top axis); *transitions with* $\Delta K = \pm 1$ *occur only for vibrational transitions for which* $[\alpha_{xz}]^{nm}$ *or* $[\alpha_{yz}]^{nm}$ *or both are different from zero; and transitions with* $\Delta K = \pm 2$ *occur only for vibrational transitions for which* $[\alpha_{xx} - \alpha_{yy}]^{nm}$ *or* $[\alpha_{xy}]^{nm}$ *or both are different from zero.* From this general rule and the vibrational selection rules in Table 55 it follows immediately that for all transitions between totally symmetric vibrational states of molecules of point groups $\boldsymbol{C_{2v}}$, $\boldsymbol{D_2}$, $\boldsymbol{V_h}$, and all axial molecules with a more than two-fold axis, only $\Delta K = 0$ occurs; whereas for all transitions between a totally symmetric and a non-totally symmetric vibrational state only $\Delta K = \pm 1$ or $\Delta K = \pm 2$ or both occur, but never $\Delta K = 0$. In particular, for E—A_1 transitions of molecules of point group $\boldsymbol{C_{3v}}$, both $\Delta K = \pm 1$ and $\Delta K = \pm 2$ occur, whereas for the similar transition E'—A_1' of $\boldsymbol{D_{3h}}$ only $\Delta K = \pm 2$ occurs. For E—A_1 of $\boldsymbol{C_{4v}}$ only transitions with $\Delta K = \pm 1$ occur, whereas for B_1—A_1 and B_2—A_1 of $\boldsymbol{C_{4v}}$ (and similarly B_{1g}—A_{1g} and B_{2g}—A_{1g} of $\boldsymbol{D_{4h}}$) only $\Delta K = \pm 2$ occurs.

If the upper state is a degenerate vibrational state with Coriolis splitting ($\zeta_i \neq 0$) and the lower state totally symmetric, the additional rule applies that *for* $\Delta K = +1$ *and* $\Delta K = -2$ *only the* $+l$ *sublevels, for* $\Delta K = -1$ *and* $\Delta K = +2$ *only the* $-l$ *sublevels of the degenerate state combine with the lower non-degenerate state.* If the degenerate state is the lower the reverse rule holds.

Just as for the infrared spectrum, *only rotational levels of the same species can combine with one another* (see p. 415). This rule does not lead, however, to any additional restriction of the possible ΔK values, since it is implicitly taken into account in the above rules for ΔK. However, this symmetry rule does lead to restrictions of the combinations of the sublevels for a given K and J.

Totally symmetric Raman bands. If the molecule has no symmetry but is accidentally a symmetric top, all transitions allowed by (IV, 74) would occur; that is, there would be in each Raman band five series of sub-bands with five branches each. Since no example is known of such a case, we shall not discuss it in detail. The structure, however, can easily be visualized by superimposing appropriate bands from among those discussed in the following paragraphs.

If the molecule is a (nearly or genuine) symmetric top on account of its symmetry and belongs to the point groups $\boldsymbol{C_{2v}}$, $\boldsymbol{D_2}$, $\boldsymbol{V_h}$, or any of the axial point groups with a

[41] This restriction corresponds to the rule $J = 0 \leftrightarrow J = 0$ for dipole radiation.

more-than-two-fold axis, for a transition between two totally symmetric vibrational states *only* $\Delta K = 0$ *occurs* (see above). We have, therefore, *one series of sub-bands* of much the same type as for a || infrared band (Fig. 122), except that (1) there are now in each sub-band—in addition to the three branches P, Q, and R—an O and an S branch with double the spacing of the P and R branches ($\Delta J = \pm 2$; see p. 20), and that (2) the Q branches are relatively much stronger (there is a strong Q branch also for $K = 0$). Since for Raman bands the difference between the rotational constants in the upper and lower states is always very small, the *complete band also has five branches*, each line of which arises by the superposition of a number of lines of different sub-bands, very similar to the three branches of a || infrared band for which $A' \approx A''$ and $B' \approx B''$. Alternate lines of the P and R branches coincide with the lines of the O and S branches. The resulting appearance of the band is much the same as that of the rotational Raman spectrum (see Fig. 13), although here there would be a slight convergence of the lines as in a 1—0 Raman band of a diatomic molecule. Since all the line-like Q branches of the sub-bands coincide, they will form in general a *very strong central line-like Q branch of the whole band*. This is usually the only part of the band that is observed. The relative intensity of the five branches in its dependence on I_A/I_B is indicated in Fig. 134. It is seen that the intensity of P and R branches decreases with decreasing I_A/I_B and is zero for $I_A/I_B = 0$, the case of the linear molecule (see above). The formulae for the intensities of the individual lines in the branches are the same as for the pure rotation spectrum [see equations (I, 49) and (I, 50), and Placzek and Teller (701)].

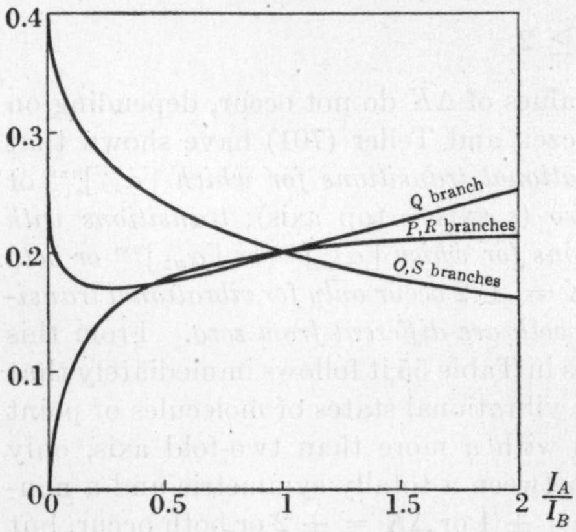

FIG. 134. Relative intensity of the branches in a totally symmetric Raman band of a symmetric top molecule as a function of I_A/I_B [after Placzek and Teller (701)].—See caption of Fig. 125.

For molecules of point groups C_2 and C_{2h} that are nearly symmetric tops, in addition to α_{zz} and $\alpha_{xx} + \alpha_{yy}$, also α_{xy} is totally symmetric (see Table 55); therefore, in addition to $\Delta K = 0$, also $\Delta K = \pm 2$ occurs for totally symmetric Raman bands. A discussion of bands with $\Delta K = \pm 2$ will be given below. It is clear that as before the line-like Q branch will be the predominant feature of the Raman bands.

Non-totally symmetric non-degenerate Raman bands. For nearly or accidentally symmetric top molecules of point groups C_{2v},[42] D_2, V_h, and for genuine symmetric top molecules with a four-fold axis, the non-totally symmetric non-degenerate Raman bands A_2—A_1, B_1—A, B_{1g}—A_g, B—A respectively have $\Delta K = \pm 2$ only, according to the selection rule on p. 441 and Table 55. The resulting fine structure is very similar to that of a ⊥ infrared band as given in Fig. 128, except that each sub-band now has, in addition to P, Q, and R branches, an O and an S branch, that the spacing of the sub-bands, that is, of the Q branches, is equal to $4(A—B)$ rather than $2(A—B)$ (if the interaction of rotation and vibration is neglected) and finally that,

[42] Assuming that the C_2 coincides with the top axis.

since for $\Delta K = -2$ the smallest K value is 2, one sub-band (with $K = 1$) is missing which would occur at ν_0. For medium resolution, therefore, such a Raman band should consist of a *series of lines of nearly equal spacing*, $4(A-B)$, *with a zero-gap and without a strong central line.* No such structure has as yet been resolved. For the intensity alternation in such bands see Chapter I, section 2, and Placzek and Teller (701).

In the case of nearly symmetric top molecules of point groups C_s, C_2, C_{2h}, C_{2v}, D_2, V_h, the non-totally symmetric non-degenerate Raman bands $A''-A'$, $B-A$, B_g-A_g, $B_{1,2}-A_1$, $B_{2,3}-A$, $B_{2g,3g}-A_g$ respectively have $\Delta K = \pm 1$ instead of $\Delta K = \pm 2$, since for them only $[\alpha_{xz}]^{nm}$ and $[\alpha_{yz}]^{nm}$ are different from zero (see Table 55). Raman bands with $\Delta K = \pm 1$ would be still more similar to $\perp$ infrared bands in that the spacing of the Q branches (sub-bands) would be $2(A-B)$.

Degenerate Raman bands. For Raman transitions from the (totally symmetric) ground state to a degenerate upper state, we may have $\Delta K = \pm 1$ or $\Delta K = \pm 2$ or both.

The first case, $\Delta K = \pm 1$, as can be seen from Table 55 in conjunction with the general selection rule on p. 441, applies to $E''-A_1'$ transitions of D_{3h} molecules, $E-A_1$ transitions of C_{4v}, D_{4h} molecules, E_1-A_1 transitions of C_{5v}, C_{6v}, D_{5h}, D_{6h}, and a few other molecules. In this case, since the selection rule for the Coriolis sub-levels is also the same as for a $\perp$ infrared band of the same molecules, we have the same fine structure (see Fig. 128), except that again each sub-band has an O and an S branch in addition to the P, Q, and R branches. The line-like Q branches which form the main feature of the band can be represented by the same formula (IV, 60) as for the $\perp$ infrared band; that is, they form a *series of nearly equidistant lines* of spacing $2[A(1-\zeta_i) - B]$. No such fine structure has as yet been resolved.

The second case, $\Delta K = \pm 2$, applies to $E'-A_1'$ transitions of D_{3h} molecules, E_2-A_1 transitions of C_{5v}, C_{6v}, D_{5h}, D_{6h} molecules, and a few others. In this case we have again a series of sub-bands (see Fig. 128) whose line-like Q branches form the most prominent feature. But since ΔK is different (± 2), and also since the selection rule for the Coriolis sublevels is different, a different formula holds. We obtain from (IV, 41) and (IV, 42), taking account of the selection rules,

$$\nu_0^{\text{sub}} = \nu_0 + 4[A'_{[v]}(1+\zeta_i) - B'_{[v]}] \pm 4\left[A'_{[v]}\left(1+\frac{\zeta_i}{2}\right) - B'_{[v]}\right]K \\ + [(A'_{[v]} - B'_{[v]}) - (A''_{[v]} - B''_{[v]})]K^2, \quad \text{(IV, 75)}$$

where the + sign holds for $\Delta K = +2$ and the − sign for $\Delta K = -2$ and where in the former case $K = 0, 1, 2, \cdots$, in the latter $K = 2, 3, 4, \cdots$. It follows that if, as usual, we put $A' = A''$, $B' = B''$, the Q branches form *a series of equidistant "lines" of spacing* $4\left[A\left(1+\frac{\zeta_i}{2}\right) - B\right]$ which, if ζ_i is small, is double the spacing in a band with $\Delta K = \pm 1$ (just as for the previously considered non-degenerate band with $\Delta K = \pm 2$). There is again a line missing in this series, now at $\nu_0 + 2A\zeta_i$. It is important to note that ζ_i does not enter into equation (IV, 75) in the same way as it does into equation (IV, 60). If, therefore, a degenerate state combines with the ground state both in the infrared and in the Raman spectrum, and if the Raman band has $\Delta K = \pm 2$, its ζ_i may be determined from the observed spacing in the infrared and in the Raman effect without any assumption about the force constants, lack of anharmonicity, and so on; and therefore I_A may also be directly determined. For example, the transition E'—A_1' of D_{3h} is possible both in the infrared and Raman spectrum and might thus be used for a determination of I_A. Actually, however, no Raman band of this type has as yet been sufficiently resolved.

The third case ($\Delta K = \pm 1$ and $\Delta K = \pm 2$) applies to E—A_1 Raman bands of C_{3v} and D_3 molecules, to E_g—A_{1g} Raman bands of D_{3d} molecules, and to a few others, since for them both $[\alpha_{xy}]^{nm}$ and $[\alpha_{xz}]^{nm}$, $[\alpha_{yz}]^{nm}$ are different from zero (see Table 55). In this case we have, of course, simply a *superposition of two series of line-like Q branches*, one with a constant spacing $2[A(1 - \zeta_i) - B]$, the other with a constant spacing $4[A(1 + \zeta_i/2) - B]$, assuming again that $A' = A''$ and $B' = B''$. The former has no missing line, the latter has a line missing near the center. Fig. 135 shows such a structure schematically. Naturally for three-fold symmetry there is an

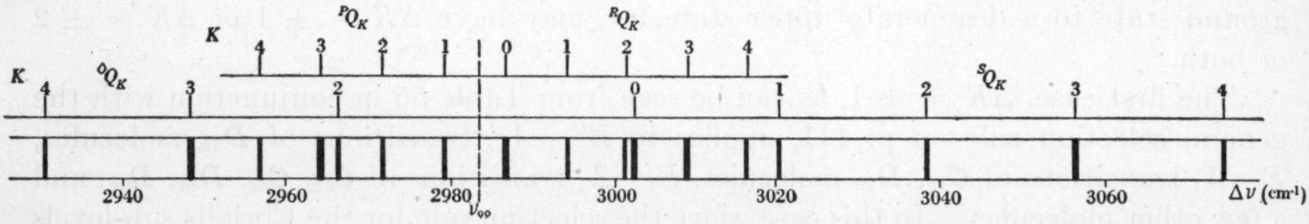

FIG. 135. **Fine structure of a degenerate Raman band with $\Delta K = \pm 1$ and ± 2.**—The figure has been drawn to scale for ν_4 of CH_3F for which the necessary data are known from the infrared spectrum. No Raman observations of sufficient dispersion have as yet been made. The intensity alternation is indicated by the weight of the lines.

intensity alternation of the type strong, weak, weak, strong, $\cdots$ in each of the two series of Q branches. From the spacing in the two series, if B is known from $\|$ infrared (or Raman) bands, A and ζ_i can be determined, and from the former the moment of inertia, I_A, about the symmetry axis. Unfortunately, for NH_3, the only molecule of this type for which the Raman spectrum of the gas has been investigated under sufficiently high dispersion, the degenerate Raman bands were too weak for observation.

Unresolved Raman bands. According to the above discussion, Raman bands of symmetric top molecules corresponding to totally symmetric vibrations have a strong central Q branch resulting from the superposition of the (strong) line-like Q branches of all sub-bands, whereas Raman bands corresponding to non-totally symmetric vibrations (degenerate or non-degenerate) have a series of Q branches which are widely spaced if the moment of inertia about the top axis is small. In the first case, therefore, if the Raman spectrum is investigated with the usual low dispersion (that is, when the fine structure is not resolved), *very sharp Raman lines* will be observed. They represent the Q branches of the bands, the other branches usually not being recorded since their lines do not all coincide. On the other hand, in the second case the unresolved Raman bands appear as broad lines (bands) with a flat maximum, whose height is much smaller than in the first case, since the various Q branches do not coincide. This is one of the main reasons why non-totally symmetric Raman lines are usually much weaker than totally symmetric ones. With medium dispersion Nielsen and Ward (670) have indeed found that in the gaseous state the *non-totally symmetric Raman lines are broad.* This is well illustrated by

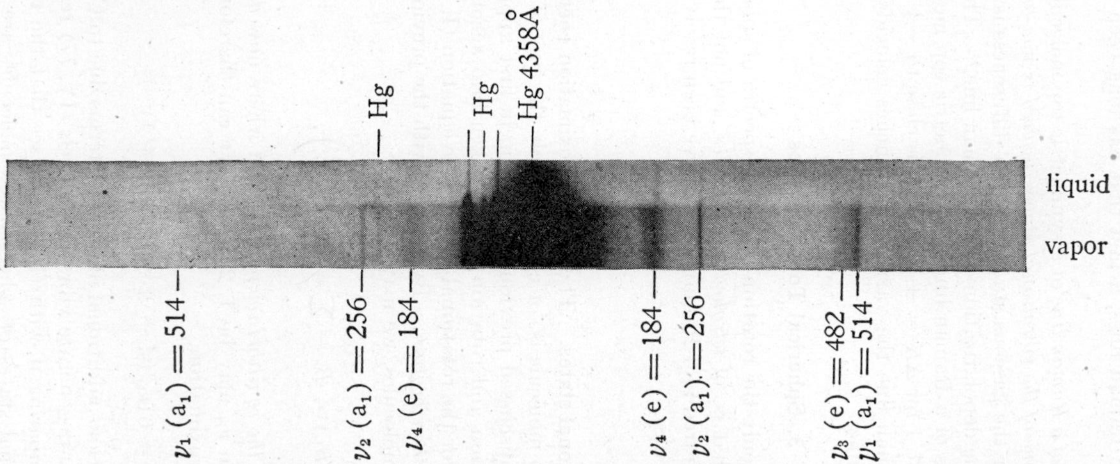

FIG. 136. **Raman spectrum of PCl_3 in the liquid and vapor state showing broadness of degenerate Raman lines** [after Nielsen and Ward (670)].*—The spectrum is excited by the mercury line 4358 Å. The figures given at the bottom are the shifts in wave number units for the vapor (which differ slightly from the figures for the liquid in Table 38.)

* The author is greatly indebted to Professor J. R. Nielsen for this spectrogram.

the Raman spectrum of PCl_3 in Fig. 136, due to Nielsen and Ward. While they found that in the liquid state, on account of quenching of molecular rotation (see Chapter V, section 2) the lines are sharper, even in the liquid state they are usually not as sharp as totally symmetric Raman lines, a fact that has long been recognized [see Kohlrausch (14)].

Conversely, *diffuseness of a Raman line* of a symmetric top molecule may be taken as a *strong indication that one of the vibrational states involved is non-totally symmetric.* If definitely established for the gaseous state,[43] such a diffuseness is in fact at least as good an indication as the depolarization of the Raman line.[44] The opposite conclusion, from the sharpness of a Raman line of a symmetric top molecule, is not as certain, since if ζ_i is close to 1 for $\Delta K = \pm 1$, or if ζ_i is close to $-\frac{1}{2}$ for $\Delta K = \pm 2$, even for a degenerate Raman line the various Q branches coincide approximately and would form a sharp "line."

3. Spherical Top Molecules

We shall consider here only the rotation-vibration spectra of spherical top molecules of point group T_d, that is, of *tetrahedral molecules*, and not those of lower or higher symmetry, since infrared and Raman bands have been resolved only for this type of spherical top molecules.

(a) Energy levels

Non-degenerate vibrational states. The zero-approximation energy of a rotating and vibrating spherical top molecule is, of course, simply the sum of the vibrational and the rotational energy discussed previously. Again, in first approximation, the interaction between vibration and rotation can be taken into account by using in the expression $BJ(J+1)$ for the rotational energy [see equation (I, 51)] an *effective B value, $B_{[v]}$, averaged over the vibration.* By analogy with the previous formulae for linear and symmetric top molecules, we may put

$$B_{[v]} = B_e - \sum \alpha_i \left(v_i + \frac{d_i}{2} \right), \qquad \text{(IV, 76)}$$

where B_e is the *B value of the equilibrium position* and where now d_i, the degree of degeneracy of the vibration ν_i, may be 1, 2, or 3. We can therefore write for the total energy, in a first approximation,

$$T = G(v_1, v_2, \cdots) + B_{[v]}J(J+1). \qquad \text{(IV, 77)}$$

While for degenerate vibrations additional interaction terms due to Coriolis coupling appear (see below), for non-degenerate vibrational levels (IV, 77) represents a very good approximation. Comparing it with (IV, 6), we see that the rotational levels in non-degenerate vibrational states of spherical top molecules are very similar to those of linear molecules, except that the *statistical weight* is $(2J+1)^2$ instead of $(2J+1)$.

[43] In the liquid state there may also be other reasons for a broadening.

[44] It will be remembered that a totally symmetric Raman line may have a degree of depolarization up to $\frac{6}{7}$, that is, may in exceptional cases be completely depolarized, so that it could not be distinguished by the polarization from a non-totally symmetric Raman line.

As for linear molecules, the finer interaction of rotation and vibration leads also to *centrifugal distortion* represented by a term $D_{[v]}J^2(J+1)^2$ in the energy formula. But up to the present time sufficiently accurate data are not available to require taking account of this effect.

Degenerate vibrational states. As in the case of symmetric top molecules, the *Coriolis forces* that occur in the rotating molecule may produce an interaction between mutually degenerate vibrations, which in its turn will lead to an appreciable *splitting of the degeneracy.*

There are three species of degenerate vibrational levels in tetrahedral molecules, E, F_1, and F_2. The fundamentals of the molecules Y_4 and XY_4 have only species E and F_2 (see p. 140). Now it can fairly easily be seen by a consideration of the vibrations in Fig. 41 that if one component of the doubly degenerate vibration ν_2 is excited the Coriolis force does *not* tend to excite the other component, no matter what the direction of the axis of rotation. Therefore *no Coriolis splitting arises for the doubly degenerate vibrational states.* The rotational energy levels are the same as for the non-degenerate vibrational states (see equation IV, 77).

The absence of Coriolis splitting follows quite generally for any state of species E of point group T_d from Jahn's general rule (p. 376). The product $E \times E$ of the species of the two interacting vibrations, according to Table 33, is $A_1 + A_2 + E$, that is, does not contain the species of the rotation, which is F_1 in this case (see Table 28).

However, *for the triply degenerate vibrational states the Coriolis interaction does cause a splitting.* This is most easily seen by considering the vibration ν_3 of XY_4 in Fig. 41. For a rotation about the z axis, if the component ν_{3a} is excited, the Coriolis force tends to excite ν_{3c}, whereas ν_{3b} is uninfluenced. Therefore a splitting occurs in this case into three components, one of which has the "original" frequency. The other two, as for symmetric top molecules, are not simply ν_{3a} and ν_{3c}, but linear combinations of these vibrations which no longer tend to go over into each other in consequence of the Coriolis force. These two linear combinations are, as previously, the two circular oscillations, the clockwise and the counter-clockwise, whose angular momentum is $\boldsymbol{p}$. Actually, since the force on the Y nuclei is not the same in all directions, the motion is not circular but elliptic. $\boldsymbol{p}$ is parallel or antiparallel to the total angular momentum $\boldsymbol{J}$.

The *rotational energy values for the three sublevels* are given by a formula very similar to that for symmetric top molecules [see Teller (836), Shaffer, Nielsen, and Thomas (781), and Dennison (280)], namely [45]

$$\begin{aligned} F^{(+)}(J) &= B_{[v]}J(J+1) + 2B_{[v]}\zeta_i(J+1), \\ F^{(0)}(J) &= B_{[v]}J(J+1), \\ F^{(-)}(J) &= B_{[v]}J(J+1) - 2B_{[v]}\zeta_i J, \end{aligned} \qquad \text{(IV, 78)}$$

where ζ_i is the magnitude of $\boldsymbol{p}$ in units $\dfrac{h}{2\pi}$ for the particular vibrational state.[46] Fig.

[45] Here a small additive term that is independent of J and is the same for all three component levels has been omitted.

[46] Tindal, Straley anh Nielsen (866) use B_e instead of $B_{[v]}$ in the second term on the right, but Dennison (280) states explicitly that $B_{[v]}$ should be used.

137 gives at the top a graphical representation of the three sets of levels while at the bottom the levels of a totally symmetric state (ground state) are given.

The formulae (IV, 78) hold for vibrational states of species F_1 as well as of F_2 (the former occurring only as upper states of certain overtone and combination bands of the XY_4 molecule). The ζ_i values can be expressed in terms of the potential

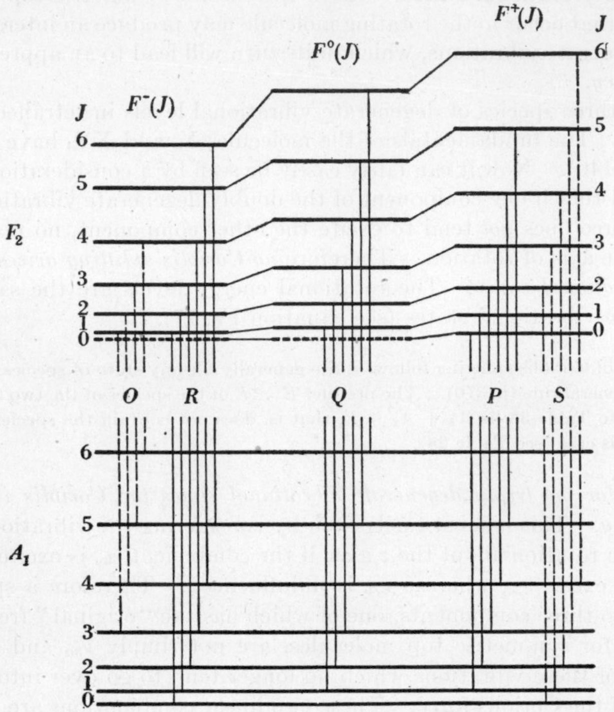

FIG. 137. Rotational energy levels of a spherical top molecule in a triply degenerate (F_2) and totally symmetric (A_1) vibrational state.—The transitions indicated will be discussed on p. 454 and p. 459. The broken-line levels do not occur (see Fig. 138).

constants and the masses [see Johnston and Dennison (476) and Shaffer, Nielsen, and Thomas (781)]. But, as for symmetric top molecules, the *sum of the* ζ_i *for all* $v_i = 1$ *states of the same species is independent of the potential constants.* For the two F_2 vibrations ν_3 and ν_4 of XY_4 molecules, it follows immediately from the previous formula (IV, 47) for an XYZ_3 molecule that

$$\zeta_3 + \zeta_4 = \tfrac{1}{2}, \qquad \text{(IV, 79)}$$

since here $B = A$ and since one of the degenerate vibrations of the XYZ_3 molecule goes over into the E vibration of XY_4, which has $\zeta_2 = 0$ (see above). In forming the sum, and also in (IV, 78), as previously, ζ_i is to be taken positive or negative depending on whether or not the direction of rotation of the electric dipole moment of the molecule coincides with the direction of the angular momentum of vibration $\mathbf{p}$.

Johnston and Dennison (476) also worked out the ζ *values for some of the overtone and combination levels.* They found for the F_2 sublevels of $2\nu_3$ and $2\nu_4$: $\zeta = -\zeta_3$ and $-\zeta_4$ respectively, while for $\nu_3 + \nu_4$ they found $\zeta = -\frac{1}{2}(\zeta_3 + \zeta_4) = -\frac{1}{4}$. However, they did not take into account the Coriolis interaction with the other sublevels, which leads to rather more complicated relations [see Shaffer, Nielsen, and Thomas (782)]. For the combination levels $v_1\nu_1 + \nu_3$ and $v_2\nu_2 + \nu_3$, $v_1\nu_1 + \nu_4$ and $v_2\nu_2 + \nu_4$ the ζ values are the same as for ν_3 and ν_4 respectively [Shaffer, Nielsen, and Thomas (781)].

It may be mentioned that while E vibrational levels, apart from second-order effects (see below), do not show any Coriolis splitting, the E sublevels of overtone and combination levels that also have F_1 or F_2 sublevels do split, because of the interaction of E with F_1 or F_2, which in zero approximation have the same energy. Thus the state $2\nu_3$ (or $2\nu_4$) has sublevels A_1, E, and F_2. In consequence of anharmonicity these three sublevels (even without rotation) have slightly different energy. In consequence of Coriolis interaction both the sublevels E and F_2 split linearly with increasing J [see Shaffer, Nielsen, and Thomas (782)].

Symmetry properties of the rotational levels. As for symmetric top molecules, the rotational eigenfunctions of the spherical top molecules have certain symmetry properties which correspond to the symmetry types (species) of the rotational subgroup to which the molecule belongs. For the tetrahedral molecules of point group T_d (the only ones we are considering here), the *rotational subgroup* (that is, the point group that has as symmetry elements only the symmetry axes of T_d) is T (see Table 30). The species of this group are A, E, and F. It is obvious that both A_1 and A_2 of T_d belong to A, and both F_1 and F_2 of T_d belong to F of T. Depending on the behavior of the total eigenfunction $\psi \approx \psi_e\psi_v\psi_r$ with respect to the symmetry elements of T, we have *three over-all species of rotational levels, A, E, and F.*

If $\psi_e\psi_v$ is totally symmetric, the species of the rotational level depends on the symmetry of ψ_r only. A closer study of the rotational eigenfunctions of the spherical top shows [see Wilson (933)] that the species are those given in Fig. 138a for the first twelve rotational levels.[47] For all but the first few rotational levels we have one or more sublevels of each of the three species. They are drawn separately in Fig. 138a, but it should be understood that without taking into account second-order interactions of vibration, rotation, and electronic motion (see below), all levels of a given J ($2J + 1$ in all) coincide. For large J values, as can be seen qualitatively from Fig. 138a, there are for each J value equally many A and E sublevels and three times as many F levels as A levels.

If $\psi_e\psi_v$ is not totally symmetric with respect to the rotational subgroup T, we have to "multiply" the species of $\psi_e\psi_v$ by the species of ψ_r given in Fig. 138a, according to the rules of Table 33. For example, for $J = 4$ we have for totally symmetric $\psi_e\psi_v$ the species $A + E + 2F$ of the rotational levels (see Fig. 138a). Therefore, if $\psi_e\psi_v$ has species E we have for the rotational sublevels the species $E \times (A + E + 2F) = E + 2A + E + 4F$. The over-all species of the levels up to $J = 12$ obtained in this way for $\psi_e\psi_v$ of species E and of species F (that is, F_1 or F_2) are given in Fig. 138b and 138c. In the latter it is also indicated, according to Jahn (468), which rotational sublevels belong to $F^{(+)}$, which to $F^{(0)}$, and which to $F^{(-)}$. It may be noted that the $F^{(0)}$ levels of F_1 or F_2 have the same species as the levels of a totally symmetric vibrational state, except that the level with $J = 0$ is missing. The species of the $F^{(-)}$ and $F^{(+)}$ levels are obtained by shifting those of Fig. 138a up or down respectively by one unit.

If the *influence of the nuclear spin* is disregarded, it is easily verified from Fig. 138 that the statistical weight of a set of levels with a given J (apart from the ordinary space degeneracy) is $2J + 1$, $2(2J + 1)$, and $3(2J + 1)$ for vibrational levels of species A, E, and F respectively. However, in order to obtain the actual statistical

[47] The species of the higher levels can easily be obtained from formulae given by Wilson (933).

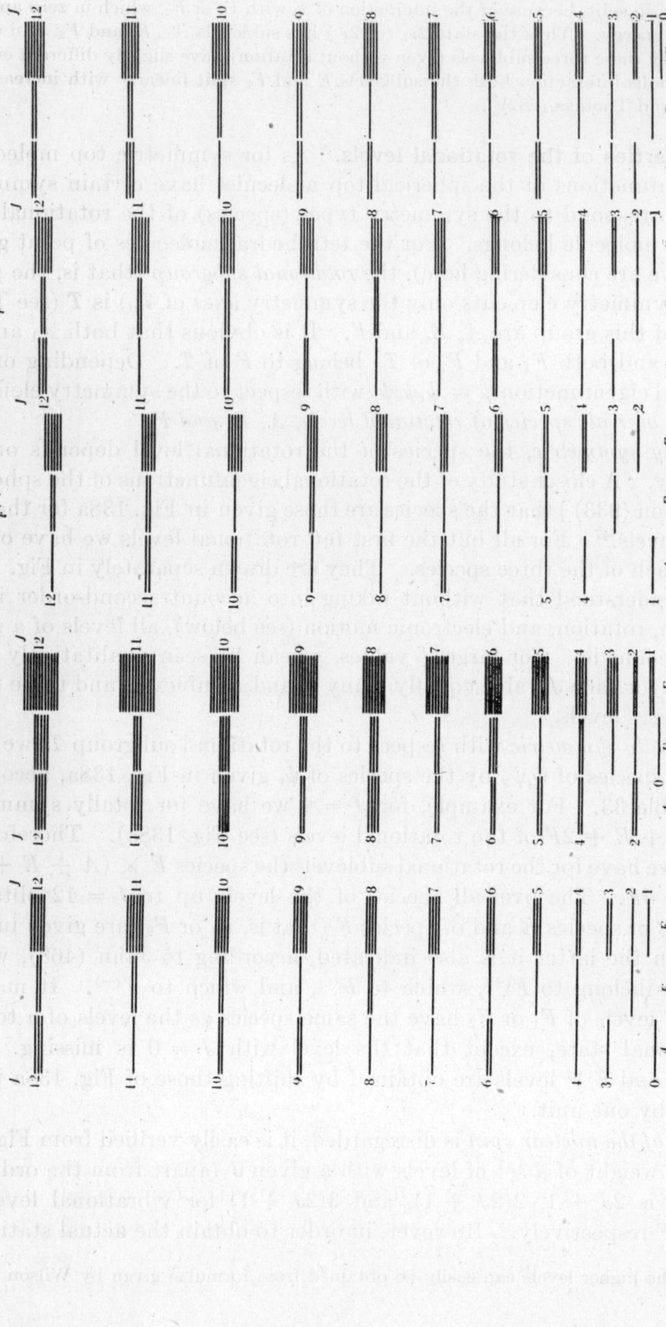

FIG. 138. Species of the rotational levels of a spherical top molecule of point group T_d (a) in an A, (b) in an E, (c) in an F vibrational state.—A doubly or triply degenerate level is drawn as one line. Thus in (a) for $J = 6$ there are three F levels, one E and one A level. Levels drawn close together coincide when there is no Coriolis interaction (see text).

weights, we have to take the nuclear spin I into account. *If the spin of the identical nuclei is zero*, only those rotational levels can occur whose eigenfunctions remain unchanged for any rotations that lead to an exchange of identical nuclei; that is, *only the A rotational levels* of Fig. 138 *can occur*. This would be the case for the molecules SO_4^{--}, ClO_4^{-}, $Ni(CO)_4$ if they were regular tetrahedrons. For them, for example in a vibrational state of species A_1 or A_2, the rotational levels $J = 1, 2$, and 5 would not occur (see Fig. 138a).

If the nuclear spin I of the identical atoms is not zero, the inclusion of the spin function may cause the total eigenfunction of all rotational sublevels to be of species A. *All rotational sublevels may occur but with different statistical weights.* For a tetrahedral XY_4 molecule with $I(Y) = \frac{1}{2}$, as Wilson (933) has shown, in a way similar to that indicated previously for XY_3 molecules, the species of the spin functions are $5A + E + 3F$. For the A rotational levels we have to use a spin function of species A, of which there are five; for the E rotational levels we have to use the one spin function of species E in order to make the total eigenfunction of species A; and for the F rotational levels we have to use a spin function of species F, of which there are three. Since $E \times E$ gives two functions of species A while $A \times A$ and $F \times F$ give only one each, it follows that the statistical weights of the A, E, and F rotational levels are 5, 2, and 3 respectively. From this the total statistical weights for each J can be obtained. For a vibrational state of species A (A_1 or A_2), they have been given in the previous Table 7, p. 39. For the other vibrational states they are easily obtained on the basis of Fig. 138b and 138c. For example, for $J = 4$ the three sublevels 4^-, 4^0, and 4^+ have statistical weights (apart from the usual factor $2J + 1$ for space degeneracy) $(5 + 2 \times 3) = 11$, $(5 + 2 + 2 \times 3) = 13$ and $(2 + 3 \times 3) = 11$ respectively. These statistical weights apply, for example, to the molecules CH_4 and SiH_4. For $I(Y) = 1$ according to Wilson (933), the symmetry of the spin function is $15A + 6E + 18F$, and therefore the statistical weights of the A, E, and F rotational levels are 15, 12, and 18 respectively, leading to the total statistical weights given in Table 7. This table contains also the weights for $I(Y) = \frac{3}{2}$.

Inversion doubling. In all tetrahedral molecules the inversion doubling is unobservably small since the potential hill separating the "left" from the "right" configuration is very high. But, strictly speaking, each of the rotational sublevels discussed above is double, one component being "positive," the other "negative" with respect to an inversion. The potential field is then of point group $\boldsymbol{O_h}$ (compare the treatment of $\boldsymbol{C_{3v}}$ molecules p. 413), and the rotational subgroup is $\boldsymbol{O}$, which has the same species (A_1, A_2, E, F_1, F_2) as $\boldsymbol{T_d}$ (see Table 28). The A rotational sublevels in Fig. 138 are therefore split into A_1 and A_2, the F levels into F_1 and F_2, the E levels into two E levels. For $I = 0$ and Bose statistics of the identical nuclei, only the A_1 levels occur; for $I = \frac{1}{2}$ and Fermi statistics, only the A_2, E, and F_2 levels occur, with weights 5, 1, and 3 respectively. Thus for $I = \frac{1}{2}$ only the E levels are actually doubled by the possibility of inversion. Only for larger I values will all levels be doubled. But in any case, just as for $\boldsymbol{C_{3v}}$ molecules, the total statistical weight of each sublevel is not changed by the doubling. There is hardly a chance that the doubling will ever be observed in the rotation-vibration spectra of tetrahedral molecules. For all practical purposes we can therefore draw our conclusions *as though there were only one nuclear configuration;* we can therefore omit a detailed discussion of which levels are shifted up and which down [see Jahn (468)].

Coriolis splitting of the rotational levels. We have seen above that each rotational level of a given J value consists of a number of sublevels ($2J + 1$ in all) which coincide in the approximation in which (IV, 77) and (IV, 78) hold. But if the finer interaction of vibration and rotation is taken into account, a splitting of these sublevels occurs, for reasons similar to those for the l-type doubling of linear molecules (see p. 377). However, a splitting can occur only into as many levels of slightly different energy as there are different lines in Fig. 138. A rotational sublevel of species E or F, which is doubly or triply degenerate respectively, cannot be split into two or three components by the finer interactions, since these interactions have all a tetrahedral symmetry. Their degeneracy could only be removed by an external field.

The *splitting of levels with the same J is largely due to the Coriolis interaction of different vibrations.* The splitting will be the larger the closer together are the two interacting vibrational levels, and will be proportional to $J(J+1)$. This is in contrast to the Coriolis splitting of a triply degenerate vibrational state which is proportional to J, since it is due to the Coriolis interaction of mutually degenerate vibrations, not of different vibrations of different frequency.

Since the rotation of a tetrahedral molecule (considered as a non-genuine vibration) has species F_1 (see Table 28) there are, according to Jahn's rule (p. 376), seven types of Coriolis perturbations, namely those between the following pairs of vibrational levels (compare Tables 31 and 33):

$$A_1—F_1, \quad A_2—F_2, \quad E—F_1, \quad E—F_2, \quad F_1—F_1, \quad F_1—F_2, \quad F_2—F_2.$$

Jahn (469) has shown that for A_2-F_2 perturbations in second order approximation only the $F^{(0)}$ component of the F_2 vibrational level is perturbed, and that for $F_2—F_2$ perturbations only the $F^{(+)}$ and $F^{(-)}$ components are affected. Also he showed that in both cases the shift is the same for each group of levels of a given J, so that there is no splitting but only a slight change of the effective $B_{[v]}$ values. The same applies to $A_1—F_1$ and $F_1—F_1$ perturbations not explicitly treated by Jahn. These conclusions may also be obtained qualitatively from Fig. 138 if it is remembered that only rotational levels of the same J and the same symmetry (including the behavior with respect to inversion) can perturb one another. Thus the rotational levels of the vibrational ground state as well as of any other non-degenerate vibrational state are not split by Coriolis interaction with any other vibrational state.[48] However, in the case of the three remaining pairs $E—F_1$, $E—F_2$, and $F_1—F_2$, different sublevels are affected differently, and therefore a splitting of the rotational levels of vibrational states of species E, F_1 and F_2 into as many sublevels as there are lines in Fig. 138 does take place. For example, consider the interaction between an E and an F_2 vibrational level. The A sublevels of F_2 for $J = 1$ and 3 are unperturbed since there are no A sublevels for $J = 1$ and 3 in the vibrational state E. But the F sublevels will be shifted, since they occur in both vibrational states for these J values. Similarly, for other J values, the number of sublevels of the three species is not the same in the two vibrational levels, and therefore the shift will be different for the different sublevels.

While in general the Coriolis splitting here considered will be small, as is the l-doubling for linear molecules, it will be *very appreciable when the two interacting vibrational levels are close together*. This occurs, for example, in CH_4 and similar molecules for the $v_i = 1$ levels of $\nu_2(e)$ and $\nu_4(f_2)$, which for CH_4 occur at 1526 and 1306.2 cm^{-1} respectively. The Coriolis splitting will naturally become of importance also for the higher vibrational levels. Jahn (468) (469) has given detailed formulae for the perturbations of the individual levels of an F_2 vibrational level interacting

[48] They may be slightly split by higher-order interactions and by centrifugal stretching terms, but the effect of these has not as yet been calculated [see Wilson (934)].

with an E and an F_1 vibrational level. He has calculated the shifts numerically for the rotational levels of ν_4 of CH_4 up to $J = 10$ [see also Shaffer, Nielsen, and Thomas (781) (782) and Murphy (646)].

In addition to the above-discussed perturbations, we have to expect also the occurrence of more irregular perturbations, for which a shift and splitting occurs only for a few J values, similar to the ordinary perturbations of diatomic and linear polyatomic molecules. It is, however, premature to discuss these since no relevant experimental data are available.

(*b*) *Infrared spectrum*

Selection rules. Just as for linear and symmetric top molecules, as long as the interaction of vibration and rotation is not too large, the vibrational selection rules are the same for the rotation-vibration spectrum as they are for the pure vibration spectrum (Table 55). In particular, the *ground state can combine* (in infrared absorption) *only with vibrational states of species* F_2. The selection rule for the rotational quantum number J is, as always,

$$\Delta J = 0, \pm 1. \qquad \text{(IV, 80)}$$

In F_2 vibrational states we have the $F^{(+)}$, $F^{(0)}$, and $F^{(-)}$ sublevels (see p. 447). According to Teller (836), *for* $\Delta J = +1$ *only the* $F^{(-)}$ *levels combine with the* (A_1) *ground state, for* $\Delta J = 0$ *only the* $F^{(0)}$ *levels, and for* $\Delta J = -1$ *only the* $F^{(+)}$ *levels.*

As for symmetric top molecules, since rotational levels of different species have different nuclear spin functions and since the coupling of the nuclear spin with the rest of the molecule is extremely slight, *rotational levels of a given species can only combine with rotational levels of the same species;* that is,

$$A \leftrightarrow A, \qquad E \leftrightarrow E, \qquad F \leftrightarrow F. \qquad \text{(IV, 81)}$$

This rule holds very strictly not only for infrared transitions but also for transitions brought about in any other way. Thus, as mentioned before, there are three non-combining modifications (analogous to ortho and para H_2) for any tetrahedral molecule. In the ground state their rotational levels are given by the three columns in Fig. 138a.

As always, *only states of opposite symmetry with respect to inversion can combine with one another* $(+ \leftrightarrow -)$. However, this rule does not lead to any additional restrictions unless the inversion doubling is resolved. Up to the present time no such case has been observed.

$F_2 - A_1$ transitions. The only rotation-vibration bands of tetrahedral molecules whose fine structures have as yet been studied in any detail are those whose lower state is the ground state (species A_1) and whose upper state is an F_2 state. If for the moment we neglect the Coriolis splitting of the F_2 state into F^+, F^0, F^- components, it is clear from the energy formula (IV, 77) and the selection rule (IV, 80) that we obtain a simple P, Q, and R branch, with the same formulae as for linear molecules. The spacing of the lines in the P and R branches would be $2B$ (apart from the slight convergence introduced by the fact that $B' \neq B''$), whereas the lines of the Q branch would almost coincide. In this approximation, unlike the case of || bands of symmetric top molecules, the individual lines would be single, even for a large difference between B' and B'', since here the band is not a superposition of sub-bands. But, as for a || band of a symmetric top molecule, the intensity of the lines with low J is relatively much smaller than that for bands of linear molecules, since the statistical weight is $(2J + 1)^2$ rather than $(2J + 1)$.

Actual investigation of the infrared bands of CH_4, SiH_4, and others with medium dispersion shows indeed such a simple structure (see Fig. 139). However, the spacing in different bands is rather different. The reason for this is, as for symmetric top molecules, the Coriolis coupling between rotation and vibration in the upper de-

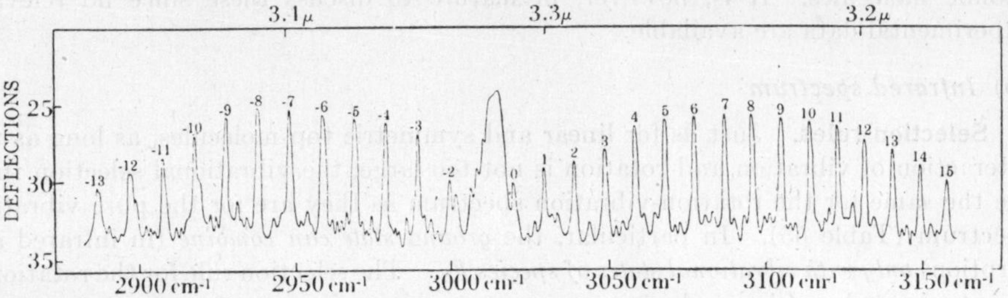

FIG. 139. Fine structure of the fundamental band ν_3 of CH_4 at 3.31μ [after Nielsen and Nielsen (656)].—The length of the absorbing path was 2 cm at atmospheric pressure. The numbers written on the maxima are *m* values.

generate vibrational state. In consequence of this the upper states of corresponding lines of the *P*, *Q*, and *R* branches are somewhat different (see Fig. 137). Using the formulae (IV, 77) and (IV, 78) for the rotational levels of the lower and upper vibrational state respectively, and remembering the above selection rule for the F^+, F^0, F^- sublevels, we obtain for the *R branch*,

$$R(J) = \nu_0 + 2B'_{[v]} - 2B'_{[v]}\zeta_i + (3B'_{[v]} - B''_{[v]} - 2B_{[v]}\zeta_i)J + (B'_{[v]} - B''_{[v]})J^2; \quad \text{(IV, 82)}$$

for the *Q branch*,

$$Q(J) = \nu_0 + (B'_{[v]} - B''_{[v]})J + (B'_{[v]} - B''_{[v]})J^2; \quad \text{(IV, 83)}$$

and for the *P branch*,

$$P(J) = \nu_0 - (B'_{[v]} + B''_{[v]} - 2B'_{[v]}\zeta_i)J + (B'_{[v]} - B''_{[v]})J^2. \quad \text{(IV, 84)}$$

As for diatomic molecules, the *P* branch and *R* branch form *one series of lines* represented by the one formula

$$\nu = \nu_0 + (B'_{[v]} + B''_{[v]} - 2B'_{[v]}\zeta_i)m + (B'_{[v]} - B''_{[v]})m^2, \quad \text{(IV, 85)}$$

where $m = J + 1$ for the *R* and $m = -J$ for the *P* branch, and where one line, $m = 0$, is missing. If $B' - B'' \approx 0$, as is always the case for fundamentals, the separation of successive lines in this series is very nearly constant and is equal to $2B(1 - \zeta_i)$, while all lines of the *Q* branch coincide at the zero line ν_0. Since ζ_i is different for different vibrations (within the limits +1 and −1), it is clear that the *spacing of the lines in different infrared bands may differ widely.* The moment of inertia of a tetrahedral molecule can therefore not be determined from the fine structure of one infrared band alone.

However, because of the sum rule for ζ_i it is possible to *determine the moment of inertia if the spacing in all infrared-active fundamentals is known.* For XY_4 molecules there are two such fundamentals, ν_3 and ν_4, for which $\zeta_3 + \zeta_4 = \frac{1}{2}$ [see (IV, 79)]. Therefore the sum of the spacings in ν_3 and ν_4 is $3B$, and thus B, I_B, and the X—Y distance in XY_4 can be determined. For CH_4, according to Childs (205), the spacing of the lines in the band ν_3 at 3020 cm^{-1} (Fig. 139) near the origin [where they are least disturbed by the quadratic term in (IV, 85)] is 9.93 cm^{-1}, and in the band ν_4 at 1306 cm^{-1} it is 5.74 cm^{-1}. Therefore $3B = 15.67$ cm^{-1} and $B = 5.223$ cm^{-1}. This is, of course, an average of the B values of the upper and lower states. Correcting with the $(B' - B'')$ values known from the convergence of the bands, Childs obtained $B_{[0]} = 5.252$ cm^{-1}. From this and the observed spacing the individual ζ_i values can be obtained. They are $\zeta_3 = 0.05$ and $\zeta_4 = 0.45$.

The above $B_{[0]}$ value cannot, of course, claim the accuracy of the B values of linear molecules since for the ζ_i sum rule harmonic vibrations have been assumed. The corresponding moment of inertia is $I_B{}^0 = 5.330 \times 10^{-40}$ g cm^2, and from this, since $I = \frac{8}{3}m_Y r^2_{XY}$, the C—H *distance in* CH_4 is found to be 1.0929×10^{-8} cm, which refers of course to an average internuclear distance in the lowest vibrational state. This value is in excellent agreement with the value 1.0936×10^{-8} cm that follows from the CH_3D spectrum (see p. 439). The latter value is probably the more accurate one, since it is obtained from bands that are unaffected by Coriolis coupling.

It is easily seen that, just as for $\perp$ bands of symmetric top molecules, the use of *combination differences* is not of much value in the analysis of infrared bands of tetrahedral molecules, since no two lines in the band have the same upper state (see Fig. 137). From (IV, 82) and (IV, 84) it follows immediately that

$$R(J) - P(J) = 4B'(1 - \zeta')(J + \tfrac{1}{2}) \qquad \text{(IV, 86)}$$

and

$$R(J-1) - P(J+1) = (4B'' - 4B'\zeta_i')(J + \tfrac{1}{2}). \qquad \text{(IV, 87)}$$

The initial and final levels are therefore not separated. However, at any rate, the differences $R(J) - P(J)$ and $R(J-1) - P(J+1)$ plotted against J should yield very nearly straight lines. Actually, in the case of CH_4 this is not very well fulfilled, as there is a slight curvature. This is apparently due to the fact that the higher rotational levels of the upper vibrational state are perturbed by Coriolis interaction with other vibrational levels (see p. 452).

Indeed, with higher dispersion the *higher rotational lines* in the CH_4 bands are *resolved into two or more components.* Such a splitting is just indicated for the last lines of the P branch of the fundamental ν_3 of CH_4 in Fig. 139. The splitting is very large for ν_4 of CH_4, as seen in Fig. 140, and clearly disturbs the regularity of the P and R branches. The splitting is also very clear in the photographic infrared band of CH_4 at 11060 Å, reproduced in Fig. 141a.

In CH_4 the frequency of the active fundamental $\nu_4(f_2)$ is fairly close to that of the inactive fundamental $\nu_2(e)$ (1306 compared to 1526 cm^{-1}). Therefore the Coriolis interaction (which according to the above is possible between E and F_2 levels) will be strong, leading to a considerable splitting of the higher rotational levels into sublevels (see Fig. 138). Childs and Jahn (208) have carried out detailed calculations of the perturbations of ν_4 by ν_2, and have obtained striking agreement between the calculated and observed spectra (Fig. 140). Such an agreement can, of course, be obtained only when the perturbing state is known, as well as fairly accurate values of the moment of inertia and of ζ_i (that is, of the unperturbed energy levels). It would be very difficult to use the perturbation formulae for an exact evaluation of the moment of inertia. Most of the combination and overtone bands of CH_4 (see for example Fig. 141b), have a still more complicated structure than the fundamental ν_4, obviously because the perturbations are still larger. None of these, with the exception of the simple band at 9042 cm^{-1} (Fig. 141a), has as yet been analyzed.

The only tetrahedral molecules other than CH_4 whose infrared bands have been resolved are CD_4 [Nielsen and Nielsen (658)], SiH_4 [Steward and Nielsen (807),

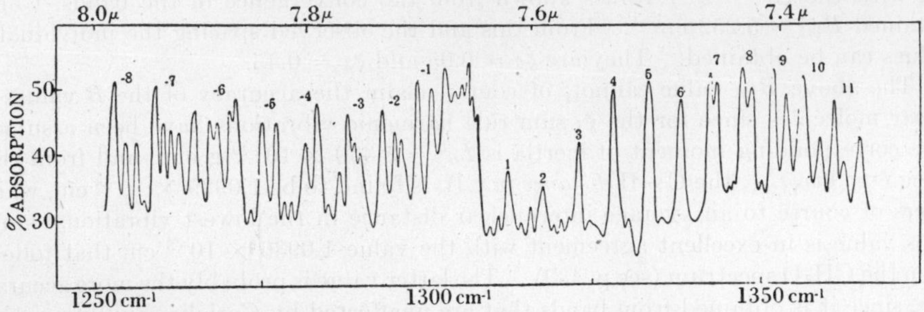

Fig. 140. Fine structure of the fundamental band ν_4 of CH_4 at 7.65μ [after Nielsen and Nielsen (656)].—The length of the absorbing path was 2 cm at atmospheric pressure. The numbers written on the maxima are *m* values.

Tindal, Straley, and Nielsen (865) (866)], and GeH_4 [Steward and Nielsen (808) Straley, Tindal, and Nielsen (865) (818)]. For them the situation is very similar to that in CH_4, that is, while ν_3 shows a simple structure, only the lines with large J being split, ν_4 is very strongly perturbed, and so are most of the combination and overtone bands. The values for the rotational constants $B_{[0]}$ obtained for these molecules are less accurate than for CH_4. They are given together with the resulting moments of inertia and internuclear distances in Table 133.

TABLE 133. ROTATIONAL CONSTANTS, MOMENTS OF INERTIA, AND INTERNUCLEAR DISTANCES OF TETRAHEDRAL XY_4 MOLECULES.

Molecule	$B_{[0]}$ (cm^{-1})	$I_B{}^0$ (10^{-40} gm cm^2)	r(X—Y)(10^{-8} cm)
CH_4	5.252	5.330	1.0929[49]
CD_4	2.64_7	10.57_6	1.089
SiH_4	2.96[50]	9.46	1.45_6[51]
GeH_4	2.87[50]	9.75	1.47_8

Forbidden vibrational transitions. The Coriolis perturbation causes a mixing of the eigenfunctions of the two levels concerned (as do all perturbations). If the perturbation is sufficiently strong this may lead to the breakdown of the vibrational selection rules which hold for rotation-vibration spectra only under the assumption of small interaction of vibration and rotation. If according to the vibrational selection rules one of two interacting states can combine with the ground state but not the other, with increasing rotation the second will assume to some extent properties of the first and therefore will be enabled to combine with the ground state. *Thus the interaction of vibration and rotation may lead to the occurrence of forbidden vibrational transitions, particularly for the higher rotational levels* (for $J = 0$ the vibrational selection rules hold rigorously).

---

[49] A more accurate value is probably 1.0936, obtained from the CH_3D spectrum (see p. 439).

[50] These are the B'' values given by Tindal, Straley, and Nielsen (866) (818). Their B_0 is our B_e, which is not given here since Tindal, Straley, and Nielsen's method of evaluation does not appear to be without objections (for example, they assume B to be independent of v_1).

[51] Tindal, Straley, and Nielsen (866), due to an arithmetical error, give 1.55.

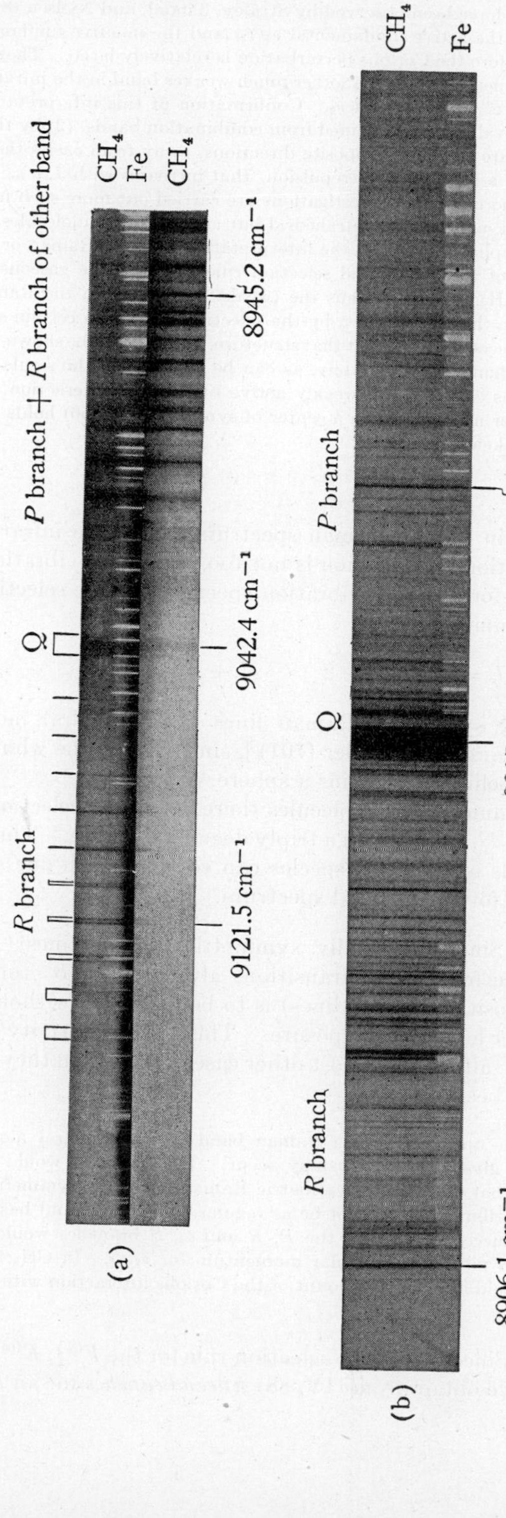

Fig. 141. Photographic infrared bands of CH_4 (a) at 1.106μ and (b) at 1.135μ.—For the band 1.106μ two spectrograms of different density are given. The length of the absorbing path was in (a) for the upper spectrogram 1650 cm at 2 atm. pressure, for the lower 1100 cm at 1 atm. pressure, in (b) 400 cm at a pressure of 890 mm and a temperature of −80° C.

Such forbidden transitions have been observed by Straley, Tindal, and Nielsen (817) (866) (818) for SiH_4 and GeH_4, for which the active fundamental $\nu_4(f_2)$ and the inactive fundamental $\nu_2(e)$ are fairly close together, and therefore the Coriolis perturbation is relatively large. They found in both cases, in addition to the fundamental band ν_4, another much weaker band in the infrared which must be interpreted as the "inactive" fundamental ν_2. Confirmation of this interpretation is supplied (1) by the agreement with the value for ν_2 obtained from combination bands, (2) by the fact that the Q branches of the two bands are shaded in opposite directions, away from each other, as would be expected on the assumption of a perturbation (repulsion) that increases with J.

It appears likely that as more detailed investigations are carried out more such forbidden vibrational transitions will be found, not only for tetrahedral but also for other molecules. Their actual occurrence for SiH_4, GeH_4, emphasizes that in the interpretation of weak Raman or infrared bands the possibility of a violation of the vibrational selection rules even in the gaseous state must be considered (see the case of C_2H_4, p. 328). Thus the (weak) occurrence in the Raman or infrared spectrum of a certain vibration that is forbidden by the selection rules for a certain structure (point group) of a molecule does not necessarily rule out that structure, unless it can be shown that no Coriolis coupling could produce these bands. Fortunately, as can be seen from Jahn's rule (p. 376) by no means all forbidden transitions can be made weakly active by Coriolis interaction. For example, the rule of mutual exclusion for molecules with a center of symmetry (p. 256) holds rigorously even when Coriolis interaction is taken into account.

(c) *Raman spectrum*

Selection rules. Again for the Raman spectrum as for the infrared spectrum, if the interaction of vibration and rotation is not too strong the vibrational selection rules remain the same as for the pure vibration spectrum. The selection rule for J is the same as for the symmetric top:

$$\Delta J = 0, \pm 1, \pm 2; \qquad J' + J'' \geq 2. \tag{IV, 88}$$

However, for the totally symmetric Raman lines of tetrahedral molecules, only $\Delta J = 0$ can occur [see Placzek and Teller (701)], since during the whole vibrational motion the polarizability ellipsoid remains a sphere.

Unlike the case of symmetric top molecules there is here no selection rule for the three Coriolis sublevels $F^{(+)}$, $F^{(0)}$, $F^{(-)}$ of a triply degenerate state. The rule (IV, 81) that only rotational levels of the same species can combine with one another holds of course here, as well as for the infrared spectrum.

A_1—A_1 transitions. Since for totally symmetric Raman lines (A_1—A_1) only $\Delta J = 0$ occurs, and since for Raman transitions always $B' \approx B''$, only *one strong sharp "line"* (the superposition of all Q lines) is to be expected, without any accompanying branches even for heavy overexposure. This is in conformity with observation, for example for CH_4, although in most other cases, even when they should occur, other branches have not been observed.

E—A_1 transitions. If the upper state of a Raman band of a tetrahedral molecule is doubly degenerate, all five branches given by (IV, 88) may occur. The structure would therefore be expected to be very similar to that of a totally symmetric Raman band of a symmetric top molecule, except that the intensity distribution would not be as regular. No such band has as yet been observed. The separation of successive lines in the P, R and O, S branches would be $2B$ and $4B$ respectively, since there is no vibrational angular momentum for $\nu_2(e)$. In CH_4, SiH_4, and GeH_4 the higher rotational lines would be split on account of the Coriolis interaction with $\nu_4(f_2)$ which has a similar frequency.

F_2—A_1 transitions. Since there is no selection rule for the $F^{(+)}$, $F^{(0)}$, $F^{(-)}$ levels of an F_2 vibrational level, we obtain from (IV, 88) *fifteen branches* for an F_2—A_1 transition, five for each sub-band $F_2^+—A_1$, $F_2^0—A_1$, and $F_2^-—A_1$. These fifteen branches are shown schematically in Fig. 142. The superscripts +, 0, − indicate the upper vibrational sublevel concerned. According to (IV, 77) and (IV, 78) the spacings in the three pairs of branches S^+O^-, S^0O^0, S^-O^+ would be (neglecting the difference between B' and B'') $2B(2 + \zeta_i)$, $4B$, and $2B(2 - \zeta_i)$ respectively; in the three pairs R^+P^-, R^0P^0, R^-P^+ the spacing would be $2B(1 + \zeta_i)$, $2B$, and $2B(1 - \zeta_i)$ respectively; in the three branches Q^+, Q^0, Q^- it would be $2B\zeta_i$, 0, and $-2B\zeta_i$ respectively.

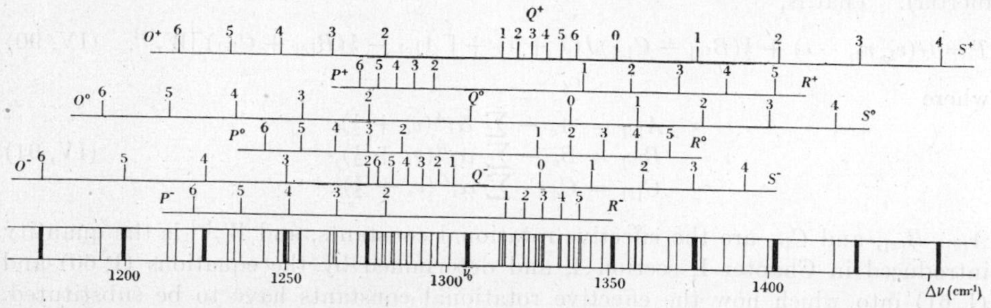

FIG. 142. **Branches of an $F_2—A_1$ Raman band of a tetrahedral molecule.**—The figure is drawn approximately to scale for ν_4 of CH_4 with $B = 5.2$ cm^{-1} and $\zeta = 0.45$. The difference of the B values in the upper and lower state is neglected. The three predominating branches are indicated by heavy lines. For small ζ values, as in the only observed case (ν_3 of CH_4), many of the branches here separated will almost coincide.

Fortunately, as has been shown theoretically by Teller (836), the S^+ series with spacing $2B(2 + \zeta_i)$ and the Q^0 branch with spacing 0 and the O^- series with spacing $2B(2 + \zeta_i)$ (that is, the transitions $F_2^+—A_1$ for $\Delta J = +2$, $F_2^0—A_1$ for $\Delta J = 0$, and $F_2^-—A_1$ for $\Delta J = -2$) have predominant intensity.

The Raman band ν_3 of CH_4 observed by Dickinson, Dillon, and Rasetti (287) in the gaseous state does indeed show these three branches. The observed spacing of the lines, 21.5 cm^{-1}, is therefore $2B(2 + \zeta_3)$, whereas the line spacing 9.93 cm^{-1} of the infrared band ν_3 is $2B(1 - \zeta_3)$. The sum of the two spacings yields $6B = 31.43$ or $B = 5.24$ cm^{-1}. This result is independent of the sum rule for ζ_i and the somewhat doubtful spacing in the infrared band ν_4, but is in very satisfactory agreement with the B value for CH_4 obtained above. If this Raman band could be measured more accurately with larger dispersion, and particularly if also the other branches could be observed, it would appear to supply the best method of obtaining a really precise value for the moment of inertia of CH_4 (and similar molecules).

Since the upper states of the S^+ and O^- branches of an $F_2—A_1$ Raman band are the same as those of the P^+ and R^- branches of the corresponding $F_2—A_1$ infrared band, *combination differences* may be used for the evaluation of the constants. From Fig. 137, in which the S^+ and O^- branches are shown by dotted lines, one can see easily that the following relation must hold:

$$S^+(J) - P^+(J + 3) = R^-(J) - O^-(J + 3) = F''(J + 3) - F''(J) = 6B''(J + 2). \quad \text{(IV, 89)}$$

When the data of Dickinson, Dillon, and Rasetti (287) for the Raman band and of Nielsen and Nielsen (656) for the infrared band ν_3 of CH_4 are used, it is found that the agreement of $S^+(J) - P^+(J + 3)$ with $R^-(J) - O^-(J + 3)$ is not very satisfactory. This may be due to insufficient accuracy of the Raman measurements, or to the second-order Coriolis splitting of the levels with larger J values in the upper vibrational state.

4. Asymmetric Top Molecules

(a) *Energy levels*

Unperturbed energy levels. As one would expect, just as in the case of linear, symmetric top, and spherical top molecules, so for asymmetric top molecules a good approximation to the energy of a vibrating and rotating molecule is obtained by taking the *sum of the pure vibrational energy* (see Chapter II) *and the rotational energy* (see Chapter I) *calculated with effective values of the rotational constants* (moments of inertia). That is,

$$T = G(v_1, v_2, \cdots) + \tfrac{1}{2}(B_{[v]} + C_{[v]})J(J+1) + [A_{[v]} - \tfrac{1}{2}(B_{[v]} + C_{[v]})]W_\tau^{[v]}, \quad \text{(IV, 90)}$$

where

$$\begin{aligned} A_{[v]} &= A_e - \sum \alpha_i^A(v_i + \tfrac{1}{2}), \\ B_{[v]} &= B_e - \sum \alpha_i^B(v_i + \tfrac{1}{2}), \\ C_{[v]} &= C_e - \sum \alpha_i^C(v_i + \tfrac{1}{2}). \end{aligned} \quad \text{(IV, 91)}$$

$A_{[v]}$, $B_{[v]}$, and $C_{[v]}$ are the effective rotational constants, and $W_\tau^{[v]}$ is the quantity introduced in Chapter I, section 4, and determined by the equations (I, 60) and (I, 61) into which now the effective rotational constants have to be substituted. A_e, B_e, C_e are the values of the rotational constants referring to the *equilibrium position*, that is, apart from the factor $\dfrac{h}{8\pi^2 c}$, the reciprocal equilibrium moments of inertia:

$$A_e = \frac{h}{8\pi^2 c I_A{}^e}, \quad B_e = \frac{h}{8\pi^2 c I_B{}^e}, \quad C_e = \frac{h}{8\pi^2 c I_C{}^e}, \quad \text{(IV, 92)}$$

where, usually, it is assumed that

$$A_e > B_e > C_e. \quad \text{(IV, 93)}$$

The correctness of the energy formula (IV, 90)—that is, the legitimacy of using the Wang formulae (I, 58) and (I, 60) for the rigid asymmetric top with effective rotational constants—has been proven in detail by Wilson and Howard (944) [see also Shaffer and Nielsen (780), Darling and Dennison (263), and Nielsen (666)]. However, this proof is valid only on the assumption that there are no nearby vibrational states that may perturb strongly the one considered. If the latter is the case we have to expect deviations from (IV, 90) (see below).

According to (IV, 90) and the discussion in Chapter I, *for each value of J we have $2J + 1$ different energy levels,* whose positions are given by the equations (I, 60) for W_τ. But these equations are slightly different for different vibrational levels on account of the dependence of $A_{[v]}$, $B_{[v]}$, and $C_{[v]}$ on the v_i according to (IV, 91). For practical calculations, the use of Ray's equation (I, 59) rather than Wang's equation (I, 58) with effective rotational constants may be more convenient, on account of the tables given by King, Hainer, and Cross (504) (see Chapter I, section 4). Naturally the approximations discussed in Chapter I, particularly for the case of a nearly symmetric top, may also be used here with effective rotational constants.

As in the case of linear molecules, the rotational constants α_i^A, α_i^B, α_i^C are usually small compared to the A_e, B_e, C_e respectively. As there, the α_i can be resolved into three parts [equation (IV, 11)] $\alpha_i^{(\text{harm.})} + \alpha_i^{(\text{anharm.})} + \alpha_i^{(\text{Cor.})}$. The main contribution is $\alpha_i^{(\text{anharm.})}$, corresponding to the fact that in consequence of the anharmonicity of the vibrations the moments of inertia and therefore the rotational constants change. The term $\alpha_i^{(\text{Cor.})}$ is due to the Coriolis interaction of different vibrations, and may become appreciable when two vibrations that can interact are close together (see further below). But even if all anharmonic terms in the potential energy were zero, and if the Coriolis interaction were negligible, there would still be the (small) contribution $\alpha_i^{(\text{harm.})}$, which is due to the fact that $\frac{h}{8\pi^2 cI}$ averaged over a harmonic vibration is not exactly (though nearly) equal to $\frac{h}{8\pi^2 cI^e}$. For the case of the nonlinear triatomic XY_2 molecule, Shaffer and Nielsen (780) and Darling and Dennison (263) have given explicit expressions for the α_i (or their equivalents) in terms of the potential constants and dimensions of the molecule.

It should be understood that the *effective moments of inertia* $I_A^{[v]}$, $I_B^{[v]}$, $I_C^{[v]}$ obtained from $A_{[v]}$, $B_{[v]}$, $C_{[v]}$ *are not simply the average moments of inertia in the vibrational state* $[v]$. It is the average A, B, C rather than the average I_A, I_B, I_C that matters for the rotational energy levels of the vibrating molecules. This distinction has a striking consequence in the case of non-linear triatomic molecules. For these as for any plane body, we have, for every instantaneous position of the nuclei,

$$I_C = I_A + I_B. \tag{IV, 94}$$

This relation would, of course, also hold for the average moments of inertia. But it does not in general hold for the effective moments of inertia, since the latter are $\frac{h}{8\pi^2 cA} = \frac{1}{(1/I_A)_{\text{average}}}$, and similarly for B and C, and since $(I)_{\text{average}} \neq \frac{1}{(1/I)_{\text{average}}}$. Therefore there is a *defect*,

$$\Delta = I_C^{[v]} - I_A^{[v]} - I_B^{[v]}, \tag{IV, 95}$$

in the relation (IV, 94) if it is applied to the effective moments of inertia. Such a defect was first found from the experimental data for H_2O by Mecke and his coworkers (612) (130). Darling and Dennison (263) have given explicit theoretical formulae for Δ as a function of v. It is found, as is plausible from the above, that in a first approximation Δ does not depend on the anharmonic terms of the potential energy. For the equilibrium values I^e of the moments of inertia, the relation (IV, 94) must of course be fulfilled, although even for the lowest vibrational level a slight defect Δ exists, and has been observed for H_2O. Similar considerations apply for more-than-triatomic planar molecules; but for them vibrations perpendicular to the plane of the molecule are possible, leading to a non-planar configuration in which (IV, 94) would not be fulfilled in any case, so that the effect is not as striking.

The *influence of centrifugal stretching* has already been considered briefly in Chapter I. In a first approximation the correction introduced will be the same for different vibrational levels. (This amounts to the assumption that $D_v = D_e$ for diatomic molecules, which is usually a fair approximation.) In the sum rules of Table 8 (p. 50), which serve to determine the $A_{[v]}$, $B_{[v]}$, and $C_{[v]}$, these corrections have to be taken into account for accurate evaluations [see Darling and Dennison (263)]. Detailed equations, which replace the Wang equations (I, 58) and (I, 60) when both the dependence of the rotational constants on the v_i and the centrifugal stretching terms (as well as the dependence of the latter on the v_i) are taken into account, have been given by Nielsen (665) for J values up to and including $J = 6$.

Symmetry properties. As we have seen in Chapter I, section 4, each energy level of an asymmetric top has one of the symmetries $++$, $+-$, $-+$, $--$ of its rotational eigenfunctions [or A, B_c, B_a, and B_b respectively, after Mulliken (645); see Table 9]. Here the first sign refers to the behavior with respect to a rotation by 180° about the axis of largest moment of inertia (c axis), and the second sign refers to that with respect to rotation by 180° about the axis of smallest moment of inertia (a axis). An asymmetric top molecule may have no, one, or three two-fold axes of symmetry which coincide with the principal axes of inertia. It is therefore easy to see what the *species of the rotational eigenfunctions* is *with respect to the rotational subgroup of the point group of the molecule.*

For example, for *point group* $\boldsymbol{C_{2v}}$ the rotational subgroup is $\boldsymbol{C_2}$. If the C_2 coincides with the axis of least moment of inertia it is clear that the rotational eigenfunctions have species A or B (see Table 12) for $++$, $-+$ or $+-$, $--$ levels respectively; if the C_2 coincides with the axis of largest moment of inertia they have species A or B for $++$, $+-$ or $-+$, $--$ respectively; if the C_2 coincides with the axis of intermediate moment of inertia they have species A or B for $++$, $--$, or $+-$, $-+$ respectively. The *species of the total eigenfunction* (*over-all species*) with respect to the rotational subgroup is immediately obtained if it is remembered that for $\boldsymbol{C_2}$ $A \times A = A$, $A \times B = B$, $B \times B = A$ (see Chapter II, section 3e). Assuming a totally symmetric electronic state, Fig. 143a and b give the resultant species of the lowest rotational levels for A_1 or A_2 and B_1 or B_2 vibrational states of a $\boldsymbol{C_{2v}}$ molecule when the C_2 coincides with the axis of least moment of inertia. Fig. 144a and b give the species when the C_2 coincides with the axis of intermediate moment of inertia. Fig. 143 would apply to a molecule like H_2CO, Fig. 144 to a molecule like H_2O. The same figures would also apply to molecules of point groups $\boldsymbol{C_2}$ or $\boldsymbol{C_{2h}}$.

If there are three mutually perpendicular two-fold axes, as in the *point groups* $\boldsymbol{V}$ *and* $\boldsymbol{V_h}$, the rotational subgroup is $\boldsymbol{V}$, which has the four species A, B_1, B_2, B_3 (see Table 13). As is easily seen, these four species are the species of the rotational eigenfunctions of the levels $++$, $+-$, $-+$, and $--$ respectively, if the z and the y axes are the axes of largest and least moment of inertia respectively. If the z and x axes are the axes of largest and least moment of inertia respectively, A, B_1, B_2, B_3 correspond to $++$, $+-$, $--$, and $-+$ respectively, and similarly in other cases (see Table 13). The *over-all species* of the rotational levels for non-totally symmetric vibrational states are again obtained by the multiplication rules, which for point group $\boldsymbol{V}$ are $A \times A = A$, $A \times B_i = B_i$, $B_i \times B_i = A$, $B_1 \times B_2 = B_3$, $B_2 \times B_3 = B_1$, $B_1 \times B_3 = B_2$. Fig. 145 gives, for the various vibrational levels of a molecule of point group $\boldsymbol{V_h}$, the species of the lowest rotational levels, assuming that the z and x axes are the axes of largest and smallest moment of inertia respectively, as is the case for C_2H_4 with our previous choice of axes (see p. 108).

If the *spins of the identical nuclei* are *zero* (in which case they follow Bose statistics, and the total eigenfunction must be symmetric with respect to an exchange of any two of them) only the A rotational levels occur, both when the rotational sub-group is $\boldsymbol{C_2}$ and when it is $\boldsymbol{V}$. This would, for example, be the case for NO_2 and for N_2O_4 if they had the plane symmetrical structure. If the identical nuclei have *non-zero spin*, we have to multiply by the nuclear spin function in order to get the total eigenfunction, and it is this total eigenfunction that must be of the same species for all occurring levels. As previously, by a suitable choice of the spin function the total eigenfunction may be made symmetric (or antisymmetric) with respect to any ex-

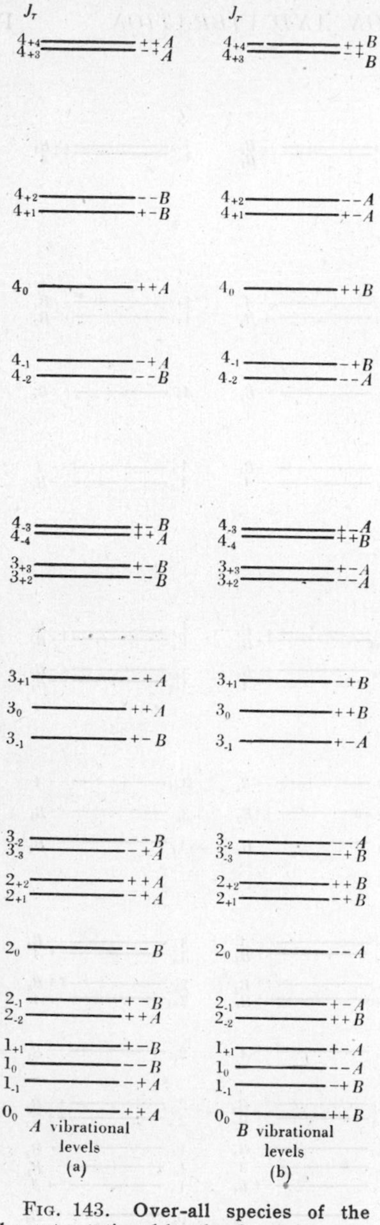

FIG. 143. **Over-all species of the lowest rotational levels of a C_{2v} (or C_{2h} or C_2) molecule when the C_2 coincides with the axis of *least* moment of inertia (a) for A_1 and A_2 (or A, A_g, A_u) vibrational levels, (b) for B_1 and B_2 (or B, B_g, B_u) vibrational levels.**—The signs $++$, $+-$ $\cdots$ have the same meaning as previously (for example Fig. 19). At the same time the first of these signs gives for planar molecules the "parity," that is, the behavior with respect to inversion, for vibrational states that are symmetric with respect to the plane of the molecule (A_1 and B_1 for C_{2v}). For vibrational states that are antisymmetric with respect to this plane (A_2 and B_2 of C_{2v}) the parities are reversed (see also p. 465).

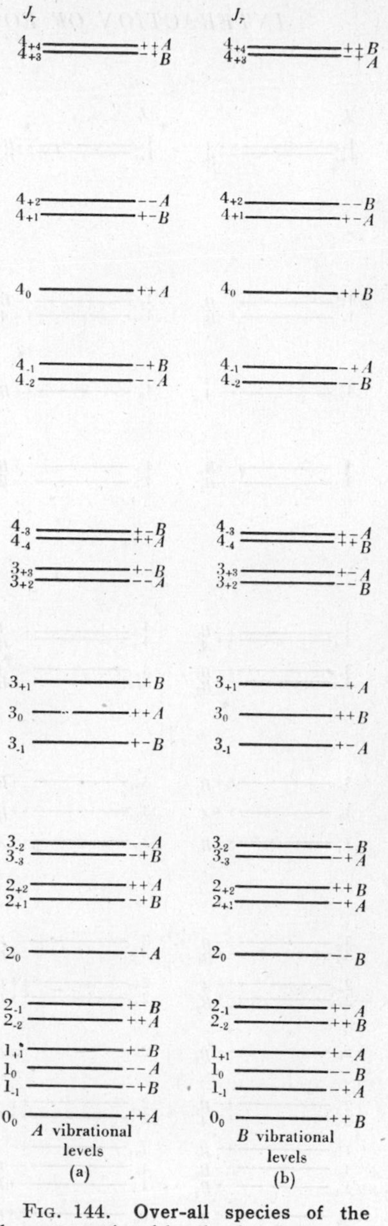

FIG. 144. **Over-all species of the lowest rotational levels of a C_{2v} (or C_{2h} or C_2) molecule when the C_2 coincides with the axis of *intermediate* moment of inertia (a) for A_1 and A_2 (or A, A_g, A_u) vibrational levels, (b) for B_1 and B_2 (or B, B_g, B_u) vibrational levels.**—See caption of Fig. 143.

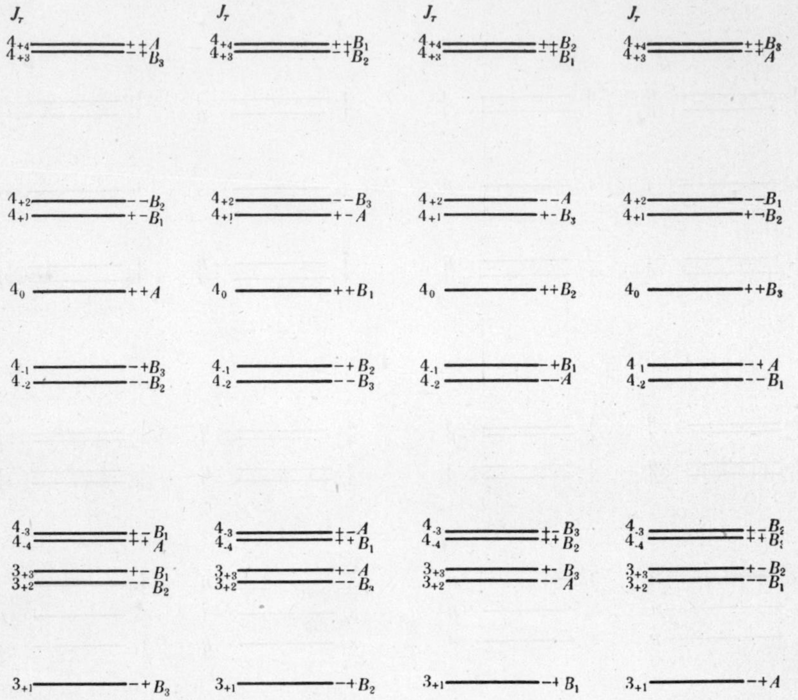

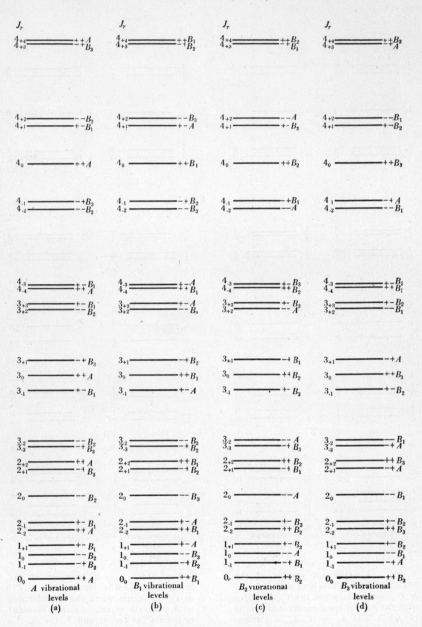

FIG. 145. **Over-all species of the lowest rotational levels of a V_h (or V) molecule when the z and x axes are the axes of largest and smallest moment of inertia (a) for A_g and A_u (or A) vibrational levels, (b) for B_{1g} and B_{1u} (or B_1) vibrational levels, (c) for B_{2g} and B_{2u} (or B_2) vibrational levels, (d) for B_{3g} and B_{3u} (or B_3) vibrational levels.**— See caption of Fig. 143.

change of identical nuclei for all rotational levels; that is, in general *all rotational levels can occur.*

When the rotational sub-group is $\mathbf{C}_2$, since a rotation by 180° exchanges two identical nuclei, the total eigenfunction must be of species A if the nuclei follow Bose statistics, and of species B if they follow Fermi statistics. If there are several pairs of identical nuclei it is the "resultant" statistics that matters (see p. 54). For one pair of identical nuclei of spin $\frac{1}{2}$ (for example in H_2CO and H_2O) there are three nuclear spin functions of species A and one of species B(for the same reasons that there are three symmetric functions and one antisymmetric function for diatomic molecules; see Molecular Structure I, p. 146). In order to make the total eigenfunction of species B we have to combine the A rotational levels with B spin functions and the B levels with A spin functions. Thus the B levels in Fig. 143 and 144 have three times the statistical weight of the A levels. For other spin values, and in particular for the case of several pairs of identical nuclei, the weights of the A and B levels have been given in the previous Table 10 (Chapter I). They are determined by the formulae (I, 8) and (I, 9).

When the rotational subgroup is V there is at least one set of four identical nuclei (for example in C_2H_4). A rotation by 180° about one of the two-fold axes corresponds to two exchanges of identical atoms. Therefore the total eigenfunction must be of species A no matter what is the statistics of the nuclei. If the four identical nuclei have $I = \frac{1}{2}$ and all other nuclei present have $I = 0$ (as in C_2H_4) there are sixteen different spin functions, of which seven are of species A and three each of species B_1, B_2, and B_3 [the proof for this is similar to that given on p. 410 for an XY_3 molecule; see also Wilson (933)]. If the total eigenfunction is to be of species A the rotational levels of species A in Fig. 145 must be taken with spin functions of species A, the rotational levels of species B_1 with spin functions of species B_1, and similarly for B_2 and B_3. Therefore the rotational levels A, B_1, B_2, and B_3 have statistical weights in the ratio 7 : 3 : 3 : 3. The results for some other similar cases have already been given in the previous Table 11, p. 54.

In addition to the above symmetry properties, we have again the property "*positive* or *negative*" *with respect to an inversion.* For non-planar molecules each rotational level is double (inversion doubling), one component being positive, the other negative. For plane molecules no such doubling exists; the rotational levels are either "positive" or "negative." Since for a plane molecule a rotation by 180° about the axis of largest moment of inertia followed by a reflection at the plane of the molecule is equivalent to an inversion, therefore in a totally symmetric vibrational state the $+ +$ and $+ -$ rotational levels (see p. 51) are "positive" the $- +$ and $- -$ levels are "negative." These are the first signs on the right in Figs. 143a, 144a, and 145a. As has been shown in more detail by Mulliken (645), the "parities" shown in these figures apply to all vibrational levels that are symmetric with respect to the plane of the molecule—that is, A_1 and B_1 of $\mathbf{C}_{2v}$ and A_g, B_{1g}, B_{2u}, and B_{3u} of V_h—while for vibrational levels that are antisymmetric with respect to the plane of the molecules the reverse parities apply—that is, for A_2 and B_2 of $\mathbf{C}_{2v}$ and for A_u, B_{1u}, B_{2g}, and B_{3g} of V_h.[52]

In the case of non-planar asymmetric top molecules the two inversion doublet sublevels will in general not have the same statistical weight (except for a molecule without symmetry). However,

[52] Molecules of point group V cannot be planar.

the sum of the weights of the two sublevels is again the same as the one obtained without consideration of the inversion doubling. Thus for CH_2F_2 the ratio of the statistical weights of the A and B rotational levels according to Table 10 (p. 53) is 10 : 6. The weights of the positive and negative sublevels of an A level can be shown to be 7 and 3 respectively, of a B level 3 and 3 respectively. In this case, a rotation about the two-fold axis (which leads to the distinction of A and B levels) always produces a simultaneous exchange of both pairs of identical nuclei, while a twisting of the molecule by 180° (which is equivalent to an inversion) leads to an exchange of one pair only.

A somewhat similar situation arises in the case of plane molecules of the type C_2H_4, where a twisting of the molecule about the C—C axis leads to an exchange of nuclei that cannot be brought about by simple rotations. The resulting doubling of the energy levels (which is not an inversion doubling; see p. 225f.) again does not lead to an increase in the statistical weight (see also section 5 of this Chapter).

Perturbations. Just as for linear, symmetric top, and spherical top molecules, so for asymmetric top molecules perturbations may occur between vibrational levels that lie close together. These perturbations may again be of the *Fermi resonance* or the *Coriolis type*, and in either case we may have *regular (vibrational) perturbations and irregular (rotational) perturbations*. In all cases, only rotational levels of the same over-all species and the same J value can perturb one another. For perturbations of the Fermi resonance type, that is, between vibrational levels of the same species, the same considerations apply as for linear molecules (see p. 378f.).

However, for perturbations of the Coriolis type, that is, perturbations between vibrational levels of different (vibrational) species which take place in consequence of the rotation of the molecule, some special considerations are necessary. As previously, such perturbations vanish for the rotationless state ($J = 0$), and increase quadratically with J. If the two vibrational levels that perturb each other are fairly far apart we obtain simply a change of the rotational constants $A_{[v]}$, $B_{[v]}$, and $C_{[v]}$ as compared to the unperturbed values. Thus we obtain the contributions $\alpha_i^{(\text{Cor.})}$ to the rotational constants α_i. If the two vibrational levels are very close together we obtain rotational perturbations, that is, somewhat irregular deviations of the rotational levels from those given by the Wang formulae.

For molecules of point group $\mathbf{C_{2v}}$ the three rotations have species A_2, B_1, and B_2 (see Table 13). According to Jahn's rule for Coriolis perturbations (p. 376) and the multiplication rules for the species ($A_i \times A_i = A_1, B_i \times B_i = A_1, A_1 \times A_2 = A_2, A_1 \times B_i = B_i, A_2 \times B_1 = B_2, A_2 \times B_2 = B_1, B_1 \times B_2 = A_2$) we find immediately that the following pairs of vibrational levels may perturb one another:

$$(A_1, A_2),\ (A_1, B_1),\ (A_1, B_2),\ (A_2, B_1),\ (A_2, B_2),\ (B_1, B_2).$$

Thus, in the general case (when there are vibrational levels of each species) there are *six kinds of Coriolis perturbations*.

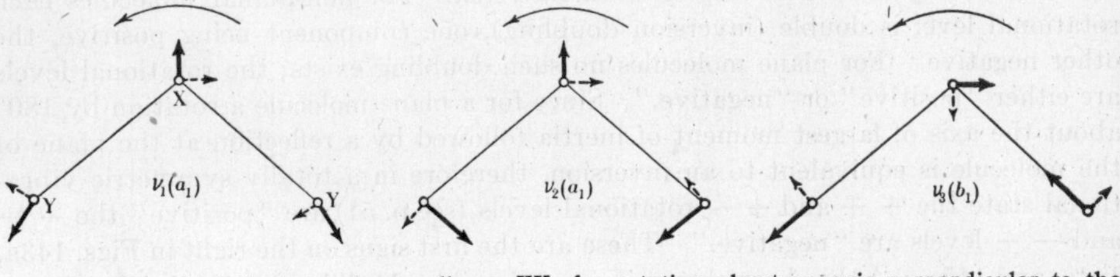

FIG. 146. **Coriolis forces in non-linear XY₂ for rotation about an axis perpendicular to the plane of the molecule** (qualitative).—The solid arrows represent the velocities of the nuclei on account of the vibration, the broken arrows represent the Coriolis forces.

For *non-linear* XY_2 *molecules* since they have only vibrational levels of species A_1 and B_1 only one type of Coriolis perturbation (A_1, B_1) occurs. Fig. 146 shows the direction of the Coriolis forces for the three normal vibrations for a rotation about an axis perpendicular to the plane of the molecule. It is seen that in $\nu_3(b_1)$ the Coriolis forces tend to excite mainly $\nu_2(a_1)$ but to a slight extent also $\nu_1(a_1)$, and conversely in $\nu_1(a_1)$ and $\nu_2(a_1)$ they tend to excite $\nu_3(b_1)$. Actually, for example for H_2O and H_2S, since ν_1 is very close to ν_3 (see Table 37), the Coriolis interaction of ν_1 and ν_3 is much stronger than that between ν_2 and ν_3. From Fig. 143 or 144 it can be seen which particular rotational levels will perturb each other, if account is taken of the rule that they have to have the same over-all species and the same J. If it is furthermore considered that the shift produced by the perturbation is inversely proportional to the separation of the unperturbed levels, and is in general different for different over-all species, it is immediately realized that the perturbation does not merely consist in an equal shift of all levels of a given J. As long as the unperturbed vibrational levels are fairly far apart, the perturbed rotational levels may be obtained from the Wang formula by using appropriately changed rotational constants. However if the unperturbed levels are close together, more irregular perturbations arise, since the energy differences of different pairs of perturbing levels may be widely different. Formulae for this case have been given by Wilson (935).

For *plane* XYZ_2 *molecules* all the above types of Coriolis interactions are possible, although for the fundamentals only (A_1, B_1), (A_1, B_2), and (B_1, B_2) interactions are possible, since no A_2 fundamentals exist [see Jahn (470)]. Fig. 147 shows as an example the Coriolis forces for the vibrations

FIG. 147. **Coriolis forces in the vibrations ν_5 and ν_6 of a plane XYZ_2 molecule for a rotation about the axis of symmetry.**—The solid arrows represent the vibrational velocities, the broken arrows the Coriolis forces. The molecule is shown in oblique projection. The figure is approximately drawn to scale for D_2CO. (Compare also the normal vibrations in Fig. 24.)

$\nu_5(b_1)$ and $\nu_6(b_2)$ during a rotation about the symmetry axis. It is seen that in $\nu_5(b_1)$ they tend to excite $\nu_6(b_2)$, and vice versa. In the case of H_2CO, since the two vibrations have very similar magnitudes (see Table 76), a strong interaction results. Similarly a strong interaction is produced between $\nu_3(a_1)$ and $\nu_5(b_1)$, and between $\nu_1(a_1)$ and $\nu_4(b_1)$ by a rotation about the axis of largest moment of inertia (perpendicular to the plane of the molecule), and between $\nu_3(a_1)$ and $\nu_6(b_2)$ by a rotation about the axis of intermediate moment of inertia. There are also some other pairs that can interact, but they have widely different frequencies in H_2CO and therefore the interaction is very slight.

For a *molecule of point group* $V_h(\equiv D_{2h})$ the three rotations about the z, y, and x axes have the species B_{1g}, B_{2g}, and B_{3g} respectively (see Table 14). Therefore, using the same procedure as before, the following pairs of vibrational levels may interact in consequence of Coriolis forces:

$$(A_g, B_{1g}),\quad (A_g, B_{2g}),\quad (A_g, B_{3g}),\quad (A_u, B_{1u}),\quad (A_u, B_{2u}),\quad (A_u, B_{3u}),$$
$$(B_{1g}, B_{2g}),\quad (B_{1g}, B_{3g}),\quad (B_{2g}, B_{3g}),\quad (B_{1u}, B_{2u}),\quad (B_{1u}, B_{3u}),\quad (B_{2u}, B_{3u});$$

that is, there are twelve kinds of Coriolis perturbations. Of these, for the fundamentals of plane X_2Y_4 molecules, the perturbations (A_g, B_{3g}), (B_{1g}, B_{3g}), and (B_{2g}, B_{3g}) do not occur since there are no fundamentals of species B_{3g}; the perturbation (A_u, B_{1u}) does not occur since for the only fundamentals of species A_u and B_{1u} (ν_4 and ν_7 in Fig. 44) all nuclei move parallel to the axis of rotation. As an example, in Fig. 148, the Coriolis forces for the vibration $\nu_{10}(b_{2u})$ of X_2Y_4 are shown when it

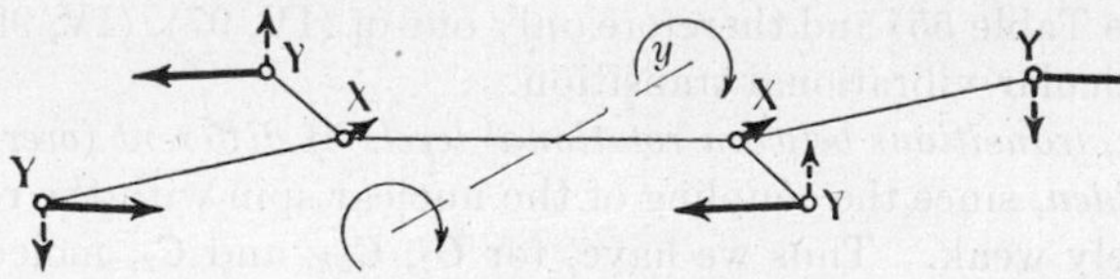

FIG. 148. **Coriolis forces in the vibration $\nu_{10}(b_{2u})$ of plane X_2Y_4 for a rotation about the y axis.**—See caption of Fig. 147. The Coriolis force on the X atoms is zero since they are moving parallel to the axis of rotation.

is rotating about the y axis. It is seen that $\nu_{10}(b_{2u})$ will interact with $\nu_4(a_u)$, the twisting vibration. For C_2H_4, ν_{10} and ν_4 have fairly similar frequencies and therefore the interaction may well be strong enough to cause a sufficient admixture of $\nu_{10}(b_{2u})$ to $\nu_4(a_u)$ so that the latter may occur weakly in the infrared in spite of the fact that without rotation such a transition is rigorously forbidden (see p. 458 and p. 328).

In the actual calculation of the influence of the Coriolis interaction on the energy levels, it is necessary to work out, just as for linear molecules, the vibrational angular momenta p_x, p_y, p_z produced by the various pairs of normal vibrations that can enter into Coriolis interaction [equation (IV, 10)], for example (ν_1, ν_3) and (ν_2, ν_3) for non-linear XY_2, and then introduce them into the general Hamiltonian (II, 276) [see Wilson (935) and Jahn (470)]. Such calculations have been carried out for the case of the three vibrations ν_3, ν_5, and ν_6 of D_2CO (see above) by Nielsen (664). In this case the formulae are considerably simplified by the fact that the molecule is nearly a symmetric top.

(*b*) *Infrared spectrum*

Selection rules. If, as is usually the case, the interaction of vibration and rotation is not too large, the selection rules for the infrared rotation-vibration spectrum are again the same as those discussed previously for the rotation spectrum and the vibration spectrum separately, except that it is now the direction of the *change* of dipole moment during the vibration (see Table 55) that matters for the rotational selection rules. Thus, as always, we have

$$\Delta J = 0, \pm 1; \quad J = 0 \leftrightarrow J = 0. \qquad \text{(IV, 96)}$$

Furthermore, *if the alternating dipole moment lies in the axis of least moment of inertia* (I_A) we have the selection rule (see p. 55) that only the transitions

$$+\,+ \leftrightarrow -\,+ \quad \text{and} \quad +\,- \leftrightarrow -\,- \qquad \text{(IV, 97)}$$

can take place, where the $+\,+$, $+\,-$, $\cdots$ refers to the symmetry of the rotational eigenfunction (see p. 51f.). *If the alternating dipole moment lies in the axis of intermediate moment of inertia*, we have

$$+\,+ \leftrightarrow -\,- \quad \text{and} \quad +\,- \leftrightarrow -\,+; \qquad \text{(IV, 98)}$$

and finally, *if the alternating dipole moment lies in the axis of largest moment of inertia*, we have

$$+\,+ \leftrightarrow +\,- \quad \text{and} \quad -\,+ \leftrightarrow -\,-. \qquad \text{(IV, 99)}$$

If, as is usually the case in an entirely unsymmetrical molecule, the alternating dipole moment has none of these directions, all the above transitions can occur; that is, all but the transitions between states of the same $+$, $-$ symmetry are allowed. If the alternating dipole moment is in the plane of two principal axes, only the transitions corresponding to these two axes are allowed. For the more symmetric molecules of point groups $\boldsymbol{C_{2v}}$, $\boldsymbol{V}$, and $\boldsymbol{V_h}$ the changing dipole moment always lies in one of the principal axes (see Table 55) and therefore only one of (IV, 97), (IV, 98), and (IV, 99) applies for a particular vibrational transition.

As previously, *transitions between rotational levels of different (over-all) species are very strictly forbidden*, since the coupling of the nuclear spin with the rest of the molecule is so extremely weak. Thus we have, for $\boldsymbol{C_2}$, $\boldsymbol{C_{2h}}$, and $\boldsymbol{C_{2v}}$ molecules,

$$A \leftrightarrow A, \qquad B \leftrightarrow B; \qquad \text{(IV, 100)}$$

and for $\boldsymbol{V}$ and $\boldsymbol{V_h}$ molecules,

$$A \leftrightarrow A, \qquad B_1 \leftrightarrow B_1, \qquad B_2 \leftrightarrow B_2, \qquad B_3 \leftrightarrow B_3. \qquad \text{(IV, 101)}$$

These selection rules hold even for collisions, and therefore any particular gas consists of *as many almost non-convertible modifications as there are rotational species of its molecules*. Thus for H_2O, H_2CO, and all other $\boldsymbol{C_{2v}}$ molecules with non-zero spin of at least one pair of identical nuclei there are two modifications A and B (para and ortho), whose statistical weights are given by the previous considerations (1 : 3 for H_2O and H_2CO; see Figs. 143 and 144). For C_2H_4 and similar $\boldsymbol{V_h}$ molecules with four identical nuclei of non-zero spin there are four practically non-convertible modifications A, B_1, B_2, B_3 (for C_2H_4 in the ratio 7 : 3 : 3 : 3), while for $N_2{}^{14}O_4{}^{16}$ there are only two such modifications on account of the zero spin of O^{16} (see Table 11 and Fig. 145). As can easily be seen from Figs. 143–145, the selection rules (IV, 100) and (IV, 101) are automatically fulfilled when one of (IV, 97)–(IV, 99) is fulfilled.

Finally, as always, we have the selection rule for the symmetry property *positive-negative with respect to inversion:*

$$+ \leftrightarrow -. \qquad \text{(IV, 102)}$$

For non-planar molecules this rule is of importance only when the inversion doubling is resolved (see p. 465). For planar molecules, for which no inversion doubling occurs, it is automatically fulfilled when one of (IV, 97)–(IV, 99) is fulfilled.

It must be realized that the selection rules (IV, 100)–(IV, 102) hold for any strength of coupling between vibration and rotation, whereas (IV, 97)–(IV, 99) hold only when this coupling is weak. But it is easily seen that for allowed vibrational transitions of plane molecules of point groups $\boldsymbol{C_{2h}}$ and $\boldsymbol{C_{2v}}$ and of any molecules of point groups $\boldsymbol{V}$ and $\boldsymbol{V_h}$ the rules (IV, 100)–(IV, 102) lead to the same restrictions as (IV, 96)–(IV, 98). However, for forbidden vibrational transitions of these molecules made possible by strong Coriolis interaction, the rules (IV, 96)–(IV, 98) will be violated while (IV, 100)–(IV, 102) will still hold.

Since the selection rules are different for the different orientations of the alternating dipole moment, we obtain in a molecule of point group $\boldsymbol{C_{2v}}$, $\boldsymbol{V}$, or $\boldsymbol{V_h}$ *three types of infrared bands* which we call *type A,*[53] *type B,*[53] and *type C bands depending on whether the change of dipole moment is in the direction of the axis of least, intermediate, or largest moment of inertia.* The infrared bands of molecules of lower symmetry consists of a superposition of two or all three of these types (hybrid bands).

Type *A* bands. Fig. 149 gives schematically the possible transitions allowed by (IV, 96) and (IV, 97) for a type A band. Except for the species designations A, B, · · ·, this figure holds for any asymmetric top molecule, as long as the change of dipole moment is along the axis of least moment of inertia. The over-all species designations A, B that are added to the levels hold for a B_1—A_1 transition of a $\boldsymbol{C_{2v}}$ molecule whose C_2 axis coincides with the axis of intermediate moment of inertia, as in H_2O and H_2S. For other directions of the C_2 axis the designations have to be changed according to the previous rules (p. 462). In brackets in Fig. 149 the species designations of the levels for a B_{3u}—A_g transition of a $\boldsymbol{V_h}$ molecule such as C_2H_4 are added, assuming that the x axis (C═C axis) is the axis of least moment of inertia. The only importance of these species designations for the spectrum is that they determine the missing levels (see p. 462), and therefore the missing lines, when the spin of the identical nuclei is zero; and, when the nuclear spin is not zero (see p. 465), the intensity alternation of successive lines in the branches. At the bottom of Fig. 149 the actual spectrum that is produced is shown. (The figure is drawn to scale for the H_2O band ν_3). It is seen that a rather complicated pattern arises, even though only the levels up to $J = 3$ have been used.

[53] These designations should not be confused with the species designations A and B.

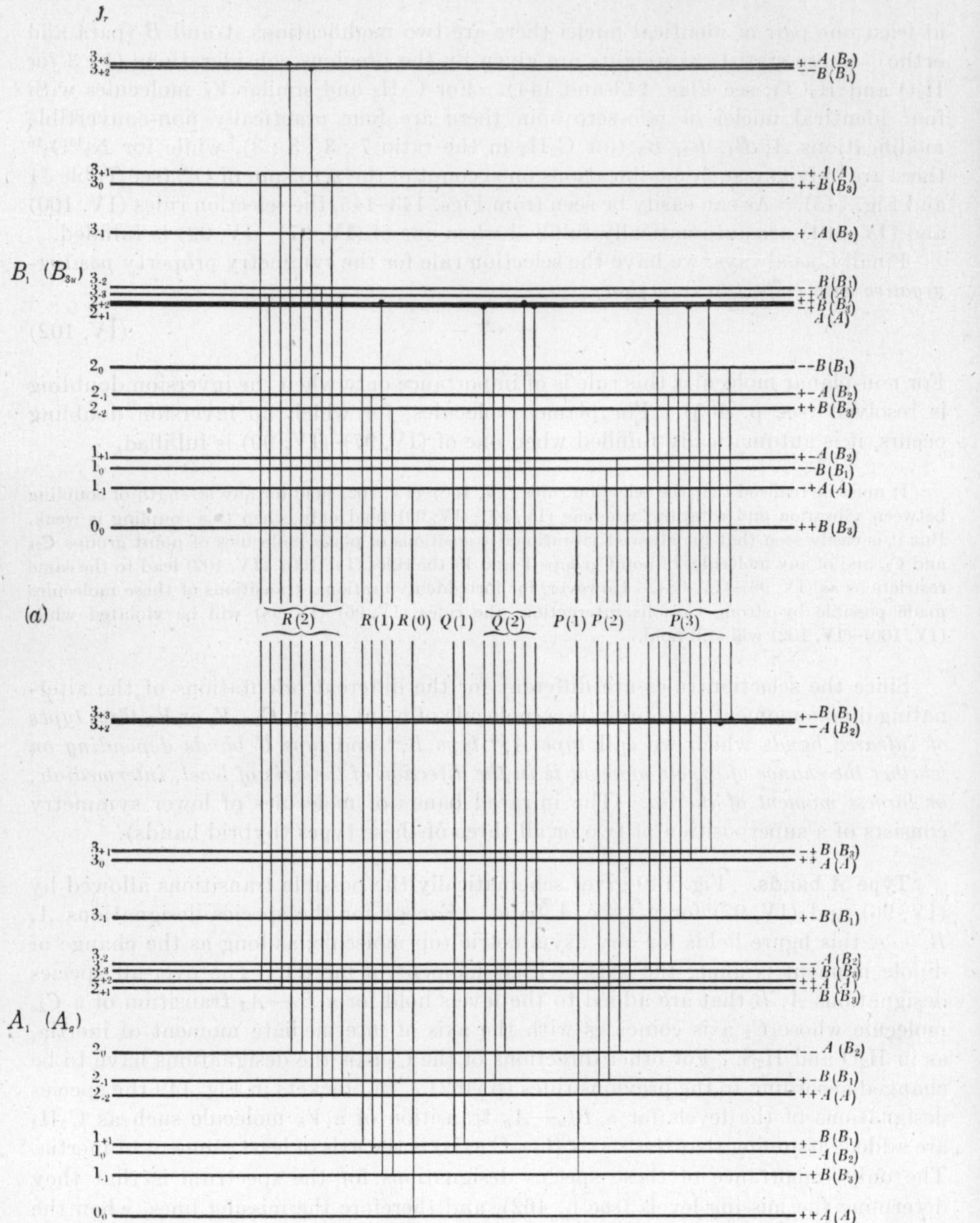
J_τ
B_1 (B_{3u})
A_1 (A_g)
(a)
(b)
R(2) R(1) R(0) Q(1) Q(2) P(1) P(2) P(3)
+−A(B₂)
−−B(B₁)
++A(A)
++B(B₃)
+−B(B₁)
−−A(B₂)
3900
3800
3700
3600
cm⁻¹

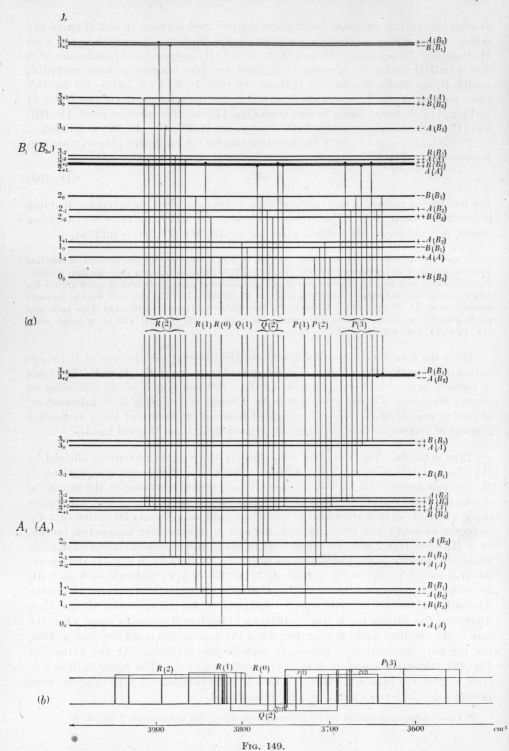

FIG. 149.

The *structure of the band* obtained depends greatly, of course, on the relative values of the moments of inertia. Only near the limiting cases of the symmetrical top or the linear molecule can we expect any simple regularity. Dennison (279) has calculated the energy levels with $J = 0, 1, 2, 3$, and 4 for the ten different ratios $\rho = B/A = I_A/I_B = 0.1, 0.2, \cdots 1.0$ of a plane molecule for which $I_A + I_B = I_C$, and has plotted the spectrum assuming that $A' = A''$, $B' = B''$, $C' = C''$. His figure is reproduced in Fig. 150. Corresponding lines in the ten cases are connected by the thin curves onto which also the designations of the transitions are written. For the limiting case $\rho = 1$ we have $I_C = 2I_A$ and $C = \frac{1}{2}A$. In the resulting $\perp$ band of the symmetric top (see p. 424f.) the separation of successive Q branches is $2(C - A) = -2C\ (= -A)$, whereas the separation of the lines in the P and R branches of each sub-band is $2A = 4C$. Thus only one series of equidistant lines of spacing $2C$ appears. It is assumed in the diagram that C is the same for all cases given, and it is used as the unit on the abscissa. In the limiting case $\rho = 0$ we have $A = 0$, and obtain the structure of a $||$ band of a linear molecule of moment of inertia I_C, that is, we have a simple P and R branch, again of spacing $2C$. For very small ρ we have practically a $||$ band of a symmetric top molecule ($I_C = I_A + I_B \approx I_B$), with a central Q branch and P and R branches of spacing $B + C$ (see Fig. 150 and p. 418).

Kramers and Ittmann (540) have given general formulae for the *intensities of the individual lines* of a band. But their evaluation would be extremely laborious [see also Casimir (4)].[53a] The intensities indicated in Fig. 150 by the height of the lines have been derived by Dennison on the assumption of the approximate validity of the symmetric top intensities. At the same time a Boltzmann factor 1 has been assumed. Therefore all the lines in the limiting case $\rho = 1$ have the same intensity. Naturally, to obtain a more accurate picture it would be necessary to include higher J values as well as the effect of the Boltzmann factor. This has been done for J up to 6 and $\rho = 0.05, 0.10, 0.15$, and 0.20 by Nielsen (660). The general result of Fig. 150 is thereby unaltered: With decreasing ρ the lines of the Q branches ($\Delta J = 0$) shift toward the band center and for $\rho = 0$ their intensity would be zero. Under medium dispersion *for small ρ a type A band would therefore look exactly like a $||$ band of a symmetric top molecule*, consisting of a strong unresolved central maximum accompanied by a series of nearly equidistant lines on either side.

It must be realized that Fig. 150 is based on the assumption of equal rotational constants in the upper and lower states. If this assumption is not fulfilled the three branches will be shaded one way or the other. If the rotational constants in the upper and lower states are very different, as may happen for the photographic infrared bands, the crowding of the lines in the center of the band (Fig. 150) may not be at all

[53a] Quite recently Cross, Hainer, and King (249a) have given a very detailed and useful discussion of the intensities for asymmetric top molecules and in particular have given extensive tables of line strengths up to $J = 12$ based on the rigorous formulae.

FIG. 149. **(a) Energy level diagram for a type *A* band of an asymmetric top molecule and (b) spectrum for the lowest *J* values.**—Both the energy level diagram and the spectrum are drawn to scale for the fundamental $\nu_3(b_1)$ of H_2O [after the data of Nielsen (667)] for which the C_2 coincides with the axis of intermediate moment of inertia. The species designations apply to this case. In brackets the species designations for a B_{3u}—A_g transition of a V_h molecule are added assuming that the x axis is the axis of least moment of inertia (as in C_2H_4). The vertical transition lines in (a) have not been made to fit the spectrum in (b).

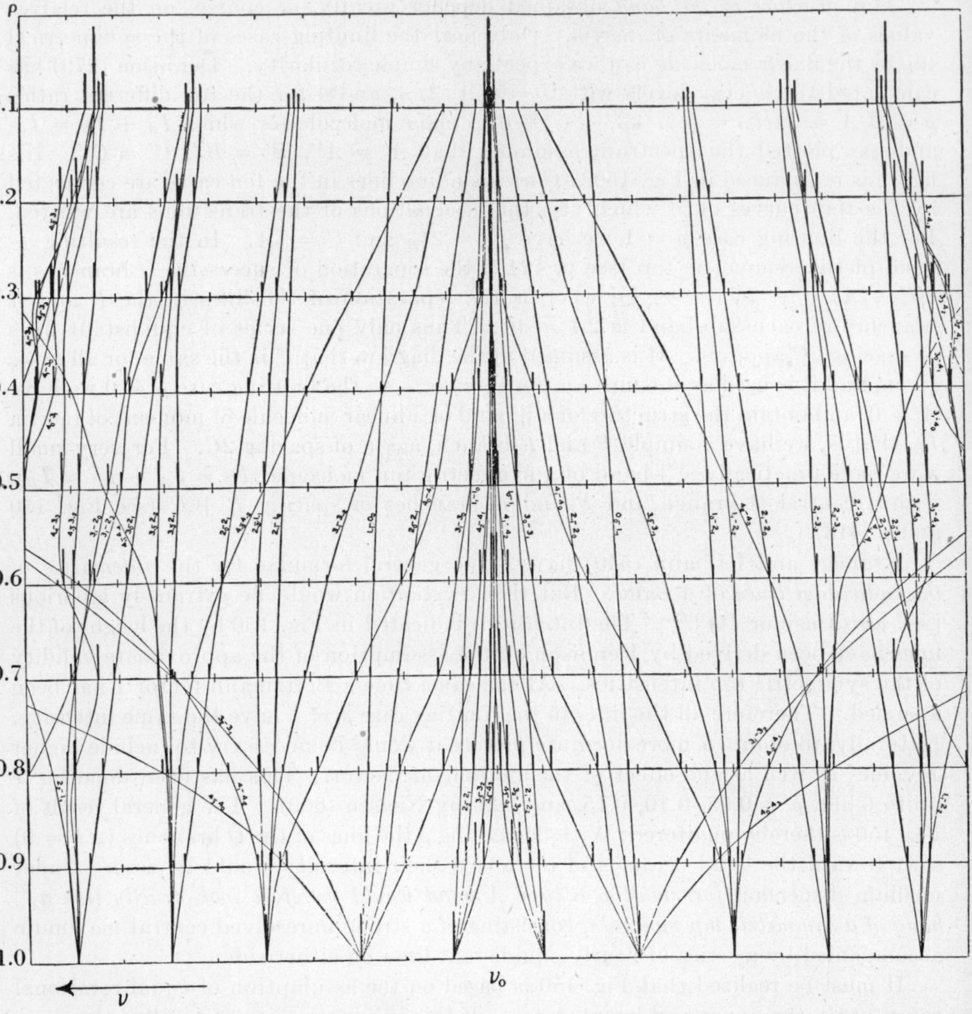

FIG. 150. **Type A bands for various values of $\rho = B/A$ of a planar molecule** [after Dennison (279)].—Corresponding lines for different ρ are connected by light curves on which the $J'_{\tau'} - J''_{\tau''}$ are indicated. The spectral lines are indicated by heavy vertical lines whose height is proportional to the intensity. The influence of the Boltzmann factor is neglected.

prominent. An example of this case is the H_2O band at 9400 Å, reproduced in Fig. 151a. On the other hand, the H_2O band at 8200 Å in Fig. 151b shows the central Q branch clearly. While these two bands are due to a molecule that is not even approximately a symmetric top, Fig. 152 and 153 give two type A bands of the nearly symmetric top molecules H_2CO and C_2H_4 respectively ($\rho = 0.13$ and 0.16 respectively). It is seen that these bands are practically identical with || bands of symmetric top molecules.

While there is no obvious regularity in the type A band of a strongly asymmetric molecule (such as H_2O), it must be remembered that even in the most general case

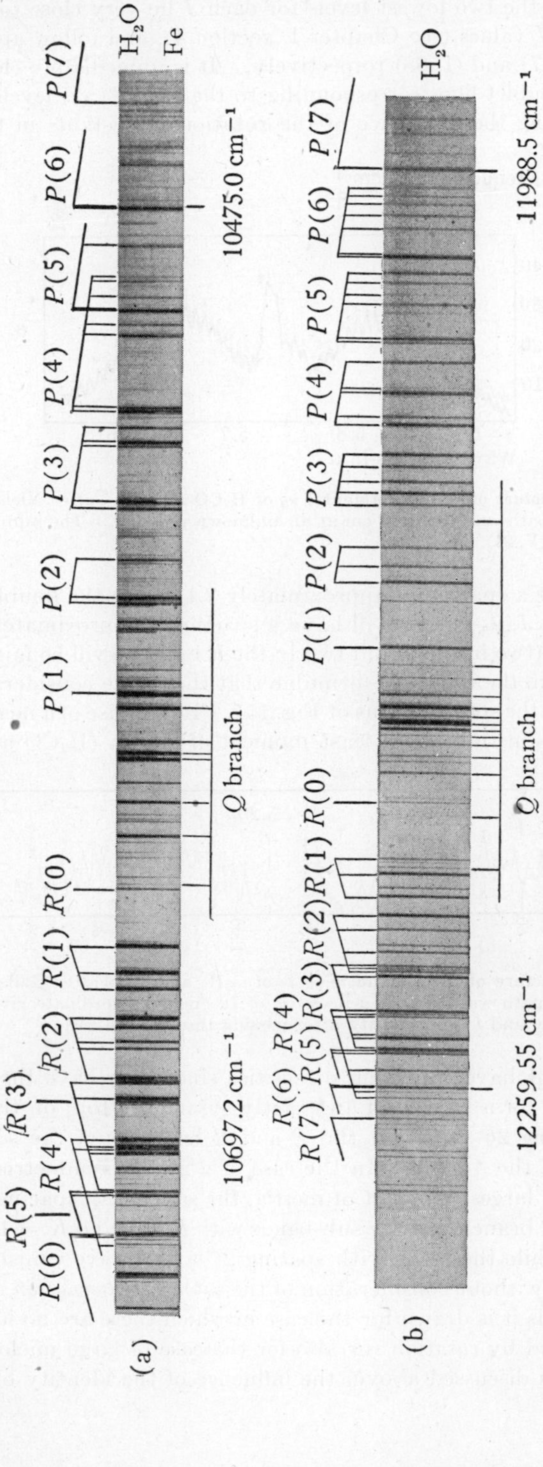

FIG. 151. **Photographic infrared bands of H_2O at (a) 9400 and (b) 8200 Å in the solar spectrum.**—Only the stronger lines (with $\Delta\tau = \pm 1$) are indicated as far as they have been analyzed (see Table 134 p. 486). The last two lines at the right and left of each group of given J are the doublets discussed in the text.

the two highest and the two lowest levels for each J lie very close together, except for the very lowest J values (see Chapter I, section 4), and follow approximately a simple formula, (I, 67) and (I, 68) respectively. It is immediately clear from these formulae that the doublet lines corresponding to the J_{+J}, J_{+J-1} levels in the R and P branches, neglecting the difference of the rotational constants in the upper and

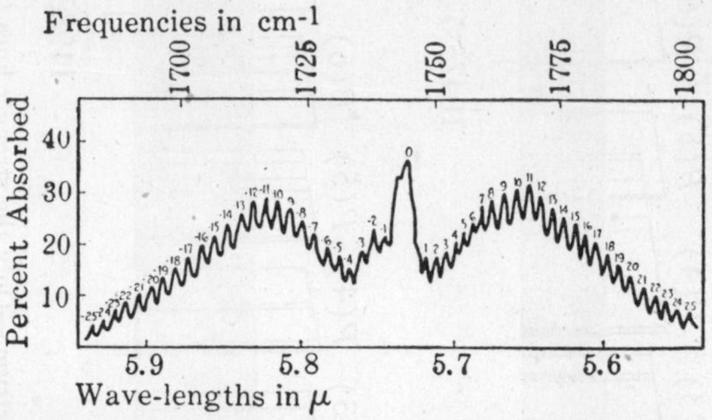

FIG. 152. **Fine structure of the fundamental ν_2 of H_2CO at 5.73μ** [after Nielsen (662)].—The length of the absorbing path was about 10 cm at an unknown pressure. The numbers on the maxima are m values as in (IV, 21).

lower state, will have a spacing of approximately $2A$, while the doublet lines corresponding to the J_{-J}, J_{-J+1} levels will have a spacing of approximately $2C$. These *four series of doublets* (two in the P and two in the R branch) will be fairly prominent, since it turns out from the intensity formulae that they have considerable intensity. They are indicated in the spectrograms of Fig. 151. In the case of a nearly symmetric top with the top axis in the axis of least moment of inertia (H_2CO and C_2H_4), the

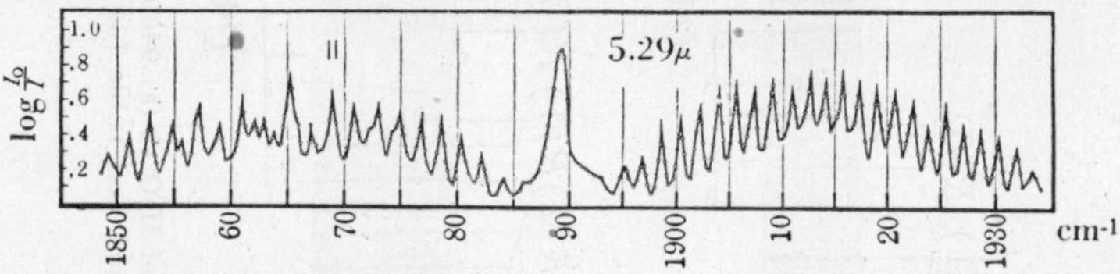

FIG. 153. **Fine structure of the || band $\nu_7 + \nu_8$ of** C_2H_4 **at 5.29μ** [after Gallaway and Barker (345)].—The absorbing path was 6 cm at a pressure of 72 cm. The ordinate gives $\log I_0/I$ where I_0 is the incident intensity and I the intensity after passing through the gas.

series with spacing $2A$ have vanishing intensities since they have then $\Delta K = \pm 1$, which does not occur for a || band of an exactly symmetric top; on the other hand, the series with spacing $2C$ represent the P and R branches of the sub-bands with $K = 0$ and $K = 1$ of the || band. In the case of a nearly symmetric top with the top axis in the axis of largest moment of inertia, the series with spacing $2A$ would go over into the P and R branches of the sub-bands with $K = 0$ and $K = 1$ of the || band (compare Fig. 17), while the series with spacing $2C$ would have vanishing intensity.

Fig. 150 is drawn without consideration of the *intensity alternation* due to nuclear spin, or in other words it is drawn for the case in which there are no identical nuclei that can be exchanged by rotation (or also for the case of large nuclear spin of the identical nuclei). As discussed above, the influence of the identity of the nuclei is different for different symmetries of the molecule and different orientations of the symmetry axes with respect to the principal axes of inertia.

For the case of a C_{2v} molecule whose two-fold axis coincides with the axis of intermediate moment of inertia (b axis), the intensity alternation can be seen immediately from Fig. 149, if it is realized that the intensity ratio of the lines connecting A levels to those connecting B levels equals the ratio of the statistical weights of these levels

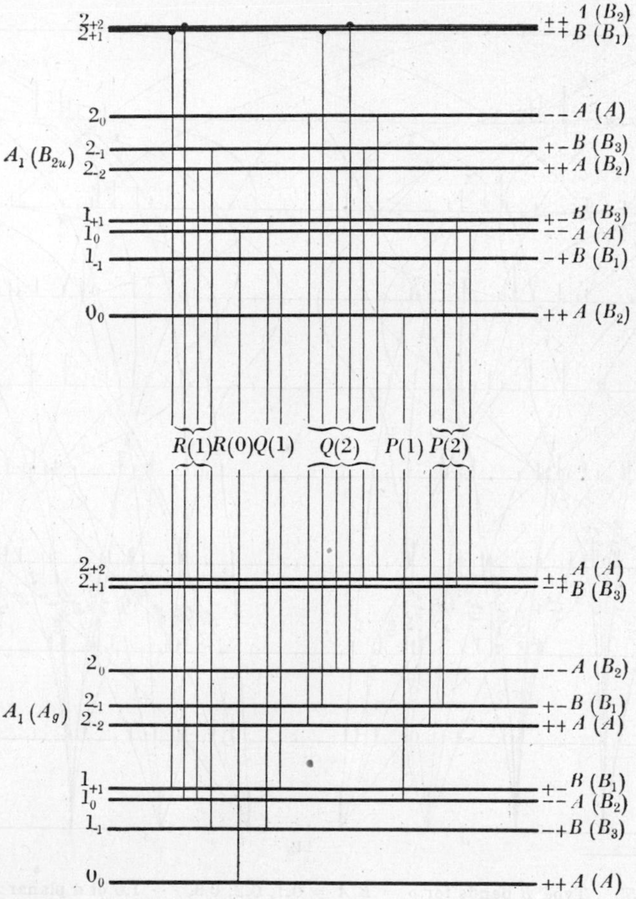

FIG. 154. **Energy level diagram for a type B band of an asymmetric top molecule.**—The energy levels are drawn to scale for the fundamental $\nu_1(a_1)$ of H_2O [after the data of Nielsen (667)] for which the C_2 coincides with the b-axis. Also the species designations apply to this case (see Fig. 144). In brackets the species for a B_{2u}—A_g transition of a V_h molecule are added assuming that the y axis is the b-axis.

(1 : 3 for H_2O, 2 : 1 for D_2O; see p. 465 and Table 10). In the series of doublets mentioned above, as can be seen immediately from Fig. 149, alternately the short- and the long-wave-length component is the stronger. This can be seen clearly in the H_2O bands reproduced in Fig. 151. In the limiting case of a nearly symmetric top with the a axis as top axis (prolate top), to each K value (for each J) belongs an A and a B level, and therefore there will be no intensity alternation in the sub-

bands of the || band into which the type A band goes over (Fig. 122), with the exception of that with $K = 0$. Thus for not too high resolution the whole band will not have an intensity alternation, except a very slight one if only small K values occur.

In the case of a C_{2v} molecule whose two-fold axis coincides with the axis of least moment of inertia (for example H_2CO), it can be seen, by changing the species in

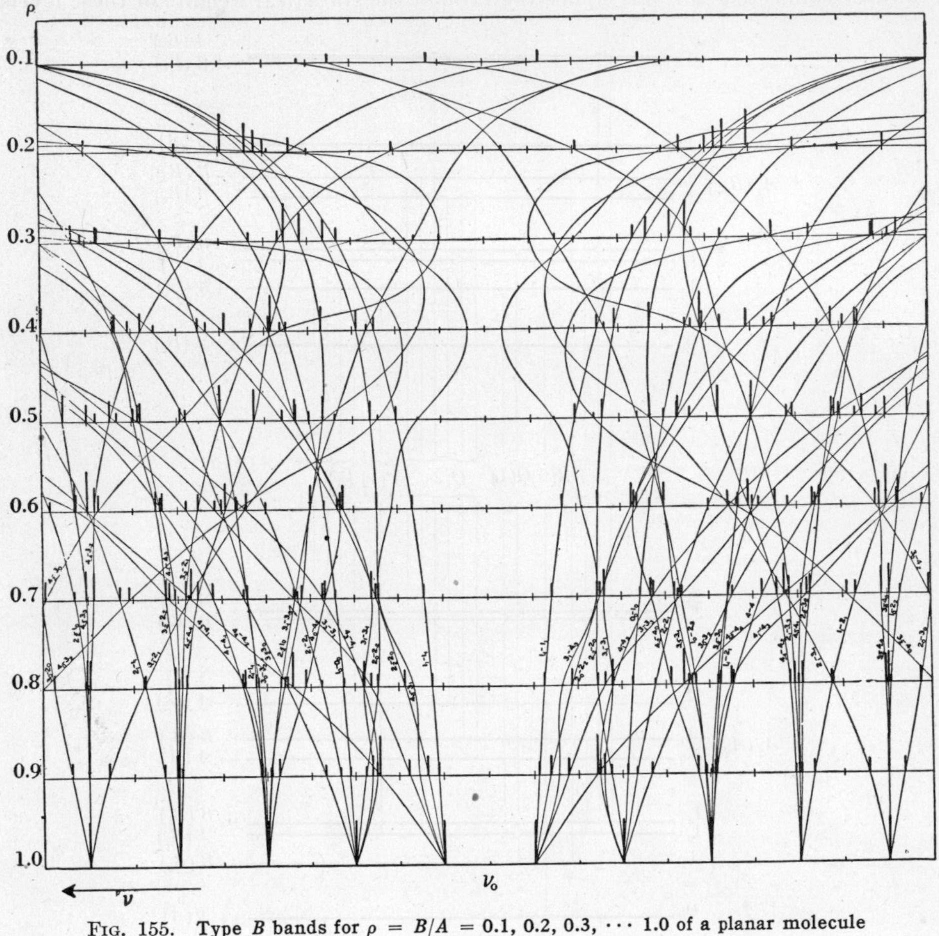

FIG. 155. **Type B bands for $\rho = B/A = 0.1, 0.2, 0.3, \cdots 1.0$ of a planar molecule** [after Dennison (279)].—See caption of Fig. 150.

Fig. 149 according to the previous rule (p. 462f.), that the doublet series with spacing $2C$ behaves in the same way as in the preceding case, while in the doublet series with spacing $2A$ both doublet components are alternately weak and strong. Again, in the limiting case of a prolate nearly symmetric top no intensity alternation should appear, in agreement with observation for the H_2CO band reproduced in Fig. 152.

In the case of a molecule of point group V_h the intensity alternation for the general case can be read from Fig. 149. In the case of a prolate nearly symmetric top of this point group it can be seen from Fig. 149, by considering which levels have the same K, that each sub-band of the || band will have an intensity alternation

[(7 + 3) : (3 + 3) = 10 : 6 for C_2H_4], but in successive sub-bands alternately the even and the odd lines are strong, and therefore the whole band (if as usual the K fine structure is not resolved) will not show an intensity alternation. This is verified by the C_2H_4 band in Fig. 153.

Type *B* bands. When the *alternating dipole moment* lies *in the axis of intermediate moment of inertia* (b axis) the symmetry selection rule is (IV, 98) rather than (IV, 97). Fig. 154 gives schematically the possible transitions for the lowest J values of a type

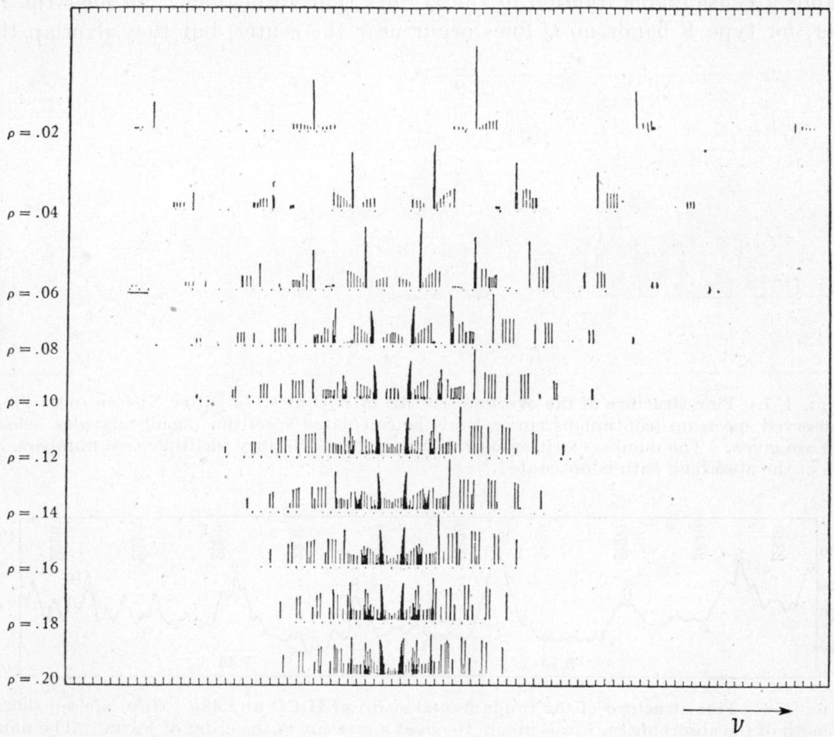

FIG. 156. **Type *B* bands for $\rho = B/A = 0.02 \cdots 0.20$** [after Nielsen (660)].—Unlike Figs. 150 and 155 here as well as in Fig. 161 the Boltzmann factor has been taken into account assuming $I_C = 20 \times 10^{-40}$ gm cm² and absorption at room temperature. The intervals between divisions at the bottom as well as at the top are $5C$.

B band. The over-all species symbols added refer to molecules of point group $\boldsymbol{C_{2v}}$, for which (as in Fig. 149) the two-fold axis coincides with the b axis (as in H_2O and H_2S). The transitions would be the same for any other orientation of the two-fold axis. Only the designations A, B of the rotational levels would have to be changed (see p. 462f.). In Fig. 154 also, in brackets, the over-all species have been added for B_{2u}—A_g transitions of molecules of point group $\boldsymbol{V_h}$, assuming that the b axis is the y axis (which is perpendicular to the C—C axis in C_2H_4 but in the plane of the molecule).

The actual structure of a type B band for a plane molecule has been calculated by Dennison (279) for $\rho = I_A/I_B = 1, 0.9, 0.8, \cdots 0.1$, and by Nielsen (660), for

$\rho = 0.20, 0.18, \cdots 0.02$. Their results are reproduced in Fig. 155 and Fig. 156. In the latter figure all levels up to $J = 6$ are taken into account, in the former only those up to $J = 4$. Also, in Fig. 156 the Boltzmann factor has been taken into account, assuming a constant $I_C = 20 \times 10^{-40}$ gm cm^2. In both cases it has been assumed that $A' = A''$, $B' = B''$, and $C' = C''$. It is seen that unlike the type A bands *the type B bands do not have a strong central branch for any value of* ρ. Rather there is a gap between the first strong group of lines on the short- and the long-wavelength side of the origin (which, however, is filled by weaker lines). While for type A bands a considerable fraction of the Q lines (not all of them) fall near the band center, for type B bands no Q lines occur near the center, but they overlap the P

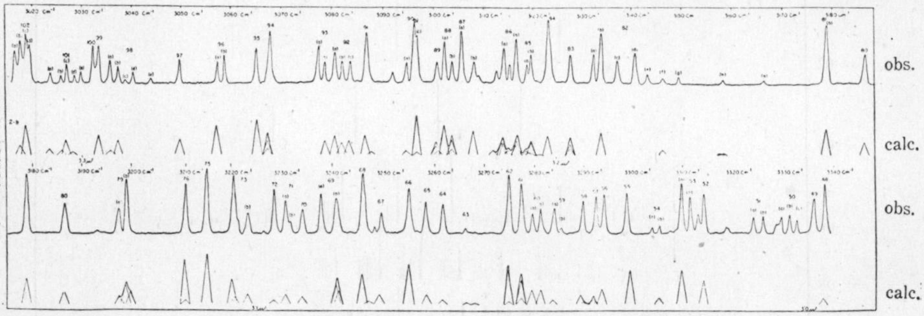

FIG. 157. **Fine structure of the overtone $2\nu_2(A_1)$ of H_2O at 3.17μ [after Nielsen (665)].—Both the observed spectrum (continuous curve) and the calculated spectrum (small triangles below the curve) are given. The numbers written on the maxima are arbitrary identification numbers. The length of the absorbing path is not stated.**

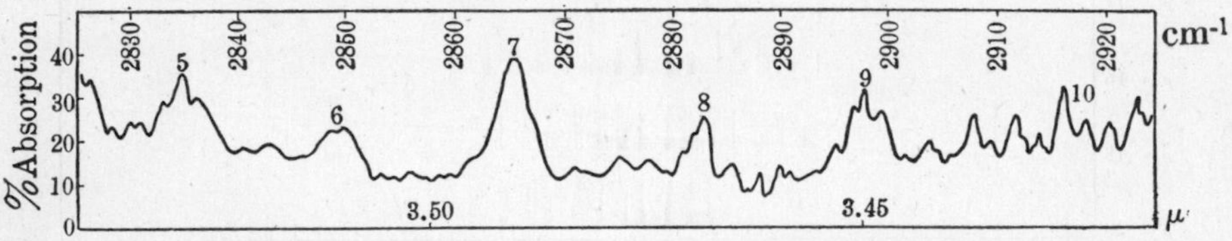

FIG. 158. **Fine structure of the fundamental $\nu_4(b_1)$ of H_2CO at 3.48μ [after Nielsen (662)].—The length of the absorbing path was about 10 cm at a pressure of the order of 1 atm. The numbers on the maxima are arbitrary running numbers.**

and R branches. The reason for this is clear from Fig. 154, since it follows from (IV, 98) that Q lines with $\Delta\tau = 0, \pm 1$, which would be close to the center, cannot occur.

For $\rho = 1$ $(I_A = I_B)$ the type B band is, of course, identical with the type A band. In the neighborhood of this limiting case the type B band under medium dispersion would still consist of a number of approximately equidistant lines. While for intermediate ρ values the structure is very complicated, as the opposite limiting case is approached (ρ small) we have again an approximately symmetric top, since $I_C = I_A + I_B \approx I_B$. But since now the change of dipole moment is perpendicular to the axis of the nearly symmetric top, *the type B band*, unlike the type A band, *approaches in structure the $\perp$ band of a symmetric top molecule.* This is clearly seen by comparing the spectra at the top of Fig. 156 with Fig. 128. In the limiting case,

$\rho = 0$, we obtain a $\perp$ band of a linear molecule; that is, only one of the sub-bands (with P, Q, and R branch) in the top row of Fig. 156 remains.

The intensities in Fig. 155 and Fig. 156 have been calculated in the same way as was briefly indicated for the type A bands.

As an example of a type B band of a strongly asymmetric top, Fig. 157 gives the fine structure of the overtone $2\nu_2(A_1)$ of H_2O, after Nielsen (665). In this figure is also given the calculated spectrum, assuming certain values for the rotational constants in the upper and lower state. Unlike the type A bands, here the series corresponding to the transitions involving the two highest and two lowest levels of each set with given J no longer stand out, and thus the structure appears even more complicated than that of type A bands. As examples of type B bands of nearly symmetric top molecules, Fig. 158 and 159 give the fine structures of the fundamentals $\nu_4(b_1)$ and $\nu_9(b_{2u})$ of H_2CO and C_2H_4 respectively. They correspond rather closely to the theoretical spectrograms near the top of Fig. 156. We have essentially a series of nearly equidistant lines which are the unresolved Q branches of the sub-bands of the $\perp$ bands. The separation of successive lines is approximately $2A$. In contrast to the $\perp$ bands of exactly symmetric tops (see Fig. 128), the C_2H_4 band in Fig. 159 shows an *intensity minimum near the center,* in agreement with Fig. 156,

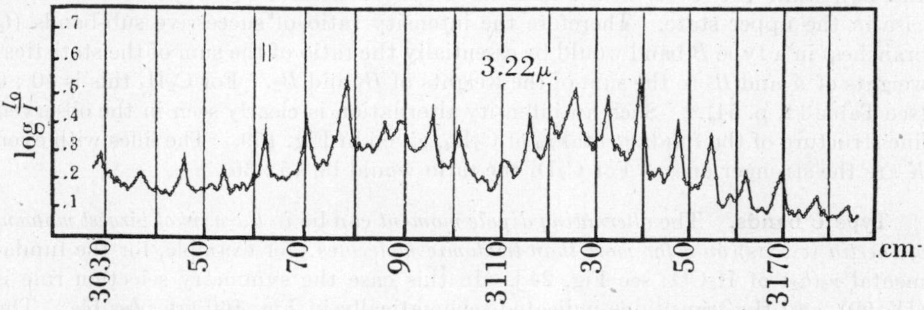

FIG. 159. **Fine structure of the fundamental $\nu_9(b_{2u})$ of C_2H_4 at 3.22μ [after Gallaway and** Barker (345)].—The length of the absorbing path was 6 cm. at a pressure of 150 mm. The ordinate gives log I_0/I (see caption of Fig. 153).

and indicating the deviation from the symmetric top. (For the H_2CO band of Fig. 158 the overlapping by other bands is too great to show this effect.)

The *intensity alternation* due to nuclear spin can again be read from Fig. 154 or similar figures. We consider only the case of nearly symmetric top molecules. If for a C_{2v} molecule the axis of least moment of inertia (a axis) coincides with the two-fold axis and is the top axis (as in H_2CO), it is easily seen that in the ground state (ψ_v of species A_1), since the $+ +$ and $- +$ levels have species A and the $+ -$ and $- -$ levels species B, the levels with even K have species A and those with odd K have species B. In the upper state (ψ_v of species B_1) the over-all species are reversed; that is, the levels with even K have species B, those with odd K species A. (This follows also from the fact that a rotation by 180° about the a axis exchanges the identical nuclei, and from the properties of the symmetric top eigenfunctions; see p. 406f.) Therefore there will be an intensity alternation of the type strong, weak, strong $\cdots$ in the series of Q branches, the intensity ratio being equal to the ratio of the statistical weights of the A and B levels. Such an intensity alternation (ratio

1 : 3) can indeed be seen in the H_2CO band reproduced in Fig. 158. According to the above, *the first "line" on the short-wave-length side of the origin* ($K' = 1 \leftarrow K'' = 0$) *will be strong or weak depending on whether the* (resultant) *statistics of the identical nuclei is Bose or Fermi, respectively.* Conversely, this condition supplies a valuable check on the correctness of the choice of the band origin.

In a similar manner it can be seen that for a nearly symmetric top molecule of point group $\boldsymbol{C_{2v}}$ no intensity alternation in the series of Q branches of a type B band occurs when the two-fold axis has the direction of the b axis while the a axis is the top axis; nor when the c axis is the top axis (oblate symmetric top) with the two-fold axis in either the a or b axis. Only when the two-fold axis is the c axis (which is not possible for plane molecules) does an intensity alternation again appear. Remembering that the type A bands of nearly symmetric top molecules do not show an intensity alternation of the above type under any circumstances, we see that the observation of an intensity alternation greatly restricts the possible interpretations of a given band.

Finally, for a nearly symmetric top molecule of point group $\boldsymbol{V_h}$ with the x axis (C—C axis in C_2H_4) as the a axis, it follows immediately from Fig. 154 (species symbols in brackets) that in the lower state the rotational levels with even K have species A and B_3 (with the exception of those with $K = 0$, which are alternately A and B_3), while the levels with odd K have species B_1 and B_2. The reverse is the case in the upper state. Therefore the intensity ratio of successive sub-bands (Q branches) in a type B band would be essentially the ratio of the sum of the statistical weights of A and B_3 to the sum of the weights of B_1 and B_2. For C_2H_4 this is 10 : 6 (see Table 11, p. 54).[54] Such an intensity alternation is clearly seen in the observed fine structure of the fundamental ν_9 of C_2H_4, given in Fig. 159. The lines with even K are the stronger ones. For C_2D_4 the ratio would be 45 : 36.

Type *C* bands. The *alternating dipole moment* can lie *in the axis of largest moment of inertia* (c axis) *only for more-than-triatomic molecules* [for example, for the fundamental $\nu_6(b_2)$ of H_2CO; see Fig. 24]. In this case the symmetry selection rule is (IV, 99), and the transitions indicated schematically in Fig. 160 are possible. The over-all species of the rotational levels given in this figure (unlike Figs. 149 and 154) refer to the case of a $\boldsymbol{C_{2v}}$ molecule with the two-fold axis in the a axis (as in H_2CO, see Fig. 143). The species in brackets refer again to a $\boldsymbol{V_h}$ molecule with the x axis as the a axis (as in C_2H_4, see Fig. 145).

The actual band structure to be expected for a type C band has not been evaluated for the general case, but Nielsen (660) has given diagrams for a number of small ρ values (assuming again a plane molecule). These diagrams are reproduced in Fig. 161. It is seen that *for very small ρ the structure is practically the same as that of a type B band*, simply because both the type C and type B bands approach $\perp$ bands of a symmetrical top ($I_B \approx I_C$). For larger ρ values the structure is increasingly different from that of a type B band. This is because the type C band for $\rho = I_A/I_B \rightarrow 1$ goes over into a $\|$ band of a symmetric top, rather than a $\perp$ band as do the type A and B bands. This $\|$ band in the limiting case has a strong Q branch and a spacing of the P and R lines of $2A = 2B = 4C$, whereas the limiting type A and B bands have no strong central branch and a spacing of $2C$. For medium resolution one

[54] The same result is also obtained by considering that a rotation by 180° about the a axis leads to an exchange of two pairs of H atoms. Just as for linear molecules (see Table 2), this gives an intensity alternation 10 : 6 with the even levels the stronger (the resultant statistics is Bose).

would therefore expect a *central maximum for not too small* ρ, but none for very small ρ.

As an example, Fig. 162 gives the fine structure of the fundamental $\nu_7(b_{1u})$ of C_2H_4, which according to Fig. 44 corresponds to a vibration perpendicular to the plane of the molecule. The central maximum is seen to be very prominent. Even

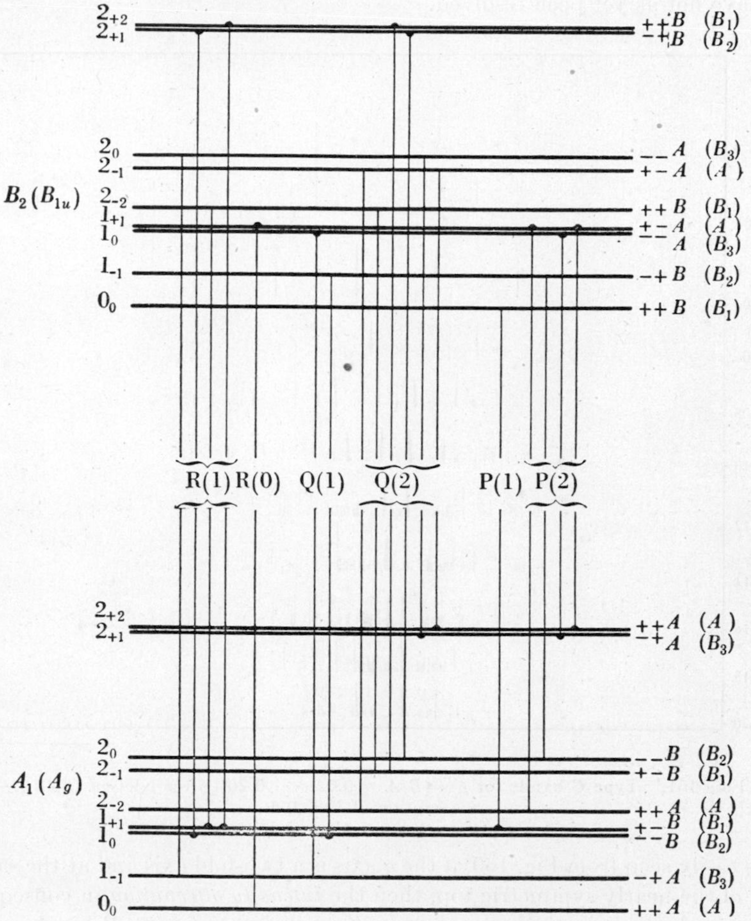

FIG. 160. **Energy level diagram for a type *C* band of an asymmetric top molecule.**—The energy levels are drawn for the case of a nearly symmetric top as in H_2CO for which the C_2 coincides with the *a*-axis. The species designations refer to this case. In brackets the species for a B_{1u}—A_g transition of a V_h molecule are added assuming that the x axis is the *a*-axis, the z axis the *c*-axis.

though here $\rho = 0.16$, the molecule is still sufficiently close to a prolate symmetric top that the Q branches of the sub-bands of the $\perp$ band which the band approaches stand out clearly. Their separation is approximately $2A$ as for the type B bands. That these Q branches do not stand out in Fig. 161 for $\rho = 0.16$ is due to the fact that in this figure only transitions with $J \leq 6$ have been included.

In H_2CO the fundamental $\nu_6(b_2)$ is a type C band. Its fine structure has been investigated for both H_2CO and D_2CO by Ebers and Nielsen (295) (296). Here a complication arises due to the fact that the fundamental $\nu_5(b_1)$ which is a type B band is very close to ν_6. This results in a strong Coriolis perturbation between the two upper vibrational levels, giving rise to a strong convergence of the series of Q branches ("lines"), which is of opposite direction in the two bands. Since here $\rho = 0.1$ the central branch is no longer prominent. Type C bands with larger ρ values have not as yet been resolved.

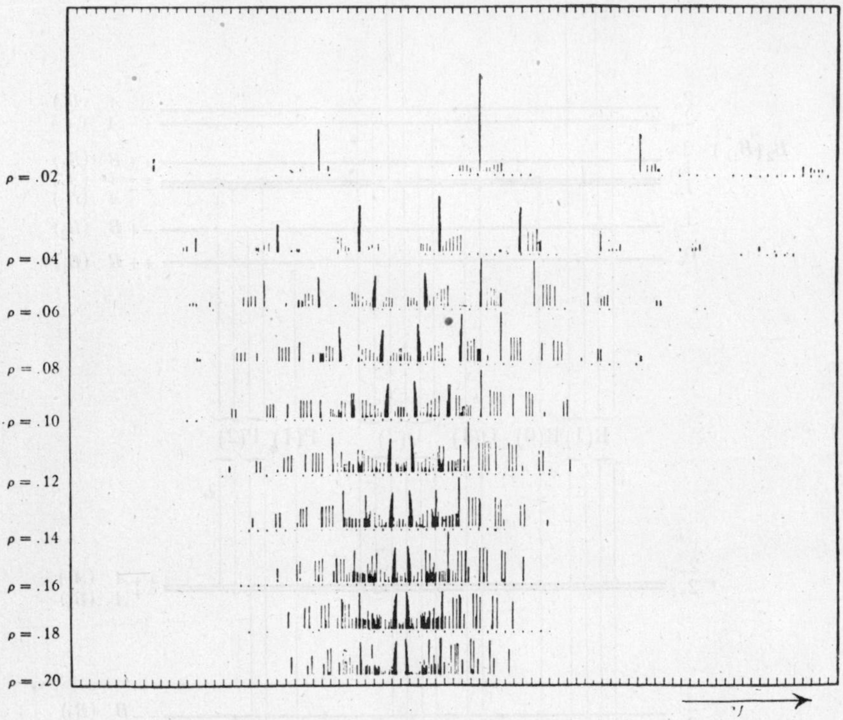

FIG. 161. **Type C bands for $\rho = B/A = 0.02, \cdots 0.20$** [after Nielsen (660)].—See caption of Fig. 156.

As is easily seen from Fig. 160, if the a axis is a two-fold axis and at the same time the axis of the nearly symmetric top, then the *intensity alternation* in consequence of nuclear spin depends in the same way on K as it does for type B bands. This intensity alternation is not very clear in the C_2H_4 band in Fig. 162. The H_2CO band, mentioned above but not reproduced here [see Ebers and Nielsen (295)], shows it clearly. Near the other limiting case, when the c axis is the symmetric top axis, an intensity alternation will occur only when the c axis is also a symmetry axis.

Unresolved infrared bands. Only too frequently the infrared bands of asymmetric top molecules are not resolved. But if the dispersion used is not too small, so that the *envelopes of the bands* can be distinguished from simple maxima, it is sometimes possible to draw conclusions as to the type of the bands, which are of use for the vibrational analysis and have been used to some extent in Chapter III.

It is clear from the above discussion of the band types and from Figs. 150, 155, 156, and 161 that *type A bands* will exhibit a *fairly strong central maximum*, corresponding to the Q lines, with two accompanying maxima on either side, corresponding essentially to the P and R lines, for any value of ρ. *Type C bands* will have similar

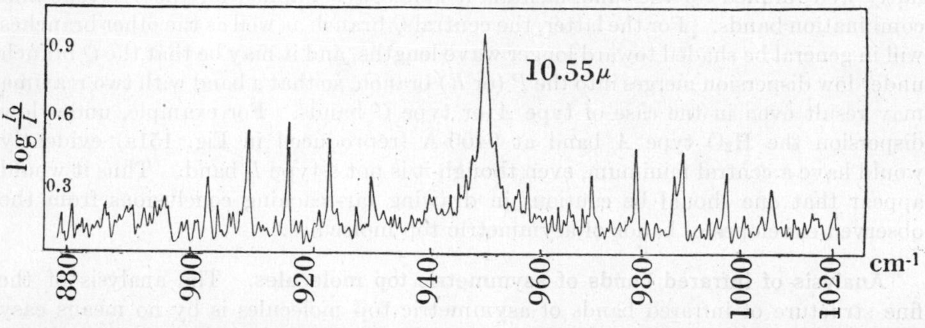

FIG. 162. Fine structure of the fundamental $\nu_7(b_{1u})$ of C_2H_4 at 10.55μ [after Gallaway and Barker (345)].—The length of the absorbing path was 6 cm at a pressure of 5 cm. The ordinate is log I_0/I (see caption of Fig. 153).

envelopes, except for small ρ values when the central branch no longer stands out, so that only one broad maximum results. *Type B bands* on the other hand have *no central maximum* but, as can be seen from Figs. 155 and 156, the *Q lines form two maxima* on either side and fairly close to the zero line, in addition to the P and R maxima. These Q maxima should be prominent, particularly for the larger ρ values. More detailed calculations of the envelopes for a considerable variety of cases have been carried out by Badger and Zumwalt (76). We give in Fig. 163 only one case

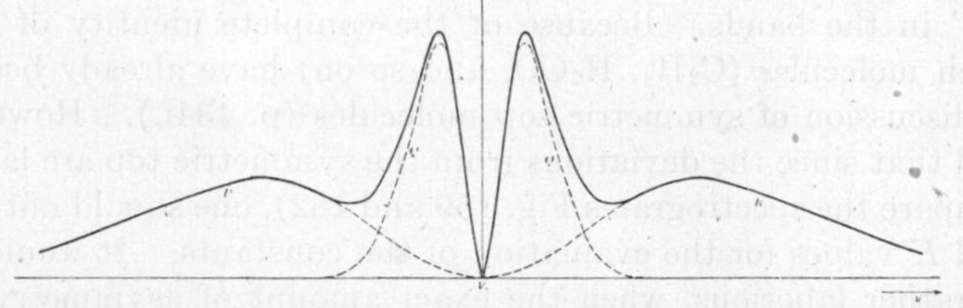

FIG. 163. Envelope of a type B band of a (non-planar) asymmetric top molecule with $\rho = B/A = 0.73$ and $C/A = 0.64$ [after Badger and Zumwalt (76)].—For small moments of inertia the intensity distribution is not quite symmetrical about the band origin.

of a type B band. With insufficient resolution it appears quite possible that the two Q maxima are observed as one and that therefore the type B band has the appearance of a type A band.

Conversely, when an unresolved band of an asymmetric top molecule is observed to have three maxima, one can conclude that it is not a type B band only if one is sure that the resolution would have been sufficient to resolve the two Q maxima of a type B band, or when one knows from other evidence that ρ is small, and that therefore the Q maxima are not prominent. If an infrared band shows four maxima, as in Fig. 163, one can be reasonably sure that it is a type B band. The same is true when it has only two maxima with a distinct minimum at the center.

However, it must be remembered that all the previous schematic spectra of asymmetric top molecules, as well as the calculation of envelopes by Badger and Zumwalt, have been carried out under the assumption that the rotational constants A, B, C are the same in the upper and the lower state. While this assumption is fairly well fulfilled for the fundamentals it is less well fulfilled for the overtone and combination bands. For the latter, the central Q branch as well as the other branches will in general be shaded toward longer wave lengths, and it may be that the Q branch under low dispersion merges into the P (or R) branch, so that a band with two maxima may result even in the case of type A or type C bands. For example, under low dispersion the H_2O type A band at 9400 Å (reproduced in Fig. 151a) evidently would have a central minimum, even though it is not a type B band. Thus it would appear that one should be cautious in drawing far-reaching conclusions from the observed envelopes of bands of asymmetric top molecules.

Analysis of infrared bands of asymmetric top molecules. The analysis of the fine structure of infrared bands of asymmetric top molecules is by no means easy except close to the limiting cases of symmetric tops. In these latter cases the procedure is, of course, exactly the same as described above for exactly symmetric top molecules. The only difference is that in the case of the near-prolate top (a axis the top axis) the energy formula is (IV, 90) with $W_\tau^{[v]} = K^2$, rather than (IV, 41); that is, $B_{[v]}$ is replaced by $\frac{1}{2}(B_{[v]} + C_{[v]})$. In the less frequent case of a near-oblate top (c axis the top axis), $B_{[v]}$ in (IV, 41) is replaced by $\frac{1}{2}(A_{[v]} + B_{[v]})$ and $A_{[v]}$ by $C_{[v]}$. Thus *the analysis of type A bands of near-prolate tops gives* $\frac{1}{2}(B_{[v]} + C_{[v]})$ in the upper and lower states and *the analysis of type B and C bands of near-prolate tops gives* $A_{[v]} - \frac{1}{2}(B_{[v]} + C_{[v]})$ in the upper and lower states. These constants are most accurately obtained from appropriate *combination differences*, which also, if formed for different bands with the same lower state, will supply a useful check on the correctness of the analysis. Approximate values of the constants are obtained from the spacing of the "lines" in the bands. Because of the complete identity of procedure the results for such molecules (C_2H_4, H_2CO, and so on) have already been included in the previous discussion of symmetric top molecules (p. 434f.). However, it should be emphasized that since the deviations from the symmetric top are largest for small K values (compare the spectrograms Fig. 159 and 162), one should not use sub-bands with too small K values for the evaluation of the constants. It would also be wise, even though rather laborious, when the exact amount of asymmetry is known, to calculate the spectrum with the more accurate asymmetric top formulae in order to see whether the simplifications introduced by the symmetric top approximations were justified.

In the case of *strongly asymmetric top molecules* the lines in a band no longer form easily recognizable branches of the usual type. However, the series of doublets in the R and P branches that correspond to the two highest and the two lowest levels of each set form fairly normal branches. If they can be picked out of the large number of apparently irregular lines, from the separation of successive doublets approximate values for $2A$ and $2C$ respectively are obtained. For a plane molecule according to (IV, 94) this leads also to an approximate value for B. In the case of type B bands a further aid in determining approximate rotational constants is afforded by the fact that, as can be seen from Fig. 155, the first strong lines on the short- and long-wave-length sides of the origin are the Q lines $1_{+1} - 1_{-1}$ and $1_{-1} - 1_{+1}$

respectively. These two lines stand out clearly, for example, in the H_2O band in Fig. 157 [the maxima marked 81(b) and 82(b)]. As can easily be seen with the help of Table 8 the separation of these two lines is $(A' - C') + (A'' - C'')$.

If in this way or from other evidence (for example, from the known internuclear distances in other molecules) rough values for the rotational constants have been obtained, they may be used for the calculation of a theoretical spectrum by calculating the rotational energy levels from the formulae (I, 58)–(I, 62). If the assumed approximate constants are not too inaccurate, it is usually possible, by comparing the theoretical with the observed spectrum, to assign many of the observed lines to specific transitions. A critical test on the correctness of this assignment can then be obtained by forming *combination differences* from the observed lines. For example, in the case of a type A band (see Fig. 149) the difference of the lines $Q(1_{+1} - 1_0)$ and $P(1_{+1} - 2_0)$, since they have the same upper state, is the difference $\Delta_1F''(2_0 - 1_0)$ between the levels 2_0 and 1_0 in the lower state. This same difference is also obtained as the difference of the lines $R(2_{-1} - 1_0)$ and $Q(2_{-1} - 2_0)$. Similar *combination relations*,

$$Q[J_{\tau_1'} - J_{\tau_1''}] - P[J_{\tau_1'} - (J+1)_{\tau_2''}] = R[(J+1)_{\tau'_2} - J_{\tau_1''}] - Q[(J+1)_{\tau_2'} - (J+1)_{\tau_2''}] \quad \text{(IV, 103)}$$

and

$$R[(J+1)_{\tau_2'} - J_{\tau_1''}] - Q[J_{\tau_1'} - J_{\tau_1''}] = Q[(J+1)_{\tau_2'} - (J+1)_{\tau_2''}] - P[J_{\tau_1'} - (J+1)_{\tau_2''}], \quad \text{(IV, 104)}$$

with all possible combinations of the τ_1', τ_2', τ_1'', τ_2'' must hold for other J values, independent of any perturbations that might occur in the rotational energy levels. In addition, for the higher J values, combination relations of the form

$$Q(J_{\tau_1'} - J_{\tau_1''}) - Q(J_{\tau_1'} - J_{\tau_2''}) = Q(J_{\tau_2'} - J_{\tau_1''}) - Q(J_{\tau_2'} - J_{\tau_2''}) \quad \text{(IV, 105)}$$

must hold, and similarly for P and R branches.

Once the correct assignment of the lines has been found and checked by the combination relations, *accurate values for the rotational constants* in the upper and lower states can be obtained *by the application of Mecke's sum rules* (see Table 8, p. 50). For example, the difference $\Delta_1F''(2_0 - 1_0)$, according to Table 8, is $4B''$. Similarly, the differences $\Delta_1F''(2_{-1} - 1_{+1})$ and $\Delta_1F''(2_{+1} - 1_{-1})$ are $4C''$ and $4A''$ respectively. In a similar manner, the other combination differences or simple combinations of them may be expressed in terms of A'', B'', C'' for the lower, and of A', B', and C' for the upper state [see Table 3 in Mecke (612)].

The rotational constants thus obtained are accurate except for the *neglect of centrifugal stretching*. If it is necessary to take this into account it is best to apply an appropriate correction to the observed combination differences, and then apply Mecke's sum rules. For a detailed discussion of the centrifugal stretching corrections, see the references quoted in Chapter I, p. 50.

Independent of any assumption about the analytic representation, the *positions of the rotational energy levels* of a given type $(+\,+, +\,-, -\,+$ or $-\,-)$ relative to the lowest of that type can be obtained directly from the spectrum simply by *adding up appropriate combination differences.* The lowest levels of the four types are $0_0(+\,+)$, $1_{-1}(-\,+)$, $1_0(-\,-)$, and $1_{+1}(+\,-)$ (see Fig. 149). Taking the energy of the lowest rotational level $0_0(+\,+)$ as zero, the energies of the levels 1_{-1}, 1_0, and 1_{+1}

TABLE 134. WAVE NUMBERS AND ASSIGNMENT OF THE LINES IN THE H_2O BAND AT 8227 Å, AFTER BAUMANN AND MECKE (130).

$J''_{\tau''}$	*P* branch		*Q* branch		*R* branch	
	$J'_{\tau'}$	ν_{vacuum}[55] (cm^{-1})	$J'_{\tau'}$	ν_{vacuum}[55] (cm^{-1})	$J'_{\tau'}$	ν_{vacuum}[55] (cm^{-1})
0	—		—		1_{-1}	12,173.77 (4)
1_{-1}	0	12,127.44 (5*)	—		2_{-2}	195.19 (8)
1_{-1}	—		—		2_{+2}	262.82
1_0	—		1_{+1}	12,156.21 (5)	2_{-1}	190.71 (3)
1_{+1}	—		1_0	145.43 (6)	2_0	202.02 (7)
2_{-2}	1_{-1}	103.66 (4)	—		3_{-3}	212.07 (5)
2_{-2}	—		2_{+1}	215.28	3_{+1}	289.92
2_{-1}	1_0	108.32 (8)	2_0	164.90 (5)	3_{-2}	207.75 (8)
2_0	1_{+1}	098.17 (3)	2_{-1}	132.71 (3)	3_{-1}	224.67 (4)
2_0	—		—		3_{+3}	341.64 ?
2_{+1}	—		2_{+2}	151.82 (9)	3_0	218.83 (6)
2_{+1}	—		2_{-2}	084.09	—	
2_{+2}	—		2_{+1}	149.16 (4)	3_{+1}	223.82 (2)
2_{+2}	1_{-1}	037.50	—		3_{-3}	145.43*
3_{-3}	2_{-2}	082.22 (8)	—		4_{-4}	226.10 (10)
3_{-3}	2_{+2}	149.83*	3_0	216.93	4_0	323.41
3_{-2}	2_{-1}	085.60 (4)	3_{-1}	177.58 (2)	4_{-3}	222.95 (4)
3_{-2}	—		3_{+3}	294.91	4_{+1}	387.64
3_{-1}	2_0	071.02 (8)	3_{-2}	113.88 (2)	4_{-2}	224.72 (10)
3_{-1}	—		—		4_{+2}	357.97
3_0	2_{+1}	079.02 (4)	3_{+1}	153.70 (2)	4_{-1}	237.45 (2)
3_0	—		3_{-3}	075.76	—	
3_{+1}	2_{+2}	074.56 (5)	3_0	141.56 (5)	4_0	247.99 (8)
3_{+1}	2_{-2}	006.82	—		4_{-4}	301.51
3_{+2}	—		3_{+3}	150.30 (6)	4_{+1}	243.54 (1)
3_{+2}	2_{-1}	11,941.05	3_{-1}	032.99	4_{-3}	078.81*
3_{+3}	—		3_{+2}	149.83 (3)	4_{+2}	244.73 (10*)
3_{+3}	2_0	957.90	3_{-2}	000.80	4_{-2}	131.45 ?
4_{-4}	3_{-3}	12,060.09 (4)	—		5_{-5}	238.31 (4)
4_{-4}	3_{+1}	138.30	4_{-1}	221.71	—	
4_{-3}	3_{-2}	062.41 (8)	4_{-2}	193.22 (1)	5_{-4}	236.53 (8)
4_{-3}	3_{+2}	212.07*	4_{+2}	305.87	—	
4_{-2}	3_{-1}	044.29 (3*)	4_{-3}	089.79 (0)	5_{-3}	261.00 (2)
4_{-2}	3_{+3}	161.55*	4_{+1}	254.53 ?	—	
4_{-1}	3_0	053.37 (4)	4_0	159.81 (4)	—	
4_{-1}	—		4_{-4}	062.41*	—	
4_0	3_{+1}	044.29 (3*)	4_{-1}	127.98 (0)	—	
4_0	3_{-3}	11,966.39	—		5_{-5}	144.82*
4_{+1}	3_{+2}	12,052.80 (2)	4_{+2}	147.69 (3)	—	
4_{+1}	3_{-2}	11,903.17	4_{-2}	033.87	5_{-4}	077.34
4_{+2}	3_{+3}	12,051.71 (0)	4_{+1}	144.82 (3)	—	
4_{+2}	3_{-1}	11,934.72 ?	4_{-3}	11,979.69	5_{-3}	150.31* ?
5_{-5}	4_{-4}	12,037.50 (5)	—		6_{-6}	249.38 (8)
5_{-4}	4_{-3}	038.65 (1)	5_{-3}	12,209.85 (−2)	6_{-5}	248.57 (3)
5_{-3}	4_{-2}	018.62 (2)	5_{-4}	062.01 (−1)	—	
6_{-6}	5_{-5}	013.65 (2)	—		7_{-7}	259.93 (1)
6_{-5}	5_{-4}	014.14 (4)	—		7_{-6}	259.55 (5)
7_{-7}	6_{-6}	11,988.51 (2)	—		8_{-8}	268.98 (3)
7_{-6}	6_{-5}	988.51 (2)	—		8_{-7}	268.98 (3)
8_{-8}	7_{-7}	962.67 (2)	—		—	
8_{-7}	7_{-6}	962.67 (2)	—		—	

[55] The numbers in parentheses give the estimated intensities. An asterisk indicates an overlapped line.

are, according to the Wang formulae (I, 58, 60, 61) (see also Table 8), $B + C$, $A + C$, and $A + B$ respectively. Since the centrifugal stretching correction for these levels is certainly negligible, and since no perturbation is likely to occur for the vibrational ground state, the positions of all rotational levels of this state with respect to the lowest can be determined. The energies of the rotational levels of the excited vibrational states can then be determined simply by adding the wave numbers of appropriate lines to the energies of the rotational levels of the ground state.

If the observed and identified lines are not sufficient to form all the combination differences necessary for an evaluation of the constants in the above-described way, the only other alternative, in order to obtain these constants, seems to be to change the original rough constants by trial and error in such a way that a complete agreement of theoretical and observed spectra is finally obtained. The complexity of the Wang equations obviously makes this an exceedingly tedious procedure, particularly since there are six rotational constants to adjust.

Examples, moments of inertia and internuclear distances. Mecke and his co-workers (612) (130) (333) were the first to analyze fully the rotation-vibration spectrum of an asymmetric top molecule, namely that of H_2O. This is still the only case of a strongly asymmetric top for which a really complete analysis is available. The advantage here is that, because of the strong water-vapor absorption in the atmosphere, a very complete spectrum is obtainable under large dispersion in the photographically accessible region of the solar spectrum. All bands in the photographic region are found to be type *A* bands. As an example, we give in Table 134 the wave numbers and assignments of the lines in the band 8227 Å of H_2O, which is reproduced in Fig. 151b. The reader may use these data for verifying that the previous combination relations are fulfilled. Table 135 illustrates the agreement of some of the

TABLE 135. SOME COMBINATION DIFFERENCES FOR THE VIBRATIONAL GROUND STATE OF THE H_2O MOLECULE, AS CALCULATED FROM SEVERAL ROTATION-VIBRATION BANDS AS WELL AS FROM THE ROTATION SPECTRUM.

Term difference	Photographic infrared spectrum				Rotation spectrum[59]
	Combination[56]	λ8227[57]	λ7957[57]	λ6994[58]	
$3_{+2}-2_0$	$R(2_0)-Q(3_2)$	191.68	191.72	191.68	$(3_0-2_0)+(3_2-3_0)=190.23$
	$Q(2_0)-P(3_2)$	191.66	191.61	191.61	
$4_{-1}-3_{-3}$	$R(3_{-3})-Q(4_{-1})$	163.60	163.60	(163.68)	$(4_{-1}-4_{-3})+(4_{-3}-3_{-3})=163.59$
	$Q(3_{-3})-P(4_{-1})$	163.56	163.59	163.60	
$5_{-5}-4_{-1}$	$Q(4_{-1})-P(5_{-5})$	(24.91)	24.95	(24.89)	$(5_{-5}-4_{-3})-(4_{-1}-4_{-3})=25.00$
$5_{-4}-3_{-2}$	$R(3_{-2})-P(5_{-4})$	184.30	184.32	(184.43)	$(5_{-4}-4_{-4})+(4_{-4}-3_{-2})=184.18$
$5_{-3}-4_{-3}$	$R(4_{-3})-Q(5_{-3})$	174.52	174.48	—	$(5_{-3}-4_{-1})+(4_{-1}-4_{-3})=174.63$
	$Q(4_{-3})-P(5_{-3})$	174.60	174.60	174.64	

[56] In brackets the J_τ of the lower state are given. For the asymmetric top (see Fig. 149, 154, 160) this does not always identify a line unambiguously, since two or more *R* lines (and similarly *P* and *Q* lines) with different τ' may have the same lower state. In the table τ' is not always the same for all differences in one row.

[57] From the data of Baumann and Mecke (130).

[58] From the data of Freudenberg and Mecke (333).

[59] From the data of Randall, Dennison, Ginsburg, and Weber (712).

combination differences for the lower state within this band as well as their agreement with corresponding combination differences in other bands and with appropriate combinations in the far infrared rotation spectrum. It is seen that with one exception ($3_2 - 2_0$) the agreement of the combination differences for a given pair of levels, as determined from different bands and from the far infrared rotation spectrum, is very satisfactory indeed. The slight disagreement for $3_2 - 2_0$ seems to be due to some erroneous assignment in the far infrared spectrum, since the different photographic bands give close agreement among themselves. Mecke (612) has also analyzed a number of H_2O bands in the ordinary infrared on the basis of the measurements of Plyler and Sleator (704). These bands have been recently remeasured under somewhat higher dispersion by Nielsen (665) (667) (see Fig. 157), who has also extended Mecke's analysis to higher J values and has, in addition, identified and analyzed the type B band ν_1 which overlaps the type A band ν_3.

From these analyses the *positions of the rotational levels* have been determined with considerable accuracy (see the papers quoted above for the excited vibrational states, and Randall, Dennison, Ginsburg and Weber for the ground state). However up to now the rotational constants have not been evaluated from these observed energy levels with a corresponding accuracy (comparable to that obtained for linear molecules).

TABLE 136. ROTATIONAL CONSTANTS OF THE H_2O MOLECULE IN THE ELECTRONIC GROUND STATE.

Rotational constants A, B, C		Moments of inertia[60]		Rotational constants α		
Lowest vibrational level (cm^{-1})	Equilibrium position (cm^{-1})	Lowest vibrational level (10^{-40} gm cm^2)	Equilibrium position (10^{-40} gm cm^2)	α_1 (cm^{-1})	α_2 (cm^{-1})	α_3 (cm^{-1})
$A_{[0]} = 27.79$	$A_e = 27.33$	$I_A^0 = 1.007_3$	$I_A^e = 1.024_3$	$\alpha_1^A = +0.49_5$	$\alpha_2^A = -2.65_9$	$\alpha_3^A = +1.23_4$
$B_{[0]} = 14.50_8$	$B_e = 14.57_5$	$I_B^0 = 1.929_6$	$I_B^e = 1.920_7$	$\alpha_1^B = +0.224$	$\alpha_2^B = -0.202$	$\alpha_3^B = +0.112$
$C_{[0]} = 9.28_9$	$C_e = 9.49_9$	$I_C^0 = 3.013_7$	$I_C^e = 2.947_0$	$\alpha_1^C = +0.145$	$\alpha_2^C = +0.105$	$\alpha_3^C = +0.169$

Mecke and his co-workers have evaluated the rotational constants in the way indicated above, neglecting the effect of centrifugal stretching. Darling and Dennison (263) have re-evaluated their data, taking account of this interaction in the manner proposed by Wilson (936). However, they only employed the levels 1_{-1}, 1_0, 1_{+1}, 2_{-1}, 2_0, 2_{+1}, making use of the relations (see Table 8 and p. 485)

$$4A = F(2_{+1}) - F(1_{-1}), \quad 4B = F(2_0) - F(1_0), \quad 4C = F(2_{-1}) - F(1_{+1}), \qquad \text{(IV, 106)}$$

with the addition of a slight correction for centrifugal stretching. Thus only a few lines of each band are actually used, so that the error of the A, B, and C values may

60 The effective moments of inertia for the lowest vibrational level are the values of Dennison and Darling (263) except for a very small change on account of a slightly different value for the conversion factor used by them. The A_e, B_e, C_e values and the moments of inertia for the equilibrium position (which were not taken over from Darling and Dennison) were obtained from the A_0, B_0, C_0 and newly determined α's. In this derivation the validity of the relation $I_C^e = I_A^e + I_B^e$ was not assumed.

be as much as ± 0.03 cm^{-1}. Darling and Dennison have only given the moments of inertia. Their data have been combined with those of Nielsen (665) (667), yielding the rotational constants and moments of inertia in Table 136. It is seen that while the relation $I_C{}^e = I_A{}^e + I_B{}^e$ is well fulfilled there is a noticeable difference even between $I_C{}^0$ and $I_A{}^0 + I_B{}^0$, and of course much larger differences between $I_C{}^{[v]}$ and $I_A{}^{[v]} + I_B{}^{[v]}$ (see above).

From the values of $I_A{}^e$ and $I_B{}^e$ one obtains, for the O—H *distance* in H_2O in the equilibrium position,

$$r_e(\text{O—H}) = 0.958_4 \times 10^{-8} \text{ cm},$$

and for the HOH *angle*,

$$\measuredangle \text{ H—O—H} = 104° \, 27'.$$

The value of r_e(O—H) is slightly smaller than the r_e value 0.9710×10^{-8} cm in the free OH radical. From the effective moments of inertia in the lowest vibrational level one obtains

$$r_0(\text{O—H}) = 0.956_8 \quad \text{and} \quad \measuredangle \text{ H—O—H} = 105° \, 3'.$$

Apart from H_2O, the only other rotation-vibration spectra of strongly asymmetric top molecules that have been studied in any detail are those of HDO [Herzberg (446)] and H_2S [Cross (248) and Crawford and Cross (242)]. In each case only one band has been analyzed. In the papers referred to further details concerning the method of analysis may be found. We give in Table 137 the rotational constants

TABLE 137. ROTATIONAL CONSTANTS AND MOMENTS OF INERTIA OF H_2S IN THE LOWEST VIBRATIONAL STATE, AFTER CRAWFORD AND CROSS (242).

Rotational constants	Moments of inertia [61]
$A_0 = 10.393$ cm^{-1}	$I_A^0 = 2.694 \times 10^{-40}$ gm cm^2
$B_0 = 9.040$ cm^{-1}	$I_B^0 = 3.097 \times 10^{-40}$ gm cm^2
$C_0 = 4.723$ cm^{-1}	$I_C^0 = 5.927 \times 10^{-40}$ gm cm^2

and moments of inertia of H_2S in the lowest vibrational state. The values for the equilibrium position cannot be evaluated from the present data. The effective dimensions in the lowest state calculated from A_0 and B_0 are

$$r_0(\text{S—H}) = 1.334 \times 10^{-8} \text{ cm}, \qquad \measuredangle \text{ H—S—H} = 92° \, 16'.$$

If the dimensions were to be derived from A_0 and C_0 or B_0 and C_0, slightly different values would be obtained, since the relation $I_C = I_A + I_B$ is not exactly fulfilled. The above values are therefore less accurate than those for the equilibrium position in H_2O. It seems significant that just as for H_2O the value of r_0(S—H) in H_2S is slightly smaller than the r_0 value 1.35_0 in the free SH radical.

(*c*) *Raman spectrum*

Selection rules. The selection rules for the rotation-vibration Raman bands have been given by Placzek and Teller (701). To a good approximation the vibrational selection rules are the same as for the pure vibration spectrum (see Table 55).

[61] These I values have been recalculated from the rotational constants given by Crawford and Cross, using the new value for the conversion factor (see appendix, p. 538).

For the total angular momentum we have, as always,

$$\Delta J = 0, \pm 1, \pm 2 \qquad (J' + J'' \geq 2). \tag{IV, 107}$$

All transitions between the various rotational levels of two vibrational states that obey this rule are possible if the molecule has no symmetry. When the molecule has symmetry we have additional *symmetry selection rules* which depend on the components of the polarizability (Table 55) that are different from zero for the particular vibrational transition [for a proof see Placzek and Teller (701)]:

If only $[\alpha_{xx}]^{nm}$, $[\alpha_{yy}]^{nm}$, *and* $[\alpha_{zz}]^{nm}$ *are different from zero* we have the selection rule

$$++ \leftrightarrow ++, \qquad +- \leftrightarrow +-, \qquad -+ \leftrightarrow -+, \qquad -- \leftrightarrow --, \tag{IV, 108}$$

as for the rotational Raman spectrum. *If only* $[\alpha_{xy}]^{nm}$ *is different from zero* and if the x, y, z axes are the a, b, c axes respectively we have

$$++ \leftrightarrow +-, \qquad -+ \leftrightarrow --. \tag{IV, 109}$$

If only $[\alpha_{xz}]^{nm}$ *is different from zero*, with the same orientation of the axes, we have

$$++ \leftrightarrow --, \qquad +- \leftrightarrow -+; \tag{IV, 110}$$

and finally, *if only* $[\alpha_{yz}]^{nm}$ *is different from zero*, we have

$$++ \leftrightarrow -+, \qquad +- \leftrightarrow --. \tag{IV, 111}$$

If several of the α_{ik} are different from zero the transitions allowed by any of the corresponding selection rules may occur.

In applying the rules one has to make sure that the α_{ik} in Table 55 are referred to the above system of axes ($a \rightarrow x$, $b \rightarrow y$, $c \rightarrow z$). Otherwise appropriate changes have to be made. For example, for H_2O we have to put the z axis perpendicular to the plane of the molecule and the y axis in the two-fold axis, unlike the choice in Table 55. If these changes of axes are also made in Table 13, it follows, according to the methods outlined in Chapter III, section 1, that the antisymmetric vibration ν_3 of species B_1 is connected with the polarizability component α_{xy} (rather than with α_{xz} as in Table 55). Consequently the selection rules (IV, 109) would hold for this Raman band. For the totally symmetric Raman bands of the same molecule only the transitions (IV, 108) can occur.

According to Table 55, the transitions (IV, 108) can occur for the totally symmetric Raman bands of any asymmetric top molecule; but for the lower symmetries $\boldsymbol{C_s}$, $\boldsymbol{C_2}$, $\boldsymbol{C_{2h}}$ in addition the transitions (IV, 109) can occur; and for $\boldsymbol{C_i}$ all transitions compatible with (IV, 107) can occur for totally symmetric Raman bands.

It is clear that also for Raman transitions only rotational levels of the *same over-all species* can combine with one another. As to the property *positive* and *negative with respect to inversion* we have the same rule as for the Raman spectra of linear and symmetric top molecules:

$$+ \leftrightarrow +, \qquad - \leftrightarrow -. \tag{IV, 112}$$

But as in the case of the infrared spectrum, the last two rules do not introduce any further restrictions of the possible transitions unless forbidden vibrational transitions are considered or unless the inversion doubling is resolved.

Unresolved Raman bands. No rotation-vibration Raman band of an asymmetric top molecule has as yet been resolved. But a few remarks may be made about the structure of unresolved bands on the basis of the above selection rules.

Since for *totally symmetric Raman bands* rotational levels of the same type (+ +, + −, − +, − −) may combine with one another, and since at the same time $\Delta J = 0$ is possible, and since finally the rotational constants in the upper and lower state are very nearly the same, it is obvious that *a large number of lines of the Q branch will coincide near the band origin.* This sharp strong central "line" will in general be the only feature that is observed. It should be realized, however, that the strong central "line" does not represent the whole Q-branch of the band, as can be seen by consideration of Fig. 149.

For *non-totally symmetric Raman bands* the transitions (IV, 108) do not occur. Only rotational levels of different + − symmetry combine with one another [(IV, 109) or (IV, 110) or (IV, 111)]. Therefore the lines of the Q branch do not, in general, coincide at the origin. Whether they lie fairly close together and form a central maximum depends on the asymmetry of the molecule, and on whether $[\alpha_{xy}]^{nm}$, $[\alpha_{xz}]^{nm}$, or $[\alpha_{yz}]^{nm}$ is different from zero. As can be seen by comparison of the selection rules (IV, 109–111) with (IV, 97–99), the possible rotational transitions for the three cases $[\alpha_{xy}]^{nm} \neq 0$, or $[\alpha_{xz}]^{nm} \neq 0$, or $[\alpha_{yz}]^{nm} \neq 0$ are the same as for type C, B, and A infrared bands respectively (see Figs. 160, 154, and 149), except for the additional transitions with $\Delta J = \pm 2$. Since the latter give rise to lines which are in general at a greater distance from the origin, the same considerations as for unresolved infrared bands can be applied. In particular, therefore, Raman bands with $[\alpha_{xz}]^{nm} \neq 0$ will in general have a central minimum. But also the other non-totally symmetric Raman bands will in general not have such a sharp central maximum of outstanding intensity as do the totally symmetric Raman bands, but will be more or less broad. Thus we come to the same conclusion as for symmetric top or nearly symmetric top molecules:

If a Raman line of an asymmetric top molecule is definitely broad (broader than other lines), *we can be certain that it corresponds to a non-totally symmetric vibration.* However, if the line is sharp it does not necessarily mean that it corresponds to a totally symmetric Raman line, although it makes this probable.

We see, therefore, that the observation of the width of Raman lines supplements in a very significant way the observation of their state of polarization, since it allows of a definite decision in just that case in which the observation of the state of polarization does not give an unambiguous answer: When a degree of depolarization of $\frac{6}{7}$ is observed for a Raman line it is probable, but not certain, that it corresponds to a non-totally symmetric vibration. Observation of a great width of the Raman line would make this certain. On the other hand, if a Raman line is quite sharp (even under fairly high dispersion) it is not certain, although probable, that the line corresponds to a totally symmetric vibration. But if a degree of depolarization $< \frac{6}{7}$ is observed it is certain to be a totally symmetric Raman line.

5. Molecules with Free or Hindered Internal Rotation

When free rotation or only slightly hindered rotation of one part of a molecule against the other is possible, further complications arise in the structure of rotation-vibration bands.

(a) Energy levels

Free rotation. The possibility of torsional oscillations exists, as we have seen previously (see Chapter II, section 5d), in molecules like C_2H_4, C_2H_6, $CH_3—C{\equiv}C—CH_3$,

CH_3OH and similar ones. It is clear that as long as the frequency of these torsional oscillations is large, or in other words as long as the potential maxima separating the different positions of equilibrium are high, the rotational energy levels in each vibrational level that come into play at ordinary temperatures are entirely similar to those of molecules without such torsional oscillations, and need not be separately discussed. However, at least for some of the above-named molecules, the possibility that a free or nearly free internal rotation occurs has to be considered.

For the limiting case of entirely free rotation (hindering potential = 0), Nielsen (661) first gave the energy formula [see also Koehler and Dennison (517)], assuming the molecule to be a symmetric top (moment of inertia about the top axis = I_A), and assuming that two of its parts (of moments of inertia $I_A^{(1)}$ and $I_A^{(2)}$) can rotate relative to each other about the top axis. He found that *the term*

$$F_t(k_1, k) = \frac{A_1A_2}{A}\left(k_1 - k\frac{A}{A_1}\right)^2 \qquad \text{(IV, 113)}$$

has to be added to the ordinary rotational energy $F(J, K)$ *of the symmetric top* [see equation (IV, 41)]. Here

$$A_1 = \frac{h}{8\pi^2 c I_A^{(1)}}, \quad A_2 = \frac{h}{8\pi^2 c I_A^{(2)}}, \quad A = \frac{h}{8\pi^2 c I_A}. \qquad \text{(IV, 114)}$$

$k(= \pm K)$, as previously, is the component of the total angular momentum $\boldsymbol{J}$ about the top axis, and k_1 is the quantum number of the angular momentum of part 1 (moment of inertia $I_A^{(1)}$) of the molecule, which can assume the values

$$k_1 = 0, \pm 1, \pm 2, \cdots. \qquad \text{(IV, 115)}$$

Formula (IV, 113) may also be written

$$F_t(k_1, k) = A_1k_1^2 + A_2(k - k_1)^2 - Ak^2, \qquad \text{(IV, 116)}$$

where $k - k_1 = k_2$ is the quantum number of the angular momentum of the part 2 of the molecule about the top axis. The expression (IV, 116), when added to (IV, 41), gives for the *total rotational energy*

$$F(J, K, k_1, k_2) = BJ(J + 1) - BK^2 + A_1k_1^2 + A_2k_2^2. \qquad \text{(IV, 117)}$$

In other words, for free rotation the term AK^2 in the ordinary symmetric top formula is replaced by $A_1k_1^2 + A_2k_2^2$.

In Fig. 164a we give for a number of K values and for $J = K$ the energy levels resulting from (IV, 113) up to $|k_1| = 5$. Actually, of course, these series of levels occur for every J value (see Fig. 8). The figure is drawn approximately to scale for the case of CH_3OH, assuming this molecule to be a symmetric top with the axis of torsion coincident with the top axis. Assuming further that part 1 of the molecule is the OH group, part 2 the CH_3 group, and that the vibrational state is totally symmetric, the *species of the rotational levels* are given [see Koehler and Dennison (517)]. They are determined by $k_2 = k - k_1$, since k_2 rather than k_1 corresponds to the rotation of the CH_3 group. If $|k_2|$ is a multiple of 3 there are two coinciding A levels (except for $k_2 = 0$, $K = 0$ when there is only one), otherwise the species is E.

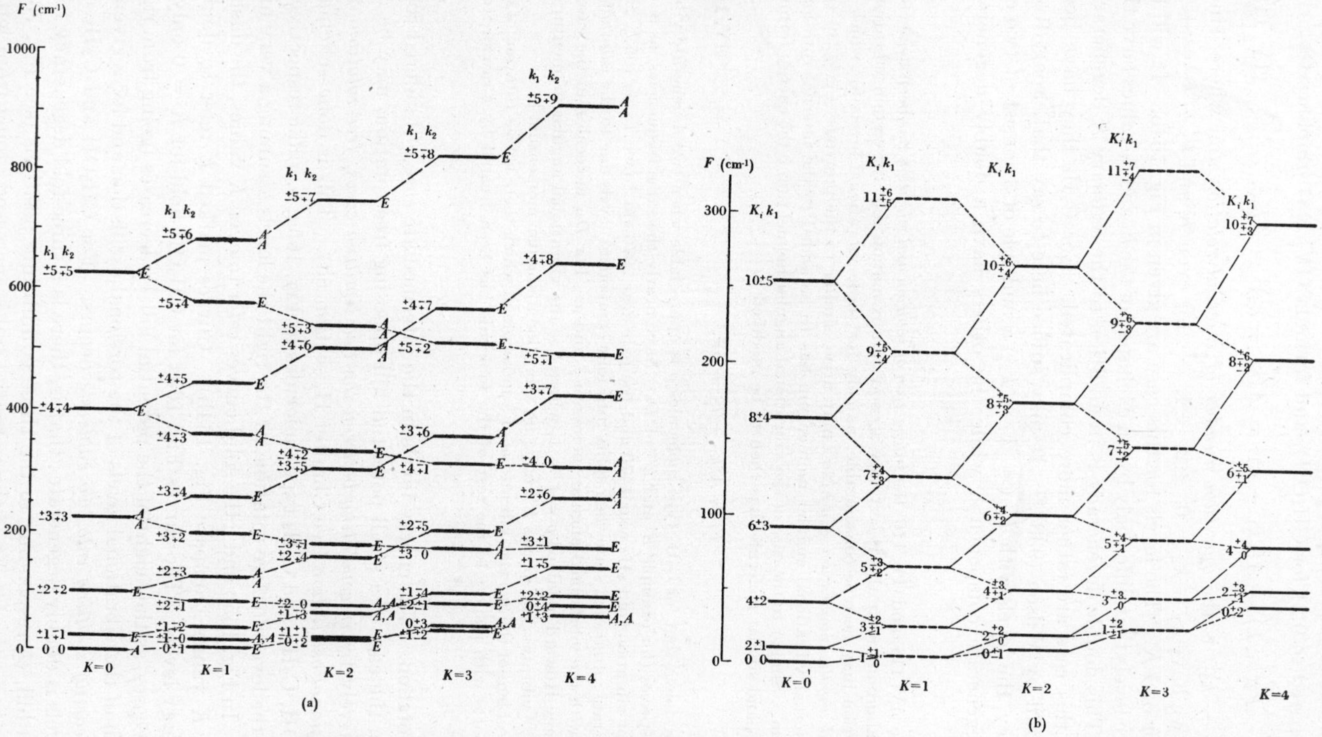

FIG. 164. **Energy levels of a molecule with free internal rotation for a number of K values and $J = K$ in a totally symmetric vibrational state: (a) molecule with two unequal parts (CH_3OH), (b) molecule with two equal parts (C_2H_6).**—The levels are drawn approximately to scale for CH_3OH and C_2H_6 assuming that there be free rotation (which actually is not present, see text). The oblique dashed lines connect levels of the same k_1 values. The energy is the total energy for $J = K$ not just the contribution of internal rotation. For other J values an appropriate shift has to be applied. In (b) the k_2 values are the same as the k_1 values. The former are therefore not given, but instead the K_i values. The levels with odd K in (b) are indicated by heavy broken lines for the sake of a comparison with Fig. 165.

For *molecules with two equal parts*, such as C_2H_6, $CH_3—C{\equiv}C—CH_3$, C_2H_4, we have $A/A_1 = \frac{1}{2}$ (compared to 0.21 for CH_3OH), and formula (IV, 113) simplifies to

$$F_t(k_1, k) = A(2k_1 - k)^2 = A(k_1 - k_2)^2 = AK_i^2, \qquad \text{(IV, 118)}$$

where $K_i(= |k_1 - k_2|)$ is the *quantum number of internal rotation* and where, since $K = |k_1 + k_2|$, we have $k_1 - k_2 = 0, \pm 2, \pm 4, \cdots$ for even K and $k_1 - k_2 = \pm 1, \pm 3, \pm 5, \cdots$ for odd K. The levels for this case are given in Fig. 164b. It will be noticed that here levels that differ only by an exchange of the k_1 and k_2 values have the same energy. This double degeneracy is in addition to the ordinary K degeneracy. The species in this case are rather more complicated. For C_2H_6 they have been discussed in detail by Wilson (938).[62] It may suffice here to say that, apart from the K degeneracy, the levels with K_i $(= |k_1 - k_2|)$ a multiple of 3 consist of two coinciding non-degenerate components, while the others have a doubly degenerate species.

In the formulae (IV, 113) and (IV, 116) the *interaction of rotation and vibration* has been neglected. For the vibrational ground state this is certainly a good approximation; but for certain vibrational levels this interaction may become considerable, namely for those vibrational levels for which an *internal vibrational angular momentum* $\pm\rho(h/2\pi)$ may arise, similar to the previous $\pm\zeta(h/2\pi)$ (see p. 402). This internal vibrational angular momentum has to be added to the internal rotational angular momentum. Similar to the case of the vibrational angular momentum ζ the energy formula (IV, 118) for C_2H_6 and similar molecules has then to be modified to

$$F_t = A(k_1 - k_2 \mp \rho)^2. \qquad \text{(IV, 119)}$$

Howard (461) has shown that for C_2H_6 type molecules ρ is appreciable when two degenerate vibrations of different species (for example E' and E'' of $\boldsymbol{D}_{3h}$) have nearly the same frequencies, as is the case for the pair of vibrations of C_2H_6 near 1470 and the pair near 2970 cm^{-1} (see Table 105). The four-foldly degenerate vibrational state then splits into four component levels due to the interaction of the rotational with the vibrational angular momenta ζ and ρ. For $\boldsymbol{D}_{3h}$ molecules, in the case of complete degeneracy Howard obtained $\rho = \zeta - 1$, where ζ is the vibrational angular momentum of the vibration $\nu(e')$ [vibrations of species E'' have no vibrational angular momentum].

The general problem of the normal vibrations of a molecule in which there is free rotation of one part or several parts with respect to the rest of the molecule has been treated by Crawford and Wilson (247).

Hindered rotation. The energy levels in the intermediate case of hindered rotation, that is, in the case of a small potential hill opposing free rotation, may be obtained qualitatively by *interpolating between the two limiting cases: free rotation* (see above) *and torsional oscillation* (see Chapter II, section 5d). This is done schematically for CH_3OH, C_2H_6, and C_2H_4 type molecules in Fig. 165a, b, and c respectively. At the left are the levels for free rotation, at the right the levels assuming a very high potential hill. In Fig. 165b and c the solid curves refer to even K values, the dashed curves to odd K values (compare Fig. 164b). Curves marked E refer to doubly degenerate energy levels. The curves Fig. 165a for CH_3OH hold for $K = 0$ only.

In all three cases, as the height of the potential hill V_0 decreases (going from right to left) for each of the vibrational levels of the torsional oscillation and for a given K value, *an increasing splitting into two sublevels* occurs. For CH_3OH and C_2H_6 one of these sublevels is doubly degenerate; that is, there is a three-fold degeneracy for a large potential hill, corresponding to the three equilibrium positions (see Fig. 73b). In all three cases, as the height of the potential hill is increased (going from left to

[62] It may be noted that Wilson's m is our $k_1 - k_2$, while Koehler and Dennison's m is our k_1.

right) the rotational levels that are not genuinely degenerate (E) split into two sublevels of different energy.

Fig. 165b and c show that, for C_2H_6 and C_2H_4 type molecules for a given vibrational level and a given height of the potential hill, the splitting due to the possibility of torsion is the same for all even K values but is different from the splitting for odd K values. For CH_3OH the dependence on K is more complicated and this is why

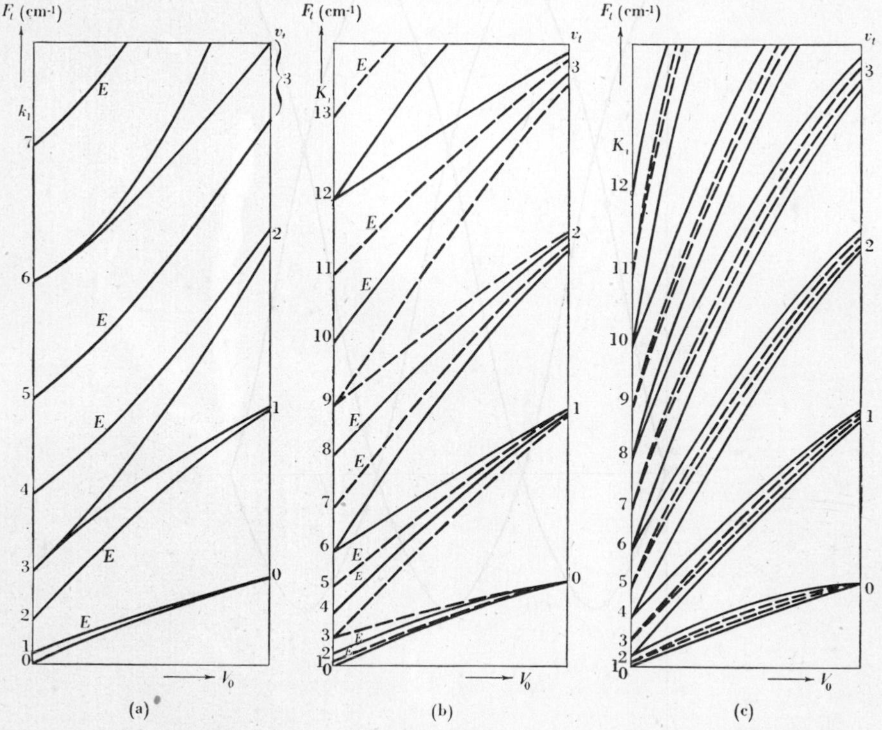

FIG. 165. Correlation between free rotation and torsional oscillation: (a) for CH_3OH, (b) for C_2H_6, (c) for C_2H_4 (qualitative).—There is a linear energy scale for each ordinate axis. Unlike Fig. 164 in this figure only the part of the energy due to internal rotation is plotted. The potential barrier V_0 is increasing from left to right. (a) Holds for $K = 0$ only, in (b) and (c) the solid curves refer to even, the broken curves to odd K values. It may be noted that for $V_0 = 0$ there is no zero-point energy whereas for large K_0 there is one half quantum of the torsional oscillation.

Fig. 165a gives only the curves for $K = 0$. Koehler and Dennison (517) have shown that here the splitting as a function of K still varies periodically but not with an integral period since A/A_1 is not simply $\frac{1}{2}$. This is shown for the first two levels of the torsional oscillation in Fig. 166. It should be noted that the degeneracy of two of the three levels for $K = 0$ does not remain for $K \neq 0$. The magnitude of the splitting in all cases, of course, increases rapidly with increasing quantum number v_t of the torsional oscillation. But on account of the zero-point vibration it exists even for $v_t = 0$ (see Fig. 166), that is, for any vibrational state, even if the torsional oscillation is not excited. However, it will be noticeable in this case only when the potential hill is very low.

Less symmetrical and more general systems than those discussed above have been treated theoretically by Crawford (236), Price (708), and Pitzer and Gwinn (698).

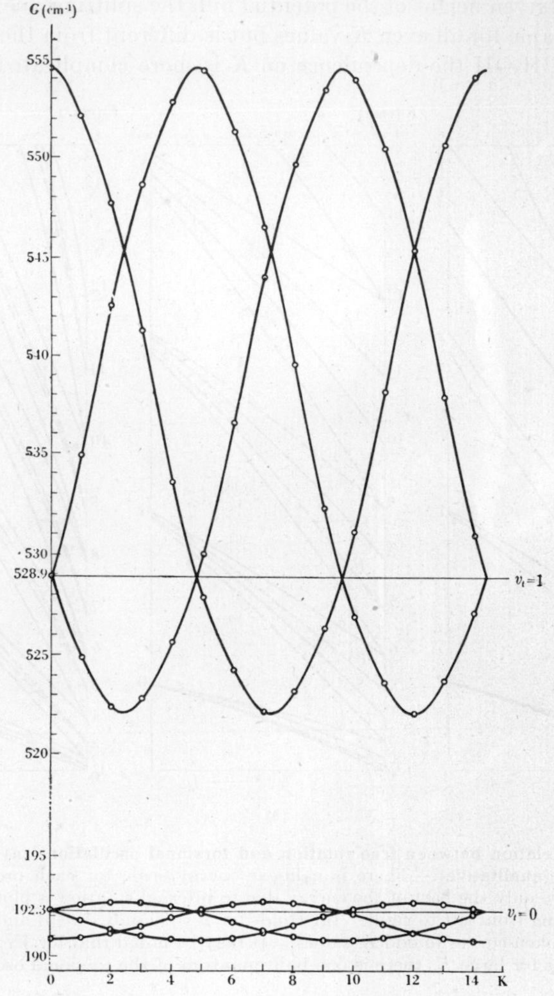

FIG. 166. **Variation with K of splitting due to possibility of passage through barrier for torsional oscillation in CH_3OH** [after Koehler and Dennison (517)].—The energy levels for each K value are indicated by circles. For both vibrational levels shown ($v_t = 0$ and $v_t = 1$) the splitting is drawn to the same scale (unlike Koehler and Dennison's figure) in order to emphasize the rapid increase with v_t.

(*b*) *Infrared spectrum*

Symmetrical molecules. The torsional oscillation in a symmetrical molecule such as C_2H_6 or C_2H_4 is infrared inactive. It is obvious that this will also hold for the free internal rotation, that is, for the limiting case when there is no potential barrier, since no oscillating dipole moment is connected with this motion. In other words, there is *no pure rotation spectrum corresponding to the free internal rotation*, just as there is no ordinary pure rotation spectrum in these molecules.

For the *rotation-vibration spectrum* we have to add to the previous selection rules for symmetric top molecules the selection rule for the quantum number $K_i = |k_1 - k_2|$ of internal rotation. Nielsen (661) has shown that

$$\Delta K_i = 0 \quad \text{for} \quad \Delta K = 0 \qquad \text{and} \qquad \Delta K_i = \pm 1 \quad \text{for} \quad \Delta K = \pm 1; \qquad \text{(IV, 120)}$$

that is K_i *does not change for* || *bands, whereas it changes by* ± 1 *for* $\perp$ *bands.*[63] Consequently, each sub-band of a || band of a symmetric top molecule with free internal rotation will consist of a number of sub-sub-bands corresponding to the different K_i values populated in the lower state. But since $\Delta K_i = 0$ all these sub-sub-bands will exactly coincide as long as the interaction of vibration and internal rotation is neglected. Even if the latter is taken into account they will almost coincide, just as do the sub-bands (see Fig. 122). Thus, except for extremely high resolution or for very large coupling of rotation and vibration, *a* || *band of a symmetric top molecule with free internal rotation will have the same structure as one without free internal rotation.*

In the case of a $\perp$ band each sub-band will also consist of a number of sub-sub-bands, two for each K_i value of the lower state (since $\Delta K_i = \pm 1$). Since the contribution of the internal rotation to the energy for molecules like C_2H_6, according to (IV, 118), is AK_i^2, the structure of a sub-band (with given K and ΔK) is entirely similar to that of a full perpendicular band without free rotation (Fig. 128), except that the separation of the line-like Q branches is $2A$ instead of $2(A - B)$. Actually, as we have seen previously (p. 429f.), the spacing of the sub-bands is $2A(1 - \zeta_i) - 2B$, on account of the interaction of rotational and vibrational angular momentum about the top axis. Also, according to Howard (see above), on account of the interaction of *internal* rotational and vibrational angular momentum (if the upper state of species E' is accidentally degenerate with one of species E'', as is frequently the case), the separation of the sub-sub-bands is $2A(1 - \zeta_i)$. Thus *in a* $\perp$ *band of a symmetric top molecule with free internal rotation each of the line-like Q branches of Fig. 128 will be split into a number of nearly equidistant "lines" of spacing* $2B$ (neglecting the dependence of A and B on v). No such structure has as yet been found.

The fact that the observed $\perp$ infrared bands of C_2H_6 do not exhibit any evidence of such a secondary structure shows, as was first pointed out by Howard (461), that there is *no free internal rotation in* C_2H_6. Howard (461) also calculated the expected structure for slightly hindered rotation, and concluded that if the potential hill preventing free rotation were lower than 700 cm^{-1} the fine structure of the $\perp$ bands should be different from what is actually observed. A molecule for which such a double fine structure due to free (or nearly free) internal rotation should be observable is $CH_3—C{\equiv}C—CH_3$ (see p. 356). But the infrared spectrum of this molecule has not as yet been investigated with sufficiently high dispersion.

Slightly asymmetric molecules, CH_3OH. In a molecule of the type of CH_3OH the torsional oscillation is infrared active and therefore also the *internal rotation*, if it is free, *is infrared active.* Since in CH_3OH only the rotation of the OH group

[63] It may be remembered that K_i is even for even K and odd for odd K; for this reason alone, $\Delta K_i = 0$ would be impossible for $\Delta K = \pm 1$, and $\Delta K_i = \pm 1$ impossible for $\Delta K = 0$.

about the top axis gives rise to a change of dipole moment, the selection rules for the pure internal rotation spectrum are

$$\Delta J = 0, \pm 1, \quad \Delta K = \pm 1, \quad \Delta K_1 = \pm 1, \quad \Delta K_2 = 0, \quad \text{(IV, 121)}$$

where (as previously) the subscripts 1 and 2 refer to the OH and CH_3 groups respectively, and where $K_1 = |k_1|$ and $K_2 = |k_2|$.

With these selection rules it is seen immediately from the energy formula (IV, 117) that the Q "lines" ($\Delta J = 0$) of the *free internal rotation spectrum* form the double series

$$\nu = A_1 - B \mp 2BK \pm 2A_1K_1, \quad \text{(IV, 122)}$$

where the upper signs hold for positive ΔK and ΔK_1, the lower signs for negative ΔK and ΔK_1. Since $A_1 \gg B$, formula (IV, 122) represents a series of bands of spacing $2A_1$ each of which consists of sub-bands whose zero lines have a spacing $2B$. Such a series of bands corresponding to large K_1 values has indeed been found by Borden and Barker (169) in the infrared spectrum of CH_3OH, in the region 600–860 cm^{-1}. The spacing is about 40 cm^{-1}, giving $A_1 = 20$ cm^{-1} and an OH distance of about 0.92 Å. The CH_3OH spectrum becomes much more complicated at smaller frequencies, indicating that for smaller K_1 values there is no longer approximately free rotation; and the spectrum seems to end in a strong band at 270 cm^{-1} [see Lawson and Randall (560)] very probably corresponding to the 1—0 transition of the torsional oscillation (see also Fig. 165). Koehler and Dennison (517), on the basis of a more detailed comparison of the observed spectrum and the theoretical spectrum assuming hindered rotation, have derived a *potential barrier of about* 470 ± 40 cm^{-1} *for the internal rotation in* CH_3OH.

For the *rotation-vibration spectrum* we have, in the case of || *bands* ($\Delta K = 0$), the selection rule

$$\Delta K_1 = 0, \quad \Delta K_2 = 0, \quad \text{(IV, 123)}$$

and for ⊥ *bands* ($\Delta K = \pm 1$) [see Borden and Barker (169)],

$$\Delta K_1 = \pm 1, \quad \Delta K_2 = 0 \quad \text{or} \quad \Delta K_1 = 0, \quad \Delta K_2 = \pm 1 \quad \text{(IV, 124)}$$

depending on whether the dipole moment of the vibrational transition is in part 1 or part 2 of the molecule; that is, for CH_3OH, in the OH or CH_3 group respectively.

From (IV, 123) in connection with the energy formula it is immediately seen that the *internal rotation does not influence the structure of a* || *band,* just as for molecules of the C_2H_6 type, as long as the interaction of vibration and internal rotation is not too strong and the dispersion not too high. Thus Borden and Barker (169) found the usual structure for the CH_3OH band ν_4^{CO} at 1033.9 cm^{-1}. The resulting B_0 value has been given in Table 132, p. 437.

For the ⊥ *bands* it follows from the energy formula (IV, 117) and the selection rules (IV, 124) that, similar to the case of molecules of the C_2H_6 type, we have a *double rotational structure.* Each sub-band with a given K and $\Delta K(= \pm 1)$ (see Fig. 128) consists of a number of sub-sub-bands corresponding to the different K_1 values and $\Delta K_1 = \pm 1$ if the oscillating dipole moment is in the OH group, or to the different K_2 values and $\Delta K_2 = \pm 1$ if it is in the CH_3 group. As is easily seen from the energy formula (IV, 117), the spacing of the sub-sub-bands is $2A_1$ or $2A_2$ respectively, while the spacing of the sub-bands is $2B$ (the same as the spacing of the

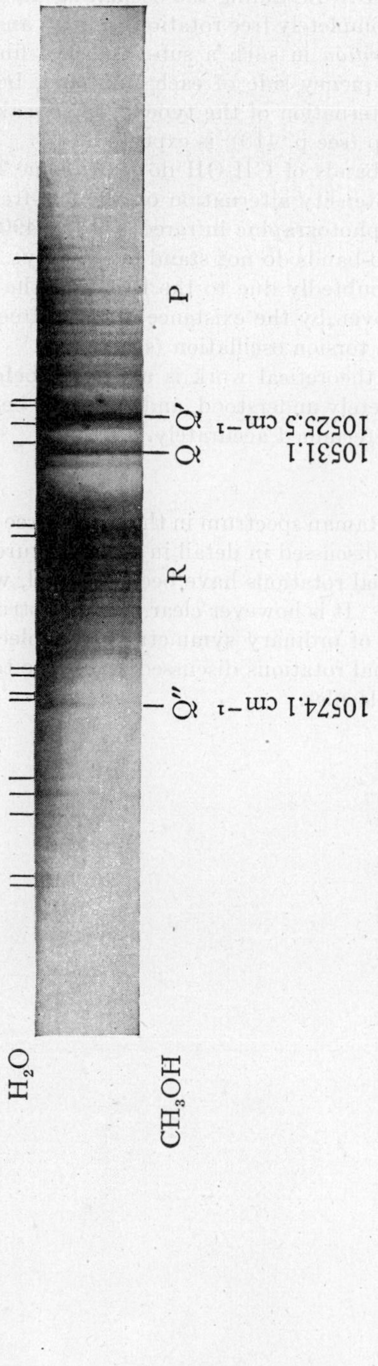

FIG. 167. **Photographic infrared band λ9490 Å of CH_3OH.** The length of the absorbing path was 1100 cm at a pressure of about 100 mm. The H_2O lines occurring in the same spectral region are indicated at the top. All other lines are due to CH_3OH. In addition to the main *Q* branch there are two further "lines" that are probably *Q* branches. They are marked *Q′* and *Q″* according to Badger and Bauer (66).

lines in the P and R branches). Since A_1 and A_2 are very much larger than B, we may also say that we have a number of sub-bands of spacing $2A_1$ or $2A_2$, each of which consists of lines with spacing $2B$. Assuming the oscillating dipole moment to be in the OH group, and assuming completely free rotation, Borden and Barker (169) have calculated the *intensity distribution* in such a sub-band and find it to be strongly *asymmetrical*. On the low-frequency side of each sub-band, but not on the high-frequency side, an intensity alternation of the type strong, weak, weak, strong, $\cdots$ characteristic of the CH_3 group (see p. 410), is expected.

The actually observed $\perp$ bands of CH_3OH do show some but not all of these features. For example, the intensity alternation on the low-frequency side of a $\perp$ band is clearly shown by the photographic infrared band $\lambda 9490$ reproduced in Fig. 167; however, the different sub-bands do not stand out clearly. This lack of agreement with expectation is undoubtedly due to the fact that the internal rotation is not entirely free, as is also proven by the existence of a low-frequency fundamental vibration corresponding to the torsion oscillation (see above).

Further experimental and theoretical work is necessary before the structure of the CH_3OH spectrum is completely understood, and therefore before all internuclear distances and angles can be determined accurately.

(c) *Raman spectrum*

The selection rules for the Raman spectrum in the case of free or hindered internal rotation have not as yet been discussed in detail in the literature. Since no Raman bands of molecules with internal rotations have been resolved, we shall not attempt a discussion of their structure. It is however clear that this structure is in the same relation to the Raman bands of ordinary symmetric top molecules as the infrared bands of molecules with internal rotations discussed above are to the infrared bands of ordinary symmetric top molecules.

CHAPTER V

APPLICATIONS

In addition to the immediate application of the study of infrared and Raman spectra of polyatomic molecules to the determination of the structure of these molecules, there are a number of other important applications. Of these we shall here discuss only the two which appear to be most important: the calculation of thermodynamic quantities and certain investigations concerning the nature of the liquid and solid state. Infrared and Raman spectra have been used also for chemical analysis, including the detection of new compounds in certain mixtures (which are not easily amenable to strictly chemical detection) and the determination of chemical equilibria, and Raman spectra have been applied to the study of electrolytic dissociation in solutions. But these and certain other applications will not be taken up here [see, for example, Kohlrausch (13) (14) and Hibben (10)].

1. Calculation of Thermodynamic Quantities

On the basis of the molecular data obtained from the spectra, as was first suggested by Urey (881) and Tolman and Badger (869), it is possible to predict with great precision the values of thermodynamic quantities, such as the heat capacity of the particular gases. This possibility is of great practical importance, particularly since the direct experimental measurement of these quantities is usually difficult and tedious and sometimes impossible. Frequently the values calculated from the spectroscopic data are more accurate than those determined by direct thermal measurements.

The partition function (state sum). According to the Maxwell-Boltzmann distribution law, in thermal equilibrium the number of atoms or molecules N_n in a state of total energy ϵ_n and of total statistical weight (degeneracy) g_n is proportional to $g_n e^{-\epsilon_n/(kT)}$, where k is Boltzmann's constant and T is the absolute temperature in degrees Kelvin. The total number of atoms or molecules, N, in a given volume is therefore proportional to

$$Q = \sum g_n e^{-\epsilon_n/(kT)}, \qquad \text{(V, 1)}$$

with the same proportionality factor. The quantity Q is the *partition function* of the gas. *All thermodynamic quantities can be expressed in terms of it.* Therefore we consider first its calculation from spectroscopic data.

Since for a perfect gas the translational and the internal energy, ϵ_{tr} and ϵ_{int}, are entirely independent of each other, the total energy ϵ_n of an atom or molecule can always be written as the sum

$$\epsilon_n = \epsilon_{\text{tr}} + \epsilon_{\text{int}} \qquad \text{(V, 2)}$$

and at the same time the total statistical weight can be written

$$g_n = g_{\text{tr}} \cdot g_{\text{int}}. \qquad \text{(V, 3)}$$

Therefore the partition function Q can be separated into a product

$$Q = Q_{\text{tr}} \cdot Q_{\text{int}} \qquad \text{(V, 4)}$$

of the *translational partition function*

$$Q_{\text{tr}} = \sum g_{\text{tr}} e^{-\epsilon_{\text{tr}}/(kT)} \qquad \text{(V, 5)}$$

and the *internal partition function*

$$Q_{\text{int}} = \sum g_{\text{int}} e^{-\epsilon_{\text{int}}/(kT)}. \qquad \text{(V, 6)}$$

It is shown in standard texts on statistical mechanics [for example (16)] that the translational partition function is given by

$$Q_{\text{tr}} = V\left(\frac{2\pi mkT}{h^2}\right)^{\frac{3}{2}} = 1.8793_6 \times 10^{20} V M^{\frac{3}{2}} T^{\frac{3}{2}}, \qquad \text{(V, 7)}$$

where V is the volume considered and m the absolute mass of the atom or molecule, M the chemical atomic or molecular weight.

The internal partition function is frequently simply called *the partition function* or also the *state sum.* It can be calculated if the g_{int} and ϵ_{int} have been determined from the spectrum.

The internal energy ϵ_{int} is the sum of three contributions, the electronic, vibrational, and rotational energy, and similarly the internal statistical weight is the product of three corresponding factors.

However, for practically all polyatomic molecules the Boltzmann factors of excited electronic states are entirely negligible compared to those of the ground state. Only for the very few polyatomic molecules with a multiplet ground state (NO_2, ClO_2, and various free radicals) does the electronic contribution to the energy have to be considered. Disregarding such cases, we can write for the *internal energy*

$$\epsilon_{\text{int}} = [G_0(v_1, v_2, \cdots) + F_v(J, \cdots)]hc, \qquad \text{(V, 8)}$$

and for the *statistical weight*

$$g_{\text{int}} = g_v \cdot g_r, \qquad \text{(V, 9)}$$

where g_v is the weight of the vibrational level (without rotation) and g_r that of the rotational sublevel (without vibration). In (V, 8) we have used $G_0(v_1, v_2, \cdots)$ rather than $G(v_1, v_2, \cdots)$, in conformity with the custom in statistical mechanics of *referring all internal energies to the lowest state of the molecule considered.* Substituting in (V, 6), we obtain

$$Q_{\text{int}} = \sum_v \left(g_v e^{-G_0(v_1, v_2, \cdots)hc/(kT)} \sum_r g_r e^{-F_v(J, \cdots)hc/(kT)}\right), \qquad \text{(V, 10)}$$

where for each vibrational level we have to sum over all rotational sublevels. For convenience, we may also write

$$Q_{\text{int}} = \sum Q^v, \qquad Q^v = g_v e^{-G_0(v_1, v_2, v_3 \cdots)hc/(kT)} \sum_r g_r e^{-F_v(J, \cdots)hc/(kT)}. \qquad \text{(V, 11)}$$

Q^v is the contribution of one vibrational level with all its rotational sublevels to the state sum. A shift of the zero of energy by $\Delta\epsilon$ would introduce a factor $e^{-\Delta\epsilon/(kT)}$ in (V, 10) in front of the first summation sign.

It is obviously very tedious to evaluate the partition function Q_{int} by direct summation according to (V, 10). However, *if we neglect the interaction between vibration and rotation*, that is, the dependence of F on the vibrational quantum numbers, a considerable simplification is obtained, since then the sum $\sum_r g_r e^{-F_v hc/(kT)}$ is the same for all vibrational levels, and *the partition function can be written as the product of two factors, the vibrational and the rotational partition function*,

$$Q_{\text{int}} = Q_v \cdot Q_r, \tag{V, 12}$$

where

$$Q_v = \sum_v g_v e^{-G_0(v_1, v_2 \cdots) hc/(kT)}, \qquad Q_r = \sum_r g_r e^{-F_v(J, \cdots) hc/(kT)}. \tag{V, 13}$$

Now each factor can be evaluated separately. (Note the difference between Q_v and Q^v.)

The vibrational partition function. The vibrational contribution to the partition function is most easily evaluated when a further simplification is introduced: the *neglect of the anharmonicities*. These can, of course, be neglected safely only for the lower vibrational levels, that is, for lower temperatures, for which the neglect of the interaction of vibration and rotation is also permissible. In this *harmonic-oscillator approximation* we have, for the vibrational energy,

$$G_0(v_1, v_2, \cdots) = \sum \omega_i v_i, \tag{V, 14}$$

rather than $G_0(v_1, v_2, \cdots)$ from (II, 284). Since the energy is thus a sum of independent terms the contribution to the partition function is a product of terms each one of which is due to one vibration only [similar to the products (V, 4) and (V, 12)]. We obtain from (V, 13) with (V, 14), considering first only non-degenerate vibrations,

$$Q_v^{(\text{harm})} = \sum_{v_1 v_2 \cdots} e^{-(\omega_1 v_1 + \omega_2 v_2 + \cdots) hc/(kT)} = \left(\sum_{v_1} e^{-\omega_1 v_1 hc/(kT)}\right)\left(\sum_{v_2} e^{-\omega_2 v_2 hc/(kT)}\right) \cdots. \tag{V, 15}$$

Each of the latter sums represents a geometric progression whose sum for $v_i = 0, 1, 2, \cdots \infty$ is given by the elementary formula

$$\sum_{v_i} e^{-\omega_i v_i hc/(kT)} = \frac{1}{1 - e^{-\omega_i hc/(kT)}}.$$

Thus we have for the vibrational partition function

$$Q_v^{(\text{harm})} = (1 - e^{-\omega_1 hc/(kT)})^{-1}(1 - e^{-\omega_2 hc/(kT)})^{-1}(1 - e^{-\omega_3 hc/(kT)})^{-1} \cdots. \tag{V, 16}$$

From the above derivation, letting two or more ω_i coincide, it is easily seen that for a degenerate vibration the appropriate factor in (V, 15) has to be repeated as many times as the degeneracy indicates, and therefore we obtain, instead of (V, 16), for the *vibrational partition function*,

$$Q_v^{(\text{harm})} = (1 - e^{-\omega_1 hc/(kT)})^{-d_1}(1 - e^{-\omega_2 hc/(kT)})^{-d_2}(1 - e^{-\omega_3 hc/(kT)})^{-d_3} \cdots, \tag{V, 17}$$

where $d_1, d_2, d_3, \cdots$ are the degrees of degeneracy of the vibrations $\omega_1, \omega_2, \omega_3, \cdots$ respectively. This includes the case of accidental degeneracy, for example, when a number of CH vibrations have very nearly the same frequency.

Thus, in this approximation, it is easy to calculate the vibrational partition function if the frequencies of the normal vibrations and their degrees of degeneracy are known. It is not necessary to know anything about the form of the oscillations or about their species, except that the degree of degeneracy must be known. One should, of course, use for the ω_i in (V, 17) the wave numbers ν_i of the fundamental bands as observed in the infrared or Raman spectrum (that is the $\Delta G_{\frac{1}{2}}$) and not the zero-order frequencies.

As examples, in Table 138 we give for three temperatures the values of the vibrational partition function of HCN and CH_4 on the basis of the harmonic-oscillator

TABLE 138. VIBRATIONAL PARTITION FUNCTIONS OF HCN AND CH_4 (HARMONIC-OSCILLATOR APPROXIMATION).

Molecule		$T = 300°$ K.	$T = 1000°$ K.	$T = 2000°$ K.
HCN	$(1 - e^{-\omega_1 hc/(kT)})^{-1}$	1.00004	1.05207	1.28611
	$(1 - e^{-\omega_2 hc/(kT)})^{-2}$	1.06918	2.43229	6.22323
	$(1 - e^{-\omega_3 hc/(kT)})^{-1}$	1.00000	1.00860	1.10177
	$Q_v^{(\text{harm})}$	1.06922	2.58095	8.81830
CH_4	$(1 - e^{-\omega_1 hc/(kT)})^{-1}$	1.00000	1.01533	1.14018
	$(1 - e^{-\omega_2 hc/(kT)})^{-2}$	1.00132	1.26542	2.25138
	$(1 - e^{-\omega_3 hc/(kT)})^{-3}$	1.00000	1.03991	1.43700
	$(1 - e^{-\omega_4 hc/(kT)})^{-3}$	1.00573	1.64420	4.41926
	$Q_v^{(\text{harm})}$	1.00706	2.19681	16.30153

approximation. The contributions of the different vibrations are also given separately. The fundamental frequencies are those given in Table 59 and Table 80 respectively.

For numerical calculations it is convenient to note that

$$\frac{\omega_i hc}{kT} = \frac{\omega_i}{0.6951T} = 1.439\,\frac{\omega_i}{T}, \qquad \text{(V, 18)}$$

where ω_i is in cm^{-1}.

If *anharmonicity* of the vibrations is not to be neglected (and its neglect appears to be the greatest source of error in all these statistical calculations), the most straightforward and, for not too high temperatures, simplest way of obtaining the partition function is the *direct summation* according to (V, 13) over all vibrational levels for which $e^{-G_0(v_1, v_2, \cdots)hc/(kT)}$ is not negligibly small. This method has the advantage that any vibrational perturbations (Fermi resonance) can easily be taken into account, and that the degree of degeneracy of the various vibrational levels can be substituted without difficulty. It must, of course, be understood that all vibrational levels, not only the observed ones, have to be used in the summation. Since for high temperatures the direct summation becomes very awkward, Gordon (388) (389) and Kassel (491) have developed expansions for three- and four-atomic molecules which take the anharmonicities into account and are somewhat easier to handle. We shall not, however, reproduce these formulae. If, as frequently happens, Fermi resonances occur for the molecule, the formulae become even more complicated than without them. In either method, the direct summation or the use of Gordon's and Kassel's formulae, it is necessary to know all the anharmonic constants x_{ik} and g_{ik} for the molecule considered. As we have seen in section 3 of Chapter III, there are only very few molecules for which these anharmonic constants are all known. Even for the molecules for which they are known, the calculation of the partition function at high temperatures involves rather uncertain extrapolations: for example, for HCN, while the levels $v_3\nu_3$ have actually been observed up to 15000 cm^{-1}, the levels $v_2\nu_2$ have been observed only up to 2800 cm^{-1}. But at high temperatures higher $v_2\nu_2$ levels, say up to 9000 cm^{-1}, would become of importance, and it is very doubtful whether they can be represented by the x_{22} value obtained from the observed $v_2\nu_2$ levels.

Conversely, these considerations emphasize the need for more detailed investigations of overtone and combination vibrations of all thermodynamically important molecules.

The rotational partition function. The formulae for the rotational partition function are of course different for linear, symmetric top, and asymmetric top molecules. For *diatomic and linear polyatomic molecules* we have, in very good approximation (rigid rotator),

$$F(J, \cdots) = BJ(J+1); \tag{V, 19}$$

and therefore

$$Q_r = \sum_J (2J+1)e^{-BJ(J+1)hc/(kT)}. \tag{V, 20}$$

In this formula the influence of the nuclear spin is neglected (see below).

If the temperature is very low it is best to form Q_r according to (V, 20) by *direct summation*, since only comparatively few terms matter. However, for ordinary temperatures, the number of rotational levels involved is usually very great. In that case it is easier to use an *asymptotic expansion* first given by Mulholland (640) [see also Kassel (491)]:

$$Q_r = \frac{kT}{hcB} + \frac{1}{3} + \frac{1}{15}\frac{hcB}{kT} + \frac{4}{315}\left(\frac{hcB}{kT}\right)^2 + \frac{1}{315}\left(\frac{hcB}{kT}\right)^3 + \cdots \tag{V, 21}$$

For small B and large T this formula goes over into the (classical) result

$$Q_r{}^{(\text{class})} = \frac{kT}{hcB} = 0.6951\,\frac{T}{B}, \tag{V, 22}$$

which is also obtained by replacing the summation in (V, 20) by an integration (see Molecular Structure I, p. 132). Even for $T/B = 5$ (that is, for example, for CO at as low a temperature as 10° K.) the first term in (V, 21) gives 98 per cent, the first two give 99 per cent, and the first three give 99.95 per cent of the correct result.

The expansion (V, 21) can of course be used with appropriate $B_{[v]}$ values for each vibrational level, and thus the interaction of vibration and rotation can be taken in account according to (V, 11). If the latter formula is applied for linear polyatomic molecules it is necessary also to take into account the fact that for Π, Δ, $\cdots$ vibrational levels the energy formula is not (V, 19) but

$$F(J, \cdots) = B[J(J+1) - l^2], \quad J \geq l \tag{V, 23}$$

(see p. 371). This introduces an additional factor for the particular vibrational levels. However, the influence on the complete internal partition function is very slight, much smaller than the influence of the neglect of the interaction of vibration and rotation [using (V, 12) instead of (V, 11)].

For *rigid symmetric top molecules* the rotational energy is given by [see formula (I, 20)]

$$F(J, \cdots) = BJ(J+1) + (A-B)K^2, \tag{V, 24}$$

and therefore the rotational partition function is [1]

$$Q_r = \sum_{J=0}^{\infty} \sum_{K=-J}^{+J} (2J+1)e^{-[BJ(J+1)+(A-B)K^2]hc/(kT)}. \tag{V, 25}$$

[1] In order to avoid confusion with the Boltzmann constant we are here using K for the previous k that is, $K = J, J-1, \cdots -J$.

For all actual cases, except for extremely low temperatures, this formula may be replaced by the expansion [see Viney (902) Kassel (486) (491)]

$$Q_r = e^{Bhc/(4kT)} \sqrt{\frac{\pi}{B^2 A}\left(\frac{kT}{hc}\right)^3} \left[1 + \frac{1}{12}\left(1 - \frac{B}{A}\right)\frac{Bhc}{kT} + \frac{7}{480}\left(1 - \frac{B}{A}\right)^2 \left(\frac{Bhc}{kT}\right)^2 + \cdots\right]. \quad \text{(V, 26)}$$

For small values of B/T this expression approaches the *classical value*

$$Q_r^{(\text{class})} = (Q_r)_{B/T \to 0} = \sqrt{\frac{\pi}{B^2 A}\left(\frac{kT}{hc}\right)^3} = 1.02718 \sqrt{\frac{T^3}{B^2 A}}, \quad \text{(V, 27)}$$

which represents a fairly good approximation even for comparatively low temperatures.

If the interaction of vibration and rotation is to be taken into account, one may again use appropriate $B_{[v]}$ and $A_{[v]}$ values in (V, 26) and then sum over the various vibrational levels according to (V, 11). While in this summation no account is taken of the Coriolis splitting of degenerate vibrational levels, since (V, 26) is based on (V, 24), Wilson (939) has shown that the influence of this splitting is negligible. For all practically important temperatures the partition function is close to the classical value (V, 27) in any case.

For *spherical top molecules* one has simply to substitute $A = B$ in the above formulae for symmetric top molecules.

Since there is no explicit formula for the rotational levels of an *asymmetric top molecule*, it is impossible to derive a rigorous asymptotic expansion for Q_r in this case. However, it may be expected [see Gordon (388)] that the formula for the symmetric top with rotational constants A and $\sqrt{BC}$ instead of A and B (if A, B, C are the rotational constants of the asymmetric top) will give a good approximation to the rotational partition function of the asymmetric top, if B and C are not too different.[2] We have then, according to (V, 26), for the asymmetric top,

$$Q_r = e^{\sqrt{BC}hc/(4kT)} \sqrt{\frac{\pi}{ABC}\left(\frac{kT}{hc}\right)^3} \left[1 + \frac{1}{12}\left(1 - \frac{\sqrt{BC}}{A}\right)\frac{\sqrt{BC}hc}{kT} + \cdots\right]. \quad \text{(V, 28)}$$

Gordon (388) has shown for a particularly unfavorable case that the equation (V, 28) approximates the state sum obtained by direct summation with an error that is only 0.1 per cent at 100° K. and much smaller at higher temperatures.

For sufficiently high temperatures (or small rotational constants) (V, 28) goes over into

$$Q_r^{(\text{class})} = \sqrt{\frac{\pi}{ABC}\left(\frac{kT}{hc}\right)^3} = 1.02718 \sqrt{\frac{T^3}{ABC}} = 0.006935 \times 10^{60} \sqrt{T^3 I_A I_B I_C}. \quad \text{(V, 29)}$$

This corresponds to the classical equipartition and is frequently a sufficiently good approximation.[3] Even in the above-mentioned unfavorable case the difference

[2] If A and B are more nearly alike than B and C one has to use C and $\sqrt{AB}$ for the rotational constants A and B of the symmetric top.

[3] It should be noted that some authors use A, B, C for the moments of inertia. Here they are, as always in this book, the rotational constants $h/(8\pi^2 c I_A)$, and so on.

between (V, 29) and the exact value is only 0.3 per cent. Equation (V, 29) rather than (V, 28) has almost always been used in calculations of thermodynamic quantities of asymmetric top molecules. As examples we give in Table 139 the values of the rotational partition functions at the temperatures 100°, 300°, and 1000° K. for a

TABLE 139. ROTATIONAL PARTITION FUNCTIONS OF HCN, CH_3Cl, CH_4, AND C_2H_4 AT THREE DIFFERENT TEMPERATURES (NEGLECTING THE IDENTITY OF THE NUCLEI AND NUCLEAR SPIN).

Molecule	$T = 100°$ K.		$T = 300°$ K.	$T = 1000°$ K.
	Classical	Exact		
HCN	47.02	47.35	141.05	470.2
CH_3Cl	928.4	930.1	4824.4	29362
CH_4	85.50	87.13	444.31	2704.1
C_2H_4	512.8	514.5	2664.5	16216

linear, a symmetric top, a spherical top, and an asymmetric top molecule. For 100° K. both the classical value $Q_r^{(class)}$ and the exact value are given, for the other temperatures only the classical value, which however is indistinguishable from the exact value.

In discussing the rotational partition function we have thus far considered the molecule as rigid. The stretching of the molecule in consequence of centrifugal forces causes a shift of the higher rotational levels and therefore a change of the rotational partition function. Since the stretching terms in the energy formula are known only for very few polyatomic molecules we shall not give the formulae for this correction to the partition function, but refer for linear molecules to Giauque and Overstreet (360), Johnston and Davis (473), Gordon (389), and Kassel (491); for non-linear molecules to Kassel (486) (491) and Wilson (936). It should be stressed that these corrections for higher temperatures may be of the same order as the higher terms in (V, 21) and (V, 26).

Up to now we have neglected entirely the *influence of the identity of nuclei and of the nuclear spin.* For low temperatures, when a direct summation is used, this influence is simply taken into account by using the proper statistical weight for each rotational level (see Chapter I and IV). For high temperatures, when the asymptotic expansions are applicable for molecules without identical nuclei, these expansions can also be applied to symmetrical molecules if an appropriate change is made as follows.

In the case of linear molecules with two or more identical nuclei (CO_2 and others), the even rotational levels (assuming a ${}^1\Sigma_g^+$ ground state) are symmetric, the odd antisymmetric. It is clear that at high temperatures the partition function for the even levels equals the partition function for the odd levels. Therefore if only one set of levels occurs, as in the case of *zero nuclear spin*, the rotational partition function is one half of the previously derived value (V, 21) or (V, 22). Similarly, in the case of symmetric top molecules of symmetry C_{3v}, only the rotational levels with K divisible by 3 are of species A, and they are the only ones that occur if the nuclear spin of the identical atoms is zero (see Chapter IV, section 2a). Therefore the partition function in this case, for high temperatures, is one-third of what it would be without identical nuclei [equation (V, 26) or (V, 27)]. Similarly, in the case of point group V_h (molecules such as C_2H_4) we have rotational levels of species A, B_1, B_2, B_3 (see p. 462), which occur equally often if the identity of the nuclei is disregarded, but of which only the A levels actually occur for zero nuclear spin of the identical atoms. Therefore the partition function, at high temperatures, is only one-quarter of the value that follows from (V, 28) or (V, 29). Finally, in the case of tetrahedral molecules (point group T_d) we have the three species of rotational levels A, E, and F. For high J values there are three times as many F sublevels as there are A sublevels, and as many E sublevels as A sublevels (see Chapter IV, section 3a). Considering the degree of degeneracy of the E and F levels, it is immediately seen that for zero nuclear spin (when only the A sublevels occur) the rotational partition function is only one-twelfth [that is, $1/(1 + 2 + 3 \times 3)$] of what it would be without considering the identity of the nuclei [equation (V, 26) or (V, 27) with $B = A$].

The number 2, 3, 4, 12, as the case may be, by which the rotational partition function has to be divided in the above cases when the spin of the identical nuclei is zero, is called the *symmetry number*. It was first introduced by Ehrenfest and is frequently designated by σ. It is characteristic for each point group, and can be shown to be equal to "*the number of indistinguishable positions into which the molecule can be turned by simple rigid rotations*" [Wilson (941)]. The reader may easily verify this for the above examples. Table 140 gives the symmetry numbers for the more important point groups.

TABLE 140. SYMMETRY NUMBERS (σ) FOR VARIOUS POINT GROUPS.

Point group	Symmetry number	Point group	Symmetry number	Point group	Symmetry number
C_1, C_i, C_s	1	D_2, D_{2d}, $D_{2h} \equiv V_h$	4	$C_{\infty v}$	1
C_2, C_{2v}, C_{2h}	2	D_3, D_{3d}, D_{3h}	6	$D_{\infty h}$	2
C_3, C_{3v}, C_{3h}	3	D_4, D_{4d}, D_{4h}	8	T, T_d	12
C_4, C_{4v}, C_{4h}	4	D_6, D_{6d}, D_{6h}	12	O_h	24
C_6, C_{6v}, C_{6h}	6	S_6	3		

If the identical nuclei have *non-zero spin* ($I \neq 0$), in general, all rotational levels occur, but with different weights. Therefore the rotational partition function is larger than for zero nuclear spin by a certain factor. The general rule for this factor is perhaps most easily understood if we consider first a simple example. In the case of a molecule of symmetry number $\sigma = 2$ with two identical nuclei of spin $I = \frac{1}{2}$ (for example H_2, C_2H_2, H_2O, H_2CO) the two modifications (ortho and para) have weights 3 and 1 respectively. Thus, at not too low temperatures the rotational partition function is $(3 + 1)$ times the partition function for zero nuclear spin, that is, $\frac{4}{2}$ times the rotational partition function obtained when the identity of nuclei and the nuclear spin are neglected. This figure may also be obtained by considering the space quantization of the nuclear spins independently of the rotational motion. Each spin $I = \frac{1}{2}$ can have two orientations in a magnetic field and therefore with two nuclei of spin $\frac{1}{2}$ the statistical weight is $2 \times 2 = 4$ times as large as for $I = 0$; that is, we have the factor $\frac{4}{2}$ for the rotational partition function. If the spin of the two identical nuclei were I, the factor would be $(2I + 1)^2/2$ since there are $2I + 1$ possible orientations for each nucleus. If an atom that does not belong to a set of identical atoms (such as O in H_2O, C or O in H_2CO, C in CO_2 and CH_4, and so on) has non-zero nuclear spin I, this also introduces a factor $2I + 1$ in the partition function, but in this case for high as well as low temperatures (both ortho and para modifications would have the same factor). It is easy to see that in the general case *the nuclear spins introduce*, at not too low temperatures, *a factor* $(2I_1 + 1)(2I_2 + 1)(2I_3 + 1)\cdots$ *into the partition function for zero spin*, where the product is formed over all nuclei of the molecule (see p. 16). For example, for $B^{11}Cl_3^{35}$ with $I(B^{11}) = \frac{5}{2}$, $I(Cl^{35}) = \frac{3}{2}$, this nuclear spin factor would be $6 \times 4 \times 4 \times 4 = 384$; for DCN with $I(D) = 1$, $I(C) = 0$, $I(N) = 1$ it would be $3 \times 1 \times 3 = 9$.

Usually the nuclear spin factor can be and *is entirely neglected in statistical calculations,* since for all molecules but H_2 and D_2 it causes a detectable effect on measurable quantities only at extremely low temperatures. For these low temperatures (for H_2 up to room temperature) it is necessary to use the direct summation with the proper statistical weights, and at the same time it is necessary to take account of the fact that the two (or more) modifications do not readily go over into one another.

The exact calculation of the vibrational and rotational partition functions, taking all refinements into account (particularly anharmonicity and centrifugal stretching), is exceedingly tedious and requires molecular data that are available only for very few molecules. Fortunately the harmonic-oscillator approximation and the (classical) rigid-rotator approximation are very satisfactory as long as the temperature is not too high. Since in most practical applications it is this *harmonic-oscillator rigid-rotator approximation* that is used, we summarize the result: *Neglecting the spin contribution, anharmonicity, and non-rigidity, the internal partition function of linear molecules is given by*

$$Q_{\text{int}} = \frac{\dfrac{kT}{\sigma hcB}}{(1 - e^{-\omega_1 hc/(kT)})^{d_1}(1 - e^{-\omega_2 hc/(kT)})^{d_2}(1 - e^{-\omega_3 hc/(kT)})^{d_3}\cdots}, \quad (V, 30)$$

where σ is 2 or 1 depending on whether the molecule has point group $D_{\infty h}$ or $C_{\infty v}$ respectively. For *non-linear molecules* the partition function, under the above conditions, is given by

$$Q_{\text{int}} = \frac{\dfrac{1}{\sigma}\sqrt{\dfrac{\pi}{ABC}\left(\dfrac{kT}{hc}\right)^3}}{(1 - e^{-\omega_1 hc/(kT)})^{d_1}(1 - e^{-\omega_2 hc/(kT)})^{d_2}(1 - e^{-\omega_3 hc/(kT)})^{d_3}\cdots}, \quad (V, 31)$$

where the symmetry number σ is given by Table 140 and where $k/(hc) = 0.6951$. For ω_i the frequencies of the fundamentals should be substituted; the d_i are their degrees of degeneracy. For the numerical factor in (V, 31), when either ABC or $I_A I_B I_C$ are used, see equation (V, 29) p. 506.

If the nuclear spin contribution is to be taken into account in the partition function, the right-hand side of (V, 30) or (V, 31) has to be multiplied simply by $(2I_1 + 1)(2I_2 + 1)(2I_3 + 1)\cdots$ unless the temperature is very low. But in that event (V, 30) or (V, 31) could not be applied in any case.

According to Wilson (941) and Hirschfelder (454), the product $I_A I_B I_C$ which occurs in (V, 31) through ABC can be evaluated for molecules for which the position of the principal axes is not obvious, by means of

$$I_A I_B I_C = \begin{vmatrix} +I_{xx} & -I_{xy} & -I_{xz} \\ -I_{xy} & +I_{yy} & -I_{yz} \\ -I_{xz} & -I_{yz} & +I_{zz} \end{vmatrix}. \quad (V, 32)$$

Here the I_{xx}, I_{xy}, $\cdots$ are the moments and products of inertia with respect to any convenient coordinate system having the center of mass as origin; that is,

$$\begin{aligned} I_{xx} &= \sum m_i(y_i^2 + z_i^2), \quad \cdots, \quad \cdots, \\ I_{xy} &= \sum m_i x_i y_i, \quad \cdots, \quad \cdots, \end{aligned} \tag{V, 33}$$

where m_i is the mass of atom i whose coordinates are x_i, y_i, z_i. If a coordinate system is chosen whose origin is not the center of mass and with respect to which the coordinates of atom i are x_i', y_i', z_i', the relations (V, 33) have to be replaced (from the theorem of parallel axes) by

$$\begin{aligned} I_{xx} &= \sum m_i(y_i'^2 + z_i'^2) - \frac{1}{M}(\sum m_i y_i')^2 - \frac{1}{M}(\sum m_i z_i')^2, \\ I_{xy} &= \sum m_i x_i' y_i' - \frac{1}{M}(\sum m_i x_i')(\sum m_i y_i'), \end{aligned} \tag{V, 34}$$

where $M = \sum m_i$.

Partition function for molecules with internal rotations. Up to now we have implicitly assumed the molecule to be semirigid; that is, that the amplitudes of the oscillations are small compared to the internuclear distances, and that any centrifugal stretchings are small. While this assumption is well fulfilled for most of the simpler molecules, there are molecules for which it is not fulfilled, namely those in which free internal rotations or slow torsional oscillations are possible.

If the potential hill preventing free internal rotation is very high, as for example in C_2H_4 and similar molecules, so that the vibrational levels of the torsional oscillation can be represented by the ordinary vibrational formula for all energies of importance for the temperatures considered, the torsional oscillation may simply be included in the previous formula for the vibrational partition function. However, *if the internal rotation is entirely free*, this "vibrational" degree of freedom has to be omitted from the vibrational partition function Q_v, and instead an appropriate term has to be added to the rotational partition function Q_r. The accurate expression for Q_r in the case of a symmetric top molecule in which just two parts can rotate with respect to each other (C_2H_6, CH_3OH, $\cdots$) is obtained if in (V, 25) the term $F_t(k_1, k)$ of (IV, 113) or (IV, 118) is added in the exponent, and a summation over k_1 or K_i, the quantum number of internal rotation, is included. In the case of *symmetric top molecules consisting of two equal parts* which carry out an internal rotation with respect to each other about the top axis and which are also symmetric tops (for example C_2H_6, if there were free rotation, $CH_3—C{\equiv}C—CH_3$, and others), one obtains for the rotational partition function at sufficiently high temperatures [see Eidinoff and Aston (301) and Kassel (488)],

$$Q_r = \frac{1}{\sigma}\frac{\pi}{A_1 B}\left(\frac{kT}{hc}\right)^2, \tag{V, 35}$$

where A_1 is the rotational constant corresponding to the moment of inertia of one-half of the molecule (the CH_3 group in the examples) about the top axis, and where B corresponds to the moment of inertia of the whole molecule about an axis perpendicular to the top axis. The symmetry number σ for a molecule with free rotation is different from that for the same molecule without free rotation. Thus, in the case of rigid C_2H_6 (point group D_{3h} or D_{3d}) the symmetry number is 6 (see Table 140), but with free rotation there are three indistinguishable positions of the two CH_3 groups with respect to each other, and for each of these we have the six positions as for rigid C_2H_6; that is, in all there are $3 \times 6 = 18$ indistinguishable positions, and thus $\sigma = 18$.

Wilson (938) has discussed in detail the partition function of molecules like C_2H_6 at very low temperatures under the assumption of free rotation when the summation has to be carried out separately for the different species of rotational levels.

Formulae for the complete rotational partition functions of a number of other non-rigid molecules—such as propane, diphenyl, toluene, isobutane, tetramethylmethane—assuming free internal rotation, have been given by Eidinoff and Aston (301) and Kassel (488) (489) (490). For *molecules with a number of symmetric tops* attached to an essentially rigid frame, according to Pitzer and Gwinn (698), in a good approximation *each top, assuming free rotation, contributes a factor*

$$Q_f = \frac{(8\pi^3 I_m kT)^{\frac{1}{2}}}{hn} = 0.27930\,\frac{(I_m 10^{40}T)^{\frac{1}{2}}}{n} \qquad \text{(V, 36)}$$

to the partition function in addition to Q_{int} from (V, 31). Here n is the number of indistinguishable positions of the attached top considered and I_m is its "reduced" moment of inertia. The latter is given by

$$I_m = I_m^0\left[1 - I_m^0\left(\frac{\lambda_{mA}^2}{I_A} + \frac{\lambda_{mB}^2}{I_B} + \frac{\lambda_{mC}^2}{I_C}\right)\right], \qquad \text{(V, 37)}$$

a formula that holds accurately for a single attached top but only approximately if there are several such tops. In this formula $I_m{}^0$ is the moment of inertia of the m'th top, λ_{mA} is the cosine of the angle between the axis of the top and the axis of the least moment of inertia I_A of the whole molecule and similarly for λ_{mB} and λ_{mC}. It is easily seen that if a molecule with two identical tops is considered, such as C_2H_6 (where one top then serves as the rigid framework), then (V, 36) and (V, 37), combined with the partition function for overall rotation (V, 31), lead to (V, 35). The symmetry number appears then as the product of n and the symmetry number σ_0 for over-all rotation (for ethane $n = 3$ and $\sigma_0 = 6$).

As has been shown mainly by a comparison of calculated and observed thermodynamical quantities (see below), the *internal rotation is in general not free but more*

TABLE 141. PARTITION FUNCTIONS FOR FREE AND HINDERED ROTATION IN ETHANE OR DIMETHYL ACETYLENE.

V_0 (cal)	$T = 100°$ K.	$T = 300°$ K.	$T = 500°$ K.	$T = 1000°$ K.
0	1.548	2.682	3.462	4.896
500	1.32_6	2.41_1	3.21_9	4.69_3
1000	1.13_4	2.03_0	2.82_9	4.35_3
2000	1.04_1	1.58_3	2.25_5	3.74_3
3000	1.01_8	1.38_0	1.91_0	3.26_2
5000		1.21_0	1.56_7	2.62_0
10000		1.08_4	1.27_8	1.91_9

or less hindered. Wilson (941), Crawford (236), Price (708), and Pitzer and Gwinn (698) have given detailed discussions of this intermediate case for one or more attached tops. The formulae for the energy levels, which were illustrated qualitatively for three simple cases by the previous Fig. 165, and likewise the formulae for the partition functions are rather complicated and will not be given here. Instead we give, in Table 141, for several values (including zero) of the height of the potential barrier (V_0) and for several temperatures, the factor contributed to the partition function by the hindered internal rotation in C_2H_6 or $CH_3—C{\equiv}C—CH_3$, assuming a cosine-like potential of the form of equation (II, 301) with $n = 3$. The values in the first row of the table (for $V_0 = 0$) are those obtained from (V, 36), which applies to free internal rotation.[3a] It is seen from the table that with increasing height of the potential barrier the partition function tends to approach the value 1, particularly at low temperatures. The reason for this tendency is, of course, that for a high barrier the contribution to the partition function approaches that of a torsional oscillation $(1 - e^{-\omega hc/(kT)})^{-1}$, which goes to 1 for large ω and for not too high temperatures. On the other hand, for barriers less than say 500 cal (175 cm^{-1}), the partition function is close to the free rotation value.

A situation somewhat similar to the above arises for molecules in which a potential barrier is separating the two equilibrium positions corresponding to inversion, as in NH_3 (see p. 221f.). The thermodynamic functions for this case have been discussed in some detail by Pitzer (693b).

Heat content and heat capacity. The total internal energy E^0 of one mole of a perfect gas (including translational as well as inner degrees of freedom) is

$$E^0 = E_0{}^0 + N_1\epsilon_1 + N_2\epsilon_2 + N_3\epsilon_3 + \cdots, \qquad \text{(V, 38)}$$

where $E_0{}^0$ is the energy at absolute zero (zero-point energy) and $N_1, N_2, \cdots$ the number of molecules having energies $\epsilon_1, \epsilon_2, \cdots$ above the lowest energy. The numbers N_n, according to the Maxwell-Boltzmann distribution law, are given by

$$N_n = N\frac{g_n e^{-\epsilon_n/(kT)}}{Q},$$

where Q is the total partition function (V, 1) and N the Avogadro number. Substituting in (V, 38), one obtains[4]

$$E^0 = E_0{}^0 + N\sum\frac{g_n\epsilon_n e^{-\epsilon_n/(kT)}}{Q} = E_0{}^0 + Nk\frac{T^2\,dQ/dT}{Q} = E_0{}^0 + RT^2\frac{d(\ln Q)}{dT}. \qquad \text{(V, 39)}$$

Here $R = Nk$ is the gas constant per mole.

The *heat content* H^0 *of one mole of a perfect gas* is the sum of the total internal energy E^0 and the external energy $pV = RT$. That is, we have

$$H^0 = E_0{}^0 + RT + RT^2\frac{d(\ln Q)}{dT}, \qquad \text{(V, 40)}$$

The *molar heat capacity at constant pressure* is given by

$$C_p{}^0 = \frac{dH^0}{dT} = R + R\frac{d}{dT}\left[T^2\frac{d(\ln Q)}{dT}\right]. \qquad \text{(V, 41)}$$

[3a] The moment of inertia of the CH_3 group was assumed to be 5.53×10^{-40} gm cm^2 which is one half of an older value for $I_A(C_2H_6)$ but is only insignificantly different from the new value given in Table 132. Most authors until recently have used $I(CH_4) = 5.33 \times 10^{-40}$ gm cm^2 for the moment of inertia of a CH_3 group.

[4] Compare the similar derivation in Molecular Spectra I, p. 507, where, however, the zero-point energy $E_0{}^0$ was omitted and where the translational part was separated off first.

It is thus seen that the *heat content and heat capacity can be calculated by simple differentiations if the partition function Q is known.*

Since Q occurs in H^0 and $C_p{}^0$ only as $\ln Q$, if Q is a *product* of a number of factors the heat content as well as the heat capacity are the *sums* of a number of corresponding contributions. In particular, since according to (V, 4) Q can always be written as the product of the translational and internal partition function Q_{tr} and Q_{int}, we can write, taking account of the expression (V, 7) for Q_{tr},

$$H^0 = E_0^0 + \tfrac{5}{2}RT + H_{int}^0, \tag{V, 42}$$

$$C_p^0 = \tfrac{5}{2}R + C_{p,\,int}^0, \tag{V, 43}$$

where

$$H_{int}^0 = RT^2 \frac{d(\ln Q_{int})}{dT} \tag{V, 44}$$

and

$$C_{p,\,int}^0 = R\frac{d}{dT}\left[T^2 \frac{d(\ln Q_{int})}{dT}\right] \tag{V, 45}$$

are the contributions of the internal degrees of freedom to heat content and heat capacity respectively.

Again, as we have seen previously, Q_{int} can be written as the product of a number of factors if certain approximations are made. In particular, if the interaction of vibration and rotation is neglected, as is almost always done in practical calculations, H_{int}^0 and $C_{p,\,int}^0$ can be written as the *sums of a rotational and a vibrational term;* thus

$$H_{int}^0 = E_{int}^0 = H_r^0 + H_v^0 \tag{V, 46}$$

$$C_{p,\,int}^0 = C_{v,\,int}^0 = C_{pr}^0 + C_{pv}^0, \tag{V, 47}$$

where

$$H_r^0 = RT^2 \frac{d(\ln Q_r)}{dT}, \qquad C_{pr}^0 = R\frac{d}{dT}\left[T^2\frac{d(\ln Q_r)}{dT}\right], \tag{V, 48}$$

$$H_v^0 = RT^2 \frac{d(\ln Q_v)}{dT}, \qquad C_{pv}^0 = R\frac{d}{dT}\left[T^2\frac{d(\ln Q_v)}{dT}\right]. \tag{V, 49}$$

We consider first the *rotational contribution to H^0 and $C_p{}^0$ for molecules without free or hindered internal rotations.* For high temperatures, assuming a rigid molecule, the rotational partition function Q_r is given by (V, 22) or (V, 29). Since it occurs both in H^0 and $C_p{}^0$ in the form $d(\ln Q_r)/dT$, all constant (that is, temperature-independent) factors drop out, and we obtain for linear molecules, from (V, 22),

$$H_r^0 = RT, \qquad C_{pr}^0 = R, \tag{V, 50}$$

and for other rigid molecules, from (V, 29),

$$H_r^0 = \tfrac{3}{2}RT, \qquad C_{pr}^0 = \tfrac{3}{2}R. \tag{V, 51}$$

Thus at high temperatures we have for each rotational degree of freedom the classical equipartition values $\tfrac{1}{2}RT$ and $\tfrac{1}{2}R$ of H^0 and $C_p{}^0$ respectively. For all gases except H_2 these classical values are practically reached at room temperature or even lower temperatures.

It should be noted that the moments of inertia do not enter the formulae (V, 50) and (V, 51) for heat capacity and heat content at not too low temperatures. Similarly the symmetry number and the nuclear spin factor drop out. For lower temperatures, when the asymptotic expansions (V, 21), (V, 26), and (V, 28) have to be used, the moments of inertia do enter, while symmetry number and nuclear spin factor can still be neglected. However, for *very low temperatures*, when direct summation has to be carried out for Q_r in (V, 13), the *identity of nuclei and the nuclear spin* do produce a noticeable effect on the heat capacity except when the symmetry number is 1. We have then also (for $\sigma > 1$) to realize that the equilibrium values for H^0 and $C_p{}^0$ calculated by using in Q_r all rotational levels with their proper statistical weights (inclusive of nuclear spin contribution) do not in general coincide with the actual, observed values, because the *different modifications* having different rotational species do not go over into one another within the time of an experiment. It is therefore necessary to calculate the rotational contribution to the heat content and heat capacity of each modification separately and add them in the proportion of the statistical weights of the modifications.

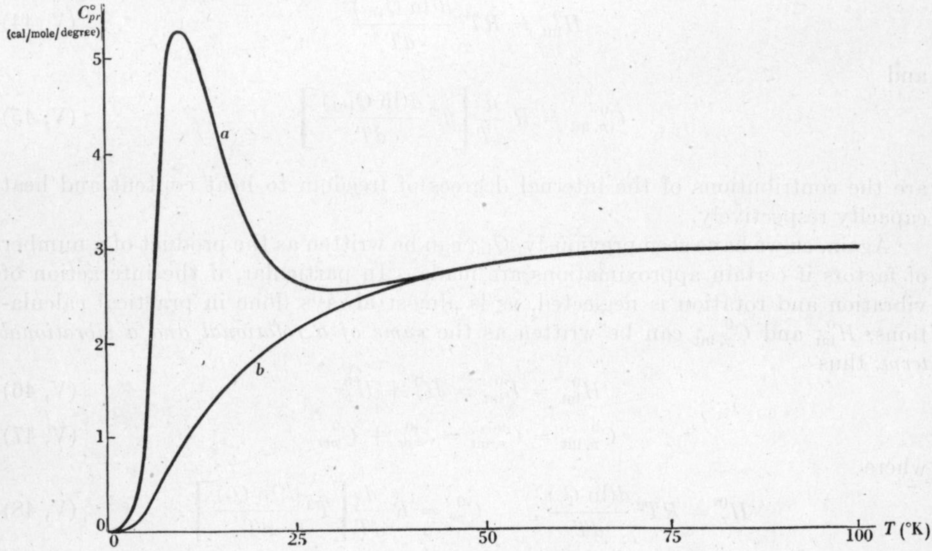

FIG. 168. **Calculated rotational heat capacity of gaseous H_2O at low temperatures (a) for equilibrium and (b) for no equilibrium between the two modifications** [after Stephenson and McMahon (805)].

For example, for H_2, H_2O, H_2CO, and similar molecules: $C_p{}^0 = \frac{1}{4}C_p{}^0$ (para) $+ \frac{3}{4}C_p{}^0$ (ortho); for NH_3, CH_3X, and others: $C_p{}^0 = \frac{1}{3}C_p{}^0$ (para) $+ \frac{2}{3}C_p{}^0$ (ortho), where para and ortho stand for the less and the more abundant modification. Fig. 168 gives as an example the rotational heat capacity of H_2O for equilibrium (curve *a*) and no equilibrium (curve *b*) between the two modifications.[5]

At low temperatures the corrections due to centrifugal stretching are always negligibly small. But at high temperatures they may become noticeable even though still small. According to Wilson (936), for H_2O, for example, at a temperature of 1000° K. an amount 0.08 cal/degree has to be added to the C^0_{pr} value obtained from (V, 51). The correction is smaller for almost all other molecules, particularly for heavier molecules.

The exact *vibrational contribution to the heat content and heat capacity* is obtained by substituting the vibrational partition function Q_v from (V, 13) into (V, 49). In

[5] It may be noted that while the partition function for the equilibrium mixture of the two (or more) modifications is the sum of two (or more) parts corresponding to the two (or more) modifications, the heat content and heat capacity are not simply the sums of two (or more) contributions because of the occurrence of ln Q in the formulae (V, 44) and (V, 45).

the *harmonic-oscillator approximation* [equation (V, 17)], Q_v is a product of terms due to different normal vibrations. In this approximation, therefore, the vibrational heat content and heat capacity are sums of terms due to the different normal vibrations. Substituting (V, 17) in (V, 49), we obtain

$$H_v^0 = R\frac{hc}{k}\sum_i \frac{d_i\omega_i e^{-\omega_i hc/(kT)}}{1 - e^{-\omega_i hc/(kT)}}, \quad R\frac{hc}{k} = 2.858 \text{ cal/cm}^{-1}, \qquad (V, 52)$$

$$C_{pv}^0 = R\left(\frac{hc}{kT}\right)^2 \sum_i \frac{d_i\omega_i^2 e^{-\omega_i hc/(kT)}}{(1 - e^{-\omega_i hc/(kT)})^2},$$

$$R\left(\frac{hc}{kT}\right)^2 = \frac{4.111}{T^2} \text{ cal/degree/cm}^{-2}, \qquad (V, 53)$$

where ω_i is in cm^{-1} and the summation is over all fundamentals of the molecule. Thus if the fundamental frequencies of a molecule and their degrees of degeneracy are known, the evaluation of the vibrational heat content and heat capacity in this approximation is a simple matter. The calculations are further simplified by tables prepared by Johnston and published in Wilson's review (941). They have also been reproduced, corrected for the new value of h, by Aston (60). These tables give

$$\frac{H_{v_i}^0}{d_i T} = R\frac{hc\omega_i}{kT}\frac{e^{-\omega_i hc/(kT)}}{1 - e^{-\omega_i hc/(kT)}} = R\frac{\omega_i hc/(kT)}{e^{+\omega_i hc/(kT)} - 1} \qquad (V, 54)$$

and

$$\frac{C_{pv_i}^0}{d_i} = R\left(\frac{hc\omega_i}{kT}\right)^2 \frac{e^{-\omega_i hc/(kT)}}{(1 - e^{-\omega_i hc/(kT)})^2} \qquad (V, 55)$$

as functions of ω_i/T. Hull and Hull (464) have given tables of the functions (V, 54) and (V, 55) divided by R, using $\omega_i hc/(kT) = \omega_i/(0.6951T)$ as the independent variable. This has the advantage that the latter tables are independent of the value of h. The functions on the right in (V, 54) and (V, 55) were first introduced by Einstein and are often called *Einstein functions.*

For high temperatures, $e^{-\omega_i hc/(kT)}$ can be replaced by $1 - \omega_i hc/(kT)$. If this value is substituted in (V, 52) and (V, 53) it is seen that, *asymptotically,*

$$H_v^0 \rightarrow RT\sum d_i, \qquad C_{pv}^0 \rightarrow R\sum d_i. \qquad (V, 56)$$

These are the classical values: for each vibrational degree of freedom the contribution to H^0 and $C_p{}^0$ is RT and R respectively. It should, however, be realized that these classical values are approached only at very much higher temperatures than the classical value for the rotational contributions to H^0 and $C_p{}^0$ [$\omega_i/(0.6951T)$ must be very small compared to 1]. At such high temperatures the harmonic-oscillator approximation on which (V, 56) is based is a poor one, since for the higher vibrational levels the influence of anharmonicity is large.

If *anharmonicity* is to be taken into account, the evaluation of the partition function and correspondingly of H_v^0 and C_{pv}^0 are much more complicated (see above). Instead of giving any explicit formulae [see Gordon (388) (389), Kassel (487) (491)] we illustrate in Fig. 169 the dependence of the specific heat of N_2O on the temperature, first according to the harmonic-oscillator formula (V, 35) (broken-line curve), and second if anharmonicities are taken into account (full-line curve). The circles refer to observed values. It is seen that for not too high temperatures and unless very high accuracy is required the harmonic-oscillator approximation is quite satisfactory. As mentioned before, this approximation together with the rigid-rotator approximation is almost always used in practical calculations.

As *examples* of this approximation, in Table 142 the rotational and vibrational as well as the total heat capacities of a number of gases at various temperatures are given as calculated from (V, 50 or V, 51), (V, 53) and $C_p^0 = \frac{5}{2}R + C_{pr}^0 + C_{pv}^0$.

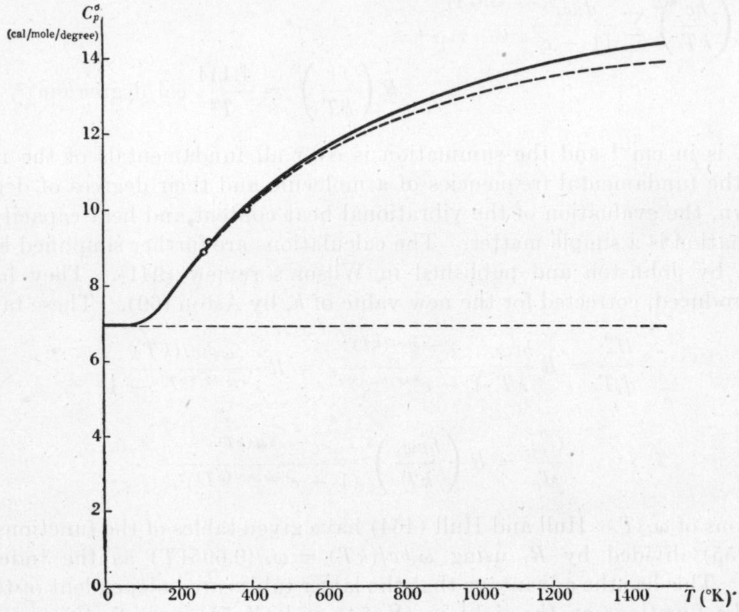

FIG. 169. **Calculated heat capacity of N_2O [after Kassel (487)].—The solid curve represents the data calculated by Kassel taking anharmonic terms into account. The broken line curve is based on the harmonic oscillator approximation. The three circles represent values observed by Eucken and Lüde (312). The drop of $C_p{}^0$ at very low temperatures when the classical value for the rotational contribution no longer applies is not shown. The vibrational contribution is the part above the light horizontal line. The anharmonic constants used in Kassel's calculations were substantially the same as those given in (III, 55) with the exception of x_{22} for which he used −3.1 instead of −2.28. Unfortunately this constant gives the greatest contribution to the difference between the harmonic and anharmonic oscillator approximation for $C_p{}^0$.**

We consider the case of CH_4 in a little more detail. For it $C_{pr}^0 = \frac{3}{2}R$. The vibrational frequencies (in cm^{-1}) are (see Table 80) 2914.2 (1), 1526 (2), 3020.3 (3), 1306.2 (3); the numbers in parentheses indicate the degeneracies. For $T = 481.2°$ K. one obtains from Johnston's tables the contributions 0.025, 0.881, 0.059, and 1.908 cal/degree/mole respectively, giving a vibrational heat capacity of $C_{pv}^0 = 2.873$ cal/degree/mole, as given in Table 142. It is seen that most of the vibrational contribution is due to the low-frequency fundamentals.

In Table 142 the observed heat capacities are also given for comparison. In every case the agreement with the calculated value is very satisfactory. Therefore one may be confident that for other molecules without internal rotations the calculated values are likewise reliable even if no experimental determination is available.

A large number of such calculations have been carried out and are summarized up to 1940 in Wilson's review (941).

It must be realized that in order to compare the observed values for heat content and heat capacity with the calculated ones, the former must first be corrected for *deviations from the perfect gas* to which the latter refer. The differences between the real-gas and perfect-gas values of H and C_p are given by

$$H - H^0 = \int_0^p \left[V - T\left(\frac{\partial V}{\partial T}\right)_p\right] dp, \qquad C_p - C_p{}^0 = -T\int_0^p \left(\frac{\partial^2 V}{\partial T^2}\right)_p dp, \qquad \text{(V, 57)}$$

where V is the molal volume and p the pressure in atmospheres, and where the derivatives of V are taken at constant pressure. Application of these formulae requires a knowledge of the equation of state. Fortunately, the corrections are usually small, particularly if the measurements have been made far above the boiling point. The observed $C_p{}^0$ values in Table 142 have been corrected in this way.

TABLE 142. CALCULATED AND OBSERVED MOLAR HEAT CAPACITIES OF SEVERAL GASES.

Molecule	T (°K)	C^0_{pr} cal/degree/mole	C^0_{pv} cal/degree/mole	C^0_p cal/degree/mole calculated	C^0_p cal/degree/mole observed	References
CH_4	297.7	2.980	0.572	8.52	8.57	(312)
	481.2	2.980	2.873	10.82	11.20	
	1000	2.980	9.221	17.17	—	
C_2H_2	288	1.986	3.393	10.37	9.97	(569a)
	500	1.986	6.124	13.09	—	
C_2H_4	270.7	2.980	1.838	9.78	9.74	(183)
	320.7	2.980	3.016	10.96	10.99	
$CH_3—C{\equiv}C—H$	272.28	2.980	5.82	13.77	13.76	(512)
	369.21	2.980	8.61	16.56	16.52	
Cyclo C_3H_6	272.15	2.980	4.18	12.13	12.10	(512)
	368.46	2.980	8.87	16.82	16.77	
$CH_3—C{\equiv}C—CH_3$	336.07	3.973[5a]	11.20	20.13	20.21	(513)

For *molecules with possible internal rotations* the situation with regard to a prediction of heat content and heat capacity is not as favorable as for those without such rotations, since up to now only in one case (CH_3OH) has the spectrum yielded a value for the potential barrier hindering free rotation. However, conversely, the observed heat-capacity data may be used to determine this potential barrier. As long as the interaction of the (hindered) internal rotations with the other motions is disregarded (as is almost always done) these internal rotations contribute a factor to the partition function, which may be considered separately, and therefore they contribute an additive term to heat content and heat capacity. Here it is assumed, of course, that, in place of these terms, the terms corresponding to the torsional oscillations are omitted from the vibrational contribution.

If the internal rotation is entirely free we obtain from the partition function (V, 36), when substituted in (V, 44) and (V, 45) for the contribution to H^0 and $C_p{}^0$ at not too low temperatures

$$H^0_{\text{f.i.r.}} = \tfrac{1}{2}RT, \qquad C^0_{p,\,\text{f.i.r.}} = \tfrac{1}{2}R. \qquad \text{(V, 58)}$$

[5a] This value includes the contribution of free internal rotation.

If there are several free internal rotations, a corresponding number of terms as in (V, 58) has to be added. If instead there is a torsional oscillation, the contributions would be given by one of the terms in (V, 52) and (V, 53). For a torsional oscillation of 800 cm^{-1} the contribution to C_p^0 at $T = 200°$ K. would be 0.21 cal/degree/mole as compared to 0.99 from (V, 58) for free rotation.

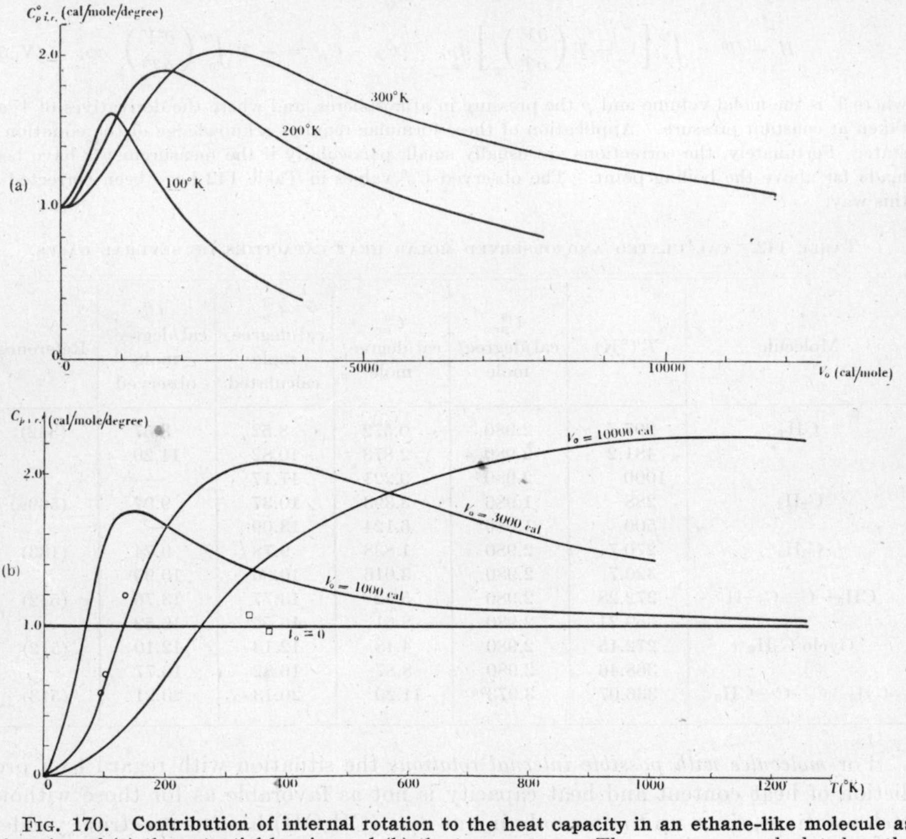

Fig. 170. Contribution of internal rotation to the heat capacity in an ethane-like molecule as a function of (a) the potential barrier and (b) the temperature.—The curves were calculated on the basis of the tables given by Pitzer and Gwinn (698) assuming a moment of inertia of the CH_3 group of 5.53×10^{-40} gm cm^2 (which is double the reduced moment occurring in the above mentioned tables)[3a]. The circles represent observed values for ethane [Kistiakowsky, Lacher and Stitt (510)], the squares represent observed values for dimethyl acetylene [Kistiakowsky and Rice (513)]. New values for C_2H_6 at higher temperatures have recently been obtained by Dailey and Felsing (261b).

An example of the case of free internal rotation seems to be dimethyl acetylene, for which the observed heat capacity agrees very well with that calculated on the assumption of free internal rotation. The data for this case are included in Table 142.

For the intermediate case of hindered internal rotation the formulae for heat content and heat capacity are too complicated to be reproduced here (see the references given in the discussion of the partition function, p. 511). We give instead in Fig. 170 a graphical representation of the dependence of the contribution to the heat capacity $C^0_{p,\text{i.r.}}$ on the height of the barrier and on the temperature for an ethane-like molecule (see Table 141 for the partition function). These curves are based on the tables given by Pitzer and Gwinn (698). It is seen from these curves that $C^0_{p,\,\mathrm{i.r.}}$ approaches zero for a very large barrier at any given temperature, and that for a given barrier $C^0_{p,\,\mathrm{i.r.}}$ approaches $\frac{1}{2}R$ for a sufficiently high temperature. It is important to realize that $C^0_{p,\,\mathrm{i.r.}}$ for small barriers increases first above the free rotation value before the decrease sets in. The reason for this initial increase and the maximum is the fact that the spacing of the lowest energy levels decreases with increasing V_0 as shown by Fig. 165.

It is clear from Fig. 170a that *if the contribution of internal rotation to the heat capacity of a gas is determined experimentally* (as the difference of the observed total heat capacity and the calculated translational, vibrational, and rotational contributions exclusive of internal rotation), *it may serve to determine the height of the potential barrier.* This was one of the ways in which the potential barrier in ethane was established (see Chapter III, section 3f.). As shown by Fig. 170a, if a $C^0_{p,\,\mathrm{i.r.}}$ value greater than $\frac{1}{2}R$ is observed, two values for the potential barrier will account for it. A decision between these is possible if $C^0_{p,\,\mathrm{i.r.}}$ values at different temperatures are available,

In Table 143 the potential barriers obtained by the above-described method, as well as by the two other methods to be described later, are summarized. It must be emphasized that all values are based on the *assumption of a cosine-like hindering potential function* [see equation (II, 301)]. For other potential functions other barrier heights would be obtained. Since as yet no independent evidence bearing on the form of the potential function is available, the values given must be considered as *equivalent* barrier heights and cannot claim to be the true heights [see also Charlesby (196a) and Pitzer and Gwinn (698)].

According to a very recent paper by Aston, Isserow, Szasz, and Kennedy (61a′) the very high values of V_0 for OH in ethyl and isopropyl alcohol given in Table 143 are very probably only apparent and due to the neglect of the fact that one of the three minima of the potential function has a different energy from the two others (compare the similar situation in the dichloroethanes, p. 347 and Fig. 98) leading to an equilibrium between two molecular forms with different barrier heights. The same authors have also developed an empirical method of calculating barrier heights from the assumption of a repulsion of hydrogen atoms according to an inverse fifth power law.

Entropy and free energy. According to statistical mechanics the entropy S^0 and the free energy F^0 of one mole of a perfect gas, in terms of the total partition function, are given by

$$S^0 = R(1 - \ln N) + RT\,\frac{d(\ln Q)}{dT} + R \ln Q, \qquad \text{(V, 59)}$$

$$F^0 = E_0{}^0 + RT \ln N - RT \ln Q, \qquad \text{(V, 60)}$$

where N is the Avogadro number and $E_0{}^0$ the zero-point energy. As for heat content and heat capacity, if Q is a product of factors, S^0 and F^0 contain sums of corresponding terms. Thus, since in all cases $Q = Q_{\mathrm{tr}} \cdot Q_{\mathrm{int}}$, we have

$$S^0 = S^0_{\mathrm{tr}} + S^0_{\mathrm{int}}, \qquad \text{(V, 61)}$$

$$F^0 = F^0_{\mathrm{tr}} + F^0_{\mathrm{int}}. \qquad \text{(V, 62)}$$

Here, introducing Q_{tr} from (V, 7), the mol volume $V = RT/p$ (p = pressure) and

TABLE 143. POTENTIAL BARRIERS HINDERING FREE ROTATION AS OBTAINED FROM HEAT CAPACITIES, ENTROPIES, OR EQUILIBRIUM DATA COMBINED WITH SPECTROSCOPIC DATA.

Molecule	Barrier [5b] cal/mole	Method	References
C_2H_6	2750	entropy, heat capacity, equilibrium	(496) (510) (511)
CH_3CCl_3	2700	entropy	(750b)
CH_3CF_3	3450	entropy	(752a)
$CH_3CH_2CH_3$	3300	entropy, heat capacity, equilibrium	(495a) (512) (508) (511)
$CH_3(CH_2)_2CH_3$	3600	entropy, heat capacity	(695a) (61b) (61d) (261b)
$(CH_3)_3CH$	3870	entropy, heat capacity	(61b) (61d) (261b)
$CH_3(CH_2)_3CH_3$	3600	entropy, heat capacity	(623a) (695a)
$(CH_3)_2CHCH_2CH_3$	8000	entropy	(774a)
$C(CH_3)_4$	4200	entropy	(693a) (61c)
$CH_3CH{=}CH_2$	2100	entropy, heat capacity, equilibrium	(245a) (236) (513) (508) (835) (511)
$CH_3CH_2CH{=}CH_2$	<800	entropy	(693a)
$CH_3CH{=}CHCH_3$	<800	entropy	(693a)
$(CH_3)_2C{=}CH_2$	1800	entropy	(693a)
$CH_3C{\equiv}CCH_3$	<500	heat capacity, entropy	(513) (972)
CH_3OH	1300	spectrum	(517)
	3400	entropy	(238)
CH_3SH	1460	entropy	(752)
CH_3CH_2OH	3000	entropy	(773)
	10000(OH)	entropy	
$CH_3CHOHCH_3$	3400	entropy	(774)
	5000(OH)	entropy	
CH_3NH_2	3000	entropy	(62a) (60a)
$(CH_3)_2NH$	3460	entropy	(60b)
$(CH_3)_3N$	4270	spectrum using (II, 303)	(61e)
$(CH_3)_2O$	3100	entropy	(498)
	2500	heat capacity	(513)
$(CH_3)_2S$	2000	entropy	(677) (851)
CH_3NO_2	≤800	heat capacity	(697) (286a)
$(CH_3)_2CO$	1000	entropy	(774)
CH_3CHO	2100	equilibrium	(635a)
o $C_6H_4(CH_3)_2$	2000	entropy	(699)
m, p $C_6H_4(CH_3)_2$	≤500	entropy	(699)
$Si(CH_3)_4$	1280	entropy	(61a)
HNO_3	7000(OH)	entropy	(326a)

the (chemical) molecular weight $M = mN$,

$$S_{\text{tr}}^0 = \tfrac{5}{2}R \ln T + \tfrac{3}{2}R \ln M + R \ln \left[\left(\frac{2\pi}{N} \right)^{\frac{3}{2}} \frac{k^{\frac{5}{2}}}{h^3} \right] + \tfrac{5}{2}R - R \ln p, \quad \text{(V, 63)}$$

$$S_{\text{int}}^0 = RT \frac{d(\ln Q_{\text{int}})}{dT} + R \ln Q_{\text{int}}, \quad \text{(V, 64)}$$

$$F_{\text{tr}}^0 = E_{0,\,\text{tr}}^0 + \tfrac{5}{2}RT - S_{\text{tr}}^0 T, \quad \text{(V, 65)}$$

$$F_{\text{int}}^0 = E_{0,\,\text{int}}^0 - RT \ln Q_{\text{int}}. \quad \text{(V, 66)}$$

The quantity usually tabulated and used for equilibrium calculations is not F^0 itself

[5b] All barriers correspond to the hindered rotation of CH_3 groups except those marked (OH) which correspond to the hindered rotation of a hydroxyl group.

but $-(F^0 - E_0{}^0)/T$, which according to (V, 62), (V, 63), (V, 65), (V, 66) is given by

$$-\frac{F^0 - E_0{}^0}{T} = S^0_{\mathrm{tr}} - \tfrac{5}{2}R + R \ln Q_{\mathrm{int}}. \qquad \text{(V, 67)}$$

For convenience in numerical calculations we give equations (V, 63) and (V, 67) also in numerical form. Substituting $R = 1.9863$ cal/degree/mole, $N = 6.0224 \times 10^{23}$, $k = 1.3807 \times 10^{-16}$ erg/degree, $h = 6.626 \times 10^{-27}$ erg sec, $p = 1$ atm. $= 1.0132 \times 10^6$ dynes/cm², and using ordinary logarithms we obtain (in cal/degree/mole):

$$S^0_{\mathrm{tr}} = 2.2868(5 \log_{10} T + 3 \log_{10} M) - 2.3135, \qquad \text{(V, 68)}$$

$$-\frac{F^0 - E_0{}^0}{T} = 2.2868(5 \log_{10} T + 3 \log_{10} M + 2 \log_{10} Q_{\mathrm{int}}) - 7.2793. \qquad \text{(V, 69)}$$

Again, as long as the interaction of vibration and rotation can be neglected, the *internal entropy and free energy are sums of independent contributions due to rotation and vibration:*

$$S^0_{\mathrm{int}} = S^0_r + S^0_v, \qquad F^0_{\mathrm{int}} = F^0_r + F^0_v, \qquad \text{(V, 70)}$$

where

$$S^0_r = RT\frac{d(\ln Q_r)}{dT} + R \ln Q_r, \qquad S^0_v = RT\frac{d(\ln Q_v)}{dT} + R \ln Q_v, \qquad \text{(V, 71)}$$

$$F^0_r = E^0_{0,r} - RT \ln Q_r, \qquad F^0 = E^0_{0,v} - RT \ln Q_v, \qquad \text{(V, 72)}$$

$$E^0_0 = E^0_{0,\mathrm{tr}} + E^0_{0,r} + E^0_{0,v}. \qquad \text{(V, 73)}$$

Unlike the case of heat content and heat capacity, since $\ln Q_{\mathrm{int}}$ occurs and not only its derivative, constant factors in Q_{int} do not now drop out. Thus, at not too low temperatures we obtain for the rotational contributions for *linear molecules* from (V, 22) (but including the symmetry number):

$$S^0_r = R\left(\ln T + \ln\frac{k}{hc} - \ln B - \ln \sigma + 1\right)$$
$$= 4.5736(\log_{10} T - \log_{10} B - \log_{10} \sigma) + 1.2639, \qquad \text{(V, 74)}$$

$$-\frac{F_r - E^0_{0,r}}{T} = S^0_r - R = S^0_r - 1.9863, \qquad \text{(V, 75)}$$

and for *other rigid molecules* from (V, 29),

$$S^0_r = \frac{R}{2}\left(3 \ln T - \ln ABC - 2 \ln \sigma + \ln\left[\pi\left(\frac{k}{hc}\right)^3\right] + 3\right)$$
$$= 2.2868(3 \log_{10} T - \log_{10} ABC - 2 \log_{10} \sigma) + 3.0327$$
$$= 2.2868(3 \log_{10} T + \log_{10} I_A I_B I_C - 2 \log_{10} \sigma) + 267.5213;^{6} \qquad \text{(V, 76)}$$

$$-\frac{F^0_r - E^0_{0,r}}{T} = S^0_r - \tfrac{3}{2}R = S^0_r - 2.9795. \qquad \text{(V, 77)}$$

---

[6] If the moments of inertia are expressed in (chemical) atomic weight units and Ångstrom units, the last constant must be replaced by −5.3838. These units are used by Wilson (941). It should be noted that Wilson uses the symbols A, B, C for these moments of inertia while here they are used for the rotational constants.

As before, these formulae are good approximations for all practically important temperatures except for very light molecules. For practical calculations it is convenient to take the translational and rotational contributions together. Then, *for linear molecules* at atmospheric pressure,[7]

$$S^0_{tr} + S^0_r = 2.2868(7 \log_{10} T + 3 \log_{10} M - 2 \log_{10} B - 2 \log_{10} \sigma) - 1.0496, \quad (V, 78)$$

$$-\frac{(F^0_{tr} + F^0_r - E^0_{0,tr} - E^0_{0,r})}{T} = S^0_{tr} + S^0_r - \tfrac{7}{2}R = S^0_{tr} + S^0_r - 6.9521; \quad (V, 79)$$

for other rigid molecules,[7]

$$S^0_{tr} + S^0_r = 2.2868(8 \log_{10} T + 3 \log_{10} M - \log_{10} ABC - 2 \log_{10} \sigma) + 0.7192, \quad (V, 80)$$

$$-\frac{(F^0_{tr} + F^0_r - E^0_{0,tr} - E^0_{0,r})}{T} = S^0_{tr} + S^0_r - 4R = S^0_{tr} + S^0_r - 7.9452. \quad (V, 81)$$

In the above formulae the *contribution of the nuclear spins* has been omitted. Since the nuclear spins cause a factor $(2I_1 + 1)(2I_2 + 1) \cdots$ in the partition function (see p. 509), we would obtain an additional term $R \ln (2I_1 + 1)(2I_2 + 1) \cdots$ to be added to S^0_r as well as to $-(F^0_r - E^0_{0,r})/T$. The entropy that includes this term is called the *absolute entropy,* whereas the entropy given above is called the *virtual entropy.* Usually only the latter is considered. It is only at extremely low temperatures that the absolute entropy must be considered, and then it cannot be obtained simply by adding the above constant term except when the symmetry number is 1; rather it is necessary to form the partition function with the proper total statistical weights for each level, and form S^0 and F^0 separately for the different (ortho, para, $\cdots$) modifications.

The *vibrational contributions,* in the harmonic-oscillator approximation, are again sums of terms due to the different vibrations. Substituting (V, 17) in (V, 71) and (V, 72), we obtain

$$S^0_v = -R \sum_i d_i \ln (1 - e^{-\omega_i hc/(kT)}) + R \frac{hc}{kT} \sum_i \frac{d_i \omega_i e^{-\omega_i hc/(kT)}}{1 - e^{-\omega_i hc/(kT)}}, \quad (V, 82)$$

$$-\frac{F^0_v - E^0_{0,v}}{T} = -R \sum_i d_i \ln (1 - e^{-\omega_i hc/(kT)}). \quad (V, 83)$$

In Johnston's tables [see Wilson (941) and Aston (60)], the contribution of a non-degenerate vibration ($d_i = 1$) to $-(F^0_v - E^0_{0,v})/T$, that is, $-R \ln (1 - e^{-\omega_i hc/(kT)})$, is given as a function of ω_i/T, thus making a calculation of the free energy very simple for any molecule for which all the ω_i are known Since the first term in the expression (V, 82) for S^0_v is identical with $-(F^0_v - E^0_{0,v})/T$ and the second term is identical with H^0_v/T, which is also given in Johnston's tables (see above), the entropy also can easily be calculated. At very high temperatures, as in the case of heat content and heat capacity, the influence of *anharmonicity* will make itself felt and (V, 82) and (V, 83) will no longer give an accurate representation. However, even

[7] If in the formulae (V, 78–81) the moments of inertia in atomic weight and Ångstrom units are used rather than the rotational constants, it is necessary to replace $-2 \log_{10} B$ and $-\log_{10} ABC$ by $+2 \log_{10} I_B$ and $+\log I_A I_B I_C$ respectively, and -1.0496 and $+0.7192$ by -6.6607 and -7.6973 respectively.

at temperatures as high as 1000° K. the effect is quite small. Thus Gordon (389) found for N_2O a difference of only 0.016 cal/degree/mole between the value of $(F^0 - E_0^0)/T$ obtained from (V, 83) and that obtained if anharmonicity is taken into account, this difference being only 0.18 per cent of the total $(F^0 - E_0^0)/T$.

As an illustration we give in Fig. 171 the variation of the partial entropies S_{tr}^0, S_r^0, S_v^0 as well as of the total entropy of methyl chloride as a function of the temperature according to the formulae (V, 68), (V, 76), and (V, 82). The necessary molecular data are taken from the previous Tables 84 and 132. The exact values for the partial and total entropies at 298.16° K. are

$$S_{tr}^0 = 37.66, \qquad S_r^0 = 17.63, \qquad S_v^0 = 0.50, \qquad S^0 = 55.79 \text{ cal/degree/mole.}$$

93 per cent of $S_v{}^0$ is due to the two vibrations of lowest frequency, $\nu_3(a_1)$ and $\nu_6(e)$. It is seen from these data and from Fig. 171 that except for very high temperatures

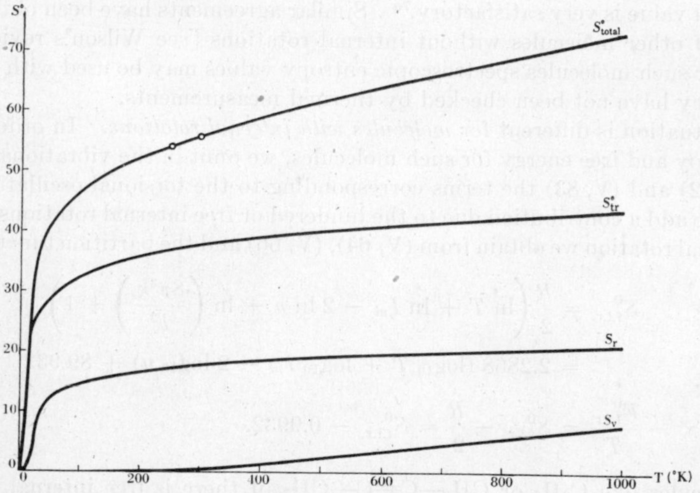

FIG. 171. **Total entropy and partial entropies (in cal/degree/mole) of methyl chloride as a function of the temperature.**—The two circles represent points observed by Messerly and Aston (623).

the vibrational contribution to the entropy is very small compared to the other contributions. This is in contrast to the heat capacity, for which in general at not too high temperatures the vibrational contribution is relatively much larger (in the above example $C_{pv}^0 = 1.81$, $C_p^0 = 9.74$). Thus a statistical calculation of the entropy is much less dependent upon precise vibrational data than that of the heat capacity.

"Observed" values of the entropy are usually obtained from the observed low-temperature heat capacities $C_p{}^s$, $C_p{}^l$, $C_p{}^g$ of the solid, liquid, and gas, and the heat of fusion L_f and heat of vaporization L_v, according to the formula

$$S = \int_0^{T_f} \frac{C_p{}^s}{T}\,dT + \frac{L_f}{T_f} + \int_{T_f}^{T_v} \frac{C_p{}^l}{T}\,dT + \frac{L_v}{T_v} + \int_{T_v}^{T} \frac{C_p{}^g}{T}\,dT, \qquad \text{(V, 84)}$$

where it is known from the third law of thermodynamics that $\int_0^T \frac{C_p{}^s}{T}\,dT \to 0$ for

$T \to 0$. The observed entropy values are therefore sometimes also called *"third-law" values.* A Debye function is used in evaluating the first integral in (V, 84) between $T = 0$ and the lowest temperature for which C_p^s is observed.

For a comparison of the observed entropy values with those calculated from spectroscopic data the former have to be *corrected for gas imperfections.* This correction is approximately given by

$$S^0 - S = \frac{27RT_c^3p}{32T^3p_c}, \qquad \text{(V, 85)}$$

where T_c is the critical temperature and p and p_c the pressure and critical pressure respectively.

In this way, for example, Messerly and Aston (623), for the case of CH_3Cl considered above, obtained an entropy of 55.94 cal/degree/mole at 298.16° K. from low-temperature heat-capacity measurements. The agreement with the previous theoretical value is very satisfactory.[7a] Similar agreements have been obtained for a number of other molecules without internal rotations [see Wilson's review (941)], so that for such molecules spectroscopic entropy values may be used with confidence even if they have not been checked by thermal measurements.

The situation is different for *molecules with internal rotations.* In order to calculate entropy and free energy for such molecules, we omit in the vibrational contribution (V, 82) and (V, 83) the terms corresponding to the torsional oscillations and in their place add a contribution due to the hindered or free internal rotations. For one *free* internal rotation we obtain from (V, 64), (V, 66) and the partition function (V, 36)

$$S^0_{\text{f.i.r.}} = \frac{R}{2}\left(\ln T + \ln I_m - 2\ln n + \ln\left(\frac{8\pi^3k}{h^2}\right) + 1\right)$$

$$= 2.2868\,(\log_{10} T + \log_{10} I_m - 2\log_{10} n) + 89.932 \qquad \text{(V, 86)}$$

$$-\frac{F^0_{\text{f.i.r.}}}{T} = S^0_{\text{f.i.r.}} - \frac{R}{2} = S^0_{\text{f.i.r.}} - 0.9932. \qquad \text{(V, 87)}$$

For molecules like C_2H_6 or $CH_3{-}C{\equiv}C{-}CH_3$, if there is free internal rotation of one CH_3 group against the rest of the molecule about the top axis [$n = 3$ and $I_m = I_A/4 = 2.759 \times 10^{-40}$ gm cm² (see p. 511)], we obtain from (V, 86)

$$S^0_{\text{f.i.r.}} = 2.2868 \log_{10} T - 2.714. \qquad \text{(V, 88)}$$

This variation is represented graphically in Fig. 172b (curve $V_0 = 0$). It is seen that $S^0_{\text{f.i.r.}}$ is of the order of several entropy units. On the other hand, for a very high potential barrier when the torsional oscillation has a high frequency, the contribution to the entropy (and free energy) according to (V, 82) is obviously quite small, at least for low temperatures. Fig. 172b shows, in addition to the free rotation curve, curves representing the dependence of the internal-rotation part of the entropy $S^0_{\text{i.r.}}$ on the temperature for a few intermediate barrier heights as obtained from Pitzer and Gwinn's tables. Fig. 172a shows for three temperatures the dependence of $S^0_{\text{i.r.}}$ on the barrier height. It is clear from these curves that conversely *the measurement of the entropy may serve to determine the barrier hindering the rotation* if all other

[7a] Messerly and Aston used slightly different values for the moments of inertia and obtained an even better agreement.

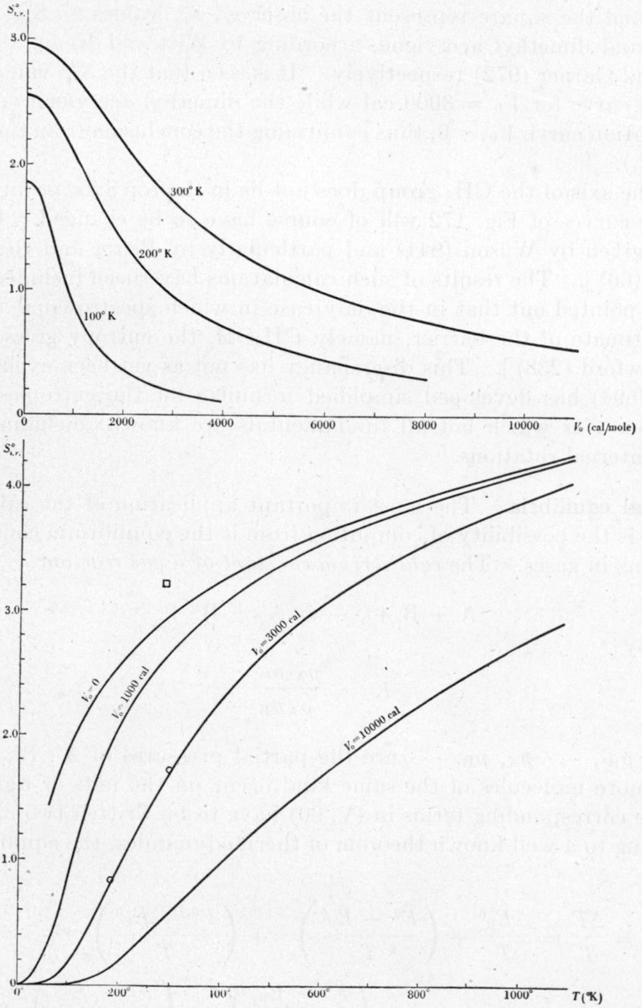

Fig. 172. Contribution of internal rotation to the entropy in an ethane-like molecule as a function, (a) of the barrier height and (b) of the temperature.—In (b) the curve for $V_0 = 0$ has not been drawn down to $T = 0$ since for very low temperatures the effect of the different modifications which has not as yet been calculated will enter. The two circles represent observations of the entropy by Witt and Kemp (947) for C_2H_6 at 184.1° and 298.1° K subtracting the contributions of translation, vibration, and over-all rotation (48.70 and 53.13 cal/degree/mol. respectively).[7b] The small square represents the observation by Yost, Osborne and Garner (972) for dimethyl acetylene at 291.0° K (subtracting 64.26 cal/degree/mol.).[7c]

[7b] These are values calculated by the writer. There seems to be a slight error in the values given by Kemp and Pitzer (496).

[7c] This value is slightly higher than that given by Yost, Osborne and Garner since they have assumed the CH_3 group to have the same dimensions as in CH_4 whereas here it is assumed that the dimensions are the same as in C_2H_6 (see footnote 3a).

contributions to the entropy can be calculated from spectroscopic data. In Fig. 172b the circles and the square represent the observed $S^0_{\text{i.r.}}$ values ($= S^0_{\text{obs.}} - S^0_{\text{tr}} - S^0_r - S^0_v$) for ethane and dimethyl acetylene, according to Witt and Kemp (947) and Yost, Osborne, and Garner (972) respectively. It is seen that the $S^0_{\text{i.r.}}$ values of ethane lie close to the curve for $V_0 = 3000$ cal while the dimethyl acetylene value is close to the free rotation curve $V_0 = 0$, thus confirming the conclusion from the heat capacity (see p. 518f.).

When the axis of the CH_3 group does not lie in the top axis, or for other rotating groups, the curves of Fig. 172 will of course have to be changed. Detailed tables have been given by Wilson (941) and particularly by Pitzer and Gwinn (698) [see also Aston (60)]. The results of such calculations have been included in Table 143. It must be pointed out that in the only case in which spectroscopic data have supplied an estimate of the barrier, namely CH_3OH, the entropy gives a much larger value [Crawford (238)]. This discrepancy has not as yet been explained.

Pitzer (694) has developed simplified formulae for the entropies of long chain hydrocarbons (for which not all fundamentals are known) including the effect of restricted internal rotations.

Chemical equilibria. The most important application of the calculation of the free energy is the possibility of computing from it the equilibrium constants of chemical reactions in gases. The *equilibrium constant of a gas reaction*

$$\mathrm{A} + \mathrm{B} + \cdots \rightleftarrows \mathrm{A'} + \mathrm{B'} + \cdots \tag{V, 89}$$

is defined by

$$K_p = \frac{p_{\mathrm{A'}} p_{\mathrm{B'}} \cdots}{p_{\mathrm{A}} p_{\mathrm{B}} \cdots} \tag{V, 90}$$

where $p_{\mathrm{A'}}, p_{\mathrm{B'}}, \cdots p_{\mathrm{A}}, p_{\mathrm{B}}, \cdots$ are the partial pressures of A′, B′, ⋯ A, B, ⋯. If two or more molecules of the same kind occur on the left- or right-hand side of (V, 89), the corresponding terms in (V, 90) have to be written two or more times.

According to a well-known theorem of thermodynamics, the equilibrium constant is given by

$$-R \ln K_p = \frac{\Delta F}{T} = \frac{\Delta E_0{}^0}{T} + \left(\frac{F^0 - E_0{}^0}{T}\right)_{\mathrm{A'}} + \left(\frac{F^0 - E_0{}^0}{T}\right)_{\mathrm{B'}} + \cdots - \left(\frac{F^0 - E_0{}^0}{T}\right)_{\mathrm{A}} - \left(\frac{F^0 - E_0{}^0}{T}\right)_{\mathrm{B}} - \cdots, \tag{V, 91}$$

where ΔF is the total standard molar free energy change and $\Delta E_0{}^0$ the total standard molar zero-point energy change, which is positive when energy is absorbed in going from A + B + ⋯ to A′ + B′ + ⋯.

Instead of expressing the equilibrium constant in terms of the free energies one may also express it directly in terms of the partition functions by substituting (V, 60) into (V, 91), obtaining

$$K_p = \frac{\dfrac{Q_{\mathrm{A'}}}{N} \cdot \dfrac{Q_{\mathrm{B'}}}{N} \cdots e^{-\Delta E_0^0/(RT)}}{\dfrac{Q_{\mathrm{A}}}{N} \cdot \dfrac{Q_{\mathrm{B}}}{N} \cdots}. \tag{V, 92}$$

Here the Q are the total partition functions referred to the lowest energy level of each molecule. If we use instead the partition functions Q^0 referred to a zero point of energy that is the same for all molecules taking part in the reaction, we can also write

$$K_p = \frac{\dfrac{Q^0_{A'}}{N} \cdot \dfrac{Q^0_{B'}}{N} \cdots}{\dfrac{Q^0_A}{N} \cdot \dfrac{Q^0_B}{N} \cdots}. \qquad (V, 93)$$

In the case of reactions in which the produced molecules are equal in number to the reactant molecules, the N's drop out in both (V, 92) and (V, 93), and a very simple formula results.

According to either (V, 91) or (V, 92), the equilibrium constant of a chemical gas reaction can be predicted if both $(F^0 - E_0{}^0)/T$ (or Q) for all reaction partners and $\Delta E_0{}^0$ are known.

We discuss first in a little more detail the *influence of* $(F^0 - E_0{}^0)/T$ *or* Q *on the equilibrium.* These quantities can be determined according to the previous formulae from spectroscopic molecular data except when there are internal rotations (for which as yet no satisfactory spectroscopic data are available). $(F^0 - E_0{}^0)/T$ has been tabulated for a number of molecules by various authors [see Wilson (941) and Aston (60)].

It is easily seen from (V, 92) (or V, 91) that the *nuclear spin* has no influence on chemical equilibria except at very low temperatures. As we have seen previously, the nuclear spin, at not too low temperatures, contributes a constant factor $(2I_1 + 1)(2I_2 + 1) \cdots$ to the partition function of each molecule and, since the nuclei are the same for the reactant and produced molecules, these factors cancel. For this reason the nuclear spin is usually omitted in tabulations of the free energy.

From the formulae for the rotational partition function (V, 22), (V, 27), (V, 29), and from (V, 92), it is seen that the equilibrium constant at a given temperature is the larger, the larger the product of the moments of inertia of the produced molecules (and the smaller that of the reactant molecules). Similarly, it is seen from the vibrational partition function (V, 17) that the equilibrium constant is the larger, the smaller the vibrational frequencies of the produced molecules (and the larger those of the reactant molecules). Quite generally it can be said that the side of a gas reaction is favored that has the lower and more closely spaced energy levels of its molecules. This is, of course, only a secondary influence if $\Delta E_0{}^0$ is large.

It is also interesting to consider the *influence of internal rotations on the equilibrium.* If one of the produced molecules has a free internal rotation but none of the reactant molecules has such (as might conceivably have been the case for the reaction $C_2H_4 + H_2 \rightarrow C_2H_6$), it is immediately seen from Table 141 and equation (V, 93) that the equilibrium constant may be several times larger than in the case of no free rotation, all other factors being the same. For intermediate potential barriers, intermediate values of the equilibrium constant arise (see Table 141). Conversely, therefore, the measurement of the equilibrium constant of an appropriate reaction may serve as a (third) method for the determination of barrier heights (see the example below and Table 143).

Even though the equations (V, 92) and (V, 93) are convenient for the discussion of the influence of a particular term on the equilibrium constant, for practical calculations (V, 91) is usually used. At any rate, most authors prefer to tabulate $(F^0 - E_0^0)/T$ rather than Q.

The second datum that enters the equilibrium constant, *the zero-point energy change* ΔE_0^0, may in certain cases also be obtained from spectroscopic data. If the atomic heats of formation, $D(\mathrm{A}), D(\mathrm{B}), \cdots, D(\mathrm{A'}), D(\mathrm{B'}), \cdots$ of all reaction partners are known [that is, the energies required to dissociate the molecules A, B, $\cdots$, A′, B′, $\cdots$ from their lowest states into free atoms (see Molecular Spectra I)] the zero-point energy change is simply given by

$$\Delta E_0^0 = D(\mathrm{A}) + D(\mathrm{B}) + \cdots - D(\mathrm{A'}) - D(\mathrm{B'}) - \cdots. \quad \text{(V, 94)}$$

As an example, for the thermal dissociation of H_2O into OH and H according to the equation $H_2O \rightarrow OH + H$, we have $\Delta E_0^0 = D(H_2O) - D(OH) = 218.9 - 99.4 = 119.5$ kcal. To be sure, the atomic heats of formation of most polyatomic molecules are not of purely spectroscopic origin, since use is made in their derivation of thermochemical heats of formation (in the example the heat of formation of H_2O).

In the case of *isotopic exchange reactions* such as

$$H_2O + HD \rightleftarrows HDO + H_2, \quad \text{(V, 95)}$$

$$H_2S + D_2 \rightleftarrows D_2S + H_2, \quad \text{(V, 96)}$$

$$C_2H_2 + C_2D_2 \rightleftarrows 2C_2HD, \quad \text{(V, 97)}$$

the zero-point energy change ΔE_0^0 and therefore the equilibrium constant can be obtained entirely from infrared and Raman data. For these reactions, in (V, 94) the differences of the atomic heats of formation of isotopic molecules occur. But these differences are equal to the *differences of the zero-point energies* $G(0, 0, \cdots)$ which can be obtained from the observed vibration spectra. We have in this case (in cm^{-1})

$$\Delta E_0^0 = G^{A'}(0, 0, \cdots) + G^{B'}(0, 0, \cdots) + \cdots - G^{A}(0, 0, \cdots) - G^{B}(0, 0, \cdots) \cdots, \quad \text{(V, 98)}$$

or if anharmonicities are neglected (see p. 78),

$$\Delta E_0^0 = \tfrac{1}{2}\sum \nu_i^{A'} + \tfrac{1}{2}\sum \nu_i^{B'} + \cdots - \tfrac{1}{2}\sum \nu_i^{A} - \tfrac{1}{2}\sum \nu_i^{B} - \cdots, \quad \text{(V, 99)}$$

where in the sums d_i-foldly degenerate vibrations have to be added d_i times.

In this way Glockler and Morrell (377) have calculated ΔE_0^0 for the reaction (V, 97), Grafe, Clusius, and Kruis (398) for the reaction (V, 96), and Libby (577) for the reaction (V, 95) and similar ones, taking account of anharmonicity [see also Black and Taylor (155a)]. Other similar reactions have also been studied.

In many cases ΔE_0^0 cannot as yet be obtained entirely from spectroscopic data, but it is obtained from the observed *heat of reaction* ΔH [8] with the help of spectroscopic data. This heat of reaction, apart from the correction for deviation from the perfect gas, according to (V, 42), is given by

$$\Delta H^0 = \Delta E_0^0 + \Delta \sum \left(\tfrac{5}{2}RT + H_{\mathrm{int}}^0\right) = \Delta E_0^0 + \Delta \sum (H^0 - E_0^0). \quad \text{(V, 100)}$$

[8] The heat of reaction may be determined by direct calorimeter measurements or indirectly from the temperature dependence of the equilibrium constant (van't Hoff's equation).

From this equation $\Delta E_0{}^0$ can easily be evaluated according to the previous discussion of the heat content H if ΔH^0 is measured.

It may be noted that when the number of molecules is not changed by the reaction the translational term $\Delta(\sum \frac{5}{2}RT)$ in (V, 100) is zero, since $\frac{5}{2}RT$ is the same for all gases. In such cases, therefore, the dependence of the heat of reaction ΔH^0 on the temperature is determined solely by the difference in temperature dependence of $H_{\text{int}} = RT^2[d(\ln Q_{\text{int}})/dT]$, that is, of the internal partition function of the gases concerned, or qualitatively, by the difference in the vibrational and rotational energy levels which, on account of the Boltzmann factor, are differently populated at a given temperature. In a similar manner it is seen from (V, 92) that the deviation of the equilibrium constant K_p from the value $e^{-\Delta E_0{}^0/(RT)}$ is determined only by the different temperature variation of the internal partition functions, if the number of molecules is not changed by the reaction.

As a first illustration of the calculation of chemical equilibria let us consider the industrially important *water-gas equilibrium* [see Kassel (487)]:

$$CO_2 + H_2 \rightleftarrows CO + H_2O. \tag{V, 101}$$

The zero-point energy change $\Delta E_0{}^0$ in this case is not known from spectroscopic data, but can be calculated from the observed heat of reaction,[9] which is at 300° K.: $\Delta H_{300} = 9808$ cal. In order to obtain $\Delta E_0{}^0$ from it we have to know the quantities $H^0 - E_0{}^0$ at 300° for the four gases involved. According to calculations, based on the previous formulae, by Kassel (487) for CO_2, by Davis and Johnston (270)[10] for H_2, by Johnston and Davis (473)[10] for CO, and by Gordon (388)[11] for H_2O, the values of $H^0 - E_0{}^0$ for 300° K. are 2256.1, 2036.2, 2085.1, and 2376.3 cal/mole respectively. From these values it follows that for the reaction (V, 101)[12]

$$\Delta E_0{}^0 = \Delta H^0 - \Delta\Sigma(H^0 - E^0) = 9808 - (2085.1 + 2376.3) + (2256.1 + 2036.2) = 9639 \text{ cal/mole.}$$

The quantities $-(F^0 - E_0{}^0)/T$ for five temperatures as derived from spectroscopic data by Kassel (487) for CO_2, Giauque (357)[13] for H_2, Clayton and Giauque (212) for CO, and Gordon (388) for H_2O

TABLE 144. CALCULATION OF THE EQUILIBRIUM CONSTANT FOR THE WATER-GAS REACTION.[13a]

T, °K.	$-\frac{F_0-E_0{}^0}{T}$ (cal/mole/degree)				$\frac{\Delta E_0{}^0}{T}$	$R \ln K_p$	K_p (calculated)	K_p (observed)
	CO_2	H_2	CO	H_2O				
300	43.620	24.480	40.408	37.230	32.130	−22.59	1.15×10^{-5}	
600	49.261	29.218	45.238	42.765	16.065	−6.541	3.71×10^{-2}	
900	53.074	32.020	48.114	46.106	10.710	−1.584	0.451	0.46
1200	56.049	34.027	50.210	48.579	8.033	+0.680	1.41	1.37
1500	58.513	35.605	51.880	50.586	6.426	+1.922	2.63	

[9] It is the difference of the heats of combustion of H_2 and of CO.

[10] These authors give the "total energy" which is $H^0 - E_0{}^0$ apart from the term $+\frac{5}{2}RT$. This term has to be added to their data to make them consistent with the others.

[11] Gordon gives only $-(F^0 - E_0{}^0)/T$ and S^0; but we have $H^0 - E_0{}^0 = (S^0 + (F^0 - E_0{}^0)/T)T$.

[12] Here a slight correction of ΔH_{300} to the value ΔH_{300} for the perfect gas has been neglected. According to Kassel (487) it amounts to only 0.4 cal/mole, which is within the accuracy of ΔH_{300}.

[13] Giauque gives $-(F^0 - E_0{}^0)/T$ including nuclear spin. To make the data consistent with the others $R \ln 4 = 2.755$ had to be subtracted from his values.

[13a] This table, unlike all others in this book, is based on the older set of physical constants as given in the International Critical Tables, since the free energies in the references quoted are based on these constants.

are given in Table 144. We obtain now $-R \ln K_p$ according to (V, 91) by adding $\Delta E_0^0/T$, given in the sixth column of Table 144, to $-(F_0 - E_0^0)/T$ for CO_2 and H_2 in the second and third columns and subtracting $-(F_0 - E_0^0)/T$ for CO and H_2O in the fourth and fifth columns. The resulting values of $R \ln K_p$ are given in the seventh column while the values of $K_p = (p_{CO} \cdot p_{H_2O})/(p_{CO_2} \cdot p_{H_2})$ itself are given in the eighth column. The last column gives, for two temperatures, values observed by Neumann and Köhler (652) which agree very satisfactorily with the calculated ones.[14] The latter are believed to be more accurate than the former.

As a second example we consider the *ethylene-ethane equilibrium*

$$C_2H_6 \rightleftarrows C_2H_4 + H_2.$$

This equilibrium has been studied by various investigators, both theoretically and experimentally, most recently by Guggenheim (403) (where references to the earlier work may be found) and by Kistiakowsky and Nickle (511). The calculations for the three temperatures 653°, 723°, and 863° K. are summarized in Table 145 (similar to Table 144) and compared to the observed values of Kistiakowsky and Nickle (511) for the two lower temperatures and of Travers and Pearce (869a) for the highest temperature. The calculations have been carried out with the most recent values of the molecular constants (see Tables 92, 105, and 132), for the alternative assumptions of free rotation in C_2H_6 and of a potential barrier of 2750 cal. The zero-point energy change $\Delta E_0^0 = 30{,}890$ cal was obtained from the heat of reaction $\Delta H_{298}^0 = 32{,}575$ cal observed by Kistiakowsky, Romeyn, Ruhoff, Smith, and Vaughan (514). It is seen that the agreement of the observed values with those calculated under the assumption of a barrier of 2750 cal is very satisfactory while the values calculated under the assumption of free rotation fall far outside the limits of error of the observed ones.

TABLE 145. CALCULATION OF THE EQUILIBRIUM CONSTANT OF THE REACTION $C_2H_6 \rightleftarrows C_2H_4 + H_2$.

T, °K.	$-\frac{F_0 - E_0^0}{T}$ (cal/mole/degree): C_2H_6 (with $V_0 = 2750$ cal)	C_2H_4	H_2	$\frac{\Delta E_0^0}{T}$ (cal/mole/degree)	K_p (cal): $V_0 = 0$	$V_0 = 2750$ cal	K_p (obs) (p in atm)
653	54.31	51.65	29.80	47.30	2.40×10^{-5}	3.91×10^{-5}	4.04×10^{-5}
723	55.79	52.88	30.50	42.72	3.08×10^{-4}	4.90×10^{-4}	5.16×10^{-4}
863	58.59	55.18	31.73	35.79	1.53×10^{-2}	2.30×10^{-2}	2.44×10^{-2}

Similar statistical calculations of equilibrium constants have been carried out for numerous other reactions. We mention only some of them. Extensive data for the reactions $2CO_2 \rightleftarrows 2CO + O_2$ and $CO_2 + C \rightleftarrows 2CO$ are given by Kassel (487), and for the reactions $H_2 + \frac{1}{2}O_2 \rightleftarrows H_2O$, $OH + \frac{1}{2}H_2 \rightleftarrows H_2O$, $O_3 \rightleftarrows 1\frac{1}{2}O_2$, $O_2 + O \rightleftarrows O_3$ (in addition to a number of diatomic dissociation equilibria) by Lewis and von Elbe (575). Furthermore, equilibrium constants for a number of organic reactions have been calculated on the same basis by Kassel (489) (490), Schumann and Aston (773) (774), Pitzer (694), and Thompson (852) (853). In some of these reactions, just as for the ethane-ethylene equilibrium, agreement between observed and calculated equilibrium constants is obtained only when restricting potentials hindering internal rotations are assumed (see the previous Table 143).

[14] The "observed" values given are interpolated between the actually observed values, which do not happen to fall exactly at the temperatures given.

2. Nature of Liquid and Solid States: Intermolecular Forces

From an investigation of the changes occurring in the spectrum in going from the gaseous to the liquid and solid states, together with a knowledge of the structure of the free molecule concerned (as obtained from the spectrum of the gas), it is possible to draw important conclusions about the nature of the liquid and solid states and of the intermolecular forces. This field of investigation has been developing rapidly in recent years. It is not possible to give a complete summary here; we shall limit the discussion to some of the important points. In the case of solids we shall discuss only molecular lattices, and not metallic, atomic, or ionic lattices, in which the individual molecules can no longer be considered as separate units.

Rotation of molecules in liquids and solids. When the pressure of a gas is increased the individual lines in the fine structure of infrared and Raman bands become broader and broader, because of the frequent collisions of the absorbing molecules with others, and on account of the increasing interaction of the molecules with decreasing average distance from one another. It is therefore not surprising that in liquids, in which the molecules are even closer together, Raman and infrared bands do not in general show a rotational fine structure. A notable exception is liquid hydrogen (see Molecular Spectra I), for which a well-resolved rotational Raman spectrum has been observed. The only two cases of polyatomic molecules that form exceptions do not refer to pure liquids but to solutions:

NH_3 in aqueous solution, according to Langseth (548), shows a vibrational Raman band at 3311.8 cm^{-1}, with a well-developed though diffuse fine structure consisting of S, R, Q, P, and O branches. This must indicate, as for H_2, that in the solution a quantized rotation of the NH_3 molecules takes place such that the rotational levels, though broad, do not merge into one another. H_2O in an inert solvent such as CCl_4 or CS_2 shows, according to Kinsey and Ellis (505) and Borst, Buswell, and Rodebush (171), a fairly well-developed fine structure of its bands,[15] which seems to be correlated to that of H_2O vapor and therefore also indicates a *nearly free, quantized rotation.* It seems at first somewhat difficult to understand why this fine structure should occur in these two cases and not in many other cases. However, on second thought, it is realized that there are not many cases suitable for such investigations, and only very few of these have actually been investigated. Only a molecule that gives widely spaced rotational lines would be expected to show bands with fine structure in the liquid state. The only molecules fulfilling this condition are HF, H_2O, NH_3, CH_4, and their homologues. Liquid HCl and H_2O definitely do not show any rotational structure of their infrared and Raman bands [for HCl see particularly West (918)]. This is easily understandable on the basis of their large dipole moments and their shape, which does not lend itself easily to undisturbed rotation in the liquid state. On the other hand, NH_3 and CH_4, which would be expected to rotate with less disturbance in the liquid state, have not been investigated in any detail. For NH_3 there is only the above-mentioned investigation of NH_3 dissolved in water and for CH_4 there is only an early investigation of the Raman spectrum of the liquid by McLennan, Smith and Wilhelm (608b) which does show indications of a fine structure of the band ν_3. A further study of the spectrum of liquid methane would appear to be most promising.

[15] This was, however, not found in similar investigations by Fox and Martin (328).

It is important to realize that in the case of HCl, H_2O, and molecules with smaller rotational spacings, the spectrum of the liquid is not simply a duplicate of the spectrum of the gas with each fine structure line so broad that the whole band appears diffuse. There is a drastic *change of the intensity distribution* as well. While, for example, in gaseous HCl at ordinary pressure we have in the infrared bands the two branches P and R separated by the zero gap, and at higher pressures at least two maxima, in the liquid there is only one comparatively sharp maximum (at least as sharp as the P and R maxima in the gas). This is shown in Fig. 173, taken from West (918). Another example is C_6H_6, for which in the gaseous state many bands are found to have three maxima, very probably corresponding to P, Q, and R branches (see p. 365), while in the liquid (and solid) state as well as in solution only one much sharper maximum occurs for each band [see Leberknight (561) and Fox and Martin (328)]. Similarly, in the Raman spectra of most liquids we do not observe simply an unresolved rotation band, but instead of there being a maximum on either side of the undisplaced line, as in the gas at low pressure, the intensity drops continuously from the undisplaced line. This is observed for the liquid as well as the gas at very high pressure. Fig. 174 shows as an illustration the intensity distribution in the rotational Raman spectrum of CO_2 at 15 and 60 atmospheres after Weiler (914). Also, in the vibrational Raman spectrum, lines that are broad in the gaseous state because they correspond to non-totally symmetric vibrations are found to be sharper in the liquid state [see Fig. 136 and Nielsen and Ward (670)].

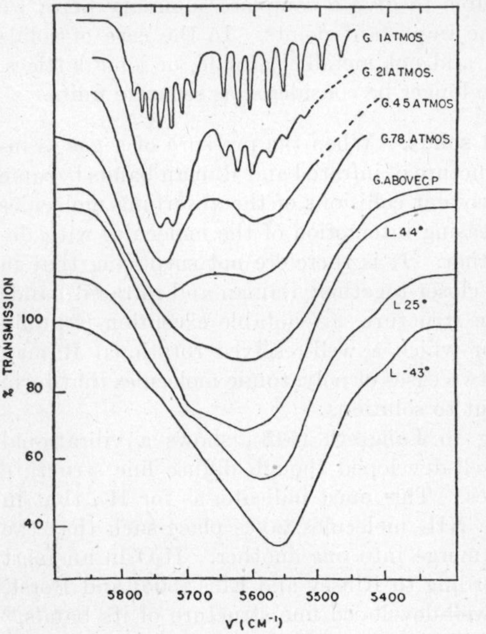

Fig. 173. First overtone of HCl at 1.76μ in the gas and in the liquid at various pressures and temperatures [after West (918)].

On the other hand, this difference between liquid and vapor does not always seem to occur between *solution* and vapor. The two previously discussed cases of NH_3 in H_2O, and H_2O in CCl_4 and CS_2 are examples of this. In many other cases the situation is made more complicated by polymerization and formation of hydrogen bonds.

The situation is again somewhat different in *solids*. In no case has quantized rotation been detected spectroscopically.[15a] The observation of a fine structure in solid HCl by Shearin mentioned in Molecular Spectra I has been disproved by Lee, Sutherland, and Wu (571). However, the existence of ortho hydrogen in solid hydrogen proves that quantized rotation takes place in solid hydrogen just as in the liquid, if in no other case. The infrared and Raman fundamental of solid HCl does

[15a] Note added in proof: Very recently Beck (132a) has reported fairly well resolved rotational fine structures in the infrared absorption bands of solid ammonium chloride and bromide.

show two maxima [see Hettner (448) and Lee, Sutherland, and Wu (571)]. But it is rather doubtful whether this means incipient quantized rotation, particularly since the two maxima merge into one for temperatures above 98° K. where a thermal transition point has been observed. If free quantized rotation occurs at all, it is more likely to occur above the transition point than below it. For solid HBr and HI, Zunino (978) found only one maximum. Again it would appear to be promising to investigate solid CH_4, particularly since thermal measurements for CH_4 as well as SiH_4 and CD_4 seem to indicate free rotation above a certain transition point. But from these thermal measurements it cannot be decided whether or not the rotation is quantized [see Clusius (220), Eucken and Veith (315), and others]. By means of thermal measurements as well as electrical measurements (dielectric constant, piezoelectric effect), free rotation in crystals has been shown to occur in a number of other cases [for example, for NH_4Cl by Hettich and Hendricks (447), for H_2O by Giauque and Ashley (358)].

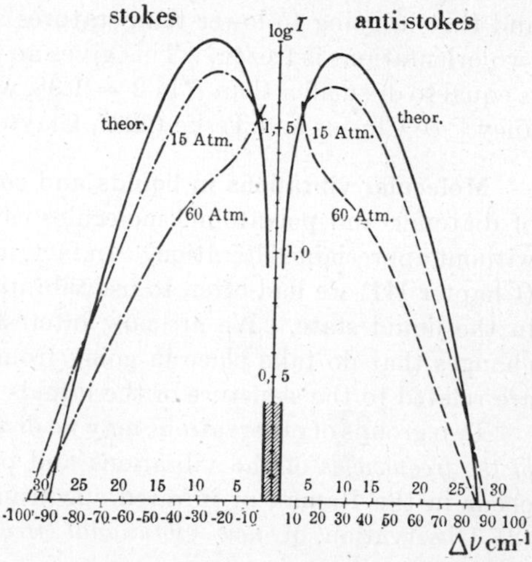

FIG. 174. **Intensity distribution in the unresolved rotational Raman band of CO_2 at 15 and 60 atm.** [after Weiler (914)].—The full curve represents the theoretical distribution at low pressure. The numbers above the abscissa axis are the running numbers of the rotational lines which are $(J/2) + 1$.

A *theory* of molecular rotation in *solids* was first given by Pauling (685), and was later developed by Fowler (327), Nielsen (663), Devonshire (286), and Cundy (251). These authors have shown, among other things, that with increasing strength of the interaction of the molecules a transition takes place from free quantized rotation to oscillation about an equilibrium orientation, and conversely that with increasing temperature a transition takes place from oscillation to rotation, explaining the observed transition points that are not accompanied by change of crystal structure.

Debye (271) has given a theoretical treatment of molecular rotation in *liquids.*

Intimately connected with the question of molecular rotation in crystals are certain *apparent contradictions to the third law of thermodynamics.* For example, if the absolute entropy of light or heavy hydrogen is determined from thermal measurements by calculating $\int_0^T \frac{C_p}{T}\,dT$, assuming that the integral vanishes at absolute zero, a value is obtained that is appreciably smaller than the entropy calculated statistically in the manner described in section 1 of this chapter. The explanation is that in the crystal the equilibrium between ortho and para molecules—that is, rotating and non-rotating molecules—freezes, and the C_p measurements at low temperature refer to the non-equilibrium mixture of the two modifications. By adding the entropy of mixing to the thermal entropy, agreement with the statistical values is obtained [see Giauque and Johnston (359) (357), Johnston and Long (474) and Clusius and Bartholomé (222)]. A similar situation seems to exist in H_2O [see Giauque and Ashley (358)], but such an effect does not occur for N_2, indicating that the interaction with the lattice prevents a free rotation at any temperature [see Clusius and Bartholomé (222)]. On the other hand, for slightly unsymmetrical molecules such as CO and N_2O, which apparently do not rotate freely in the solid state, a discrepancy between thermal and statistical entropies still occurs, and is explained by the fact that two almost equivalent orientations (CO and OC) are possible in the lattice and that in going to lower temperatures an almost 1 : 1 equilibrium between these two orientations is frozen. This gives an addition to the entropy of an amount that is equal to or smaller than $R \ln 2 = 1.38$, which agrees well with the observed discrepancy [see Clusius and Teske (223), Clayton and Giauque (211), and Clusius (221)].

Molecular vibrations in liquids and solids. It is well known that the *vibrations* of diatomic and polyatomic molecules can take place in the liquid and solid state without appreciable alteration. In fact, in the discussion of the vibrational structure (Chapter III) we had often to use vibrational frequencies of molecules as measured in the liquid state. We are now interested in determining how the usually small changes that do take place in going from the vapor to the liquid and solid phases are related to the structure of the liquids and solids.

Two groups of observations may be distinguished: (1) Observation of *small changes of the frequencies* of the vibrations and possibly of the intensities with which they occur in the Raman or infrared spectrum in the gaseous, liquid, and solid states: (2) Observation of *new vibrational frequencies* which are absent in the gaseous state.

In order to give an idea of the *magnitude of the frequency changes* we list in Table 146 the fundamental frequencies of some molecules in the gaseous, liquid, and solid states. The changes vary from 0 to 5 per cent and are practically always toward smaller values. It is noteworthy that in some cases different vibrations of one and the same molecule show very different relative frequency shifts. Furthermore, it must be noted that at least for the hydrogen halides [Zunino (978), West (918)], and probably also for the other molecules, the shift decreases with increasing temperature both in the liquid and solid states. This is, of course, what one would expect, since with increasing temperature the average separation of the molecules increases. Frequency shifts have also been investigated for many solutions, and have been found to be of the same order of magnitude, although frequently somewhat larger than for the corresponding pure liquids [see, for example, Ellis and Kinsey (303) and West (918)]. Kirkwood [quoted by West and Edwards (919)] and Bauer and Magat (127) have developed theoretical formulae for the shift, assuming a simple electrostatic interaction of an oscillating dipole with its surroundings of dielectric constant D. They find that the relative shift should be proportional to $(D - 1)/(2D + 1)$, which seems roughly to be the case for HCl in different non-ionizing solvents. Moreover, the order of magnitude of the effect comes out correctly. The theory implies essentially free rotation of the oscillating dipole.

In the second group of observations *new bands* are found which do not occur in the gas; or, in other words, for some vibrations the shifts are so great that they can no longer be considered as due to the influence of the average interaction with all the neighboring molecules. Thus in solid H_2O at low temperatures, in place of the

TABLE 146. VIBRATION FREQUENCIES (FUNDAMENTALS) IN THE GASEOUS, LIQUID, AND SOLID STATES.

Molecule	Vibrations	ν (gas) (cm^{-1})	ν (liquid) (cm^{-1})	ν (solid) (cm^{-1})	References
HCl		2886	2785	2768 (100° K.)	(448)
CO_2	$\nu_1(\sigma_g^+)$	1285.5	1285.5	1285	(608a)
	$2\nu_2(\Sigma_g^+)$	1388.3	1387.5	1388	
	ν_2	667.3		656[16] (87° K.)	(261a)
	ν_3	2349.3		2288[16] (87° K.)	
CS_2	$\nu_1(\sigma_g^+)$	655	656.5		(466)
H_2S				2545.8	(647)
	$\nu_1(a_1)$	2610.8	2573.6	2553.7	
				2520.8	
SO_2	$\nu_1(a_1)$	1151.2	1144.3		See Table 65
	$\nu_2(a_1)$	519	524.5		
	$\nu_3(b_1)$	1361	1336.0		
HCN	$\nu_1(\sigma^+)$	2089.0	2094		See Table 59
	$\nu_3(\sigma^+)$	3312.0	3213		
C_2H_2	$\nu_1(\sigma_g^+)$	1973.8	1959		(377)
	$\nu_2(\sigma_g^+)$	3373.7	3338		(380)
CH_4	$\nu_1(a_1)$	2914.2	2909	2906 (83° K.)	(823) (608b)
CH_3Cl	$\nu_1(a_1)$	2966.2	2955		
	$\nu_3(a_1)$	732.1	709		See Table 84
	$\nu_4(e)$	3041.8	3036		
$CHCl_3$	$\nu_1(a_1)$	3033	3018.9		See Table 86
	$\nu_2(a_1)$	672	668.3		
	$\nu_3(a_1)$	363	365.9		
	$\nu_6(e)$	262	262		
C_2H_4	$\nu_1(a_g)$	3019.3	3009		(378)
	$\nu_2(a_g)$	1623.3	1621		
	$\nu_3(a_g)$	1342.4	1340		
$CH_3—C\equiv CH$	$\nu_1(a_1)$	3429	3305		See Table 100
	$\nu_2(a_1)$	2941	2926.2		(386) (368)
	$\nu_3(a_1)$	2142	2123.5		
	$\nu_5(a_1)$	930	929.5		
C_2H_6	$\nu_1(a_1')$	2899.2	2884		(379)
	$\nu_3(a_1')$	993.0	994		
	$\nu_6(a_2'')$	1379.0	1370	1370	(63)
	$\nu_8(e')$	1486.0	1462	1462	
	$\nu_{11}(e'')$	1460	1463		
C_6H_6	$\nu_1(a_{1g})$	3069	3061.9		(152)
	$\nu_2(a_{1g})$	992	992		
	$\nu_{12}(e_{1u})$	3099	3090	3089 (193° K.)	(561)
	$\nu_{13}+\nu_{16}(E_{1u})$	3045	3035	3034 (193° K.)	

Raman line 3654 cm^{-1} of the vapor, two lines appear at 3090 and 3135 cm^{-1} [Sutherland (823)]. The generally accepted explanation of the new lines in this and other cases is that they are due to *polymers or associated molecules.*

Perhaps the most clear-cut case is that of formic acid (HCOOH). Here the vapor shows the characteristic O—H vibration (3570 cm^{-1}) only at higher temperatures, when it certainly is monomeric. At lower temperatures and in the liquid state this O—H vibration does not occur, but instead a new band at 3080 cm^{-1} is observed

[16] These are the main maxima.

[Bonner and Hofstadter (167)],[17] which must be ascribed to the O—H vibration in the *dimer*. The reason for the great change of the O—H vibration frequency in the dimer is now generally agreed to be *hydrogen bonding*, which is also responsible for the considerable stability of the dimer [binding energy = 12.6 kcal, see Hermann (428a)]. If the structure of the dimer is

$$
\begin{array}{ccc}
& \text{O—H}\cdots\text{O} & \\
& \diagup \qquad \diagdown\!\!\diagdown & \\
\text{H—C} & & \text{C—H} \\
& \diagdown\!\!\diagdown \qquad \diagup & \\
& \text{O}\cdots\text{H—O} &
\end{array}
$$

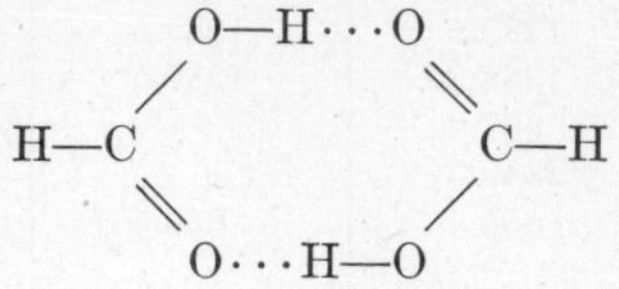

the H atoms may jump from one molecule to the other, leading to a configuration which differs from the above only by a reflection in the mid-plane. This gives rise to what is now usually called hydrogen bonding. On this basis one would expect the O—H vibration of the dimer to be greatly different from that of the monomer, while the other vibrational frequencies should be only slightly different, in agreement with observation. The structure indicated above has indeed been confirmed by electron-diffraction data [Pauling and Brockway (686) and Karle and Brockway (484a)].

A somewhat similar case is that of methyl (and ethyl) alcohol. As mentioned in Chapter III, p. 334, in the vapor the O—H vibration occurs at 3682 cm^{-1}, whereas in the liquid it occurs at 3400 cm^{-1}. That this is not simply a shift of the first type is shown by investigation of solutions of alcohol in various inert solvents [see for example Errera, Gaspart, and Sack (309)]. For certain small concentrations *both* bands occur at low temperature, but at higher temperatures or still greater dilutions only (or predominantly) the band at 3682 cm^{-1} occurs. This indicates very definitely that the latter band corresponds to the monomer, whereas the 3400 band corresponds to a polymer. Errera, Gaspart, and Sack (309) have even been able to resolve this band into two bands, one of which they ascribe to the dimer and the other to a higher polymer. At any rate it follows that in pure liquid alcohol there are practically no single molecules but only associated ones. The same conclusion applies also to liquid and solid H$_2$O (see above).

Similar results have also been obtained for HF [Buswell, Maycock, and Rodebush (187)], although it has only been investigated in the gaseous state and in solutions. As for HCOOH, at room temperature and not too low pressures, an additional band appears (at 3450 cm^{-1}): this band is much stronger than the ordinary HF band, indicating almost complete association due to hydrogen-bond formation. In very dilute solutions in CCl$_4$, however, the monomer predominates. Even in liquid HCl a secondary band appears close to each of the ordinary vibrational bands [see Freymann (336), West (918)]. These secondary bands disappear with increasing temperature and in all probability are due to a dimer or polymer, whose concentration in this case is very slight and which is probably not due to hydrogen bonding.

A number of investigations have also been carried out for weak solutions of CH$_3$OH, H$_2$O, and others in *"active"* solvents, that is, in solvents whose molecules can form hydrogen bonds with the CH$_3$OH, H$_2$O, $\cdots$ molecules. This bonding results in new bands of the associated molecules different from those observed in the

[17] In HCOOH the new band is overlapped by the C—H vibration which occurs at the same place both in the monomer and dimer. But investigation of HCOOD, where this chance coincidence does not occur, has confirmed that the above interpretation for HCOOH is correct.

pure liquids. By comparing the shift of the O—H band the strength of the hydrogen bonds of these associated molecules can be compared; that is, the *proton-attracting power of the solvents* can be compared [see Gordy (390), Gordy and Stanford (391)]. Further spectroscopic work on association, hydrogen bonding, and related topics has been carried out by Freymann (337), Thompson (845), Goubeau (395), Kempter and Mecke (497), and Giguère (360a).

In a few cases in the Raman spectra of crystals, new Raman lines very close (0–50 cm^{-1}) to the exciting line have been found. These lines have to be interpreted as due to *vibrations of the lattice* rather than of the molecule [see, for example, Venkateswaran (892)]. In going over to the liquid state these lines disappear in general, but in a few cases some indication of them remains, pointing to a quasi-crystalline structure of the liquid [see Gross and Vuks (401)]. Lattice vibrations in crystals and quasi-crystalline structures of liquids have also been detected by means of the electronic spectra of rare-earth ions by Freed and Weissmann (331). Further experimental material is necessary before more detailed conclusions can be drawn about this subject [compare also the summary by Glockler (364)].

APPENDIX

PHYSICAL CONSTANTS AND CONVERSION FACTORS

The values for the physical constants used in this book are given in the second column of Table 147. While these values are essentially the new values of Birge, he has recently given slightly different values based on a more detailed discussion and on more recent work [Birge (154)]. These latter values are given in the last column of Table 147. The difference between the two sets is in each case within the probable error given by Birge. The values used here are the same as those used in Molecular Spectra I and it is mainly for the sake of consistency with the earlier book that Birge's more recent values were not used here. In almost all cases thus far investigated the difference between the molecular data calculated with the two sets of physical constants is within the accuracy of the spectroscopic data.

In Table 148 are given the conversion factors of the energy units that follow from the physical constants adopted here. The values that follow from Birge's more recent constants are given in parentheses.

The value of the numerical factor $\frac{h}{8\pi^2 c}$ in the equation for the moment of inertia $I_B = \frac{h}{8\pi^2 cB}$ (see p. 14) using the more recent constants would be $27.98_{65} \times 10^{-40}$ instead of 27.994×10^{-40} used here. The factor $4\pi^2 c^2 M_1$ in the equation for the force constant $k = 4\pi^2 M_1 \mu_A c^2 \omega^2$ (see p. 160) would be 5.8890×10^{-2} instead of 5.8894×10^{-2}.

TABLE 147. PHYSICAL CONSTANTS.

Constant		Value used in this book	Birge's new value
Electronic charge	e	4.8029×10^{-10} e.s.u.	4.8025×10^{-10}
Planck's constant	h	6.626×10^{-27} erg sec	6.624×10^{-27}
Velocity of light	c	2.99776×10^{10} cm/sec	2.99776×10^{10}
Electronic mass (rest mass)	m	9.111×10^{-28} gm	9.1066×10^{-28}
$\frac{1}{16}$ mass of the O^{16} atom	M_1	1.6600×10^{-24} gm	$1.6599_2 \times 10^{-24}$
Number of molecules in a mole:			
Referred to Aston's atomic weight scale ($O^{16} = 16$)	N_A	6.0240×10^{23}	6.0244×10^{23}
Referred to the chemical atomic weight scale	N_{ch}	6.0224×10^{23}	6.0228×10^{23}
Boltzmann's constant	k	1.3807×10^{-16} erg/degree	$1.3804_7 \times 10^{-16}$
Gas constant per mole	R	1.9863 cals/degree/mole	1.98647

TABLE 148. CONVERSION FACTORS FOR ENERGY UNITS.

Unit	cm^{-1}	erg/molecule	cal/mole chem.	electron-volts (absolute)
1 cm^{-1}	1	1.9863×10^{-16} (1.9858)	2.8581 (2.8575)	1.2398×10^{-4} (1.2395)
1 erg/molecule	5.0344×10^{15} (5.0358)	1	1.4389×10^{16} (1.4390)	6.2416×10^{11} (6.2421)
1 cal/mole chem.	0.34988 (0.34996)	6.9498×10^{-17} (6.9494)	1	4.3378×10^{-5} (4.3379)
1 electron-volt (absolute)	8066.0 (8067.5)	1.6022×10^{-12} (1.60203)	23053 ($23052._8$)	1

BIBLIOGRAPHY

I. Textbooks, Monographs and Tables

1. R. B. Barnes, R. C. Gore, U. Liddel and V. Z. Williams, Infrared Spectroscopy, Industrial Applications and Bibliography (Reinhold, New York, 1944).
1a. S. Bhagavantam, Scattering of Light and the Raman Effect (Andhra University, Waltair, 1940).
2. M. Born, Optik (J. Springer, Berlin, 1933).
3. J. Cabannes, Anisotropie des Molécules; Effet Raman (Herrmann et Cie., Paris, 1930).
4. H. B. G. Casimir, Rotation of a Rigid Body in Quantum Mechanics (J. B. Wolters, The Hague, 1931).
5. P. Frank and R. v. Mises, Die Differential und Integralgleichungen der Mechanik und Physik (F. Vieweg, Braunschweig, 1925 and 1927).
6. S. Glasstone, K. J. Laidler and H. Eyring, The Theory of Rate Processes (McGraw-Hill, New York, 1941).
7. H. Hellmann, Einführung in die Quantenchemie (Deuticke, Leipzig, 1937).
8. G. Herzberg, Atomic Spectra and Atomic Structure (Dover Publications, New York, 1944).
9. G. Herzberg, Molecular Spectra and Molecular Structure I: Diatomic Molecules (Prentice-Hall, New York, 1939).
10. J. H. Hibben, The Raman Effect and its Chemical Applications (Reinhold Publishing Corporation, New York, 1939).
11. W. Jevons, Band Spectra of Diatomic Molecules (Physical Society, London, 1932).
12. E. C. Kemble, The Fundamental Principles of Quantum Mechanics (McGraw-Hill, New York, 1937).
13. K. W. F. Kohlrausch, Der Smekal-Raman-Effekt (J. Springer, Berlin, 1931).
14. K. W. F. Kohlrausch, Der Smekal-Raman-Effekt, Ergänzungsband 1931–1937 (J. Springer, Berlin, 1938).
15. R. de L. Kronig, The Optical Basis of the Theory of Valency (Cambridge University Press, 1935).
15a. H. Margenau and G. M. Murphy, The Mathematics of Physics and Chemistry (Van Nostrand, New York, 1943).
16. J. E. Mayer and M. G. Mayer, Statistical Mechanics (Wiley, New York, 1940).
17. L. Pauling, The Nature of the Chemical Bond (Cornell University Press, 1939 and 1941).
18. L. Pauling and E. B. Wilson Jr., Introduction to Quantum Mechanics (McGraw-Hill, New York, 1935).
19. F. G. Rawlins and A. M. Taylor, Infrared Analysis of Molecular Structure (Cambridge University Press, 1929).
20. V. Rojansky, Introductory Quantum Mechanics (Prentice-Hall, New York, 1938).
21. C. Schaefer and F. Matossi, Das ultrarote Spectrum (J. Springer, Berlin, 1930).
22. H. Sponer, Molekülspektren und ihre Anwendung auf chemische Probleme I. Tabellen, II. Text (J. Springer, Berlin, 1935 and 1936).
22a. G. B. B. M. Sutherland, Infrared and Raman Spectra (Methuen, London, 1935).
23. B. L. van der Waerden, Die gruppentheoretische Methode in der Quantenmechanik (J. Springer, Berlin 1932).
24. W. Weizel, Bandenspektren (Akademische Verlagsgesellschaft, Leipzig, 1931).
25. E. T. Whittaker, Analytical Dynamics of Particles and Rigid Bodies (Dover Publications, New York, 1944).
26. T. Y. Wu, Vibration Spectra and Structure of Polyatomic Molecules (National University of Peking, Kun-ming China, 1939).

II. References to Individual Papers

30. W. S. Adams and T. Dunham, Pub. Astron. Soc. Pac. **44**, 243 (1932).
31. A. Adel, Phys. Rev. **44**, 691 (1933).
32. ———, Phys. Rev. **45**, 56 (1934).
33. ———, Phys. Rev. **46**, 222 (1934).
34. ———, Astrophys. J. **94**, 375, 379 (1941).
34a. ———, Astrophys. J. **97**, 190 (1943).

35. A. Adel and E. F. Barker, Phys. Rev. **45**, 277 (1934).
36. ——— and ———, J. Chem. Phys. **2**, 627 (1934).
37. ——— and D. M. Dennison, Phys. Rev. **43**, 716 (1933).
38. ——— and ———, Phys. Rev. **44**, 99 (1933).
39. ——— and V. M. Slipher, Phys. Rev. **46**, 902 (1934).
40. ——— and ———, and O. Fouts, Phys. Rev. **49**, 288 (1936).
41. H. Aderhold and H. E. Weiss, Z. Physik, **88**, 83 (1934).
42. E. Amaldi and G. Placzek, Z. Physik, **81**, 259 (1933).
43. R. Ananthakrishnan, Proc. Ind. Acad. Sci. **2A**, 452 (1935).
44. ———, Proc. Ind. Acad. Sci. **3A**, 52 (1936).
45. ———, Proc. Ind. Acad. Sci. **3A**, 527 (1936).
46. ———, Proc. Ind. Acad. Sci. **4A**, 82 (1936).
47. ———, Proc. Ind. Acad. Sci. **5A**, 76 (1937).
48. ———, Proc. Ind. Acad. Sci. **5A**, 285 (1937).
49. ———, Proc. Ind. Acad. Sci. **5A**, 447 (1937).
50. T. F. Anderson, J. Chem. Phys. **4**, 161 (1936).
51. ——— and A. B. Burg, J. Chem. Phys. **6**, 586 (1938).
52. ———, E. N. Lassettre and D. M. Yost, J. Chem. Phys. **4**, 703 (1936).
53. D. H. Andrews and J. W. Murray, J. Chem. Phys. **2**, 634 (1934).
54. W. R. Angus, C. R. Bailey, J. B. Hale, C. K. Ingold, A. H. Leckie, C. G. Raisin, J. W. Thompson and C. L. Wilson, J. Chem. Soc. 1936, p. 966.
55. ———, C. R. Bailey, J. B. Hale, C. K. Ingold, A. H. Leckie, C. G. Raisin, J. W. Thompson and C. L. Wilson, J. Chem. Soc. 1936, p. 971.
56. ———, C. K. Ingold and A. H. Leckie, J. Chem. Soc. 1936, p. 925.
57. ——— and A. H. Leckie, Proc. Roy. Soc. London, **149**, 327 (1935).
58. ——— and ———, Proc. Roy. Soc. London, **150**, 615 (1935).
59. ——— and ———, J. Chem. Phys. **4**, 83 (1936).
60. J. G. Aston, in Taylor—Glasstone's Treatise on Physical Chemistry, vol. I, p. 511 (Van Nostrand, 1942).
60a. ——— and P. M. Doty, J. Chem. Phys. **8**, 743 (1940).
60b. ———, M. L. Eidinoff and W. S. Forster, J. Amer. Chem. Soc. **61**, 1539 (1939).
61. ———, H. L. Fink and S. C. Schumann, J. Amer. Chem. Soc. **65**, 341 (1943).
61a′. ———, S. Isserow, G. J. Szasz and R. M. Kennedy, J. Chem. Phys. **12**, 336 (1944).
61a. ———, R. M. Kennedy and G. H. Messerly, J. Amer. Chem. Soc. **63**, 2343 (1941).
61b. ———, ——— and S. C. Schumann, J. Amer. Chem. Soc. **62**, 2059 (1940).
61c. ——— and G. H. Messerly, J. Amer. Chem. Soc. **58**, 2354 (1936).
61d. ——— and ———, J. Amer. Chem. Soc. **62**, 1917 (1940).
61e. ———, M. L. Sagenkahn, G. J. Szasz, G. W. Moessen and H. F. Zuhr, J. Amer. Chem. Soc. **66**, 1171 (1944).
62. ———, S. C. Schumann, H. L. Fink and P. M. Doty, J. Amer. Chem. Soc. **63**, 2029 (1941).
62a. ———, C. W. Siller and G. H. Messerly, J. Amer. Chem. Soc. **59**, 1743 (1937).
63. W. H. Avery and C. F. Ellis, J. Chem. Phys. **10**, 10 (1942).

64. W. Bacher and J. Wagner, Z. physik. Chem. B. **43**, 191 (1939).
65. R. M. Badger, Phys. Rev. **35**, 1038 (1930).
66. ——— and S. H. Bauer, J. Chem. Phys. **4**, 469 (1936).
67. ——— and ———, J. Chem. Phys. **4**, 711 (1936).
68. ——— and ———, J. Amer. Chem. Soc. **59**, 303 (1937).
69. ——— and ———, J. Chem. Phys. **5**, 599 (1937).
70. ——— and ———, J. Chem. Phys. **5**, 605 (1937).
71. ——— and J. L. Binder, Phys. Rev. **37**, 800 (1931).
72. ——— and ———, Phys. Rev. **38**, 1442 (1931).
73. ——— and L. G. Bonner, Phys. Rev. **43**, 305 (1933).
74. ——— and C. H. Cartwright, Phys. Rev. **33**, 692 (1929).
75. ——— and R. Mecke, Z. physik. Chem. B. **5**, 333 (1929).
76. ——— and L. R. Zumwalt, J. Chem. Phys. **6**, 711 (1938).
77. C. R. Bailey, quoted by Sutherland and Penney (831).
78. ——— and S. C. Carson, J. Chem. Phys. **7**, 859 (1939).

79. C. R. Bailey, S. C. Carson, and E. F. Daly, Proc. Roy. Soc. London, **173**, 339 (1939).
80. ——— and A. B. D. Cassie, Proc. Roy. Soc. London, **132**, 236 (1931).
81. ——— and ———, Proc. Roy. Soc. London, **135**, 375 (1932); **140**, 605 (1933).
82. ——— and ———, Proc. Roy. Soc. London, **137**, 622 (1932).
83. ——— and ———, Proc. Roy. Soc. London, **142**, 129 (1933).
84. ——— and ———, Nature, **131**, 239, 910 (1933).
85. ——— and ———, Proc. Roy. Soc. London, **145**, 336 (1934).
86. ———, ——— and W. R. Angus, Proc. Roy. Soc. London, **130**, 142 (1930).
87. ——— and R. R. Gordon, J. Chem. Phys. **6**, 225 (1938).
88. ——— and ———, Trans. Farad. Soc. **34**, 1133 (1938).
89. ——— and J. B. Hale, Phil. Mag. **25**, 98 (1938).
90. ———, ———, C. K. Ingold and J. W. Thompson, J. Chem. Soc. 1936, p. 931.
91. ———, ——— and J. W. Thompson, Proc. Roy. Soc. London, **161**, 107 (1937).
92. ———, ——— and ———, J. Chem. Phys. **5**, 274 (1937).
93. ———, ——— and ———, Proc. Roy. Soc. London, **167**, 555 (1938).
94. ———, J. W. Thompson and J. B. Hale, J. Chem. Phys. **4**, 625 (1936).
95. O. Ballaus and J. Wagner, Z. physik. Chem. B. **45**, 165 (1940).
96. G. B. Banerji and B. Mishra, Ind. J. Physics, **15**, 359 (1941).
97. D. Barca-Galateanu, Z. Physik, **117**, 589 (1941).
98. P. Barchewitz, C. R. Paris, **203**, 930 (1936).
99. ——— and G. Costeanu, C. R. Paris, **207**, 722 (1938).
100. ——— and J. Garach, C. R. Paris, **208**, 2071 (1939).
101. ——— and M. Parodi, J. de Phys. **10**, 143 (1939).
102. ——— and ———, C. R. Paris, **209**, 30 (1939).
103. E. F. Barker, Astrophys. J. **55**, 391 (1922).
104. ———, Phys. Rev. **23**, 200 (1924).
105. ———, Phys. Rev. **33**, 684 (1929).
106. ———, Phys. Rev. **55**, 657 (1939).
107. ———, J. Chem. Phys. **7**, 277 (1939).
107a. ———, Rev. Mod. Phys. **14**, 198 (1942).
108. ——— and A. Adel, Phys. Rev. **44**, 185 (1933).
109. ——— and G. Bosschieter, J. Chem. Phys. **6**, 563 (1938).
110. ——— and H. H. Nielsen, Phys. Rev. **37**, 727 (1931).
111. ——— and E. K. Plyler, J. Chem. Phys. **3**, 367 (1935).
112. ——— and W. W. Sleator, J. Chem. Phys. **3**, 660 (1935).
113. ——— and T. Y. Wu, Phys. Rev. **45**, 1 (1934).
114. R. B. Barnes, Phys. Rev. **39**, 562 (1932).
115. ———, Phys. Rev. **47**, 658 (1935).
116. ———, W. S. Benedict and C. M. Lewis, Phys. Rev. **47**, 129 (1935).
117. ——— and R. R. Brattain, J. Chem. Phys. **3**, 446 (1935).
118. J. Barriol, J. de Phys. **10**, 215 (1939).
119. E. Bartholomé, Z. physik. Chem. B. **23**, 152 (1933).
120. ———, Z. Electrochem. **42**, 341 (1936).
121. ——— and K. Clusius, Z. Electrochem. **40**, 529 (1934).
122. ——— and J. Karweil, Z. physik. Chem. B. **35**, 442 (1937).
123. ——— and ———, Z. physik. Chem. B. **39**, 1 (1938).
124. ——— and E. Teller, Z. physik. Chem. B. **19**, 366 (1932).
125. P. F. Bartunek and E. F. Barker, Phys. Rev. **48**, 516 (1935).
126. M. Battista, N. Cimento, **14**, 343 (1937).
127. E. Bauer and M. Magat, Physica **5**, 718 (1938).
128. S. H. Bauer and R. M. Badger, J. Chem. Phys. **5**, 852 (1937).
129. E. Bauermeister and W. Weizel, Physik. Z. **37**, 169 (1936).
130. W. Baumann and R. Mecke, Z. Physik, **81**, 445 (1933).
131. J. Y. Beach and K. J. Palmer, J. Chem. Phys. **6**, 639 (1938).
132. ——— and A. Turkevich, J. Amer. Chem. Soc. **61**, 303, 3127 (1939).
132a. C. Beck, J. Chem. Phys. **12**, 71 (1944).
133. C. M. Beeson and D. M. Yost, J. Chem. Phys. **7**, 44 (1939).
134. F. K. Bell, J. Amer. Chem. Soc. **57**, 1023 (1935).

134a. R. P. Bell, Trans. Farad. Soc. **38**, 422 (1942).
135. D. Bender, Phys. Rev. **47**, 252 (1935).
136. W. S. Benedict, Phys. Rev. **47**, 641 (1935).
137. ———, K. Morikawa, R. B. Barnes and H. S. Taylor, J. Chem. Phys. **5**, 1 (1937).
138. W. H. Bennett and C. F. Meyer, Phys. Rev. **32**, 888 (1928).
139. E. Bernard and C. Manneback, Ann. Brux, **59**, 113 (1939).
139a. ———, ——— and A. Verleysen, Ann. Brux. **59**, 376 (1939); **60**, 45 (1940).
140. H. J. Bernstein, J. Chem. Phys. **6**, 718 (1938).
141. ——— and W. H. Martin, Trans. Roy. Soc. Canada III, **31**, 95 (1937).
142. Best and Trampe quoted by D. M. Yost, Proc. Ind. Acad. **8A**, 333 (1938).
143. H. Bethe, Ann. Physik, **3**, 133 (1929).
144. S. Bhagavantam, Ind. J. of Phys. **5**, 66, 86 (1930).
145. ———, Ind. J. of Phys. **6**, 595 (1931).
146. ———, Nature, **129**, 830 (1932).
147. ———, Nature, **130**, 740 (1932).
148. ———, Nature, **138**, 1096 (1936).
149. ———, Phys. Rev. **53**, 1015 (1938).
150. ——— and A. V. Rao, J. Chem. Phys. **4**, 293 (1936).
151. ——— and ———, Proc. Ind. Acad. Sci. **3A**, 135 (1936).
152. ——— and ———, Nature, **139**, 114 (1937); Proc. Ind. Acad. Sci. **5A**, 18 (1937).
153. ——— and T. Venkatarayudu, Proc. Ind. Acad. Sci. **8A**, 101, 115 (1938).
154. R. T. Birge, Rev. Mod. Phys. **13**, 233 (1941).
155. N. Bjerrum, Verh. d.d. phys. Ges. **16**, 737 (1914).
155a. J. F. Black and H. S. Taylor, J. Chem. Phys. **11**, 395 (1943).
156. E. Blum and H. Verleger, Physik. Z. **38**, 776 (1937).
157. H. Boersch, Wien. Ber. (IIb) **144**, 1 (1935).
158. ———, Monatshefte f. Chem. **65**, 311 (1935).
159. G. B. Bonino, Lincei **25**, 502 (1937).
160. ——— and R. Manzoni-Ansidei, Lincei **25**, 489, 494 (1937).
161. L. G. Bonner, Phys. Rev. **43**, 305 (1933).
162. ———, Phys. Rev. **46**, 458 (1934).
163. ———, J. Amer. Chem. Soc. **58**, 34 (1936).
164. ———, J. Chem. Phys. **5**, 293 (1937).
165. ———, J. Chem. Phys. **5**, 704 (1937).
166. ——— and R. Hofstadter, Phys. Rev. **52**, 249 (1937).
167. ——— and ———, J. Chem. Phys. **6**, 531 (1938).
168. ——— and J. S. Kirby-Smith, Phys. Rev. **57**, 1078A (1940).
169. A. Borden and E. F. Barker, J. Chem. Phys. **6**, 553 (1938).
170. M. Born and E. Brody, Z. Physik, **6**, 140 (1921).
171. L. B. Borst, A. M. Buswell and W. H. Rodebush, J. Chem. Phys. **6**, 61 (1938).
172. M. van den Bossche and C. Manneback, Ann. Brux. **54**, 230 (1934).
173. M. Bourguel and L. Piaux, Bull. Soc. Chim. Fr. **2**, 1958 (1935).
174. K. Bradacs and L. Kahovec, Z. physik. Chem. B. **48**, 63 (1941).
175. C. A. Bradley, Phys. Rev. **40**, 908 (1932).
176. H. Braune and G. Engelbrecht, Z. physik. Chem. B. **19**, 303 (1932).
177. ——— and S. Knoke, Z. physik. Chem. B. **21**, 297 (1933).
178. C. J. Brester Thesis, Utrecht (1923): Kristallsymmetrie und Reststrahlen, and Z. Physik, **24**, 324 (1924).
179. L. O. Brockway, Rev. Mod. Phys. **8**, 231 (1936).
180. ——— and L. Pauling, Proc. Nat. Acad. Sci. U. S. A. **19**, 68 (1933).
181. ——— and ———, Proc. Nat. Acad. Sci. U. S. A. **19**, 860 (1933).
182. A. E. Brodskii and A. M. Sack, J. Chem. Phys. **3**, 449 (1935); Acta Physicochimica **2**, 215 (1935).
183. E. J. Burcik, E. H. Eyster and D. M. Yost, J. Chem. Phys. **9**, 118 (1941).
184. ——— and D. M. Yost, J. Chem. Phys. **7**, 1114 (1939).
185. O. Burkard, Proc. Ind. Acad. **8**, 365 (1938).
186. A. M. Buswell, V. Deitz and W. H. Rodebush, J. Chem. Phys. **5**, 501 (1937).
187. ———, R. L. Maycock and W. H. Rodebush, J. Chem. Phys. **8**, 362 (1940).
188. ———, G. W. McMillan, W. H. Rodebush and F. T. Wall, J. Amer. Chem. Soc. **61**, 2809 (1939).

189. J. Cabannes, Ann. de Phys. (10) **18**, 285 (1932).
190. ———, J. Chim. Phys. **35**, 1 (1938).
191. ——— and A. Rousset, Ann. de Phys. **19**, 229 (1933).
191a. ——— and ———, J. de Physique (8) **1**, 155 (1940).
192. D. M. Cameron and H. H. Nielsen, Phys. Rev. **53**, 246 (1938).
193. ———, W. C. Sears and H. H. Nielsen, J. Chem. Phys. **7**, 994 (1939).
194. A. Carrelli and P. Tulipano, N. Cimento, **15**, 1 (1938).
195. A. B. D. Cassie, Proc. Roy. Soc. London, **148**, 87 (1934).
196. S. H. Chao, Phys. Rev. **50**, 27 (1936).
196a. A. Charlesby, Proc. Phys. Soc. London, **54**, 471 (1942).
197. J. Chédin, C. R. Paris, **201**, 552 (1935); **203**, 1509 (1936).
198. ———, Ann. de Chim. **8**, 243 (1937).
199. ———, J. de Phys. **10**, 445 (1939).
200. ——— and J. C. Pradier, C. R. Paris, **203**, 722 (1936).
201. H. C. Cheng, Z. physik. Chem. B. **26**, 288 (1934).
202. ———, J. de Chim. Phys. **32**, 715 (1935).
203. ———, C. F. Hsueh and T. Y. Wu, J. Chem. Phys. **6**, 8 (1938).
204. ——— and J. Lecomte, J. de Phys. **6**, 477 (1935).
205. W. H. J. Childs, Proc. Roy. Soc. London, **153**, 555 (1936).
206. ——— and H. A. Jahn, Z. Physik, **104**, 804 (1937).
207. ——— and ———, Proc. Roy. Soc. London, **169**, 428 (1939).
208. ——— and ———, Proc. Roy. Soc. London, **169**, 451 (1939).
209. ——— and R. Mecke, Z. Physik, **64**, 162 (1930); **68**, 344 (1931).
210. K. N. Choi and E. F. Barker, Phys. Rev. **42**, 777 (1932).
211. J. O. Clayton and W. F. Giauque, J. Amer. Chem. Soc. **54**, 2610 (1932).
212. ——— and ———, J. Amer. Chem. Soc. **55**, 5071 (1933).
213. A. P. Cleaves and E. K. Plyler, J. Chem. Phys. **7**, 563 (1939).
214. ———, H. Sponer and L. G. Bonner, J. Chem. Phys. **8**, 784 (1940).
215. C. E. Cleeton and N. H. Williams, Phys. Rev. **45**, 234 (1934).
216. F. F. Cleveland, J. Amer. Chem. Soc. **63**, 622 (1941).
217. ———, D. T. Hamilton and M. J. Murray, Phys. Rev. **61**, 735A (1942).
218. ——— and M. J. Murray, J. Chem. Phys. **9**, 390 (1941).
218a. ——— and ———, J. Chem. Phys. **11**, 450 (1943); **12**, 320 (1944).
218b. ———, ——— and H. J. Taufen, J. Chem. Phys. **10**, 172 (1942).
219. ———, ———, J. R. Coley and V. I. Komarewsky, J. Chem. Phys. **10**, 18 (1942).
220. K. Clusius, Z. physik. Chem. B. **23**, 213 (1933).
221. ———, Nature, **130**, 775 (1932).
222. ——— and E. Bartholomé, Z. physik. Chem. B. **30**, 258 (1935).
223. ——— and W. Teske, Z. physik. Chem. B. **6**, 135 (1929).
224. W. W. Coblentz, Publ. Carnegie Inst. of Wash. D. C. No. 35 Part I (1905).
225. W. F. Colby, Phys. Rev. **47**, 388 (1935).
226. G. K. T. Conn and G. B. B. M. Sutherland, Proc. Roy. Soc. London, **172**, 172 (1939).
227. ——— and C. K. Wu, Trans. Farad. Soc. **34**, 1483 (1938).
228. J. B. Conn, G. B. Kistiakowsky and E. A. Smith, J. Amer. Chem. Soc. **61**, 1868 (1939).
229. J. P. Cooley, Astrophys. J. **62**, 73 (1925).
230. A. S. Coolidge, J. Amer. Chem. Soc. **50**, 2166 (1928).
231. C. Corin, Bull. Soc. Roy. Sci. Liège, (1938), p. 243.
232. ———, J. de Chim. Phys. **32**, 241 (1935).
233. ——— and G. B. B. M. Sutherland, Proc. Roy. Soc. London, **165**, 43 (1938).
234. G. Costeanu, C. R. Paris, **207**, 285 (1938).
234a. T. G. Cowling, Nature, **152**, 694 (1943).
235. B. L. Crawford, Jr., J. Chem. Phys. **7**, 555 (1939).
236. ———, J. Chem. Phys. **8**, 273 (1940).
237. ———, J. Chem. Phys. **8**, 526 (1940).
238. ———, J. Chem. Phys. **8**, 744 (1940).
239. ———, W. H. Avery and J. W. Linnett, J. Chem. Phys. **6**, 682 (1938).
240. ——— and S. R. Brinkley, Jr., J. Chem. Phys. **9**, 69 (1941).
241. ——— and P. C. Cross, J. Chem. Phys. **5**, 371 (1937).
242. ——— and ———, J. Chem. Phys. **5**, 621 (1937).

243. B. L. Crawford and P. C. Cross, J. Chem. Phys. **6**, 525 (1938).
244. ——— and J. T. Edsall, J. Chem. Phys. **7**, 223 (1939).
245. ——— and L. Joyce, J. Chem. Phys. **7**, 307 (1939).
245a. ———, G. B. Kistiakowsky, W. W. Rice, A. T. Wells and E. B. Wilson, J. Amer. Chem. Soc **61**, 2980 (1939).
246. ——— and W. W. Rice, J. Chem. Phys. **7**, 437 (1939).
247. ——— and E. B. Wilson, J. Chem. Phys. **9**, 323 (1941).
248. P. C. Cross, Phys. Rev. **47**, 7 (1935); J. Chem. Phys. **5**, 370 (1937).
249. ——— and F. Daniels, J. Chem. Phys. **1**, 48 (1933).
249a. ———, R. M. Hainer and G. W. King, J. Chem. Phys. **12**, 210 (1944).
250. ——— and J. H. Van Vleck, J. Chem. Phys. **1**, 350 (1933).
251. H. M. Cundy, Proc. Roy. Soc. London, **164**, 420 (1938).

252. A. Dadieu, Naturwiss. **18**, 895 (1930); Monatshefte f. Chem. **57**, 437 (1931).
253. ——— and W. Engler, Wien. Anz. p. 128 (1935).
254. ——— and ———, Wien. Anz. p. 192 (1935).
255. ——— and K. W. F. Kohlrausch, Chem. Ber. **63**, 1657 (1930).
256. ——— and ———, Wien. Ber. **139**, 77, 165 (1930); Monatshefte f. Chem. **55**, 379; **56**, 461 (1930).
257. ——— and ———, Physik. Z. **33**, 165 (1932).
258. ———, ——— and A. Pongratz, Monatshefte f. Chem. **60**, 221 (1932).
259. ———, ——— and ———, Monatshefte f. Chem. **61**, 426 (1932).
260. ——— and H. Kopper, Wien. Anz. p. 92 (1935).
261. ———, A. Pongratz and K. W. F. Kohlrausch, Monatshefte f. Chem. **61**, 369 (1932).
261a. W. Dahlke, Z. Physik, **102**, 360 (1936).
261b. B. P. Dailey with W. A. Felsing, J. Amer. Chem. Soc. **65**, 42, 44 (1943).
262. I. Damaschun, Z. physik. Chem. B. **16**, 81 (1932).
263. B. T. Darling and D. M. Dennison, Phys. Rev. **57**, 128 (1940).
264. E. Darmois and M. Théodoresco, C. R. Paris, **208**, 1308 (1939).
265. P. Daure, Ann. de Phys. **12**, 375 (1929).
266. ——— and A. Kastler, C. R. Paris, **192**, 1721 (1931).
267. ———, ——— and H. Berry, C. R. Paris, **202**, 569 (1936).
268. M. M. Davies, Trans. Farad. Soc. **35**, 1184 (1939).
269. ———, Trans. Farad. Soc. **36**, 333, 1114 (1940).
270. C. O. Davis and H. L. Johnston, J. Amer. Chem. Soc. **56**, 1045 (1934).
271. P. Debye, Physik. Z. **36**, 100 (1935).
272. J. M. Delfosse, Ann. Brux. **55**, 114 (1935).
273. ———, Nature, **137**, 868 (1936).
274. ——— and R. Goovaerts, Bull. Belg. **21**, 410 (1935).
274a. M. Delwaulle, C. R. Paris, **206**, 1965 (1938); **208**, 999 (1939).
275. ———, F. François and J. Wiemann, C. R. Paris, **208**, 1818 (1939).
276. D. M. Dennison, Astrophys. J. **62**, 84 (1925).
277. ———, Phil. Mag. **1**, 195 (1926).
278. ———, Phys. Rev. **28**, 318 (1926).
279. ———, Rev. Mod. Phys. **3**, 280 (1931).
280. ———, Rev. Mod. Phys. **12**, 175 (1940).
281. ——— and J. D. Hardy, Phys. Rev. **39**, 938 (1932).
282. ——— and S. B. Ingram, Phys. Rev. **36**, 1451 (1930).
283. ——— and M. Johnston, Phys. Rev. **47**, 93 (1935).
284. ——— and G. E. Uhlenbeck, Phys. Rev. **41**, 313 (1932).
285. ——— and N. Wright, Phys. Rev. **38**, 2077 (1931).
286. A. F. Devonshire, Proc. Roy. Soc. London, **153**, 601 (1936).
286a. T. DeVries and B. T. Collins, J. Amer. Chem. Soc. **64**, 1224 (1942).
287. R. G. Dickinson, R. T. Dillon and F. Rasetti, Phys. Rev. **34**, 582 (1929).
288. G. H. Dieke and G. B. Kistiakowsky, Phys. Rev. **45**, 4 (1934).
289. P. Donzelot, C. R. Paris, **203**, 1069 (1936).
290. J. Duchesne, C. R. Paris, **204**, 1112 (1937).
291. ———, Nature, **142**, 256 (1938).
292. ——— and M. Parodi, Nature, **144**, 382 (1939).

293. A. B. F. Duncan and J. W. Murray, J. Chem. Phys. **2**, 636 (1934).
294. Th. Dunham, Publ. Astron. Soc. Pac. **45**, 42 (1933).

295. E. S. Ebers and H. H. Nielsen, J. Chem. Phys. **5**, 822 (1937).
296. ——— and ———, J. Chem. Phys. **6**, 311 (1938).
297. W. F. Edgell and G. Glockler, J. Chem. Phys. **9**, 375 (1941).
298. J. T. Edsall, J. Chem. Phys. **5**, 225 (1937).
299. ———, J. Chem. Phys. **5**, 508 (1937).
299a. ——— and H. Scheinberg, J. Chem. Phys. **8**, 520 (1940).
300. ——— and E. B. Wilson, Jr., J. Chem. Phys. **6**, 124 (1938).
301. M. L. Eidinoff and J. G. Aston, J. Chem. Phys. **3**, 379 (1935).
302. J. W. Ellis, Phys. Rev. **23**, 48 (1924); **32**, 906 (1928).
303. ——— and E. L. Kinsey, J. Chem. Phys. **6**, 497 (1938).
304. G. Emschwiller and J. Lecomte, J. de Phys. **8**, 130 (1937).
305. W. Engler, Z. physik. Chem. B. **35**, 433 (1937).
306. ——— and K. W. F. Kohlrausch, Z. physik. Chem. B. **34**, 214 (1936).
307. J. Errera and P. Mollet, Nature, **138**, 882 (1936).
308. ——— and ———, C. R. Paris, **204**, 259 (1937).
309. ———, R. Gaspart and H. Sack, J. Chem. Phys. **8**, 63 (1940).
310. A. Eucken and H. Ahrens, Z. physik. Chem. B. **26**, 297 (1934).
311. ——— and A. Bertram, Z. physik. Chem. B. **31**, 361 (1936).
312. ——— and K. v. Lüde, Z. physik. Chem. B. **5**, 413 (1929).
313. ——— and A. Parts, Z. physik. Chem. B. **20**, 184 (1933).
314. ——— and F. Sauter, Z. physik. Chem. B. **26**, 463 (1934).
314a. ——— and K. Schäfer, Z. physik. Chem. B. **51**, 60 (1942).
315. ——— and H. Veith, Z. physik. Chem. B. **34**, 275 (1936).
316. E. H. Eyster, J. Chem. Phys. **6**, 576 (1938).
317. ———, J. Chem. Phys. **6**, 580 (1938).
318. ———, J. Chem. Phys. **8**, 135 (1940).
319. ——— and R. H. Gilette, J. Chem. Phys. **8**, 369 (1940).

320. F. Fehér, Z. Elektrochem. **43**, 663 (1937).
321. ——— and W. Kolb, Naturwiss. **27**, 615 (1939).
321a. ——— and G. Morgenstern, Z. anorg. allgem. Chem. **232**, 169 (1937).
322. E. Fermi, Z. Physik, **71**, 250 (1931).
322a. R. Fichter, Helv. Phys. Acta, **13**, 309 (1940).
323. A. M. de Ficquelmont, M. Magat and L. Ochs, C. R. Paris, **208**, 1900 (1939).
324. H. M. Foley and H. M. Randall, Phys. Rev. **59**, 171 (1941).
325. R. Fonteyne, Nature, **138**, 886 (1936); Natuurwet. Tijdschr. **20**, 112 (1938).
325a. ———, Naturwiss. **31**, 441 (1943).
326. ———, J. Chem. Phys. **8**, 60 (1940).
326a. W. R. Forsythe and W. F. Giauque, J. Amer. Chem. Soc. **64**, 48 (1942).
327. R. H. Fowler, Proc. Roy. Soc. London, **149**, 1 (1935); **151**, 1 (1935).
328. J. J. Fox and A. E. Martin, Proc. Roy. Soc. London, **167**, 257 (1938).
329. ——— and ———, Proc. Roy. Soc. London, **174**, 234 (1940).
330. ——— and ———, Proc. Roy. Soc. London, **175**, 208 (1940).
330a. F. François, C. R. Paris, **207**, 425 (1938).
331. S. Freed and S. I. Weissmann, J. Chem. Phys. **8**, 840 (1940).
332. W. Fresenius and J. Karweil, Z. physik. Chem. B. **44**, 1 (1939).
333. K. Freudenberg and R. Mecke, Z. Physik. **81**, 465 (1933).
334. M. Freymann and R. Freymann, C. R. Paris, **200**, 1043 (1935); **205**, 852 (1937).
335. ———, ——— and Y. Ta, C. R. Paris, **207**, 728 (1938).
336. R. Freymann, J. Chem. Phys. **6**, 497 (1938).
337. ———, J. de Phys. **9**, 517 (1938).
338. L. W. Fung and E. F. Barker, Phys. Rev. **45**, 238 (1934).
339. G. W. Funke, Z. Physik. **99**, 341 (1936).
340. ———, Z. Physik. **104**, 169 (1937).
341. ——— and G. Herzberg, Phys. Rev. **49**, 100 (1936).

342. G. W. Funke and E. Lindholm, Z. Physik. **106**, 518 (1937).
343. N. Fuson, H. M. Randall and D. M. Dennison, Phys. Rev. **56**, 982 (1939).

344. D. M. Gage and E. F. Barker, J. Chem. Phys. **7**, 455 (1939).
345. W. S. Gallaway and E. F. Barker, J. Chem. Phys. **10**, 88 (1942).
346. P. Gänswein and R. Mecke, Z. Physik, **99**, 189 (1936).
347. H. Gerding and J. Lecomte, Rec. Trav. Chim. **58**, 614 (1939).
348. ——— and ———, Physica **6**, 737 (1939).
349. ——— and W. J. Nijveld, Nature, **137**, 1070 (1936).
350. ———, ——— and G. J. Muller, Z. physik. Chem. B. **35**, 193 (1937).
350a. ———, ——— and G. W. A. Rijnders, Rec. Trav. Chim. **60**, 25 (1941).
350b. ——— and G. W. A. Rijnders, Rec. Trav. Chim. **58**, 603 (1939).
350c. ——— and E. Smit, Z. physik. Chem. B. **51**, 217 (1942).
351. S. L. Gerhard, Phys. Rev. **42**, 622 (1932).
352. ——— and D. M. Dennison, Phys. Rev. **43**, 197 (1933).
353. W. Gerlach, Sitz. Ber. Bay. Akad. Math. Nat. Kl. **39**, No. 1 (1932).
354. H. Gershinowitz and E. B. Wilson, Jr., J. Chem. Phys. **5**, 500 (1937).
355. ——— and ———, J. Chem. Phys. **6**, 247 (1938).
356. J. C. Ghosh and S. K. Das, J. Phys. Chem. **36**, 586 (1932).
357. W. F. Giauque, J. Amer. Chem. Soc. **52**, 4816 (1930).
358. ——— and M. F. Ashley, Phys. Rev. **43**, 81 (1933).
359. ——— and H. L. Johnston, J. Amer. Chem. Soc. **50**, 3221 (1928).
360. ——— and R. Overstreet, J. Amer. Chem. Soc. **54**, 1731 (1932).
360a. P. A. Giguère, Trans. Roy. Soc. Can. **35**, 1 (1941).
361. R. H. Gillette and F. Daniels, J. Amer. Chem. Soc. **58**, 1139, 1143 (1936).
362. N. Ginsburg and E. F. Barker, J. Chem. Phys. **3**, 668 (1935).
363. L. Giulotto and P. Caldirola, Z. physik. Chem. B. **49**, 34 (1941).
364. G. Glockler, Rev. Mod. Phys. **15**, 111 (1943).
365. ——— and J. H. Bachmann, Phys. Rev. **54**, 970 (1938).
366. ——— and ———, Phys. Rev. **55**, 669 (1939).
367. ——— and ———, Phys. Rev. **55**, 1273 (1939).
368. ——— and H. M. Davis, J. Chem. Phys. **2**, 881 (1934).
369. ——— and W. F. Edgell, J. Chem. Phys. **9**, 224 (1941).
370. ———, ——— and G. R. Leader, J. Chem. Phys. **8**, 897 (1940).
371. ——— and G. E. Evans, J. Chem. Phys. **10**, 607 (1942).
372. ——— and G. R. Leader, J. Chem. Phys. **7**, 278 (1939).
373. ——— and ———, J. Chem. Phys. **7**, 382 (1939).
374. ——— and ———, J. Chem. Phys. **7**, 553 (1939).
375. ——— and ———, J. Chem. Phys. **8**, 125 (1940).
376. ——— and ———, J. Chem. Phys. **8**, 699 (1940).
377. ——— and C. E. Morrell, J. Chem. Phys. **4**, 15 (1936).
378. ——— and M. M. Renfrew, J. Chem. Phys. **6**, 170, 409 (1938).
379. ——— and ———, J. Chem. Phys. **6**, 295, 409, 682 (1938).
380. ——— and ———, J. Chem. Phys. **6**, 340 (1938).
381. ——— and ———, J. Chem. Phys. **6**, 408 (1938).
382. ——— and C. G. Sage, J. Chem. Phys. **8**, 291 (1940).
383. ——— and ———, J. Chem. Phys. **9**, 387 (1941).
384. ——— and F. T. Wall, J. Chem. Phys. **5**, 813 (1937).
385. ——— and ———, J. Phys. Chem. **41**, 143 (1937).
386. ——— and ———, Phys. Rev. **51**, 529 (1937).
387. C. F. Goodeve, Trans. Farad. Soc. **30**, 60 (1934).
388. A. R. Gordon, J. Chem. Phys. **2**, 65 (1934).
389. ———, J. Chem. Phys. **3**, 259 (1935).
390. W. Gordy, J. Chem. Phys. **7**, 93 (1939); **9**, 215 (1941).
391. ——— and S. C. Stanford, J. Chem. Phys. **8**, 170 (1940); **9**, 204 (1941).
392. ——— and D. Williams, J. Chem. Phys. **3**, 664 (1935); **4**, 85 (1936).
393. E. Gorin, J. Walter and H. Eyring, J. Amer. Chem. Soc. **61**, 1876 (1939).
394. J. Goubeau, Ber. d.d. chem. Ges. **68**, 912 (1935).
395. ———, Z. physik. Chem. B. **45**, 237 (1940).

396. J. Goubeau and J. Karweil, Z. physik. Chem. B. **40**, 376 (1938).
397. S. Gradstein, Z. physik. Chem. B. **22**, 384 (1933).
398. D. Grafe, K. Clusius and A. Kruis, Z. physik. Chem. B. **43**, 1 (1939).
399. P. Grassmann and J. Weiler, Z. Physik. **86**, 321 (1933).
400. B. Gredy, C. R. Paris, **197**, 327 (1933).
401. E. Gross and M. Vuks, Nature, **135**, 100 (1935).
402. J. Guéron, C. R. Paris, **199**, 136, 945 (1934).
403. E. A. Guggenheim, Trans. Farad. Soc. **37**, 97, 271 (1941).
405. J. Gupta, Ind. J. Phys. **10**, 199, 465 (1936).
406. ———, Sci. and Culture, **3**, 245 (1937).

407. J. O. Halford, L. C. Anderson and G. H. Kissin, J. Chem. Phys. **5**, 927 (1937).
407a. D. T. Hamilton and F. F. Cleveland, J. Chem. Phys. **12**, 249 (1944).
408. W. Hanle, Ann. d. Physik, **15**, 345 (1932).
409. I. Hanson, Phys. Rev. **46**, 122 (1934).
410. L. Harris, A. A. Ashdown and R. T. Armstrong, J. Amer. Chem. Soc. **58**, 852 (1936).
411. ———, W. S. Benedict and G. W. King, Nature, **131**, 621 (1931).
412. ——— and G. W. King, J. Chem. Phys. **2**, 51 (1934).
413. ——— and ———, J. Chem. Phys. **8**, 775 (1940).
414. ———, ———, W. S. Benedict and R. W. B. Pearse, J. Chem. Phys. **8**, 765 (1940).
415. J. B. Hatcher and D. M. Yost, J. Chem. Phys. **5**, 992 (1937).
416. R. R. Haun and W. D. Harkins, J. Amer. Chem. Soc. **54**, 3917 (1932).
417. T. Hayashi, Scient. Pap. Inst. Phys. Chem. Res. (Tokyo), **21**, 69 (1933).
418. K. Hedfeld and P. Lueg, Z. Physik. **77**, 446 (1932).
419. ——— and R. Mecke, Z. Physik, **64**, 151 (1930).
420. F. Heidenreich, Z. Physik, **97**, 277 (1935).
421. M. de Hemptinne and J. M. Delfosse, Ann. Bruxelles, **56**, 373 (1936).
422. ———, J. Jungers and J. M. Delfosse, J. Chem. Phys. **6**, 319 (1938).
423. ——— and C. Manneback, Proc. Ind. Acad. **9A**, 286 (1939).
424. ——— and C. Velghe, Physica **5**, 958 (1938).
425. ——— and J. Wouters, Nature, **138**, 884 (1936).
426. ———, ——— and M. Fayt, Bull. Belg. **19**, 318 (1933).
427. S. B. Hendricks and L. Pauling, J. Amer. Chem. Soc. **47**, 2904 (1925).
428. V. Henri and P. Angenot, C. R. Paris, **201**, 895 (1935); J. Chim. Phys. **33**, 641 (1936).
428a. R. C. Herman, J. Chem. Phys. **8**, 252 (1940).
429. ——— and R. Hofstadter, J. Chem. Phys. **6**, 534 (1938); **7**, 460, 630 (1939).
430. ——— and V. Williams, J. Chem. Phys. **8**, 447 (1940).
430a. E. Herz, L. Kahovec, and K. W. F. Kohlrausch, Z. physik. Chem. B. **53**, 124 (1943).
431. G. Herzberg, Z. physik. Chem. B. **17**, 68 (1932).
432. ———, J. Chem. Phys. **8**, 847 (1940).
433. ———, Paper at the Columbus meeting on Molecular Structure, (1941).
434. ———, Rev. Mod. Phys. **14**, 219 (1942).
434a. ———, unpublished.
435. ——— and K. Franz, Z. Physik, **76**, 720 (1932).
436. ——— and A. E. McKay, (unpublished).
437. ———, F. Patat and J. W. T. Spinks, Z. Physik, **92**, 87 (1934).
438. ———, ——— and H. Verleger, Z. Elektrochem. **41**, 522 (1935).
439. ———, ——— and ———, Z. Physik, **102**, 1 (1936).
440. ———, ——— and ———, J. Phys. Chem. **41**, 123 (1937).
441. ——— and J. W. T. Spinks, Z. Physik, **91**, 386 (1934).
442. ——— and ———, Proc. Roy. Soc. London, **147**, 434 (1934).
443. ——— and H. Verleger, Physik, Z. **35**, 622 (1934).
444. ——— and ———, Phys. Rev. **48**, 706 (1935).
445. ——— and ———, Physik. Z. **37**, 444 (1936).
446. L. Herzberg, Z. Physik, **107**, 549 (1937).
447. A. Hettich and S. B. Hendricks, Naturwiss, **21**, 467 (1933).
448. G. Hettner, Z. Physik, **89**, 234 (1934).
449. ———, R. Pohlman and H. J. Schumacher, Z. Physik, **91**, 372 (1934).
450. ———, ——— and ———, Z. Physik, **96**, 203 (1935).

451. J. H. Hibben, J. Chem. Phys. **4**, 324 (1936).
452. J. O. Hirschfelder, Dissertation Princeton (1935).
453. ———, J. Chem. Phys. **6**, 795 (1938).
454. ———, J. Chem. Phys. **8**, 431 (1940).
455. R. Hofstadter, J. Chem. Phys. **6**, 540 (1938).
456. F. T. Holmes, J. Chem. Phys. **4**, 88 (1936).
456a. H. Hönl and F. London, Z. Physik, **33**, 803 (1925); Ann. d. Phys. **79**, 273 (1926).
457. J. Horiuti, Z. Physik, **84**, 380 (1933).
458. W. V. Houston and C. M. Lewis, Proc. Nat. Acad. Sci. Amer. **17**, 229 (1931).
459. J. B. Howard, J. Chem. Phys. **3**, 207 (1935).
460. ———, J. Chem. Phys. **5**, 442 (1937).
461. ———, J. Chem. Phys. **5**, 451 (1937).
462. ——— and E. B. Wilson, Jr., J. Chem. Phys. **2**, 620 (1934).
463. G. F. Hull, J. Chem. Phys. **3**, 534 (1935).
464. J. R. Hull and R. A. Hull, J. Chem. Phys. **9**, 465 (1941).
465. F. Hund, Z. Physik, **43**, 805 (1927).

466. S. Imanishi, Nature, **135**, 396 (1935).
467. C. K. Ingold, Proc. Roy. Soc. London, **169**, 149 (1938).

468. H. A. Jahn, Proc. Roy. Soc. London, **168**, 469, 495 (1938).
469. ———, Proc. Roy. Soc. London, **171**, 450 (1939).
470. ———, Phys. Rev. **56**, 680 (1939).
471. ——— and E. Teller, Proc. Roy. Soc. London, **161**, 220 (1937).
472. H. M. James and A. S. Coolidge, J. Chem. Phys. **1**, 834 (1933).
473. H. L. Johnston and C. O. Davis, J. Amer. Chem. Soc. **56**, 271 (1934).
474. ——— and E. A. Long, J. Chem. Phys. **2**, 389 (1934).
475. ——— and M. K. Walker, Phys. Rev. **39**, 535 (1932).
476. M. Johnston and D. M. Dennison, Phys. Rev. **48**, 868 (1935).
476a. E. J. Jones, J. Amer. Chem. Soc. **65**, 2274 (1943).
477. G. Jung and H. Gude, Z. physik. Chem. B. **18**, 380 (1932).

478. L. Kahovec, Z. physik. Chem. B. **40**, 135 (1938).
478a. ——— and K. Knollmüller, Z. physik. Chem. B. **51**, 49 (1941).
478b. ——— and K. W. F. Kohlrausch, Z. physik. Chem. B. **37**, 421 (1937).
479. ——— and ———, Z. physik. Chem. B. **38**, 96 (1937).
480. ——— and ———, Z. physik, Chem. B. **46**, 165 (1940).
481. ——— and ———, Chem. Ber. **73**, 159 (1940).
482. ———, ———, A. W. Reitz and J. Wagner, Z. physik. Chem. B. **39**, 431 (1938).
483. ——— and J. Wagner, Z. physik. Chem. B. **47**, 48 (1940).
484. ——— and ———, Z. physik. Chem. B. **48**, 188 (1941).
484a. J. Karle and L. O. Brockway, J. Amer. Chem. Soc. **66**, 574 (1944).
485. J. Karweil and K. Schäfer, Z. physik. Chem. B. **40**, 382 (1938).
486. L. S. Kassel, J. Chem. Phys. **1**, 576 (1933).
487. ———, J. Amer. Chem. Soc. **56**, 1838 (1934).
488. ———, J. Chem. Phys. **4**, 276 (1936).
489. ———, J. Chem. Phys. **4**, 435 (1936).
490. ———, J. Chem. Phys. **4**, 493 (1936).
491. ———, Chem. Rev. **18**, 277 (1936).
492. A. Kastler, C. R. Paris, **194**, 858 (1932).
493. L. Kellner, Proc. Roy. Soc. London, **157**, 100 (1936).
494. ———, Proc. Roy. Soc. London, **177**, 447 (1941).
495. ———, Proc. Roy. Soc. London, **177**, 456 (1941).
495a. J. D. Kemp and C. J. Egan, J. Amer. Chem. Soc. **60**, 1521 (1938).
496. ——— and K. S. Pitzer, J. Chem. Phys. **4**, 749 (1936); J. Amer. Chem. Soc. **59**, 276 (1937).
496a. H. Kempter, Z. Physik, **116**, 1 (1940).
497. ——— and R. Mecke, Z. physik. Chem. B. **46**, 229 (1940).
498. R. M. Kennedy, M. Sagenkahn and J. G. Aston, J. Amer. Chem. Soc. **63**, 2267 (1941).
499. J. A. A. Ketelaar, J. Chem. Phys. **9**, 775 (1941).
500. ——— and K. J. Palmer, J. Amer. Chem. Soc. **59**, 2629 (1937).

501. C. F. Kettering, L. W. Shutts and D. H. Andrews, Phys. Rev. **36**, 531 (1930).
502. G. W. King, J. Chem. Phys. **5**, 413 (1937).
503. ———, R. T. Armstrong and L. Harris, J. Amer. Chem. Soc. **58**, 1580 (1936).
504. ———, R. M. Hainer and P. C. Cross, J. Chem. Phys. **11**, 27 (1943).
505. E. L. Kinsey and J. W. Ellis, Phys. Rev. **51**, 1074 (1937).
506. J. S. Kirby-Smith and L. G. Bonner, J. Chem. Phys. **7**, 880 (1939).
507. J. G. Kirkwood, J. Chem. Phys. **7**, 506 (1939).
508. G. B. Kistiakowsky, J. R. Lacher and W. W. Ransom, J. Chem. Phys. **8**, 970 (1940).
509. ———, ——— and F. Stitt, J. Chem. Phys. **6**, 407 (1938).
510. ———, ——— and ———, J. Chem. Phys. **7**, 289 (1939).
511. ——— and A. G. Nickle, J. Chem. Phys. **10**, 78, 146 (1942).
512. ——— and W. W. Rice, J. Chem. Phys. **8**, 610 (1940).
513. ——— and ———, J. Chem. Phys. **8**, 618 (1940).
514. ———, H. Romeyn, J. R. Ruhoff, H. A. Smith and W. E. Vaughan, J. Amer. Chem. Soc. **57**, 65 (1935).
515. O. Klein, Z. Physik, **58**, 730 (1929).
515a. C. H. Kline and J. Turkevich, J. Chem. Phys. **12**, 300 (1944).
516. A. Klit and A. Langseth, J. Chem. Phys. **5**, 925 (1937).
517. J. S. Koehler and D. M. Dennison, Phys. Rev. **57**, 1006 (1940).
517a. K. W. F. Kohlrausch, Z. physik. Chem. B. **30**, 305 (1935).
518. ———, Monatshefte f. Chem. **68**, 349 (1936).
519. ———, Physik. Z. **37**, 58 (1936).
520. ——— and F. Köppl, Z. physik. Chem. B. **24**, 370 (1934).
521. ——— and ———, Z. physik. Chem. B. **26**, 209 (1934).
522. ——— and ———, Monatshefte f. Chem. **65**, 185 (1935).
522a. ——— and O. Paulsen, Monatshefte f. Chem. **72**, 268 (1939).
523. ——— and A. Pongratz, Z. physik. Chem. B. **22**, 373 (1933).
524. ——— and ———, Z. physik. Chem. B. **27**, 176 (1934).
525. ——— and ———, Chem. Ber. **67**, 976 (1934).
526. ———, ——— and R. Seka, Monatshefte f. Chem. **70**, 213 (1937).
527. ——— and A. W. Reitz, Proc. Ind. Acad. Sci. **8A**, 255 (1938).
528. ——— and R. Skrabal, Monatshefte f. Chem. **70**, 377 (1937).
529. ——— and ———, Z. Elektrochem. **43**, 282 (1937).
530. ——— and W. Stockmair, Z. physik. Chem. B. **29**, 292 (1935).
531. ——— and J. Wagner, Z. physik. Chem. B. **45**, 93 (1939).
532. ——— and ———, Z. physik. Chem. B. **45**, 229 (1940).
532a. ——— and ———, Z. physik. Chem. B. **52**, 185 (1942).
533. ——— and H. Wittek, Z. physik. Chem. B. **47**, 55 (1940).
533a. ——— and ———, Z. physik. Chem. B. **48**, 177 (1941).
534. ——— and G. P. Ypsilanti, Z. physik. Chem. B. **29**, 274 (1935).
535. V. Kondratjew and O. Ssetkina, Physik. Z. U.S.S.R. **9**, 279 (1936).
536. H. Kopper, Z. physik. Chem. B. **34**, 396 (1936).
537. ——— and A. Pongratz, Monatshefte f. Chem. **62**, 78 (1933).
538. K. Kozima and S. Mizushima, Sci. Pap. Inst. Phys. Chem. Res. (Tokyo), **31**, 296 (1937).
539. H. A. Kramers and G. P. Ittmann, Z. Physik. **53**, 553 (1929).
540. ——— and ———, Z. Physik. **58**, 217; **60**, 663 (1929).
541. P. Krishnamurti, Ind. J. Phys. **5**, 633 (1930).
542. R. de L. Kronig, Z. Physik. **50**, 347 (1928).
543. Z. W. Ku, Phys. Rev. **44**, 376 (1933).
544. T. G. Kujumzelis, Physik. Z. **39**, 665 (1938).
545. ———, Z. Physik. **109**, 586 (1938); **110**, 760 (1938).

546. R. T. Lagemann and H. H. Nielsen, J. Chem. Phys. **10**, 668 (1942).
547. A. Langseth, Z. Physik. **72**, 350 (1931).
548. ———, Z. Physik. **77**, 60 (1932).
549. ——— and H. J. Bernstein, J. Chem. Phys. **8**, 410 (1940).
550. ———, ——— and B. Bak, J. Chem. Phys. **8**, 415 (1940).
551. ———, ——— and ———, J. Chem. Phys. **8**, 430 (1940).
552. ——— and R. C. Lord, J. Chem. Phys. **6**, 203 (1938).

553. A. Langseth and R. C. Lord, Danske Vidensk Selskab. Math. Fys. **16**, 6 (1938).
554. ——— and J. R. Nielsen, Nature, **130**, 92 (1932).
555. ——— and ———, Z. physik. Chem. B. **19**, 427 (1932).
556. ——— and ———, Phys. Rev. **46**, 1057 (1934).
557. ———, ——— and J. O. Sørensen, Z. physik. Chem. B. **27**, 100 (1934).
558. ———, J. O. Sørensen and J. R. Nielsen, J. Chem. Phys. **2**, 402 (1934).
559. ——— and E. Walles, Z. physik. Chem. B. **27**, 209 (1934).
560. J. R. Lawson and H. M. Randall, unpublished; quoted by Koehler and Dennison, Phys. Rev. **57**, 1006 (1940).
561. C. E. Leberknight, Phys. Rev. **43**, 967 (1933).
562. F. Lechner, Wien. Ber. **141**, 291 (1932).
563. ———, Wien. Ber. **141**, 633 (1932).
564. ———, Wien. Anz. (1933). No. 14.
565. J. Lecomte, C. R. Paris, **196**, 1011 (1933).
566. ———, J. de Phys. **8**, 489 (1937).
567. ———, C. R. Paris, **206**, 1568 (1938).
567a. ———, C. R. Paris, **207**, 395 (1938).
568. ———, J. de Phys. **9**, 13 (1938).
568a. ——— and J. P. Mathieu, C. R. Paris, **213**, 721 (1941).
569. ———, H. Volkringer and A. Tchakirian, C. R. Paris, **204**, 1927 (1937); J. de Phys. **9**, 105 (1938).
569a. A. Leduc in Int. Crit. Tables, vol. V, 79 (1929).
570. E. Lee and G. B. B. M. Sutherland, Proc. Cambr. Phil. Soc. **35**, 341 (1939).
571. ———, ———, and C. K. Wu, Nature, **142**, 669 (1938).
572. ——— and C. K. Wu, Trans. Farad. Soc. **35**, 1366 (1939).
573. C. G. LeFèvre and J. W. LeFèvre, J. Chem. Soc. London, 1935, p. 1696.
574. A. Levin and C. F. Meyer, J. Opt. Soc. Amer. **16**, 137 (1928).
574a. H. A. Lévy and L. O. Brockway, J. Amer. Chem. Soc. **59**, 2085 (1937).
575. B. Lewis and G. von Elbe, J. Amer. Chem. Soc. **57**, 612 (1935).
576. C. M. Lewis and W. V. Houston, Phys. Rev. **44**, 903 (1933).
577. W. F. Libby, J. Chem. Phys. **11**, 101 (1943).
578. L. N. Liebermann, Phys. Rev. **60**, 496 (1941).
579. E. Lindholm, Z. Physik **108**, 454 (1938).
580. R. Linke and W. Rohrmann, Z. physik. Chem. B. **35**, 256 (1937).
581. J. W. Linnett, J. Chem. Phys. **6**, 692 (1938).
582. ———, J. Chem. Phys. **8**, 91 (1940).
583. ———, Trans. Farad, Soc. **37**, 469 (1941).
584. ——— and W. H. Avery, J. Chem. Phys. **6**, 686 (1938).
585. ——— and H. W. Thompson, Nature, **139**, 509 (1937).
586. R. C. Lord, Jr., J. E. Ahlberg and D. H. Andrews, J. Chem. Phys. **5**, 649 (1937).
587. ——— and D. H. Andrews, J. Phys. Chem. **41**, 149 (1937).
588. ——— and F. A. Miller, J. Chem. Phys. **10**, 328 (1942).
589. ——— and E. Teller, J. Chem. Soc. London 1937, p. 1728.
590. ——— and N. Wright, J. Chem. Phys. **5**, 642 (1937).
591. P. Lueg and K. Hedfeld, Z. Physik, **75**, 512 (1932).
592. ——— and ———, Z. Physik, **75**, 599 (1932).

593. D. P. MacDougall and E. B. Wilson, Jr., J. Chem. Phys. **5**, 940 (1937).
594. G. E. MacWood and H. C. Urey, J. Chem. Phys. **4**, 402 (1936).
595. A. Maione, N. Cimento, **14**, 361 (1937).
596. C. Manneback, J. Chem. Phys. **5**, 989 (1937).
597. ———, Ann. Brux. **55**, 5, 129, 237 (1935).
598. ——— and A. Verleysen, Nature, **138**, 367 (1936); Ann. Brux. **56**, 349 (1936).
599. M. F. Manning, J. Chem. Phys. **3**, 136 (1935).
600. R. Manzoni-Ansidei and M. Rolla, Lincei, **27**, 410 (1938).
601. ——— and ———, Ric. Scient. 1938, p. 363.
602. P. E. Martin and E. F. Barker, Phys. Rev. **41**, 291 (1932).
603. F. Matossi and H. Aderhold, Z. Physik, **68**, 683 (1931).
604. ——— and O. Bronder, Z. Physik, **111**, 1 (1938).
605. A. W. Maue, Ann. Physik. **30**, 555 (1937).

606. L. R. Maxwell, S. B. Hendricks and V. M. Mosley, J. Chem. Phys. **3**, 699 (1935).
607. ——— and V. M. Mosley, J. Chem. Phys. **8**, 738 (1940).
607a. C. R. McCrosky, F. W. Bergstrom and G. Waitkins, J. Amer. Chem. Soc. **64**, 722 (1942).
608. A. McKellar and C. A. Bradley, Phys. Rev. **46**, 664 (1934).
608a. J. C. McLennan and H. D. Smith, Can. J. Research, **7**, 551 (1932).
608b. ———, ——— and J. O. Wilhelm, Trans. Roy. Soc. Can. **23**, 279 (1929).
609. R. Mecke, Z. Elektrochem. **36**, 589 (1930); Z. Physik, **64**, 173 (1930).
610. ———, Leipz. Vortr. 1931, p. 23.
611. ———, Z. physik. Chem. B. **16**, 409, 421; **17**, 1 (1932).
612. ———, Z. Physik, **81**, 313 (1933).
613. ———, Z. f. Astrophys. **6**, 144 (1933).
614. ———, Hand-und Jahrbuch d. chem. Phys. vol. 9, II p. 281, Leipzig, 1934.
615. ———, Z. physik. Chem. B. **33**, 156 (1936).
616. ———, Z. Physik. **107**, 595 (1937).
617. ——— and O. Vierling, Z. Physik **96**, 559 (1935).
618. ——— and R. Ziegler, Z. Physik **101**, 405 (1936).
619. L. Médard, J. Chim. Phys. **32**, 136 (1935).
620. ——— and F. Déguillon, C. R. Paris, **203**, 1518 (1936).
620a. A. G. Meister, F. F. Cleveland and M. J. Murray, Amer. J. Phys. **11**, 239 (1943).
621. A. C. Menzies, Proc. Roy. Soc. **172**, 89 (1939).
622. ——— and H. R. Mills, Proc. Roy. Soc. **148**, 407 (1935).
623. G. H. Messerly and J. G. Aston, J. Amer. Chem. Soc. **62**, 886 (1940).
623a. ——— and R. M. Kennedy, J. Amer. Chem. Soc. **62**, 2988 (1940).
624. M. V. Migeotte and E. F. Barker, Phys. Rev. **50**, 418 (1936).
625. S. M. Mitra, Ind. J. Phys. **12**, 9 (1938).
626. ———, Ind. J. Phys. **13**, 391 (1939).
627. S. Mizushima and Y. Morino, Sci. Pap. Inst. Phys. Chem. Res. (Tokyo) **29**, 188 (1936).
628. ——— and ———, Bull. Chem. Soc. Jap. **13**, 182 (1938).
629. ———, ——— and K. Higasi, Sci. Pap. Inst. Phys. Chem. Res. (Tokyo), **25**, 159 (1934).
630. ———, ——— and K. Kozima, Scient. Pap. Inst. Phys. Chem. Res. (Tokyo) **29**, 111 (1936).
631. ———, ——— and ———, Sci. Pap. Inst. Phys. Chem. Res. (Tokyo) **29**, 188 (1936).
632. ———, ——— and S. Nakamura, Sci. Pap. Inst. Phys. Chem. Res. (Tokyo) **37**, 205 (1940).
633. J. G. Moorhead, Phys. Rev. **39**, 83 (1932).
634. ———, Phys. Rev. **39**, 788 (1932).
635. Y. Morino and S. Mizushima, Scient. Pap. Inst. Phys. Chem. Res. (Tokyo) **32**, 220 (1937).
635a. J. C. Morris, J. Chem. Phys. **11**, 230 (1943).
636. P. M. Morse and E. C. G. Stueckelberg, Helv. Phys. Acta. **4**, 335 (1931).
637. P. A. Moses, Proc. Ind. Acad. **10A**, 71 (1939).
638. H. Moureu, M. Magat and G. Wetroff, P. Ind. Acad. **8A**, 356 (1938).
639. S. Mrozowski, Phys. Rev. **62**, 270 (1942).
640. H. P. Mulholland, Proc. Cambridge Phil. Soc. **24**, 280 (1928).
640a. P. Müller, Helv. Phys. Acta, **15**, 233 (1942).
641. R. S. Mulliken, Phys. Rev. **43**, 279 (1933).
642. ———, J. Phys. Chem. **41**, 5 (1937).
643. ———, J. Phys. Chem. **41**, 159 (1937).
644. ———, J. Chem. Phys. **7**, 14 (1939).
645. ———, Phys. Rev. **59**, 873 (1941).
646. G. M. Murphy, J. Chem. Phys. **8**, 71 (1940).
647. ——— and J. E. Vance, J. Chem. Phys. **6**, 426 (1938).
648. J. W. Murray and D. H. Andrews, J. Chem. Phys. **2**, 119 (1934).
649. ———, V. Deitz and D. H. Andrews, J. Chem. Phys. **3**, 180 (1935).
649a. M. J. Murray and F. F. Cleveland, J. Amer. Chem. Soc. **65**, 2110 (1943).
649b. ——— and ———, J. Chem. Phys. **12**, 156 (1944).

650. S. M. Naudé and H. Verleger, Physik. Z. **38**, 919 (1937).
651. W. Nespital, Z. physik. Chem. B. **16**, 153 (1932).
652. B. Neumann and G. Köhler, Z. Elektrochem. **34**, 218 (1928).
653. A. H. Nielsen, Phys. Rev. **53**, 983 (1938).
654. ———, Phys. Rev. **57**, 346 (1940).

654a. A. H. Nielsen, J. Chem. Phys. **11**, 160 (1943).
655. ——— and E. F. Barker, Phys. Rev. **46**, 970 (1934).
656. ——— and H. H. Nielsen, Phys. Rev. **48**, 864 (1935).
657. ——— and ———, J. Chem. Phys. **5**, 277 (1937).
658. ——— and ———, Phys. Rev. **54**, 118 (1938).
659. ——— and ———, Phys. Rev. **56**, 274 (1939).
660. H. H. Nielsen, Phys. Rev. **38**, 1432 (1931).
661. ———, Phys. Rev. **40**, 445 (1932).
662. ———, Phys. Rev. **46**, 117 (1934).
663. ———, J. Chem. Phys. **3**, 189 (1935).
664. ———, Phys. Rev. **55**, 289 (1939).
665. ———, Phys. Rev. **59**, 565 (1941).
666. ———, Phys. Rev. **60**, 794 (1941).
667. ———, Phys. Rev. **62**, 422 (1942).
668. ——— and W. H. Shaffer, J. Chem. Phys. **11**, 140 (1943).
669. J. R. Nielsen and N. E. Ward, J. Chem. Phys. **5**, 201 (1937).
670. ——— and ———, J. Chem. Phys. **10**, 81 (1942).
671. H. Nisi, Jap. J. Phys. **5**, 119 (1929).
672. H. D. Noether, J. Chem. Phys. **10**, 664 (1942).
673. ———, J. Chem. Phys. **10**, 693 (1942).
673a. ———, J. Chem. Phys. **11**, 97 (1943).
673b. G. Nordheim and H. Sponer, J. Chem. Phys. **11**, 253 (1943).
673c. R. Norris, Proc. Ind. Acad. **16A**, 287 (1942).
674. W. V. Norris and H. J. Unger, Phys. Rev. **43**, 467 (1933).
674a. ——— and ———, Phys. Rev. **45**, 68 (1934).
675. ———, ——— and R. E. Holmquist, Phys. Rev. **49**, 272 (1936).

676. L. S. Ornstein and J. Rekveld, Z. Physik. **61**, 593, **65**, 719 (1930); **68**, 543 (1931).
677. D. W. Osborne, R. N. Doescher and D. M. Yost, J. Amer. Chem. Soc. **64**, 169 (1942).
678. ———, C. S. Garner and D. M. Yost, J. Chem. Phys. **8**, 131 (1940).
679. J. W. Otvos and J. T. Edsall, J. Chem. Phys. **7**, 632 (1939).
680. R. G. Owens and E. F. Barker, J. Chem. Phys. **8**, 229 (1940).
681. ——— and ———, J. Chem. Phys. **10**, 146 (1942).

682. N. G. Pai, Proc. Roy. Soc. London, **149**, 29 (1935).
683. M. Parodi, C. R. Paris, **205**, 607 (1937).
684. J. R. Patty and H. H. Nielsen, Phys. Rev. **39**, 957 (1932).
685. L. Pauling, Phys. Rev. **36**, 430 (1930).
686. ——— and L. O. Brockway, Proc. Nat. Acad. Sci. Amer. **20**, 336 (1934).
687. ———, H. D. Springall and K. J. Palmer, J. Amer. Chem. Soc. **61**, 927 (1939).
688. O. Paulsen, Z. physik. Chem. B. **28**, 123 (1935).
688a. ———, Monatsh. f. Chem. **72**, 244 (1939).
689. E. Pendl, A. W. Reitz and R. Sabathy, Proc. Ind. Acad. **8A**, 508 (1938).
690. W. G. Penney, Proc. Roy. Soc. London, **144**, 166 (1934).
691. ——— and G. B. B. M. Sutherland, J. Chem. Phys. **2**, 492 (1934).
692. ——— and ———, Proc. Roy. Soc. London, **156**, 654 (1936).
693. L. W. Pickett, J. Chem. Phys. **10**, 660 (1942).
693a. K. S. Pitzer, J. Chem. Phys. **5**, 473 (1937).
693b. ———, J. Chem. Phys. **7**, 251 (1939).
694. ———, J. Chem. Phys. **8**, 711 (1940).
695. ———, Chem. Rev. **27**, 39 (1940).
695a. ———, J. Amer. Chem. Soc. **63**, 2413 (1941).
696. ———, J. Chem. Phys. **10**, 605 (1942).
696a. ———, J. Chem. Phys. **12**, 310 (1944).
697. ——— and W. D. Gwinn, J. Amer. Chem. Soc. **63**, 3313 (1941).
698. ——— and ———, J. Chem. Phys. **10**, 428 (1942).
699. ——— and D. W. Scott, J. Amer. Chem. Soc. **65**, 803 (1943).
700. G. Placzek, Marx, Handb. d. Radiologie, vol. VI, 2, p. 205 (1934).
701. ——— and E. Teller, Z. Physik, **81**, 209 (1933).

702. E. K. Plyler, Phys. Rev. **39**, 77 (1932).
703. ——— and E. F. Barker, Phys. Rev. **38**, 1827 (1931); **41**, 369 (1932).
704. ——— and W. W. Sleator, Phys. Rev. **37**, 1493 (1931).
705. R. Pohlman, Z. Physik, **79**, 394 (1932).
706. ——— and H. J. Schumacher, Z. Physik, **102**, 678 (1936).
707. D. Price, J. Chem. Phys. **9**, 725 (1941).
708. ———, J. Chem. Phys. **9**, 807 (1941).
709. E. N. Prileskajeva, J. K. Syrkin and M. W. Wolkenstein, Acta Physicochimica U.S.S.R. **12**, 176 (1940).
710. P. Pringsheim and B. Rosen, Z. Physik, **50**, 741 (1928).
711. H. M. Randall and E. F. Barker, Phys. Rev. **45**, 124 (1934).
712. ———, D. M. Dennison, N. Ginsburg and L. R. Weber, Phys. Rev. **52**, 160 (1937).
713. D. H. Rank, J. Chem. Phys. **1**, 504 (1933).
714. ———, J. Chem. Phys. **1**, 572 (1933).
715. ——— and E. R. Bordner, J. Chem. Phys. **3**, 248 (1935).
716. ———, K. D. Larsen and E. B. Bordner, J. Chem. Phys. **2**, 464 (1934).
717. A. V. Rao, Z. Physik, **97**, 154 (1935).
718. B. P. Rao, Proc. Ind. Acad. Sci. **11A**, 1 (1940).
719. B. S. R. Rao, J. Chem. Phys. **6**, 343 (1938).
720. ———, Proc. Ind. Acad. Sci. **10A**, 167 (1939).
721. C. S. S. Rao, Z. Physik, **94**, 536 (1935).
722. I. R. Rao and P. Koteswaran, J. Chem. Phys. **5**, 667 (1937).
723. K. V. K. Rao, Proc. Ind. Acad. Sci. **14A**, 521 (1941).
723a. Y. P. Rao, Ind. J. Phys. **16**, 205 (1942).
724. R. S. Rasmussen, Phys. Rev. **62**, 301A, (1942).
724a. ———, J. Chem. Phys. **11**, 249 (1943).
724b. ———, D. D. Tunnicliff and R. R. Brattain, J. Chem. Phys. **11**, 432 (1943).
725. B. S. Ray, Z. Physik, **78**, 74 (1932).
726. O. Redlich, Z. physik. Chem. B. **28**, 371 (1935).
727. ———, J. Chem. Phys. **9**, 298 (1941).
727a. ———, E. K. Holt and J. Bigeleisen, J. Amer. Chem. Soc. **66**, 13 (1944).
728. ———, T. Kurz and P. Rosenfeld, Z. physik. Chem. B. **19**, 231 (1932).
729. ———, ——— and ———, J. Chem. Phys. **2**, 619 (1934).
730. ———, ——— and W. Stricks, Monatshefte f. Chem. **71**, 1 (1937).
731. ——— and L. E. Nielsen, J. Amer. Chem. Soc. **65**, 654 (1943).
731a. ——— and F. Pordes, Monatshefte f. Chem. **67**, 203 (1936).
731b. ——— and W. Stricks, Monatshefte f. Chem. **67**, 213; **68**, 47, 374 (1936).
732. ——— and ———, Monatshefte f. Chem. **67**, 328 (1936).
733. ——— and H. Tompa, J. Chem. Phys. **5**, 529 (1937).
734. F. Reiche and H. Rademaker, Z. Physik, **39**, 444 (1926); **41**, 453 (1927).
735. A. W. Reitz, Z. physik. Chem. B. **33**, 179 (1936); **35**, 363 (1937).
736. ———, Z. physik. Chem. B. **38**, 275, 381 (1937).
737. ——— and R. Sabathy, Wien. Ber. **146**, 577 (1938).
738. ——— and R. Skrabal, Monatshefte f. Chem. **70**, 398 (1937).
739. ——— and J. Wagner, Z. physik. Chem. B. **43**, 339 (1939).
740. R. Robertson and J. J. Fox, Proc. Roy. Soc. London, **120**, 128 (1928).
741. R. Rollefson and R. Havens, Phys. Rev. **57**, 710 (1940).
742. N. Rosen and P. M. Morse, Phys. Rev. **42**, 210 (1932).
743. E. J. Rosenbaum, J. Chem. Phys. **8**, 643 (1940).
744. ——— and T. A. Ashford, J. Chem. Phys. **7**, 554 (1939).
745. ———, D. J. Rubin and C. R. Sandburg, J. Chem. Phys. **8**, 366 (1940).
746. J. E. Rosenthal, Phys. Rev. **45**, 426 (1934).
747. ———, Phys. Rev. **45**, 538 (1934).
748. ———, Phys. Rev. **46**, 730 (1934); **49**, 535 (1936).
749. ———, J. Chem. Phys. **5**, 465 (1937).
750. ——— and G. M. Murphy, Rev. Mod. Phys. **8**, 317 (1936).
750a. A. Rousset, J. Laval and R. Lochet, C. R. Paris, **216**, 886 (1943).
750b. T. R. Rubin, B. H. Levedahl and D. M. Yost, J. Amer. Chem. Soc. **66**, 279 (1944).

751. K. Rumpf and R. Mecke, Z. physik. Chem. B. **44**, 299 (1939).
752. H. Russell, Jr., D. W. Osborne and D. M. Yost, J. Amer. Chem. Soc. **64**, 165 (1942).
752a. ———, D. R. V. Golding and D. M. Yost, J. Amer. Chem. Soc. **66**, 16 (1944).

753. R. Sabathy, Z. physik. Chem. B. **41**, 183 (1938).
754. H. Sachsse and E. Bartholomé, Z. phys. Chem. B. **28**, 257 (1935).
754a. B. D. Saksena, Proc. Ind. Acad. **10A**, 449 (1939).
755. ———, Proc. Ind. Acad. Sci. **11A**, 53 (1940).
755a. ———, Proc. Ind. Acad. Sci. **12A**, 312 (1940).
756. ———, Proc. Ind. Acad. Sci. **12A**, 416 (1940).
757. E. O. Salant and J. E. Rosenthal, Phys. Rev. **42**, 812 (1932).
758. ——— and ———, Phys. Rev. **42**, 818 (1932).
759. ——— and W. West, Phys. Rev. **33**, 640 (1929).
760. E. J. Salstrom and L. Harris, J. Chem. Phys. **3**, 241 (1934).
761. J. A. Sanderson, Phys. Rev. **50**, 209 (1935).
762. C. Sannié and V. Poremski, C. R. Paris, **208**, 2073 (1939).
762a. B. S. Satyanarayana, Proc. Ind. Acad. Sci. A. **15**, 414 (1942).
762b. H. Sayvetz, J. Chem. Phys. **7**, 383 (1939).
763. C. Schaefer, Z. Physik, **60**, 586 (1930).
764. ——— and C. Bormuth, Z. Physik, **62**, 508 (1930).
765. ———, ——— and F. Matossi, Z. Physik, **39**, 648 (1926).
766. ——— and R. Kern, Z. Physik, **78**, 609 (1932).
767. ———, F. Matossi and H. Aderhold, Z. Physik, **65**, 289 (1930).
768. K. Schäfer, Z. physik. Chem. B. **40**, 357 (1938).
769. R. Schaffert, J. Chem. Phys. **1**, 507 (1933).
770. K. Schierkolk, Z. Physik, **29**, 277 (1924).
770a. V. Schomaker and R. A. Spurr, J. Amer. Chem. Soc. **64**, 1184 (1942).
771. ——— and D. P. Stevenson, J. Amer. Chem. Soc. **62**, 1267 (1940).
771a. ——— and ———, J. Chem. Phys. **8**, 637 (1940).
772. M. Schuler, in Müller-Pouillet's Lehrbuch der Physik I, 1 (11. edition) p. 731 (Braunschweig, 1929).
773. S. C. Schumann and J. G. Aston, J. Chem. Phys. **6**, 480 (1938).
774. ——— and ———, J. Chem. Phys. **6**, 485 (1938).
774a. ———, ——— and M. Sagenkahn, J. Amer. Chem. Soc. **64**, 1039 (1942).
775. M. K. Sen, Ind. J. Phys. **11**, 9 (1937).
776. W. H. Shaffer, J. Chem. Phys. **9**, 607 (1941).
777. ———, J. Chem. Phys. **10**, 1 (1942).
778. ——— and R. R. Newton, J. Chem. Phys. **10**, 405 (1942).
779. ——— and A. H. Nielsen, J. Chem. Phys. **9**, 847 (1941).
780. ——— and H. H. Nielsen, Phys. Rev. **56**, 188 (1939).
781. ———, ——— and L. H. Thomas, Phys. Rev. **56**, 895 (1939).
782. ———, ——— and ———, Phys. Rev. **56**, 1051 (1939).
783. W. Shand and R. A. Spurr, J. Amer. Chem. Soc. **65**, 179 (1943).
784. S. T. Shen, Y. T. Yao and T. Y. Wu, Phys. Rev. **51**, 235 (1937).
785. H. Y. Sheng, E. F. Barker and D. M. Dennison, Phys. Rev. **60**, 786 (1941).
786. A. A. Sidorova, Acta Physicochimica **7**, 193 (1937).
787. S. Silver, J. Chem. Phys. **7**, 1113 (1939).
788. ———, J. Chem. Phys. **8**, 919 (1940).
789. ——— and E. S. Ebers, J. Chem. Phys. **10**, 559 (1942).
790. ——— and W. H. Shaffer, J. Chem. Phys. **9**, 599 (1941).
791. A. Simon and F. Fehér, Z. Elektrochem. **41**, 290 (1935).
792. ——— and ———, Z. anorg. allgem. Chem. **230**, 289 (1936).
792a. ——— and H. Hoepper, Kolloidzeitschr. **85**, 8 (1939).
793. S. C. Sirkar, Ind. J. Phys. **6**, 131 (1931).
794. ———, Ind. J. Phys. **10**, 189 (1936).
795. Z. I. Slawsky and D. M. Dennison, J. Chem. Phys. **7**, 509 (1939).
796. ——— and ———, J. Chem. Phys. **7**, 522 (1939).
797. L. G. Smith, Phys. Rev. **59**, 924 (1941).
798. ——— and W. M. Woodward, Phys. Rev. **61**, 386 (1942).

799. H. Sponer and J. S. Kirby-Smith, J. Chem. Phys. **9**, 667 (1941).
800. ———, G. Nordheim, A. L. Sklar and E. Teller, J. Chem. Phys. **7**, 207 (1939).
801. ——— and E. Teller, J. Chem. Phys. **7**, 382 (1939).
802. ——— and ———, Rev. Mod. Phys. **13**, 75 (1941).
803. A. D. Sprague and H. H. Nielsen, Phys. Rev. **43**, 375 (1933).
804. ——— and ———, J. Chem. Phys. **5**, 85 (1937).
805. C. C. Stephenson and H. O. McMahon, J. Chem. Phys. **7**, 614 (1939).
806. D. P. Stevenson, J. Chem. Phys. **8**, 285 (1940).
807. W. B. Steward and H. H. Nielsen, Phys. Rev. **47**, 828 (1935).
808. ——— and ———, Phys. Rev. **48**, 861 (1935).
809. G. A. Stinchcomb and E. F. Barker, Phys. Rev. **33**, 305 (1929).
810. F. Stitt, J. Chem. Phys. **7**, 297 (1939).
811. ———, J. Chem. Phys. **7**, 1115 (1939).
812. ———, J. Chem. Phys. **8**, 56 (1940).
813. ———, J. Chem. Phys. **8**, 981 (1940).
814. ———, J. Chem. Phys. **9**, 780 (1941).
815. ——— and D. M. Yost, J. Chem. Phys. **4**, 82 (1936).
816. ——— and ———, J. Chem. Phys. **5**, 90 (1937).
817. J. W. Straley, C. H. Tindal and H. H. Nielsen, Phys. Rev. **58**, 1002 (1940).
818. ———, ——— and ———, Phys. Rev. **62**, 161 (1942).
819. J. Strong, Phys. Rev. **38**, 1818 (1931).
820. Y. K. Suirkin and M. V. Wolkenstein, Acta Physicochim. U.S.S.R. **2**, 308 (1935).
821. B. Susz and E. Briner, Helv. Chim. Acta **18**, 378 (1935).
822. G. B. B. M. Sutherland, Proc. Roy. Soc. London, **141**, 342 (1933).
823. ———, Proc. Roy. Soc. London, **141**, 535 (1933).
824. ———, Phys. Rev. **43**, 883 (1933).
825. ———, Proc. Roy. Soc. London, **145**, 278 (1934).
825a. ———, Trans. Farad. Soc. **34**, 325 (1937).
826. ———, Phys. Rev. **56**, 836 (1939).
827. ———, J. Chem. Phys. **7**, 1066 (1939).
828. ——— and D. M. Dennison, Proc. Roy. Soc. London, **148**, 250 (1935).
829. ——— and S. L. Gerhard, Nature, **130**, 241 (1932).
830. ———, E. Lee and C. K. Wu, Trans. Farad. Soc. **35**, 1373 (1939).
831. ——— and W. G. Penney, Proc. Roy. Soc. London, **156**, 678 (1936).

832. A. Tchakirian and H. Volkringer, C. R. Paris, **200**, 1758 (1935).
833. D. E. Teets and D. H. Andrews, J. Chem. Phys. **3**, 175 (1935).
834. D. Telfair, J. Chem. Phys. **10**, 167 (1942).
835. ——— and W. H. Pielemeier, J. Chem. Phys. **9**, 571 (1941).
836. E. Teller, Hand- und Jahrb. d. Chem. Phys. vol. 9, II, 43 (1934).
837. ——— and L. Tisza, Z. Physik, **73**, 791 (1932).
838. ——— and K. Weigert, Gött. Nachr. (1933), p. 218.
839. V. N. Thatte, Nature, **138**, 468 (1936).
840. ——— and M. S. Joglekar, Phil. Mag. **19**, 1116 (1935).
841. ——— and ———, Phil. Mag. **23**, 1067 (1937).
842. M. Théodoresco, C. R. Paris, **210**, 175 (1940); **216**, 56, 117 (1943).
843. L. H. Thomas and S. E. Whitcomb, Phys. Rev. **56**, 383 (1939).
844. H. W. Thompson, J. Chem. Phys. **6**, 748 (1938).
845. ———, J. Amer. Chem. Soc. **61**, 1396 (1939).
846. ———, J. Chem. Phys. **7**, 441 (1939).
847. ———, J. Chem. Phys. **7**, 448 (1939).
848. ———, J. Chem. Phys. **7**, 453 (1939).
849. ———, Trans. Farad. Soc. **35**, 697 (1939).
850. ———, Trans. Farad. Soc. **36**, 988 (1940).
851. ———, Trans. Farad. Soc. **37**, 38 (1941).
852. ———, Trans. Farad. Soc. **37**, 251 (1941).
853. ———, Trans. Farad. Soc. **37**, 344 (1941).
854. ——— and D. J. Dupré, Trans. Farad. Soc. **36**, 805 (1940).
855. ——— and G. P. Harris, Trans. Farad. Soc. **38**, 37 (1942).

855a. H. W. Thompson and G. P. Harris, Trans. Farad. Soc. **40**, 295 (1944).
856. ——— and J. W. Linnett, J. Chem. Soc. London (1937), p. 1376.
857. ——— and ———, J. Chem. Soc. London (1937), p. 1384.
858. ——— and ———, Proc. Roy. Soc. London, **160**, 539 (1937).
859. ———, ——— and F. J. Wagstaffe, Trans. Farad. Soc. **36**, 797 (1940).
860. ——— and N. P. Skerrett, Trans. Farad. Soc. **36**, 812 (1940).
861. ——— and H. A. Skinner, J. Chem. Phys. **6**, 775 (1938).
862. B. Timm and R. Mecke, Z. Physik, **94**, 1 (1935).
863. ——— and ———, Z. Physik, **97**, 221 (1935).
864. ——— and ———, Z. Physik, **98**, 363 (1935).
865. C. H. Tindal, J. W. Straley and H. H. Nielsen, Proc. Nat. Acad. Sci. Amer. **27**, 208 (1941).
866. ———, ——— and ———, Phys. Rev. **62**, 151 (1942).
867. L. Tisza, Z. Physik, **82**, 48 (1933).
868. R. Titeica, Ann. de Phys. (11) **1**, 533 (1934).
869. R. C. Tolman and R. M. Badger, J. Amer. Chem. Soc. **45**, 2277 (1923).
869a. M. W. Travers and T. J. P. Pearce, J. Soc. Chem. Ind. **53**, 321 (1934).
870. F. Trenkler, Physik. Z. **36**, 162, 423 (1935).
871. ———, Physik. Z. **37**, 338 (1936).
872. ———, Proc. Ind. Acad. Sci. **8A**, 383 (1938).
873. B. Trumpy, Z. Physik, **66**, 790 (1930).
874. ———, Z. Physik, **68**, 675 (1931).
875. ———, Z. Physik, **88**, 226 (1934).
876. ———, Z. Physik, **90**, 133 (1934).
877. ———, Norsk. Videns, Selsk. (1934), No. 9.
878. ———, Z. Physik, **93**, 624 (1935).
879. ———, Z. Physik, **98**, 672 (1936).
880. ———, Z. Physik, **100**, 250 (1936).
880a. J. Turkevich and P. C. Stevenson, J. Chem. Phys. **11**, 328 (1943).

881. H. C. Urey, J. Amer. Chem. Soc. **45**, 1445 (1923).
882. ——— and C. A. Bradley, Phys. Rev. **38**, 1969 (1931).
883. H. J. Unger, Phys. Rev. **43**, 123 (1933).

884. J. H. Van Vleck, J. Chem. Phys. **1**, 177, 219 (1933).
885. ——— and P. C. Cross, J. Chem. Phys. **1**, 357 (1933).
886. H. Vedder and R. Mecke, Z. Physik, **86**, 137 (1933).
887. S. Vencov and D. Stefanescu, Bull. Soc. Roumaine de Phys. **39**, 13 (1938).
888. ——— and ———, Bull. Soc. Roumaine de Phys. **40**, 45 (1939).
889. C. S. Venkateswaran, Proc. Ind. Acad. Sci. **1A**, 850 (1935).
890. ———, Proc. Ind. Acad. Sci. **2A**, 260 (1935); **4A**, 345 (1936).
891. ———, Proc. Ind. Acad. Sci. **3A**, 533 (1936).
891a. ———, Proc. Ind. Acad. Sci. **4A**, 345 (1936).
892. ———, Proc. Ind. Acad. Sci. **4A**, 414 (1936).
893. S. Venkateswaran, Nature, **127**, 406 (1931).
894. ———, Ind. J. Phys. **5**, 145 (1930).
895. ———, Phil. Mag. **14**, 258, (1932); **15**, 263 (1933).
896. K. Venkateswarlu, Proc. Ind. Acad. Sci. **10A**, 156 (1939).
897. ———, Proc. Ind. Acad. Sci. **12A**, 453 (1940).
898. H. Verleger, Z. Physik, **98**, 342 (1935).
899. ———, Naturwiss. **24**, 237 (1936).
900. A. Verleysen and C. Manneback, Ann. Brux. **57**, 31 (1937).
901. O. Vierling and R. Mecke, Z. Physik. **99**, 204 (1935).
902. I. E. Viney, Proc. Cambr. Phil. Soc. **29**, 142, 407 (1933).
903. H. H. Voge and J. E. Rosenthal, J. Chem. Phys. **4**, 137 (1936).
904. H. Volkringer, A. Tchakirian and M. Freymann, C. R. Paris, **199**, 292 (1934).
905. ———, J. Lecomte and A. Tchakirian, J. de Phys. **9**, 105 (1938).

906. J. Wagner, Z. physik. Chem. B. **40**, 36 (1938).
907. ———, Z. physik. Chem. B. **40**, 439 (1938).

908. J. Wagner, Z. physik. Chem. B. **45**, 69 (1939).
909. ———, Z. physik. Chem. B. **48**, 309 (1941).
910. F. T. Wall and C. R. Eddy, J. Chem. Phys. **6**, 107 (1938).
911. ——— and G. Glockler, J. Chem. Phys. **5**, 314 (1937).
912. S. C. Wang, Phys. Rev. **34**, 243 (1929).
912a. H. E. Watson, G. P. Kane and K. L. Ramaswany, Proc. Roy. Soc. **152**, 130 (1936).
913. M. Wehrli, Helv. Phys. Acta, **11**, 339 (1938); **13**, 153 (1940).
914. J. Weiler, Ann. Physik, **23**, 493 (1935).
915. A. J. Wells and E. B. Wilson, Jr., J. Chem. Phys. **9**, 314 (1941).
916. ——— and ———, J. Chem. Phys. **9**, 659 (1941).
917. M. Werth, Phys. Rev. **39**, 299 (1932).
918. W. West, J. Chem. Phys. **7**, 795 (1939).
919. ——— and R. T. Edwards, J. Chem. Phys. **5**, 14 (1937).
920. ——— and M. Farnsworth, J. Chem. Phys. **1**, 402 (1933).
921. ——— and R. B. Killingsworth, J. Chem. Phys. **6**, 1 (1938).
922. S. E. Whitcomb, H. H. Nielsen and L. H. Thomas, J. Chem. Phys. **8**, 143 (1940).
923. E. Wigner, Gött. Nachr. 1930, p. 133.
924. R. Wildt, Naturwiss. **20**, 851 (1932).
925. D. Williams, J. Amer. Chem. Soc. **61**, 2987 (1939).
926. ———, J. Amer. Chem. Soc. **62**, 2442 (1940).
927. ———, Phys. Rev. **57**, 1077A (1940).
928. ——— and L. Decherd, J. Amer. Chem. Soc. **61**, 1382 (1939).
929. E. B. Wilson, Jr., Phys. Rev. **45**, 427 (1934).
930. ———, Phys. Rev. **45**, 706 (1934).
931. ———, Phys. Rev. **46**, 146 (1934).
932. ———, J. Chem. Phys. **3**, 59 (1935).
933. ———, J. Chem. Phys. **3**, 276 (1935).
934. ———, J. Chem. Phys. **3**, 818 (1935).
935. ———, J. Chem. Phys. **4**, 313 (1936).
936. ———, J. Chem. Phys. **4**, 526 (1936).
937. ———, J. Chem. Phys. **5**, 617 (1937).
938. ———, J. Chem. Phys. **6**, 740 (1938).
939. ———, J. Chem. Phys. **7**, 948 (1939).
940. ———, J. Chem. Phys. **7**, 1047 (1939).
941. ———, Chem. Rev. **27**, 17 (1940).
942. ———, J. Chem. Phys. **9**, 76 (1941).
943. ——— and B. L. Crawford, Jr., J. Chem. Phys. **6**, 223 (1938).
944. ——— and J. B. Howard, J. Chem. Phys. **4**, 260 (1936).
945. ——— and A. J. Wells, J. Chem. Phys. **9**, 319 (1941).
945a. T. P. Wilson, J. Chem. Phys. **11**, 361 (1943).
945b. ———, J. Chem. Phys. **11**, 369 (1943).
946. E. E. Witmer, Proc. Nat. Acad. Sci. Amer. **13**, 60 (1927).
947. R. K. Witt and J. D. Kemp, J. Amer. Chem. Soc. **59**, 273 (1937).
948. H. Wittek, Z. physik. Chem. B. **48**, 1 (1940).
948a. ———, Z. physik. Chem. B. **51**, 103, 187 (1942).
948b. ———, Z. physik. Chem. B. **52**, 153 (1942)
949. K. L. Wolf, in Müller-Pouillet's Lehrbuch der Physik IV, 3 p. 726, (1933).
949a. M. W. Wolkenstein, J. Phys. Acad. Sci. U.S.S.R. **5**, 185 (1941).
950. S. C. Woo and R. M. Badger, Phys. Rev. **39**, 932 (1932).
951. ——— and T. C. Chu, J. Chem. Phys **5**, 786 (1937).
952. R. W. Wood, J. Chem. Phys. **4**, 536 (1936).
953. ——— and G. Collins, Phys. Rev. **42**, 386 (1932).
954. ——— and D. H. Rank, Phys. Rev. **48**, 63 (1935).
955. J. Wouters, Bull. Belg. **20**, 782 (1934).
956. N. Wright and H. M. Randall, Phys. Rev. **44**, 391 (1933).
957. C. K. Wu and G. B. B. M. Sutherland, J. Chem. Phys. **6**, 114 (1938).
958. T. Y. Wu, Phys. Rev. **46**, 465 (1934).
959. ———, J. Chem. Phys. **5**, 392 (1937).

960. T. Y. Wu, J. Chem. Phys. **7**, 965 (1939).
961. ———, J. Chem. Phys. **8**, 489 (1940).
962. ———, J. Chem. Phys. **10**, 116 (1942).
963. ——— and A. T. Kiang, J. Chem. Phys. **7**, 178 (1939).
964. ——— and S. T. Shen, Chim. Journ. Phys. **2**, 128 (1936).
965. V. L. Wu and E. F. Barker, J. Chem. Phys. **9**, 487 (1941).

966. T. Yeou, C. R. Paris, **206**, 1371 (1938).
967. ———, C. R. Paris, **207**, 326 (1938).
968. D. M. Yost and T. F. Anderson, J. Chem. Phys. **2**, 624 (1934).
969. ——— and ———, J. Chem. Phys. **3**, 754 (1935).
970. ———, D. DeVault, T. F. Anderson and E. N. Lassettre, J. Chem. Phys. **6**, 424 (1938).
971. ———, E. N. Lassettre and S. T. Gross, J. Chem. Phys. **4**, 325 (1936).
972. ———, D. W. Osborne and C. S. Garner, J. Amer. Chem. Soc. **63**, 3492 (1941).
973. ——— and J. E. Sherborne, J. Chem. Phys. **2**, 125 (1934).
974. ———, C. C. Steffens and S. T. Gross, J. Chem. Phys. **2**, 311 (1934).

974a. R. Ziegler, Z. Physik, **116**, 716 (1940).
975. L. R. Zumwalt and R. M. Badger, J. Chem. Phys. **7**, 235 (1939).
976. ——— and ———, J. Chem. Phys. **7**, 629 (1939).
977. ——— and P. A. Giguère, J. Chem. Phys. **9**, 458 (1941).
978. J. Zunino, Z. Physik, **100**, 335 (1936).

AUTHOR INDEX

SUBJECT INDEX

This subject index includes in addition to the usual material of an index also all *symbols* used in the book and all *molecules* (chemical compounds) discussed or mentioned.

Italicized page numbers refer to more detailed discussions of the subjects than ordinary page numbers, or to definitions; boldface page numbers refer to figures.

The mathematical symbols and symbols for species, point groups, molecular constants, and so forth, are listed at the beginning of the section devoted to the corresponding letter. The Greek letters are arranged under the letter with which they begin when they are written in English (for example φ, π, ψ are listed under P and in this order) except Δ when used to indicate a difference in which case it is disregarded in the alphabeting. Symbols to which a word is joined are arranged under the corresponding symbol; for example, *R branch* is under *R*, not under *Rb; B rotational levels* under *B*, not under *Br*.

All individual molecules (chemical compounds) are listed under their chemical formulae considered as words; for example, $CHCl_3$ under Chcl, H_2SO_4 under Hso. If there are several molecules giving the same "word" they are listed in the order of increasing numbers of the first, second, ⋯ atom; for example, $CHCl_3$, CH_3Cl, $C_2H_2Cl_4$, $C_2H_4Cl_2$ in this order, but C_2H_4 is ahead of $CHCl_3$ since the corresponding "word" is Ch. It should be realized that this order is somewhat different from that in the formula index of Chemical Abstracts where the number of atoms has "priority" over the alphabet. This change appears to be necessary in a combined formula and subject index. For the benefit of the hurried reader, for all molecules discussed in detail, cross references are given under their chemical names.

The order of symbols in a chemical formula is the usual one for all organic compounds, including metal-organic compounds, that is, C comes first, then H, while the remaining atoms are in alphabetical order. For inorganic compounds the central atom, if any, is put first, then H and then the other atoms in alphabetical order, except in the case of acids and H_2O for which H is put first. Thus we have $SiHCl_3$, H_2SO_4, and so forth, that is substantially the conventional order. Cross references are given in all ambiguous cases.

The alphabeting of subjects has been done first on the basis of the part before the comma except in combinations of symbols and words (see above). In alphabeting the second part, after the comma, prepositions at the beginning have been disregarded.

All material referring to a particular molecule, such as fundamentals, internuclear distances, and so on, is given under that molecule. Molecular types such as linear XY_2, pyramidal XY_3, should be looked up under XY_2, XY_3, where also various items relating to these types may be found.

A

D

F

J

K

L

M

O

P

R

S

U

V